Communications
in Computer and Information Science 228

Jun Zhang (Ed.)

Applied Informatics and Communication

International Conference, ICAIC 2011
Xi'an, China, August 20-21, 2011
Proceedings, Part V

Springer

Volume Editor

Jun Zhang
Huazhong University of Science and Technology
1037 Luoyu Road, Wuhan, China
E-mail: zhangjunhust@yeah.net

ISSN 1865-0929 e-ISSN 1865-0937
ISBN 978-3-642-23222-0 ISBN 978-3-642-23223-7 (eBook)
DOI 10.1007/978-3-642-23223-7
Springer Heidelberg Dordrecht London New York

Library of Congress Control Number: 2011934039

CR Subject Classification (1998): C, D, F, H, I, J.1, J.2

Typesetting: Camera-ready by author, data conversion by Scientific Publishing Services, Chennai, India

Printed on acid-free paper

Springer is part of Springer Science+Business Media (www.springer.com)

Preface

Computers are firmly rooted in nearly all areas of life today. In company life, documents and products are produced by using computers, communication is by e-mail, companies present their commodities and services on the Internet and copious amounts of data are stored in databases. Different application programs are required for all of these processes.

The 2011 International Conference on Applied Informatics and Communication (ICAIC 2011) held during August 20–21, 2011, in Xi'an, China, provided an excellent platform for scientists and engineers of different countries working in various fields to exchange the latest developments and foster world-wide cooperation in the field of applied informatics and communication. Hundreds of experts and scholars from different countries attended ICAIC 2011.

Being crucial for the development of applied informatics and communication, our conference encompasses numerous research topics and applications: from the core fields of information systems and software engineering, manufacturing and automation engineering, computer-based signal processing and image processing to Communications Technology and other related topics. All of the papers were peer-reviewed by selected experts and 451 were selected for the proceedings of ICAIC 2011. We believe that this book provides a valuable reference for researchers in related fields. The papers describe how efficient, user-friendly, secure and upgradeable information technology systems can be designed, built and meaningfully used and operated in practice.

ICAIC 2011 was organized by the Information Engineering Research Institute, USA, and the proceeding are published by Springer. The conference was held in Xi'an. Xi'an, located in central-northwest China, records the great changes of the country just like a living history book. Called Chang'an (meaning the eternal city) in ancient times, Xi'an is one of the birthplaces of the ancient Chinese civilization in the Yellow River Basin area. It is the eastern terminal of the Silk Road and the site of the famous Terracotta Warriors of the Qin Dynasty. More than 3,000 years of history, including over 1,100 years as the capital city of ancient dynasties, have endowed Xi'an with an amazing historical heritage.

We sincerely thank all the members of the ICAIC Organizing Committee, Academic Committee and Program Committee. Moreover, we also express our sincere appreciation to all the referees, staff and volunteers for their tremendous efforts. Finally, we would like to thank all the authors who have contributed to this volume.

We hope that readers will find lots of interesting and useful information in the proceedings.

June 2011

Dehuai Zeng
Jianwei Zhang
Jun Zhang

Table of Contents – Part V

Computing and Image Processing

Image Clustering Segmentation Based on Fuzzy Mutual Information and PSO*

Wei Jian-Xiang[1], Sun Yue-Hong[2], and Tao Zhao-Ling[3]

[1] Department of Information Science,
Nanjing College for Population Program Management 210042, China
[2] School of Mathematical Sciences, Nanjing Normal University, 210097, China
[3] College of Math & Physics, Nanjing University of Information Science and Technology, 210044, China
zhsina@yeah.net

Abstract. This paper proposes a new method for image clustering based on fuzzy mutual information. A new Particle Swarm Optimization based on diversity of symmetrical distribution (sdPSO) is adopt as optimization algorithm and the object function is based upon the fuzzy mutual information and cluster distance measure optimum between the original image and the one after clustering segmentation. Appling the method into a noised medical image, the result image has clear and continuous boundary and keeps complete internal characteristics. By comparing with the other image clustering algorithms, experimental results show that the image segmentation quality of this method is improved obviously.

Keywords: image clustering, fuzzy mutual information, particle swarm optimization.

1 Introduction

Image segmentation is one of the key techniques for image procession. The common methods are the thresholding method, the clustering method and the region growth method, and so on. Mutual Information (MI) is a new image clustring segmentation rule. Since 1995, MI has been widely used in image registration field. Rigau [1] and Kim [2] first tried to apply MI in image segmentation to get better effect than that over the maximum entropy rule. Zhen-tai Lu [3] proposed a segmentation algorithm based on the combination of the K-mean value algorithm and the maximum mutual information, the segmentation speed is improved by limiting the computation rang of the mutual information to a certain interval. Cheng-mao Wu [4] and Li-xin Liu [5] introduced the fuzzy membership into the mutual information, and put forward a multi-threshold segmentation method for fuzzy mutual information. Yu-dong Zhang [6] integrated fuzzy entropy and mutual information, gave the threshoding based on fuzzy mutual information difference standard with its performance superior to the traditional mutual information.

* This work is partially supported by National Social Science Foundation of China (09CTQ022) and the Sixth Project of Six Industries of JiangSu Province of China (09-E-016).

J. Zhang (Ed.): ICAIC 2011, Part V, CCIS 228, pp. 1–12, 2011.

In this paper, a new particle swarm optimization based on diversity of symmetrical distribution (sdPSO) is applied into image clustering. We bring forward a new image segmentation method, and the optimization object function is based on cluster distance measure and the fuzzy mutual information between the original image and the one after clustering segmentation.

2 The sdPSO

Particle swarm optimization (PSO), a novel swarm intelligence algorithm, is brought forward by Eberhart and Kennedy [7]in 1995 and proved to be a valid optimization technique and has been applied in many areas successfully. In PSO algorithm, the solution is usually supposed to be a particle in the search space for the problem being treated. Every particle consists of three parts: the present position x, the travel velocity v and the particle fitness, denoted by $P(x, v, fitness)$. In the process of the iteration, the particle renews itself by renewing two "extremes": one is the best solution found by the particle itself, we call it the personal acknowledge ability of the particle, denoted by P_{best}; another is the best solution found by the whole particle swarms, we call it the social acknowledge ability of the particle, G_{best}. With the two optimum found the position and the travel velocity of each particle can be renewed by the two formulas

$$v_i^{(t+1)} = \omega v_i^{(t)} + c_1 r_1 \left(P_{best\,i}^{(t)} - x_i^{(t)}\right) + c_2 r_2 \left(G_{best}^{(t)} - x_i^{(t)}\right) \tag{1}$$

$$x_i^{(t+1)} = x_i^{(t)} + v_i^{(t+1)} \tag{2}$$

where $P_{best\,i}^{(t)}$ is the history optimum for the i-th particle till the t-th generation search; $G_{best}^{(t)}$ is optimum for the particle swarm till the present; v_i is the present travel velocity of the ith particle; c_1 is the personal acknowledge ability coefficient, c_2 the social acknowledge ability coefficient; r_1, r_2 are random numbers in the interval [0,1].

In paper [8], we brought forward a new particle swarm optimization based on diversity of symmetrical distribution (sdPSO). We think the probability to converge to the global optimum can be increased by locating the particles around the optimum as symmetry as possible, thus decrease the probability to converge to the local optimum. Hence, the location symmetry extent of the particles in the high space is used to measure the diversity.

Definition 1. Suppose the space dimension is m, the size of the particle swarm is n, $X=(x_1,x_2,\ldots,x_n)^T$, and the i-th particle is $x_i = (x_{i1},x_{i2},\ldots,x_{im})$, the present particle swarm optimum is G_{best}: $(g_1,g_2,\ldots,g_m)$, so the diversity measure function is

$$Diversity(X) = \frac{1}{\sum_{j=1}^{m} \left| \sum_{i=1}^{n} (x_{ij} - g_j) \right| + 1} \tag{3}$$

The diversity function describes the deviation extent of the entire individual compared to the local optimum in each one-dimensional space. The smaller the deviation value is the more symmetry the particles locate around the local optima; the larger the diversity measure-function is, the better the swarm diversity is.

When the number of the iteration reaches a certain value, compute the *Diversity* according to formula (3), if $Diversity < \varepsilon$, $0 < \varepsilon < 1$, then the diversity adjustment is needed for the particle swarm. First, get value with the *j*-th (*j=1,2,…,m*) entry of each particle x_{ij} (*i*=1,2,…,*n*) minus the corresponding entry of the G_{best}, g_j, the number of the positive and the negative ones are denoted respectively n_1 and n_2. If there exists $|n_1 - n_2| > \mu(n)$, ($1 < \mu(n) < \frac{n}{2}$, $\mu(n)$ is an integer related to the size of the particle swarm),then choose $\text{int}(\frac{|n_1 - n_2|}{2})$ particles randomly among $|n_1 - n_2|$ particles. And then transform the chosen particles according to the formula:

$$x_{ij} = \begin{cases} g_j + Rand*|x_{ij} - g_j| & \text{if } x_{ij} - g_j < 0 \\ g_j - Rand*|x_{ij} - g_j| & \text{if } x_{ij} - g_j > 0 \end{cases} \tag{4}$$

3 Image Clustering Based on the Mutual Information and the sdPSO

In this section, a new optimization method is proposed; the optimization object is the mutual information quantity and the distance measure, the algorithm is based on the PSO method with particle symmetry and location diversity. Every particle in the swarm $X_i = (x_{i1}, x_{i2}, \cdots, x_{iC})$ denotes a set of clustering center vector.

A. FITNESS FUNCTION CHOICE

Mutual information quantity is one of the basic concepts in the information theory, describing the statistical mutual-dependence of the two systems, which can be considered as one system includes the information of another. During the image processing, mutual information (or fuzzy mutual information) describes the mutual-dependence of the two images, linking the images before and after segmentation. So we think mutual information should try to arrive at the maximum. Considering the three evaluation parameters, different fitness functions are given for different particles.

1)Fitness function based on maximum mutual information quantity

Assume F is a grey-scale image with size $M \times N$, where the grey-scale space $G = \{g_i \mid i = 1, 2, \cdots L\}$, and the one-dimensional grey-scale histogram is h_i. Suppose the clustering center vector is $Z = (z_1, z_2, \cdots, z_C)$, cluster the grey-scale space G of F to clustering C according to minimum distance rule.

$$Index(g_i) = Arg \min_{1 \le k \le c} d(g_i, z_k), \quad i = 1, 2, \cdots, L \tag{5}$$

where $Index(g_i)$ is the clustering sort for the grey-scale value g_i, and the grey-scale change to $g_i' = g_{Index(g_i)}$ after clustering.

Let $G' = \left\{ g_i' \mid i = 1, 2, \cdots L' \right\}$ be the grey-scale space of the image F', which is the segmentation image of F, and the one-dimensional grey-scale histogram is h_i'. There is the standard function for image clustering segmentation based on maximum mutual information.

$$(z_1^*, z_2^*, \cdots, z_C^*) = \underset{z_1, z_2, \cdots, z_C \in G}{\arg\max}\ MI(F, F') \tag{6}$$

with

$$MI(F,F') = \sum_{g_i \in G} \sum_{g_i' \in G'} p_{FF'}(g_i, g_i') \log \frac{p_{FF'}(g_i, g_i')}{p_F(g_i) \cdot p_{F'}(g_i')} \tag{7}$$

$$p_{FF'}(g_i, g_i') = \frac{h_i}{M \times N} \tag{8}$$

$$p_F(g_i) = \frac{h_i}{M \times N}, \quad p_{F'}(g_i') = \frac{h_i'}{M \times N} \tag{9}$$

Formula (7-9) respectively denotes the mutual information, joint probability and marginal probability about F and F'. Therefore, the particle fitness function is assumed as

$$fitness1 = \frac{1}{MIM(F, F')}. \tag{10}$$

2)Fitness function based on maximum fuzzy mutual information

In essence, fuzziness exists in each image, and fuzzy set theory can better describe the image fuzziness. Thence combine the fuzzy set theory with the mutual information. Fuzzy membership is introduced to define the fuzzy mutual information.For the image thresholding standard, the noise perturbation or non-uniform illumination may result in unsatisfactory result. So the fuzzy mutual information is applied to the image clustering aim to get more satisfactory segmentation result.

Define the fuzzy membership function $\mu(g_i, g_j')$ for the grey-scale g_i of image F and the grey-scale g_j' of image F' as

$$\mu(g_i, g_j') = \begin{cases} 1 - 0.5\left[\dfrac{|g_i - g_j'|}{L}\right]^\beta, & g_j' = g_i \\ \dfrac{1}{2(L-1)}\left[\dfrac{|g_i - g_j'|}{L}\right]^\beta, & g_j' \neq g_i \end{cases} \tag{11}$$

with parameter $\beta > 0$. It is easy to verify that the membership function satisfies

$$\sum_{g_j' \in G'} \mu(g_i, g_j') = 1, \quad g_i \in G \tag{12}$$

Hence, the standard function for image clustering segmentation based on fuzzy maximum mutual information is

$$(z_1^*, z_2^*, \cdots, z_C^*) = \underset{z_1, z_2, \cdots, z_C \in G}{\arg\max}\ FMI(F, F') \tag{13}$$

where the fuzzy mutual information is

$$FMI(F,F')=\sum_{g_i\in G}\sum_{g_i'\in G'}[\mu(g_i,g_i')]^m\, p_{FF'}(g_i,g_i')\log\frac{p_{FF'}(g_i,g_i')}{p_F(g_i)\cdot p_{F'}(g_i')} \tag{14}$$

where m expresses the fuzzy index, $m>0$, when $m=0$ the fuzzy mutual information degenerates to the mutual information.So the particle fitness function is assumed as

$$fitness2=\frac{1}{FMIM(F,F')} \tag{15}$$

3)Fitness function based on mutual information and optimum distance measure

Assume J_e is mean discretization degree, $d_{\max}$ is maximum intra-clustering distance,and $d_{\min}$ is minimum inter-distance[9], ie

$$J_e=\frac{\sum_{i=1}^{C}\left(\frac{\sum_{\forall x_k\in D_i}d(x_k,z_i)}{n_i}\right)}{C} \tag{16}$$

$$d_{\max}=\max_{1\le i\le C}\left\{\frac{\sum_{\forall x_k\in D_i}d(x_k,z_i)}{n_i}\right\} \tag{17}$$

$$d_{\min}=\min_{\substack{1\le i,j\le C\\ i\ne j}}\left\{d(z_i,z_j)\right\} \tag{18}$$

where $X=(x_1,x_2,\cdots,x_n)$ is the eigenvector of the image pixels (for the grey-scale image, the one dimensional eigenvector refers to the grey-scale value of the image), C denotes the clustering number, Z_i represents the i-th clustering centre vector, D_i the i-th clustering region, $d(x_k,z_i)=d_{ik}$ is the distance between the pixel eigenvector and the corresponding clustering centre, usually Euclidean distance is utilized. n_i is the sample number of the pixel eigenvector in the clustering region D_i .

The bigger the mutual information quantity between the original image and the obtained one after clustering segmentation, the smaller the J_e and $d_{\max}$, and the bigger the $d_{\min}$, the better the clustering effect. Thus the particle fitness function based on mutual information and cluster distance measure optimum can be defined as

$$fitness3=w_1d_{\max}+w_2(L-d_{\min})+w_3J_e+\frac{w_4}{MIM(F,F')} \tag{19}$$

In the same way the particle fitness function based on fuzzy mutual information and cluster distance measure optimum can be defined as

$$fitness4=w_1d_{\max}+w_2(L-d_{\min})+w_3J_e+\frac{w_4}{FMIM(F,F')} \tag{20}$$

B. FMIM-DIS-SDPSO ALGORITHM

Next we give steps for the fast image clustering algorithm based on fuzzy mutual information, the distance measure, and the Particle Swarm Optimization with Diversity of Symmetrical Distribution (FMIM-DIS- sdPSO).

Step 1. Given the original image F and the grey-scale space $G=\{g_i\}$, compute the one-dimensional grey-scale histogram $h_i, i=1,2,\cdots L$. Input the clustering classifications C , the particle number M , randomly choose the initial clustering center vector $X_i=(x_{i1},x_{i2},\cdots,x_{iC})$, $i=1,2,\cdots,M$. Then input the maximum iteration times G and the diversity adjustment threshold ε .

Step 2. With formula (20), compute the fitness of every particle, according to the fitness value, take the present best position P_{best_i} for each particle and the global best position G_{best} of the particle swarms.

Step 3. According to formula (1), update the speed V_i for each particle, and restrain the entries in interval $(-v_{\max}, v_{\max})$, if the updated speed is bigger than $v_{\max}$, let it be $v_{\max}$, if the updated is less than $-v_{\max}$, replace it with $-v_{\max}$.

Step 4. Update the location X_i for each particle according to formula (2), where the entries are restrained in $[0, L-1]$. For the particles beyond the range, randomly choose values within the interval for them.

Step 5. Update the fitness. For each particle, compare the fitness with that of the best experienced position P_{best_i} , if better; choose the particle position as the present best location. For every particle, compare the fitness with that of the global best experienced position G_{best} , if better; choose the particle position as the present global best location.

Step 6. When the iteration times reach a certain number, get the *Diversity* value according to formula (3); if $Diversity<\varepsilon$ $(0<\varepsilon<1)$, adjust the position of the particle according to formula (4).

Step 7. When the iteration times is G , stop the particle swarm algorithm, and segment the image F by the clustering center vector expressed as the global optimum G_{best} . Or else return to step2.

4 Simulation and Analysis

Select the brain MRI image (see figure 1(a)), add the salt and pepper noise with noise density 0.05 to the brain image (see figure 5(a)). Let the clustering number be 8. We simulate with various algorithms next.

A. Setting parameters

Assume the particle size is 50, the maximum iteration number $G = 100$. The linear decreasing weight is chosen as

$$\omega = \omega_{\max} - T * \frac{\omega_{\max} - \omega_{\min}}{G} \tag{21}$$

where T is the present iteration number. $w_{\max} = 2.0, w_{\min} = 0.8$, $c_1 = c_2 = 1.2$, $v_{\max} = 5$.

Considering MIM-DIS-sdPSO, for formula (19), select $w_1 = 0.04, w_2 = 0.01, w_3 = 0.95, w_4 = 0.19$. Let the diversity adjustment threshold $\varepsilon = 0.0002$, diversity adjustment start from the 5-th iteration.

Algorithms comparisons are done to describe the algorithm performance. MIM-DIS-sdPSO is compared with FCM, MIM-PSO, MIM-sdPSO and MIM-DIS-PSO. Formula (10) is utilized to compute particle fitness for those in MIM-PSO and MIM-sdPSO, no diversity adjustment for the former, the latter starts diversity adjustment from the 5-th iteration. Compared with MIM-DIS-sdPSO, MIM-DIS-PSO has not the step for the particle diversity adjustment. The rest parameters in the algorithms are the same as those in MIM-DIS-sdPSO.

B. Results comparisons and performance analysis

The brain MRI image with noise is turned into images illustrated in figure 1 (b)-(j) after clustering segmentation by FCM, MIM-PSO, MIM-sdPSO, FMIM-PSO, FMIM-sdPSO, MIM-DIS-PSO, MIM-DIS-sdPSO, FMIM-DIS-PSO and FMIM-DIS-sdPSO, respectively. Table 1 lists the corresponding result comparison. Fuzzy mutual

Table 1.

Brain MRI image with noise	J_e	d_{max}	d_{min}	**MI**	**Clustering center vector**
FCM	7.1561	13.7615	13.9592	0.4473	(1, 24, 85 , 161, 249, 17, 71, 14)
MIM-PSO	11.4314	51.0103	1.6241	0.4843	(0, 2,75,76, 88, 106, 120, 172)
MIM-sdPSO	9.3457	34.9564	0.7198	0.5078	(0, 1, 2, 74, 84, 97, 127, 196.)
FMIM-PSO	8.0544	20.4618	2.9492	0.4833	(0,3, 77, 80,99, 115, 129, 220)
FMIM-sdPSO	7.8777	17.1661	1.2248	0.5066	(0, 1, 2, 71, 88, 99, 128, 237)
MIM-DIS-PSO	5.9934	7.8698	24.2025	0.3956	(0,81,111,135,162, 190, 225, 255)
MIM-DIS-sdPSO	5.7708	9.9172	19.7227	0.3985	(0,80,112,135,174, 214, 235,255)
FMIM-DIS-PSO	5.9697	7.8284	24.2059	0.2878	(0, 80,111, 136,164, 190, 225, 255)
FMIM-DIS-sdPSO	6.0079	7.4584	24.4485	0.2883	(0, 80,111, 136,167, 194, 225,255)

Result comparison for brain MRI image with noise after various clustering segmentations information is utilized in the fitness computation for the four algorithms: FMIM-PSO, FMIM-sdPSO, FMIM-DIS-PSO and FMIM-DIS-sdPSO. As to FMIM-PSO and FMIM-sdPSO, the membership parameter is $\beta = 1.5$, the fuzzy index $m = 0.95$. The diversity adjustment threshold is selected $\varepsilon = 0.00013$ for MIM-sdPSO and FMIM-sdPSO, and diversity adjustment starts from the first iteration. MIM-DIS-sdPSO starts diversity adjustment from the fifth iteration with $\varepsilon = 0.00013$. FMIM-DIS-PSO and FMIM-DIS-sdPSO start to adjust the diversity from the 25-th iteration with parameters $\beta = 0.1$, $m = 0.9$ and $\varepsilon = 0.00012$.

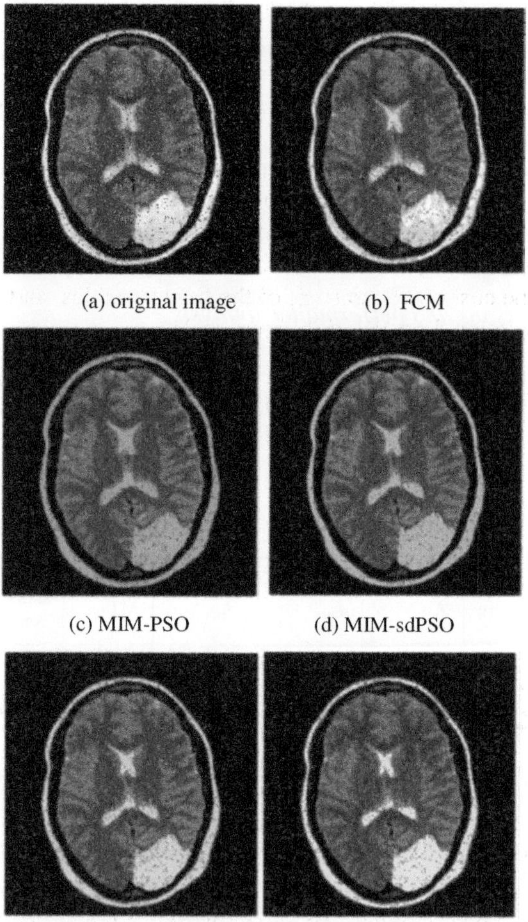

(a) original image (b) FCM

(c) MIM-PSO (d) MIM-sdPSO

(e) FMIM-PSO (f) FMIM-sdPSO

Fig. 1. Clustering segmentation for the brain MRI image with the salt and pepper noise: (a) original image; (b) FCM; (c) MIM-PSO; (d) MIM-sdPSO; (e) FMIM-PSO; (f) FMIM-sdPSO; (g) MIM-DIS-PSO; (h) MIM-DIS-sdPSO; (i) MIM-DIS-PSO; (j) MIM-DIS-sdPSO

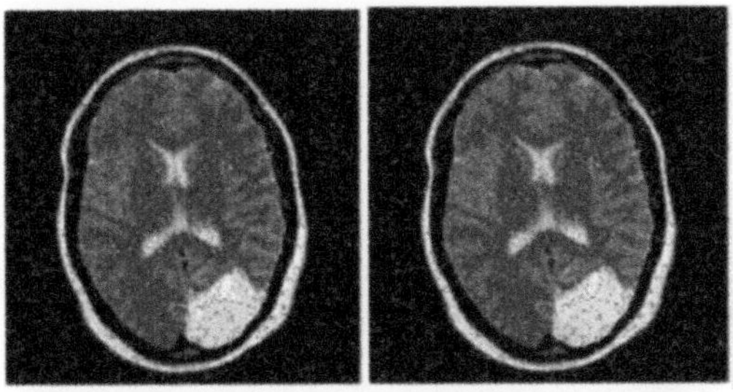

(g) MIM-DIS-PSO (h) MIM-DIS-sdPSO

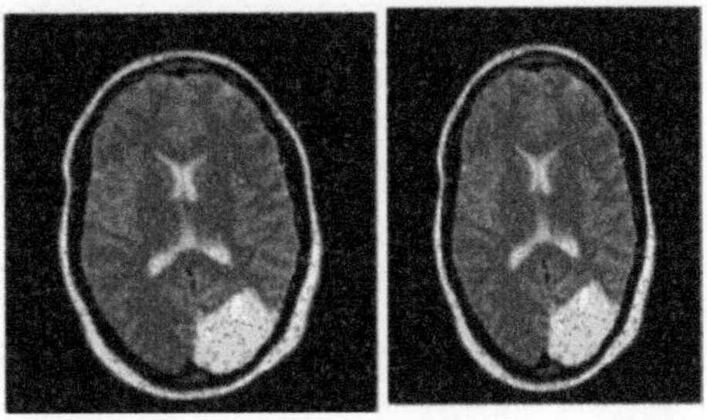

(i)FMIM-DIS-PSO (j) FMIM-DIS-sdPSO

Fig. 1. (*Continued*)

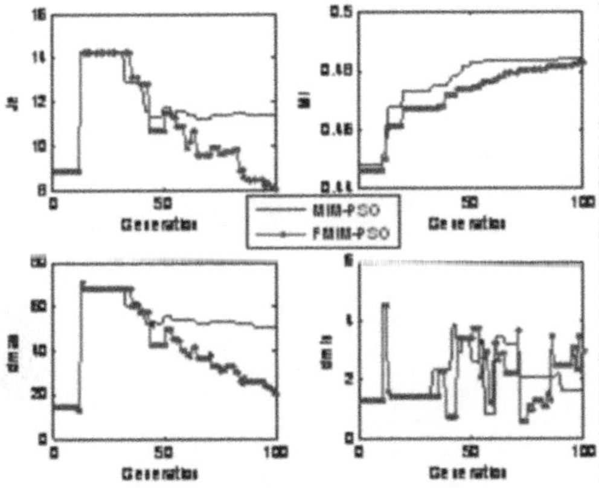

Fig. 2. Segmentation performance comparisons between MIM-PSO and FMIM-PSO for the image with noise

From table 1 , we know the particle fitness is measured only with mutual information or fuzzy mutual information for the four algorithms: MIM-PSO, MIM-sdPSO, FMIM-PSO and FMIM-sdPSO; and the resulted image have very small Mini-inter-distance after the clustering segmentation. The segmentation effect for the brain MRI image with noise is far better by FMIM-PSO and FMIM-sdPSO than those by MIM-PSO and MIM-sdPSO as shown in figures 2 and 3. Therefore, fuzzy mutual information is better suitable for the image with noise.

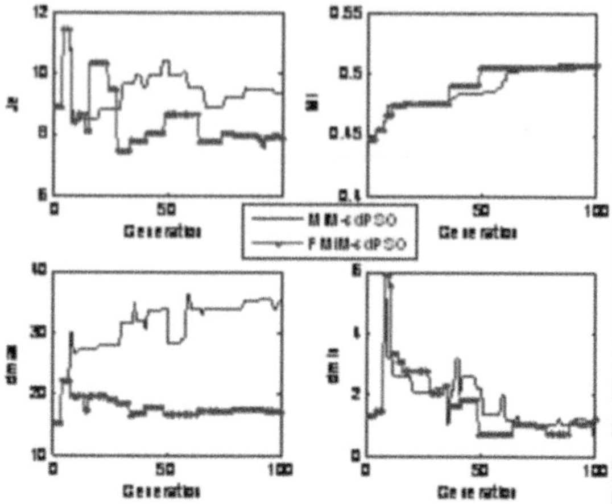

Fig. 3. Segmentation performance comparisons between MIM-sdPSO and FMIM-sdPSO for the brain MRI image with noise

As shown in table1, the evaluation indexes are good after clustering segmentation for the four algorithms: MIM-DIS-PSO, MIM-DIS-sdPSO, FMIM-DIS-PSO and FMIM-DIS-sdPSO, since not only mutual information or fuzzy mutual information but also the cluster distance measure are integrated into the particle fitness. Compared with MIM-DIS--PSO, when FMIM-DIS-PSO is applied to the brain MRI image with noise, the result has less mutual information while the three evaluation indexes have no distinct change, see figure 4. d_{max} and d_{min} are improved if FMIM-DIS-sdPSO is used for the brain MRI image with noise in comparison with those by MIM-DIS-sdPSO, refer to figure 5. Thus, as for noised image clustering segmentation, when the cluster distance measure is utilized for the particle fitness, there is not much

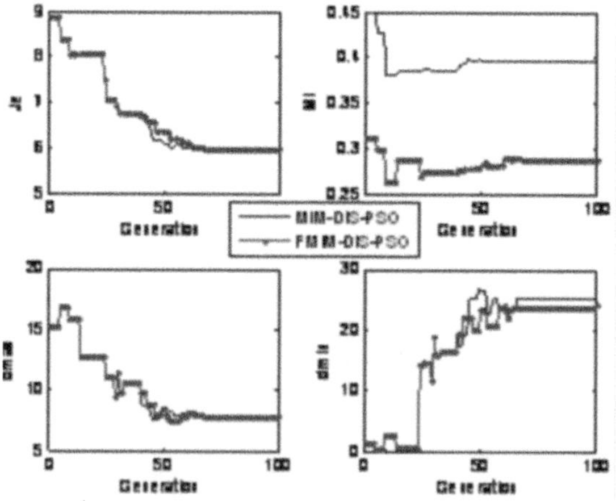

Fig. 4. Segmentation performance comparisons between MIM-DIS-PSO and FMIM-DIS-PSO for the brain MRI image with noise

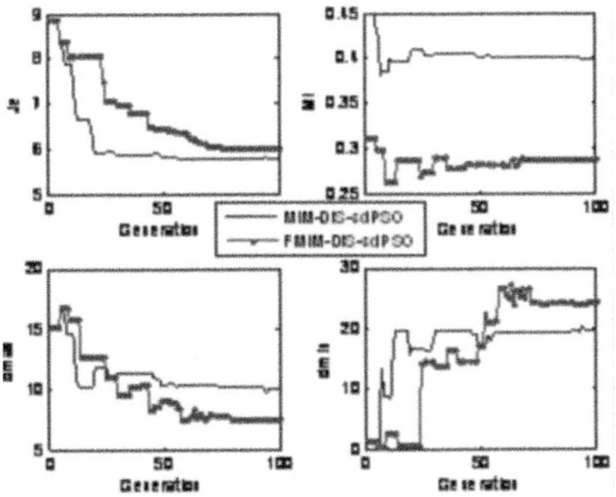

Fig. 5. Segmentation performance comparisons between MIM-DIS- sdPSO and FMIM-DIS-sdPSO for the brain MRI image with noise

difference whether mutual information or fuzzy mutual information is adopted. Comprehensive consideration on the evaluation parameters and mutual information, MIM-DIS-sdPSO also has nice effect with the noised image.

More medical images are involved in the segmentation experiment and more contrasts are done, satisfactory results are obtained.

5 Conclusions

FMIM-DIS-sdPSO proposed here is an image clustering algorithm based on fuzzy mutual information and cluster distance measure optimum, considering not only the grey-scale information, the spatial information and fuzzy mutual information before and after segmentation for the image but also the object functions optimization with the help of the particle swarm based on diversity of symmetrical distribution. A great quantity of contrast experiment and simulation discloses that the clustering performance of FMIM-DIS-sdPSO is improved obviously comparing with those of FCM, MIM-PSO or others. The image after clustering segmentation with the new algorithm has accurate object information, owns continuous and clear image boundary, and keeps complete internal image characteristics. So the new algorithm is suitable for noised image segmentation.

References

1. Rigau, J., Feixas, M., Sbert, M., Bardera, A., Boada, I.: Medical image segmentation based on mutual information maximization. In: Barillot, C., Haynor, D.R., Hellier, P. (eds.) MICCAI 2004. LNCS, vol. 3216, pp. 135–142. Springer, Heidelberg (2004)
2. Kim, J.: A Nonparametric Statistical Method for Image Segmentation Using Information Theory and Curve Evolution. IEEE Trans. on Image Process 14(10), 1486–1502 (2005)

3. Lu, Z.-t., Lü, Q.-w., Chen, W.-f.: Unsupervised segmentation of medical image based on maximizing mutual information. Journal of Image and Graphics 13(4), 658–661 (2008)
4. Wu, C.-m.: Fuzzy mutual information and its application in image segmentation. Computer Engineering 34(7), 218–220 (2008)
5. Liu, L.-x.: Image multi-threshold method based on fuzzy mutual information. Computer Engineering and Applications 45(25), 166–168 (2009)
6. Zhang, Y.-d., Wu, L.-n., Wei, G., Wu, H.-q., Guo, Y.-l.: Maximum fuzzy mutual information used for image segmentation. Computer Engineering and Applications 45(20), 1–5 (2009)
7. Kennedy, J., Eberhart, R.C.: Particle Swarm Optimisation. In: Proceedings of IEEE International Conference on Neural Networks (ICNN), Perth, Australia, vol. (4) (1948)
8. Sun, Y.-H., Wei, J.-X., Xia, D.-S.: An improved PSO based on diversity of particle symmetrical distribution. Pattern Recognition and Artificial Intelligence 23(2), 137–143 (2010)
9. Omran, M., Engelbrecht, A.P.: Particle Optimization Method for Image Clustering. International Journal of Pattern Recognition and Artificial Intelligence 19(3), 297–321 (2005)

Practical OFDM Synchronization Algorithm Based on Special Preamble Structure

Kai Zhong, Guangxi Zhu, and Desheng Wang

Department of electronic and information engineering,
Huazhong University of Science and Technology
Wuhan, 430074, China
KaiZhong119@tom.com

Abstract. Synchronization for both time and frequency is a important technology in OFDM systems. In this paper, we design a novel synchronization algorithm based on special preamble structure which includes timing synchronization, fraction frequency offset estimation and integer frequency offset estimation. All synchronization procedure can be completed in time domain, so the algorithm is simple and practical. Simulation results show that the timing synchronization has a high performance even at low SNR in multipath channel and the frequency offset synchronization have a dynamical estimation range and a satisfactory performance.

Keywords: synchronization, OFDM, special preamble structure， practical.

1 Introduction

Orthogonal frequency division multiplexing(OFDM) which is one of the most promising techniques for the next-generation wireless communication system has recently received a lot of attention due to its high spectral efficiency and its robustness to combating frequency-selective fading[1]. However, the performance of OFDM receivers is very sensitive to timing synchronization and frequency offset estimation, because loss of timing synchronization and frequency offset estimation introduces Inter Symbol Interference(ISI) and Inter Carrier Interference(ICI) and can severely degrade the performance of OFDM receivers[2][3].

Various techniques have been proposed in the literature for OFDM system. Preamble-aided methods exploit the periodic nature of the time domain preamble symbols, which are suitable for burst-mode OFDM applications [4]-[6]. CP methods exploit the correlation between the cyclic prefix and its data replica [7], but their performance is degraded in multipath environments. Blind techniques have also been studied which do not rely on cyclic prefix or preamble information[8][9], however, the processing delay and computational complexity of these algorithms is high, making them unfeasible for high delay spread environments.

In this paper, we propose a novel technique that uses only one preamble symbol with a special structure, to achieve robust, low-complexity and dynamical estimation range time-frequency OFDM synchronization in the time-domain. It combines

J. Zhang (Ed.): ICAIC 2011, Part V, CCIS 228, pp. 13–20, 2011.

autocorrelation, symmetrical correlation and PN scramble to achieve reliable synchronization accuracy in both

AWGN and multipath fading channels. It is also able to achieve faster convergence than conventional techniques for all synchronization procedure can be completed in time domain.

This work is supported by the National High Technology Research and Development Program of China under Grant 2008AA01Z204, by National key project under grant No. 2010ZX03003-001-002 & No.2009ZX03003-003-002 and by the China National Science Foundation under Grant No.60802009.

The rest of this paper is organized as follows. In Section 2, we briefly introduces OFDM system model. The proposed synchronization algorithm based on special preamble structure is presented in Section 3 and the simulation results are shown in Section 4. Finally, we conclude the paper in Section 5.

2 OFDM Signal Model

In an OFDM system, the complex baseband samples of transmission signal can be expressed as:

$$x(n)=\frac{1}{\sqrt{N}}\sum_{k=0}^{N-1}x(k)\exp\left(j\frac{2\pi kn}{N}\right) \quad -N_g\leq n\leq N-1 \tag{1}$$

where $x(k)$ represents data in the frequency domain before inverse fast Fourier transform(IFFT), $x(n)$ represents data in the time domain after inverse fast Fourier transform, N is the numbers of the subcarrier, and N_g denotes the length of the cyclic prefix(CP).

The channel between the transmitter and the receiver can be modeled as a wide-sense stationary, uncorrelated scattering, fading Rayleigh channel. Therefore, the baseband samples of received signal can be represented as:

$$r(n)=e^{\frac{j2\pi\varepsilon n}{N}}\sum_{l=0}^{L-1}x(n-l-\tau)h_l+w(n) \tag{2}$$

where L is the path number; h_l is the channel impulse response of the *lth* path; τ is the integer-valued unknown timing offset; ε is the frequency offset; and $w(n)$ is the complex Guassian noise with zero mean and variance σ_n^2.

3 Synchronization Algorithm

1. Preamble structure

In this section, we describe a special preamble structure apply to OFDM timing and frequency synchronization, as depicted in Fig.1. The whole preamble is divided into four parts, serially denoted as X, Y, X, Z.

X is a chu sequence with the length N_c used in this paper have the following form:

$$X(n) = \exp(j\pi\gamma n^2 / N_c) \tag{2}$$

where $N_c = N/4$ and $\gamma = N_c - 1$, $0 \le n < N_c$.

Let sequence M denotes the reversed and conjugated version of sequence X, which can be expressed as:

$$M(n) = X(N_c - n - 1) \qquad 0 \le n < N_c \tag{3}$$

In order to improve the accuracy of timing synchronization algorithm in multipath channels, we introduce a PN sequence to scramble the phase of M. It is assumed that P is a pseudo-random sequence with the length N_c, the element of P is randomly selected from the collection:

$$1/\sqrt{2}\{1+j, 1-j, -1+j, -1-j\}$$

So sequence Y has the following form:

$$Y(n) = M(n) * P(n) \qquad 0 \le n < N_c \tag{4}$$

Z is a geometric sequence which is adopted here to execute integer frequency offset estimation.

$$Z(n) = \exp(j2\pi n / N_c) \quad 0 \le n < N_c \tag{5}$$

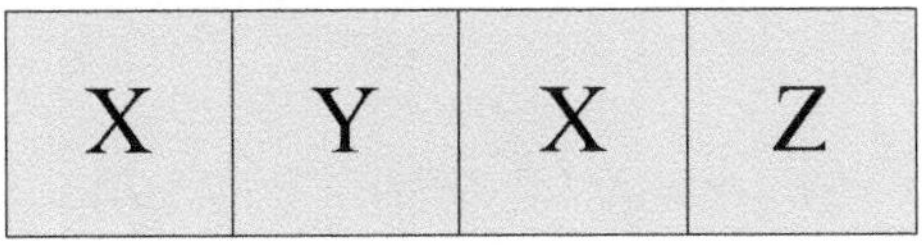

Fig. 1. The proposed preamble structure

2. Timing synchronization

Based on the special preamble structure, we can propose our timing synchronization metric as follows:

$$C(d) = \sum_{n=0}^{N/4-1} r(n+d) * r(d+N/2-1-n) * P^*(N/4-1-n) \tag{6}$$

$$E(d) = \sum_{n=0}^{N/4-1} |r(d+N/2-1-n)|^2 \tag{7}$$

$$T(d) = \frac{C(d)}{E(d)} \tag{8}$$

$$\tau = \arg\max_{d}(T(d)) \tag{9}$$

Here, P^* is the conjugated version of PN sequence P.

3. Fraction frequency offset estimation

Since the timing synchronization is completed, the fraction frequency offset can be calculated by

$$\varepsilon_f = -\frac{1}{\pi}\angle A(\tau) \tag{10}$$

$$A(\tau) = \sum_{n=0}^{N/4-1} r(\tau+n) * r^*(\tau+n+N/2) \tag{11}$$

where ε_f is estimated fractional frequency offset, $\angle$ denotes the argument of a complex number.

After the estimation of fraction frequency offset, the received signal must be compensated as follows

$$r'(n) = r(n) * \exp(-2\pi\varepsilon_f n/N) \tag{12}$$

4. Integer frequency offset estimation

Consider sequence Z which is a geometric sequence, then integer frequency offset can be estimated as

$$\varepsilon_i = round(-\frac{N}{2\pi d}\angle B(\tau)) \tag{13}$$

$$B(\tau) = \sum_{n=0}^{N/4-d-1} r'(n+\tau+3N/4) * r'^*(n+\tau+3N/4+d) * q \tag{14}$$

$$q = \exp(j2\pi/N_c) \tag{15}$$

where d is the distance of two correlated blocks, $round(n)$ denotes round n to the nearest integers.

From (13), it is obvious that the integer frequency offset estimation range is $[-N/2d, N/2d]$. When d is small, the estimation range is large, whereas the estimation range is small.

As a result, the whole CFO estimate is $\varepsilon = \varepsilon_i + \varepsilon_f$.

4 Simulation Results

This subsection shows a number of simulation results to evaluate the performance of the timing and frequency synchronization schemes described in Section 3. Simulation parameters are given in Table 1.

Table 1. Simulation Parameters

System Parameter	Parameter Value
Modulation	QPSK
Number of subcarriers, N	1024
Length of cyclic prefix, G	64
Number of multipath	6
Path gains	[-3, 0, -2, -6, -8, -10]dB
Symbol timing error	20
The normalized frequency offset	3.2 and 15.4
Simulation cycles nloop	5000

Figure 2 shows the timing metric of the proposed estimator under AWGN channel, when SNR is 0dB. In contrast with Schmidl algorithm which has a platform phenomenon, the figure obviously shows an impulse-shaped peak at the start of preamble(except CP), which makes the proposed estimator to be more accurate in time offset estimation.

The mean square error(MSE) reflect the variance of the estimation. Figure 3 shows the MSE of the timing offset estimation in multipath fading channels. It can be seen from Figure 3 that the MSE of proposed method is 0 when SNR is 1dB, which will result in prominent enhancement in OFDM receiver performance.

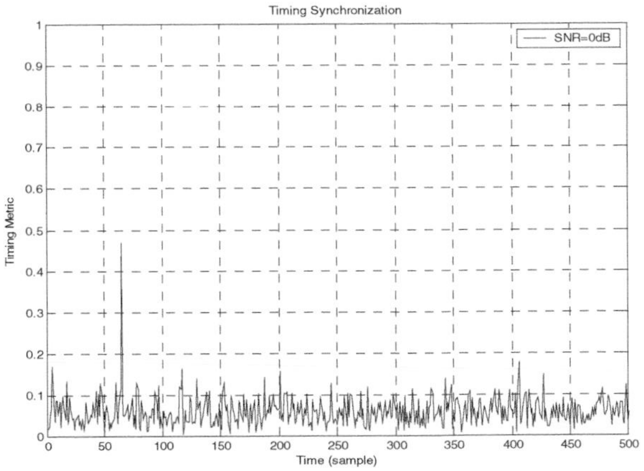

Fig. 2. Timing metric of the proposed estimator under AWGN channel

Figure 4 and Figure 5 show the MSE of the frequency offset estimation in multipath fading channels, when the normalized frequency offset is 3.2 and 15.4, respectively. The simulation results make is clear that the performance of smaller frequency offset estimation is similar to the performance of larger frequency offset estimation.

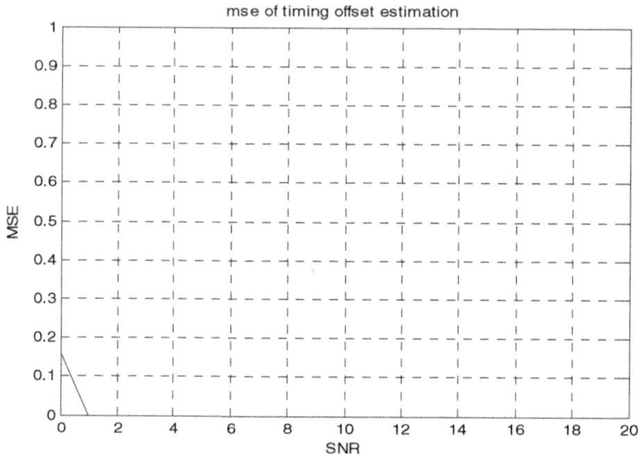

Fig. 3. MSE of the timing offset estimation under multipath channels

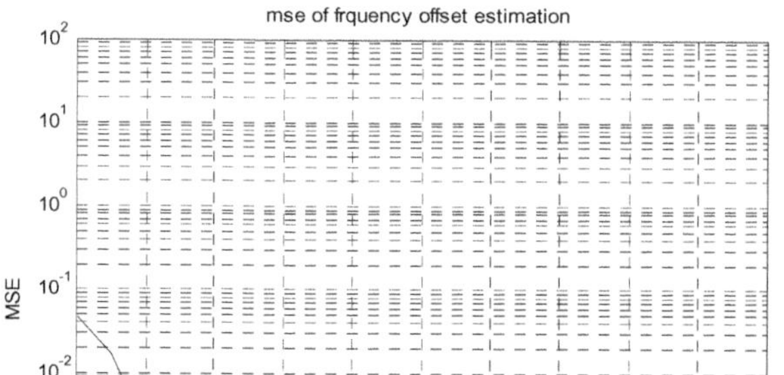

Fig. 4. MSE of the frequency offset estimation under multipath channels, ε =3.2

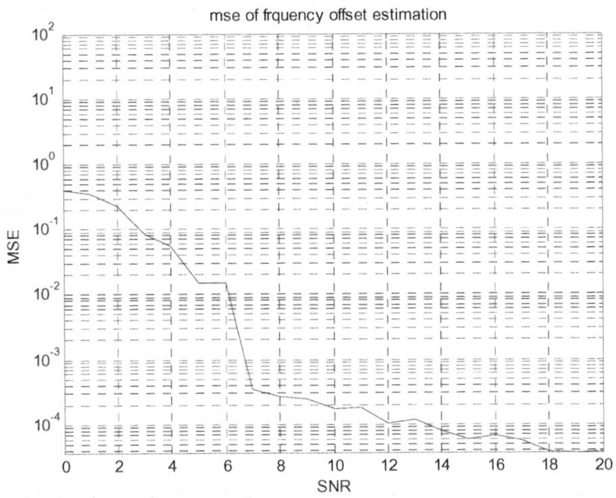

Fig. 5. MSE of the frequency offset estimation under multipath channels, ε =15.4

5 Conclusion

This paper puts forward a novel timing synchronization and frequency offset estimation algorithm for OFDM systems.

It combines autocorrelation, symmetrical correlation and PN scramble to achieve reliable synchronization accuracy in both AWGN and multipath fading channels. It is also able to achieve faster convergence than conventional techniques for all synchronization procedure can be completed in time domain. Simulation results show that the timing synchronization has a high performance even at low SNR in multipath channel and the frequency offset synchronization have a dynamical estimation range and a satisfactory performance.

References

1. van Nee, R., Prasad, R.: OFDM for multimedia communications. Artech House, Prentice Hall (2000)
2. Speth, M., Fechtel, S.A., Fock, G., Meyr, H.: Optimum Receiver Design for Wireless Broad-Band Systems using OFDM Part-I. IEEE Tran. Commun. 47, 1668–1677 (1999)
3. Mostofi, Y., Cox, D.C.: A Robust Timing Synchronization Design in OFDM Systems – Part I: Low Mobility Case. IEEE Trans. Wireless Commun. 6, 4329–4339 (2007)
4. Schmidl, T.M., Cox, D.C.: Robust Frequency and Timing Synchronization for OFDM. IEEE Trans. Commun. 45, 1613–1621 (1997)
5. Yi, G., Gang, L., Jianhua, G.: A Novel Timing and Frequency Synchronization Scheme for OFDM Systems. In: Wicom, pp. 393–397 (September 2007)
6. Wu, M., Zhu, W.-P.: A Preamble-Aided Symbol and Frequency Synchronization Scheme for OFDM systems. In: IEEE International Symposium on Circuits and Systems, ISCAS 2005, vol. 3, pp. 2627–2630 (2005)

7. Hsieh, M., Wei, C.: A low-complexity frame synchronization and frequency offset compensation scheme for OFDM systems over fading channels. IEEE Trans. Veh. Technol. 48, 1596–1609 (1999)
8. Park, B., Chenon, H., Ko, E., Kang, C., Hong, D.: A blind OFDM synchronization algorithm based on cycle correlation. IEEE Signal Process. Lett. 11, 83–85 (2004)
9. Bolcskei, H.: Blind estimation of symbol timing and carrier frequency offset in wireless OFDM systems. IEEE Trans. Commun. 49, 988–999 (2001)

Crime Location Prediction Based on the Maximum-Likelihood Theory

Su Mingche[1], Li Hanyu[2], Qin Yiming[3], and Zhang Xiaohang[4]

[1] School of Computer Science and Technology, Jilin University
[2] School of economics, Jilin University
[3] School of mathematics, Jilin University
[4] School of physics, Jilin University
sumingche@sina.cn

Abstract. In this paper, we are required to construct models to predict the offender's location. Initially, we take a rough anchor point locating based on psychological analysis of offenders. Then we construct a mathematical model based on the maximum-likelihood theory to make accurate predictions about the anchor point and next crime locations. Then we choose an actual crime cases to test the reliability and utility of our model. Finally, a conclusion that only when those crime sites are evenly distributed around a point can be drawn and accurate mathematical predictions can be realized;

Keywords: crime-prediction maximum-likelihood, crime psychology.

1 Introduction

As it is said by Dr. Kim Rossmo, who is a detective inspector in charge of the Vancouver Police, the three most important elements are location, location, location. Crime scene locations can give us some important insights into where an offender may be based, and help us focus the criminal investigative process. That's what geographic profiling is all about.

Geographic profiling is an investigative aid that predicts the serial offender's most likely location including home, work, social venues, and travel routes. Using information from a series of related crimes, a geographic profiler uses a mathematical model to analyze the locations of the crimes and the characteristics of the local neighborhoods in order to produce a map showing the areas in which the offender most likely lives and works. It helps focus investigative efforts on the probable location of the suspect, within an area defined by the 'geoprofile'.

2 Predictions of Home Location Based on Psychological Analysis

A. The Center Theory and Application

The "median centre", also known as the centre of minimum travel, is a measure of central tendency in point patterns and is found by locating the position from which travel to all points in a spatial distribution is minimized.

J. Zhang (Ed.): ICAIC 2011, Part V, CCIS 228, pp. 21–29, 2011.

With the increasing popularity of geographic profiling, it attracts great attention. So, there are various strategies of measuring the accuracy of the predictions. The minimum center is the basis for calculating the standard distance of a point pattern, a measure of spatial dispersion analogous to the standard deviation. When used with the mean centre it can help describe two-dimensional distributions, and through the concept of relative dispersion, allow for comparisons of spread between different sets of points.

The center of minimum distance of a given set of points in two dimensional Euclidean spaces is the location at which the sum of the distances to all other points is the smallest.

And

$$W(\bar{x},\bar{y}) = \sum_{i=1}^{n} dist\left((x_i,y_i),(\bar{x},\bar{y})\right)$$

$$i = 0, 1, 2, 3 \ldots$$

Obtaining a precise value for the complexity of this strategy was more difficult than other spatial distribution strategies. The problem of minimizing W is not a trivial task, and it has a long history. The function, W, is convex therefore it has a unique minimum in the convex hull of the set of given points.

B. The Basic and Developed Circle Theory

At the most basic level, a geographic profile is a prediction of the most likely location at which the search for a serial offender's home should commence. Traditionally, the basic circle theory states that the offenders probably live within an area circumscribed by their offenses. However, in the south of England for example, using advanced technology, scientists Canter and Larkin used a theory to show that 87% of the 45 serial rapists they studied within a circle defined by a diameter drawn between that offender's farthest offenses. Just basing on this theory, considering the most common situation, we may estimate the offender's living area roughly based on the theory as the following graph.

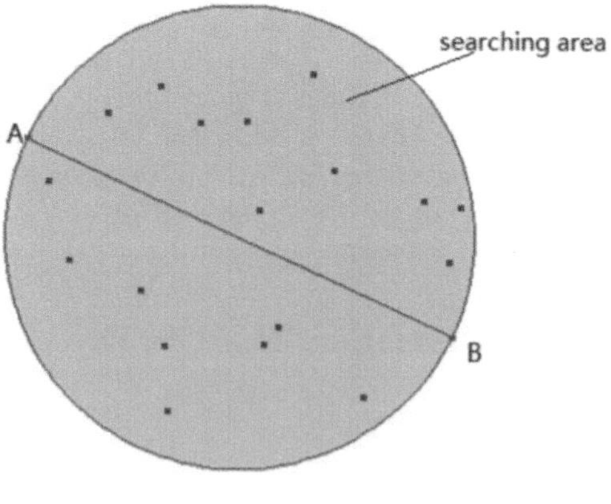

Fig. 1. Searching area depending on basic circle theory

Considering the accuracy of the model, we may revise the circle hypothesis to include all the circles surrounded by any of the two offense sites. We can define the large area covered by all the circles as awareness space, and it indeed enlarges the awareness area on the basis of the need for police forces with limited resources and technological capabilities to invest in geographic profiling systems. However, it would prove more accurate and reasonable. Now we may set the boundaries and delineate the awareness area. All the area painted with the color would be useful.

Then we may consider the area surrounded by the center of all the circles as the crucial searching area. It would be more effective and meaningful to take precedence to search this area when it comes to the special location.

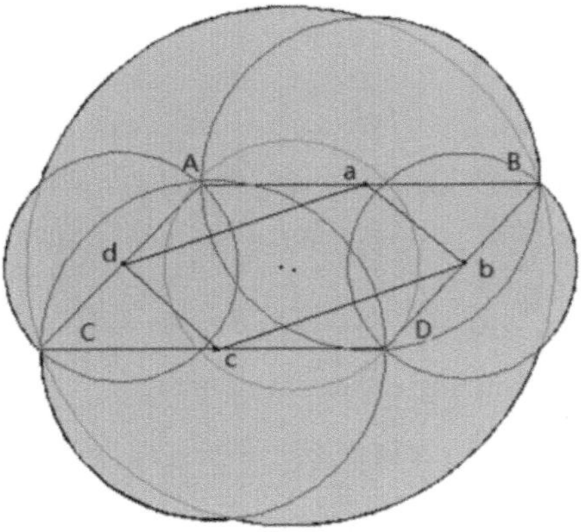

Fig. 2. Searching area based on developed circle theory

As the graph above, the dots A, B, C, D, are the four offense sites. The largest circle-like area painted with the color is the awareness area. And the quadrilateral surrounded by a, b, c, d is the crucial searching area only the half of the space surrounded by A, B, C, D. It could save more the limited force of police just to search the quadrilateral surrounded by a, b, c, and d rather than awareness area. In other words, the offenders would more probably live within the crucial searching area than awareness area.

3 Crime Predictions Based on Maximum-Likelihood Theory

A. Basic Analysis and Symbols

First we may consider the correlation between the former offense sites and next crime location. It can be inferred that at a certain distance d_0, the greater distance from former offense sites to the future crime locations are, the probability of crime occurrence would decrease. Therefore it is the same with the factor of time.

This analysis could be quite reasonable and reliable when it comes to the reason of this correlation. The probability of repeating offenses at the same place in a relatively short time interval would be very tiny.

s_i: different crime offense sites

d: distance

a, b: coefficients of the formulas

d_0: a special but crucial value determined by the former crime dat.

B. The mathematics prediction model

Suppose only that the offender chooses to offend at the new site *d* with probability density $Q(s_{n+1}|s_i, d_0)$ determined by t_i, d_0 and s_i.

$$Q(s_{n+1}|s_i, d_0)\begin{cases} ade^{b(t_i - t_1)} & d \le d_0 \\ \frac{1}{d - d_0} & d > d_0 \end{cases} \quad (i \le n) \tag{1}$$

Then we propose the following formulas to estimate the value of d_0 on the basis of the data of former crimes. Given a sample $s_1, s_2, \ldots, s_n$, (the crime sites) from a probability distribution $Q(s_{n+1}|s_i, d_0)$ and construct likelihood function

$$L(d_0) = \prod_{i=1}^{n-1} Q(s_n|s_i, d_0) \tag{2}$$

This is equivalent to maximizing the log-likelihood

$$\Lambda(d_0) = \ln \prod_{i=1}^{n-1} Q(s_n|s_i, d_0) \tag{3}$$

Then the best choice of d_0 that makes the likelihood as large as possible. Ultimately, we may get the value of d_0.

Based on the former probability model, we may imply that $P(s|z, \alpha)$:

$$P(s_i|z, \alpha) = \frac{1}{4\alpha^2} e^{-\frac{\pi}{4\alpha^2}|s_i - z|^2} \quad (i \le n) \tag{4}$$

Where an offender with anchor point z and mean offense distance α commits an offense at the location *x* with probability density.

Considering the probability function of the selection of new crime location $R(d)$, it is determined by the two factors, offender's anchor point and the former crime data. Neglecting some tiny, non-quantifiable and too complicated factors, we just construct the function as follows

$$R(d) = P(s|z, \alpha) \cdot \prod_{i=1}^{n} Q(s_{n+1}|s_i, d_0) \tag{5}$$

And *d* represents the location of the new crimes occurrence. Then we would seek the value of *d* which makes $R(d)$ reaches the maximum.

C. The Solution Process and Result

$d^2 = (x_{n+1} - x_i)^2 + (y_{n+1} - y_i)^2$,

$$z = (x_0, y_0) \tag{6}$$

$$R(x,y) = \frac{1}{4\alpha^2}\sum_{i=1}^{n}\phi(i)\cdot\prod_{i=1}^{n}\alpha\varphi(i) \tag{7}$$

$$\phi(i) = e^{-\frac{\pi}{4\alpha^2}[(x_i-x_0)^2+(y_i-y_0)^2]}$$

$$\varphi(i) = \sqrt{(x_{n+1}-x_i)^2+(y_{n+1}-y_i)^2}\,e^{b(t_i-t_0)}$$

$$\begin{cases} R_x(x,y) = 0 \\ R_y(x,y) = 0 \end{cases} \tag{8}$$

The result of *(x, y)* are the location where the next crime would most probably happen.

D. Anchor Point Prediction

Base on the assumption that an offender with anchor point z and mean offense distance α commits an offense at the location x with probability density $p(s_i|z,\alpha)$, we get as follows

$$P(s_i|z,\alpha) = \frac{1}{4\alpha^2}e^{-\frac{\pi}{4\alpha^2}[(x_i-x_0)^2+(y_i-y_0)^2]} \tag{9}$$

$$\alpha^2 = \sum_{i=1}^{3}\frac{(x_i-x_0)^2+(y_i-y_0)^2}{3}$$

$$= x_0^2 + y_0^2 - \frac{2}{3}x_0 - \frac{2}{3}y_0 + 2 \tag{10}$$

Then we construct the likelihood function

$$f(x,y) = \prod_{i=1}^{3}P(s_i|z,\alpha)$$

$$= \prod_{i=1}^{3}\frac{1}{4\alpha^2}e^{-\frac{\pi}{4\alpha^2}[(x_i-x_0)^2+(y_i-y_0)^2]} \tag{11}$$

This is equivalent to maximizing the log-likelihood

$$g(x,y) = \ln f(x,y) = \ln\frac{1}{4\alpha^2}\cdot\sum_{i=1}^{3}\{\psi(i)\} \tag{12}$$

$$\psi(i) = -\frac{\pi}{4\alpha^2}[(x_i-x_0)^2+(y_i-y_0)^2]$$

To get the maximum of $g(x,y)$, we seek the partial derivative of x and y

$$\begin{cases} g_x(x,y) = 0 \\ g_y(x,y) = 0 \end{cases} \tag{13}$$

Then we solve the equation and get the final conclusions as follows

$$\begin{cases} x_0 = \frac{1}{3} \\ y_0 = \frac{1}{3} \end{cases} \tag{14}$$

When $z = \left(\frac{1}{3}, \frac{1}{3}\right)$, $g(x,y)$ would reach the maximum.

E. Offense Locations Prediction

Considering the mean distance α, we put the $(x_0, y_0) = \left(\frac{1}{3}, \frac{1}{3}\right)$ into the following equation and get the results as follows

$$\alpha^2 = \sum_{i=1}^{3} \frac{(x_i - x_0)^2 + (y_i - y_0)^2}{3}$$

$$= x_0^2 + y_0^2 - \frac{2}{3}x_0 - \frac{2}{3}y_0 + 2$$

$$= \frac{16}{9} \tag{15}$$

At the same time, we simply suppose the crime time as follows

$$t_1 = 1, t_2 = 2, t_3 = 3,$$

Considering the probability function of the selection of new crime location $R(d)$,$d = d(x,y)$, we get the following solution process as follows:

$$R(x_4, y_4) = \frac{1}{4\alpha^2} \sum_{i=1}^{3} \phi(i) \cdot \prod_{i=1}^{3} \alpha\varphi(i) \tag{16}$$

$$= \frac{9}{16} e^{-\frac{9}{16}\pi\left[\left(x_4 - \frac{1}{3}\right)^2 + \left(y_4 - \frac{1}{3}\right)^2\right]} \cdot e^3 \cdot \omega$$

$$\phi(i) = e^{-\frac{\pi}{4\alpha^2}[(x_i - x_0)^2 + (y_i - y_0)^2]}$$

$$\varphi(i) = \sqrt{(x_{n+1} - x_i)^2 + (y_{n+1} - y_i)^2}\, e^{b(t_i - t_0)}$$

ω represents:

$$\sqrt{x_4^2+(y_4-1)^2}\sqrt{x_4^2+y_4^2}\sqrt{(x_4-1)^2+y_4^2}$$

$$R(x,y)=\frac{9}{16}e^{-\frac{9}{16\sigma}\left[\left(x-\frac{1}{3}\right)^2+\left(y-\frac{1}{3}\right)^2\right]},\sigma$$

$$\sigma=\sqrt{x^2+(y-1)^2}\sqrt{x^2+y^2}\sqrt{(x-1)^2+y^2}$$

In order to get the maximum of $R(x,y)$, we seek the partial derivative of x and y as follow

$$\begin{cases} R_x(x,y)=0 \\ R_y(x,y)=0 \end{cases} \tag{17}$$

Ultimately we get the solution through the MATLAB as follows:

$$\begin{cases} x=0.8378 \\ y=0.8378 \end{cases}$$

While $s_4=(0.8378,0.8378)$ is the location that we need to predict through a series of complicated calculations.

4 Model Instantiation

In this part, two actual crime cases should be used to analyze the sensitivity and reliability of our model. As the following graph shows, it is a series of attack and murder crime case.

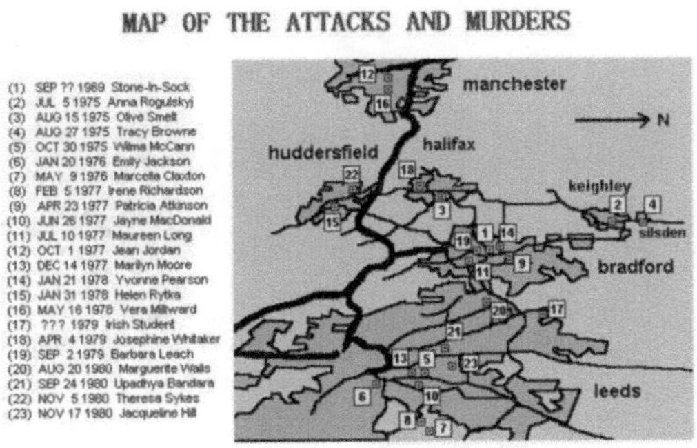

Fig. 3. Map of the attacks and murders

First we may extract these crime points and map them into a plane rectangular coordinate system in accordance with a certain proposition.

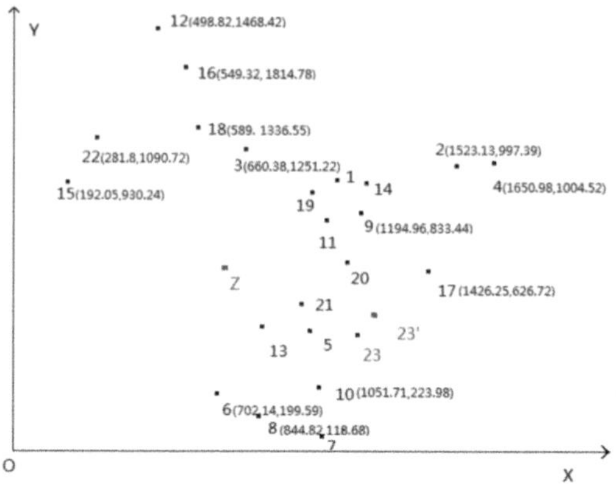

Fig. 4. Abstract rectangular coordinate based on the map

Then due to the former mathematics models, the predictions of anchor point and the 23th crime location will be proposed through the complex calculations

$$P(s_i|z,\alpha)=\frac{1}{4\alpha^2}e^{-\frac{\pi}{4\alpha^2}[(x_i-x_0)^2+(y_i-y_0)^2]}$$

$$\alpha^2=\frac{1}{22}\sum_{i=1}^{22}[(x_i-x_0)^2+(y_i-y_0)^2]$$

$$=21409.45$$

The solution of mean distance $\alpha = 146.3$

$$f(x,y)=\prod_{i=1}^{22}P(s_i|z,\alpha)$$

$$=\prod_{i=1}^{22}\frac{1}{4\alpha^2}e^{-\frac{\pi}{4\alpha^2}[(x_i-x_0)^2+(y_i-y_0)^2]}$$

$$g(x,y)=\ln f=\ln\frac{1}{4\alpha^2}\cdot\sum_{i=1}^{22}\{\psi(i)\}$$

$$\psi(i)=-\frac{\pi}{4\alpha^2}[(x_i-x_0)^2+(y_i-y_0)^2]$$

To seek the partial derivative of x and y

$$\begin{cases} g_x(x,y)=0 \\ g_y(x,y)=0 \end{cases}$$

We get the solution

$$\begin{cases} x_0 = 606.68 \\ y_0 = 698.98 \end{cases}$$

The offender's anchor point $z = (606.68, 698.98)$
Then we predict the 23th crime locations

$$R(x,y) = \frac{1}{4a^2}\sum_{i=1}^{22}\phi(i)\cdot\prod_{i=1}^{22}a\varphi(i)$$

$$\phi(i) = e^{-\frac{\pi}{4a^2}[(x_i-x_0)^2+(y_i-y_0)^2]}$$

$$\varphi(i) = \sqrt{(x_{n+1}-x_i)^2+(y_{n+1}-y_i)^2}\,e^{b(t_i-t_0)}$$

$$\begin{cases} R_x(x,y) = 0 \\ R_y(x,y) = 0 \end{cases}$$

$$\begin{cases} x = 1134.54 \\ y = 470.11 \end{cases}$$

The 23th crime location due to the predictions

$$S_{23}'(1134.53, 470.11)$$

While the actual location

$$S_{23}(1115.15, 371.2)$$

As it is shown, they are quite near and our results based on the mathematics predictions could be reliable and effective.

References

1. Brantingham, P.J., Brantingham, P.L.: Patterns in Crime. Macmillan, New York (1984)
2. Snook, B., Taylor, P.J., Bennell, C.: Geographic Profiling: The Fast, Frugal, and Accurate Way. Appl. Cognit. Psychol. 18, 105–121 (2004)
3. Snook, B., Zito, M., Bennell, C., Taylor, P.J.: On the Complexity and Accuracy of Geographic Profiling Strategies
4. Canter, D., Coffey, T., Huntly, M., Missen, C.: Predicting serial killers' home base using a decision support system. Journal of Quantitative Criminology 16(4) (2000)
5. Stile, J., Brown, D.: Geographic Profiling with Event Prediction, `http://ieeexplore.ieee.org/stamp/stamp.jsp?arnumber=01244466`
6. O'Leary, M.: Determining the Optimal Search Area for a Serial Criminal. In: Joint Mathematics Meetings, Washington DC (2009)
7. Rossmo, D.K.: A Methodological Model. American Journal of Criminal Justice XVII(2) (1993)
8. Rossmo, D.K.: Geographic Profiling. CRC Press, New York (2000)
9. `http://en.wikipedia.org/wiki/Peter_Sutcliffe.2010/2/22`

Research on Business Risk Assessment Model of City Commercial Banks Based on Grey System Theory*

Shuming Ren, Dandan Wang, and Hongjing Wang

Department of Economics, Dalian University of Technology,
Dalian, Liaoning Province, China
shumingren@yahoo.cn

Abstract. A new evaluation index system of business risk is set up after grey relational clustering. And assessment model of city commercial banks' business risk is set up on factor analysis and gray relational analysis. The innovation lies in four aspects. First, the gray relational clustering is used to remove highly relevant indicators. Second, it uses factor analysis to determine index weight, avoiding more subjective weakness of former research. Third, asset risk, liquidity risk, capital risk and profitability risk are considered in the index system.

Keywords: City commercial bank, Gray relational clustering, Gray relational analysis, Factor analysis, Gray System Theory.

1 Introduction

Domestic and foreign commercial bank risk assessment model mainly include three types:

Qualitative analysis model. It's a model that makes qualitative judgments on the risk of commercial banks with expert scoring method and the Delphi method. The greatest disadvantage is that the model sometimes is so subjective that large deviations may appear.

The regression model. The typical application is American FIMS model. It constructs econometric model to evaluate bank risk on the basis of empirical data. However, the application of the model needs a prerequisite for the survival of a large number of sample banks and failed banks. Chinese city commercial banks have a relatively short history, seldom go bankrupt. Therefore, this method isn't reasonable[1].

The comprehensive evaluation model. Methods commonly used are comprehensive scoring method, fuzzy comprehensive evaluation and multivariate statistical analysis, etc[2, 3].The first two methods are inevitably influenced by subjective factors of

* Supported by Scientific Research Foundation for Liaoning Provincial PHD "Research on market structure evolution and local enterprises' entering mode in major equipment manufacturing under owners' risk-averse"(Grant No. 20101015)and Fundamental Research Funds for the Central Universities "Research on system innovation and policy innovation of high-end service in the view of social capital"(Grant No. DUT10RW310).

J. Zhang (Ed.): ICAIC 2011, Part V, CCIS 228, pp. 30–37, 2011.

weight distribution. Although the fuzzy comprehensive evaluation method can overcome defects of the two methods, but in the application process, the method which uses the principle of maximum membership degree will lead to substantial loss of original information, and even sometimes draw abnormal conclusions.

In this paper, the gray system theory is used to construct risk assessment model of city commercial banks, which can overcome the shortcomings of the models.

2 Index System

A. Preliminary selection

Based on simple principle, representative principle and comparability principle, we reference previous studies [4, 5], combine with its own characteristics of city commercial banks to establish index system which includes asset security, liquidity, capital adequacy, profitability. As shown in table 1. Provision coverage, liquidity ratio, excess reserves ratio, capital adequacy ratio, core capital adequacy ratio, interest rate, return on capital and return on assets are positive indicators, the rest are negative indicators.

Table 1. Summary of commercial bank risk

Index classifi -cation	Index
Asset securit y index	non-performing loan rate, credit loss rate, loss rate of doubtful loans, subprime loan loss rate, estimated loan loss rate, the single largest proportion of customer loans, the maximum ratio of ten loans, rate of provision coverage
Liquidi ty index	deposit-loan ratio, liquidity ratio, the proportion of borrowing from other banks,cash reserve ratio
Capital adequa cy	Capital adequacy ratio, core capital adequacy ratio
Profita bil-ity capacit y	cost to income rate, interest collection rate, return on capital rate, return on assets rate

B. Final confirmation

For reverse indicators, we will start with the countdown method to turn them into obverse indicators.

Because the dimension of indicators is not uniform, we need to standardize the value of evaluation indicators. It chooses means method. The formula is as follows,

$$Y_{ij} = \frac{X_{ij}}{\overline{X_j}} \,. \tag{1}$$

Where $\overline{X_j}$ corresponds to mean of index X_j,

We have selected 16 city commercial banks whose total properties attain 500 million Yuan or above and data can be obtained. All of the data come from the annual report statement that each bank official site announces.

Grey clustering is used to screen indicators; results are shown in table 2, 3, 4.

Table 2. Grey correlation absolutely results of asset security index

	X1	X2	X3	X4	X5	X6	X7	X8
X1	1	0.51	0.57	0.76	0.52	0.96	0.86	0.97
X2		1	0.52	0.52	0.64	0.51	0.52	0.51
X3			1	0.64	0.57	0.58	0.60	0.57
X4				1	0.54	0.78	0.86	0.74
X5					1	0.53	0.53	0.52
X6						1	0.89	0.93
X7							1	0.84
X8								1

Table 3. Grey correlation absolutely results of liquidity risk index

	X9	X10	X11	X12
X9	1	0.52	0.70	0.90
X10		1	0.53	0.52
X11			1	0.62
X12				1

Table 4. Grey correlation absolutely results of profitability index

	X1	X166	X17	X18
X15	1	0.53	0.76	0.90
X16		1	0.53	0.53
X17			1	0.83
X18				1

Table 5. Management risk evaluation index system of city commercial bank

Index classifi -cation	Index
Asset securit y index	Non-performing loan rate, credit loss rate, loss rate of doubtful loans, estimated loan loss rate
Liquidi ty index	Deposit-loan ratio, liquidity ratio, the proportion of borrowing from other banks
Capital adequa cy	Capital adequacy ratio
Profita bil-ity capacit y	interest rate, return on capital, return on assets

In order to identify correlation among indicators strictly, the paper adopts the standard $\rho\rangle 0.85$ to cluster. We should pay special attention to the gray absolute correlation of the capital adequacy. The result is 0.97. There is a high correlation. So we can choose either one on behalf of capital adequacy. Eventually, risk assessment system of city commercial banks is established, as shown in table 5.

3 City Commercial Banks Business Risk Assessment Model Based on Grey System Theory

A. The Theoretical Basis

Gray system theory focuses on the study of such unascertained that are known with small samples or "poor information". It is through the development of "part" known information to achieve the exact description of reality and understand the world [6]. City commercial banks is still at a primary stage of development, risk management techniques and methods are still not perfect, and our statistics are not complete, resulting some evaluation indicators appear "grey" feature. The gray system theory can overcome the defects. By analyzing Chinese commercial banks reports published in recent years, it's not difficult to find that business risk assessment system of city commercial banks is a gray system.

B. Construct Model

First, determine the comparison and reference series.

Suppose there are m evaluation objects, n evaluation indicators, and then X_{ki} represents evaluation value of k of bank i.

Comparison series, $X_i = \{x_{ik} | i = 1,2,\cdots,m; k = 1,2,\cdots,n\}$.

Reference series, $X_0 = \{x_{0k} | k = 1,2,\cdots,n\}$.

Where $X_i = (x_{1i}, x_{2i}, \cdots, x_{ni})$ represents vector of bank i that consists of all evaluation indicators.

$$X_0 = (x_{10}, x_{20}, \cdots, x_{n0}). \tag{2}$$

There will be a matrix composed of m banks evaluated, n indicators evaluated.

$$X = (X_{ki})_{n\times m} = \begin{bmatrix} x_{11} & x_{12} & \cdots & x_{1m} \\ x_{21} & x_{22} & \cdots & x_{2m} \\ \vdots & \vdots & & \vdots \\ x_{n1} & x_{n2} & \cdots & x_{nm} \end{bmatrix}. \tag{3}$$

Because the indicators dimension is not uniform, we need to standardize the value of evaluation index. With reference to Formula (1).

After standardizing, vector of all indicators of bank i is as follows,

$$Y_i = (Y_{1i}, Y_{2i}, \cdots, Y_{ni}), i = 0,1,\cdots,m. \tag{4}$$

Matrix (3) is converted into Matrix (5) after standardizing.

$$Y = (Y_{ki})_{n\times m} = \begin{bmatrix} Y_{11} & Y_{12} & \cdots & Y_{1m} \\ Y_{21} & Y_{22} & \cdots & Y_{2m} \\ \vdots & \vdots & & \vdots \\ Y_{n1} & Y_{n2} & \cdots & Y_{nm} \end{bmatrix}. \quad (5)$$

Using this method above, the original data were normalized to eliminate the different dimensions of data which may cause an unreasonable impact on the result.

Next, calculate correlation coefficient.

Regard the ideal value of each indicators as a point in space, we call it the reference point, and all other indicators' values are considered as comparison points. The correlation coefficient is a function of distance between the reference point and the comparison point. The association analysis is essentially based on the distance between reference point and comparison point, by the way of finding out the difference and proximity from the multi-factor. In this paper, we use the method of Deng correlation calculation. Take the standardized vector $Y_0 = (Y_{10}, Y_{20}, \cdots Y_{n0})$ as a new reference series that consist of the ideal banks' indicators date. $Y_i (i = 1,2,\cdots,m)$ is a comparison series, and the gray correlation is $\gamma(Y_0, Y_i) = \frac{1}{n}\sum_{k=1}^{n} \gamma(Y_{k0}, Y_{ki})$.

And, the correlation coefficient between the appraised bank and ideal banks is:

$$\gamma_{ki} = \frac{\min_i \min_k |Y_{k0} - Y_{ki}| + \rho \max_i \max_k |Y_{k0} - Y_{ki}|}{\left| Y_{k0} - Y_{ki} + \rho \max_i \max_k |Y_{k0} - Y_{ki}| \right|}.$$

Where ρ is discrimination coefficient, usually $\rho \in [0,1]$, ρ =0.5[6].All the sequences of the gray relational grade obtained are arranged in accordance with descending order. According to it, we can determine the sequence of the correlation.

The correlation uses displacement difference ($\Delta Y_{ki} = |Y_{k0} - Y_{ki}|$) to reflect the development of the two sequences which are similar to each other.

According to formula (6), we can draw the following correlation coefficient matrix,

$$M = (\gamma_{ki})_{nm} = \begin{bmatrix} \gamma_{11} & \gamma_{12} & \cdots & \gamma_{1m} \\ \gamma_{21} & \gamma_{22} & \cdots & \gamma_{2m} \\ \vdots & \vdots & & \vdots \\ \gamma_{n1} & \gamma_{n2} & \cdots & \gamma_{nm} \end{bmatrix}. \quad (6)$$

Again, determine the index weight.

Sort it according to the size of $\rho_{0k}(k=1,2,\cdots,n)$. The more forward the position of indicators is, the greater the impact is, and the degree of weight should also be greater. After treating comprehensive correlation, we could get the weight of each index, that is.

$$W=[w_1,w_2,\cdots,w_n]. \tag{7}$$

Specific methods are as follows,

$$w_i=\rho_{0i}\div(\rho_{01}+\rho_{02}+\cdots+\rho_{0n}),i=1,2,\cdots,n.$$

Finally, make a model.

Using correlation coefficient Matrix M and weight Matrix W , find the correlation R ,

$$R=(r_i)_{1\times m}=(r_1,r_2,\cdots,r_m)=WM .$$

4 Model Application

We have selected 16 city commercial banks whose total properties attain 500 million Yuan or above and data can be obtained. All of the data come from the annual report statement that each bank official site announces. Through analysis, Hangzhou bank developed stably in the last few years, so we choose Hangzhou bank to carry on connection analysis and make sure the index power weight.

The method of setting up the model which is introduced according to chapter 2 is used to carry on an evaluation to the management risk of city commercial bank. The power weighs a certain result such as table 6 showing.

Table 6. Correlation and weight

	Gray Absolute Correlation degree	Gray relative correlation degree	Comprehensive Correlation degree	Weight
Z1	0.6075	0.5507	0.5791	0.0836
Z2	0.6907	0.7200	0.7054	0.1018
Z3	0.5551	0.5328	0.5440	0.0785
Z4	0.5349	0.5341	0.5345	0.0771
Z5	0.5532	0.5363	0.5440	0.0786
Z6	0I.5713	0.5402	0.5558	0.0802
Z7	0.5719	0.5387	0.5558	0.0801
Z8	0.6981	0.8562	0.7771	0.1121
Z9	0.5560	0.5369	0.5464	0.0788
Z10	0.5871	0.5442	0.5657	0.0816
Z11	0.5689	0.5394	0.5541	0.0800

Evaluation results are shown in Table 7.

Table 7. Rank of city commercial bank

City Commercial Bank	correlation degree	Rank
Bank of Beijing	0.9471	9
Bank of Tianjin	0.9426	10
Bank of Hebei	0.9506	6
Baoshang Bank	0.9550	5
Bank of Dalian	0.8749	15
Harbin Bank	0.9502	7
Bank of Jinzhou	0.9420	11
Bank of Nanjing	0.9152	13
Bank of Shanghai	0.9496	8
Bank of Ningbo	0.9618	3
Qilu Bank	0.8808	14
Bank of Hangzhou	0.9938	1
Huishang Bank	0.9579	4
Fudian Bank	0.9328	12
Bank of Chongqing	0.9777	2
Bank of Xian	0.8685	16

5 Conclusions

Banks that rank in front including Bank of Hangzhou, Bank of Chongqing, Bank of Ningbo, Huishang Bank, Bank of Hebei, Harbin Bank, Bank of Shanghai and so on. These banks are mainly distributed in north, northeast, east, south and southwest regions of China. Among them, a larger number are in North China and East China, especially in East China. This result is consistent with the development of city commercial banks regional differences in characteristics. East China is economically developed area. There exist lots of small and medium-sized enterprises. They have strong demand for funds, therefore, the situation of city commercial banks is significantly better than other regions.

Bank of Xian ranks last. It is located in the northwest, the economy is relatively backward. The scope of business is also limited, lacking of effective regional expansion.

References

1. Cole, R.A., Cornyn, B.G., Gunther, J.W.: FIMS a new monitoring system for banking institutions. Federal Reserve Bulletin (1), 1–15 (1995)
2. Fan, M.: A Study of the Quantitative Supervision system of Operating Risks of Banks. Journal of South China University of Technology (Natural Science Edition) 29(11), 99–101 (2001)
3. Guan, Q., Feng, Z.: A Study of Measurement and Warning of Non-system Financial risks of Chinese commercial bank. Economic Science (1), 35–46 (2001)

4. Chi, G.-t., Feng, X., Zhao, Z.-h.: Warning model of operating risk for commercial banks and its empirical study. Journal of Systems Engineering 24(4), 408–416 (2009)
5. Hu, W.-h., Li, M.-g.: The Design and Demonstration of the city Commercial Bank of Non-scene supervision Risk Index. Journal of Harbin University of Commerce 68(1), 61–69 (2003)
6. Liu, S.-f., Xie, N.-m.: Gray Information: Theory and Practical Applications, 4th edn., pp. 45–52. Science Press, Beijing (2008)

Study on the Application of RFID Technology in the Management System of Paper Machine's Disassembly and Assembly

Xiuqing Wang, Xian Zhang, and Chunxia Zhang

College of Electronic Information and Automation
Tianjin University of Science and Technology
Tianjin, China 300222
XiuqingWang111@foxmail.com

Abstract. This paper takes the management system of paper machine's disassembly and assembly as research objects, and introduces RFID technology as well as the application of RFID. Then it establishes basic information management of paper machine, marking information management, installation information management, serial communication and other functional modules. The monitoring system adopts VC6.0 and SQL as the original developing tools. At first, it needs to analyse and construct system functional modules of paper machine, then it designs engineering framework-a kind of visual user management interface- in VC platform. Finally, it does the mapping coefficient and data transmitting from the engineering framework to database by ADO database access.

Keywords: paper machine, RFID, disassembly and assembly, management information, serial communication.

1 Introduction

In the new extension to the paper industry, it's a great way to introduce secondhand machines from European &American countries where their paper and pulp industry are much more developed, this can save the project funds and expand business at low cost. The secondhand machines have characteristics of complex and varied components, pipeline material, electrical system and electrical instruments, so the quality of its disassembly and assembly, the scientificity and integrity of mark will have a direct impact on the quality of paper machine's overhaul, installation, project schedule, normal operation.

At present, paper machine's disassembly and assembly is mostly done and marked manually, with much ruinous disassembly, unscientific and incomplete marks, which leads to the increase of maintenance and great restrict to the process and quality of project. And in actual operation the traditional disassembly and assembly mode will isolate marking information, production video and drawing from each other, this prevents cross search and cross query about information. It is a very complex manual job with high intention and low efficiency [1].

J. Zhang (Ed.): ICAIC 2011, Part V, CCIS 228, pp. 38–45, 2011.

Considering the problems of managing paper machine's disassembly and assembly, it had better fix RFID tags on each of its parts and store the appropriate information in the tags. Then, it needs to manage and detect each of parts in the terminal computer by RFID. Finally, it can manage the appropriate data installation or maintenance information according to the mark information.

At present, the research and application on RFID in our country are less developed than the developed countries, especial in the field of the UHF RFID, because of lackness of key and core technology. In our country, the products and solution based on RFID are rarely combined with industry application. Using RFID technology to manage and mark the components of paper machine you can visually operate to add, delete, query, revise and so on, and especially implement the automation of equipment installation management [7]. It can not only reduce the intensity of labor, but also improve the efficiency and accuracy of equipment installation, and then it improves the working efficiency of paper machine. These are meaningful to improve the paper machine's quality of management and installation or maintenance.

2 Introduction to RFID Technology

Radio Frequency Identification (RFID) is an automatic identification method that utilizes wireless communication between readers and small silicon chips embedded in items to be tracked. The system of RFID generally includes two parts at least, they are: (1) tag: it consists of coupling components and chips, each of tags attached to an object has the sole electronic code, this is used to identify the object;(2) reader: it is used to read data from and write data onto the right tag, and can be designed to a hand-held one or a fixed one [2]. In practice, the tag is attached to a recognized-object, and the reader can identify the electronic data stored in the tag, so as to realize the recognition of the object automatically. Then, it can achieve the management function of data acquisition, data processing, the remote transmission and so on by computer and Internet.

As a basis for rapid, real-time, accurate data collection or processing's Hi-tech and information standardization, compared with other auto identification, such as the traditional Bar Code, magnetic card, IC card, the greatest advantages of RFID technology is that the communication between RFID tag and reader can be realized with little contact and without man-made disturbing, so the automation of system is easy to be realized. As a data medium, RFID tag plays an important role in marking objects, tracking objects and collecting information. The RFID system composed of RFID tag, reader, antenna and application software can be in direct connection with the right management information system, making every object accurate to trace. This comprehensive management information system can bring the benefit to customers, including the real-time data acquisition, secure access data channel, acquiring all products' information off line. In addition, RFID system can identify several tags at the same time, it is convenient to operate, moreover, RFID tag is dustproof, waterproof, oil proof and not easy to damage, it also has high security and high safety. So RFID tag is the best information carrier for the management system of paper machine's disassembly and assembly.

3 The Management System of Paper Machine's Assembly-Disassembly Based on RFID Technology

A. Implementation Plans of the Management System

The management system consists of the monitoring computer, the data-acquiring terminal, RFID tag and others. The design flow is as shown in Fig.1.

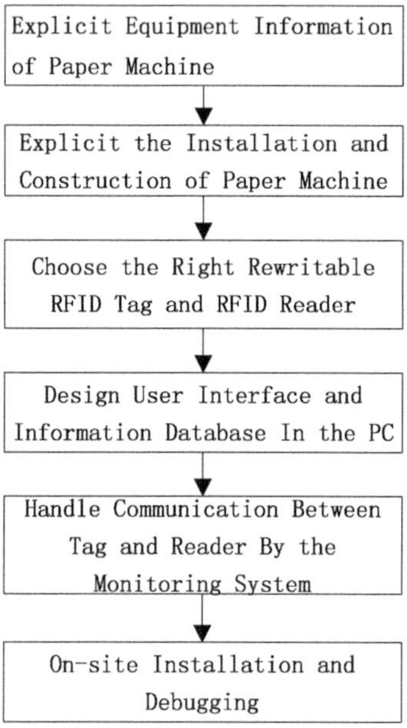

Fig. 1. Design of system flow chart

Integrating the demand of paper machine's disassembly and assembly and RFID technology, it designed the management system of paper machine's disassembly and assembly based on RFID technology. It is a shared management system for each component of paper machine. Every component can use the data resources related to components.

Following the working environment of this system, it will use anti-jamming, anti-corroding, anti-metallic and high frequency RFID tags, every component is fitted with RFID tags, the information relative to components, such as ID number, manufacturing codes, name, producers and price, is kept in the tags [7].

VC6. 0 and SQL are used as the original developing tools in the monitoring system. At first, it needs to analyze and construct system functional modules of paper machine, then design engineering framework-a kind of visual management user interface- in VC platform. At last, it needs to do the mapping coefficient and data transmitting from the engineering framework to database by ADO database access.

On the management modules of this system, it creates the index in the ways of component name, producer, time, marking ID number, manufacturing codes and others, it can call up all the information of every component from computer in detail and immediately, thus establishing the connection among components.

This system chooses the SP-S100 portable RFID reader as data-acquiring terminal, it can provide long-haul and two-way communication for RFID tags, and can monitor hundreds of objects equipped with tags in its range of 100 meters. This portable RFID reader integrates intelligent signal processing, the most advanced microchip technology, the most advanced RFID technology, high speed data transmission, massive information-storage capacity, the embedded real-time system and other technologies. After scanning the tags, the upper-management system can display the marking information of the tags relative to components. The user can click the marking information to add, query, delete and other operations on all of information of paper machine.

B. Marking Schemes of Tag Information

The present marking scheme is barcode label which is based on infrared scanners. Although this operation is very easy, barcodes have great limitations, low reliability and poor information security, so this system adopts marking scheme of RFID tags based on RFID technology [10].

The marking scheme includes the following fields:

- The overall component is marked by WNO and PNO (WNO-PNO) [9].
- Each component of paper machine is marked by the production line including approach system, molding section, press section, drying section and reeling section (WNO-PNO-Serial Number of Section) [9].
- The composition of each section is marked by serial number of single component (WNO-PNO-Serial Number of Section-Serial Number of Single Component) [9].
- The process pipe line is marked by pipeline number, caliber size and valve number (WNO-PNO- Serial Number of Section-Serial Number of Single Equipment-Drawing Number-Pipeline Number) [9].
- The control instrument is marked by WNO and serial number of component, the control cable is marked by serial number of cable [9].
- The operation platform is marked by WNO, drawing number and production video (WNO-PNO-Serial Number of Section-Serial Number of Single Equipment-Drawing Number-Production Video Number) [9].

The above marking schemes are typed into the computer before dismantling the equipment to form the first draft, and the operators can guide marking works by print files. The operators can also add or delete the marking schemes depending on the local conditions, and the finally marking schemes are formed.

C. Design of PC Monitoring Software

VC6. 0 and SQL are used as the original developing tools in the monitoring system.

1) The main interface of PC monitoring software: The main interface of PC monitoring software is shown in "Fig.2", it is divided into five modules: basic information management of paper machine, marking information management, installation information management, query information management and serial

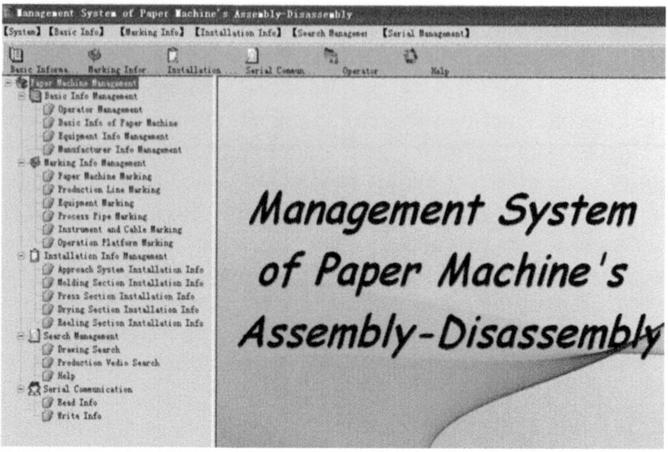

Fig. 2. The main interface of PC monitoring software

communication. Information among modules is mutually linked, it can realize adding data, deleting data, updating data and querying data and other operations on information, drawing, production video, marking information and installation information of each part of paper machine[4].

2) Serial communication interface: After reading the right ID number of RFID tags, the VC interface can reflect the marking information of the right equipment from database. For example, the marking scheme of headbox is marked by WNO-PNO-Serial Number of Section-Serial Number of Single Equipment, when RFID reader scans the tag of headbox, the operator can click the "Read Information" key to read the information of headbox, the information is as shown in "Fig.3", then the operator can link the corresponding installation information by clicking the "Browse Information" key. Similarly, the operator could click the "Write Information" key to make a mark on the right tag of equipment [8].

3) Installation information management: In "Fig.3", when the operator clicks the "Browse Information" key, the dialog as shown in "Fig.4" will be popped up, the dialog covers the installation information of equipments in approach system, molding section, pressing section, drying section and reeling section [9]. When the operator clicks the "Installation Information" key in "Fig.4", the dialog of installation information as shown in "Fig.5" will be popped up, in this dialog the operator can browse Word documents and print the installation information out [4].

When the operator scans the RFID tags, the RFID reader immediately reads the marking information from the tags, and transmits the marking information to the computer. From the linked database, the operator can acquire the equipment information corresponding to the marking information and the operator requires, thus realizing equipment management automation of paper machine.

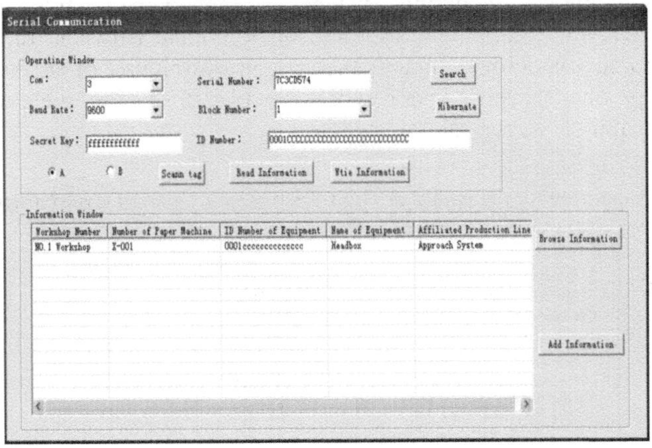

Fig. 3. Serial communication interface

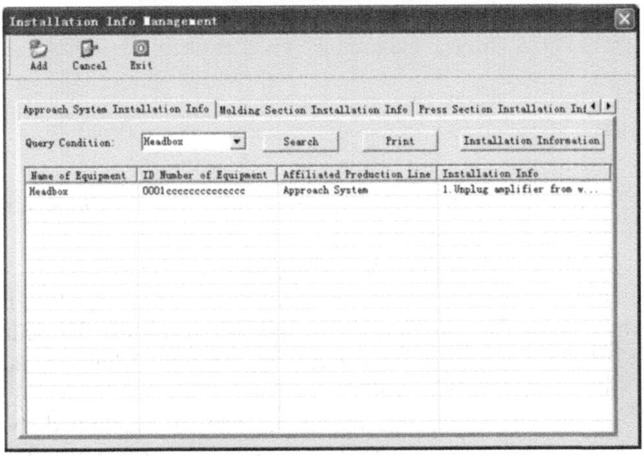

Fig. 4. assembly information management

C. Database Management

In this management system, the marking information should meet the requirement of management process with data storage and processing, it means to confirm the code structure of RFID tags in order to meet the requirement of enterprise. The data structures processing of database is a is relatively complicated problem, design of the marking scheme need to establish contacts in each other, then it need to do the mapping coefficient and data transmitting from the engineering framework to database by ADO database access. The back-end database of our system adopts SQL server database, the middleware adopts ADO method, and it is encapsulated using classes [3]. The name of the database is Paper Machine, including five tables, basic information of paper machine table-BasicInfo, operator information table-OperatorInfo, marking information table-MarkInfo, assembly information table-AssemblyInfo, drawing and video information table-DrawingVideoInfo [4].

Fig. 5. Browse and print the installation information

4 Conclusion

The management system of paper machine' disassembly and assembly based on RFID realizes the centralized management on each component of paper machine, the operator can query the information of component by different need anytime; When installing the paper machine, it can manage the installation project intelligently and improve the accuracy; When the paper machine is moving, it can query the instructions of equipment, thus improving the working efficiency. In this set of scheme, the information that the management site requires is stored in RFID tags, then using the portable reader to scan the tags can acquire the information conveniently, at last, it improves the management level of paper machine.

The introduction of RFID technology is a stage of improving process in enterprise, in order to realize the essential change and innovation, excepting for effective data acquisition, it also needs to upgrade the enterprise system software. Gathering the superiority resources, optimizing the process, only in this way can the value of RFID be played to the extreme, and it can bring more benefits to enterprise. In future, the application of RFID will play an important role in improving the efficiency of economic operation and industrial information construction [5].

Acknowledgment. Project supported by the National Natural Science Foundation of China (No. 61071207).

References

1. Yu, B.: Special Equipment Management and Testing System based on RFID. Xiamen University, doi:CNKI: CDMD: 2.2009.079888
2. Fu, J.: Study of RFID. Shanxi Science and Technology (01) (2009)

3. Shao, C., Zhang, B., Zhang, Q.: A Practical Course of Database. Tsinghua University Press, Beijing (2009)
4. Qi, M.W.: The database application example of Visual C++ + SQL Server. People's Posts and Telecommunications Publishing House (2006)
5. Samall, D.: The potential of RFID is finally a reality. Industrial Engineering 24(10), 46–47 (1992)
6. Jiang, H., Zang, C., Lin, J.: The development trend and application of RFID. Application of Electronic Technique, doi:cnki: ISSN: 0258-7998.0.2005-05-001
7. Ji, Z., Li, H., Jiang, L.: Principles and Application of RFID Tags. Xi Dian University Press (2006)
8. Zhao, J.: Technology and Application of RFID. Mechanic Industry Press (2008)
9. Bian, G.: Assembly and Maintenance of Paper Machine. China Light Industry Press (2006)
10. Association of Automatic Identification Technology of China. The Application Guide of RFID Tags and Bar Codes. Mechanic industry Press (2003)

A Fast and Accurate Approach to the Computation of Zernike Moments

Hongli Tian, Huiqiang Yan, and Hongdong Zhao

Hebei University of Technology, TianJin, China
tanhongli123@yeah.net

Abstract. This paper presents a fast and accurate approach to the computation of Zernike moments from a digital image. Exact Zernike moments are computed with these fixed exact coefficients. The digital image is transformed to be inside the unit circle, where the transformed image is divided into eight parts. Based on the specific symmetry or anti-symmetry about the x-axis, the y-axis, the origin, and the straight line y = x, we can generate the Zernike basis functions by only computing one eighth of the image. Inside the unit circle, the image information is not lost. Experimental results show that the proposed method can reduce significantly the computation operations, so the time can be reduced to just one eighth of the original computation methods.

Keywords: Zernike moments, symmetry, anti-symmetry, Zernike basis functions.

1 Introduction

Orthogonal moments, due to their orthogonality property, can describe an image fully with the minimum amount of information redundancy. In the orthogonal moments' family, the most widely used moments are Zernike moments, which were firstly proposed as digital image descriptors by Teague [1] in 1980. In the recent years, Zernike moments have been successfully utilized in image analysis, such as invariant pattern recognition [2-4], content-based image retrieval [5], and other image systems [6,7].

Nevertheless, one of the main concerns of the application of Zernike moments is how to speed up the computation of Zernike moments, the other is how to compute correctly Zernike moments. To speed up the Zernike moments, several methods have been proposed. some[8,9] of them lost the accuracy, and some[10-12] of them removed the redundancy.

In this paper, a fast and accurate approach to the computation of Zernike moments is presented. The approach computes the Zernike basis functions with the fixed exact coefficients. As the Zernike basis functions have specific symmetric or anti-symmetric properties about the x-axis, the y-axis, the origin and the straight line y=x. The proposed method shows the more performance on computation time.

The approach also can get an accurate result on Zernike moments. The exact form is presented in section 4.

J. Zhang (Ed.): ICAIC 2011, Part V, CCIS 228, pp. 46–53, 2011.

The organization of this paper is as follow. The basic theory of Zernike function is shown in section 2. The fast computation is shown section 3. The accurate form is explored section 4. Numerical experiments on Zernike moments are shown in section 5. Section 6 concludes the study.

2 Zernike Moments

The computation of Zernike moments from an input image consists of three steps: the computation of the Zernike real-valued radial polynomials; the computation of Zernike polynomials; the computation of the two-dimensional Zernike moments of order n with repetition m of an image intensity function f(x,y). In this section, each procedure for the exact form of Zernike moments for a discrete image function is described [13-16].

The Zernike moment of order n with repetition m are finally defined as

$$Z_{nm} = \frac{n+1}{\pi} \int_0^{2\pi} \int_0^1 f(\rho,\theta) V_{nm}^*(\rho,\theta) \rho d\rho d\theta \tag{1}$$

Where V_{nm}^* denotes a complex conjugate V_{nm}, $f(x,y)$ is the image function. As can be seen from the definition, the procedure for computing Zernike moments can be seen as an inner product between the image function and the Zernike basis function.

$$\int_0^{2\pi} \int_0^1 V^*(\rho,\theta) V(\rho,\theta) \rho d\rho d\theta = \begin{cases} \frac{\pi}{n+1} & if n = p, m = q \\ 0 & otherwise \end{cases} \tag{2}$$

The orthogonality implies no redundancy or overlap of information between the moments with different orders and repetitions. This property enables the contribution of each moment to be unique and independent of the information in an image.

Using the radial polynomial, complex-valued 2-D Zernike basis functions, which are defined within a unit circle, are formed by

$$V_{nm}(\rho,\theta) = R_{nm}(\rho)\exp(jm\theta), \qquad |\rho| \leq 1 \tag{3}$$

Where $j = \sqrt{-1}$ Zernike basis functions are orthogonal and satisfy.

The procedure for obtaining Zernike moments from an input image begins with the computation of Zernike radial polynomials. The real-valued 1-d radial polynomial $R_{nm}(\rho)$ is defined as

$$R_{nm}(\rho) = \sum_{k=|m|, n-k=even}^{n} \frac{(-1)^{\frac{n-k}{2}} \frac{n-k}{2}!(n-s)!}{\frac{n-k}{2}!\frac{k+m}{2}!\frac{k-m}{2}!} \rho^k$$
$$= \sum_{k=|m|, n-k=even}^{n} \beta_{n,m,k} \rho^k \quad (4)$$

In (4), n and m are generally called order and repetition, respectively. The order n is a non-negative integer, and the repetition m is an integer satisfying n-|m|=(even) and $|m| \leq$ n. The radial polynomials satisfy the orthogonal properties for the same repetition.

$$\int_0^{2\pi}\int_0^1 R_{nm}(\rho,\theta)R_{n'm}(\rho,\theta)\rho d\rho d\theta$$
$$= \begin{cases} \frac{1}{2(n+1)} & if n = n' \\ 0 & otherwise \end{cases} \quad (5)$$

3 Fast and Accurate Method

A. Limitation of Existing Fast Methods

Most of the time taken for the computation of Zernike moments is due to the computation of Zernike 1-D radial polynomials which contain many factorial terms. Therefore, the proposed faster methods by utilizing the recurrence relations, which had Prata's method, Kintner's method, Chong's method, Sun-Kyoo Hwang's method and so on, reduced the factorial terms on the radial polynomials. Through the analysis of these methods, they have some limitations. Even if only a single Zernike moment if required, these methods require the computation of extra Zernike moments because they use recurrence relations.

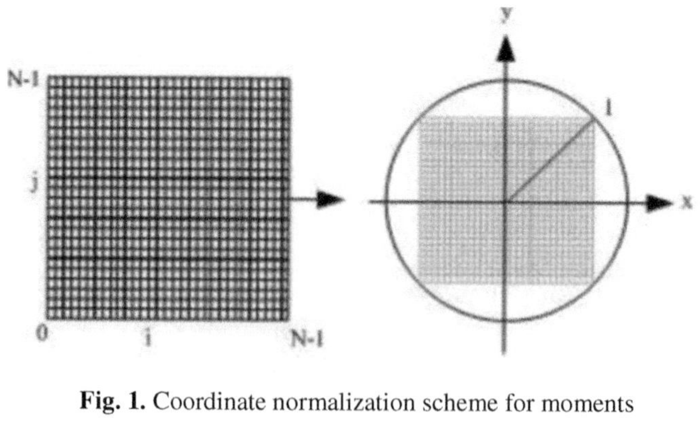

Fig. 1. Coordinate normalization scheme for moments

A. Fast and Accurate Computation of Zernike Moments

To compute Zernike moments from a N×N square digital image, a square is mapped into the unit disk of Zernike polynomials, Figure 1 shows a general case of the mapping transform. This approach is proposed by Chong.

The discrete form of the Zernike moments of an image size N×N is expressed as follows:

$$\begin{aligned} Z_{nm} &= \frac{n+1}{\lambda_N}\sum_{i=0}^{N-1}\sum_{j=0}^{N-1} f(x_i, y_j)V_{nm}^{*}((x_i, y_j)) \\ &= \frac{n+1}{\lambda_N}\sum_{x^2+y^2\le 1}\sum_{k=|m|}(\sum^{n}\beta_{n,m,k}\rho^k)e^{-jm\theta}f(x,y) \\ &= \frac{n+1}{\lambda_N}\sum_{k=|m|}^{n}\beta_{n,m,k}(\sum_{x^2+y^2\le 1}\sum \rho^k e^{-jm\theta}f(x,y)) \end{aligned} \tag{6}$$

In (6), the normalization factor λ_N must be the number of pixels located in the unit circle by the mapping transform, which corresponds to the area of a unit circle π in the continuous domain. The new axe x and y with the original indices i and j are given as

$$x = -1 + \frac{2(i+0.5)}{N}, \quad y = -1 + \frac{2(j+0.5)}{N} \tag{7}$$

In (7),i, j={0,1,2,···N-1}.

Based on the specific symmetry or anti-symmetry about the x-axis, the y-axis, the origin, and the straight line y = x, A fast computation of Zernike moments has presented in [17]. All symmetric or anti-symmetric points with respect to the point (x, y) are showed in Figure 2. r is the vector that connecting from the center of the unit disk to each pixels (x, y) of the square imgae. θ is the angle between r vector and the principle x-axis. ρ, which is the length of r vector, can be computed as

$$\rho = x^2 + y^2 \tag{8}$$

The θ can be computed as

$$\theta = \tan^{-1}(\frac{y}{x}) \tag{9}$$

The reduction in operations for Zernike moments embodied in making used of the symmetry or anti-symmetry. The symmetrical or anti-symmetrical points of *(x, y)* are in Figure 2. The propertys of the symmetric or anti-symmetric points of the point *(x, y)* are in Table 1. Zernike moments given in (5) can be rewritten as

$$Z_{nm} = \frac{n+1}{\lambda_N}\sum_{k=|m|}^{n}\beta_{n,m,k}\sum_{\substack{x^2+y^2\le 1 \\ 0\le x\le 1/\sqrt{2},0\le y\le x}}\sum \rho^k\ \psi(x,y) \tag{10}$$

In (10), the coefficients of $\beta_{n,m,k}$ have been calculated exactly and stored in advance. The $\psi(x,y)$ is written as

$$\begin{aligned}\psi(x,y) &= f(x,y)e^{-jm\theta} + f(y,x)e^{-jm(\frac{\pi}{2}-\theta)} \\ &+ f(-y,x)e^{-jm(\frac{\pi}{2}+\theta)} + f(-x,y)e^{-jm(\pi-\theta)} \\ &+ f(-x,-y)e^{-jm(\pi+\theta)} + f(-y,-x)e^{-jm(\frac{3\pi}{2}-\theta)} \\ &+ f(y,-x)e^{-jm(\frac{3\pi}{2}+\theta)} + f(x,-y)e^{-jm(2\pi-\theta)}\end{aligned} \tag{11}$$

With simple mathematics, the value of $\psi(x,y)$ can be calculated with *sin(mθ)* and *cos(mθ)* in Table 2.

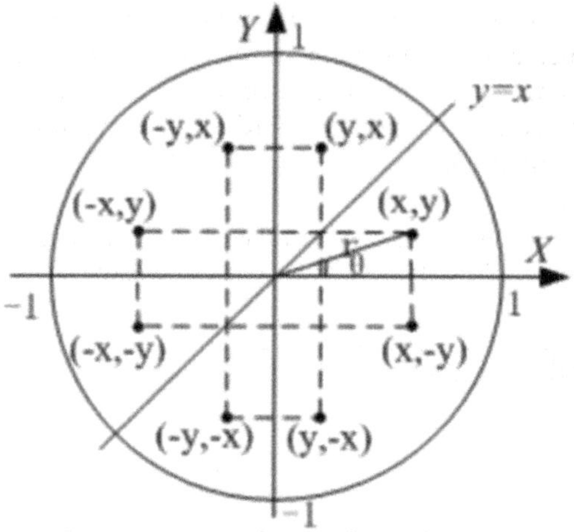

Fig. 2. The symmetric or anti-symmetric points of corresponding to the point (x, y)

Table 1. Properties of the symmetric or anti-symmetric points of (x, y)

Symmetric or anti -symmetric points	Distance(r)	Phase
(x,y)	ρ	θ
(y, x)	ρ	π/2-θ
(-y, x)	ρ	π/2+θ
(-x, y)	ρ	π-θ
(-x, -y)	ρ	π+θ
(-y, -x)	ρ	3π/2-θ
(-y, x)	ρ	3π/2+θ
(x, -y)	ρ	2π-θ

In Table2, where

$$\begin{aligned} a_1 &= \eta_m(x, y) \\ b_1 &= (-1)^k \eta_m(-y, x) \\ a_2 &= \eta_m(x, -y) \\ b_2 &= (-1)^k \eta_m(y, x) \\ k &= \begin{cases} m/2 & m is even \\ (m-1)/2 & m is odd \end{cases} \end{aligned} \quad (12)$$

$$\eta_m(x, y) = f(x, y) + (-1)^m f(-x, -y) = \begin{cases} f(x, y) + f(-x, -y) & m is even \\ f(x, y) - f(-x, -y) & m is odd \end{cases}$$

Table 2. The real and imaginary parts of $\psi(x, y)$

m	Real parts of $\Psi(x, y)$	Imaginary parts of $\Psi(x, y)$
Even	$\cos(m\theta)[a_1 + a_2 + b_1 + b_2]$	$-\sin(m\theta)[a_1 - a_2 + b_1 - b_2]$
Odd	$\cos(m\theta)[a_1 + a_2]$ $+ \sin(m\theta)[b_1 - b_2]$	$\cos(m\theta)[b_1 + b_2]$ $- \sin(m\theta)[a_1 - a_2]$

The computation of the function $\psi(x, y, \theta)$ is very simple. But it must be noted the points at $y = 0$ and $y = x$ have four points and not eight points. After obtaining the Zernike moments of one chain from *[0,0]* to $[1/\sqrt{2}, 1/\sqrt{2}]$, the entire set of Zernike moments can be obtained. The time can be reduced to just one eighth of the original computation methods to Zernike moments.

The coefficients $\beta_{n,m,k}$ without respect to the point (x, y) can be calculated with the q-recursive algorithm, but an overflow error is generated and accumulated [18]. The proposed fast method can calculate accurately the coefficients $\beta_{n,m,k}$ in advance and store them in the array, thus the cumulative error can be avoided. The array a is written as

$$a = \begin{bmatrix} a[0][0] & 0 & 0 & \cdots & 0 \\ 0 & a[1][1] & 0 & \cdots & 0 \\ a[2][0] & 0 & a[2][2] & \cdots & 0 \\ \vdots & \vdots & \vdots & \cdots & \vdots \\ a[n][0] & a[n][1] & & & a[n][m] \end{bmatrix}$$

4 Experimental Study

In this section, a comparison is performed between the several computation methods, which include the direct ZOA, Prata's method, Kintner's method, Chong's method, Sun-Kyoo Hwang's method, the proposed fast and accurate method. the test image of a 64×64 black and white Chinese character is shown in Figure 3. the entire set of Zernike moments of this Chinese character image is calculated.

Fig. 3. The 64 x 64 black and white image

The CPU elapsed time is shown in Table 3. It is obvious that the proposed method tremendously reduces the computational time.

Table 3. Cpu Elapsed Time in Milliseconds for the Zernike Moments for a 64×64 Image

Order	ZOA	Prata's	Kintner's	Chong's	Proposed
4	1.215	0.402	0.194	0.198	0.174
8	4.478	1.603	0.676	0.546	0.453
12	11.342	2.482	0.864	0.739	0.647
16	13.534	6.846	3.976	2.693	1.958
20	26.593	15.487	13.587	8.970	4.879

5 Conclusion

In this paper we proposed a fast and accurate method for computing Zernike moments from a digital image. The proposed method computed accurately Zernike moments with these fixed exact coefficients and reduced the computational time with the property of symmetry or anti-symmetry.

In this paper was evaluated using several images, and revealed that our approach can be useful in many applications of object recognition.

The proposed improved method can be used with either the computation of a single Zernike moment or a set of Zernike moments. When computing a set of Zernike moments, the proposed method can be used with other existing fast methods such as Prata's method, Kintner's method and Chong's method, and this mixed method shows the best performance. From the experimental results, we confirm that the proposed method successfully reduces the computation time in every case.

Acknowledgment. It is a project supported by Natural Science Foundation of Hebei Province (F2007000096).

References

1. Teague, M.R.: Image analysis via the general theory of moments. J. Opt. Soc. Amer. 70, 920–930 (1980)
2. Khotanzad, A., Hong, Y.H.: Invariant image recognition by Zernike moments. IEEE Trans. Pattern Anal. Mach. Intell. 12(5), 489–497 (1990)
3. Kim, W.Y., Yuan, P.: A practical pattern recognition system for translation, scale and rotation invariance. In: Proceedings of IEEE International Conference on Computer Vision and Pattern Recognition, pp. 391–396 (June 1994)
4. Khotanzad, A., Hong, Y.H.: Invariant image recognition by Zernike moments. IEEE Trans. Pattern Anal. Machine Intell. 12, 489–497 (1990)
5. Kim, Y.S., Kim, W.Y.: Content-based trademark retrieval system using visually salient feature. J. Image Vision Comput. 16, 931–939 (1998)
6. Wang, L., Healey, G.: Using Zernike moments for the illumination and geometry invariant classification of multispectral texture. IEEE Trans. Image Process. 7(2), 196–203 (1998)
7. Kim, W.Y., Kim, Y.S.: Robust rotation angle estimator. IEEE Trans. Pattern Anal. Mach. Intell. 21(8), 768–773 (1999)
8. Mukundan, R., Ramakrishnan, K.R.: Fast computation of Legendre and Zernike moments. Pattern Recognition 28(9), 1433–1442 (1995)
9. Gu, J., Shu, H.Z., Toumoulin, C., Luo, L.M.: A novel algorithm for fast computation of Zernike moments. Pattern Recognition 35(12), 2905–2911 (2002)
10. Prata, A., Rusche, W.V.T.: Algorithm for computation of Zernike polynomials expansion coefficients. Appl. Opt. 28, 749–754 (1989)
11. Kintner, E.C.: On the mathematical properties of the Zernike polynomials. Opt. Acta. 23(8), 679–680 (1976)
12. Belkasim, S.O.: Efficient algorithm for fast computation of Zernike moments. In: IEEE 39th Midwest Symposium on Circuits and Systems, vol. 3, pp. 1401–1404 (1996)
13. Brlksdim, S.O., Shridhar, M., Ahmadi, M.: Pattern recognition with moment invariants: a comparative study and new results. Pattern Recognition 24(12), 1117–1138 (1991)
14. Nukundan, R., Ramakrishnan, K.R.: Moment Functions in Image Analysis. World Scientific Publishing, Singapore (1998)
15. Chong, C.-W.: A formulation of a new class of continuous orthogonal moment invariants, and the analysis of their computational aspects, PhD thesis, Faculty of Engineering, University of Malaya, 50603 Kuala Lumpur, Malaysia (May 2003)
16. Bin, Y., Jia-Xiong, P.: Invariance analysis of improved Zernike moments. J. Opt. A: Pure Appl. Opt. 4, 606–614
17. Hwang, S.-K., Kim, W.-Y.: A novel approach to the fast computation of Zernike moments. Pattern Recognition 39, 2065–2076 (2006)
18. Papakostas, G.A., Boutalis, Y.S.: Numerical error analysis in Zernike moments computation. Image and Vision Computing 24, 960–969 (2006)

Study on Analysis and Sharing of ArcGIS Symbol and SLD Symbol

Taisheng Chen[1], Menglin Chen[2], Shanshan Tan[1], Qi Luo[1], and Mingguang Wu[1]

[1] Key Laboratory of Virtual Geographic Environment of Ministry of Education, Nanjing Normal University, Nanjing, China
[2] Land Information Engineering Department, Chuzhou University, Chuzhou, China
taishengchen@126.com

Abstract. Symbol is one of the cores of GIS and mapping system. However, difference of design styles, data structures, storage methods and render strategies among the systems leads to no synchronous sharing of map symbols when sharing geographic data. Taking map symbols as a breakth-rough, the paper analyzes and compares the ArcGIS symbol model with the SLD and provides the mapping mechanism between ArcGIS symbols and SLD symbols to study the symbol sharing between them. The analytic result shows that the SLD symbols all can be converted to the ArcGIS symbols, but only small part of ArcGIS symbols can be converted to the SLD symbols. The ArcGIS symbol system is more consummate but not perfect and the closed symbol format goes against to symbol sharing. And the SLD symbol format is open, but it is too unitary to meet the needs of Cartography. Therefore, a common symbol model is required to synthesize ArcGIS symbols and SLD symbols to achieve the map symbol sharing.

Keywords: Symbol, ArcGIS, SLD, Symbol Sharing.

1 Introduction

Map symbol is a graphic language of map, a basic means to represent map content, and could be used to communicate objective world with map makers and map users [1]. It stimulates people's left brain by pictographic and knowing graphics to generate heart images and then to convey characteristics and distribution laws of geographic entities. Symbol is one of the cores of GIS and mapping systems. But the difference of design styles, data structures, storage methods and render strategies among the systems leads to no synchronous sharing of map symbols when sharing geographic data. Therefore, it is important to implement one map symbol library with multi-usage via how to solve map symbols sharing. ArcGIS, of which the symbol system is rich and the structure is integrity, is the most widely used GIS software nowadays and the open structure of the symbol description specification, which is set by SLD for OGC, is conducive to share symbols among different software. But SLD still has some limitations [2-6]. Therefore, studying on these two kinds of symbol systems is very important for improving the map symbol model and promoting the map symbol sharing. Taking map symbols as a breakthrough, the article analyzes GIS, SLD symbol model and contrasts the structures between them to study the map symbol sharing. The article will take ArcGIS as a typical study.

J. Zhang (Ed.): ICAIC 2011, Part V, CCIS 228, pp. 54–61, 2011.

2 Contrast and Sharing Models of Point Symbol

A. Models of Point Symbol Contrast

1) ArcGIS Point symbol : ArcGIS point symbols contain SimpleMarkerSymbol, CharacterMarkerSymbol, PictureMarkerSymbol, ArrowMarkerSymbol, MultiLayer MarkerSymbol, etc. which are inherited from MarkerSymbol [7]. SimpleMarkerSymbol and ArrowMarkerSymbol are composed of simple graphics. The former including Circle, Square, Cross, X and Diamond. The latter is constituted by arrows in the shape of triangle, which is adjustable in height and width. CharacterMarkerSymbol is established by using TrueType Font. Point symbol can also be plotted by PictureMarkerSymbol which based on images. All these four kinds of point symbols can constitute MultiLayerMarkerSymbol(Fig. 1).

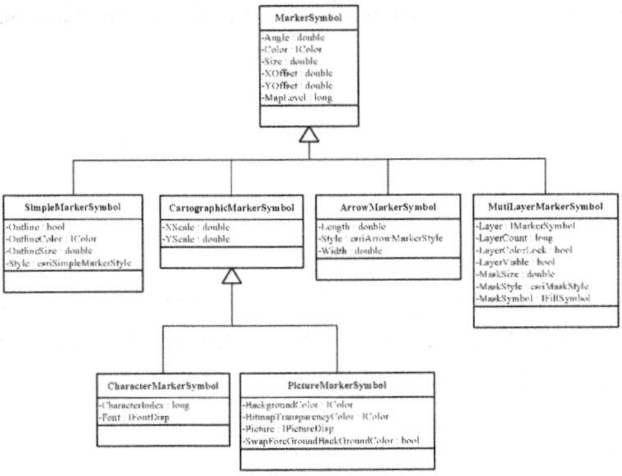

Fig. 1. Model of marker symbol in ArcGIS

2) SLD point symbol: It contains simple point symbol and External Graphic. Simple point symbol including square, circle, triangle, star, cross, x, etc. [8]. External Graphic can be divided into raster pictures and SVG, which is adjustable in attribute such as size, angle and transparency(Fig. 2).

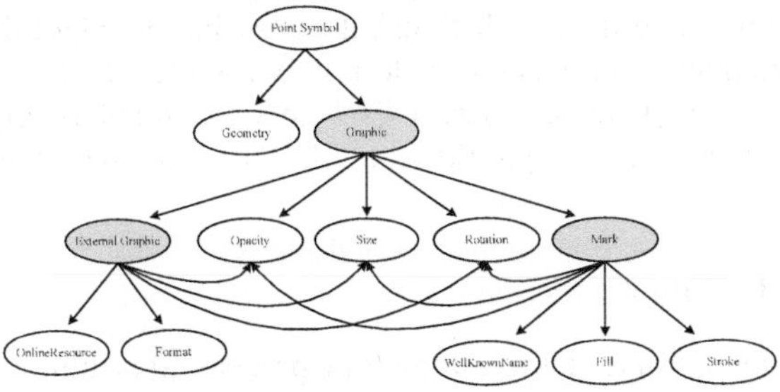

Fig. 2. Model of point symbol in SLD

B. Sharing Point Symbol

Comparative the model of point symbol in two kinds of types, SimpleMarkerSymbol can convert to Mark, Char-acterMarkerSymbol can convert to picture format symbol in External Graphic. Otherwise Mark,External Graphic can also convert to SimpleMarkerSymbol, CharacterMar-kerSymbol and PictureMarkerSymbol.we see that conversion between those two types can be achieved, and may even without damage the model(Fig. 3).

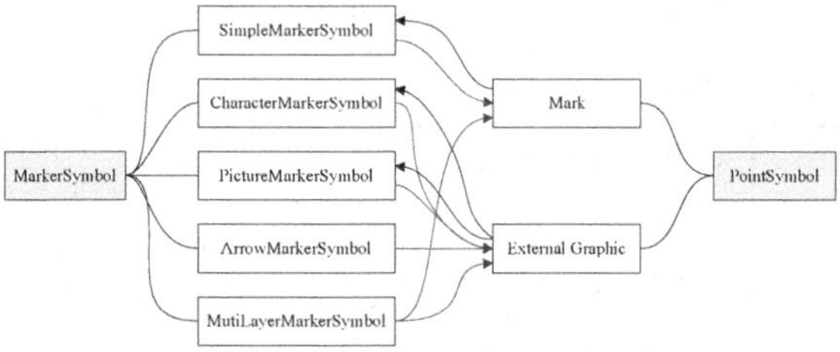

Fig. 3. Point symbol exchange between ArcGIS and SLD

C. Point Symbol Model Analysis

Point symbol in ArcGIS is mainly established by using TrueType font which is design for describe the outline of western language.TrueType font only contains lineseg-ment and quadratic bezier Lines,but most of map symbols are constitutive of regular graphs, consequently many regular graphs need fit with linesegment and bezier Lines.Because of this,the fitting method makes data redundancy and it is also mismatch our habit of mapping.Moreover, TrueType font have nothing but graph,while using TextOut which is a function of windows API to plot characters,every character only have one color,hence the point symbol need to split to several characters.Although SLD point symbol is constituted by SVG,but if symbol get a little bit more complicated as using complex line and fill adjust to point symbol,and It is hard to describe by SVG itself. Consequently, aforementioned two kinds of point symbol model can meet all demands of common GIS and the needs of mapping but it is hard to describe complex point symbol.Otherwise SLD can meet all demands of needs in the symbol sharing.

3 Contrast and Sharing Models of Line Symbol

A. Models of Line Symbol Contrast

1) ArcGIS line symbol: It contains SimpleLineSymb-ol, CartographicLineSymbol, HashLineSymbol, MarkerLineSymbol, PictureLineSymbol and MutiLayerLineSy-mbol. SimpleLineSymbol including Solid, Dashed, Dotted, Dash-Dot and Dash-Dot-Dot, etc. CartographicLi-neSymbol is established by using Mark-Gap mode, HashLineSymbol and MarkerLineSymbol are inherited from CartographicLineSymbol [7], HashSymbol which is a line symbol can be redirected to HashLineSymbol which is established by using seriation of MarkerSymbol. PictureLineSymbol can be construct by raster pictures, which is a rotative course of adjustable in length of centre line. All these five kinds of line symbol constitute MutiLayerLineSymbol(Fig. 4).

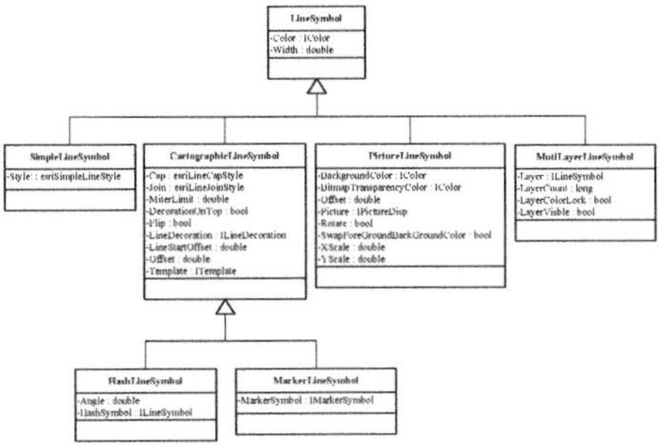

Fig. 4. Model of line symbol in ArcGIS

2) SLD line Symbol: It is constructed on stroke which contains GraphicFill and GraphicStroke [8]. Dash array can be configured in stroke as well. It correspond with Mark-Gap mode in symbol of ArcGIS Cartographic-LineSymbol,and stroke includes several line symbol attribute as opacity, width, linejoin, etc(Fig. 5).

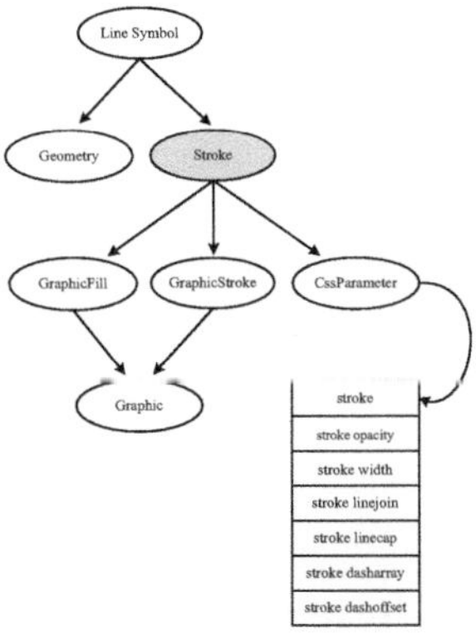

Fig. 5. Model of line symbol in SLD

B. **Sharing Line Symbol**

Comparative the model of line symbol in two kinds of types,All line symbol in ArcGIS can convert to SLD Line Symbol except HashLineSymbol and MarkerLine-Symbol.In those symbol, SimpleLineSymbol convert to simple Stroke, Template in CartographicLineSymbol convert to dasharray which is in Stroke, PictureLine-Symbol convert to GraphicStroke(Fig. 6). Otherwise the conversion between line symbol in SLD and Line Symbol of ArcGIS can be achieved.

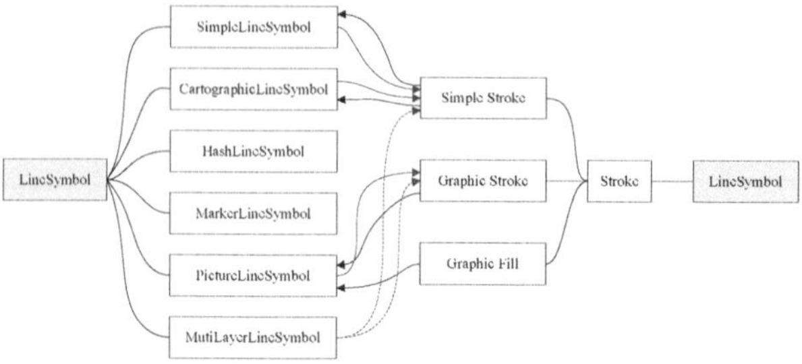

Fig. 6. Line symbol exchange between ArcGIS and SLD

C. **Line Symbol Model Analysis**

Data models of line symbol in ArcGIS can coverage most area of line model,and it is also meet most demands of needs for mapping or GIS,but It is hard to describe some complex line symbol as the symbol of power line(which needs to be adjustable in every location of telegraph poles with the corresponding point symbol), dynamic symbol(Fig. 7). Although we can achieve the upper symbols editing in Representation,but it is Consider things from the pure graphical design and ignore the relationship between symbol and geographic object,while geographic object changed but symbol cannot. SLD can only meet demands of solid line, mark-gap line and graphic line.Those line models are a little part of line symbol,just as the symbol in Fig.7 that SLD cannot describe. Consequently, both two models of line symbol can hardly satisfy the needs of mapping.And it is hard to achieve the SLD sharing because of the simple type of line symbol in SLD.

Fig. 7. Symbols of power line and arrow line

4 Contrast and Sharing Models of Fill Symbol

A. **Models of Fill Symbol Contrast**

1) ArcGIS Fill symbol: It contains SimpleFillSymbol, PictureFillSymbol, MarkerFillSymbol, GradientFillSymbol, LineFillSymbol, DotDensityFillSymbol and MultiLayerFillSymbol [7]. SimpleFillSymbol and GradientFillSymbol are color fill of object, the former is design to be filled by monochromatic color, the latter uses the gradual manner to be filled. MarkerFillSymbol means filled in fill symbol with point symbol, and the LineFillSymbol with line symbol. All these several kinds of symbol group constitute MultiLayerFillSymbol(Fig. 8).

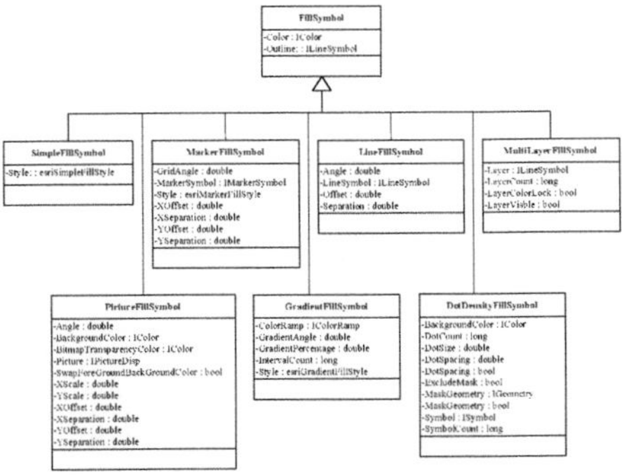

Fig. 8. Model of fill symbol in ArcGIS

2) SLD fill symbol: SLD fill symbol is constituted by Fill and Stroke.the former, contains color fill and graphic fill [8]. Color fill is adjustable in transparency,and graphic fill uses raster pictureto filled in(Fig. 9).

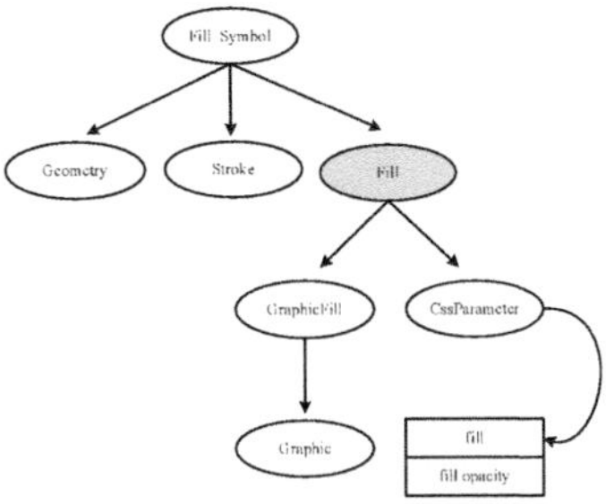

Fig. 9. Model of Polygon symbol in SLD

B. Sharing Fill Symbol

Comparative the model of fill symbol in aforementioned two kinds of types,we can see that only SimpleFillSymbol and PictureFillSymbol in ArcGIS can convert to SLD Polygon Symbol. While all of the SLD Polygon Symbols can convert to FillSymbols in ArcGIS(Fig. 10).

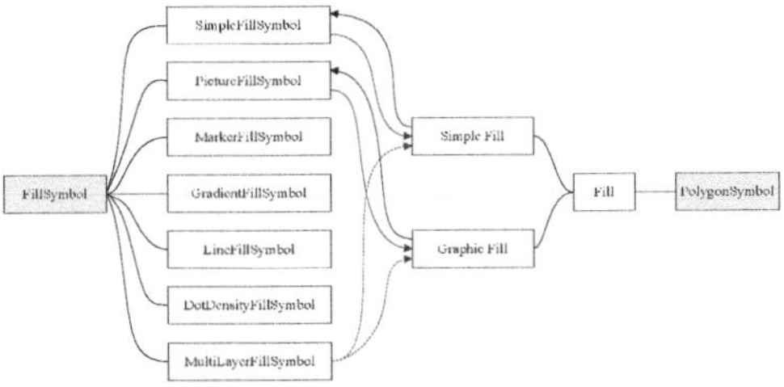

Fig. 10. Fill symbol exchange between ArcGIS and SLD

C. **Fill Symbol Model Analysis**

The model of fill symbol in ArcGIS has a well-defined support structure, and it coverage most area of fill model, which is meet most demands of needs for GIS. But there are lots of fill effects cannot be satisfied with mapping, like we find that interleave and overlap phenomenon to occur when using MarkerSymbol in MarkerFillSymbol, also find out of bounds phenomenon near the area border yet. Although we can achieve parts of problems using Representation, but Representation ignore the relationship between symbol and geographic object. Fill symbol in SLD only contains solid fill and graphic fill. Both of them limited model of fill function. On the other hand, we can use method of raster to adjustable for fill symbol, but this method cannot guarantee precision of mapping and support its qualification job. Consequently, both two models of fill symbol can hardly satisfy the needs of mapping. And it is hard to achieve the SLD sharing because of the simple type of fill symbol in SLD.

5 Conclusion

By comparison, in terms of symbol model, the ArcGIS symbol system is more perfect to meet the demands of GIS and together with the support of Representation, it can basically meet the needs of the general cartography. But, the relationship between geographical object and symbol is ignored by Representation and it only focus on the symbol syntax instead of the symbol semantic. Therefore, the ArcGIS symbol system is not at all perfect. And the SLD symbol system is too unitary to meet the needs of GIS and Cartography. In terms of symbol sharing, the file format of ArcGIS, as ESRI's commercial secret, is not exposed and it’s go against to symbol sharing. The SLD symbol system is positioned on symbol standard in order to share symbols, especially the share basing on WebService. But its own structure is so unitary that it is difficult to meet the demands of professional graphics and the map symbol sharing. In summary, neither ArcGIS nor SLD symbol model can interpret map symbols well and they are difficult for cartographers to design map and share map symbol. In consequence, there is an urgent need for a common map symbol data model, which can contain all the regular graphs completely and learn the advantages of both to achieve the symbol sharing among CAD, GIS, professional graphics software and other cartographic systems.

Acknowledgment. This work was supported by a grant from Funded by Key Laboratory of Geo-informatics of State Bureau of Surveying and Mapping(NO. 201027).

References

1. Wang, J.Y.: Cartology in Information ERA. Engineering of Surveying and Mapping 9, 1–5 (2000)
2. Brinkhoff, T.: Towards a Declarative Portrayal and Interaction Model for GIS and LBS. In: Proc. 8th Conference on Geographic Information Science (AGILE 2005), pp. 449–458. IEEE Press, Los Alamitos (2005)

3. Duarte Teixeira, M., De Melo Cuba, R., Mizuta Weiss, G.: Creating Thematic Maps with OGC Standards through the Web. In: Proc. GML & Geo-Spatial Web Services Conference (July 2005)
4. Weiser, A., Zipf, A.: A visual editor for OGC SLD files for automating the configuration of WMS and mobile map applications. In: Proc. 3rd Symposium on Location Based Services and TeleCartography. Springer, Heidelberg (2005), doi:10.1.1.70.3776
5. Dietze, L., Zipf, A.: Extending OGC Styled Layer Descriptor (SLD) for Thematic Cartography. In: Proc. 4th Int. Symp. on LBS and Telecartography. Springer, Heidelberg (2007), doi:10.1.1.71.6896
6. Neubauer, S., Zipf, A.: Suggestions for Extending the OGC Styled Layer Descriptor(SLD) Specification into 3D – Towards Visualization Rules for 3D City Models. In: Proc. of. Urban Data Management Symposium, UDMS 2007 (October 2007), doi:10.1.1.71.9606
7. ESRI, DisplayObjectModel Display, Symbols, Color and Feedbacks
8. Open GIS Consortium, Styled Layer Descriptor Implementation Specification, version 1.0.0, document 02-070

A Three-Dimensional Modeling Method for Irregular Columnar Joints Based on Voronoi Graphics Theory*

Wen-tang Zheng[1], Wei-ya Xu[2], Dong-xu Yan[2], and Hua Ji[2]

[1] Guangdong Electric Power Design Institute Guangdong Power Grid Corporation Guangzhou, Guangdong Province, China
[2] Geotechnical Engineering Research Institute University of Hohai Nanjing, Jiangsu Province, China
wen-tangzheng@qq.com

Abstract. Irregular columnar joints are generally developed in basalt. This special rock mass structure has brought some new engineering problems in geotechnical engineering, water conservancy and Hydro Power Station. Geometrical characteristics of columnar jointed structures are described and simulated by introducing the Voronoi graphics theory. Based on geometric transformation method, the visualization model of AutoCAD and numerical model of discrete element method are built with AutoLISP and Fish languages. The three-dimensional modeling method for columnar joints is applied for the Baihetan Hydro Power Station in China. The simulation results show that, Voronoi graphic has great significance for studying the nonlinear mechanical properties of irregular columnar jointed basalt.

Keywords: Irregular columnar joints, Voronoi graphics, visualization model, discrete element model, 3DEC.

1 Introduction

The irregular columnar jointed structures are primary columnar morphological tensile fractures. As shown in Fig 1, they are common in volcanic lava, dried flour, clotted ice, epidermal cells under leaf, quenched metal and quenched glass. Reports of columnar joint in volcanic lava can be traced back to 1693 while Bulkeley described the Giant's Causeway in Ireland. Müller(1998)[1]. Goehring and Morris(2007)[2] found the mixture of flour and water while drying into a certain humidity had the same process with the diffusion and extract of heat in lava. Their result showed that flour in the appropriate temperature would form a stable columnar fracture which was analogous with columnar jointed basalt. Cole (1988)[3] found columnar fracture often occurred in clotted ice. French (1925)[4], Hull and Caddock (1999)[5] found columnar fracture also existed in quenched metal and quenched glass. Reports about columnar joint in geological rock mass are generally in basic hypabyssal rock or eruptive rock[6-7].

* This work is partially supported by the National Natural Science Foundation of China (50911130366, 50979030).

J. Zhang (Ed.): ICAIC 2011, Part V, CCIS 228, pp. 62–69, 2011.

Widely distributed columnar basalt lava gave birth to the environment of columnar joints, which brought up many well-known landscapes of columnar joints at home and abroad. Fig 1(f) shows the famous Giant's Causeway in Ireland which is a popular tourist attraction. On the other hand, this special jointed rock mass has also led to new engineering problems in geotechnical engineering and hydraulic engineering. Such as the Permian continental flood basalt in southwest China, which the Emeishan basalt is one of the main foundation rock in this region. For instance, tongjiezi hydropower station at Dadu River, ertan hydropower station at Yalongjing River, xiluodu hydropower station and baihetan hydropower station planned to be built at Jinshajiang River and so on. They are all based on basalt. Therefore, the study of engineering mechanical characteristics of rock structure of basalt, especially the columnar jointed structure, has great significance in water-power engineering.

Currently, the studies on columnar jointed structure are mainly about the geometric shape and formation mechanism of columnar joints in terms of geology and mechanics. However, few researches are on numerical model and engineering application. In this article, irregular columnar joints are regarded as Voronoi graphic in the perspective of geometry, and the methods how to construct the 3D visualized numerical model of columnar jointed rock with the math theory of Voronoi are detailed introduced.

2 Voronoi Graphics

Voronoi graphics, as shown in Fig1(a), can be traced back to the book *Principles of Philosophy* written by Descartes[8] in the 17th century. Descartes considered that stars would hang in the balance and develop many swirls fields. The swirls fields coincide with each other, restrict to balance and separate the stellar system into many polygon systems containing one star respectively.

This thought was subsequently adopted in many domains of science. It has different names in different fields, for example, the Medial axis transform in biology and physiology, Wigner-Seitz field in chemistry and physics, Domains of action in crystallography, Thiessen polygons in meteorology and geography. Dirichlet(1850)[9] and Voronoi(1908)[10] are the earliest mathematician who gave a systemic research on the mathematic features of this graphics. They considered this graphics was a geometry of plane domain measured by Euclidean distance. The graphics was respectively named Dirichlet Tessellation and Voronoi Diagram and they have been the long-term use. Voronoi diagram is similar to some natural structure and possess amazing mathematic features. With the popularization of computer technology and development, Voronoi diagram has been extensively researched and applied in meteorology, geology, topography, archeology, molecular chemistry, ecology, computer science and so on. It has become a powerful tool of resolving interrelated geometry problems. Detailed introductions of the Voronoi graphics can be found in Aurenhammer (1991)[11] and Kabe(1992)[12]. They had given a detailed review on the history, definition, characteristic, arithmetic, extension and application of Voronoi graphics. Fortune(1992)[13] analyzed and compared every arithmetics of Voronoi graphics. The monograph about Voronoi graphics can be referred to Okable(2000)[14] and Mark de Berg(2000)[15].

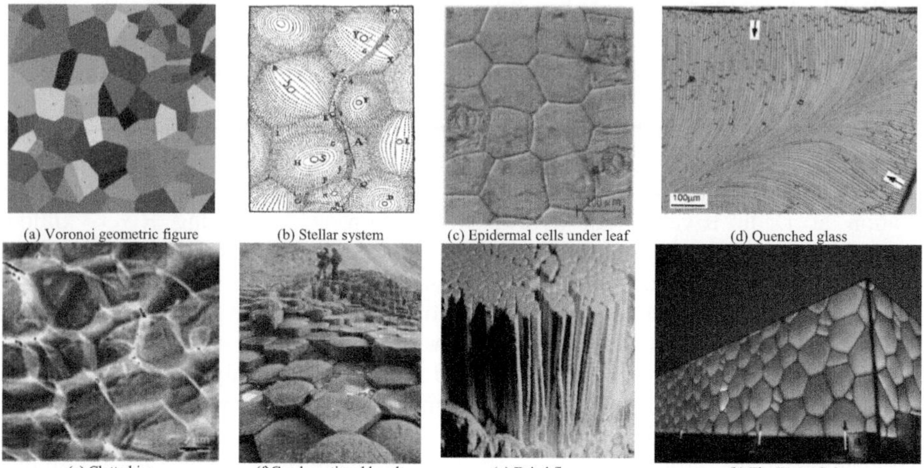

(a) Voronoi geometric figure (b) Stellar system (c) Epidermal cells under leaf (d) Quenched glass

(e) Clotted ice (f)Condensational basalt (g) Dried flour (h) The Water Cube

Fig. 1. Columnar jointed structure in nature

The basic concept of Voronoi graphics can be simply described as follows: there are n ignition sources within the region. These ignition sources are ignited at the same time and spread to all directions at the same speed, and then the figure formed by burning extinguished place is the Voronoi graphic. This process has the similarity with the forming process of columnar joints. When solidified, lava contracts with different points as the center at the surface. When the rock mass around this point is separated from the rock mass around other points, the columnar joint is formed. Describing the columnar joints with Voronoi graphics is of certain basis.

3 Mathematic Principle of Voronoi Graphics

Assume p_1 and p_2 are two points of the plane, L is the perpendicular bisector of the line segment p_1p_2. L separates the plane into two parts L_L and L_R. The point p_i in L_L has the characteristic: $d(p_l,p_1) < d(p_l,p_2)$, where $d(p_l,p_1)$ represents the Euclidean distance between p_1 and p_i (i=1,2). Points in plane L_L, defined as $V(p_i)$, are nearer to locus of point p_i than any other points in the plane, seen in fig. 2.

Applying $H(p_1, p_2)$ as the half plane L_L, while $L_R = H(p_2,p_1)$, then we have $V(p_1)=H(p_1,p_2)$, $V(p_2)=H(p_2,p_1)$. A set of points S is given with n points in the plane, $S=\{ p_1,p_2,\ldots,p_n\}$. Define the equation as follows:

$$V(p_i)=\bigcap_{i\neq j} H(p_1,p_2) \qquad (1)$$

$V(p_i)$ represents the intersection of n-1 half planes formed by locus of points nearer to p_i than any other points. It is a convex polygon zone with less than n-1 boundary called Voronoi polygon or Voronoi domain of the register of relevance p_i. As to every point of S, a certain Voronoi polygon can be made; these n different polygons construct the Voronoi graphics.

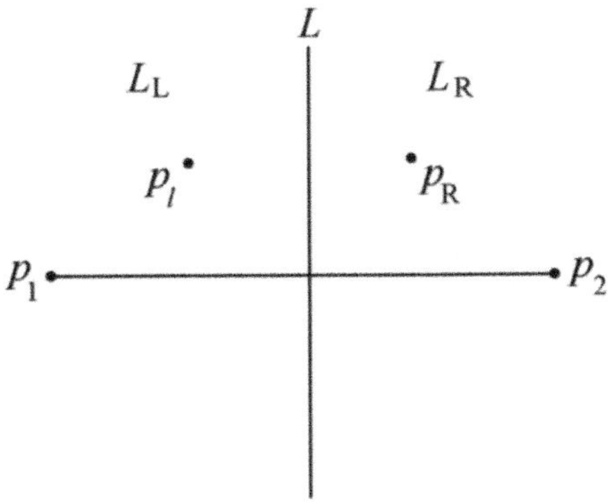

Fig. 2. Sketch of V(p1) and V(p2)

Formula (1) is used to construct the intersection of n-1 half planes, and this way can get the Voronoi polygon of point pi, then get the Voronoi graphics of S with these points. This disquisition applied increment construction method to form Voronoi graphics. The details refer to Peide Zhou's research[16]. Note the boundary and point information is written as a input dxf format, while the plane Voronoi graphics in AutoCAD are drawn as the output interface. Boundary intercepting of polygon zone can be realized to form the diagram as show in fig. 3. The maximum cylindrical diameter is 27mm and the minimum is 15mm, which is close to the geometric properties of columnar joints of level I at Baihetan hydropower station in China .

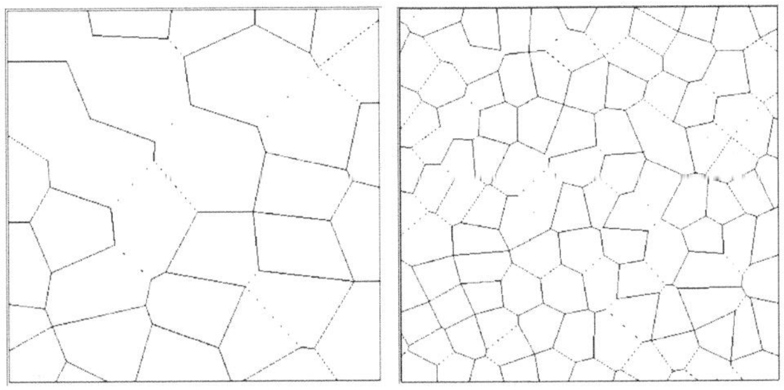

(a)1 m×1 m (b) 2 m×2 m

Fig. 3. Plane Voronoi graphic

4 Simulation of Irregular Columnar Joints

Hart and Cundall (1985)[17] used a simplified hexagonal cylindrical model to simulate the columnar joints of BWIP. But in fact, influenced by occurrence environment and geological structure and so on, the cylindrical surface is not just hexagonal. Statistics of cylindrical surface types of famous columnar joints suggests that even the well-developed hexagonal cylindrical columnar joints from Irish Giant's Causeway still contain a lot of pentagonal and quadrangular cylindrical surfaces.

The cylindrical surface of columnar joints at Baihetan hydropower station is mainly quadrangle and pentagon, whereas the hexagon cylindrical surface only takes a very small proportion, seen in fig 4. Therefore, the rock mass at Baihetan belongs to irregular columnar jointed rock mass. It is not reasonable to simulate with hexagon cylindrical surface. It can be known from the in situ test that columnar jointed rock is influenced not only by geometric shape of cylinder but also by the whole mechanical characteristics effect of hidden joints, bedding and in layers shearing band. Therefore, irregular tilted columnar jointed rock mass belongs to 3D problem. Not only the construction of cylindrical surface should be investigated, but also the transform of 3-dimensional graphics. Meanwhile, the interface of 3D visualized AutoCAD model and 3D discrete element numerical model of 3DEC should also be developed. Here the 3DEC Voronoi columnar joints pre-processer is developed based on AutoLisp language and Fish language and applied in the numerical analysis.

Columnar jointed strata have certain dip angle, and 3DEC uses the left-handed coordinate system. Thus, geometric transformation is required to the 3D graphics of geodetic coordinate system. There are 3D translation, 3D scaling, 3D reflection, and 3D rotation etc.

It is also a good way to form Voronoi graphics based on Delaunay triangle. Black points in the diagram are the vertexes of Delaunay triangle. Voronoi vertex is corresponding to the centroid of Delaunay triangle. The details can be seen in the reference [18, 19].

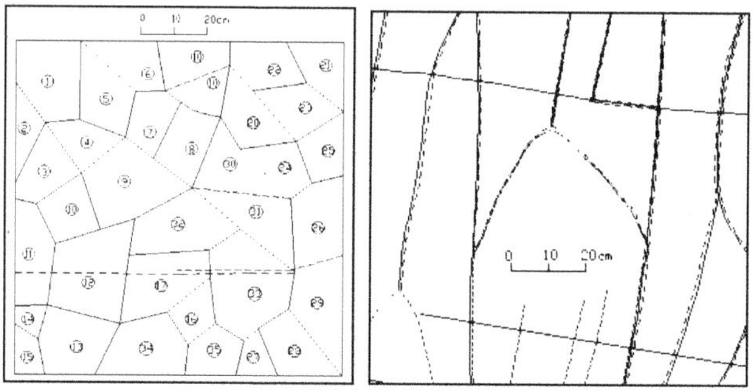

Fig. 4. The geologic sketch map of irregular columnar jointed basalt

Assume a point before transformation is O(x,y,z), and the point after transformation is O'(x,y,z). Then the geometric transformation [T] of P→P' is as follows:

$$O'=[x' \quad y' \quad z' \quad 1]=O\cdot T=[x \quad y \quad z \quad 1]\cdot\begin{bmatrix} a & b & c & 0 \\ d & e & f & 0 \\ h & i & j & 0 \\ l & m & n & 1 \end{bmatrix} \tag{2}$$

$$(x'\ y'\ z'\ 1)=(x\ y\ z\ 1)\cdot P(t_x\ t_y\ t_z)=(x\ y\ z\ 1)\begin{bmatrix}1&0&0&0\\0&1&0&0\\0&0&1&0\\t_x&t_y&t_z&1\end{bmatrix} \quad (3)$$

$$(x'\ y'\ z'\ 1)=(x\ y\ z\ 1)\cdot R(\alpha,\beta,\gamma)=(x\ y\ z\ 1)\begin{bmatrix}l_1&m_1&n_1&0\\l_2&m_2&n_2&0\\l_3&m_3&n_3&0\\0&0&0&1\end{bmatrix} \quad (4)$$

Geometric transformation matrix [T] can be decomposed to the combination of several basic transformation matrixes: 3D translational matrix [P], 3D scaling matrix [S], and rotation of coordinate matrix [R].

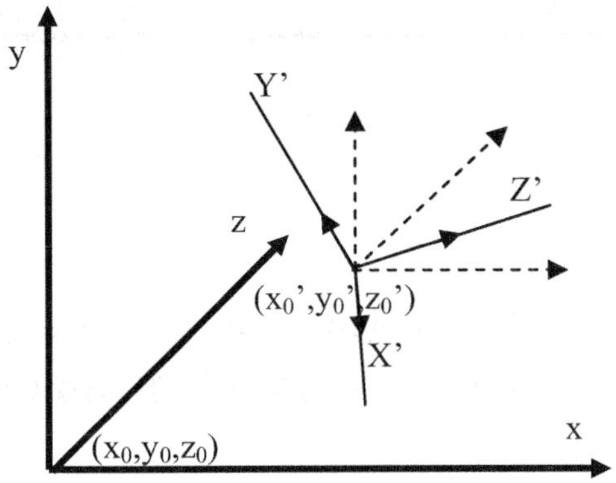

Fig. 5. Sketch of geometric transformation coordinate system of 3D figures in computer graphics

The transformation of coordinates from oxyz to o'x'y'z' is:

$$(x',y',z',1)=(x,y,z,1)\cdot P(x_0,y_0,z_0)\cdot S(s_x\ s_y\ s_z)\cdot R(\alpha,\beta,\gamma) \quad (5)$$

Table 1. The program structure of Voronoi visualization model and 3DEC discrete simulated model

Function	Grammatical structure of AutoLisp and Fish
Generate AutoCAD visualization model	(entmake (list '(0. "3DFACE") '(8. "columnar joint") '(62 . "color") (cons 10 (list x1 y1 z1)) (cons 11 (list x2 y2 z2)) (cons 12 (list x3 y3 z3)))
Generate 3DEC Irregular columnar joints	POLY PRISM mat n region o & A (x1,y1,z1) (x2,y2,z2) (x3,y3,z3) …& B (x1,y1,z1) (x2,y2,z2) (x3,y3,z3) …
Generate 3DEC oblique structural plane	JSET id n dd a dip b persistence p & Origin (x1,y1,z1)

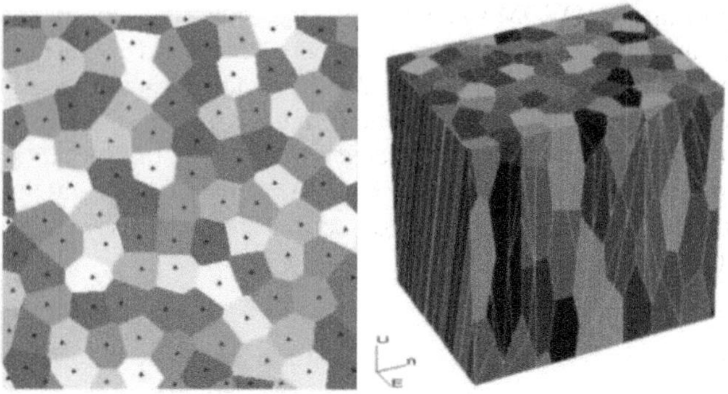

Fig. 6. Visualization model in AutoCAD and simulated model in 3DEC of irregular columnar jointed basalt

Based on geometric transformation method, visualization model of columnar joints and grammatical structure and model of discrete element are formed with AutoLISP language of AutoCAD and FISH language of 3DEC, as shown in table 1 and fig 6. Although 3DEC is the 3D version of UDEC, and due to the difficulties of simulating the 3D irregular Voronoi joints net, it does not have the inner installation Voronoi chessboard mosaic joints generator.

4 Conclusions

This article analyzed the basic geometrical characteristics of columnar jointed structure and introduced Voronoi graphics to simulate irregular columnar jointed structure. According to the characteristics of irregularities and cylinder deflection of columnar jointed rock at Baihetan hydropower station and based on the spatial 3D graphical transform technology of computer graphics, this article investigated the visualized of irregular columnar joints and generation method of numerical model. Combining with AutoLisp language and FISH language, Voronoi irregular deflection columnar joints pre-processer of 3DEC was developed. Studies have shown that the Voronoi graphics method has great significance for studying the nonlinear mechanical properties of irregular columnar jointed basalt.

Acknowledgment. Thanks to project (50911130366) and project (50979030) supported by the National Natural Science Foundation of China.

References

1. Mueller, G.: Experimental simulation of basalt columns. Journal of Volcanology and Geothermal Research 86(4), 93–96 (1998)
2. Goehring, L., Mahadevan, L., Morris, S.W.: The dynamic similarity between columnar joints in basalt and starch. American Geophysical Union Fall Meeting (2007)
3. Cole, D.M.: Crack nucleation in polycrystalline ice. Cold Regions Science and Technology 15(1), 71–79 (1988)

4. French, J.W.: The fracture of homogeneous media. Trans. Geol. Soc. Glasgow 17(1), 50–58 (1925)
5. Hul, D., Caddock, B.D.: Simulation of prismatic cracking of cooling basalt lava flows by the drying of sol-gels. Journal of Materials Science 34(23), 57–67 (1999)
6. Li, J.-l., Wang, P.-j., Zheng, C.-q., et al.: Characteristics and Genesis of Rhyolite with Columnar Joints from the Yingcheng Formation in South- eastern Uplift of the Songliao Basin. Journal of Jilin University (Earth Science Edition) 37(6), 56–62 (2007) (in Chinese)
7. Shan, Z.-Q.: In Search of pillars around China. Chinese National Geography 586(8), 32–64 (2009) (in Chinese)
8. Descartes, R.: Principia Philosophiae. Ludovicus Elzevirius, Amsterdam (1664)
9. Dirichlet, G.L.: Uber die Reduktion der positiven quadratischen For- men mit drei unbestimmten ganzen Zahlen. J. Reine. Angew. Math. 40, 209–227 (1850)
10. Voronoi, G.M.: Nouvelles applications des parametres continus a la theorie des formes quadratiques. Deuxieme Memoire: Recherches sur less parall ello edres primitifs. J. Reine. Angew. Math. 134, 198–287 (1908)
11. Aurenhammer, F.: Voronoi diagrams: A survey of a fundamental geometric data structure. ACM Comput. Surv. 23(3), 345–405 (1991)
12. Okabe, A., Boots, B., Sugihara, K.: Spatial Tessellations: Concepts and Applications of Voronoi Diagrams. John Wiley & Sons, Chichester (1992)
13. Fortune, S.: Voronoi diagrams and Delaunay triangulations. In: Du, D.Z., Hwang, F.K. (eds.) Computing in Euclidean Geometry, 1st edn. LNSC, vol. 1, pp. 193–233. World Scientic, Singapore (1992)
14. Okabe, A., Boots, B., Sugihara, K., et al.: Spatial Tessellations: Concepts and Applications of Voronoi Diagrams, 2nd edn. John Wiley, Chichester (2000)
15. de Berg, M., van Kreveld, M., Overmars, M., et al.: Chapter 7: Voronoi Diagrams. In: Computational Geometry, 2nd edn., pp. 147–163. Springer, Heidelberg (2000) ISBN 3-540- 65620-0; Includes a description of Fortune's algorithm
16. Zhou, P.: Computational geometry—design and analysis of algorithm, 2nd edn. Tsinghua university press, Beijing (2005)
17. Hart, R.D., Cundall, P.A., Cramer, M.L.: Analysis of a loading test on a large basalt block. In: Proceedings of the 26th US Symposium on Rock Mechanics Research & Engineering Applications in Rock Masses, vol. (2), pp. 759–768 (1985)
18. Riedinger, R., Habar, M., Oelhafen, P., et al.: About the Delaunay-Voronoi tesselation. Journal of Computational Physics 74(1), 61–72 (1988)
19. Franco, P.P., Michael, L.S.: Computational Geometry: An Introduction. Springer, Heidelberg (1985)

Key Technology Research on Remote Sensing Monitoring of Methane Emission in Coal Mine Area*

Hui Ci, Yong Qin, Hui Yang, Guoqiang Li, and Gefei Feng

Key Laboratory of CBM Resources and Pooling Process, MOE,
China University of Mining and Technology Xuzhou, Jiangsu Province, China
College of Kewen & College of Linguistic Sciences Xuzhou Normal University Xuzhou, Jiangsu Province, China
ci_hui@yahoo.cn

Abstract. Methane is a kind of greenhouse gas that strongly absorbs Earth outward long-wave radiation, a gas whose greenhouse effect is 21 times worse than carbon dioxide. Coal mine area is one major source of industrial methane emission and the emission sources are principally coal mines and coal bed methane (CBM) ground development zones. One of the type areas at present is southern Qinshui Basin in Shanxi province. With the support from the advantages of remote sensing monitoring technology in real time monitoring of mine area environment and on the basis of the optimal waveband from the sieving process of remote sensing monitoring of methane, this paper takes southern Qinshui Basin as the focus of research in order to perfect and progress the theoretical model of methane remote sensing response, to develop the hyperspectral monitoring method of methane concentrations, with which furthermore, to make a preliminary analysis of the temporal and spatial distribution regularity of atmospheric concentrations of methane so as to provide groundwork for further research and development of real time and high sensitive monitoring method of the atmosphere environment in mine areas.

Keywords: coalmining region, remote sensing, atmospheric monitoring, hyper spectral.

1 Research Background

Methane is a kind of greenhouse gas that strongly absorbs Earth outward long-wave radiation and its global warming potential (GWP, the total energy that 1000g gases can absorb within a certain period of time) is 21(assuming the GWP of CO2 is 1.0), that is, the greenhouse effect of every Moore equivalent weight of CH4 is 21 times as that of CO2, which means CH4 has a far worse contribution than CO2 does to greenhouse effect.

Although the normal atmospheric concentration of CO2 is 300 times as the concentration of CH4, the latter exerts a tremendous influence on global climate.

* This work is partially supported by the National Natural Science Foundation of China (Grant No. 41001230) to Hui Yang, and Important National Science & Technology Specific Projects (Project number: 2008ZX05034) to Yong Qin.

J. Zhang (Ed.): ICAIC 2011, Part V, CCIS 228, pp. 70–76, 2011.

Moreover, a series of chemical and photochemical reactions in troposphere and stratosphere also involve CH4, and the reaction products also possess the nature of greenhouse gases.

Human industrialization activity causes the continuous increase of atmospheric concentrations of CH4 in the last 100 years, and the current average atmospheric concentration, now being CH4 is 1.72×10^{-6}, is still increasing with a speed of 0.8% to 1.0% every year. It is estimated that with the generating speed of methane at present, CH4 will play a major role in the greenhouse effect in decades. Therefore, climate change caused by greenhouse gases, including CH4, has already become the theme of many international conferences.

Coal mine area is one major source of industrial methane emission and the emission sources are principally coal mines and CBM ground development zones. The main emission sources consist of coal mine exhaust wellhead, coal bunker from coal washery, coal gangue dump in coal mine area and ground CBM (gas drainage) hole.

According to the estimation from Chongqing Branch of China Coal Research Institute, in China, an average of 10 m^3 CH4 is vented to atmosphere in producing 1 ton raw coal. In 2000, CH4 emission of coal mine in China reached 9.625 billion m^3 which accounted for some 60 percent of the global emissions and is equivalent to138 million tons of CO2.

The accumulated national coal-mine gas drainage, in 2009, was 5.6 billion m^3, while the utilization rate was only 32%. 3.8 billion m^3 drainage CH4 of the total was injected into the atmosphere, and CH4 discharged into the atmosphere through ventilation air methane (VAM) was far greater than the previous number.

At present, the daily ground CBM drainage is about 2million m^3 with a utilization amount being 800,000 m^3, which means that, each year, CH4 that is discharged into the atmosphere through this source is 430 million m^3 around.

Therefore, the State Environmental Protection Administration promulgated national standard on April 25th, 2008—"the Coalbed Methane (Coal-mine gas) Emission Standard (interim)" (GB, 21522-2008), in order to curb the mine methane emission, to improve the utilization rate of coal mine methane and to alleviate global greenhouse effect.

For the time being, however, the insufficiency of real-time, effective and high sensitive monitoring method, the incomplete relevant principles, approaches and technology are still serious problems in alleviating methane emission in coal mine area.

Therefore, With the support from the incomparable advantages of remote sensing monitoring technology in real time monitoring of mine area environment and the basis foundation of the optimal waveband from the sieving process of remote sensing monitoring of methane based on ultimate remote sensing principle.

This paper takes southern Qinshui Basin as the focus of the research to perfect and progress the theoretical model of methane remote sensing response, to develop the hyperspectral monitoring method of methane concentrations, with which furthermore, to make a preliminary analysis of the temporal and spatial distribution regularity of atmospheric concentrations of methane so as to provide groundwork for the further research and development of real time and high sensitivity monitoring method of the atmosphere environment in mine areas, see Fig. 1.

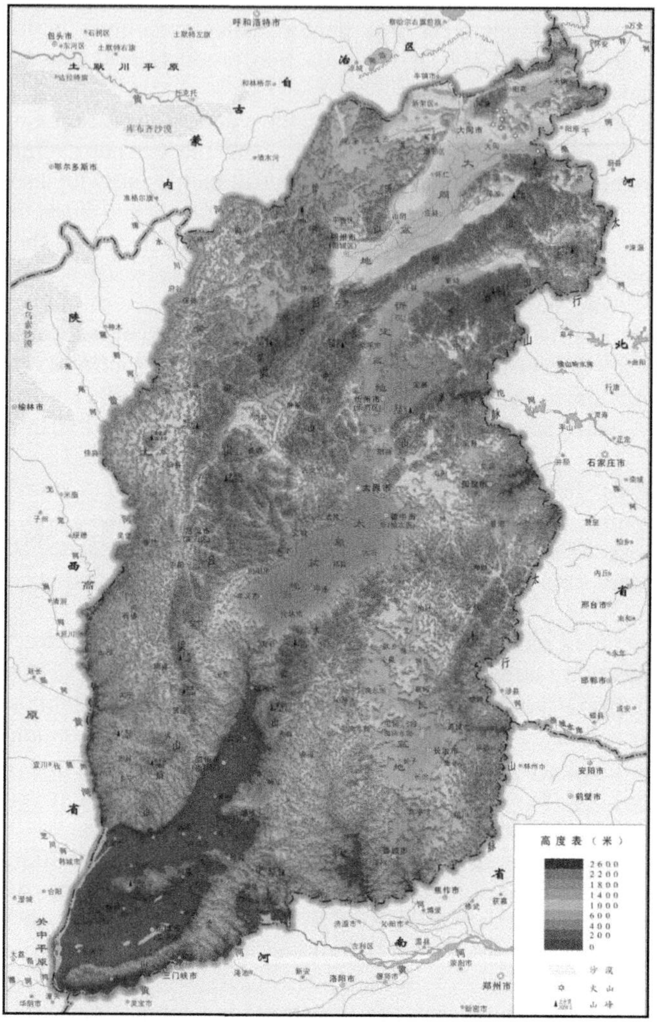

Fig. 1. Topographic map of Shanxi province

2 Primary Research Contents

A. Screening Optimal Waveband from Remote Sensing Monitoring of Methane

With the above conceivable band as the main research object, laboratory study on methane absorption/radiation characteristics is carried out to screen waveband with high absorbability and sensitivity for atmospheric methane gas and to establish foundation for selecting remote sensing images that is appropriate for monitoring methane gas.

B. Theoretical Model of Methane Remote Sensing Response

Laboratory spectrum analysis, system measurement and analysis of methane of different concentrations are conducted to determine the quantitative relation between specific band and methane concentrations response.

Meanwhile, spectral method is further employed to measure and analyze the characteristics of remote sensing images response of atmospheric methane concentrations. On the basis of that, different functions are chosen to conduct simulation to establish the quantitative response model between atmospheric methane concentrations and the spectrum value of remote sensing images.

C. Remote Sensing Monitoring Model of Atmospheric Methane Concentrations in Coal Mine Area

With southern coal mine zone/CBM development base of Qinshui Basin as the main object, spectral analysis of near-surface atmospheric samples is carried out. Based on this, further contrast analysis of the coupling relation between ground objects spectrum in mine area and remote sensing response is implemented in order to build remote sensing monitoring model of atmospheric methane concentrations in coal mine area and apply this model to the preliminary analysis of temporal and spatial distribution of the atmospheric methane concentrations.

3 Technique Procedures

Based on the processing of actual measurement spectral data of surface, the analysis of satellite hyper spectral image and spectral analysis of methane in laboratory and so on, this paper use hyper spectral data mining technology, weak information extraction technology, spectral modeling and classification techniques, to establish the correlation between the methane concentration in mining area and spectrum in remote sensing image, in order to monitor the concentration of methane gas in mining area by using remote sensing. The procedures of research and technical methods are shown in Fig. 2.

Site selection study in southern Qinshui Basin, the typical conditions of the main source of methane emissions is diverse, larger and specific types of emissions sources are concentrated. Area Jincheng and Luan existing two million-ton mines are large domestic coal mining area, the distribution of a large number of high-gas coal mine, a large number of mine drainage gas methane into the atmosphere and the lack of wind.

The area is China's largest CBM surface development base, has more than 3,000 CBM wells construction of mouth, more than 1,100 ports in operation, but currently only 40% utilization, most of the coal bed methane (gas) vent and into the atmosphere.

In practice, based on hyper spectral measurements on ground and satellite hyper spectral data processing and analysis technology, this paper systematically study the spatial distribution of methane in the mining area and the spectral variation discipline of different concentrations of methane gas.

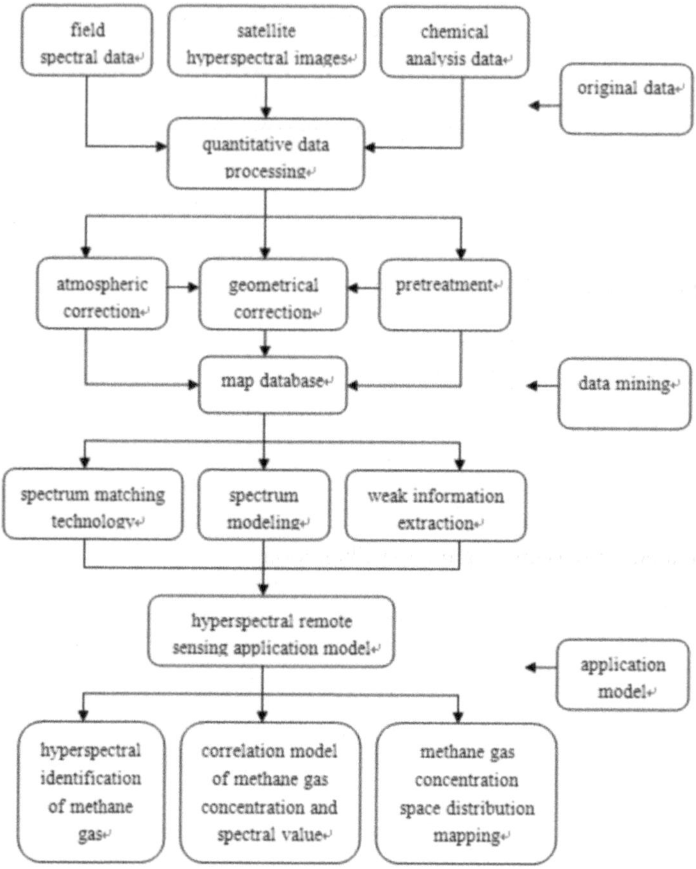

Fig. 2. Technical processes and methodology

By analyzing the relationship between the different concentrations of methane gas and spectral of remote sensing image, this paper presents a remote sensing monitoring model for methane gas, in order to offer the possibility to monitor the methane emissions of mining area in real-time quickly and high sensitively.

4 Key Technologies

A. Technique of High-Precision Radiation Correction

The accuracy of the spectrum information is the premise condition to apply hyperspectral data to directly identify the ground feature, and the radiation correction technique is one of the key technologies is to realize high precision spectra data recovery and an important segment for the quantification of hyperspectral data.

Radiation correction technique mainly studies the transmission of atmospheric radiation, in other words, this technique inverts real ground objects reflectance through the radiometric calibration of sensor and calculation of atmospheric correction model.

At the present stage, the atmospheric correction technique mainly includes two categories: the empirical approach and the model method. The empirical line method, image internal average relative reflectance, flat field and other methods compromise the empirical approach. The model method is mainly atmosphere correction model on the grounds of radiative transfer theory, including Modtran, Lowtran, 5S and 6S, ATREM and ACORN, etc. This study was to make a combination of atmospheric radiative transfer model and the measurement of selected ground targets' spectral radiance to conduct reflectivity inversion calculation.

B. Formation Mechanism and Parameter Selection of Characteristics of Methane Gas Spectral Absorption

In remote sensing physics research fields, due to the differences among earth's surface objects such as chemical composition, molecular structure, the surface status, etc. The typical, unique and diagnostic spectrum absorption characteristics are formed, which are the theoretical basis of employing hyperspectral remote sensing technology to directly identify ground features.

Numerous studies show that the spectrum absorption characteristics presented in the interaction between ground features and sun radiation energy depend on factors such as the atom amount that ground feature themselves contain, bonding force, geometric features of atom distribution (molecules and crystal structure) and the formed crystal field, electric field and magnetic field, etc.

Thanks to the extreme complexity of the interaction mechanism of the two, many dimensions have an effect on the final characteristics of spectrometric behaviors, which requires us to apply a variety of complementary theory and take the experimental data as basis in such a way to implement research and analysis on the regularity of spectrum characteristic change and variation of ground objects (methane gas), to probe into the formation mechanism of the absorption feature, and to solve the problem of fuzziness exists in remote sensing data expression. Based on this, research on quantifying parameter selection and distinguishing of spectrum absorption characteristics of methane gas can be conducted.

C. Technique of Hyperspectral Data Mining and Weak Information Extraction

Data mining and the knowledge discovery is to extract effective information from the mass data and to form multi-stage data processing of knowledge scheme that people can understand.

Specific to application and research field of remote sensing monitoring of mine area methane, hyperspectral data mining and the knowledge discovery means to extract methane' information such as spectral information, radiation information, space information and phase information by means of ground hyperspectral data measurement and spectral data processing analysis, satellite hyperspectral imaging acquirement and classification and recognition technology of weak information.

With this foundation and combination of traditional geological analysis and laboratory testing technology, we can get a full access to knowledge on the concentration changes and distribution range of methane gas.

Acknowledgment. The authors would like to thank the financial support from the Important National Science & Technology Specific Projects (No. 2008ZX05034) and the National Natural Science Foundation of China (Grant No. 41001230).

References

1. Zhang, F., Xu, L.: Effect of Methane on Global Warming and Mitigating Measures. Mining Safety & Environmrntal Protection 31(5), 6–10 (2004)
2. Yang, Z.: Atmospheric trace gases and global warming. Environment Herald (1), 3–6 (1996)
3. Zheng, S., Wang, Y., Wang, Z.: Amount of Methane Exhausting to Atomosphere in Coal Mines of China. Safety in Coal Mines 36(2), 29–33 (2005)
4. Zhang, X.: Greenhouse Gas and Global Environment Change. Journal of Shenyang College of Education 3(4), 109–111 (2001)
5. Chen, H., Wang, M., Wen, Y., Wang, G.: Background Concentration of Atmospheric CO2, CH4 and N2O at MT. Waliguan and Xinglong in China, Quarterly Journal of Applied meteorology 14(14), 402–409 (2003)
6. Huang, J., Wan, Y., Liu, L., et al.: The Character and Application of Modis. Geospatial Information 1(4), 20–28 (2003)
7. Zhou, L., Tang, J., Wen, Y., et al.: Characteristics of Atmospheric Methane Concentration Variation at MT. Waliguan, Quarterly Journal of Applied meteorology 9(4), 385–391 (1998)
8. Yin, D., Nan, Z., Jin, C.: Current Status of Study of Ecology in Mining Areas and Its Perspectives. Scientia Geographica Sinica 24(2), 38–44 (2004)
9. Plaza, A., Martinez, P., Perez, R., et al.: Spatial/Spectral endmember extraction by multidimensional morphological operations. IEEE Trans. Geosci. Remote Sensing 40(9), 2025–2041 (2002)
10. Lillesand, T.M., Kiefer, R.W.: Remote Sensing and Image Interpretation, 50th edn. John Wiley and Sons Inc., New York (2004)

Characterization of Attributes in Consistent Decision Formal Contexts

Hong Wang[1] and Minggang Du[2]

[1] College of Mathematics and Computer Science Shanxi Normal University, Linfen, 041004, P.R. China

[2] School of Urban and Environment Science Shanxi Normal University, Linfen, 041004, P.R. China

hongwang1980@sina.cn

Abstract. Formal concept analysis of binary relations has proved to be useful in resolve of many problems of theoretical or interest. In this paper, we discuss characterizations of three different types of attribute sets in consistent decision formal contex. Characteristics of three different types of attribute are examined, and the necessary and sufficient conditions to judge them are also obtained.

Keywords: Formal Context, Concept Lattice, Consistent Set, Attribute Characterization.

1 Introduction

Formal concept analysis is a mathematical framework developed by Wille[1], that is useful for representation and analysis of data. Formal concept analysis is formulated based on the notion of a formal context, which is a binary relation between a set of objects and a set of attributes. It reflects the relationship of generalization and the specialization among concepts, it thereby is more intuitional and more effective to research on reducing and discovering knowledge. Formal concept analysis as a kind of very effective methods for data analysis have been wildly applied in the field of machine studying, artificial intelligence and knowledge discovery, and so on.

Rough set theory and formal concept analysis have strong connection. Many efforts have been made to compare and combine the two theories [2-5]. The notions of rough set approximations have been introduced into formal concept analysis, and the notions of formal concept and concept lattice have also been introduced into rough set theory by considering different types of formal concepts[4]. Compared with the rough set theory, there is less effort investigated on the issue in concept lattice theory. In [8], concepts of consistent decision formal context and properties of consistent decision formal context were defined, and approaches to knowledge reduction with two requirements in these decision formal contexts were proposed. We try to investigate characteristics of three different types of attribute sets based on consistent decision formal context.

This paper is organized as follows. In Section 2, we briefly review some basic concepts and notations for formal context and formal concept. Section 3 introduce the definitions and properties about decision formal contexts. Section 4 gives

J. Zhang (Ed.): ICAIC 2011, Part V, CCIS 228, pp. 77–85, 2011.

characterizations of three different types of attribute, and provide the conditions for determining the necessary and unnecessary attribute sets. We then conclude the paper with a summary in Section 5.

2 Preliminaries

A formal context is a triplet (U, A, I), where $U = \{x_1, x_2, \cdots, x_n\}$ is a non-empty, finite set of objects called the universe of discourse, $A = \{a_1, a_2, \cdots, a_m\}$ is a non-empty, finite set of attributes, and $I \subseteq U \times A$ is a binary relation between U and A. In a formal context (U, A, I), for each $(x, a) \in U \times A$, if $(x, a) \in I$, we write xIa, we say that x has attribute a, or the attribute a ispossessed by object x. For $X \subseteq U$ and $B \subseteq A$, we denote

$$X^* = \{a \in A : \forall x \in X, (x, a) \in I\}$$
$$B^* = \{x \in U : \forall a \in B, (x, a) \in I\},$$

X^* is the maximal set of attributes shared by all objects in X. Similarly, B^* is the maximal set of objects that have all attributes in B. For $x \in U$ and $a \in A$, we denote $x^* = \{x\}^*, a^* = \{a\}^*$. Thus x^* is the set of attributes possessed by x, and a^* is the set of objects having attribute a.

Definition 2.1. A pair $(X, B), X \subseteq U, B \subseteq A$, is called a formal concept of the context (U, A, I) if $X^* = B$ and $B^* = A$. X and B are respectively referred to as the extent and the intent of the concept (X, B).

Definition 2.2. [1] Let (U, A, I) be a formal context,

$X, X_1, X_2 \subseteq U$, $B, B_1, B_2 \subseteq A$. Then

$(1) X_1 \subseteq X_2 \Rightarrow X_2^* \subseteq X_1^*$;

$(2) B_1 \subseteq B_2 \Rightarrow B_2^* \subseteq B_1^*$;

$(3) X \subseteq X^{**}, B \subseteq B^{**}$;

$(4) X^* = X^{***}, B^* = B^{***}$;

$(5) (X_1 \cup X_2)^* = X_1^* \cap X_2^*, (B_1 \cup B_2)^* = B_1^* \cap B_2^*$

Let $L(U, A, I)$ denote all concepts of formal context (U, A, I), denote $(X_1, B_1) \leq (X_2, B_2) \Leftrightarrow X_1 \subseteq X_2 \Leftrightarrow B_1 \supseteq B_2$, then $\leq$ is an order relation on $L(U, A, I)$.

Theorem 2.2.[1] Let (U,A,I) be a formal context, $(X_1,B_1),(X_2,B_2)\in L(U,A,I)$, then

$$(X_1,B_1)\wedge(X_2,B_2)=\left(X_1\cap X_2,(B_1\cup B_2)^{**}\right),$$
$$(X_1,B_1)\vee(X_2,B_2)=\left((X_1\cup X_2)^{**},B_1\cap B_2\right)$$

are concepts. Thus $L(U,A,I)$ is a complete lattice called the concept lattice of the context (U,A,I).

3 Consistent Decision Formal Context and Their Conclusion

In this section, we recall some of the basic results about concept lattice which have been established [5] .

Let (U,A,I) be a formal context, $L(U,A,I)$ denote all concepts of formal context (U,A,I), denote

$$\mathcal{L}=\{L(U,A,I)\mid I\subseteq U\times A,|A|<\infty\},$$

If for $(Y,C)\in L(U,A_2,I_2)$, there exists $(X,B)\in L(U,A_1,I_1)$, such that $X=Y$, then $L(U,A_1,I_1)$ is said to be finer then $L(U,A_2,I_2)$, denote by $L(U,A_1,I_1)\le L(U,A_2,I_2)$, then $(\mathcal{L},\le)$ is a partial ordered set.

Let U be the universe of discourse and $\mathrm{P}(U)$ the power set of U, denote

$$\mathcal{H}=\{H\subseteq \mathrm{P}(U):\varnothing,U\in H,X,Y\in H\Rightarrow X\cap Y\in H\}.$$

Definition 3.1. Let $H_1,H_2\in \mathrm{H}$. If $H_2\subseteq H_1$, then $H_1\le H_2$.

Theorem 3.1. Let (U,A,I) be a formal context, then

$$L_u(U,A,I)=\{X:X\subseteq U,X^{**}=X\}\in\mathcal{H}.$$

Proof. If $X^{**}=X$, $Y^{**}=Y$, then

$$X\cap Y\subseteq(X\cap Y)^{**}\subseteq\left(X^*\cup Y^*\right)^*=X^{**}\cap Y^{**}=X\cap Y.$$

Conversely, $(X\cap Y)^{**}=X\cap Y$.

Definition 3.2.[5] $DT=(U,A,I,D,J)$ is said to be a decision formal context, where $A\cap D=\varnothing$, $I\subseteq U\times A$ and $J\subseteq U\times D$, A and D are called the conditional attribute set and decision attribute set respectively.

Definition 3.3.[5] Let $DT=(U,A,I,D,J)$ be a decision formal context, if $L(U,A,I)\le L(U,D,J)$, we say that DT is a generalized consistent decision formal context. For any $C\subseteq A$, if $L(U,A,I)\le L(U,D,J)$, we say that D is a consistent set of DT . If C is consistent set and no proper subset of C is a consistent set of DT , then C is referred to as a reduct of DT .

Theorem 3.2. Let $DT=(U,A,I,D,J)$ be a consistent decision formal context, $C\subseteq A$, then C is a consistent set of DT iff $(X^*\cap C)^*\subseteq(X^*\cap(A-C))^*$.

Proof. Assume that $DT=(U,A,I,D,J)$ is a consistent decision formal context, then $L_u(U,D,J)\subseteq L_u(U,A,I)$. If C is a consistent set, then $L_u(U,D,J)\subseteq L_u(U,C,I_C)$. Therefore, for any $X\in L_u(U,D,J)$, we have $X\in L_u(U,A,I), X\in L_u(U,C,I_C)$, i.e., $X=(X^*\cap C)^*=X^{**}$, hence

$$X=(X^*\cap C)^*=((X^*\cap C)\cup(X^*\cap(A-C)))^*$$
$$=(X^*\cap C)^*\cap(X^*\cap(A-C))^*.$$

Therefore $(X^*\cap C)^*\subseteq(X^*\cap(A-C))^*$.

Conversely, if for any $X\in L_u(U,D,J)$, $(X^*\cap C)^*\subseteq(X^*\cap(A-C))^*$, then $(X^*\cap C)^*=X^{**}$. since $X\in L_u(U,A,I)$, then $X^{**}=X$, so $X=(X^*\cap C)^*$, i.e., $(X,X^*\cap C)\in L_u(U,C,I_C)$, then for any $X\in L_u(U,D,J)$, we have $X\in L_u(U,C,I_C)$. Thus we obtain that C is a consistent set of DT .

Let $DT=(U,A,I,D,J)$ be a decision formal context, $B\subseteq A$. Denote

$$H_B=L_u(U,B,I_B), H=L_u(U,D,J)$$
$$R_B=\{(X_i,X_j)\in H_B^2: X_i\supseteq Y\Leftrightarrow X_j\supseteq Y\,(Y\in H)\},$$
$$H_B/R_B=\{L_B(Y)\neq\varnothing: Y\in H\}$$
$$L_B(Y)=\{X\in H_B: X\supseteq Y, \forall Y'\in H, Y'\supset Y, X\not\supseteq Y'\}.$$

Theorem 3.3.[5] Let $DT=(U,A,I,D,J)$ be a consistent decision formal context, $B\subseteq A$ for any $(X_i,B_i),(X_i,B_i)\in L(U,A,I)$, we denote

$$D(X_i,X_j)=\begin{cases} B_i\Delta B_j, X_j\subset X_i, X_i, X_j\in L_u(U,D,J) \\ \qquad X_i\in L_u(U,D,J), X_j\in L_A(X_i), \\ \qquad X_j\neq X_i \\ \varnothing \qquad otherwise. \end{cases}$$

where $B_i\Delta B_j=B_i\cup B_j-B_i\cap B_j$, then B is a consistent set iff $B\cap D(X_i,X_j)\neq\varnothing.(D(X_i,X_j)\neq\varnothing)$.

Example 1. Table 1 depicts an example of a consistent decision formal context $DT=(U,A,I,D,J)$, where $U=\{x_1,x_2,x_3,x_4,x_5\}$, $A=\{a,b,c,d,e,f\}$ is the conditional attribute set, and $D=\{g,h,k,j\}$ is the decision attribute set.

Table 1. A consistent decision formal context

U	a	b	c	d	e	f	g	h	k	j
$x1$	0	1	0	0	1	0	1	0	0	0
$x2$	1	1	0	1	0	1	0	0	1	0
$x3$	0	0	1	0	1	0	1	0	0	1
$x4$	1	1	0	1	0	1	0	0	1	0
$x5$	1	1	0	0	1	1	1	1	1	0

By Theorem 3.3, we conclude that $C_1=\{a,c,e\},C_2=\{f,c,e\}$ are the reducts of DT .

4 Characterzations of Attribute in Consistent Decision Formal Conttexts

Definition 4.1. Let $DT=(U,A,I,D,J)$ be a consistent decision formal context, and $\{B_k:k\le r\}$ be the set of all reducts of DT . Then attribute sets A is divided into the following three parts:

(1) Core attribute set: $C=\bigcap_{k=1}^{r}B_k$;

(2) Relative necessary attribute set: $K=\bigcup_{k=1}^{r}C_k-C$;

(3)Absolute unnecessary attribute set: $I=A-\bigcup_{k=1}^{r}C_k$.

For any $a\in A$, denote $B_a=\{b\in A:b^*=a^*\}$.

Therorem 4.1. Let $DT=(\mathrm{U,A,I,D,J})$ be a consistent decision formal context, if $\mathrm{a}\in \mathrm{C}$, then $|B_\mathrm{a}|=1$.

Proof. If $|B_\mathrm{a}|>1$.Without any loss of generality,we suppose suppose $B_\mathrm{a}=\{\mathrm{a,b}\}$ For any $X\in L_\mathrm{u}(\mathrm{U,D,J})$,then $\mathrm{b}^*=\mathrm{a}^*$,we obtain that $\mathrm{a}\in X^*\Leftrightarrow b\in X^*$, therefore

$$X=X^{**}$$
$$=(X^*\cap((A-\{a\})\cup\{a\}))^*$$
$$=(X^*\cap(A-\{a\}))^*\cap(X^*\cap\{a\})^*$$

$$=\left(X^{*}\cap\left(A-\{a\}\right)\right)^{*}\cap\left(X^{*}\cap\{b\}\right)^{*}$$
$$=\left(X^{*}\cap\left(\left(A-\{a\}\right)\cup\{b\}\right)\right)^{*},$$

Then by Theorem 3.2, $\left(A-\{a\}\right)\cup\{b\}=A-\{a\}$ is a consistent set, but $a\notin\left(A-\{a\}\right)\cup\{b\}$, this is contrary to the fact that $a\in C$. Therefore $|B_a|=1$.

Theorem 4.2. Let $DT=(\mathrm{U,A,I,D,J})$ be a consistent decision formal context, If $|B_a|>1$, then $B_a\subseteq K$ or $B_a\subseteq I$.

Proof. If $a\in K$, then there exists a reduct set C_k ($k\leq r$) such that $a\in C_k$. If $b^{*}=a^{*}$ for any $b\in A$, then $a\in X^{*}\Leftrightarrow b\in X^{*}$. For any $X\in L_u(\mathrm{U,D,J})$, we have $\left(X^{*}\cap \mathrm{a}\right)^{*}=\left(X^{*}\cap \mathrm{b}\right)^{*}$. Therefore

$$\left(X^{*}\cap\left(\left(C_k-\{a\}\right)\cup\{b\}\right)\right)^{*}$$
$$=\left(X^{*}\cap\left(C_K-\{a\}\right)\right)^{*}\cap\left(X^{*}\cap\{b\}\right)^{*}$$
$$=\left(X^{*}\cap\left(C_k-\{a\}\right)\right)^{*}\cap\left(X^{*}\cap\{a\}\right)^{*}$$
$$=\left(X^{*}\cap\left(\left(C_k-\{a\}\right)\cup\{a\}\right)\right)^{*}==\left(X^{*}\cap C_K\right)^{*}=X$$

Then by Theorem 3.2, $\left(C_{\mathrm{k}}-\{\mathrm{a}\}\right)\cup\{\mathrm{b}\}$ is a consistent set, but $a^{*}=b,^{*}$ hence $b\notin C$. It follows that $b\in K$. Thus $B_a\subseteq K$. Conversely, assume that $a\in I$, for any $b\in A$. If $b^{*}=a^{*}$, then $b\in I$, otherwise, $B_a\subseteq K$. therefore, the conclusion is true.

Theorem 4.3. Let $\mathrm{DT}=(\mathrm{U,A,I,D,J})$ be a consistent decision formal context, then

(1) $a\in C$ iff $|B_{\mathrm{a}}|=1$ and there exists $X\in L_{\mathrm{u}}(\mathrm{U,D,J})$, $B_a\subseteq X^{*}$, $\left(X^{*}\cap\left(A-B_a\right)\right)^{*}\not\subset B_a^{*}$.

(2) $a\in I$ iff for any $X\in L_{\mathrm{u}}(\mathrm{U,D,J})$, $B_a\not\subset X^{*}$ or for any $X\in L_{\mathrm{u}}(\mathrm{U,D,J})$, $\left(X^{*}\cap\left(A-B_a\right)\right)^{*}\subseteq B_a^{*}$.

(3) $a\in K$ iff $|B_{\mathrm{a}}|>1$ and there exists $X\in L_{\mathrm{u}}(\mathrm{U,D,J})$, $B_a\subseteq X^{*},\left(X^{*}\cap\left(A-B_a\right)\right)^{*}\not\subset B_a^{*}$

Proof. (1) $a\in C$, then by Theorem 4.1, $|B_{\mathrm{a}}|=1$, that is, $B_{\mathrm{a}}=\{\mathrm{a}\}$, hence $A-B_{\mathrm{a}}$ is not a consistent set of DT. By Theorem 3.2, there must exists $X\in L_{\mathrm{u}}(\mathrm{U,D,J})$, such that $\left(X^*\cap\left(A-B_a\right)\right)^*\not\subset\left(X^*\cap B_a\right)^*$, but $B_a\not\subset X^*$, $\left(X^*\cap B_a\right)^*=U$, hence if a is an element of the core of DT, there exists $X\in L_u\left(\text{U,D,J}\right), B_a\subseteq X^*$, such that $\left(X^*\cap\left(A-B_a\right)\right)^*\not\subset B_a^*$.

Conversely, assume that there exists $X\in L_u\left(\text{U,D,J}\right)$, $B_a\subseteq X^*$ such that $\left(X^*\cap\left(A-B_a\right)\right)^*\not\subset B_a^*$. since for any $B_a\cap B\neq\varnothing$, $\left(X^*\cap\left(A-B_a\right)\right)^*\subseteq\left(X^*\cap B_a\right)^*$, $\left(X^*\cap\left(A-B_a\right)\right)^*\subseteq B_a^*$ holds, we conclude that $\left(X^*\cap B_a\right)\not\subset\left(X^*\cap\left(A-B_a\right)\right)^*$. By Theorem 3.2, B is not a consistent set of DT ,hence B is not a reduct set of DT .Thus we have proved that a is an element of the core of DT .

(2) If $a\in \text{I}$, then $X=X^{**}=\left(X^*-B_a\right)^*$, hence $B_a\not\subset X^*$ or $B_a\subseteq X^*$. If $B_a\subseteq X^*$, then $B_a^{**}\subseteq X^*$, then $a\in I$. Hence, for any a reduct set C' , $B_a\not\subset C'$, $B_a\subseteq B_a^{**}\cap\left(A-C'\right)$, we obtain that $B_a\subseteq B_a^{**}\cap\left(A-C'\right)\subseteq \text{X}^*\cap\left(A-C'\right)$, that is $\left(X^*\cap\left(A-C'\right)\right)^*\subseteq\left(B_a^{**}\cap\left(A-C'\right)\right)^*\subseteq B_a^*$. Since C' is a consistent set of DT , by Theorem 3.2, we know that, $\left(X^*\cap C'\right)^*\subseteq\left(X^*\cap\left(A-C'\right)\right)^*\subseteq B_a^*$. For any $b\in X^*\cap C'$,we have $b\in C'$, and $a\in I$,we conclude that $b^*\neq a^*$,that is, $b\notin B_a$. Hence $b\in X^*-B_a$.

Thus $X^*\cap C'\subseteq X^*-B_a$, $\left(X^*\cap\left(A-B_a\right)\right)^*\subseteq\left(X^*\cap C'\right)^*$.

Therefore $\left(X^*\cap\left(A-B_a\right)\right)^*\subseteq B_a^*$.

Conversely, if $B_a\not\subset X^*$ for any $X\in L_u\left(\text{U,D,J}\right)$,then $X=X^{**}=\left(X^*-B_a\right)^*$, hence $\left(X^*-B_a\right)^*=X\in$ $L_u\left(U,D,J\right)$, from which we conclude that a is a absolute unnecessary attribute. If $B_a\subseteq X^*, \left(X^*\cap\left(A-B_a\right)\right)^*\subseteq B_a^*$, then $X=X^{**}=\left(\left(X^*-B_a\right)\cup B_a\right)^*=\left(X^*-B_a\right)^*\cap B_a^*=\left(X^*-B_a\right)^*$, thus B_a is a absolute unnecessary attribute set, hence a is a absolute unnecessary attribute.

(3) It can immediately be obtained from (1) and (2).

Example 2. In Example 1, we can see $L_u\left(U,D,J\right)=\left\{U,245,135,5,3,\varnothing\right\}$ Obviously, for any element of $L_u\left(U,D,J\right)$, one can obtain:

$X_1 = \{2,4,5\}$, then $X_1^* = \{a,b,f\}$
$X_2 = \{1,3,5\}$, then $X_2^* = \{e\}$
$X_3 = \{5\}$, then $X_3^* = \{a,b,e,f\}$
$X_4 = \{3\}$, then $X_4^* = \{c,e\}$

We denote $B_c = \{c\}$, then $|B_c| = 1$ $\exists$ $X_4 = \{3\}$, $B_c \subseteq X_4^* = \{c,e\}$ $(X_4^* \cap (A - B_c))^* = \{b\}^* = \{1,2,4,5\} \not\subseteq B_c^* = \{3\}$. $B_e = \{e\}$, $|B_e| = 1$, $\exists$ $X_2 = \{1,3,5\}$, $B_e \subseteq X_2^* = \{e\}$,and $(X_2^* \cap (A - B_e))^* = U \not\subseteq B_e^* = \{1,3,5\}$. Therefore, $c, e \in C$.

$B_a = B_f = \{a, f\}$, $\exists$ $X_1, X_3 \in L_u(U, D, J)$, such that $B_a \subseteq X_1^*, B_a \subseteq X_3^*$,satisfy

$$(X_1^* \cap (A - B_a))^* = \{b\} = \{1,2,4,5\} \not\subseteq B_a^* = \{2,4,5\}$$
$$(X_3^* \cap (A - B_a))^* = \{b,e\} = \{1,5\} \not\subseteq B_a^* = \{2,4,5\} .$$

Therefore, $a, f \in K$.

For any $X \in L_u(U, D, J), \{d\} \not\subseteq X^*, d \in I$, $\{b\} \subseteq X_1^*, \{b\} \subseteq X_3^*$,then

$$(X_1^* \cap (A - \{b\}))^* = \{a, f\}^* = \{2,4,5\} \subseteq B_b^* = \{1,2,4,5\}$$
$$(X_3^* \cap (A - \{b\}))^* = \{a,e,f\}^* = \{5\} \subseteq B_b^* = \{1,2,4,5\} .$$

Therefore $b \in I$.

5 Conclusion

Comparing with the studies on rough set theory, there is less results on the issue in concept lattice theory. In this paper, we have introduced characterization of three important types of attribute sets and give the judgement theorems. The numerical computational results are showed.

Acknowledgment. The authors wish to thank the anonymous referees for their very constructive comments. This work is supported by the Natural Science Foundation of Shanxi Provice in China (No. 2008011012).

References

1. Ganter, B., Wille, R.: Formal concept analysis, Mathematical Foundations. Springer, Berlin (1999)
2. Kent, R.E.: Rough concept analysis: a synthesis of rough sets and formal concept analysis. Fundamenta Informaticae 27, 169–181 (1996)
3. Saquer, J., Deogun, J.S.: Formal rough concept analysis. In: Zhong, N., Skowron, A., Ohsuga, S. (eds.) RSFDGrC 1999. LNCS (LNAI), vol. 1711, pp. 91–99. Springer, Heidelberg (1999)
4. Yao, Y.Y.: Concept lattices in rough set theory. In: Dick, S., et al. (eds.) Proceedings of 2004 Annual Meeting of the North American Fuzzy Information Processing Society (NAFIPS 2004), June 27-30, pp. 796–801 (2004)
5. Zhang, W.X., Qiu, G.F.: Uncertain Decision Making Based on Rough Sets. Qinghua Press, Beijing (2005)

Multi-scale Areal Feature Matching Based on Overlapping Degree and Graph Theory*

Zhijiang Li, Ming Zhao, and Yuxin Sun

School of Printing and Packaging Wuhan University Luoyu Road 129#,
Wuhan, Hubei Province, China
School of Computer Science and Engineering Uhan Institute of Technology
Xiongchu Street 693#, Wuhan, Hubei Province, China
ZhijiangLi2022@tom.com

Abstract. Integration and updating of spatial data from disparate source is very common in the application of GIS. Since areal feature possess much in spatial objects' presentation, areal feature matching becomes a key technology in GIS application. This paper proposes a different scale areal feature matching method considering overlapping and adjacent relation. Regarding the polygons as whole entities, a primary matching is proceed due to overlapping relationship between the same feature of different scales. After that, a secondary and a final matching are executed considering adjacency between features of the same scales as well as the adjacency between the matching sets so as to ensure the topological consistency partially and globally. In the end, a case study of land-using dataset demonstrates that the proposed method is effective and reliable.

Keywords: feature Matching, Areal Entities, Adjacent Relation, Graph Theory.

1 Introduction

In GIS, spatial data matching has been widely used in many aspects such as map conflation, quality evaluation and improvement of spatial data, maintenance and updating of multi-scale spatial database, location-based navigation services, etc[1]. According to the types of map elements, matching method can be divided into node matching (including node to node and node to area), line matching (including line to line and line to area), and area matching (area to area). There are mainly three category of spatial data matching methods: geometric matching; topological matching and semantic matching. Geometric matching calculate the geometric similarity between the elements, including distance, area, shape, angle and direction. Topological matching takes the topological relations between the elements as criterion. Semantic matching compares the semantic similarity of attribute value so as to find the same object, which is not widely used because of its great dependence on data model, attribute data types and data integrity.

* This work is supported by the National High-Tech Research and Development Plan of China Under Grant No. 2009AA12Z209 (863).

J. Zhang (Ed.): ICAIC 2011, Part V, CCIS 228, pp. 86–93, 2011.

Recent years, there are many research papers on areal feature matching. Zhang et al. proposed an areal feature matching method based on the fuzzy classification of topological relationship, which identifies candidate matching set by overlapping ratio firstly, and then the matching result is ascertained by fuzzy topological relationship between areal features[3]. Masuyama introduced a method which calculates the matching possibility by overlapping ratio[4]. Tong et al. presented a probabilistic theory-based matching method which uses a variety of criteria to calculate the matching probability by which to match the same feature[5]. Guo et al. developed matching algorithms based on spatial-direction similarity and areal similarity[6][7]. Hao et al. proposed a matching method which combines various criteria[8]. Zhao et al. developed a same-scale areal feature matching method which take both overlapping degree and matching distance as criteria[9]. In the work by Zhang, Guo and Sun, a matching method is implemented to match the same residential features in the NFGIS of the same area[10].

It is necessary to research multi-scale areal feature matching in map conflation and topology consistency checking of multi-scale areal feature. Besides, some areal datasets, which comprise mutually adjacent areal features, are widely used in various field. For example, land-using data, which is used to analyse and plan the usage of land. In this paper we define the dataset mentioned above as "Adjacent Dataset". Let M be an adjacent dataset which can be denoted as $M = \{ a_1, a_2, \ldots a_i, \ldots , a_n \}$, where ai ($1 \leqslant i \leqslant n$) is a areal feature in M. In this case, M and ai meet two conditions below:

1) Features are non-overlapping

Suppose $a_i \in M$ and $a_j \in M$, then $a_i \cap a_j = \Phi$. Every feature in *M* owns its entire interior, it overlaps with no feature in *M*.

2) No gap in M

$a_1 \cup a_2 \cup a_3 \ldots \cup a_{n-1} \cup a_n = M$. Because there is no gap, and no overlapping in *M*, union of all features equals to *M*.

Considering overlapping relation between multi-scale datasets and the adjacent relation of features in adjacent dataset, this research proposes and implements a method based on overlapping degree and graph theory in section 2 and a case study of 1:10000 and 1:50000 land-using datasets was made in order to verify the proposed method in section 3.

2 Methodology

In areal feature matching between multi-scale adjacent dataset, because that small scale dataset is less detailed than large scale dataset, the matching is 1:N ($N \geqslant 1$), which means a areal feature in small scale dataset relates with N areal features in large scale dataset. Considering the relationship between different-scale dataset of same region and the adjacent relations between features in same dataset, there are several conditions should be satisfied in this case of matching.

1) Non-Ambiguity. In accordance with the real world, matching result of large scale dataset is 1:1, which means one areal feature in large scale dataset can only be related to one feature in small scale dataset.
2) Every feature in large scale dataset should be matched. Because the two datasets represent a same region, they are equal to each other, every feature in large scale dataset would overlapped at least one feature in small scale. Thus, every feature in large scale dataset must has its related one in the other dataset.
3) Partial Connectivity. Described as below.
4) Global Connectivity. Described as below.

A. Partial and Global Connectivity

1) Graph of Areal Dataset

Regard to adjacency, r_{ij} is denoted as the adjacency of areal feature a_i and a_j, where a_i and a_j are two areal features in dataset $M = \{ a_1, a_2, \dots a_i, \dots , a_{n+1}, a_n \}$. Hence, r_{ij} can be defined by the following equation.

$$r_{ij} = \begin{cases} 1, Dis(a_i, a_j) \le \delta \\ 0, Dis(a_i, a_j) > \delta \end{cases} \tag{1}$$

Where $Dis(a_i, a_j)$ is the nearest distance of a_i and a_j, as shown in Fig.1. For that $Dis(a_i, a_j)$ equals $Dis(a_j, a_i)$, so r_{ij} also equals r_{ji}. δ is a threshold for the nearest distance, which can be defined based on actual situation.

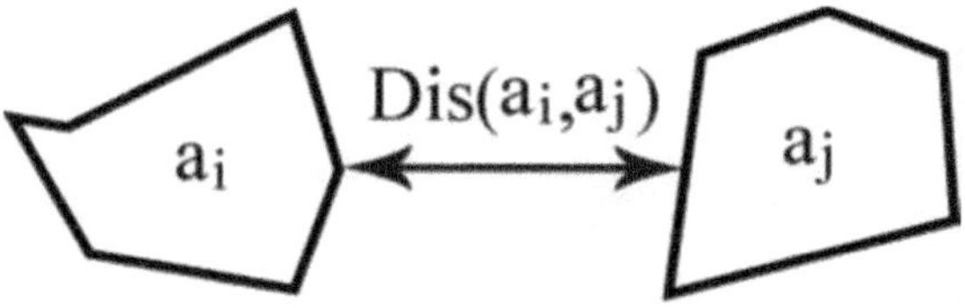

Fig. 1. Distance of two areal features

In this case, a set of adjacency is denoted as $R(M) = \{r_{11}, r_{12}, \dots, r_{ij}, \dots, r_{n-1n}, r_{nn}\}$ refers to the adjacent relations of features in *M*.

Considering the definition of graph which is consist of a set of vertexes and a set of edges which defines the connectivity of vertexes. If *M* is regarded as a set of vertexes, and *R(M)* is regarded as a set of edges, a graph of adjacent relation for areal dataset can be defined.

Suppose *M* is a areal dataset, *R(M)* is the adjacent relations of features in *M*, and *G(M, R(M))* is the graph of areal dataset M which is constructed from *M* and *R(M)*. And, the connectivity of *G(M,R(M))* represents the connectivity of the dataset. Note that *G(M,R(M))* is a undirected graph.

As shown in Fig.2, an adjacent dataset *A* consist of six features ranged from *a1* to *a6*. TABLE 1 is a matrix of adjacent relations of features in *A*, which is denoted as *T(A)*. Fig.3 shows undirected graph *G(A, T(A))*. It's obviously that *G(A, T(A))* is a connected graph, so the dataset A is connected.

Fig. 2. A adjacent dataset

Table 1. Adjacent Relations of Features in Dataset Shown in Fig.2

	a1	a2	a3	a4	a5	a6
a1	1	1	1	1	0	0
a2	1	1	1	1	1	0
a3	1	1	1	1	1	1
a4	1	1	1	1	1	1
a5	0	1	1	1	1	1
a6	0	0	1	1	1	1

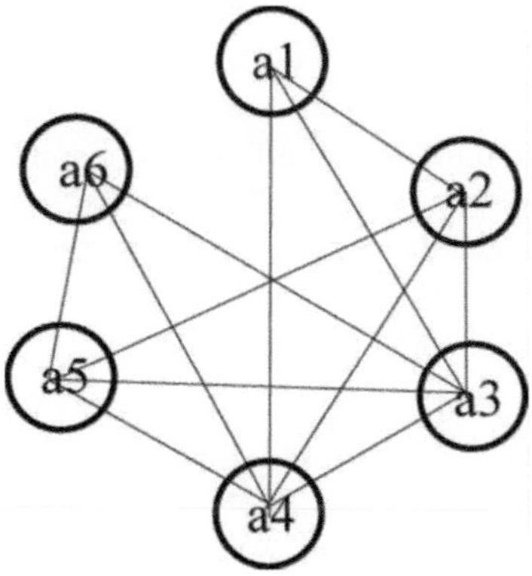

Fig. 3. Graph constructed from TABLE I and Fig.2

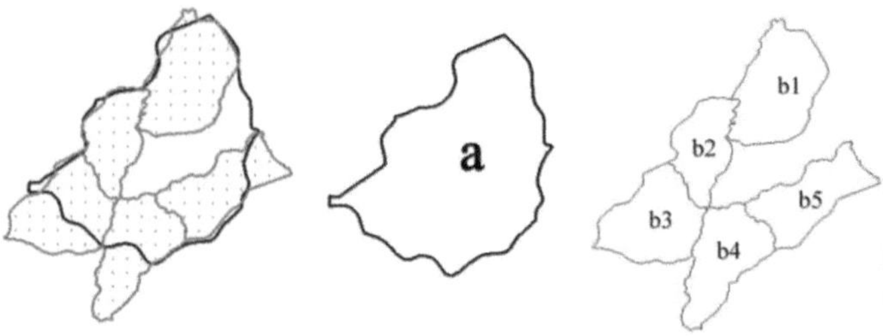

Fig. 4. Small scale feature *a* and its matching result

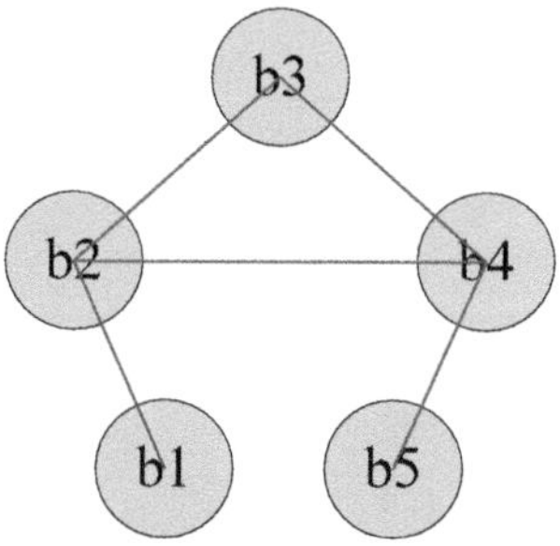

Fig. 5. Graph construct from matching dataset in Fig.4

2) Partial Connectivity
Let a be a areal feature in small scale dataset, its matching set is $B = \{ b1, b2, \ldots bi, \ldots, bn \}$ $(n \geq 1)$. Dataset B must be connected due to the real world and the definition of dataset's connectivity. Fig.4 and Fig.5 show a situation meet partial connectivity.

3) Global Connectivity
Suppose two features in small scale dataset are denoted as a_i and a_j respectively, B_i and B_j are their corresponding matching dataset. If r_{ij} equals to 1 which means a_i is adjacent to a_i and B_i and B_j is not empty. Denoting dataset C as $B_i \cup B_j$. (in this case, *C* must be connected)

As shown in Fig.6, *a1* and *a2* are areal features in small scale dataset, *b1, b2, b3,b4, b5* are areal features in large scale dataset. $B1 = \{b1, b2\}$ and $B2 = \{b3,b4\}$, the union of *B1* and *B2* is apparently connected, thus, B1 and B2 meet the global connectivity.

B. Process of the Proposed Algorithm

1) Finding candidates and matching based on overlapping degree (step 1)
After eliminating the bias of global and local coordinates between different-scale maps of the same region, overlap exists in features refer to the same areal object in the two maps. And the larger of overlapping area is, the greater of the possibility of being the same object is. Besides, the overlapping area also reflects the overall similarity between features, which is in line with the visual effects of human eyes. In addition, matching based on overlapping can also reduce blindness in matching process and will greatly improve the matching efficiency and accuracy.

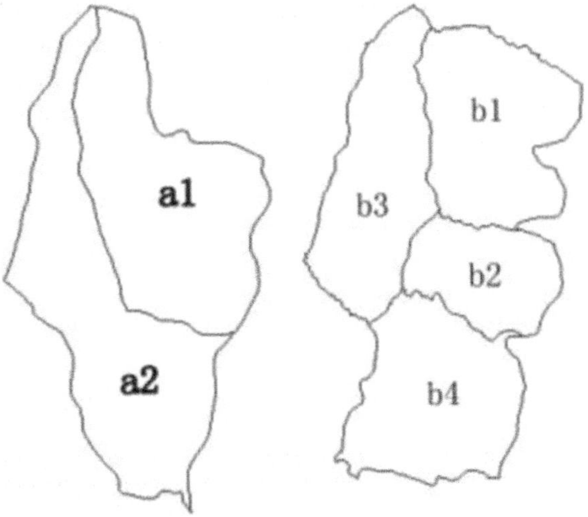

Fig. 6. Example of global connectivity

Denote *a* as an areal feature in small scale dataset, and *b* is an areal feature in large scale dataset, the overlapping degree of *a* and *b* can be defined as:

$$OVERLAP(a,b)=\frac{Area(a\cap b)}{Area(b)} \tag{2}$$

If *OVERLAP(a, b)* is larger than 0, it is potential for them to be matched. Thus, adding the ID of *a* and overlapping degree to the candidate set of *b*, correspondingly, add ID of *b* and overlapping degree to the *a*'s candidate set. Processing features with this method in both datasets iteratively. Hence, the mapping relation between features in the two datasets would be established, every feature in both datasets has a candidate set ordered by overlapping degree.

By the candidates set created in above step, a certain threshold can be set, so that an areal feature in large scale dataset can be matched with the feature in its candidate set where the overlapping degree of them is the greatest and also greater than the threshold, simultaneously, marking the large scale feature as matched. In the same way, matching and marking small scale features with large scale features in their candidates sets.

2) Matching based on partial connectivity (step 2)

After step 1, most large scale features has been matched, and most of the small scale features have non-empty matching set. However, some of the matching set may not meet the rule of partial connectivity. In this case, a secondary matching is carried out by searching unmatched large scale features in the candidates set of a small scale feature whose matching dataset is not connected so as to satisfy partial connectivity.

3) Matching based on global connectivity (step 3)

After secondary matching, although rare matching dataset does not meet partial connectivity, two issues still lies:

First, some matching dataset does not meet global connectivity.

Second, matching omission occurred in secondary matching. When the matching dataset of a small scale feature satisfy partial connectivity, the searching process would be stopped, however, there still may be unmatched large scale feature in its candidates set, thus, omission occurred.

Experiment made in this research also proved problem mentioned above. Therefore, it is necessary to carry out a third matching based on global connectivity. In this step, global connectivity of matching datasets of every two small scale feature is verified to test the unmatched large scale features in their candidates set at the same time. If adding unmatched feature to matching datasets does not break global connectivity or even makes them satisfy the rule, the unmatched feature would be added to one of the matching datasets based on overlapping degree and partial connectivity.

Here is a case handled by third matching. Small scale features *a1* and *a2* shown in Fig.7 are adjacent, and their corresponding matching sets are *B1* = { *b1*, *b2* } and *B2* = { *c1* }, while *d1* is a unmatched feature in both candidates set of *a1* and *a2*. It is obvious that both *B1* and *B2* meet the rule of partial connectivity, but they do not meet the rule of global connectivity. After processing of this step, *d1* is matched with *a1* due to overlapping degree.

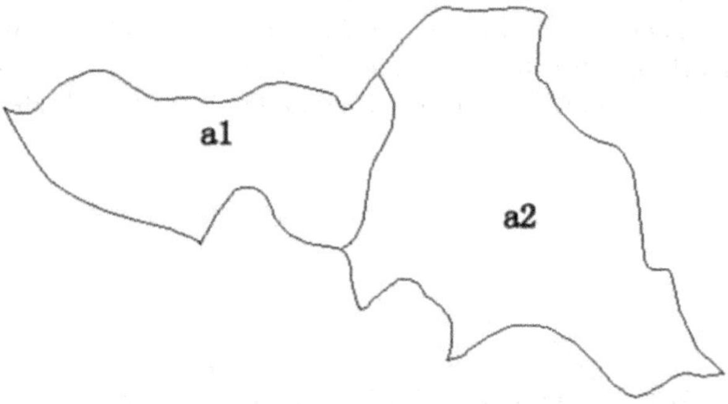

Fig. 7. Two adjacent small scale features

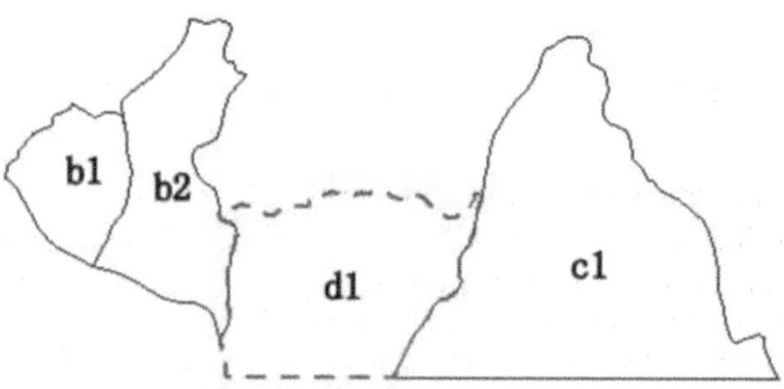

Fig. 8. Unmatched largescale feature and matching sets of features in Fig.7

3 Conclusion

The experiment was done on a notepad computer with 1.66GHz CPU and 1.5G memory in Windows XP operation system. In the matching, a 1:50000 dataset with 638 features, a 1:10000 dataset with 103 features which is a part of the small dataset were used. The matching took 2.5 minutes, all large scale features were matched successfully, without a miss. And there were 60 matching sets, with 17 1:1 matchings, 17 1:2 matchings, rest of them are 1:N ($N \geq 3$) matchings.

Test results show that the areal feature matching algorithm proposed in this paper is simple and efficient with high successful matching rate. It can serve map updating and conflation and consistency checking and other fields well. Besides, it has high practical significance in areal feature matching of multi-scale land-using dataset.

References

1. Xu, F., Deng, M., Zhao, B., Chen, J.: A Detailed Investigation on the Methods of Object Matching. Journal of Geo-Information Science 8(5), 657–663 (2009)
2. Li, D., Gong, J., Zhang, Q.: Discussion on Technology of Map Conflation. Science of Surveying and Mapping 29(1), 1–4 (2004)
3. Zhang, Q., Li, D., Gong, J.: Areal Feature Matching among Urban Geographic Databases. Journal of Remote Sensing 8(2), 107–112 (2004)
4. Masuyama, A.: Methods for Detecting Apparent Differences Between Spatial Tessellations at Different Time Points. International Journal of Geographical Information Science 20(6), 633–648 (2006)
5. Tong, X., Deng, S., Shi, W.: A Probabilistic Theory-based Matching Method. ACTA Geodaetica et Cartographica Sinica 36(2), 210–217 (2007)
6. Guo, L., Cui, T., Zheng, H., Zhang, X.: Arithmetic for Area Vetor Spatial Data Matching on Spatial Direction Similarity. Journal of Geomatics Science and Technology 25(5), 380–382 (2008)
7. Guo, L., Zheng, H., Wang, H.: Hydrographic Surveying and Charting 29(3), 12–15 (May 2009)
8. Hao, Y., Tang, W., Zhao, Y., Li, N.: Areal Feature Matching Algorithm Based on Spatial Similarity. ACTA Geodaetica et Cartographica Sinica 37(4), 501–506 (2008)
9. Zhao, B., Li, D.: Entity Matching Method Based on Overlap Degree and Matching Distance. In: The Tenth Symposium of Institute of Surveying and Mapping of Six Provinces and One City of East China, pp. 9–12 (July 2007)
10. Zhang, L., Guo, Q., Sun, Y.: Geomatics and Information Science of Wuhan University 33(6), 604–607 (2008)

Access Control Using Trusted Virtual Machine Based on Xen

Xi Wang[1] and Chuan Cheng[2]

[1] Wuhan Polytechnic Wuhan 430074, China
[2] Wuhan Maritime Communication Research Institute Wuhan 430079, China
xiwang8012@126.com

Abstract. Xen is an open-source para-virtualizing virtual machine monitor (VMM), or "hypervisor", for a variety of processor architectures including x86.Xen has impressive performance and easy to connect to LAN.

This paper proposes a method of access control using trusted virtual machine based on Xen. Combined with trusted computing technique and virtualization technique, this model is mainly concerned about safe data access. The key idea is using the isolation of virtual machine to provide an exclusive environment to sensitive data and use trusted computing to ensure the virtual environment is trusted.

Keywords: Virtualization, Xen, Access Control, TPM, Virtual Machine Monitor.

1 Introduction

It is vital for us to use security techniques when we surfing on the internet or access sensitive data. There are many security strategies that we commonly use.

The most common ones may be the network security protocols, such as the TLS, IPsec etc. However, these network security protocols can not provide the assurance of integrity and reliability about the communication terminals. Due to the vulnerability of the terminals, malicious software and hacking attack usually aim at the terminals. That leads to personal information is in great danger and distrust with the security transfer protocols.

Trusted computing technique is a new strategy for information security, whose main security measure is through constructing trusted chains to defend any attacks. The BIOS Boot loader is the start of the system. Then constructing the trusted chains from lower layer to upper layer until it reaches the application layer.

It is a complicated work to build the trusted chains on the personal computers. The reason is that a lot of application software been installed on the personal computers for the users' personal purpose and it is hard to verify every application software. However, applying the VMs and vTPM to build a virtual trusted platform could deal with such issues[1].Virtual trusted platform is usually used to process some specific tasks, which could be defined by the users. It is easy to build the trusted chains on the virtual trusted platform. Every virtual machine can build a TPM of its own by the vTPM. Moreover, the good isolation between the guest virtual machines makes the

J. Zhang (Ed.): ICAIC 2011, Part V, CCIS 228, pp. 94–101, 2011.

security and stability of the platform. Once a guest OS is being attacked, it will not affect other guest OS and the whole system platform as well.

2 Background: Xen and Virtualization, TPM

2.1 Xen and Virtualization

Computing virtualization is a technique of low cost and effective method for hardware and resource sharing. Virtual Machine Monitor (VMM) runs on the middle layer which is between the virtual machines and hardware devices. VMM is in charge of the management of computer hardware and guest operating system. It provides the isolative virtual hardware environment for the operating systems that runs on the guests. The guest operating system can not "feel" the existence of the VMM. It runs the same as that runs on the traditional operating system platform.

In a traditional VMM the virtual hardware exposed is functionally identical to the underlying machine. Although full virtualization has the obvious benefit of allowing unmodified operating systems to be hosted, it also has a number of drawbacks. This is particularly true for the prevalent x86 architecture. Efficiently virtualizing the x86 MMU is also difficult. These problems can be solved, but only at the cost of increased complexity and reduced performance[2]. In order to avoid the drawbacks of full virtualization, a virtual machine abstraction is applied which is similar but not identical to the underlying hardware. This approach is been called paravirtualization[3]. It is important to make sure that changes to the application binary interface(ABI) are not required. Neither are the modifications to guest applications. Fig. 1 gives an overview of Xen Virtual Machine Architecture.

2.2 Trusted Platform Module

TPM is the acronym for Trusted Platform Module. It is a name both of security encryption processor and data security rules which is given by TCG(Trusted Computing Group)[4]. Trusted computing platform could provide the computer hardware and software for trusted computing service. It also provides the system reliability, feasibility and security of information and behavior.

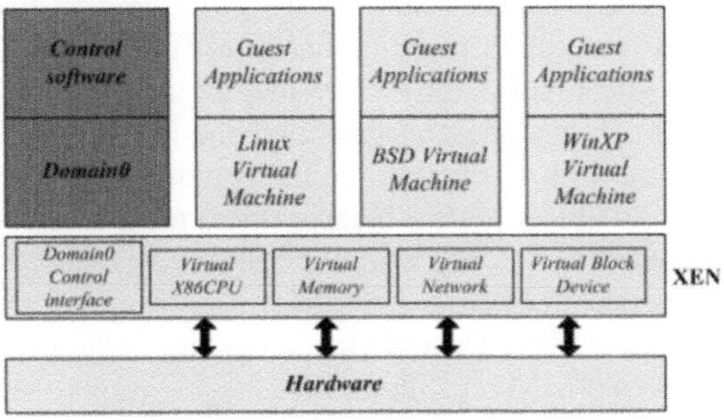

Fig. 1. Xen Virtual Machine Architecture

A mechanism is required to protect the security and integrity of the terminals for both local and remote terminals. Trusted computing technique is developed to deal with such problems. TPM offers the functionalities such as the key generation, remote attestation etc. Attestation based information about the device hardware, firmware, operating system, and applications can all be dynamically assessed to determine if the system should be trusted prior to granting a privilege. Unlike secure boot, which loads only signed and verified software, the TCG trusted boot process only takes measurements up to the bootstrap loader and leaves it up to the remote party to determine the system's trustworthiness. Thus, when the system is powered on it transfers control to an immutable base. This base will measure the next part of BIOS by computing a SHA1 secure has over its content and protect the result by using the TPM. This procedure is then applied recursively to the next portion of code until the OS has been bootstrapped[5].

3 Access Control Using Trusted Virtual Machine Based on Xen

3.1 Trusted Virtual Machine Access Control Base on Xen

Trusted virtual machine monitor(VMM) can provide the isolative virtual hardware platform for the upper layer guest virtual machines, which is similar to the traditional Virtual Machine Monitor. It enables several guest virtual machines running at the same time.

Trusted VMM has the following characteristics:

1) Isolation
Different applications can run on different guest virtual machines with the support of virtual machine monitors. Every guest virtual machine runs on the relative independent "hardware". The VMM ensures every guest virtual machine run separately.

2) Compatibility
Trusted VMM based on Xen supports the current operation systems such as Windows, Linux, BSD etc. It is easy to move current applications to trusted virtual platform without the needs to change the ABIs.

3) Security
Trusted VMM not only has the characteristics of the traditional security virtual machines, but also has some other security characteristics.

Trusted VMM has the functionality of managing and reusing TPM. As the only idenfier in the platform, every guest virtual machine has its own virtual TPM device and the trusted root is built on its corresponding virtual TPM.

Trusted VMM supports the integrity measurement mechanism of trusted computing to make sure the trustiness of the whole platform. It offers the basic of remote attestation for upper applications

The access control method for trusted virtual machine based on Xen is composed of guest machines with Xen VMM on Linux OS[8].[9].[10], remote server and verification server(in Fig. 2).

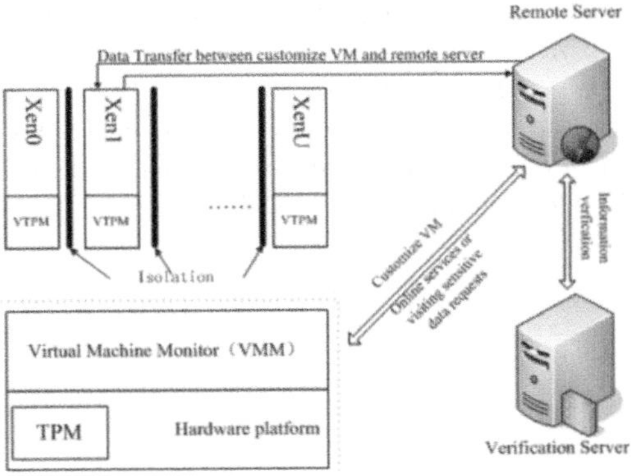

Fig. 2. Trusted VM access control mechanism based on Xen

The procedures of the access control based on Xen:
Guest = {Guest 1,Guest 2,...Guest n}
RS: Remote Server
VS: Validation Server
Cv: measurement integrity value
VSv: Validation data
IF Guest Request (Online Services) OR Request (Access Sensitive Data)
  Guest Send(Request) to RS
  RS response(Request)
  RS return(Configuration File)
  Guest Create(DomainU)
  Guest Send(Cv)
  RS Validate(VSv , Cv)
  IF Validate(VSv , Cv) is True
Continue;
Else
Error;
End IF

From Fig. 2 we can make a mirror of Xen guest virtual machine which has some basic functionalities. Modifications can be made according to the specific requirements. The guest virtual machine has the vTPM function. Moreover we could add other configuration parameters in xm command, e.g. applying the parameters of banning USB interface could prevent the copy of sensitive data at the guest terminal.

3.2 Virtualization in TPM

We implemented a TPM frontend and backend driver through which all guest VMs can transparently communicate with a virtual TPM-hosting VM where our own implementation of TPM Version1.2[6] . The virtual TPMs are managed through the new TPM commands.

Fig. 2 gives an overview of the process followed in our experiment. In order to realize our design of a virtual TPM, we apply TPM 1.2 command set with virtual commands in the following categories. TPM instance 0 is always present, part of Xen. The owner of TPM 0 has sole authority for creating further virtual TPM's These commands allow a virtual TPM instance to be created, deleted.

Virtual TPM Management commands:

1) CreateInstance

TPM_CreateInstance allows the TPM 0 owner to create an additional virtual TPM instance. After this command, the created instance state is that of a new TPM that has received TPM_Init.

TPM_CreateInstance pseudocode :

```
If ( Validate command from instance 0 unsuccessful )
{
    return TPM_AUTHFAIL;
}
else if(Validate owner authorization unsuccessful)
{
    return TPM_AUTHFAIL;
}
else if(no enough space)
{
    return TPM_RESOURCES;
}
else
{
    initialization the new TPM instance;
    return TPM_SUCCESS;
}
End IF
```

2) DeleteInstance

TPM_DeleteInstance allows the TPM 0 owner to delete a virtual TPM created by TPM_CreateInstance.

TPM_DeleteInstance pseudocode :

```
If ( Validate command from instance 0 unsuccessful )
{
    return TPM_AUTHFAIL;
}
else if(Validate owner authorization unsuccessful)
{
    return TPM_AUTHFAIL;
}
else if(instance handle == 0 || instance handle does not point a created instance)
{
    TPM_BAD_PARAMETER
}
else if(no enough space)
{
    return TPM_RESOURCES;
}
else
{
    delete the TPM instance;
free all NVRAM non-volatile memory and volatile memory;
    return TPM_SUCCESS;
}
```

3.3 Virtual TPM in Xen

To implement virtual TPM based on Xen is the key technology to realize the trusted virtual machine security access mechanism based on Xen. In relate to the physical TPM, virtual TPM has these functionalities[7].[11].The physical TPM provides the mode and functionality which the operating system runs on the real hardware .So does the virtual TPM.When the VMM creates a guest virtual machine, it must create the virtual TPM as well. When the guest virtual machine is shut down, the virtual TPM is closed too.

Software running in any Xen domain remains unaware of the fact that it is communicating with a software implementation of a TPM. Due to this transparent TPM usage model, software that has previously been written to work with a hardware TPM will simply continue to work.

Virtual TPM should keep close contact with the Trusted Computing Base(TCB).

In terms of Xen, all Xen virtual machines are called Domain. Domain0 is the starting point of all virtual machines as the system boots. Domain0 is so called virtual machine manager, while the others named DomianU are guest virtual machines. Domain0 is in charge of the control and management of physical hardware and guest virtual machines. By the frontend device drivers, DomainU connects the backend device drivers in Domain0, then through the backend device drivers visit the physical hardware. Fig. 3 is the architecture of virtual TPM based on Xen.

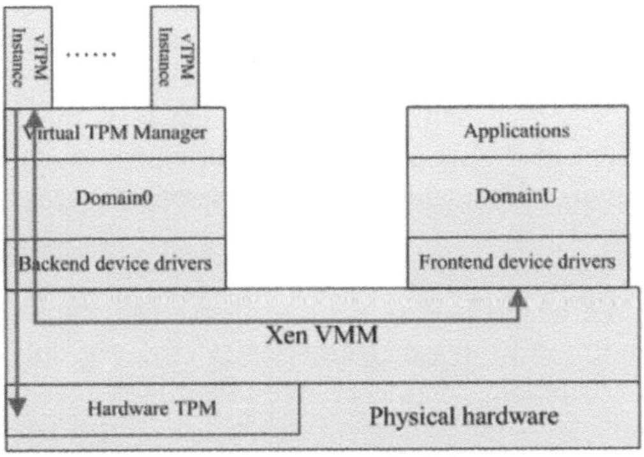

Fig. 3. Virtual TPM based on Xen Achitecture

We apply two methods for creating guest virtual machines with virtual TPM in Xen .One is using xm command, which could add self-defined parameters when creating the Xen virtual machines. The xm command is demonstrated in Fig. 4.

```
xm create –c virtual OS boot config file –vtpm 0
```

Fig. 4. xm command

Xm command is written by language Python under the directory of /xen/tools/python/xen/xm. Vtpm represents the option for using TPM. Parameter 0 means backend device drivers is in Domain0.

Another method is by adding the information of virtual TPM in configuration file of guest virtual machines. The configuration file is displayed in Fig. 5.

```
kernel = "/boot/vmlinuz-2.6.18.8-xen"
memory = 256
name = "fedora.11.x86.img"
vif = [ ' ' ]
disk = ['file:/root/fedora11/fedora.11.x86.img, xvdb, w']
root = "/dev/xvdb"
extra = "fastboot"
vTPM = [backend:0]
```

Fig. 5. The Xen virtual machine configuration file with virtual TPM option

4 Conclusions and Future Work

With the fast development of virtual techniques and trusted computing, We proposed the method of access control using trusted virtual machine based on Xen. We virtualized the TPM by TPM command set Version 1.2 and uncovered the most difficulties that arise when virtualizing the TPM and How the virtual TPMs work on Xen. It is not sufficient in the case of the access control model. We would research more in the virtual environment and make progress of the access control using trusted virtual machines based on Xen.

Acknowledgment. We would like to thank Professor Ke Zhou (Huazhong university of science and technology),Professor Dan Feng (Huazhong university of science and technology) for their valuable comments. We also thank the Xen community for its support on our model.

References

1. Chen, P.M., Noble, B.D.: When virtual is better than real. In: Proceedings of the 8th Workshop on Hot Topics in Operating Systems, p. 133. IEEE Computer Society, Washington, DC (2001)
2. Barham, P., Dragovic, B., Fraser, K., Hand, S., Harris, T., Ho, A., Neugebauer, R., Pratt, I., War_eld, A.: Xen and the Art of Virtualization. In: Proceedings of the 19th ACM SOSP (October 2000)
3. Whitaker, A., Shaw, M., Gribble, S.D.: Denali: Light weight Virtual Machines for Distributed and Networked Applications. Technical Report 02-02-01, University of Washington (2002)

4. Trusted Computing Group.TCG TPM Main Specification Level 2 Version 1.2, Revision 103-Part 2 TPM Structures
5. Sailer, R., van Doorn, L., Ward, J.P.: The role of TPM in Enterprise Security. IBM Research Report. Computer Science. RC 23363 (W0410-029), October 6 (2004)
6. Trusted Computing Group.TCG TPM Main Specification Level 2 Version 1.2. Revision 103-Part 3 Commands (2005)
7. Murray, D.G., Milos, G., Hand, S.: Improving Xen Security through Disaggregation. In: ACM VEE 2008, March 5-7 (2008)
8. The Linux Kernal Sources version 2.2.5
9. Rubini, A.: Linux Device Drivers. O'Reilly, Sebastopol (1998)
10. Siddha, S.B., Gopinath, K.: A persistent Snapshot Device Driver for Linux
11. Clark, C., Fraser, K., Hand, S., Hansen, J.G., Jul, E., Limpach, C., Pratt, I., Warfield, A.: Live Migration of Virtual Machines. In: Proceedings of the 2nd Syposium on Networked Systems Design and Implementation, NSDI 2005 (2005)

A Research on Improved Canny Edge Detection Algorithm

Jun Li[1] and Sheng Ding[2]

[1] School of Computer, Wuhan University, Wuhan, China College of Computer Science and Technology, Wuhan University of Science and Technology, Wuhan, China
[2] College of Computer Science and Technology, Wuhan University of Science and Technology, Wuhan, China
Jun_Li1980@foxmail.com

Abstract. Classical Canny Operator plays an important role in the image edge detection. The paper analyses the theory of the traditional Canny edge algorithm and does some improvements on the parts of smoothing filter selection, point amplitude calculation, and high or low threshold selection. The improved Canny algorithm uses B-spline function instead of Gaussian function; calculates gradient amplitude in 3×3 neighborhoods; and selects thresholds on the basis of gradient histogram. The experiment proves that the new algorithm improves the accuracy of positioning and provides a better and evident de-noising effect.

Keywords: Edge Detection, Improved Canny algorithm, B-spline function, Threshold Selecting.

1 Introduction

Edge, as the basic feature of target in image, indicates the sudden change of signals. It contains huge information, and is treated as important basis for image extraction. The quality of the edge map is directly related to the amount of supportive information it carries into the subsequent processing stages[1].Typical first-order derivative operators include Roberts, Prewitt, Sobel; second-order derivative operators include Laplace, log, etc. These operators are easily realized, but they're very sensitive to noise for their poor anti-interference ability. What's more, the edge is not delicate enough. Canny proposed three criteria of edge detection in 1986 which are universally acknowledged as the most strictly defined criteria so far[2]. The three criteria are signal-to-noise ratio(SNR), location accuracy criterion, and single-edge response criterion. He derived the optimal edge detection operator by adopting numerical optimization methods. Canny operator can generate information from two aspects—edge gradient direction and strength, with good SNR and edge localization performance. The algorithm is relatively easy to achieve in short time, so it becomes the evaluation standard of other edge algorithms.

However, the traditional Canny operator involves complicated calculation, and the accuracy of detection can still not reach the edge of a single pixel. In the actual

J. Zhang (Ed.): ICAIC 2011, Part V, CCIS 228, pp. 102–108, 2011.

practice, it's easily interfered by various factors, and there are still some false edges. So the specific applications are still met with some limitations. As to these issues, researchers have carried out considerable research and made many improvements. This paper, based on the in-depth study on edge detection algorithm, makes some improvement of the Canny operator. By introducing B-spline function to replace the Gaussian function, 3x3 neighborhoods of the gradient magnitude calculation, and the gradient histogram basis thresholding, it has achieved the precision and accuracy of image edge detection.

2 Canny Theory and Its Limitations

A. Canny Edge Detection Criteria

Canny's three edge detection criteria are as follows [2][3]:

1) Good detection. There should be a low probability of failing to mark true edge points, and low probalility of falsely marking nonedge points. The signal-to-noise ratio should be as high as possible.
2) Good localization. The points marked as edge points by the operator should approximate to the center of the real edge as much aspossible.
3) Low spurious response. Single edge produces fewer multiple responses, and false boundary responses are suppressed to the maximum.

B. The Four-Process Algorithm

(1) Image smoothing
In the two-dimensional case, Canny operator adopts two-dimensional Gaussian function derivatives as an image smoothing filter to suppress image noise.

(2) The calculation of gradient magnitude
The operator uses first-order partial derivative finite difference in 2x2 neighborhoods to calculate the gradient amplitude and direction of smooth data array I(x, y). $P_x[i, j]$ and $P_y[i, j]$ respectively refer to array of x and y directions partial derivative.

$$P_x[i, j]= (I[i+1, j]-I[i, j]+ I[i+1, j+1]-I[i, j+1])/2, \quad (1)$$

$$P_y[i, j]= (I[i, j+1]-I[i, j]+ I[i+1, j+1]- I[i+1, j])/2. \quad (2)$$

Pixel gradient magnitude and gradient direction are obtained with the coordinate conversion formula involving Cartesian coordinates to polar coordinates. The gradient amplitude calculated with the second order norm is

$$M(i, j) = \sqrt{f_x(i, j)^2 + f_y(i, j)^2} . \quad (3)$$

And gradient direction is

$$\theta(i, j) = \arctan(f_y(i, j) / f_x(i, j)) . \quad (4)$$

When the gradient direction θ[i, j] has been determined, the edge direction can be divided into eight directions with 45 degrees. Through the gradient of direction, the adjacent pixels of this pixel gradient direction can be found out.

(3) The non-maxima suppression of Gradient image
For each pixel with none-zero gradient strength, two adjacent pixels are searched along its gradient direction. If the gradient intensity value is less than the corresponding value of adjacent pixels in its gradient direction, then this can't be the edge point, and then set its edge strength to 0; if not, it's chosen to be candidate edge point.

(4) double threshold processs and edge connection
After non-maxima suppression, the images still have false edge.The pixels undergone non-maxima suppression are further treated with high and low threshold.The one with greater gradient amplitude than high threshold must be the edge; the one with lower gradient amplitude than low threshold is definitely not the edge; if gradient amplitude is in-between, then see whether there exists an edge pixel of eight neighborhood pixels which is greater than high threhold. If there is one, it is the edge pixel. Otherwise, it's not.

C. Processing Problems

Some problems can be found in the Canny four-process algorithm.

1. In the selection of smoothing filter. The use of Gaussian function as image smoothing filter will lead to excessive smooth. The main purpose of image smooth filter is to improve SNR, and eliminate noise. However, when images are smoothed by Gaussian, edges are treated as high frequency components and smoothed out.
2. In the Point amplitude calculation. For each pixel amplitude calculation, the traditional Canny algorithm uses center in 2x2 neighborhoods for finite difference mean value to calculate gradient amplitude. It's more sensitive to noise and easy to detect the fake edges or lose some real edges.
3. In the selection of high threshold and low thresholds. Canny algorithm adopts fixed high and low thresholds to extract image edges, and the set of high and low thresholds depends entirely on artificially gain, which lacks adaptability to different images.

These problems affect the application effect of traditional Canny algorithm. This paper aims at solving these problems, and proposing an improved scheme.

3 THE Improved Canny Algorithm

A. Using of B-Spline Function Instead of Gaussian Function

Define 1: $N_m(x) = (N_{m-1} * N_1)(x) = \int_{-0.5}^{0.5} N_{m-1}(x-t)dt$

is defined as m-order center B-spline function, among them $N_1(x) = \chi_{[-0.5,0.5]}$, * indicates convolution.

Interval [-m/2, m/2] is defined as M-order center B-spline function support set, even-order center B-spline function $N_{2m}(x)$ contains integer nodes, with Fourier transform $\hat{N}_{2m}(x) = [\sin(x/2)/(x/2)]^{2m}$.

The following theorem 1 is demostrated in [4].

Theorem 1: For the centered even-order B-spline function $N_{2m}(x)$ with integer locations, both it and its Fourier transform $\hat{N}_{2m}(x)$ would converge to a Gaussian as n tends to infinity:

$$\lim_{m\to+\infty} \hat{N}_{2m}(\sqrt{6}x/\sqrt{m}) = \exp(-x^2/2), \tag{5}$$

$$\lim_{m\to+\infty} \sqrt{m/6}N_{2m}(\sqrt{m/6}x) = \exp(-x^2/2)/\sqrt{2\pi}. \tag{6}$$

The above theorem has one corollary, which states that the derivative of centered B-spline function would converge to a Gaussian's derivative.

Theorem 2: Given a contaminated function f(x) ∈ [a,b] and a partition $a = x_1 < x_2 < ... < x_{n-1} < x_n = b$ on the interval [a,b] with the evaluation of f at each node, the centered fourth order B-spline function $N_4(x)$ is the unique function at which the below functional J(y) attains its minimum,

$$J(y) = \int_a^b |y''(x)|^2 dx + \sum_{i=1}^{n} \rho_i |f(x_i) - y(x_i)|^2, \tag{7}$$

where $\rho_i > 0$ and $y(x) = f(x)*\theta(x)$, $\theta(x)$ refers to low pass filter.

Theorem 2 is strictly testified by [5].

This theorem shows that: when only the second-order derivation after smoothing are needed, the best convolution kernels is fourth-order center B-spline function, instead of gaussian function[6].

Gaussian convolution kernels are equally smooth no matter how the signal-to-noise ratio of the signal. But if the center B-spline function is used as convolution kernels, it can choose different order number of spline function according to the noise intensity in the original image. When noise is stronger, higher order spline functions can be chosen to srengthen the smooth effect with poorer approximation; When noise is weaker, low rank of the spline functions are chosen to remove a small amount of noise, and meanwhile, to approach the original data as much as possible.

B. Calculation Gradient Amplitude in 3 * 3 Neighborhoods

The paper adopts an algorithm that calculates gradient amplitude in 3 * 3 neighborhoods[7].

Firstly, algorithm calculates X direction partial derivative

$$P_x[i,j]= I[i, j+1]-I[i, j-1], \tag{8}$$

Y direction partial derivative

$$P_y[i,j]= I[i, j+1]-I[i, j-1], \tag{9}$$

45^0 direction partial derivative

$$P_{45}[i,j]= I[i-1, j+1]-I[i+1, j-1], \tag{10}$$

135^0 direction partial derivative

$$P_{135}[i,j]= I[i+1, j+1]-I[i-1, j-1], \tag{11}$$

the horizontal difference

$$f_x[i,j]= P_x[i,j]+(P_{45}[i,j]+P_{135}[i,j])/2, \tag{12}$$

the vertical difference

$$f_y[i,j]= P_y[i,j]+(P_{45}[i,j]-P_{135}[i,j])/2. \tag{13}$$

Then gradient amplitude is

$$M(i, j) = \sqrt{f_x(i, j)^2 + f_y(i, j)^2}\ , \tag{14}$$

gradient direction is

$$\theta(i, j) = \arctan(f_y(i, j) / f_x(i, j))\ . \tag{15}$$

This method takes into consideration the pixel diagonal direction. It introduces differential mean value calculation, improves the accuracy of edge detection, and inhibits the noise.

C. Gradient Histogram—Basis Threshold Value Selection

This paper uses gradient histogram to select high threshold. High and low thresholds are respectively represented as Hth and Lth.

In an image, the ratio of non-edge points to the total points is Hratio. Image pixel numbers are gradually accumulated according to the gradient distogram from low to high. When the accumulative number reaches to the one made by total number multiplying image Hratio, the corresponding image gradient value is high threshold Hth. In this paper Hratio is set as 0.8.

Low threshold Lth can be represented as Lratio×Hth. Lratio is the corresponding scaling factor. In this paper Lratio is 0.4.

4 Experimental Result

Four figures are used in this experiment for analysis. Image 1(a) is a standard cameraman diagram. The classical Canny operator detects too many textures in 1(b), especially on the background. In 1(c), the improved Canny Operator detects fewer textures than 1(b).Image 2(a) is a standard peppers diagram. 2(b) detects fewer textures but more false edges than 2(c). In general, they vary slightly. Image 3(a) is a cameraman image with random noise. Being added to random noise, image 3(b) obviously receives interference with more false edges. The edge positioning in image 3(c) shows precision and continuity. Image 4(a) is peppers image with salt-pepper noise. It can be seen from image 4(b) that de-noising effect is very poor, but de-noising effect in image4(c) is evident with accurate edge.

This experiment demonstrates that improved canny algorithm improves the accuracy of positioning and provides a better and evident de-noising effect.

(a) Cameraman (b) classical Canny (c)improved Canny

Fig. 1. Cameraman image and edge images

(a) Peppers (b) classical Canny (c) improved Canny

Fig. 2. Peppers image and edge images

(a) Cameraman with random noise (b) classical Canny (c) improved Canny

Fig. 3. Cameraman image with random noise and edge images

(a) Peppers with salt-pepper noise (b) classical Canny (c) improved Canny

Fig. 4. Peppers image with salt-pepper noise and edge images

5 Conclusion

The paper focuses on the principle of Canny algorithm as well as its existing problems, provides some improvements to the traditional Canny algorithm in three aspects: smoothing filter function selection, point amplitude calculation and the selection of high and low threshold. The result of experiment shows that the algorithm in this paper provides better positioning accuracy and precision, thus greatly strengthens the edge detection effect.

References

1. Meer, P., Georgescu, B.: Edge Detection with Embedded Confidence. IEEE Trans. on PAMI 23(12), 1351–1365 (2001)
2. Canny, J.: A Computational Approach to Edge Detection. IEEE Trans. on PAMI 8(6), 679–698 (1986)
3. Bao, P., Zhang, L., Wu, X.: Canny Edge Detection Enhancement by Scale Multiplication. IEEE Trans. on PAMI 27(9), 1485–1490 (2005)
4. Unser, M., Aldroubi, A., Eden, M.: On the Asymptotic Convergence of B-Spline Wavelets to Gabor Functions. IEEE Trans. on Information Theory 38(2), 864–872 (1992)
5. Cheng, Z.X.: Data fitting. Xian Jiaotong University Press, Xian (1986)
6. Jia, T.-x., Zheng, N.-n., Zhang, Y.-l.: Center B-spline Dyadic Wavelet Multiscale Edge Detection. Acta Automation Sinica 24(2), 192–199 (1998) (in chinese)
7. Wang, N., Li, X.: An improved edge detection algorithm based on the Canny operator. Journal of Shenzhen University Science and Engineering 22(2), 149–153 (2005) (in chinese)

Fragile Watermarking Scheme with Pixel-Level Localization*

Jie Guo, Weidong Qiu, and Ping Li

School of Information Security Engineering Shanghai Jiao Tong University Shanghai, 200240, China
Shanghai Antzone Information Technology Ltd Shanghai, 200240, China
jieguo29@yahoo.cn

Abstract. A new fragile watermarking scheme is proposed for image authentication. The advantageous properties of both block-wise schemes and pixel-wise schemes are combined in our schem. A block-wise mechanism is exploited so as to obtain a high detection probability, while a pixel-wise mechanism is used to achieve accurate tamper localization. Both mechanisms can be adjusted and optimized based on application requirements. Meanwhile, forgery of watermark is prevented by using keyed hashes during watermark generation. Experimental results show that the proposed scheme achieves both accurate tamper localization and strong security.

Keywords: fragile watermarking, image authentication, tamper detection.

1 Introduction

Multimedia authentication is a technology to check authenticity and integrity of multimedia signals. It is often desirable to localize tampered pixels or samples for a tampered signal so unmodified parts can still be used. Technologies to fulfill this goal have been actively studied in recent years. A class of proposed technologies, called fragile watermark, is to detect any modifications to a multimedia signal.

Fragile watermarking technologies can be classified into pixel-wise schemes and block-wise schemes [1][2][3][4]. A pixel-wise scheme is designed to localize tampered pixels in additional to verify authenticity for the whole signal. A block-wise scheme, on the other hand, is designed to localize tampered blocks. A block-wise scheme is securer in general than a pixel-wise one, but has much coarser tamper localization capability.

In this paper, we propose a new fragile watermarking scheme, which combines the advantageous properties of both block-wise schemes and pixel-wise schemes. Tampered areas in a watermarked image can be localized to the pixel level by cross localization with row and column signatures. According to different requirements, one can adjust the block size to obtain a proper detectability and localization

* This work is supported by NSFC (60703032), NSFC (61073157) and the National 863 Projects (2009AA01Z426).

J. Zhang (Ed.): ICAIC 2011, Part V, CCIS 228, pp. 109–115, 2011.

accuracy. Our scheme does not require any codebook or reference in authentication process. Experimental results indicate that our scheme has adequate performance to be used in real world applications.

The rest of this paper is organized as follows. Section 2 presents our fragile watermarking algorithms. Section 3 evaluates our watermarking scheme using real data. Section 4 concludes this paper.

2 Proposed Watermarking Algorithm

We propose a new fragile watermarking scheme, which provides a graceful trade-off between security and tamper localization. A block-wise mechanism is exploited so as to obtain a high detection probability, while a pixel-wise mechanism is used to achieve accurate tamper localization.

A. Watermark Embedding

Take a grayscale image $X_{m,n}$ of M by N pixels for example. Our watermark-embedding process involves the following steps.

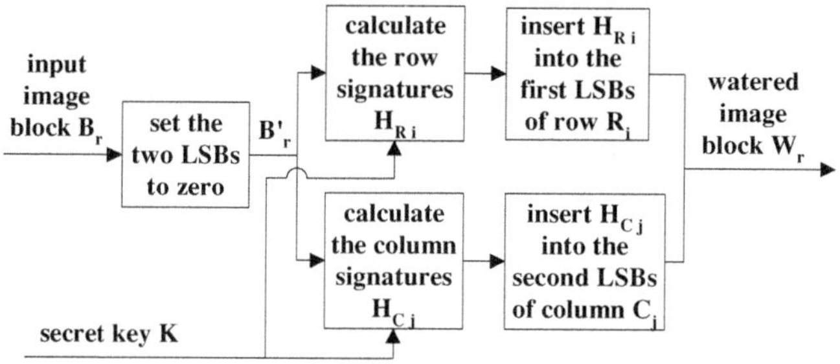

Fig. 1. Watermark embedding process

Step 1) Divide the image into blocks of I by J pixels. Each block will be watermarked independently. Let B_r denote the rth block in the image. The watermark embedding procedure for block B_r is shown in Fig. 1.

Step 2) Create a corresponding block $B_r^{'}$ from B_r by setting the two low-order least significant bits (LSBs) to zero. Let R_i and C_j denote the ith row and the jth column of pixels in the block $B_r^{'}$, respectively.

Step 3) Let $H(\cdot)$ be a keyed-hash function. Similar to the row-column hash function mentioned in paper [5], hash calculation is performed on each row R_i and each column C_j. To resist against the famous quantization attack [6], the position information (m, n) of each pixel is included in the hash calculation:

$$H_{R_i} = H(K,m,n,R_i) = (hr_1, hr_2, \cdots, hr_S) \tag{1}$$

$$H_{C_j} = H(K,m,n,C_j) = (hc_1, hc_2, \cdots, hc_S) \tag{2}$$

where K is a secret key, hr_w and hc_w denote output bits from the hash function, and S is the number of output bits.

Step 4) Embed the output H_{R_i}, called row signature, into the first LSBs of row R_i. Meanwhile, embed the output H_{C_j}, called column signature, into the second LSBs of column C_j. Since row signatures and column signatures are embedded into different LSBs, they form a grid which will be used to localize tampered pixels during image authentication process. In the case that S is larger than the block row length I or block column length J, H_{R_i} and H_{C_j} are truncated to the block row length and the block column length, respectively.

Step 5) Repeat the steps 2-4 for each block. Then, all the blocks in the original image are processed to form a watermarked image.

B. Watermark Detection

As shown in Fig. 2, the watermark verification process for each block consists of three basic steps: i) Calculate row signatures and column signatures for the block; ii) Extract the embedded block signatures from the block; iii) Verify the image integrity by comparing the calculated signature with the embedded signature. The main steps of our watermark-detecting process are as followed.

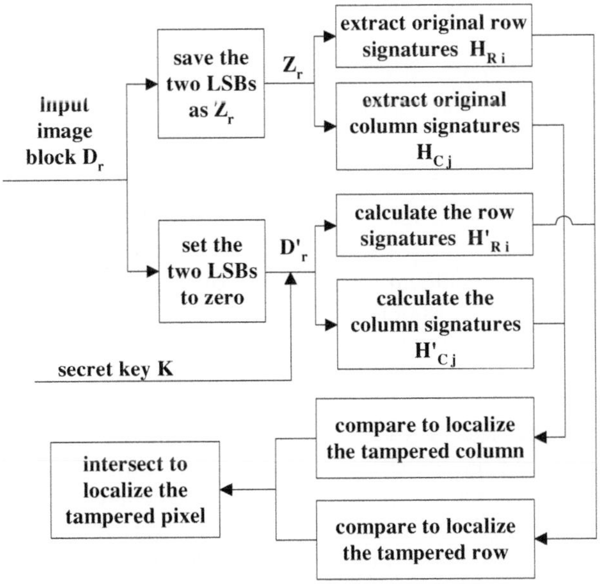

Fig. 2. Watermark verification process

Step 1) Divide the suspect image $X'_{m,n}$ into blocks of I by J pixels. Let D_r denote the rth block in the image $X'_{m,n}$.

Step 2) Split each image block D_r into two parts. The first part Z_r consists of the two LSBs of each pixel value. It contains both row signatures and column signatures of the original image. The other part D_r' consists of the pixel values whose two LSBs are zeroed out.

Step 3) Let R_i' and C_j' denote the ith row and the jth column of data in D_r', respectively. Similar to the hash calculation in watermark embedding, the hash values $H_{R_i'}$ and $H_{C_j'}$ are computed for each row R_i' and column C_j', respectively.

Step 4) Comparing the calculated hash values with the embedded hash values, tampered pixels in block D_r can be localized. For example, supposing that pixel (i, j) in a block has been modified, the calculated hash values $H_{R_i'}$ and $H_{C_j'}$ will be different from the original hash values H_{R_i} and H_{C_j}. The crossed point of the row R_i' and column C_j' pinpoints the tampered pixel (i, j).

Step 5) Repeat the steps 2-4 for each block in image $X'_{m,n}$. Then, all the blocks in the image $X'_{m,n}$ are processed to localize the tampered areas in the protected image.

3 Experimental Results

To investigate the characteristics of the proposed scheme, we conducted a series of simulation experiments in Visual C++ Version 6. All experiments are run on a HP Compaq computer with a Pentium(R) 4 CPU of clock rate 3.00GHz, 1.0 GB of RAM, and 40 GB hard-disk running Microsoft Windows XP. The standard test images are 256 × 256-pixel grayscale images.

A. Fragility

The fragility illustrates the authentication degree of fragile watermarking scheme. A fragile watermarking scheme should be sensitive to any modifications or malicious attacks to the watermarked image. In this section, we test the detectability of the proposed scheme to malicious attacks. The test image Cameraman, with 256×256 pixels, is shown in Fig. 3(a).

By setting the block size to 8 by 8 pixels, we obtain the watermarked image as shown in Fig. 3(b). The PSNR (peak signal-to-noise ratio) is 42.13 dB; thus, the imperceptibility of our scheme is high. To verify the accuracy of pixel-level tamper localization, we add the salt-and-pepper noise into the watermarked image obtain the tampered image shown in Fig. 3(c), and the detected image in Fig. 3(d). We also manipulate the watermarked image by moving the “high building” from the right to the left, removing the “low building”, and adding a “SMU” logo in the upper-right

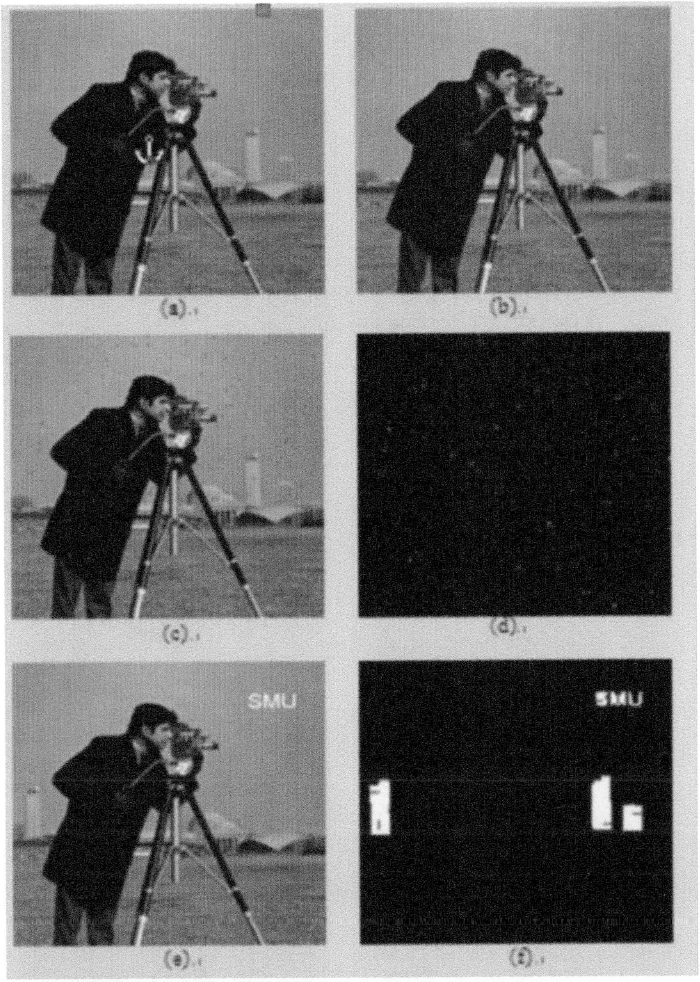

Fig. 3. Experimental results

corner. The modified image is shown in Fig. 3(e), and the detected image in Fig. 3(f). We have conducted many other experiments to other test images. All the experimental results indicate that the proposed scheme is able to localize tampered area correctly.

B. Tradeoff on Selecting Block Size

As mentioned above, how to choose the block size is a tradeoff problem for detectability and localization accuracy. In this section, we test the effect of block size on the performance of the proposed scheme. The test image F16, with 256×256 pixels, is shown in Fig. 4(a).

We first set the block size to be 8 by 8 pixels. According to embedding steps, a watermarked image is obtained, as shown in Fig. 4(b). The PSNR (peak signal-to-noise ratio) is 40.57 dB. We manipulate the watermarked image by erasing the words

"U.S.AIR Force" from the middle of the plane, changing the logo "F16" to be "F18", and adding a "SMU" logo in the upper-left corner of the plane. The modified image is shown in Fig. 4(c), and the detected image in Fig. 4(d).

Then, we set the block size to be 2 by 2 pixels. By performing embedding steps to original image, a new watermarked image is obtained. The second watermarked image is manipulated with the same modifications. That is, we erase the words "U.S.AIR Force" from the middle of the plane, change the logo "F16" to be "F18", and add a "SMU" logo in the upper-left corner of the plane. The modified image is shown in Fig. 4(e), and the detected image in Fig. 4(f). By comparing the two results, it is found that the scheme with 8×8 blocks has better detectability and the scheme with 2×2 blocks has better localization accuracy.

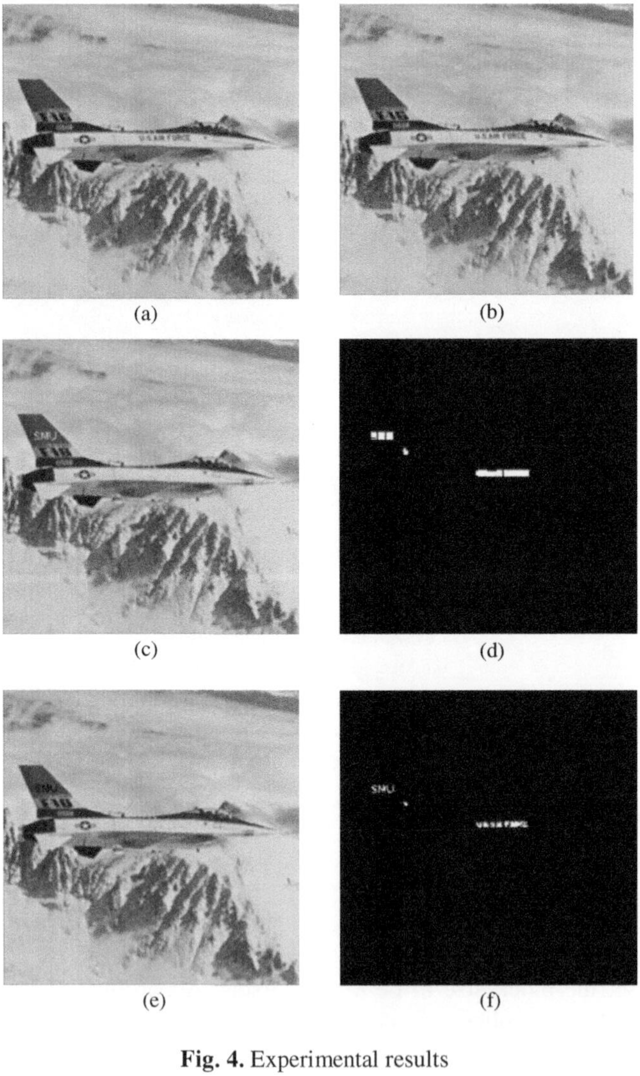

Fig. 4. Experimental results

4 Conclusion

In this paper, a new fragile watermarking scheme is presented for image tamper detection. To balance the detectability and the localization accuracy, both block-wise schemes and pixel-wise schemes are employed to embed the authentication information. A block-wise mechanism is exploited so as to obtain a high detection probability, while a pixel-wise mechanism is used to achieve accurate tamper localization. Both mechanisms can be adjusted and optimized based on application requirements. Meanwhile, keyed-hash function is employed to prevent the forgery of watermark during watermark generation. Experiments demonstrate the effectiveness and validity of our scheme.

References

1. Zhang, X., Wang, X.: Fragile watermarking scheme using a hierarchical mechanism. Signal Processing 89(4), 675–679 (2009)
2. He, Y., Han, Z.: A fragile watermarking scheme with pixel-wise alteration localization. In: 9th International Conference on Signal Processing, pp. 2201–2204 (October 2008)
3. Ho, A.T.S., Zhu, X., Shen, J., Marziliano, P.: Fragile Watermarking Based on Encoding of the Zeroes of the z-Transform. IEEE Transactions on Information Forensics and Security 3(3), 567–569 (2008)
4. Yuan, H., Zhang, X.: Multiscale Fragile Watermarking Based on the Gaussian Mixture Model. IEEE Transactions on Image Processing 15(10), 3189–3200 (2006)
5. Wolfgang, R., Delp, E.: Fragile watermarking using the vw2d watermark. In: Proc. SPIE Security and Watermarking of Multimedia Contents, vol. 3657, pp. 204–213 (January 1999)
6. Holliman, M., Memon, N.: Counterfeiting attacks on oblivious blockwise independent invisible watermarking schemes. IEEE Transaction on Image Processing 9(3), 432–441 (2000)

Study on Simulation Software of Personnel Safety Evacuation in Large Public Building Fire*

Wang Yun

Department of Fire Engineering
Chinese People's Armed Police Force Academy
Langfang , Hebei Province, China
wangyun222@yeah.net

Abstract. It is very necessary to develop an evacuation model for improve fire protection design of public building. New evacuation model and new application software was developed in this paper. Four occupant types were considered in the model, which have there own personnel characters and can reflect their influence on evacuation. Behavior rules was established in the model, which can describe the normal purposeful movement of the people and can reflect the influence of crowd density. The fire data which was simulated by fire simulation software FDS can be loaded into the evacuation model in every compute step, which can describe the evading movement and the following movement of the people. By using this software, the user can obtain the personnel position at every compute step, the people quantities at every compute step and the dynamic escaping course of the people. The result of this paper will provide the theory basis for evaluating the escaping design, implementing the performance-based fire design, and guiding the fire rescue in the fire fighting.

Keywords: public building fire, personnel evacuation, simulation software.

1 Introduction

With the development of computer technology, the simulation of evacuation has made rapid progress since the 1980s. At present the scholars of this field have developed 28 evacuation models and computation software, including BuildingExodus of English Greenwich University, Simulex of English Edinburgh University's, and EXITT of American National Institute of Standards and Technology and so on[1]. In China, the research of evacuation model and the simulation of escape still at starting stage, also has obtained some achievements. Lu Zhaomingof Hong Kong City University, Fangzheng of the Wuhan University, Chen Baozhi of Northeastern University, Wang Zhigang of Tianjin Fire Research Institute, Yang Lizhong of University of Science and Technology of China and the researchers of Sichuan fire research institute have already done a great deal of research and studies on simulating the personnel escape in buildings with different function [2-5]. The Spatial-Grid Evacuation Model, which

* Distinguishing Specialty Development Project on Fire Protection Engineering Sponsored by the Ministry of Education of China for Colleges and Universities (TS10113).

J. Zhang (Ed.): ICAIC 2011, Part V, CCIS 228, pp. 116–123, 2011.

developed by Hong Kong City University and the Wuhan University together, has been already applied in some fire evaluation work in Hong Kong [2]. The behaviors which occur during evacuation should be simulated in the model. If the personnel behavior in fire which the model can simulate to be more, then the simulation is more approaches to the real fire scene. Therefore the evacuation model in the future should contain the more behavior details and describe the escape scene more comprehensively.

A new evacuation model of large public building is established in this paper. The smoke data which were simulated by fire simulation software FDS (Fire Dynamics Simulator) can be loaded into this model in every compute step, by which the influence of smoke to personnel escape can be described and more people's behavior in fire can be simulated.

2 The Establishment of Evacuation Model

The thin network was used in this model to describe the building structure, which can reflect the outer figure and the inner obstruction of the building more correctly, can describe the people's exact position in every moment and the movement track during whole evacuation. For describe the personnel egress movement more accurately, personnel's pre-operating time and the smoke detector operating time are all considered in the model.

A. The Hypothesis of Evacuation Model

1) Grid division: The researchers of Tianjin Fire Research Institute measure the body thickness and shoulder width of Chinese, and compute the body projective area [6]. The body thickness and shoulder width is both not more than 0.5 meters. The body projective area regarded as an ellipse is 0.146 square meters, yet, is 0.197 square meters when regarded as a rectangle. Considering these metrical data, a person is expressed by a round which diameter is 0.5m, and the grid distance is 0.5m too. The grid division and personnel arrangement are showed in Fig. 1.

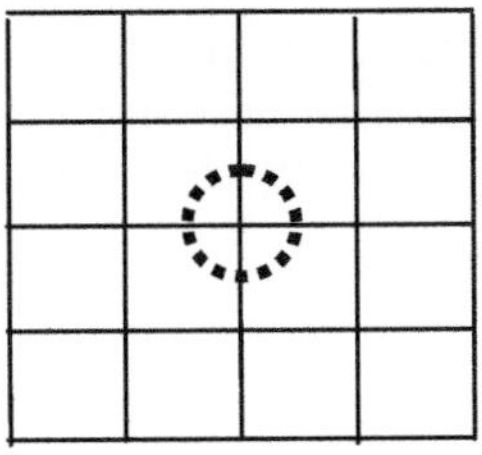

Fig. 1. Grid division and personnel arrangement in model

2) Grid point states: Every grid has different state, including wall, obstruction, exit, person and vacancy. If the grid is occupied by person, obstruction or wall, then its state is "1". Moreover, if the grid is vacancy or occupied by exit, then its state is "0".

3) Personnel character: Different types of people have different move velocity. In order to express individual character of different type of people correctly, four types of people are considered in this model, including male, female, elderly and child. The move velocities of these four types of occupant are listed in table1. The user of this model can change personnel proportion of various occupant types according actual condition.

Table 1. Move velocity of various occupant types in model

Body type	Unimpeded mean velocity （m/s）
Adult male	1.35
Adult female	1.15
Child	0.9
Elderly	0.8

B. The Movement Rules of Personnel

Supposing a person occupy the grid point (i, j), there are four directions which this people can move towards in next computer step. These four directions are showed in Fig. 2. These rules which people should comply with during movement are as follows.

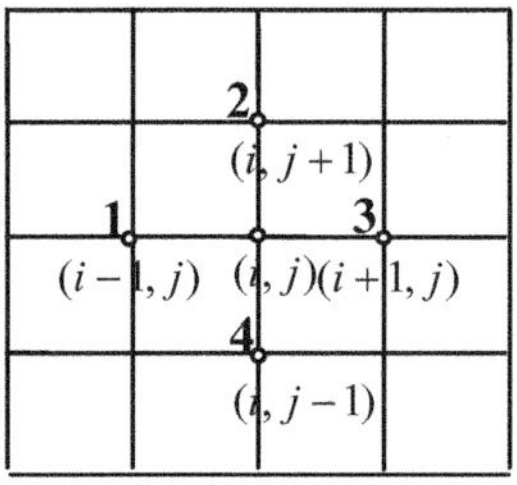

Fig. 2. The expected move direction of occupant

1) The time of preparing: Considering the recognize time and response time of people in fire, the time of preparing is simulated in this model. In order to reflect the individual character of Chinese, the research result of Zhang Shuping is used [7]. The conclusion is listed in table 2.

Table 2. The preparing time of Chinese

The time of preparing (min)	The proportion (%)
0-2	26%
3-5	39%
6-10	28%
11-15	7%

2) Move towards the nearest exit: Comparing the distances between the nearest exit and four grid points which surrounds person, the maximal distance is defined as $D_{e\max}$. The probability which grid point should be selected as the next move direction can be calculated by

$$P_e = \frac{D_{e\max} - D_e(m,n)}{D_{e\max}} \tag{1}$$

Where, P_e means the grid point selection probability considering the position of exit; $D_e(m,n)$ is the distance between grid point and the nearest exit; m, n represents the coordinate of the grid point.

The grid point with the biggest P_e is the aiming grid point of occupant movement.

3) Move keeps away from the fire: Comparing the distances between the location of fire and four grid points which surrounds person, the maximal distance is defined as $D_{f\max}$ .Considering the influence of the fire, the probability which grid point should be selected as the next move direction can be calculated by

$$P_f = \frac{D_f(m,n)}{D_{f\max}} \tag{2}$$

Where, P_f stands for selection probability of the grid point considering the influence of the fire; $D_f(m,n)$ is the distance between grid point and the location of fire; m, n represents the coordinates of the grid point.

The bigger P_f is, the bigger the distance between grid point and the location of fire is. The grid point with the biggest P_f is the aiming grid point of occupant movement.

4) Effect of the crowd density: If the surrounding grid points are both occupied by person, the crowd density should be considered in simulation. The movement of one will be influenced by other occupants, and the choices of most people are regarded as the move direction in next computer step.

5) Effect of the smoke density: Jin had done a great deal of experiments on visibility in smoke [8]. The connection between visibilities in smoke and the extinction coefficient of smoke are found.

$$\text{For lumination object } KS = 8 \tag{3}$$

$$\text{For reflection object } KS = 3 \tag{4}$$

Where, S is visibilities in smoke; K means extinction coefficient of smoke.

The extinction coefficient can be represents by the product of ratio extinction coefficient and smoke mass concentration, as follows:

$$K = K_m m_s \tag{5}$$

Where, K_m is ratio extinction coefficient; m_s means smoke mass concentration, which can be computed by fire simulation software FDS (Fire Dynamics Simulator).

The smoke mass concentration which is simulated by FDS can be loaded into the model in every compute step, and the visibilities of every grid point can be calculated. If the distance between grid point and exit is less than the visibility of this grid point, then the occupant in this grid point can find the exit in simulation.

3 The Simulation Software

A. The Input of Data

There are three types of parameters should be input into software, including the building structure, individual character, fire scene. The plan of building can be established by the input of corresponding data about building structure. In dialog box of individual character, the number of occupants, the proportion of various occupant types and the time of preparing are considered, which can reflect the influence of individual characteristics in evacuation simulation. The locations of fire, the intensity of fire, the type of combustion and simulation result of FDS are input to represent the influence of fire in evacuation simulation. The main interface and each input interface are showed in Fig. 3.

B. The Simulate Process

After the initialization of various types' parameters, the movement of person is simulated according to the flow chat showed in Fig. 4. The process of loading fire data into the model is showed in the broken line frame.

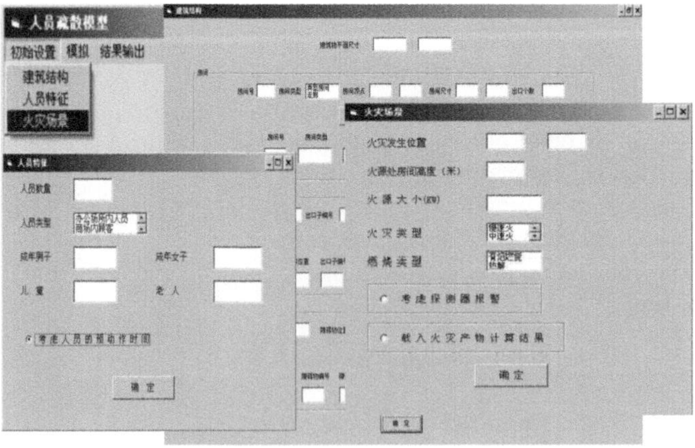

Fig. 3. The main interface of software

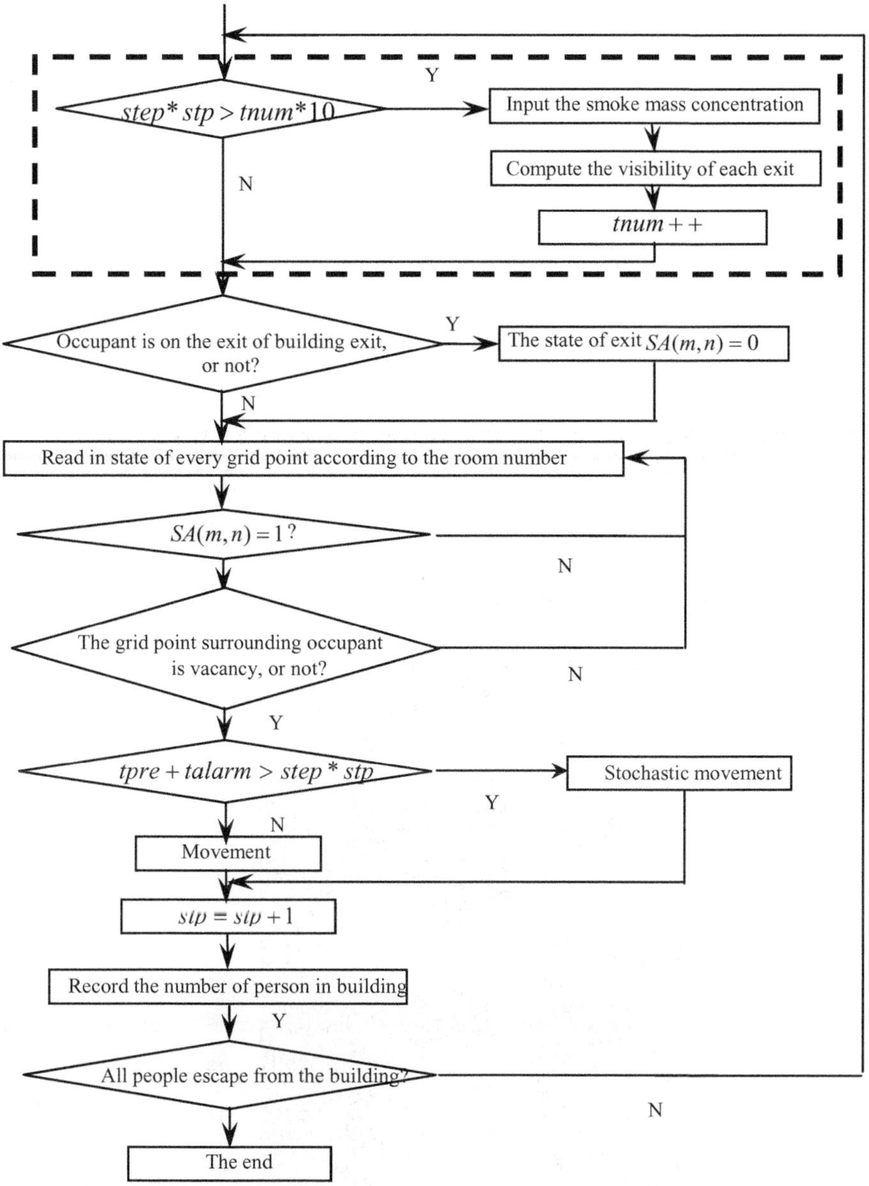

Fig. 4. The flow chats of evacuation simulation

C The Output of Simulate

The time of evacuation movement can be calculated, and the dynamic escaping course of the people in every compute step can be displayed in this software, as showed in Fig. 5.

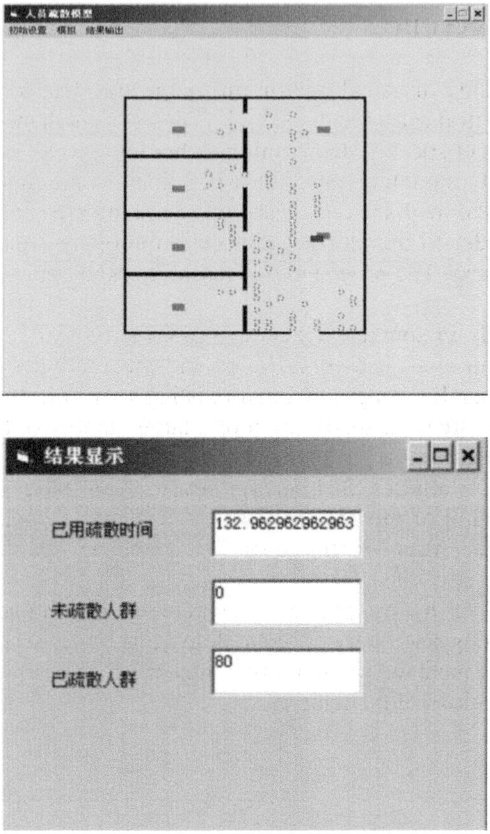

Fig. 5. The output of simulate

5 The Conclusion

In this paper, the existing domestic and foreign models were reviewed, and a new evacuation model was established, and a corresponding simulate software was compiled.

(1) A new evacuation model was developed in this paper. In the model, the people were classified to four categories and every category have there own personnel characters, which can reflect their influence on evacuation. Behavior rules was established in the model, which can describe the normal purposeful movement of the people.

(2) The fire data which was simulated by fire simulation software FDS can be loaded into the evacuation model in every compute step, which can reflect the influence of the smoke visibility in escape. The behavior rules in the fire were also established in the model, which can describe the evading movement and the following movement of the people.

(3) The application software that can calculate the time of evacuation movement was developed. By using this software, the user can obtain the personnel position at every compute step, the people quantities at every compute step and the dynamic escaping course of the people.

References

1. Kuligowski, E.D., Peacock, R.D.: A Review of Building Evacuation Models. National Institute of Standards and Technology Technical Note 1471 (2005)
2. Lo, S.M., Fang, Z.: A Spatial-Grid Evacuation Model for Buildings. Journal of Fire Sciences 18, 376–394 (2000)
3. Wen, L., Chen, Q., Chen, B.: Human Evacuation Scenarios in Case of Fire and Simulation by Computers. Journal of Northeastern University (Natural Science) 19, 445–447 (1998)
4. Wang, Z.: An occupant evacuation model for the underground shopping mall during fire incidents. Fire Safety Science 10, 57–62 (2001)
5. Yang, L., Fang, W., Huang, R., Deng, Z.: Evacuation Model of people in Fire Based on Cellular Automaton. Chinese Science Bulletin 47, 896–901 (2002)
6. Guo, Y.: The Research on the Character of Safe Evacuation in High-rise Building Fires. Zhongqing University (2001)
7. Zhang, S.: Study on Human Behaviour in Building Fire. Xi'an University of Architecture & Technology (2004)
8. Jin, T.: Studies on Human Behavior and Tenability. In: Proceedings of the Fifth International Symposium Fire Safety Science, pp. 3–21 (1997)

Short Term Load Forecasting with Least Square Support Vector Regression and PSO

Zou Min and Tao Huanqi

College of Electronics and Information Engineering
Wuhan Textile University
Wuhan, Hubei Province, 430073, P.R. China
ZouMin22222@163.com

Abstract. Load forecasting has been an important topic in power system research. Short term forecasting is important for power dispatch, especially in the modern electricity market. In this paper, an approach based on least square support vector regression (LSSVR) is proposed to short term load forecasting. An effective forecasting model can only be built under optimal parameters. The algorithm of particle swarm optimization is applied to search optimal parameters of the above forecasting model. The experimental results based on above model for a sample load series are shown that the model proposed in this paper outperforms the BP neural network approaches and the simple LSSVR methods on the mean absolute percent error criterion.

Keywords: short term load forecast, least support vector regression, particle swarm optimization.

1 Introduction

Short term load forecasting plays an important role in all aspects of reliable, economic and secure strategies for power systems. The power load series are influenced by many factors such as climate, economic policy, price and etc, thus it's difficult to get accurate prediction. Many load forecasting models are constructed[1, 2], of which neural network is often used.

Neural network method is generally used in nonlinear time series prediction[3]. But the over-fitting problems in neural network method retard the performance improvement to more high accuracy. With the development of statistic learning theory (SLT) and support vector machines (SVM) algorithm, SVM and its extended versions, such as least squares support vector machines (LS-SVM) are gradually applied to nonlinear time series prediction research with good performance[4, 5]. However, parameters determination in time series prediction based on SVM is generally a complicated and important problem to be solved carefully, which is directly related to the model forecast performance.

Particle swarm optimization (PSO) is a computational method that optimizes a problem by iteratively trying to improve a candidate solution with regard to a given measure of quality. After kernel function of SVM methods is determined, the kernel

J. Zhang (Ed.): ICAIC 2011, Part V, CCIS 228, pp. 124–132, 2011.

function parameter and penalty parameter need to be set properly. In the regression prediction research based on SVM, the model parameters are generally determined by the trail-and-error method, which determines the final optimal set of parameters with best performance from sets of selected parameters through analyzing the influences to the model by these parameters. With the development of SVM research, the research of optimal parameters determination for SVM is gradually becoming a common problem in SVM research areas. With particle swarm optimization, a novel forecasting model based on least squares support vector regression (LSSVR) with parameters optimized through PSO, also named as PSO-LSSVR model, is proposed in this paper, which searches for the optimal parameters of LSSVR model using PSO algorithm and adopts the optimal parameters to construct the LSSVR model. This paper is organized as follows. In section 2, LSSVR is introduced. Then the PSO-LSSVR forecasting model based on particle swarm optimization and LSSVR is discussed in section 3. Experimental data applied in PSO-LSSVR model is proposed in section 4. Finally, the conclusions are given in section 5.

2 Least Square Support Vector Regression

Support vector machines[6](SVM) algorithm is a machine learning algorithm, proposed by Vapnik based on statistical learning theory (SLT). LS-SVM algorithm is proposed by Suykens[7]based on standard SVM theory, which is a least squares version of standard SVM algorithm and involves equality instead of inequality constraints and works with a least squares object function. Owing to the excellent performance in nonlinear forecasting problems, support vector machines prediction models, including models based on its extension, have been generally applied for forecasting model.

Given data $\{(x_1, y_1), \cdots, (x_N, y_N)\}$, where $x_i \in R^n$ denotes the inputs of sample data and has a corresponding target value $y_i \in R$ for $i = 1, \cdots, N$, where *N* is the size of the sample data. SVM regressive algorithm maps the input space into high dimension feature space through nonlinear function and constructs a linear regressive function as follows in this feature space.

$$y(x) = \omega^{\mathrm{T}} \phi(x) + b \tag{1}$$

where $\phi(x)$ is a nonlinear function, which maps the input space into the feature space, ω^T is an m-dimensional vector, and *b* is a bias.

To solve the above regressive function, in LS-SVM, one defines the optimization problem[7]

$$\begin{cases} \min\limits_{\omega,b,e} J_2(\omega, b, \xi) = \dfrac{1}{2}\left(\omega^{\mathrm{T}} \omega\right) + \gamma \dfrac{1}{2} \sum\limits_{i=1}^{n} \xi_k^{\,2} \\ y_k\left[\omega^{\mathrm{T}} \phi(x_k) + b\right] = 1 - \xi_k, \quad k = 1, \cdots, n \end{cases} \tag{2}$$

Where ξ is a slack variable, which is necessary to allow regression error, and γ is the penalty parameter also named as regularization parameter, which compromises the effect between maximum margin and the minimum regression error.

The corresponding Lagrange dual problem is

$$L(\omega,b,\xi;\alpha)=J_2(\omega,b,\xi)-\sum_{k=1}^{n}\alpha_k\left\{y_k\left(\omega^{\mathrm{T}}\phi(x)+b\right)-1+\xi_k\right\} \tag{3}$$

where $\alpha_i \left(i=1,\cdots,n\right)$ is Lagrange multiplier.

According to Karush-Kuhn-Tucker (KKT) conditions, the equations as follows can be obtained.

$$\begin{cases} \frac{\partial L}{\partial \omega}=0\rightarrow \omega=\sum_{k=1}^{n}\alpha_k y_k \phi(x_k) \\ \frac{\partial L}{\partial b}=0\rightarrow \sum_{k=1}^{n}\alpha_k y_k=0 \\ \frac{\partial L}{\partial \xi_k}=0\rightarrow \alpha_k=\gamma\xi_k \\ \frac{\partial L}{\partial \alpha_k}=0\rightarrow y_k\left[\omega^{\mathrm{T}}\phi(x_k)+b\right]-1+\xi_k=0 \end{cases} \tag{4}$$

The above equations can be written as the solution to the following set linear equations

$$\begin{bmatrix} 0 & -y^{\mathrm{T}} \\ y & \Omega+\gamma^{-1}I \end{bmatrix}\begin{bmatrix} b \\ \alpha \end{bmatrix}=\begin{bmatrix} 0 \\ \bar{1} \end{bmatrix} \tag{5}$$

where $y=[y_1,\cdots y_n]^T$, $\bar{1}=[1,\cdots 1]^T$, $\alpha=[\alpha_1,\cdots\alpha_n]^T$, $\Omega=ZZ^T$, $Z=\left[\phi(x_1)^T y_1,\cdots,\phi(x_n)^T y_n\right]$.

In the matrix Ω, each element of the matrix is in the form:

$$\Omega_{i,j}=y_i y_j \phi(x_i)^{\mathrm{T}}\phi(x_j)=y_i y_j K(x_i,x_j) \tag{6}$$

Where $K(x_i,x_j)$ is defined as kernel function. The value of a kernel function equals to the inner product of two vectors x_i and x_j in the feature space $\phi(x_i)$ and $\phi(x_j)$, that is the kernel function $K(x_i,x_j)=\phi(x_i)^T\phi(x_j)$.

The typical examples of kernel function are polynomial kernel and radius basis function (RBF) kernels:

Polynomial: $K(x,x_k)=\left(x_k^T x+1\right)^d \ (d=1,\cdots,n)$

RBF: $K(x,x_k)=\exp\left(-\|x-x_k\|^2/2\sigma^2\right)$

The LS-SVM model for regression can be expressed as follows

$$y(x) = \sum_{k=1}^{n} \alpha_k K(x, x_k) + b \tag{7}$$

3 Pso-lssvr

A. PSO-LSSVR Optimization Procedure

To build an efficient LSSVR prediction model, LSSVR's parameters must be determined carefully. These parameters include:

(1) *Kernel function*: Kernel function is important in LSSVR model and is used to construct a nonlinear decision hyper-surface in the LSSVR input space. RBF kernel is the most typical one of all kernel functions and has the same performance under some conditions with other kernel functions, as liner polynomial and sigmoid kernel[7].

(2) *Regularization parameter:* γ determines the trade-off cost between minimizing the training error and minimizing the model's complexity.

(3) *Bandwidth of kernel function* (σ^2): represents the variance of the Gaussian kernel function.

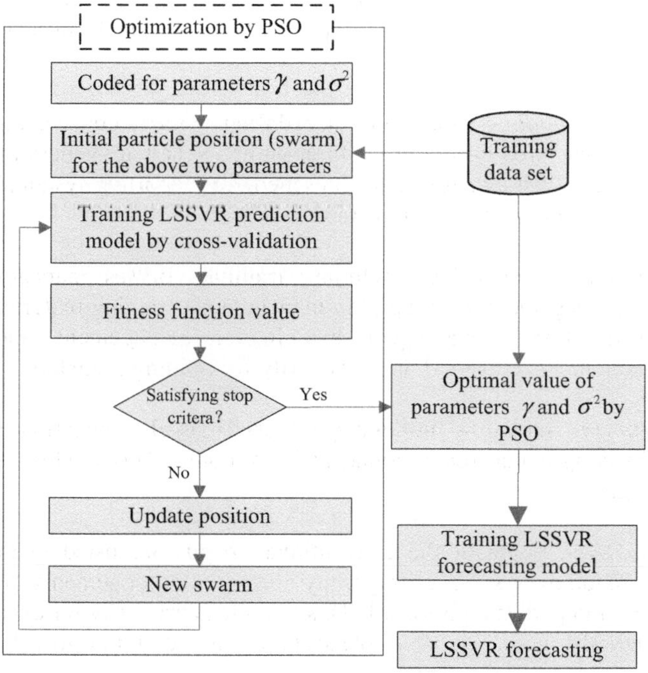

Fig. 1. PSO-LSSVR model

In contrast with the trial-and-error method, this paper proposes a new method known as PSO-LSSVR. This model uses particle swarm optimization to seek the optimal value of LSSVR's parameters and improve the model prediction performance. The proposed PSO-LSSVR forecasting model dynamically optimizes the optimal value of LSSVR's parameters during the PSO iteration process to construct the optimized LSSVR model.

Fig. 1 illustrates the process of the PSO-LSSVR model. Details of the proposed model are presented as follows:

(1) *Representation*: When particle swarm optimization is used for optimization, parameters can be directly real-value coded. The two LSSVR parameters γ and σ^2 are directly coded to construct the particle in the proposed model. The particle X can be represented as X=[p1,p2], which p1,p2 represent the above parameters γ and σ^2 respectively.

(2) *Swarm initialization*: In this model, initial swarm is composed of 30 random particles. The swarm size of 30 is trade-off between model complexity and population diversity.

(3) *Fitness function*: Fitness function is directly related to the performance of particle swarm optimization. If fitness value of training data is calculated directly, PSO may lead to over-fitting. In order to solve this problem, k-fold cross-validation[7] is concerned. In k-fold cross-validation, the training data is randomly split into k mutually exclusive subsets (the folds) of approximately equal size. The regression model is obtained with a given set of parameters $\{\gamma, \sigma^2\}$, using k-1 subsets as the training set and the performance of the parameter set is measured by the mean absolute percent error (MAPE) on the subset left out. The above procedure is repeated k times and in this fashion, each subset is used for testing once. Averaging the MAPE over the k trials (MAPE_{CV}) gives an estimate of the expected generalization error. Conventionally, the training error of k-fold cross-validation is applied to estimate the generalization error (k=5 is suggested in [8]). The fitness function of the proposed model is defined as the MAPE_{CV} of 5-fold cross-validation method on the training data set, as follows:

$$\min\ f = \text{MAPE}_{\text{CV}} \tag{8}$$

$$\text{MAPE}_{\text{CV}} = \frac{1}{k}\sum_{i=1}^{k}\text{MAPE}_i \tag{9}$$

$$\text{MAPE} = \frac{1}{n}\sum_{i=1}^{n}\left|\frac{x_i - \hat{x}_i}{x_i}\right| \tag{10}$$

where n is the number of training data samples, x_i is the actual value and $\hat{x}_i$ is the predicted value.

(4) *Particle range and max velocity*: These two parameters of PSO are related to the optimization convergence, which are decided by the optimal problem, in this

paper, particle range is not beyond γ and σ^2 allowed values and max velocity is not beyond above optimal parameters range.

(5) *Stopping criteria*: The process is repeated as shown in Fig. 1 until the max iteration number is equal to 50 or the fitness function value variation is not beyond 0.005.

B. Embedded Dimension

In time series prediction problems, the time series are generally expanded into three-dimension or higher dimensions space to explicit the implicit information of the series, which is called state reconstruction[9]. Embedding dimension m is an important factor in this method. The given series $\vec{x} = \{x_1, x_2, \cdots, x_N\}$ are transformed as follows by the embedding dimension, where X is the input matrix and Y is the corresponding output matrix.

$$X = \begin{bmatrix} \vec{x}_1 \\ \vec{x}_2 \\ \vdots \\ \vec{x}_{N-m} \end{bmatrix} = \begin{bmatrix} x_1 & x_2 & \cdots & x_m \\ x_2 & x_3 & \cdots & x_{m+1} \\ \vdots & \vdots & \ddots & \vdots \\ x_{N-m} & x_{N-m+1} & \cdots & x_{N+1} \end{bmatrix}, Y = \begin{bmatrix} \vec{y}_1 \\ \vec{y}_2 \\ \vdots \\ \vec{y}_N \end{bmatrix} = \begin{bmatrix} x_{m+1} \\ x_{m+2} \\ \vdots \\ x_N \end{bmatrix} \tag{11}$$

According to the load series, false nearest neighbours[10] (FNN) method is adopted in this paper to estimate the embedding dimension of the given series.

C. Error Criterion

There are many criterions to evaluate the performance of prediction models; here the following root mean square error (RMSE) criterion and the MAPE criterion shown in formula (10) are adopted in this paper to evaluate the performance of the hybrid model for prediction.

$$RMSE = \sqrt{\frac{1}{n}\sum_{t=1}^{n}\left(x(t) - \hat{x}(t)\right)^2} \tag{12}$$

where n is the number of training data samples, x_i is the actual value and , $\hat{x}_i$ is the predicted value.

4 Pso-Lssvr Model Application

Week max load series of some area are used in this paper to testify the proposed forecasting model performance. The week max load series are shown in Fig. 2.

A. Estimation of Embedding Dimension

Embedding dimension is very important in time series prediction with state reconstruction. Fig. 3 shows embedding dimension of load series As can be seen from the figure, embedding dimension is estimated while FNN percentage (FNNP) reduces to 0 and the dimension is 4.

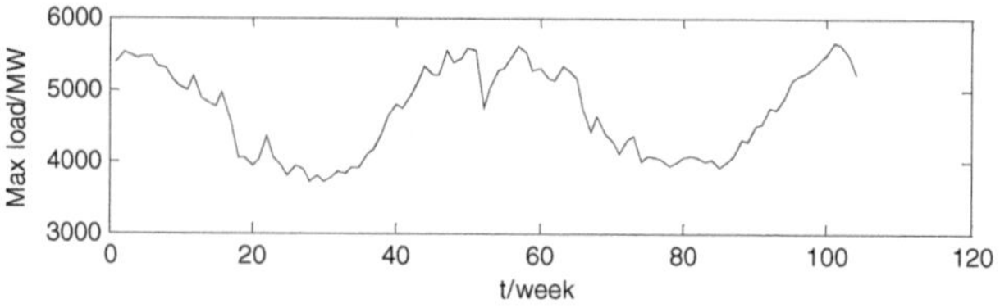

Fig. 2. Load series of some area

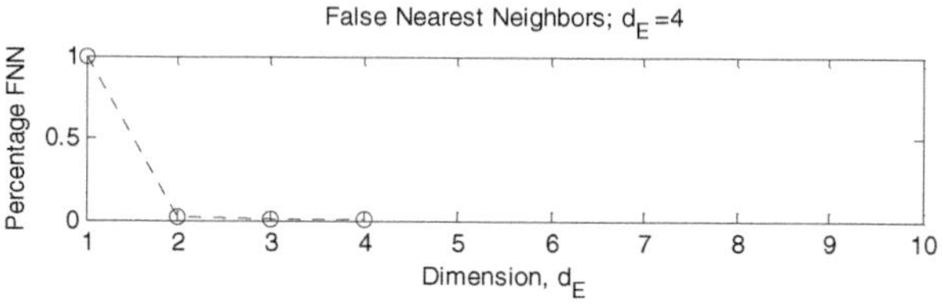

Fig. 3. FNNP value of load series

B. Result Analysis

In order to contrast the performance of PSO-LSSVR with other existing prediction methods, a BP neural network prediction model and a simple LSSVR prediction model are concerned, which outputs are shown in Fig. 4 and Fig. 5, in which, solid-line is for original data and dash-line is for regression and prediction data. In BP neural network prediction model, the sample data are normalized before used in the model. In LSSVR prediction model, a grid search and cross-validation method is adopted to find the optimal value of the model parameters.

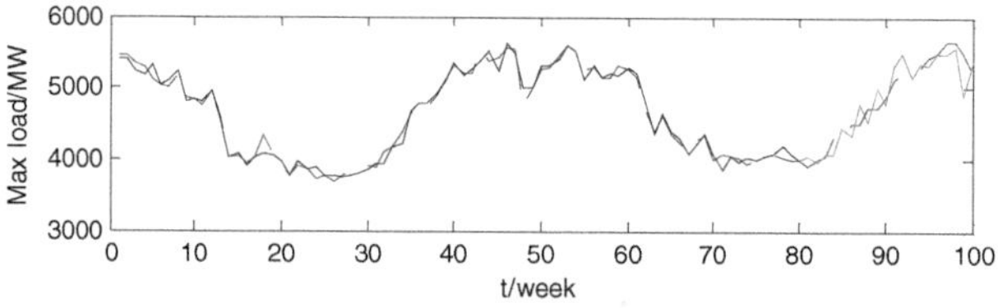

Fig. 4. Prediction result of BP neural network

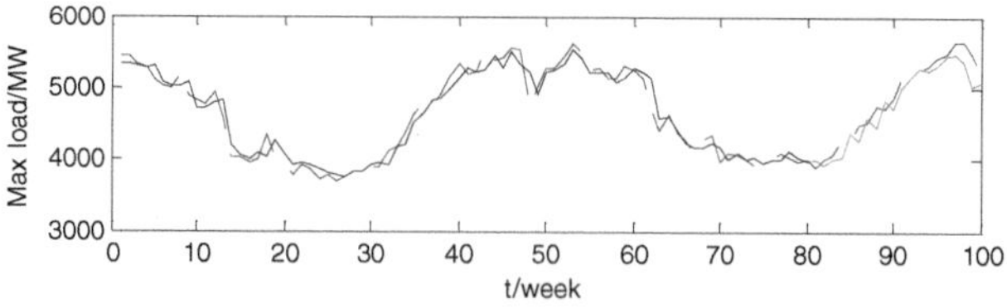

Fig. 5. Prediction result of LSSVR

In PSO-LSSVR model, the search domains of parameters and should be given at the beginning. After optimal searching in the above domains, the optimal value of parameters and are found, which are used to construct the LSSVR forecasting odel. The performance of PSO-LSSVR is shown in Fig. 6.

In all of the above three models, the regression and prediction error indices of above three models are shown in Table 1.

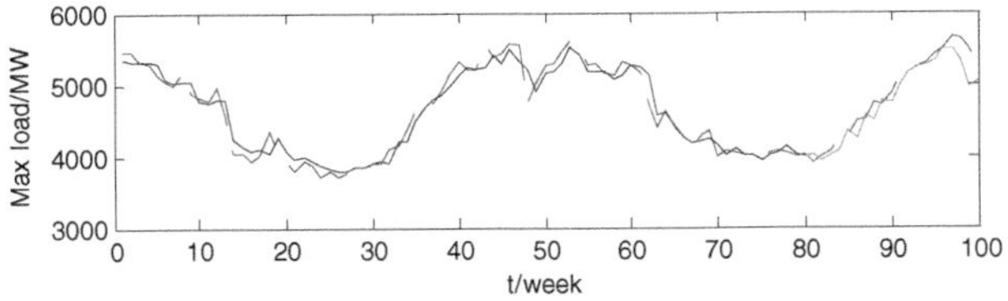

Fig. 6. Prediction result of PSO-LSSVR

Table 1. Regression And Prediction Error Of Load Series

Forecasting model	Regression error	Prediction error
	MAPE	MAPE
BP NNs	0.1623	0.1815
LSSVR	0.0219	0.0288
PSO-LSSVR	0.0196	0.0202

As can be seen from the above three figures, the BP neural network model has better performance in the regression curve but worse performance in the prediction curve, which indicates the over-fitting problems in BP neural network model, while the PSO-LSSVR model has approximate performance to the simple LSSVR model with parameters optimized by grid search and cross-validation. But the performance of both the simple LSSVR model and PSO-LSSVR model is better than that of BP neural network model. The over-fitting problems are resolved better in the last two models. From error indices in TABLE 1, we can see the PSO-LSSVR performance is slightly better than that of the simple LSSVR model.

Upon getting desirable regression and prediction performance, PSO-LSSVR model has better fitting performance to the original load series curve.

5 Conclusion

Short term load forecasting is one of important works in the power dispatch automation. Combining LSSVR with PSO, a model named as PSO-LSSVR is proposed in this paper. K-fold cross-validation fitness function is used to search the optimal parameters to construct the LSSVR forecasting model. With comparison to the BP neural network model and simple LSSVR with grid search and cross-validation, the PSO-LSSVR shows its great performance in regression and prediction. The PSO-LSSVR model is an adoptable technique for the short term load forecasting.

References

1. Senjyu, T., Mandal, P., Uezato, K., Funabashi, T.: Next day load curve forecasting using recurrent neural network structure. IEE Proceedings Generation, Transmission and Distribution 151, 388–394 (2004)
2. Zhang, Y., Zhou, Q., Sun, C., Lei, S., Liu, Y., Song, Y.: RBF Neural Network and ANFIS-Based Short-Term Load Forecasting Approach in Real-Time Price Environment. IEEE Transactions on Power Systems 23, 853–858 (2008)
3. Huang, R., Xi, L., Li, X., Richard Liu, C., Qiu, H., Lee, J.: Residual life predictions for ball bearings based on self-organizing map and back propagation neural network methods. Mechanical Systems and Signal Processing 21, 193 (2007)
4. Youshen, X., Henry, L., Hing, C.: A prediction fusion method for reconstructing spatial temporal dynamics using support vector machines. IEEE Transactions on Circuits and Systems II: Express Briefs 53, 62–66 (2006)
5. Vong, C.-M., Wong, P.-K., Li, Y.-P.: Prediction of automotive engine power and torque using least squares support vector machines and Bayesian inference. Engineering Applications of Artificial Intelligence 19, 277–287 (2006)
6. Vapnik, V.N.: An overview of statistical learning theory. IEEE Transactions on Neural Networks 10, 988–999 (1999)
7. Suykens, J.A.K.: Nonlinear modelling and support vector machines. In: Proceedings of the 18th IEEE Instrumentation and Measurement Technology Conference, IMTC 2001, Budapest, pp. 287–294 (2001)
8. Kim, K.-j.: Financial time series forecasting using support vector machines. Neurocomputing 55, 307–319 (2003)
9. Yang, H.-Z., Jiao, X.-N., Zhang, L.-Q., Li, F.-C.: Parameter Optimization for SVM using Sequential Number Theoretic for Optimization. In: 2006 International Conference on Machine Learning and Cybernetics, Dalian, China, pp. 3461–3464 (2006)
10. Kennel, M.B., Brown, R., Abarbanel, H.D.I.: Determining embedding dimension for phase-space reconstruction using a geometrical construction. Physical Review A (General Physics) 46, 3404–3411 (1992)

GIS-Based Emergency Management on Abrupt Air Pollution Accidents in Counties, China*

Hui Zhang and Mao Liu

College of Environmental Science and Engineering, Urban Public Safety Research Center
University of Nankai
Tianjin, Weijin Road 97, China
huizhang73@yahoo.cn

Abstract. Nowadays, with the expansion of industrial activities and the increase of chemical categories and products in counties of China, more and more abrupt air pollution accidents have occurred and seriously influenced the ecological security. It is important to make systematic studies on the emergency management of abrupt air pollution caused by industrial accidents in counties. Limited researches have been made on it in China and it is almost impossible to make early warning and emergency management in time on abrupt air pollution accidents in counties. This paper established an efficient emergency management system on abrupt air pollution caused by industrial accidents in counties of China based on the technology of Geographic Information System (GIS). The design and application of the system was illustrated in this paper. This system will help the companies and governments to inspect abrupt air pollution caused by industrial accidents, make effective emergency response measures, guarantee the public security and develop emergency management and prevention systems on abrupt air pollution accidents in counties of China.

Keywords: GIS, county, abrupt air pollution, industrial accidents, emergency management system.

1 Introduction

During recent years, many enterprises causing serous pollution problems have been moved into counties of China. The expansion of industrial activities and the increase of chemical categories and products in counties make the possibility for the occurrence of abrupt air pollution problems caused by industrial accidents increase, the incidence and losses caused by these accidents increase and the ecological security seriously destroyed[1-3]. Enterprises abroad pay much attention on abrupt air pollution problems caused by industrial accidents and systematic studies have been made on abrupt air pollution accidents abroad [4-7]. Enterprises in China only pay

---

* This work is partially supported by the key Technologies R&D Project of 11th 5-Year Plan (Nos. 2006Bak08B01, 200603746006) and National Natural Science Foundation of China (Nos. 70833003) to M. Liu.

J. Zhang (Ed.): ICAIC 2011, Part V, CCIS 228, pp. 133–141, 2011.

little attention on these problems and many studies have been made on urban abrupt air pollution accidents, especially on the development of emergency plan systems [8-12]. Besides, it is in lack of sequential inspection methods on air pollution problems in China and it is almost impossible to make early warning and emergency management in time on abrupt air pollution accidents in counties. The study on emergency management of abrupt air pollution accidents in counties has just begun, so there is no support of key technologies, emergency management system, emergency response plan or emergency response organization on abrupt air pollution accidents [13-15]. It is necessary for the enterprises and governments to make effective emergency response and inspection measures to reduce the negative influence caused by industrial accidents.

This paper developed an emergency management system on abrupt air pollution accidents in counties based on the Geographic Information System (GIS) and from the aspect of ecological security on counties. This system will support the establishment of emergency response and prevention system on abrupt air pollution accidents in counties of China, and it can help the enterprises and the governments to inspect abrupt air pollution problems caused by industrial accidents, make accidents simulation and emergency management, reduce the losses caused by these accidents and guarantee the public security.

2 Methods

A. Geographic Information System Technology

Emergency management system of air pollution problems caused by industrial accidents in counties needs great amounts of information. These data include spatial geographical data and social-economic data and they all have spatial and temporal characteristics. Geographic Information System (GIS) has great functions of collection, management, analysis and output of spatial information and it can blend figures and database. The framework of emergency management system on abrupt air pollution accidents can be established based on GIS. The spatial data management and analysis functions of GIS can provide good technical support platform for the implement of the system. The accidents information can be seamlessly integrated and well managed.

Based on computer technology and GIS, this paper developed an emergency management system on abrupt air pollution accidents, including basic database (including historical database, background database and so on), methods and models database, graphics database, emergency plans database, results display and query database. With the supports of GIS technology, database technology, multimedia technology and virtual reality technology, the research results can be visualized. The information on simulation results and emergency response measures can be shared and applied to the related departments and governments timely, exactly, authoritatively and dramatically.

B. Platform of the System

This system is developed based on the Window XP operation system, and uses the products of ESRI Company, such as ArcGIS Engine, ArcSDE and so on. The database system is built upon SQL Server2000, and the developing tool is Microsoft C#.

C. Design and Implementation of the System

Emergency management system on abrupt air pollution accidents in counties was developed using three-tier architecture model (Fig.1). Client program was second integrated developed using C# and ArcGIS Engine and it can access remote sensing images, vector data and attribute Data stored in SQL Server by ArcGIS Engine components. The middle layer uses ArcSDE to access vector data and grid data between client program and spatial data server. Using three-tier architecture pattern can make security limit on the access of data at the server side, central backup and restore data to prevent the loss of data caused by any contretemps and some business rules can be realized at database server to promise the access efficiency.

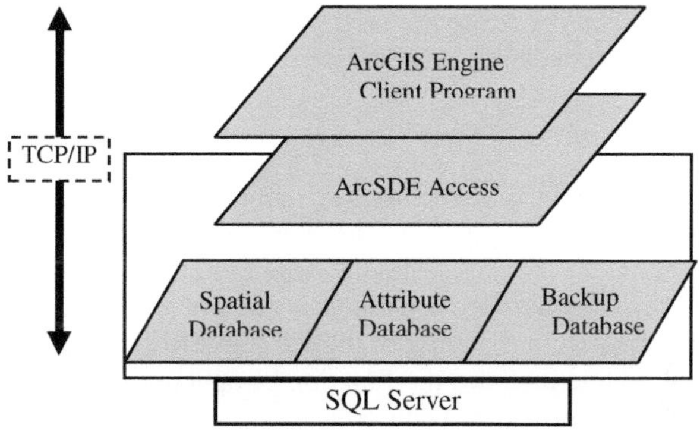

Fig. 1. Three-tier architecture of emergency management system on abrupt air pollution accidents in counties, China

D. Design of Database

When the system is operated, layered data can be overlay to form the background map, and based on monitoring results, scene of accidents is displayed on the map. Then emergency management and spatial analysis such as distance analysis, area analysis, route analysis and buffer analysis on abrupt air pollution accidents in counties are made.

Layers include: 1) residential areas, enterprises and other point objects; 2) medical treatment organizations, rescue centers and other point objects; 3) roads, rivers and other line objects; 4) buildings, factories areas and other polygon objects. Because of fast development of counties, the locations of some objects may be changed, so short-term update must be promised.

Besides, non-spatial data, such as the county's population density, property allocation, emergency rescue vehicles and instruments, and accidents simulation models of abrupt air pollution accidents should also be input in the system.

Based on the characteristics of the data needed in the system, this system uses database management system such as SQL Sever 2000 as the background database. This system has the function of storing and managing attribute and spatial data, sharing data among the departments, dealing with data, maintaining the system, outputting results, and so on. The database for this system is divided into four sub-databases: spatial database, non-spatial database, models database and temporal database. The spatial database can store figures, images and maps, such as remote sensing images, thematic maps, regionalized maps, risk assessment maps on abrupt air pollution accidents, and so forth. Non-spatial database stores historical accidents data, environmental background data and regional social-economic data (such as the property, population density, the amounts and categories of emergency response resources, and so on). Models database stores accidents assessment models, resource optimizing allocation models, accident simulation software models and so on. Temporal database is used to store data produced when operating the system and the users can not see them. Spatial data and non-spatial data can associate with each other. With the organization of programs, they can be integrated with the models and the functions of the system can then be carried out.

E. Function Modules of the System

The system is divided into seven modules, including files operation module, data collection module, layered display module, thematic maps query module, abrupt air pollution accidents model module, accidents simulation module and emergency plans query module . These modules are separated to each other on the screen, but they are connected to each other from the point of logical framework and data base.

Files operation module includes the function of opening files, database connection, printing, print review, and so on. Data collection module can support the system to quickly collect the basic data on the study area in time. Layered display module includes the function of adding layers, decreasing layers, zooming, legend display, overlay display, eagle eye, map output, and so on. Thematic maps query module supports the query on thematic maps, e.g. influence area caused by abrupt environmental pollution accidents, accidents simulation maps. Abrupt environmental pollution accidents models module can use existing models to calculate the results caused by abrupt environmental pollution accidents. With this module, it can easily get the pollution concentration in different area. Accidents simulation module mainly uses the abrupt environmental pollution accidents models to simulate pollution situation and display the results visualized on electronic maps. Accident status under any possible situation can be simulated. With this module, the clients can know about the accident visually. It can support the governments to make emergency response as early as possible. Emergency plans query module is mainly used to make emergency plans query on any environmental pollution accidents and emergency management on abrupt environmental pollution accidents in counties.

3 Mathematic Models of the System

The mathematic model used in the emergency management system on abrupt air pollution accident in counties can choose the Gaussian Diffusion Model. Comparing the simulation results with practical results, it can prove that the simulation results are reasonable. Its calculation equation on concentration is:

$$C(x,y,z,He)=\frac{Q}{2\pi u\sigma_y\sigma_s}\exp\left\{-\frac{y^2}{2\sigma_y^2}\right\}\left[\exp\left\{-\frac{(z-He)^2}{2\sigma_s^2}\right\}+\exp\left\{-\frac{(z+He)^2}{2\sigma_s^2}\right\}\right] \quad (1)$$

Where r is the distance to the point source; u is the wind speed at the point r meters away from the point source; C is the concentration of the contamination; Q is the emission velocity of the contamination; u is the average wind speed at the direction of X; y is the distance at the cross direction; He is the valid height of the point source; H is the height of the gas; ΔH is the rising height of the gas and it can be get according to the 《GB/T 3840-91》; σ_y is the diffusion parameter at the cross direction and can be get according to the 《GB/T 3840-91》; σ_s is the diffusion parameter at the vertical direction and can be get according to the 《GB/T 3840-91》.

When it is winding, calculating the pollution concentration at any point in the polluted area, define z=0, and then the pollution concentration at the point is calculated in the following equation:

$$C(x,y,0,He)=\frac{Q}{\pi u\sigma_y\sigma_s}\exp\left\{-\frac{y^2}{2\sigma_y^2}\right\}\exp\left\{-\frac{He^2}{2\sigma_s^2}\right\} \quad (2)$$

When it is breezing or not winding, assuming that the wind never changes, horizontal diffusion of the contamination has no obvious direction, the pollution concentration at the point is calculated in the following equation: (annual average wind speed at the height of 10m is: $u_{10} \leq 1.5m/s$)

$$C_r=\left(\frac{2}{\pi}\right)^{1/2}\frac{Q}{2\pi ur\sigma_s}\exp\left(-\frac{He^2}{2\sigma_s{}^2}\right) \quad (3)$$

4 Practical Usage

In order to show the practical usage of the emergency management system on abrupt air pollution accidents in counties, this paper chooses the abrupt leakage of ammonia storages in a factory of Xinzhuang Town as an example.The procedure is as follows:

A. Query on Dangerous Source

Get the basic information on the dangerous source in the emergency management system on abrupt air pollution accidents in counties (Fig. 2).

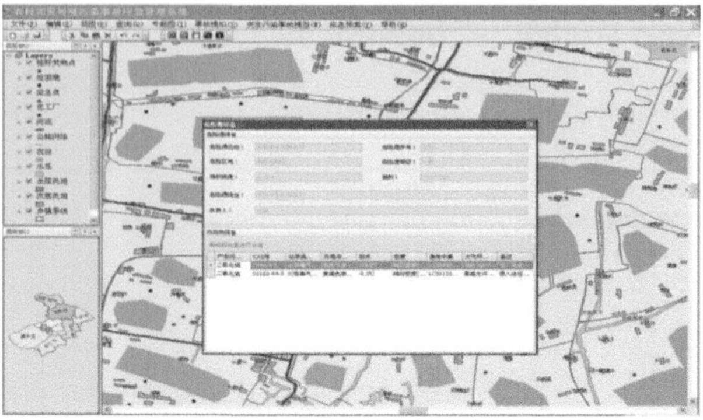

Fig. 2. Query on the dangerous source

B. Choose a Simulation Method

Choose simulation method of abrupt air pollution accidents, input data into the model, and press the button "begin" (Fig. 3).

Fig. 3. A simulation on abrupt air pollution accidents

C. Simulation Result

Show the simulation result on the map (Fig.4). From the simulation result, it is easy to find that the ammonia spreads from the chemical factory to the northwest of Xinzhuang Town. Ammonia concentration near the chemical plant is at its maximum and the maximum concentration is 148.8 ~ 170.1mg/m^3. According to the hazardous standard of Ammonia (table 1) [16], people near the chemical factory will be moderately affected within 30min.Their breath and pulse will speed up, eyes and noses will be stimulated and these people will feel obviously uncomfortable. When ammonia spreads to the edge of Xinzhuang Town, its concentration reaches 127.5 mg/m^3, people around this area will be lightly affected within 30min. Their breath will slow down, eyes and upper respiratory tract will be stimulated and they will be obviously uncomfortable. People stay at the area between the chemical factory and the first street of Xinzhuang Town that is about 800m away from the chemical factory will be obviously affected by ammonia.

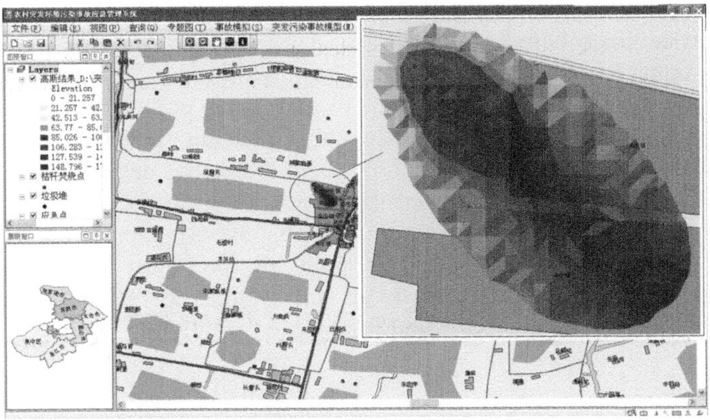

Fig. 4. Simulation result on abrupt air pollution accidents

Table 1. Dangerous Concentration of Ammonia

Concentration of ammonia in the air (mg/m³)	Contact Time (min)	Short Term Exposure Limit (STEL)
30	30	Almost no effect
70-140	28	People around the area will feel sick, their breath will slow down, and their eyes and upper respiratory tract will not uncomfortable (Light)
210-350	30	People around the area will feel obviously uncomfortable, their breath and pulse will speed up, and eyes and noses will be stimulated (Moderate)
700	30	People around the area will cough immediately, and will be strongly stimulated (Moderate)
1750-4500	30	People around the area will die immediately (Severe)

So after querying the emergency plan of abrupt air pollution in the system emergency management measures will be made: a) people around the pollution area should be evacuated and the area must be isolated immediately; b) People is not allowed in and all the fire source must be cut off; c) emergency personnel should take on the gas masks and protective clothing and cut off the leakage source if possible; d) keep ventilation and accelerate the diffusing ammonia; e) spray low concentration hydrochloric acid at the high concentration area to counteract, dilute and dissolve ammonia, f) built banks or dig trenches to store the wastewater; and g) the ammonia container should be used after reparation and severe proof.

5 Conclusions

In order to reduce the negative influence caused by the expansion of industrial activities and the increase of chemical categories and products in counties of China, this paper established an emergency management system of abrupt air pollution accidents on counties based on GIS. The system mainly has the function of data collection, thematic maps query, abrupt air pollution accidents model, accidents simulation and emergency plans query. An abrupt air pollution accident caused by leakage of ammonia storages in a factory of Xinzhuang Town was analysed in the system to show the practical use of the system. With the development of accidents model and simulation technology on abrupt air pollution accidents, It will provide great supports to the enterprises and governments for timely, quick and exact emergency management on abrupt air pollution problems caused by industrial accidents and reduce unnecessary losses in counties of China. This system is established on practical application and can be used widely.

References

1. Wu, Z., Sun, M.: Statistic analysis and countermeasure study on 200 road transportation accidents of dangerous chemicals. Journal of Safety Science and Technology 2(2), 3–8 (2006) (in Chinese)
2. Zhang, C.Y., Sun, L., Jin, L.J.: Developing Environment Risk Evaluation and Emergency Surveillance System to Prevent and Control Accidental Environment Pollution. Journal of Occupational Health and Damage 23(2), 65–67 (2008) (in Chinese)
3. Wei, G., Yang, Z.F., Li, Y.H.: Non-explosive dangerous chemical accidents in major Chinese cities. Environmental Pollution & Control 28(9), 711–714 (2006) (in Chinese)
4. Kneese, A.V.: Measuring the benefits of clean air and water. China Zhanwang Press, Beijing (1989) (in Chinese)
5. Eric, M., Sieglein, W.: Security Planning and Disaster Recovery. Posts and Telecom Press, Beijing (2003) (in Chinese)
6. John, A.D., Scylla, L.F., Carpenter, R.A.: Economic Analysis on Environmental Influence. China Environmental Science Press, Beijing (2001) (in Chinese)
7. Gradinscak, M., Beck, V., Brennan, P.: 3D Computer Modeling and Human Response. Fire and Material 23, 389–393 (1999)
8. Zhan, K.X., Chen, G.H.: Discussion on the Establishment of Inter-city Disaster Emergency Management System and its Key Issues. China Safety Science Journal 19(9), 172–176 (2009) (in Chinese)
9. Liu, D.H., Liu, M., Ren, C.X.: Estimating Water Pollution Risk Arising from Road Accidents. Journal of Safety and Environment 8(6), 140–143 (2008) (in Chinese)
10. Li, Z.M., Lv, D.: Research and Development of Assessment System for Calculating the Risk Area of Toxic Gas Leaking. Petrochemical Safety and Environmental Protection Technology 23(2), 27–28 (2007)
11. Zhang, B., Wang, Q., Li, S.: The system dynamics based research of water quality simulation in the Songhua River water pollution accident. In: Proceedings of the 2007 Conference on Systems Science, Management Science and System Dynamics: Sustainable Development and Complex Systems, pp. 2797–2804 (2007)

12. Liu, Y., Yin, P., Shi, C.: The development and research of Geographical Information System of Marine Oil Spill Emergency Response. In: Proceedings of the 2001 International Conference on Science and Engineering, pp. 425–429 (2001)
13. Chen, H.M.: GIS-based Urban Atmospheric Dispersion Model -A Case Study in Fuzhou. Geo-information Science 7(4), 101–106 (2005) (in Chinese)
14. Shu, L.Y., Wang, Y.J.: Atmospheric pollution dispersion prediction system based on Mapinfo and Surfer. Journal of Pingdingshan Institute of Technology 17(3), 26–31 (2008) (in Chinese)
15. Zhangand, M.G., Jiang, J.C.: Simulation of toxic gas dispersion based on GIS and real-time weather data. Journal of Nanjing University of Technology (Natural Science Edition) 31(3), 23–28 (2009) (in Chinese)
16. Wang, Y., Gu, Z.W., Zhang, S.: Modern Occupational Medicin, pp. 642–645. People's Medical Publishing House, Beijing (1996) (in Chinese)

A New Granular Computing Model Based on Qualitative Criterion Topological Space

Zhou Ru Qi and Xu Ning

Department of Computer Science Guangdong University of Education Guangzhou, China
Department of Computer Science and Information Engineering Shanghai
Institute of Technology hanghai, China
zhouruqi@sina.cn

Abstract. It was discussed that the way to reflect the internal relations between judgment and identification, the two most fundamental ways of thinking or cognition operations, during the course of granular information processing. Although it is conducive for the establishment of classical logic to emphasize the certainty of judgment while excluding uncertain identification from the logical category, from the objective point of view, such practice cut off the internal relations between judgment and identification, which is very unfavorable for the study of thinking or cognitive science and artificial intelligence. Matters reflect their state, movement and change process by means of their attributes and change process. Taking attributes and their change processes as the center, the method defines the criterion change operation in the criterion topological space, and further defines the basic information granules composed by attributes and criterions of matters, and provides the computing rules. The results show that this method can not only reflect the thinking process of the judgment with uncertain identification, but also express the transformation process amongst granules through the definition of qualitative mapping, so as to endow granular information processing with general sense in the process of thought expressing.

Keywords: Granular Computing, Qualitative Criterion, Criterion Topological Space, Qualitative Mapping.

1 Introduction

Although great achievements have been made by artificial intelligence in respect of logic thinking simulation, it is still far from the intended purpose of simulating human's low-level intelligence, such as the ability of imaginal thinking. The human cognition process is the transformation process of objects and attributes. People obtain a better cognition of objects through attributes understanding, or make further attributes judgment through cognition of objects [1].

Information granules play an important role in the cognition process of human. Granular structure and granular computing are two basic elements of granular computing. Granular structure involves the formal representation and interpretation of granules, mainly including the definition of granule; measurement of the approximation

J. Zhang (Ed.): ICAIC 2011, Part V, CCIS 228, pp. 142–149, 2011.

degree among granules and the law for the composition of a more complex granule by a number of granules. Granular computing refers to solving or reasoning the problem with granules as the objects.

However, the existing granular computing methods [2-4] even fail to give enough description to a simple thinking judgment process. Usually, regarding a simple judgment, much attention is paid to the results rather than the process. For example, this liquid is water, this liquid is steam, and these are two simple propositions. The evaluation process of these two propositions is actually that: this liquid is water because its temperature is between 0℃ and 100℃ and that this liquid is steam because its temperature is more than 100℃. In general logic symbolic computation, temperature is not involved in the simple judgment process, but we clearly understand that, in the reasoning process, temperature as a criterion for judgment is helpful for the correct evaluation of these two propositions.

The user interest model is mainly obtained by the data clustering technology. Data clustering, regarded as a branch of statistical learning and a guideless machine learning method, is widely used in various fields of e-commerce, image processing, pattern recognition, textual study and database. Clustering intrinsically is a knowledge discovery method of finding implicit latent useful information by dividing or covering object sets in accordance with some rule.

This paper aims to propose a new granular computing model based on qualitative criterion topological space. This model takes attributes and the attribute change process as the center and reflects the formal thinking process with judgment criteria in the criterion transformation operation process. By this way, cognitive thinking process of people can be further reflected so as to endow granular information processing with general sense.

2 Cognitive Thinking and Attributes Processing

Information granule plays a major role in the cognitive processing of human. During the cognition, reasoning and decision making process of human, a large amount of complex information is divided into a number of simple blocks, classes, groups or sets according to respective characteristics and performance. Such block, class, group or set is call as a granule. The information processing process is known as information granulation.

During the cognition process of human, new concepts are continuously generated while old concepts are continuously refined. Namely, the cognition is a process of making judgment constantly through the adequacy or necessity of the objects and attributes. When we get an object that is sufficient and necessary to some attributes, we will get to know a concept. Therefore, we can say that attributes are necessary or sufficient when the object and attributes are inconsistent with each other; attributes are necessary and sufficient when the object and attributes are consistent with each other. A concept can be formed when object and attributes are unified, which is the nature of people's cognition process [1].

In fact, the so-called inconsistence between the object and its attributes is just the judgment of attributes. This judgment is in close relation to the membership degree of the nature of attributes to some judgment standard. In the entire cognitive judgment process, attributes are the most basic information granules for cognitive thinking. Since the state, movement and change rules of a matter can be considered as some attributes of a matter while a variety of sensory organs as brain portals can only respond to various perceptive attributes to which they are sensitive, it is indicated that,

Proposition 1 [5]: Information received by people's brain is, and can only be a variety of perceptive attributes.

Namely, even if the information is defined as the state, movement and change rules of a matter, it can only be delivered to people's sensory organs in the form of perceptive attributes.

According to the philosophic definition that "natures of a matter demonstrated when having relationship with another matter is termed attributes", the following proposition can be equivalently deduced.

Proposition 2 [5]: Matters represent or demonstrate their natures through their attributes.

Philosophically, all matters have quality and quantity characteristics. Quality is "the essential characteristics of a matter to be different from other matters" while "quantity refers to the grade, size and scope of the quality that can be expressed by number and shape". Since attributes are quality demonstrated by a matter, attributes not only demonstrate the quality (characteristics or nature) they want to demonstrate but also demonstrate the quantity to be defined and standardized (characteristics). In other words, matters have two characteristic values, quality (nature) and quantity, wherein the quality characteristic can standardize the change scope of quantity characteristic by means of qualitative criterion. In addition, quality characteristic and quantity characteristic can not only change with external conditions, but also can convert to each other.

As we know, both a matter's nature and relation are considered as its attributes in formal logic. If the relation is considered as a nature reflecting the interrelation among multiple objects, the relation can also be considered as an attribute among multiple objects, or an attribute of a system formed by the relation by multiple objects through such interrelation.

That is, the integration of attributes of a plurality of members can not only generate the member relation determined by attributes of all members, but also induce a system based on the relationship and the overall attributes of the system.

Thus, if different systems can be distinguished by their different states, the state can then be considered as a kind of "quality" which is an attribute can be demonstrated (including quantity characteristics, quality characteristics and relation). In addition, the state, movement and change process and law of matters can also be attributed to the change process and law of the system attributes. Thus, the following basic assumptions can be proposed:

Assumption 1[5]: A matter demonstrates its state, movement and change process by its attributes (including quantity characteristics, quality characteristics and relation) and the change process.

Inference 1[5]: the processing of perceptive attributes and the law of their change is one of key problems for the information processing of human brain.

If we see characteristic extraction as the first step for human brain to decipher perceptive attributes of a matter, perception image of the characteristic will be formed or will emerge in human brain. By integrating different perception characteristic images, a perceptive or conscious image of an object can be built in human brain. Once the memory pattern of a matter is formed, it can be used for model identification.

In short, we believe that only if the change law of a matter is understand can a method be proposed which describes that the matter reflects its change law by its attributes, and can stimulate human brain to interpret the change of the matter's attribution, understand matters and the word and handle various matters intelligently. Here, the attributes of matters are the most basic information granule.

3 Qualitative Criterion Topological Space

From a psychological perspective, human feelings reflect the (simple) attributes of matters while the perception is a process integrating various simple attributes into the complex attribute of a matter. Assume $a(x)$ as a perceptive attribute of a matter x, $a(x)$ generally has both quantitative attribute value, namely, value (data) $d_a(x)$ and qualitative attribute value $P_a(x)$. When x has the qualitative attribute value $P_a(x)$, it also can be said that x has the "nature" reflected by $P_a(x)$. Therefore, the so-called "nature $P_a(x)$" is in fact a qualitative attribute value of the matter's attribute. In particular, when $P_a(x)$ is a nature which can distinguish x from other matter y, $P_a(x)$ is known as a property or characteristic of x.

Definition 1: Assume $d(x)$ as a quantitative value of nature $p(x)$ and $(\alpha_i,\beta_i]$ as a qualitative reference field of $p(x)$, if $d(x)\in(\alpha_i,\beta_i]$, it can be said that $(\alpha_i,\beta_i]$ is a neighboring field of $d(x)$.

If assume $(\alpha_i,\beta_i]$ and $(\alpha_j,\beta_j]$ as qualitative criterions of $p_i(x)$ and $p_j(x)$ respectively, since their intersection $(\alpha_i,\beta_i]\cap(\alpha_j,\beta_j]$ and union $(\alpha_i,\beta_i]\cup(\alpha_j,\beta_j]$ can be considered as qualitative criterion of conjunctive nature $q(x)=p_i(x)\wedge p_j(x)$ and disjunctive nature $r(x)= p_i(x)\vee P_j(x)$ of $p_i(x)$ and $p_j(x)$ respectively, it can be deduced that

Theorem 1: Numeric field $d_a(x)$ and criterion field cluster $\Gamma=\{(\alpha_i,\beta_i]\subseteq X|(\alpha_i,\beta_i]$ is qualitative criterion of $p_i(x)$ $\}$ and constitutes the criterion topological space $T(d_a(x),\Gamma)$.

Proof: since "x has attributes" and "x does not have attributes" are also two (ordinary) natures of x (recorded as $a(x)$ and $\neg a(x)$), $d_a(x)$ and $\varnothing$ are the qualitative criterions of $a(x)$ and $\neg a(x)$, so $\varnothing$, $d_a(x)\in\Gamma$ is tenable. In addition, Γ is a covering of $d_a(x)$, namely, $d_a(x)=\cup_{i\in N}[d_{ik}(x)]=\cup_{i\in N}(\alpha_i,\beta_i]$, so $(\alpha_i,\beta_i]\cap(\alpha_j,\beta_j]$, $(\alpha_i,\beta_i]\cup(\alpha_j,\beta_j]$ and $\cup(\alpha_j,\beta_j]$ all belong to Γ. Therefore, $(d_a(x),\Gamma)$ constitutes a (qualitative criterion) topological space $T(d_a(x),\Gamma)$.

Definition 2: Assume $\Gamma=\{N_i \mid i\in I\}$ as the cluster of all qualitative criterions N_i of proposition p, and say T: $\Gamma\rightarrow\Gamma$ is the criterion mapping of proposition p. If there is $N_j\in\Gamma$ for any $N_i\in\Gamma$, which makes

$$T(N_i)= N_j$$

tenable. In particular, if the following equation is tenable

$$T_{ij}(N_i)= N_j$$

for two criterion fields N_i and N_j, we say that T_{ij} is the qualitative criterion transformation from N_i to N_j.

Definition 3: Assume that criterion field $N_i(o_i,r_i)$ is the kernel of a fuzzy set A, namely, $N_i(o_i,r_i)=A_1 = \{x \mid \mu_{A(x)}=1\}$. Where, $\mu_{A(x)}$ refers to the membership grade of $x\in A$, and λ cut set of A is $A_\lambda=\{x \mid \mu_{A(x)}\geq\lambda\}$, $\lambda\in(0,1]$, which is composed of element x with membership grade $\mu_{A(x)}$ more than or equal to λ. It is known that T_λ is the λ criterion transformation of fuzzy set A with $N_i(o_i,r_i)$ as the kernel. If it is tenable that

$$T_\lambda(N_i)=A_\lambda$$

Theorem 2: If criterion field $N_i(o_i,r_i)$ is the kernel of fuzzy set A and its λ cut set A_λ has the form of $N_\lambda(o_i,r_\lambda)$, then its λ criterion transformation can be indicated by a topological equation T_λ satisfying the following equation:

$$T_\lambda(N_i)=N(T_\lambda(o_i), T_\lambda(r_i))=N_\lambda(o_i,r_\lambda)=A_\lambda$$

Namely, $T_\lambda(o_i)=o_i$ and $T_\lambda(N_i)=k_{i\lambda}\times r_i=r_\lambda$

And the fuzzy set A can be generated by all transformations $T_\lambda(N_i)$, i.e.,

$$A=\cup_{\lambda\in[0,1]}\lambda\bullet T_\lambda(N_i)= \cup_{\lambda\in[0,1]}\lambda A_\lambda$$

Where $\lambda A\lambda$ is the multiple fuzzy set with the membership degree as below,

$$\mu_{(\lambda\cdot A\lambda)}(x)=\lambda\wedge_{xA\lambda}(X)$$

Proof: since the kernel of fuzzy set A is $A_1=N_i$ and λ cut set $A_\lambda=N_\lambda$ is also a topological neighboring field with O_i as the center of sphere (n dimensions), assume the radius as r_λ (note: different values for different directions), and then, the transformation from N_i to N_λ is the topological transformation between two topological neighboring fields. As for parameter λ, this is a topological transformation keeping fuzzy set kernel $A_i=N_i$ unchanged and just change the neighboring field radius r_i. Topological transformation is flexible; therefore, it can be easily achieved through transformation T_λ in line with the equation below.

$$T_\lambda(N_i)=N(T_\lambda(o_i), T_\lambda(r_i))=N_\lambda(o_i,r_\lambda)=A_\lambda$$

Namely, T_λ will not change the center of sphere but amplify radius ri flexibly. It meets the two equations as below.

$$T_\lambda(o_i)=o_i \quad\text{and}\quad T_\lambda(N_i)=k_{i\lambda}\times r_i=r_\lambda$$

Therefore, when λ takes all values in [0,1], all multiple fuzzy sets λA_λ of fuzzy set A with $N_i(o_i,r_i)$ as the kernel can be obtained. That means the following equation is tenable:

$$A = \cup_{\lambda \in [0,1]} \lambda \bullet T_\lambda(N_i) = \cup_{\lambda \in [0,1]} \lambda\ A_\lambda$$

Definition 4: Assume A and B as two fuzzy sets with criterion fields $N_i(o_i,r_i)$ and $N_\alpha(o_\alpha,r_\alpha)$ as kernels respectively, say T_α is the criterion transformation form $N_i(o_i,r_i)$ to $N_\alpha(o_\alpha,r_\alpha)$. If $T\alpha$ meets the equation below:

$$T_\alpha(N_i) = N(T_\alpha (o_i), T_\alpha(r_i)) = N_\alpha(o_\alpha,r_\alpha)$$

Definition 5: Say criterion transformation $T_{\alpha\lambda}$ is the complex transformation of T_α and λ–criterion transformation T_λ, and if it meets the equation below:

$$T_{\alpha\cdot\lambda}(N_i) = T_\lambda \cdot T_\alpha(N_i) = T_\lambda(N_\alpha) = N_\lambda(o_\alpha,r_\lambda)$$

Obviously, it can be deduced through definitions 3, 4 and 5 and theorem 2 above that the theorem below is tenable:

Theorem 3: Starting from the kernel N_i of fuzzy set A, λ cut set $B_\lambda = N_\lambda(o_\alpha,r_\lambda) = \{x \mid \mu_{A(x)} \geq \lambda\}$ of fuzzy set B with N_α as the kernel can be obtained through complex criterion transformation $T_{\alpha\cdot\lambda} = T_\lambda \cdot T_\alpha$.

All the abovementioned demonstrates that ambiguity occurred during the course of qualitative judgment can be explained by criterion transformation (non-ambiguous criterion transformation) and others.

4 Representation and Computing Model of Granule

Definition 6: An information granule in attribute topological space can be represented as bivariate function below.

$$(P, \Gamma) = (p_i(x), (\alpha_i,\beta_i]), \text{ or } (d_{a(x)}, (\alpha_i,\beta_i])$$

Where, $a(x)$ is a perceptive attribute of matter x, $p_i(x)$ is the nature of $a(x)$ and $(\alpha_i,\beta_i]$ is a qualitative criterion field of $p_i(x)$.

Definition 7: Assume $(p_i(x), (\alpha_i,\beta_i])$ and $(p_j(x), (\alpha_j,\beta_j])$ as two basic granule. Then, the operation of their connectives (~ negation, $\wedge$ conjunction, $\vee$ disjunction, $\rightarrow$ inclusion, $\leftrightarrow$ equivalence) is defined as below.

(1) $\sim (p_i(x), (\alpha_i,\beta_i]) = (\sim pi(x), \Gamma - (\alpha_i,\beta_i])$

(2) $(p_i(x), (\alpha_i,\beta_i]) \wedge (p_j(x), (\alpha_j,\beta_j]) = (p_i(x) \Delta p_j(x), (\alpha_i,\beta_i] \cap (\alpha_j,\beta_j])$

where, Δ is the conjunction of attributes, also known as attributes integration.

(3) $(p_i(x), (\alpha_i,\beta_i]) \vee (p_j(x), (\alpha_j,\beta_j]) = (p_i(x) \nabla p_j(x), (\alpha_i,\beta_i] \cup (\alpha_j,\beta_j])$

where, ∇ is the disjunction of attributes.

(4) $(p_i(x), (\alpha_i,\beta_i]) \rightarrow (p_j(x), (\alpha_j,\beta_j]) = (p_i(x) \Rightarrow p_j(x), (\alpha_j,\beta_j] = T((\alpha_i,\beta_i])$

where, $\Rightarrow$ is the attribute transformation or attribute reasoning; T is the criterion transformation operation in definition 2.

(5) $(p_i(x), (\alpha_i,\beta_i]) \leftrightarrow (p_j(x), (\alpha_j,\beta_j])=(p_i(x) \leftrightarrow p_j(x), (\alpha_i,\beta_i] =(\alpha_j,\beta_j])$

In the above definition (4), $(p_i(x), (\alpha_i,\beta_i]) \rightarrow (p_j(x), (\alpha_j,\beta_j])$ on the left is known as granular information reasoning. As a result, a new granule or the original granule can be obtained. When the original granule is obtained, we say that the transformation is self-reciprocal; when a new granule is obtained, the following qualitative mapping can be used to demonstrate its essential meaning. The qualitative mapping model of granule transformation operation is defined as below:

Definition 8[6]: Assume $a(o)=\overset{n}{\underset{i=1}{\wedge}} a_i(o)$ as the integrated attribute of the n basic attributes $a_i(o)$, $i=1,\ldots,n$, of object o and assume $x=(x_1,\ldots,x_n)$ as the quantity value of attribute $a(o)$, where x_i is the quantitive characteristic value of $a_i(o)$, $p_i(o), \in P_o$ is a nature of $a_i(o)$, $\Gamma=\{(\alpha_i,\beta_i] \mid (\alpha_i,\beta_i]$ is the qualitative criterion of nature $p_i(o)$ and parallelepiped $[\alpha,\beta]=[\alpha_1,\beta_1]\times\ldots\times[\alpha_n,\beta_n]$ is the qualitative criterion of the integrated nature $p(o)=\overset{n}{\underset{i=1}{\wedge}} p_i(o)$, we say that mapping τ: $X\times\Gamma\rightarrow\{0,1\}\times P_o$ is the qualitative (judgment) mapping (QM) of $x=(x_1,\ldots,x_n)$ with n dimentional parallelepiped $[\alpha,\beta]$. If there is $[\alpha,\beta]\in\Gamma$ and a nature $p(o)=\overset{n}{\underset{i=1}{\wedge}} p_i(o) \in P_o$ with $[\alpha,\beta]$as qualitative criterion for any $x\in X$, which makes

$$\tau(x,[\alpha,\beta]) = x \underset{?}{\in} [\alpha,\beta] = \overset{n}{\underset{i=1}{\wedge}} (x_i \underset{?}{\in} [\alpha_i,\beta_i]) = \overset{n}{\underset{i=1}{\wedge}} \tau_{p_i(o)}(x_i)$$

where, $\tau_{p_i(o)}(x_i) = \begin{cases} 1 & x_i \in [\alpha_i,\beta_i] \\ 0 & x_i \notin [\alpha_i,\beta_i] \end{cases}$ is the true value of nature proposition $p_i(o)$.

Therefore, τ is the computing form of granular transformation in criterion topological space.

5 Conclusion

It is discussed above how the judgment criterion is realized in thinking judgment process of granular computing. Granular computing rules are defined in the topological space composed of the criterions and its transformation; moreover, the granular transformation among granules is reflected by qualitative mapping. The results show that this new method will no longer split the certain judgment and the uncertain identification. This model is a preliminary model, its computing rules are to be further improved. The specific application of the model will be one of the key topics for further study.

References

1. Zhang, W.x., Xu, W.-h.: Cognitive MODEL based on granular computing. Chinese Journal of Engineering Mathematics 24(6) (2007)
2. Li, H.: Theory of granular set:A new model of granular computing. Journal of Chongqing University of Posts and Telecommunications (Natural science) 19(4), 397–404 (2007)

3. Yan, L., Zhang, X.-D., Wei, Y.-T., He, J.-C.: Granularly semantic reasoning based on granular computing in granular space. PR & AI 21(4), 462–468 (2008)
4. Qu, T.-R., Lu, Q., Huang, H.-K.: Granular computing based hierarchical concept capture Algorithm in multi-valued information system. PR & AI 22(1), 22–27 (2009)
5. Feng, J.L.: Thought, Intelligence and Attribute theory method. Chinese Journal of Guangxi Normal University 15(3), 1–6 (1997)
6. Feng, J.L.: A Mathematical Theory and Method for Thinking Construction and Intelligence Modeling—An Introduction of Attribute Theory and Method. Chinese Journal of Guangxi Normal University 17(2), 1–7 (1999)

Performance Study of Tactical Ad-Hoc Network Based on Self-similarity Traffic Model

Lu Ying[1], Zhong Lian-jiong[1], Kang Feng-ju[2], and liang Xiang-yang[2]

[1] Computer Science and Engineering College i'an Technological University
Xi'an, Shaanxi Province, China
[2] College of Marine Engineering Northwestern Polytechnical University
Xi'an, Shaanxi Province, China
LuYing215@yeah.net

Abstract. This paper analyzed self-similarity of Ad-hoc network traffic, and how the network performance was influenced by the self-similarity. Based on Fractional Brownian Motion model, paper proposed a self-similar traffic sequence generation method of RMD algorithm. RMD algorithm is one of the self-similar statistic models. Furthermore, simulation and modeling for Ad-hoc tactic network was implemented based on this kind of traffic model. According to the each performance indicator of curve contrast chart generated by simulating experiment, conclusion is made as follow: self-similarity has great influence to Ad-hoc network indicator and has great contrast to traditional flow models' short-range dependency character. The research result of this thesis has significant meaning for precise modeling of Ad-hoc network flow, resource optimization and improvement of network performance.

Keywords: Ad-hoc, self-similarity, RMD, traffic.

1 Introduction

All the message arriving models of the wireless network emulation under the traditional model are on the assumption that in the queuing system, customers' arrival submit to Markovian characteristic, that is a kind of short-range dependence process, such as Poisson model. With the deep research on network, data and experimental results indicate that within a few milliseconds to several hours even many days, the traffic of wireless network shows a strong bursting characteristic. This characteristic shows that the arrival of messages is of long-range dependency. Carlos Oliveria et al have researched on the characteristic of wireless network traffic and pointed out that the burstiness of traffic in WLAN is associated with self-similar and long-range dependency; QilianLiang point out that Ad-hoc network traffic also exists such characteristic and predicted the Ad-hoc traffic through self-similarity model[1];Jumpei Taketsugu et al researched on the traffic of wireless cellular networks, and pointed out that self-similarity also exists in DCA-TDMA system. Therefore, simulation model of Poisson's arrival has not yet accord with the actual conditions of the tactical Ad-hoc network. This paper takes into consideration the arrival of message submitted to long-range dependence process, makes use of self-similar model which can describe

J. Zhang (Ed.): ICAIC 2011, Part V, CCIS 228, pp. 150–156, 2011.

long-range dependence characteristic better, simulates tactical Ad-hoc network in virtue of the computer simulation technology ,and analysis and evaluates simulation results to study the influence of the self-similar model on tactical Ad-hoc network performance.

2 Model of Self-similarity Traffic

The model of self-similarity traffic is given followed:

Let $X = \{Xt;t = 1,2,...\}$ be a time series which, for example, represents the trace of data flow at a bottleneck link measured at some fixed time granularity. We define the aggregated series $x_i^{(m)}= (X_{im-m+1}+...+X_{im})/m$. That is, X_t is partitioned into blocks of size m, their values are averaged, and i is used to index these blocks. Let $r(k)$ and $r^{(m)}(k)$ denote the autocorrelation functions of X_t, and $X_i^{(m)}$, respectively. X_t is self-similar more precisely, asymptotically second-order self-similar—if the following conditions hold[2,3]:

$$r(k) \sim ck\text{-}\beta, \tag{1}$$

$$r(m)\ (k) \sim r(k), \tag{2}$$

for k and m large where $0 <\beta< 1$. That is, X_t is "self-similar" in the sense that the correlation structure is preserved with respect to time aggregation—relation (2) and $r(k)$ behaves hyperbolically with $\sum_{k=0}^{\infty} r(k) = \infty$ as implied by (1). The latter property is referred to as long-range dependence.

Let Hurst parameter H=1-β/2,and by the range of β, 1/2<H<1. It follows from (1) that the farther H is away from 1/2 the more long-range dependent X_t is, and vice versa. Thus the H value acts as an indicator of the degree of self-similarity. The FBN is such a process of self-similarity.

3 Modeling and Stimulation

A. Conformation of Self-similar Traffic Model

At present , the most commonly used self-similar model are: FBN,F-ARIMA, ON/OFF and Wavelet Transform model [4].RMD method, offset selection is the key factor of this method, can generate Fractional Brownian Motion (FBM) sequence. The main parameters of the original sequence have self-similar parameter H and self-similar function r(k).

The self-similarity of traffic is mainly described by H, so H, is chosen to construct offset function, D(n). The value of a certain point of generated sequence equals to the arithmetic average of previous point and latter point plus D(n), which is shown as X[n]=(X[n-1]+X[n+1])/2+D(n). Since offset will generate gauss distributing sequence of zero mean value, a random number set, G(n), with normal distribution has to be obtained with mean value =0 and variance=1. Offset function is $D(n) = G(n)\sqrt{\frac{1-2^{2H-2}}{2^{2Hm}}}$,

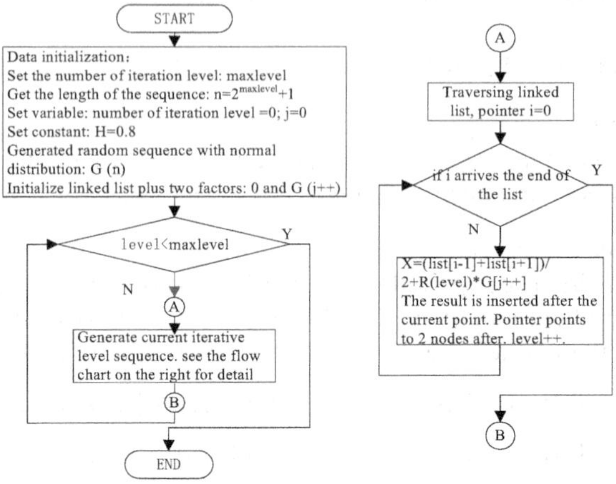

Fig. 1. Flow chart of RMD arithmetic

m is number of iteration and $\sqrt{\frac{1-2^{2H-2}}{2^{2Hm}}}$ is noted as R(m). Self-similar sequence can be generated through iterative method. Let H be the self-similar parameter of the original sequence and $\hat{H}$ be the self-similar parameter of generated sequence, $\Delta H=(\hat{H}-H)/H$ is the extent of self-similar parameter offset between generated sequence and original sequence. The smaller the ΔH value, the closer to original sequence the character of the simulated sequence is.

Assume a FBM sequence is generated within the time interval [1, 0]. Inward calculation is the basic idea of RMD algorithm, which divides interiorly within interval [0, 1]. The value of certain point of generated sequence equals to the arithmetic average of previous point and latter points plus the value of offset function of this point. The example of process of recurrent algorithm is as follow:

Initial Value: X[0] =0, X[1] = G(1)

Iteration 1: $X[1/2] = (X[0] + X[1])/2+\sqrt{\frac{1-2^{2H-2}}{2^{2H}}}\,G(2)$

Iteration 2: $X[1/4] =(X[0] + X[1/2])/2+\sqrt{\frac{1-2^{2H-2}}{2^{2H\times 2}}}\,G(3)$

$X[3/4] = (X[1/2] + X[1])/2+\sqrt{\frac{1-2^{2H-2}}{2^{2H\times 2}}}\,G(4)$

…

After n steps of recurrent, the algorithm can generate $\boldsymbol{2}^{n-1}$ data. For ensuring the accuracy of the random sequence, normal random array, G(n), has to be generated by different seed in order to insure its independency. The algorithm process is as Fig 1, linked list, which is used to store generated self-similar sequence.

The theoretical time complexity of RMD generation algorithm is O(n) and the speed of generating self-similar random sequence is faster, which can satisfy the requirement of generating self-similar sequence in high speed in practical simulation.

B. Simulation Environment

Tactical mobile Ad-hoc network is simulated by using the simulation software OPNET Modeler. The modeling of OPNET includes 3 levels of the hierarchy: network model, node model and process model[5]. Network model is the top model, which describes the network objects needed to simulate. Network model may contain any number of communicating entities called nodes and is used to specify the physical topology of a communication network, which defines the position and interconnection of communicating entities, i.e., node and link. Node models are expressed as interconnected modules. These modules can be grouped into two distinct categories. The first set is modules that have predefined characteristics and a set of built-in parameters. Examples are packet generators, point-to-point transmitters and radio receivers. The second group contains highly programmable modules. These modules referred to as processors and queues, relying on process model specifications. Process models are used to describe the logic flow and behavior of processor and queue modules. Communication between process is supported by interrupts. Process models are expressed in a language called Proto-C, which consists of state transition diagrams, a library of kernel procedures, and the standard C programming language.

C. Modeling of OPNET

1) Network model:

We set 18 wireless nodes in an Ad-hoc network. The nodes density function is of random distribution and they communicate with each other through wireless link. Fig. 2 shows the distribution of workstations' position. In this model, workstations have identical media access opportunity and move with certain speed.

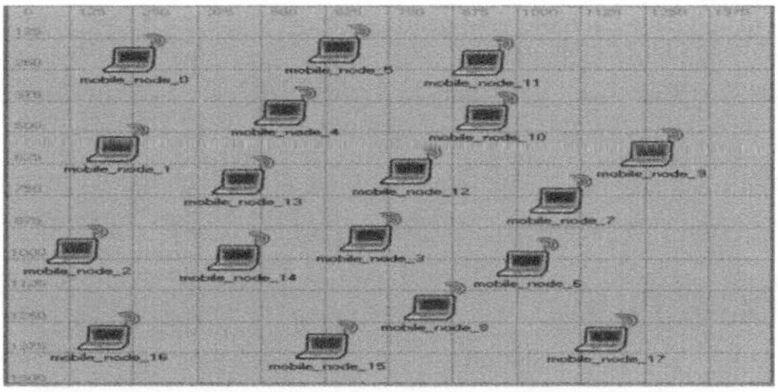

Fig. 2. Network topology model

2) Node model

All the nodes in network model are based on the same node model. Fig. 3 shows the node model which is mainly composed of some process modules followed:

(1) data source module :generate data traffic , generate packet according to its length and the arrival time intervals. Once generated, the packet will be sent to lower layer. Here the arrival time interval of the packet is in accordance with the self-similar distribution which generated by RMD algorithm.

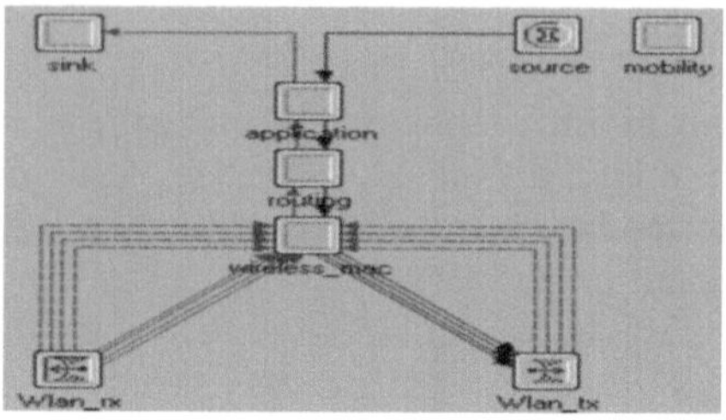

Fig. 3. Node model

(2) sink module: be responsible for statistic work concerning network average time delay and channel throughput, destroying received packets.
(3) application module: set a random destination address for upper-layer message, generate a service request to routing layer according to inner-communication interface ICI format, the message is sent to routing module together with ICI.
(4) routing modules :receive PDU from application module and execute routing arithmetic.
(5) wireless_mac module :simulate the MAC layer protocols,such as CSMA/CA,ALOHA,TDMA,IEEE802.11.
(6) transmitter module :send data frames to channel.
(7) receiver module :examine the state of channel, receive data frames from it and then send the messages to MAC module.
(8) mobility module: execute mobility schemes pre-configured.

3) Process model
OPNET Process editor uses a powerful state-transition diagram approach to support specification of any type of protocol, resource, application, algorithm, or queuing policy. States and transitions graphically define the progression of a process in response to events. Within each state, general logic can be specified by using a library of predefined functions and even the full flexibility of the C language.

4 Analysis of Simulation

A. Simulation Parameters

The wireless network operates in the Ad-hoc mode, without any central infrastructure. 18 nodes are randomly placed in a grid topology within 1200m by 1200m area and move for 0.5 hour. The distance between a node and its closest node is 200m. Each node is provided with identical traffic and can be transmitted to any of its neighbors by full-duplex work manner. Nodes in this network move towards to the destination at a constant speed uniformly chosen between min speed 0s/m and max speed 10s/m. After it reaches its destination, the node stays there for 10s.Data packet length is 1KB with exponential distribution. The protocols we adopted for MAC layer and routing layer are IEEE802.11 and DSR.

B. Analysis and Interpretation of the Simulation Results

In order to obtain ideal simulation curve, under heavy loads conditions, we modeled the different bursting characteristic network traffic via the change of self-similar parameter, the value of H=0.5/0.7/0.8/0.9 (As H=0.5, it means the traffic is Poisson distribution model). Based on the simulation model proposed above, the messages arrive as a self-similar model and then is processed accordingly, such as joining in the queue, queuing and outputting. The simulation is finished until all the packet are sent out, the result data obtained from simulation will be described and statistic.

As for simulation results, we mainly focus on the statistic characteristic of the average delay time, average throughput, packet loss ratio of the indicative network performance. The simulation data is shown in Table 1.

Table 1. Network parameters under different H VALUE

H	Average Throughput (kbit/s)	Average Delay Time (s)	Average Packet Loss Ratio (kbit/s)
H=0.9	684.246	0.675	9.352
H=0.8	679.201	0.301	3.105
H=0.7	632.912	0.048	0.307
H=0.5	619.205	0.028	0.198

According to the data above, the conclusion is:Under the same average length of message, as H value increase, average time delay, average packet loss ratio and average throughput will increase. The higher the H value, the lower the network performance.

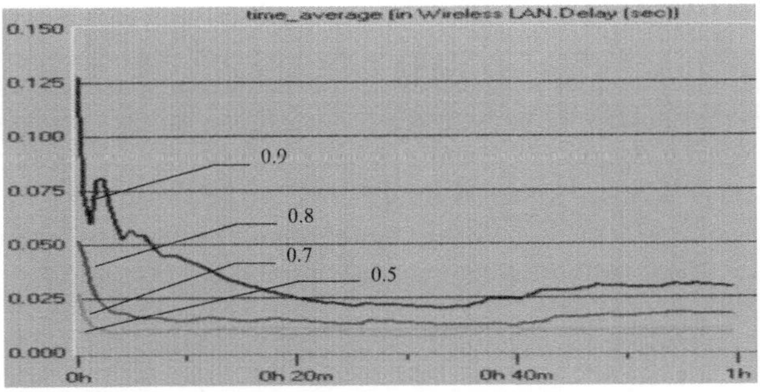

Fig. 4. Average delay time

As is shown in Fig. 4 and 5, x-coordinate is simulation time, y-coordinate in Fig.4 is average delay time, y-coordinate in Fig.5 is packet loss ratio. From the simulation and analytical results shown in Fig.4 and Fig. 5, it can be observed that under heavy network loads conditions, for different H, the higher the value of H is, the tenser the network traffic burst degree is and the more the performance of packet loss ratio and average delay time degrade. It is because of the increase of the collision probability of wireless channel, the movement station cannot compete for limited channel resources in time, so message packet queues overflow, resulting in the loss of message packet, in

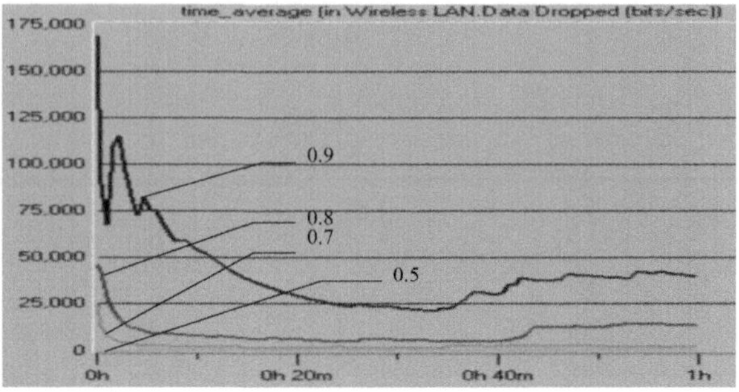

Fig. 5. Packet loss ratio

the mean time, causing the service time of packet in the queue is prolonged, which leads to the prolong of end to end delay. Trough the simulation of Ad-hoc network under self-similar traffic, we can fairly model the reaching law realistically in wireless network traffic. According to Poisson distribution which is of short-range dependence characteristic, the queuing delay of high burst traffic link is always underestimated, but through the adoption of self-similar distribution as the traffic model, we can accurately estimate the conflict of the network and queuing delay with the increase of network link utilization and accordingly increase the buffer capacity and improve the performance of network.

5 Conclusions

To conclude, according to our experience, the self-similar characteristics of Ad-hoc network traffic have an impact on the network performance and the various indexes are changed as self-similar degree of network traffic changes; the changing of network performance is particularly obvious when the traffic burst degree is high. In particular ,under heavy traffic load, the conflict degree of data frame is more serious and transmission delay of data frame is longer and buffer occupancy rate is larger than analysis results of traditional classical queuing analysis, which is worth emphasizing during tactical Ad-hoc network planning and optimal configuration.

References

1. Qilian, L.: Ad Hoc Wireless Network Traffic-Self-Similarity and Forecasting. IEEE Communications Letters 6(7), 297–299 (2002)
2. Li, L.-F.: An Iterative Method to Estimate Hurst Index of Self-similar Network Traffic. Electronics & Information Technology 28(12), 2371–2373 (2006)
3. Leland, W.E., Taqqu, M.S., Willinger, W., Wilson, D.V.: On the self-similar nature of ethernet Traffic. IEEE/ACM Trans. on Networking 2(1), 1–15 (1994)
4. Liang, Z.-T.: The Study of Algorithms Generating Self-similar Traffic. Control & Automation 24(2), 219–221 (2008)
5. OPNET Technologies, Inc. (EB/OL) (March 2002), `http://www.opnet.com/`
6. Pei, J.: Modeling Design of RMD Algorithm Based on Network Traffic. Jiangxi Communicaiton Science & Technology (2), 28–30 (2008)

Research on the Security Strategy for Wireless Sensor Networks

Jing Zhang

College of Electronic Engineering Huaihai Institute of Technology,
Lian yungang, China
JingZhang112@tom.com

Abstract. Wireless sensors networks (WSNs) consist of a number of sensor nodes connected through a wireless network that detect data and relay the detected data to the sink node using the multi-hop wireless transmission. Now, one of the most important issue is how to implent critical security tasks such as intrusion or tamper detection, and therefore must be protected. In this paper, we first analyse the security challenges in the wireless sensor network from the application layer, transport layer, network layer, data link layer and physical layer based on the system model. Then we give the security strategy which usually used in wireless sensor networks from the key management, security routing, security integration, intrusion detection and confidence model. This strategy can ensure the data confidentially implement, reliably generate, efficiently integrate and securely transmit.

Keywords: wireless sensor networks (WSNs), security strategy, key management, security routing, security integration, intrusion detection.

1 Introduction

A wireless sensors networks (WSNs) [1] consist of a number of sensor nodes connected through a wireless network that detect data and relay the detected data to the sink node using the multi-hop wireless transmission. The applications of a wireless sensor network can be used in monitoring disaster areas, monitoring patients, assisting disabled patients, helping the military, remote virtual measurement and so on.

Wireless sensor network as a task-based network is not only for data transmission, data,but also for data collection and integration, cooperative controln etc. It is a important issue to ensure the data for confidentially implement, reliably generate, efficiently integrate and securely transmit. In order to ensure the confidentiality and security, WSNs need implement some basic security mechanisms: confidentiality, point to point message authentication, integrity, freshness, authentication broadcasting and security management.

In recent years, more and more people are engaged in wireless sensor network security technology. Eschenauer and Gligor [2] proposed a distributed sensor network key management scheme, which including three processes of key pre-distribution, shared key discovery and path key establishment. Chan etc [3] put forward a random

J. Zhang (Ed.): ICAIC 2011, Part V, CCIS 228, pp. 157–164, 2011.

key pre-distribution scheme for a sensor network which have three security mechanisms: First is has q composite key management scheme; second, multi-path key to strengthen the program; third is random key distribution scheme. Jolly etc [4] put forward an energy-efficient key management protocol which based on the IBSK scheme. Perrig etc [5] proposed a sensor network security protocol SPINS, which consists of two parts: SNEP (secure network encryption protocol) by using the counter and the message authentication code to provide data confidentiality, authentication and freshness and so basic security mechanism, but it can not provide efficient broadcast authentication; uTESLA((micro timed efficient streaming loss - tolerant authentication protocol) is an extended form of TESLA which can provide broadcast authentication. Ren [6] given a detection methods to ensure the safety of the MACL program, but this program is only the ability to resist passively. J. Deng etc [7] proposed a new secure routing protocol, namely, INSENS (Intrusion-tolerant routing in wireless sensor networks),which is able to tolerate attacks.

In this paper,we first analyse the security challenges in the wireless sensor network from the application layer, transport layer, network layer, data link layer and physical layer based on the system model.Then we give the security strategy which usually used in wireless sensor networks from the key management, security routing, security integration , intrusion detection and confidence model.This strategy can ensure the data confidentially implement, reliably generate, efficiently integrate and securely transmit.

2 System Model

A. Network Model

In this work, we consider the wireless sensor networks where all nodes in the network are homogenous and energy constrained.In order to convenient describe,we assume that the network is a two-tier structure rather than multi-level topology,namely,the entire wireless sensor network is divided into many clusters,each cluster has a cluster head and lots of menber nodes. the communication of cluster members with cluster head, cluster heads with base station are the single-hop.the information which apperceived by menmber nodes directly send to the cluster head,then forwarded to the base station by cluster head. Base station can directly send commands and control information to all nodes which do not through the cluster head.

Cluster head implements periodic election,when a node to becomes cluster head,it not only shoulder the task of data fusion and forwarding,but also Still able to perform cognitive tasks.

We assumed that all sensor nodes are the same and have the same initial energy, base station determines the selection of cluster head and has enough storage capacity,processing power and energy,and can store all nodes information.

B. Energy Model

We adopt a simplified power model of radio communi-cation in document [8], namely, in order to send a k-bit packet information and the sending distance is d, the sending energy consumption is

$$E_{Tx}(k,d) = E_{elec} \times k + \varepsilon_{amp} \times k \times d \times d \tag{1}$$

The receiving energy consumption is

$$E_{Rx}(k) = E_{elec} \times k \tag{2}$$

where Eelec is the energy/bit consumed by the sender and receiver electronics, J/bit, Eelec=50nJj/bit, . ε_{amp} is the J/(bitxm2), ε_{amp} =100pJ/bit/m2.we commonly assume that the sending distance and d2 is directly proportional for shorter distance, while the sending distance and d4 is directly proportional for longer distance, so we can see the directly sending to long distance is consumed more energy than multi-hop sending.

But the differentiation from the document [9],we consider the processing consumption in order to proximity real scene, the energy consumption of cluster head is Ep

$$E_P(k,m) = \sum_{i=1}^{m} 1/3 \times E_{elec} \times k_i \tag{3}$$

So the residual energy of cluster head is

$$E_r(i) = E_r(i) - \sum_{n=1}^{n1} E_{Tx}(k_n, d_n) - \sum_{l=1}^{n2} E_{Rx}(k_l) - Ep(k,m), n1, n2 \in N \tag{4}$$

Where n1,n2 are the cluster head respectively sending and receiving times before time Ti.

The residual energy of ordinary node is

$$E_r(i) = E_r(i) - E_{Tx}(k_n, d_n) - E_{Rx}(k_l) \tag{5}$$

C. Security Challenges of Wireless Sensor Network

Wireless sensor networks are much different from common network at nodes and network properties.Wireless sensor networks security facing a great challenges[10].

- Limited communication capabilities:the communica-tion bandwidth of sensor is narrow and often changes,its communication scope only cover a few dozen to several hundred meters.there are frequently disconnected in communication and often resulting in communication failure.
- Limited power:the sensor power is extremely limited and often cause failure.the constraint of sensor power is a serious problem.
- Limited computing power : the sensor has embedded processor and memory,but its computing power is very limited.
- A large number of sensors and wide distribution, maintenance is very difficult,so the sensor network hardware and software must have high robustness and fault tolerance.
- dynamics:Network dynamics and strong. Sensor network is highly dynamic. Network sensors, sensing object and the observer may have mobility, and often has new node to join or have node failure. Therefore,Sensor network must be reconstructed and self-adjusting .

- Large-scale distributed triggers. Many sensor networks need to control the object , so it is very difficult to dynamic manage the thousands of flip-flop .
- Sensing data flow is huge. Each sensor in the sensor networks usually have a greater flow of data,and real-time. while each sensor has only limited computing resources, it is difficult to handle large real-time data stream.

3 Analyse of Security in Wireless Sensor Network

WSN framework has five layers ,as shown in Figure 1,such as the application layer, transport layer, network layer, data link layer and physical layer from top to bottom.The function of physical layer provides a simple robust modulation, transmission and access technology;the data link layer is responsible for controlling the use of wireless channel, reducing conflicts caused by the surrounding nodes;network layer answer for the data fusion, routing the pre-processed data to sink node;transport layer is maintain data stream cooperate with application layer;application layer has the various application-related protocols, software and so on.

Wireless sensor networks typically face four kinds of communication security threats: eavesdropping, tampering, counterfeiting, and blocking (Figure 2). Meanwhile, as a depletion-mode network, sensor nodes energy is very limited and precious,the system function are vulnerable to denial of service (DoS)attacks.

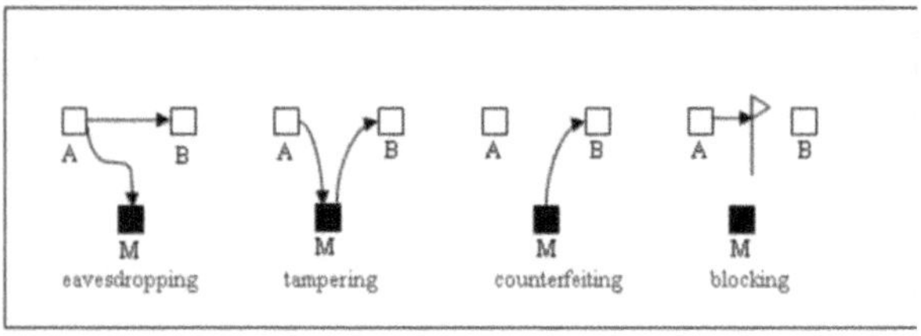

Fig. 2. The four kinds security threats in woreless sensor networks

There are faced with some security issues in each layer,the specific description as follows:

A. Scurity Analysis of Physical Layer

The main security problem of physical layer are the wireless communication interference and nodes fall,including congestion attacks and physical destruction.

B. Scurity Analysis of Data Link Layer

The traditional network security mechanism is not suitable for wireless sensor networks,the end to end security mechanisms do not apply in wireless sensor networks,because of the data is continuous processed by the intermediate nodes which need to access data packets.

There are two kinds security mechanism of the data link layer :one is use the key method to achieve information integrity, confidentiality, and authentication.These research focuses on how to optimize the security protocols for the lack energy and weak computing capabilities.

C. Scurity Analysis of Network Layer

Network layer routing protocol provide the key network routing services for the entire wireless sensor network,the attack for the routing may lead to paralysis of the entire network.Secure routing algorithm directly affects the security and usability,therefore,it is the key point.

There are various methods to attacks on routing protocols,according to the attacker's ability,it can be divided into Mote-class attack and Laptop-class attack,in the former attack,the attacker and the normal node are alike in resource; while in the latter attack, the attacker have more equipment and resources,such as energy, CPU, memory, radio transmitter, etc.and will bring a a more dangerous.

D. Scurity Analysis of Transport Layer

Transport layer is responsible for connecting wireless sensor networks with Internet or external network. The node can not save a lot of information and the node will consume a lot of energy for sending information,so there is little research on thetransport layer security protocol.

E. Scurity Analysis of Application Layer

The application layer can provides many services which based on key,such as data gathering, task distribution, target tracking, etc.So the research of the application layer security are focuse on the infrastructure,namely,the key management. Currently, the key management scheme is divided into the pre-configured key scheme and the key distribution center management.

4 Research on Security Strategy for Wireless Sensor Networks

The research of wireless sensor network security technology are focused on five areas: key management, security routing, security integration , intrusion detection and confidence model [11].

A. Key Management

The traditional key management techniques can not be effectively used for WSN. now,the most study of the key management technology is based on symmetric key mechanisms which has advantages in the calculation complexity and power consumption.There are include pre-shared key model and non-pre-shared key mode;the probabilistic key management and the deterministic key management. such as LEAP [12] is a typical deterministic key management techniques which use a variety of key mechanisms.

B. Security Routing

Common routing protocols are easily attacked which mainly consider the efficiency and energy-saving and little concern about the routing safety.Ganesan etc[13] proposed the multi-path routing mechanism which can defense the selective forwarding attacks from malicious nodes. Deng etc[14] proposed the INSENS routing protocols can limit the harm from the invasion nodes within a certain range and can continue provide routing services in the case of invasion nodes can not be excluded.

C. Data Security Integration

The goal of data security integration is to rule out the impact from the false information.in order to this goal, data security integration needs introduce various security technology based on the general data integration to ensure data integrity, secrecy ,identification and even introduce the identification mechanism for malicious node.Such as Hu and Evans proposed SA safety data aggregation techniques [15] based on μTESLA protocol,which can solve the node invasion and individual nodes captured. Proposed by Zhu and other false data filtering technology was proposed by Zhup[16] which prevent the tampered data to reach the base station.

D. Intrusion Detection

Simple password techniques can only identify the invasion of alien node, but can not identify those captured node invasion, which have the same encryption and authentication key with the normal nodes. The current study focuses on the invasion of captured nodes, such as Ye[] proposed a SEF mechanism which is a intrusion detection mechanisms to identify false data; Wang etc[] proposed a intrusion detection technology to identify malicious nodes.

E. Confidence Model

Trust model is to establish and manage the trust relationships framework which can be used to solve many other security problem, Yan etcl [19] using the trust model to solve the security in the Ad Hoc routing, data protection and other issues. Tanachaiwiwat etc [20] proposed a method to isolate malicious behavior and implement confidence-routing based on the location.

5 Conclusions

In this paper, we first analyse the security challenges in the wireless sensor network from the application layer, transport layer, network layer, data link layer and physical layer based on the system model. Then we give the security strategy which usually used in wireless sensor networks from the key management, security routing, security integration , intrusion detection and confidence model. This strategy can ensure the data confidentially implement, reliably generate, efficiently integrate and securely transmit.

Acknowledgment. We acknowledge the support of High-Tech Research plan of Jiangsu Province (No.BG2007045),we sincerely thank the anonymous reviewers for their constructive comments and suggestions.

References

1. Akylidiz, I.F., Su, W., Sankarasubramaniam, Y., Cayirci, E.: A survey on sensor networks. IEEE Personal Communications Magazine (2002)
2. Eschenauer, L., Gligor, V.D.: A key-management scheme for distributed sensor networks. In: Proc.of the 9th ACM Conf. n Computer and Communications Security (CC5 2002), pp. 41–47. ACM Press, Washington D.C (2002)
3. Chan, H., Perrig, A., Song, D.: Random key predistribution schemes for sensor networks. In: Proc. of IEEE 2003 Symposium on Research in Security and Privacy, pp. 197–213. IEEE Computer Society, Berkeley (2003)
4. Jolly, G.: A Low-energy key management protocols for wireless sensor networks. In: Eighth IEEE Symposiumon Computers and Communications, ISCC (2003)
5. Perrig, A.: SPINS: Security protocols for sensor network. Wireless Networks (5), 521–534 (2002)
6. Zhou, L., Haas, J.Z.: Securing ad hoc networks. IEEE Network Magazine 13(6), 24–30 (1999)
7. Deng, J., Han, R., Mishra, S.: Securit Support for In-Network Processing in Wireless Sensor Networks (October 2003), `http://www.cs.colorado.edu/mishras/research/paoers/sasn03.gdf`
8. Gandham, S.R., Dawande, M., Prakash, R., et al.: Energy efficient schemes for wireless sensor networks with multiple mobile base station. In: GLOBECOM 2003, pp. 377–381. IEEE Comunnications Society, San Francisco (2003)
9. Braginsky, D., Estrin, D.: Rumor routing algorithm for sensor networks. In: WSNA 2002, Atlanta, GA (September 2002)
10. Li, J., Li, J., et al.: Sensor networks and database. Sensor networks and database 10(14) (2007)
11. Zhu, Z.-j., Tan, Q.-p., Zhu, P.-d.: A Survey of WSN Security Research. Computer Engineering & Science 30(4), 102–103 (2008)
12. Zhu, S., Setia, S., Jajodia, S.: Leap:Efficient Security Mechanisms for Large-Scale Distributed Sensor Networks. In: Proc. of the 10th ACM Conf. on Computer and Communications Security, pp. 62–72 (2003)
13. Ganesan, D., Govindan, R., Shenker, S., et al.: Highly-Resilient,Energy-Efficient Multipath Routing in Wireless Sensor Networks. ACM SIGMOBILE Mobile Computing and Communications Review 5(4), 11–25 (2002)
14. Deng, J., Han, R., Mishra, S.: INSENS: Intrusion-Tolerant,Routing in Wireless Sensor Networks[R] ·Technical Report,CUCS-939-02, Department of Computer Science, University of Colorado (2002)
15. Hu, L., Evans, D.: Secure Aggregation for Wireless Networks. In: Proc. of the Workshop on Security and Assurance in Ad Hoc Networks (2003)
16. Zhu, S., Setia, S., Jajodia, S., et al.: An Interleaved Hop-by-Hop Authentication Scheme for Filtering of Injected False Data in Sensor Networks. In: Proc. of the IEEE Symp. on Security and Privacy, pp. 259–271 (2004)
17. Ye, F., Luo, H., Lu, S., et al.: Statistical En-Route,Filtering of Injected False Data sensor Networks. In: Proc. of IEEE INFOCOM 2004 (2004)

18. Wang, G., Zhang, W., Gao, G., et al.: On supporting Distributed Collaboration Sensor Networks. In: Proc. of MILCOM 2003 (2003)
19. Yan, Z., Zhang, P., Virtanen, T.: Trust Evaluation Based Security Solution in Ad Hoc Networks. In: Proc. of the 7th Nordic Workshop on Secure IT Systems (2003)
20. Tanachaiwiwat, S., Dave, P., Bhindwale, R., et al.: Location-Centric Isolation of Misbehavior and Trust Routing in Energy-Constrained Sensor Networks. In: Proc. of the 2004 IEEE Int'l Conf. on Performance, Computing, and Communications (2004)

Remote Sensing Image Automatic Classification Based on Texture Feature

Yunjun Zhan, Yujing Liang, and Jiejun Huang

Department of Resources and Environment Engineering Wuhan University of Technology, Wuhan 430070, P.R. China
yunjunzhan@hotmail.com

Abstract. Texture feature of ground object not easily change influenced by environment, so it is stable. In order to reflect object feature better, we extracted texture feature with wavelet transform method, Including contrast, correlation, energy and homogeneity. And we do the image classification based on texture feature. In order to test the method, we adopted QuickBird satellite image to experiment, and then compared with image classification based on spectral characteristics. Result suggests, in a way, that the image classification based on texture feature is able to improve the remote sensing image automatic classification precision and obtain the better classification effect.

Keywords: Texture Feature, Image Classification, Wavelet Transform, Classification Precision.

1 Introduction

Pixel attribute described by gray is more beneficial to computer processing, so it only considers single pixel gray character basically in remote sensing image computer automatic classification. Pixel gray value characterizes spectrum information of ground object, reflects the ability of electromagnetic wave reflection and is the more visual reflection of ground object feature. But due to interference and influence factors of spectrum information are too many, existing the phenomenon that different objects have same spectrum and the same objects have different spectrum, therefore, improving classification precision is not easy. In the visual interpretation, according to map spot with certain meaning composed of local many pixels, professionals analyze and reasoning, such as texture feature formed by shape, size, position, correlation and combination of the map pot. Texture feature is a spatial distribution model which does not depend on hue or brightness of object surface, reflects the image gray or color, enable to reflect the visual feature of the identical quality phenomenon in image. It includes image surface information and the relationship with environment and gives good consideration to image macrostructure and microstructure. Compared with ground object spectral features' easy change influenced by environment , the texture features of ground objects are stable. Therefore, combined spectral feature and texture feature of ground object organically is the trend to improve remote sensing image automatic classification precision.

J. Zhang (Ed.): ICAIC 2011, Part V, CCIS 228, pp. 165–172, 2011.

At present, many domestic and foreign scholars concern texture feature effect in image classification and mainly study image classification methods of texture feature analysis and texture feature extraction. Inheriting traditional image statistical methods, gray level co-occurrence matrix is usual [1]; in analysis methods based on model, the application of fractal dimension method is more, most adopt Fractal Brown function, some improve the fractal method [2]; recently, MRF method is used more, but the main difficulty is to determine parameter [3].studies have enriched texture description and feature extraction method and improved classification precision. But the problems are that features are too much, parameter is not easy to be determined, the realizing is complicated and the calculation speed is slow [4][5].

According to the characteristics of urban remote sensing high spatial resolution, this article, consideration to classification precision and implementation complexity, analyzes and extract object contrast, correlation, energy and homogeneity of texture feature in the principle of divisibility, reliability, independence and simplicity, and use discrete wavelet transform to extract image texture feature. Taking urban QuickBird data as example, experimental analysis proved that the method can improve the classification precision in high spatial resolution remote sensing image and it is simple.

2 Using Wavelet Transform to Extract Texture Feature Method

A. The Basic Principle of Texture Information Extraction

Wavelet transform has good location capability in time domain and frequency domain, and adopting gradual fine time domain and spatial domain step to high frequency component can be focused to any details of image [6]. The advantage of using wavelet transform to mergence image is that it can keep hue and saturation unchanged while improving spatial resolution. The basic principle of using two-dimensional discrete wavelet transform to extract image texture information are as follows: from $V^2_{j+1}=V^2_j\oplus W^2_j$,we know that it exists $f_j\in V^2_j$, $g_j\in W^2_j$, $f_{j+1}=f_j+g_j$。 for two-dimensional image f(x, y), dyadic wavelet of image is decomposed to:

$$f_{j+1}(x,y)=\sum_{k,m}(c_{j,k,m}\phi_{j,k,m})+\sum_{\varphi=1,2,3}\sum_{k,m}(D^{\varepsilon}_{j,k,m}\psi^{\varepsilon}_{j,k,m}) \tag{1}$$

Which :

$$C_{j,k,m}=\sum_{l,n}(h_{l-2k}h_{n-2m}c_{j+1,l,n})$$

$$D^1_{j,k,m}=\sum_{l,n}(h_{l-2k}g_{n-2m}c_{j+1,l,n})$$

$$D^2_{j,k,m}=\sum_{l,n}(g_{l-2k}h_{n-2m}c_{j+1,l,n})$$

$$D^3_{j,k,m}=\sum_{l,n}(g_{l-2k}g_{n-2m}c_{j+1,l,n})$$

$C_{j,k,m}$、$D^1_{j,k,m}$、$D^2_{j,k,m}$ and $D^3_{j,k,m}$ are image low frequency component, high frequency component in horizontal direction, vertical direction and diagonal direction respectively. $\{h_k\}_{k\in z}$ is low-pass filter coefficient, $\{g_k\}_{k\in z}$ is high-pass filter coefficient, Φ is scaling function, Ψ is orthogonal wavelets, orthonormal basis of W^2_j composed of $\{\Phi^{\varepsilon}_{j,k,m}|k,m\in z\}$, orthonormal basis of W^2_j composed of $\{\Psi^{\varepsilon}_{j,k,m}|k,m\in z\}$.

B. The Determination of Texture Measure Index

Gray level co-occurrence matrix is the statistical form of two pixel gray grade joint distribution in image, namely, statistical law of a pair of pixel gray appearing in certain direction and with a certain distance is the most representative second-order statistics texture feature calculating method [7]. This article extracts texture feature quantities again on the basic of gray level co-occurrence matrix, which is called secondary statistics. There are many statistics calculated by gray level co-occurrence matrix method and we mainly study contrast, correlation, energy and homogeneity.

① contrast: $$CON = \sum_{i=1}^{n} \sum_{j=1}^{n} p(i, j) \times (i - j)^2 \quad (2)$$

② correlation: $$COR = \left[\sum_{i=1}^{n} \sum_{j=1}^{n} p(i, j) \times (i - n) \times (j - n) \right] / \left(\sqrt{(i-n)^2} \times \sqrt{(j-n)^2} \right) \quad (3)$$

③ energy: $$ASM = \sum_{i=1}^{n} \sum_{j=1}^{n} p(i, j)^2 \quad (4)$$

④ homogeneous: $$HOM = \sum_{i} \sum_{j} \frac{1}{1 + |i - j|} p(i, j) \quad (5)$$

n: maximum pixel value, $P(i, j)$ is the element of row i and column j in gray level co-occurrence matrix .

3 Image Classification Experiment Based on Texture Information

A. Experimental Data Introduction

The experimental image is QuickBird satellite image, spatial resolution is from 0.61 to 0.72 m after fusion, and the spectrum range from 450 to 900nm, including blue(450-520nm), green(520-660nm), red(630-690nm) and near infrared(760-900nm). Due to high spatial resolution , clear veins and rich information ,QuickBird image has been the important urban remote sensing data source. Experimental image is as fig 1.

B. Using Wavelet Transform to Extract Texture Feature

We choose suitable and local square area to be the sliding window starting from left corner of the image.

Doing n layers wavelet decomposition to window region and according to texture measure formula, we calculate the texture measure in every frequency band and endow center pixel as the feature vector, and then we slip the window with step 1,repeating the steps , we can calculate the feature vector of next window center pixel.

1) Wavelet Base Selection: To decompose wavelet of image and extract the texture structure feature, we must determine the wavelet base firstly. We select Daubechies4 to decompose wavelet. Daubechies4 has good orthogonality and supportative property and is able to do continuous and discrete wavelet transformation and can reflect the continuity and mutagenicity of frequency signals better[8].

Fig. 1. Remote sensing image

2) The Determination of Decomposition Level N: Considering filtering, we hope N is more large, but in image decomposition, N is proportional to image boundary effect[9], especially when the decomposition level is high, the influence is higher. As the level N increase, the sub-image information entropy decreased gradually, and the texture sub-image information is low too. Therefore, we determine the wavelet decomposition level as 2.

3) Sliding Window Size: We try using 5x5, 7x7, 9x9 size window to do wavelet decomposition and calculate the texture feature image. Contrasting the texture image, we can find that the crude texture of texture image getting from big window is obvious and the small window is more suitable to fine texture extraction. Window has fuzzy and smoothing effect to image and if the window is too small, the texture has not strong edge effect. The result suggests that if the window (5x5) is suitable, the texture edge can be extracted well and it also can reduce the fogging effect brought by small window.

4) Texture Feature Factors Selection: Wavelet decomposition level and texture measure index determine that there are too many sub-image using wavelet transform. In order to find the most effective sub-image to urban object type, we need to compare them. In this experiment, study area is divided into water, road, grassplot, Tree, Building and other constractions. Every object is chosen 10 map spot and calculated 24 texture feature, and so we can get the texture value of every object. Calculate the dispersion of every forest type in 24 texture feature factors and select 6 factors that the dispersion is larger.

C. Image Classification Using Texture Information

Image classification adopt the Ecognition software which is object-oriented classification. We segment the image according to texture feature parameters, and the image is segmented into many map spot of different size and shape .These map spots are also called objects. Segment and classify first, and then segment and classify again, doing this until we get the satisfy classification result. The result is as fig 2 and error analysis is as table1.

Fig. 2. Classification result based on texture

D. Image Classification Based on Spectral Feature

In order to contrast the classification results based on texture feature, the experience apply supervising classification method based on spectral feature. The result is as fig 3 and the error analysis is as table 2.

In order to test the various objects recognition effect, we do the error matrix analysis to the classification results. We extracted 150 samples randomly from and different land category has different samples. In error matrix, the user precision is the proportion of land category, which represents that we extracted a random sample from classification results arbitrarily, and its types have the same condition probability with the ground practical types. Cartographic accuracy is the proportion of land category samples. It represents that extracting a random sample arbitrarily, the condition probability of classification results are consistent to the same location in the classification figure.

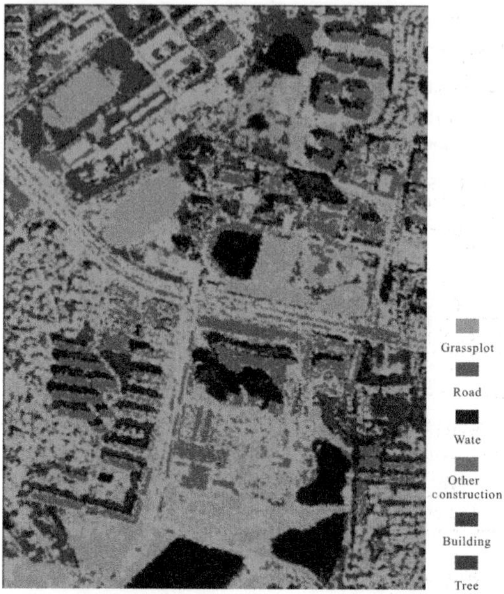

Fig. 3. Classification result based on spectral feature

E. Results Comparative Analysis

From fig 3 and table 2 we can see that Building spectra information is almost the same to the road spectra information, and the error is much. There is error phenomenon in other objects, such as the Tree in the water, and building shadow being seen as water wrongly. The holistic classification precision is 79.33% based on spectra information

Table 1. Error Matrix Based on Texture Feature Classification

Category	Water	Grassplot	Road	Other construction	Tree	Building	Others
water	22	2	1	0	2	1	0
grassplot	1	25	0	2	0	2	0
Road	0	1	18	0	0	3	0
Other construction	0	2	0	19	2	1	0
Tree	1	1	0	0	24	1	0
Building	0	0	2	0	0	14	0
Others	0	0	0	0	0	0	3
Sampling number	28	21	27	30	15	30	3
Reference number	24	20	23	33	14	38	3
User precision	91.67%	80.65%	85.71%	90.48%	85.71%	63.64%	100%
Cartographic precision	78.57%	83.33%	81.82%	79.19%	88.89%	87.5%	100%
Holistic precision				83.33%			
Kappa coefficient				80.06%			

Table 2. Error Matrix Based on Spectra Feature Classification

Category	Water	Grassplot	Road	Other construction	Tree	Building	Others
water	16	2	0	3	2	1	0
grassplot	3	15	0	2	0	1	0
Road	0	0	20	0	0	7	0
Other construction	0	2	0	26	0	2	0
Tree	0	1	0	2	12	0	0
Building	0	0	3	0	0	27	0
Others	0	0	0	0	0	0	3
Sampling number	24	21	27	30	15	30	3
Reference number	19	20	23	33	14	38	3
User precision	84.21%	75%	89.96%	78.79%	85.71%	71.05%	100%
Cartographic precision	66.67%	71.43%	74.07%	86.67%	80%	90%	100%
Holistic precision				79.33%			
Kappa coefficient				75.07%			

classification, and the Kappa statistical value is 75.07%. Therefore, we can not reach the ideal classification precision only using spectra information classification.

From fig 2 and table 1, we can see that there are less errors of building and road, and the building shadow is classified correctly. The holistic classification precision reach 83.33%, and the Kappa statistical value, which is the degree of coincidence reach 80.06%.

4 Conclusion

In remote sensing image classification, texture information can eliminate the error that the different objects have the same spectra and the same object has different spectra, which improve the classification precision and we then can obtain the better classification effect.

Acknowledgment. The work was supported by the fundamental research funds for the central universities (no.2010-ia-057), the wihan university of technology scientific research found for doctor (471-38650564), the national natural science foundation of China (41071104), the natural science found of hubei province (no.2009cda015), the national high technology research and development program of china (863 program, no.2009aa12z201).

References

1. Feng, J., Yang, Y.: Study of Texture Images Extraction Based on Gray Level Co-Occurence Matrix. Beijing Surveying (3), 19–22 (2007)
2. Qin, Q., Lu, R.: Satellite Image Classification Based on Fractal Dimension and Neural Networks. Acta Scientiarum Naturalium Universitatis Pekinensis 36(6), 858–864 (2000)
3. Zhao, Y., Zhang, L., Li, P.: Universal Markov Random Fields and Its Application in Multispectral Textured Image Classification. Journal of Remote Sensing 10(1), 123–129 (2006)
4. Xu, C.: A New Method for Analysis Image Texture and Its Application, Doctoral Dissertation. Fudan University (2005)

5. Zeng, W.: Texture Information Extraction in Remote Sensing Imageries with Gray Level Co-occurrence Matrix and Wavelet Transform, masteral dissertation. Northeast Normal University (2006)
6. Liu, H., Mo, Y.: Modified Texture Segmentation Algorithm Based on Multiresolution Model. Acta Optica Sinica 20(6), 16–20 (2000)
7. Liu, X., Shu, N.: Application of Texture Feature in Multi-Spectral Images Classification. Journal of Geomatics 31(3), 31–32 (2006)
8. Chen, S., Qin, M.: The Classification of Texture and Structure in the High Resolution Imagery Based on Wavelet Transform. Geography and Geo-Information Science 19(3), 6–9 (2003)
9. Jin, W., Yu, J.: Image Retrieval Based on Texture Features and EBP-OP Algorithm. Computer Engineering and Applications 65(8), 61–62 (2002)

The Design and Implementation of Large-Scale Call Center Management System*

Chuanliu Xie, Junfeng Wang, and Ying Mou

College of Computer Science, Sichuan University, Chengdu 610065, China
chuanliuxie211@sina.com

Abstract. With the development of call center, its business become more and more sophisticated. Its scale is also increasingly large. And its integrated subsystems become more and more. These make it very difficult to manage a large-scale call center efficiently and economically. Considering this problem, the organization of large-scale call center is combed firstly in this paper, and the functions of subsystems are analyzed. Then a scheme of large-scale call center is designed. Finally, a large-scale call center management system based on improved J2EE architecture is implemented. It is proved that the system has successfully overcome the difficulty in large-scale call center management, solved the problem of low management efficiency, and achieved intelligent management and scientific decision-making. The system can meet the requirements of large-scale call center very well.

Keywords: Index Terms - call center, information system, J2EE, intelligent scheduling.

1 Introduction

Call center is an organization that provides services to its customers via telephone by its customer service representatives[1]. A call center is generally composed of the automatic call distributor, interactive voice response systems, computer telephony integration systems, database systems, call management systems, and business processing system. With the development of technologies related to call center, its composition become more and more complex, its scale is also increasingly large, and the centralized management of large-scale call center become increasingly difficult. Therefore, it urgently needs a platform that can manage all the subsystems of call center to provide perfect management mechanism, improve management efficiency and operational capabilities, reduce management costs, achieve scientific decision-making and intelligent management, and ultimately achieve minimum costs and maximum profit.

In recent years, domestic and foreign scholars have done a variety of research on call center management system. Such as: research on optimization of business management[2], improvement of service level[3], optimization of human resource[4], and optimization of scheduling management[5,6]. However, previous research and schemes are focus on small and medium scale call centers, their research areas are

* This work was supported by the National Natural Science Foundation of China (60772105), the National High Technology Research and Development Program of China (2008AA01Z208, 2009AA01Z405), the Doctoral Fund of Ministry of Education of China (20090181110053) and the Youth Foundation of Sichuan Province (09ZQ026-028).

J. Zhang (Ed.): ICAIC 2011, Part V, CCIS 228, pp. 173–181, 2011.

confined to a particular aspect, their business is simpler, and their systems integrate fewer function modules. So their systems are not suitable for large-scale call center management. Based on the previous research, we survey the business logic of large-scale call center, and analyze the problems of large-scale call center management. We summarize the subsystems' functions of large-scale call center and design a large-scale call center management system scheme, and successfully develop the system based on improved J2EE architecture. Finally, the system is verified in a call center, and achieved good results.

2 System Implementation Process

A. Requirement Analysis

1) Departments of Call Center: According to lots of investigation and analysis, large-scale call center generally establishes system maintenance department, personnel department, project department, quality control department and training department. System maintenance department is responsible for daily maintenance of the management system. Personnel department is responsible for personnel recruitment, arrangement and management. Project department is responsible for the progress of project to ensure the project can complete punctually and guarantee the quality. Quality control department is responsible for supervising and assessing the quality of customer service. Training Department is responsible for working out the future development plan of company, and providing necessary training for employees as far as possible.

The department structure of large-scale call center is as shown in figure 1.

2) Roles of Call Center: According to staff functions and departments of call center, we can divide staff of call center into several roles, as shown in Table 1.

Table 1. System User Roles

Sequence	Name	Function
1	System Administrator	maintain the whole system
2	General Manager	CEO of the call center
3	Personnel Minister	view and manage all the personnel information and all the department
4	Personnel Administrator	maintain the personnel information
5	Remuneration Administrator	manage the remuneration of employees
6	Quality Manager	make quality control plan, arrange quality controllers' work
7	Quality Controller	execute quality control plan
8	Training Minister	make training plan according to requirement
9	Training Lecturer	execute staff training
10	Scheduling Administrator	make staff scheduling
11	Project Minister	manage all the projects
12	Project Manager	manage a project
13	Class Monitor	manage a class
14	Group Leader	manage a group
15	Employee	agent, answer the phone

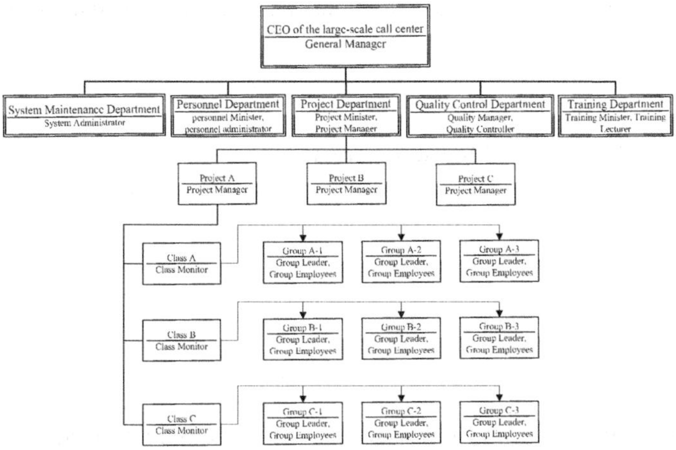

Fig. 1. Department structure of the large-scale call center

3) System Functional Requirement: The goal of large-scale call center management system is to integrate all the resources to provide perfect management mechanism. The system should integrate all the management subsystems that business requires. It is able to combine all the data perfectly via computer hardware and software technology, telecommunications, Internet and other technologies. It is able to statistics, analysis and processing all the data, to guarantee standardized management and scientific decision-making of large-scale call center. And it is able to ensure the informatization and intellectualization of call center management. Ultimately, it is able to integrate and coordinate various functions of large-scale call center, and to achieve minimum costs and maximum profit.

B. Detailed Design

According to business and functional requirements of large-scale call center, we divide large-scale call center management system into several functional modules as follows:

1) System Management: Including department management and authority management. For the convenience of using the system, we added a data dictionary sub-module. We put the conventional data into the data dictionary to make the system convenient to use. This module is primarily for system administrators.

2) Personnel Management: Including staff archives management, staff demission management, staff deployment management, and personnel reporting system. This module is primarily for personnel administrators.

3) Remuneration Management: Including attendance management, performance management, accumulation fund scheme, social insurance scheme, income tax scheme, salary formation management, payroll management, and remuneration reporting system. This module is primarily for remuneration administrators.

4) Quality Control: Including quality control templates management, quality control planning management, implementation of quality control, verification of quality control, and quality control analysis reports. This module is primarily for quality controllers.

5) Intelligent Scheduling: Including scheduling parameters management, traffic forecasting, agent forecasting, class forecasting, schedule generating, class table management, class shift management, and scheduling reporting system. This module is primarily for staff scheduling administrators.

6) Training and Examination: Including training courses management, training requirements management, training plan management, training material management, examination questions management, examination management, and examination material management. This module is primarily for training ministers.

7) KPI Monitoring: Including monitoring of service level, connection rate, line count, call count, busying count, ready count, and rest count. This module is primarily for project managers.

8) Personal Menu: Some of the conventional menus are listed in this module for individuals, including system message, my archives, my documents, my salary table, my schedules, and personal password. This module is primarily for every employee.

The functional structure of the whole system is as shown in figure 2.

C. Architecture of System Implementation

Compared to C/S (Client/Server) structure, B/S (Browser/Server) structure can greatly simplify the loads of client computer, and reduce the cost and effort for system maintenance and upgrade. So we select B/S structure, and develop the system based on the J2EE-SSH prototype which is a kind of MVC framework. MVC is a kind of the most popular software development architectures. The MVC architecture divides application programs into three core modules: model, view, and controller. Each module processes its own business separately. The relationship of model, view and controller is as shown in figure 3.

However, the SSH architecture has some shortcomings, so we made two improvements as follows:

(1) Since it is not convenient for dynamic asynchronous interaction between the foreground and background programs with Struts[8], it is used the DWR[9] framework to be supplemented.

(2) The encapsulation of Hibernate[10] is too high to manipulate the database flexibly and conveniently, especially for some low-level implementation of complex queries. In view of this, we developed a database access framework which is similar with Hibernate. It is able to support the complex queries and transactions, customize SQL statement and multi-table join query more conveniently.

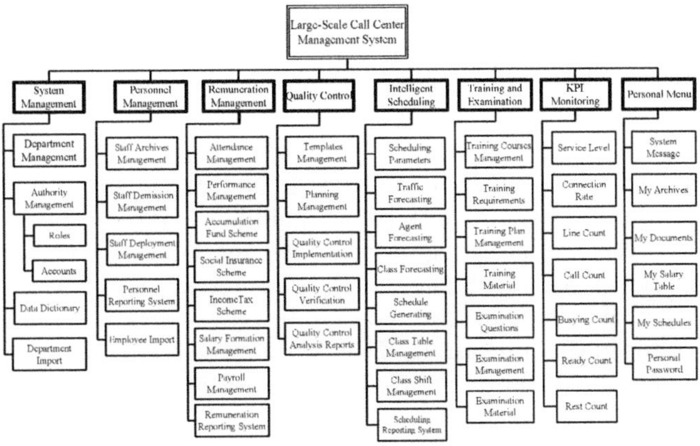

Fig. 2. Functional structure of the large-scale call center management system

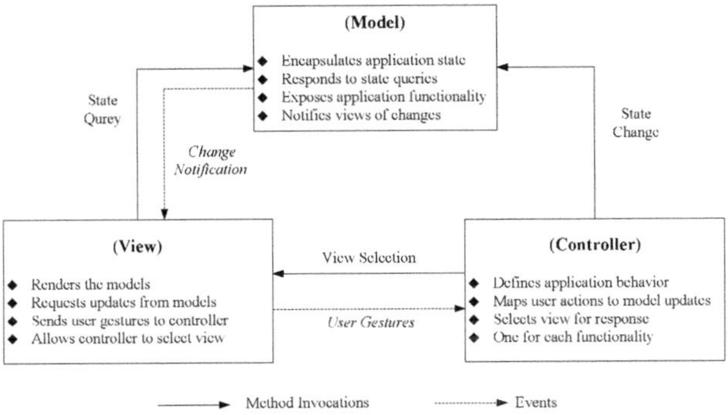

Fig. 3. Relationship of model, view and controller

Specifically, the foreground framework utilizes the combination of DWR and Struts, using the unified operation interface to access database through callback function in form of objects. The foreground operations of the system are mainly implemented by JavaScript. Background framework adopts improved light-level object-relation mapping framework, in which all the database operations are encapsulated into dao files, and the data that be queried out are encapsulated into Entities. The logic implementation in the background is achieved through operation-entity-dao. The architecture is as shown in figure 4.

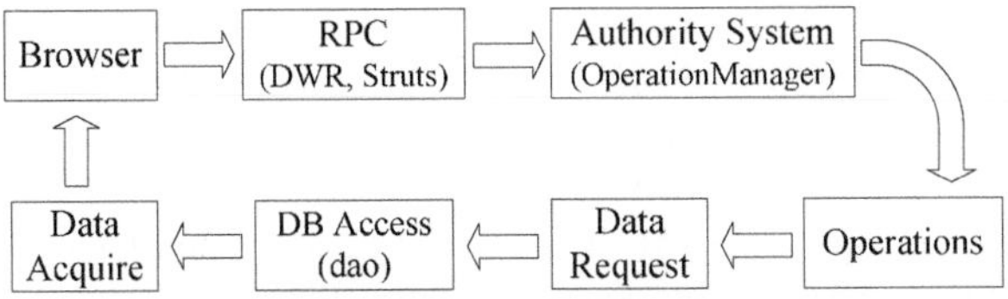

Fig. 4. Architecture of system implementation

D. Core Algorithm

Core algorithms of large-scale call center management system are mainly concentrated in scheduling module. They are as follows:

1) Back Propagation Neural Networks[11]: It is used to traffic forecasting. Back propagation neural network is a kind of multilayer forward feedback network. According to universal approximation theory: if the hidden layer nodes can be set freely by requirement, 3 layers S-shaped model is able to approximate to any function that has limited breakpoint with any precision. In view of this, we used the 3 layers BP neural network for traffic forecasting. The algorithm selects 6 parameters related to a time point as input of BP neural network to forecast the traffic of the time point, and selects 5 hidden nodes, and finally outputs the traffic of the time point. A traffic forecasting result is as shown in figure 5.

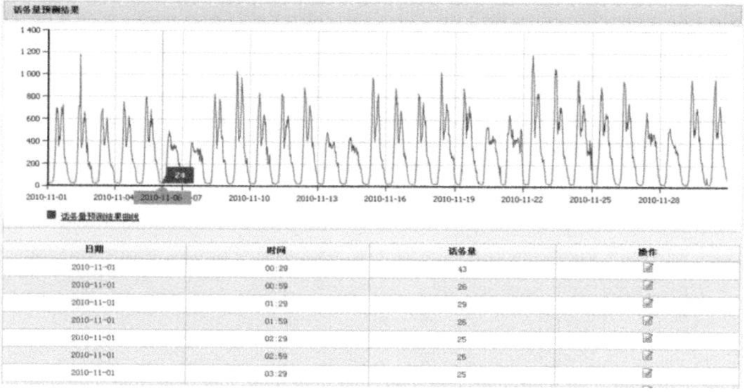

Fig. 5. A traffic forecasting result

2) Erlang-C Formula[12]: It is used to agent forecasting. The algorithm sets the service level parameters first, and the required agent count is able to be calculated. The formula is as equation (1), where W_t is the destination service level, t is target waiting time, $E_c(m,u)$ as equation (2) is the probability that a traffic is not answered immediately and has to wait, $\mu=\lambda T_s$ is the traffic intensity, λ is the average customer arrival rate, T_s is the average call duration, m is the count of agents available, and $\rho=\mu/m$ is the agent occupancy rate.

$$W_t = 1 - E_c(m,\mu)\exp(-(m-\mu)t/T_s) \tag{1}$$

$$E_c(m,\mu) = (\mu^m/m!)/(\mu^m/m! + (1-\rho)\sum_{k=0}^{m-1}\mu^k/k!) \tag{2}$$

An agent forecasting result is as shown in figure 6.

3) Exterior Point Algorithm[13,15]: It is used to abstract mathematical model of class forecasting. Exterior Point Method is actually the external punishment function algorithm. Its mathematical prototype is as equation(3), where $p(x,m_k)$ is the punishment function, $f(x)$ is the destination function, m_k is the punishment factor, and $a(x)$ is the conditions collection. The algorithm sets $a(x)$ as the result of agent forecasting and all the other conditions that must be met, and sets $f(x)$ as the sum of variances that between the result of agent forecasting result and the result that is fitted by class forecasting. Thus, class forecasting is abstracted to an optimization problem.

4) Particle Swarm Optimization [14,15]: It is used to solve the class forecasting optimization problem. The basic concept of PSO is looking for the best solution via group collaboration and information sharing between of the particles in the swarm. The algorithm sets fitness function of PSO as the punishment function $p(x,m_k)$ which is abstracted by exterior point algorithm, and then look for the best solution in a certain search space via PSO algorithm. It is proved that the class forecasting result generated by the algorithm is able to fit the agent forecasting curve in a certain precision. A class forecasting result is as shown in figure 7.

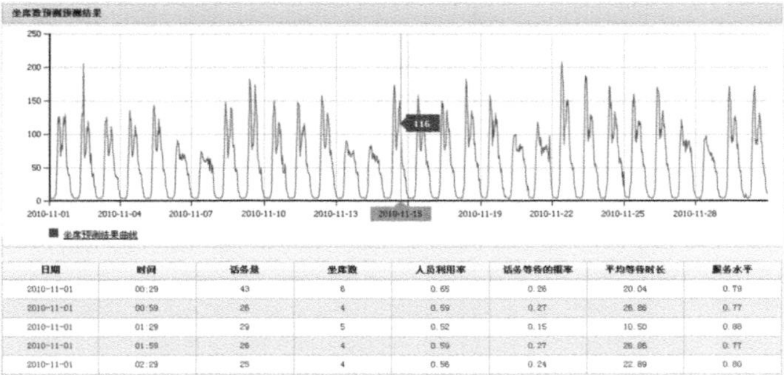

Fig. 6. An agent forecasting result

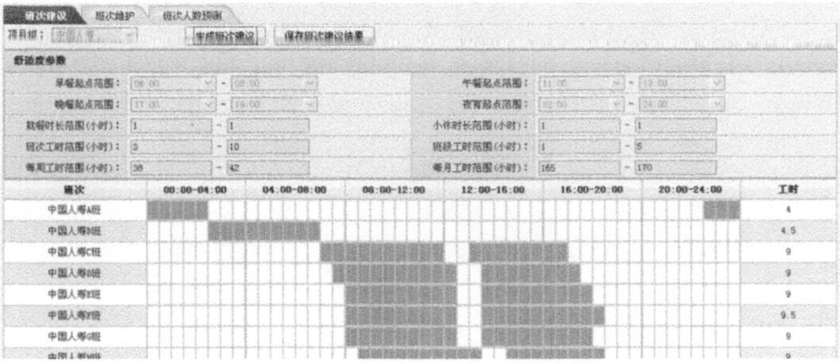

Fig. 7. A class forecasting result

5) Queue Circulating Algorithm[15]: It is used to generate the schedule. Suppose there are *n* classes (including the rest classes), the algorithm will define *n* queues and connect all the queues in series, then put the employees into the queues, and finally circulate the employees in the queues every day to get the schedule. A personal class table of scheduling result is as shown in figure 8.

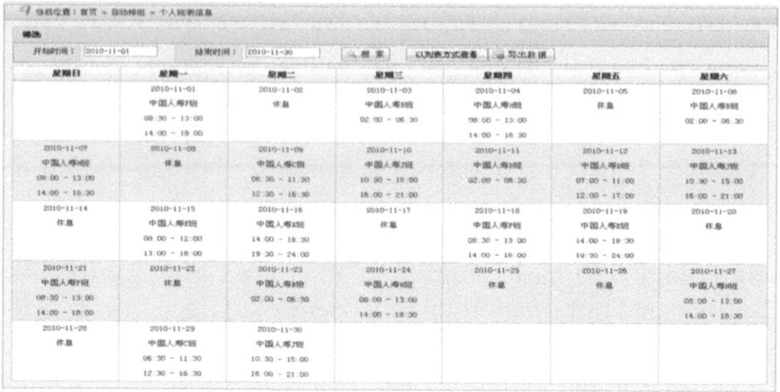

Fig. 8. A personal class table of scheduling result

3 Conclusion

In this paper, we analyze the business and requirements of the large-scale call center comprehensively and design a good scheme of the management system for large-scale call center. And the scheme is implemented based on the improved J2EE architecture. It has a good effect on the application of a call center. It is observed that some enterprises are ready to establish distributed large-scale call center. The future research will focus on how to design a management system scheme for distributed large-scale call center.

References

1. Chen, H.: Design and Application of Call Center Composition, Master Thesis. Jilin University, Changchun (2008)
2. Gurvich, I., Armony, M., Mandelbaum, A.: Service level differentiation in call centers with fully flexible servers. Management Science 54(2), 279–294 (2008)
3. Mandelbaum, A., Zeltyn, S.: Staffing many-server queues with impatient customers: constraint satisfaction in call centers. Operations Research 57(5), 1189–1205 (2009)
4. Gurvich, I., Luedtke, J., Tezcan, T.: Staffing call centers with uncertain demand forecasts: a chance-constrained optimization approach. Management Science 56(7), 1093–1115 (2010)
5. Wang, R.: Research on Accurate Scheduling Management of Call Center, Master Thesis. Huazhong University of Science and Technology, Wuhan (2008)
6. Zhang, F.: Modeling and Research of Call Center Scheduling Optimization, Master Thesis. University of Science and Technology of China, Wuhan (2010)
7. Chen, S.: The Design and Implementation of Enterprise Oriented MVC Framework, Master Thesis. Tianjin University, Tianjin (2010)
8. Chen, C., Li, L.: The analysis of Struts framework and its improvement. Computer Applications and Software 27(1), 26–28 (2010)
9. Zhang, L., Zhang, F., Wei, S.: Web application design and realization based on DWR frame. Computer Technology and Development 18(8), 84–87 (2008)
10. Wu, J., Qian, Z.: Research on web applications based on Hibernate object persistence. Computer Applications and Software 26(2), 89–91 (2009)
11. Cottrell, G.W.: New life for neural networks. Science 313(5786), 454–455 (2006)

12. Avramidis, A.N., L'Ecuyer, P.: Modeling and Simulation of Call Center. In: 2005 Proceedings of the Winter Simulation Conference, Canada, pp. 144–152 (December 2005), http://ieeexplore.ieee.org/stamp/stamp.jsp?tp=&arnumber=1574247
13. Hiroshi (Mathematical Systems Inc., JPN), Y.: Approaching large scale optimization by way of active set method, interior point method and exterior point method. Systems, Control and Information 50(9), 210–220 (2006)
14. Tripathi, P.K., Bandyopadhyay, S., Pal, S.K.: Adaptive multi-objective particle swarm optimization algorithm. In: IEEE Congress on Evolutionary Computation, CEC 2007, Singapore, pp. 2281–2288 (September 2007), http://ieeexplore.ieee.org/stamp/stamp.jsp?arnumber=04424755
15. Xie, C., Wang, J., Xia, Z., Mou, Y.: Scheduling algorithm of large-scale call center. Computer Engineering and Design (2010) (accepted)

The Design of Machine Engineering Data Management (MEDM) Based on Distributed PDM

Jing Zhang

College of Electronic Engineering Huaihai Institute of Technology
Lian yungang, China
ChengFangxiao@163.com

Abstract. Current manufacturing industry is facing an increasing challenge to satisfy customers and compete in market,and because of lacking communication among different product development stages often causes consistency problems in product lifecycle. In this paper, we first analysis the principle and the limition of distributed PDM,then we design the machine engineering data management system (M-EDMS) based on the principle of distributed PDM, the whole system is composed of two modules::the architecture design of M-EDM system,which have three layers,the first layer is the M-EDM system support level, the second layer is the core function of M-EDM object layer, the third layer is the user layer;and the function design of M-EDM system,which have the rights and organizational management module, workflow / design process management module, Project management, product structure and configuration management module, electronic collaboration module, image processing module and so on.This proposed system designment scheme is well suited for various applications for medium and small enterprises and has good prospects.

Keywords: Product data management (PDM), machine engineering data management system (M-EDMS), architecture design, system function design.

1 Introduction

Current manufacturing industry is facing an increasing challenge to satisfy customers and compete in market. To stay competitive, manufacturing companies are adopting IT solutions to facilitate collaborations and improve their product development/ production[1]· Among these solutions, product data management (PDM) systems play an essential role by managing product data electronically.

In traditional industry, because of lacking communication among different product development stages often causes consistency problems in product lifecycle. The concept of concurrent engineering, integrated product and process development and others are introduced [2].While Product data management (PDM) systems enable enterprises to conduct its business activities in a more efficient way via ingenious management of product information [3,4].Now it is a important tendency to combine the both sides to manage all product related data and provide data retrieval for product design and production. The efficiency and quality of design and manufacturing processes can be greatly improved by product information sharing and visualization in the system[5].

J. Zhang (Ed.): ICAIC 2011, Part V, CCIS 228, pp. 182–188, 2011.

There are a lot of data in the enterprise,such as personnel data, financial capital data, marketing data, technical data, all types of documents etc. and these large amounts of data has not been effectively addressed.we usually use the PDM to deal with these large amounts of data in order to effective using. But it exist the following restrictions in medium and small enterprises[6]:

- There are many confusion of drawing and data management during designment. the relationship between components is not easy, and it is difficult to update.
- It is largely change to implement a complete PDM for medium and small enterprises.
- It is not realistic to introduce a large-scale PDM because of limited fund.
- Even if the introduction of PDM, which uses only a small part of the function, the product does not play a major role .
- A large-scale implementation of PDM product cycle is long, small and medium manufacturing enterprises technical limitations make it difficult to bear fruit.
- At the same time, the domestic product of PDM is not mature enough.

So , It is necessary to develop own machine engineering data management system (MEDMS),which meets the actual needs of medium and small enterprises,based on the principle of distributed PDM and with open interfaces.

In this paper, we first analysis the principle and the limition of distributed PDM,then we design the machine engineering data management system (M-EDMS) based on the principle of distributed PDM, the whole system is composed of two modules::the architecture design of M-EDM system,which have three layers,the first layer is the M-EDM system support level, the second layer is the core function of M-EDM object layer, the third layer is the user layer;and the function design of M-EDM system,which have the rights and organizational management module, workflow / design process management module, Project management, product structure and configuration management module, electronic collaboration module, image processing module and so on.This proposed system designment scheme is well suited for various applications for medium and small enterprises and has good prospects.

2 Principles of Distributed PDM

There is no single definition for the product data management (PDM) system,, we might define it as follow: PDM is product which combine the scientific management framework and business realities to optimize the enterprises management based on IT technology. PDM system appeared to be a hot business concerns and applications, in particular the rapid development in recent years, mainly due to many large international companies are gradually accepted it as business process reengineering, concurrent engineering, and ISO9000 certification enabling technology. Currently the more mature products is client / server model,which is a object-oriented applications under the relational database management system platform[7].

With the market and international competition, he mode of enterprises operation have taken place great changes. the product data may be distributed in different regions and even different countries, each region is responsible for only a part of product data generation, maintenance and use;at same time, the region must also be mutual exchange the product data with other region ,the any region in the enterprise can get all the product data, the enterprise is develop to the distribution, groups and professional trends, and more and more closer with the outside world ,so the enterprise come forth remote collaborative desig, agile manufacturing and other new business organization and cooperation,which adopt dispersed network manufacturing, off-site product development and other new Web-based, distributed product development model.

Thus, in a distributed product development environment, PDM system should not only concern unified data management in the various departments, also consider regional requirements,any users in the enterprise can use the PDM system at anywhere to operate data within his purview,which do not know the login server and data stored in which region. including the PDM system remote configuration, data query, share, exchange, synthesis and consistency control mechanisms in order to effectively manage, publish and update product data, ensuring accuracy and consistency of product data, which is distributed PDM.

3 The Design of Machine Engineering Data Management System (M-EDMS) Based on Distributed PDM

A. The Architecture Design of M-EDM System

The M-EDM system based on distribute PDM has three layers as shown in figure 1:

- The first layer is the M-EDM system support level, including the heterogeneous hardware platform processing system, distributed database processing systems, network communication services system, and database management systems such as Oracle, Sybase, SQL Server, DB2, Informix and so on. This layer is responsible for the dynamic data convert into a few, or even hundreds of two-dimensional relational tables to achieve the object-oriented product management requirements.while this layer mainly provides database operations with the operating system, which included of adding, deleting, updating ,querying ans so on in the network conditions.

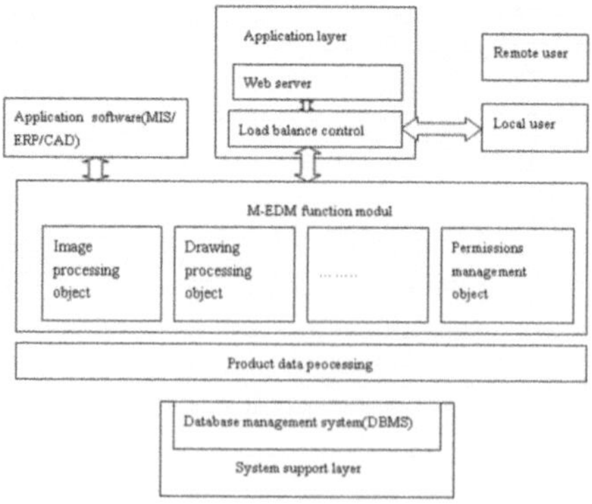

Fig. 1. The system architecture of the machine engineering data management (M-EDM)

- The second layer is the core function of M-EDM system, including the M-EDM application server (which encapsulates the business logic) and product data processing layer.According to the M-EDM system management objectives ,we establish the corresponding function modules and package the appropriate business logic code based on the object-oriented analysis and design, such as system management module objects (including systems management and work environment settings, etc.), the basic function of the module object (including the file management object, product configuration management object, project management object and workflow management object etc.), these modules can reside in the same application server,and also can be distributed in different application servers. The main tasks of product data processing layer are data transmission ,data convertion data access requests etc.it can achieve data processing among user layer,M-EDM function layer and database server,so as to form the unity of data which can be accept by M-EDM system.
- The third layer is the user layer, including the application of tools, remote access users, M-EDM local information management users. The first step is the basis for information integration,which packages the corresponding CAx systems, MIS systems, and MRP II system to achieve integrated management of the system; Second is the product information integration based on the designment and manufacturing process modeling; Meanwhile, in order to meet the needs of e-commerce, M-EDM system general use Web technology and build on the Internet / Intranet / Extranet,it can provide enterprise product development solutions,so the remote user can achieve remote off-site operations by access the web server,thus constitute the M-EDM system publishing platform.

B. The System Function of M-EDM

The main functions of M-EDM system based on distributed PDM are: project management, workflow and process management, product structure and configuration management, authority and organizational management, electronic collaboration, and view / mark for drawing ,classification / retrieval, drawing browse, and image processing functions. The functional modules shown in Figure 2.

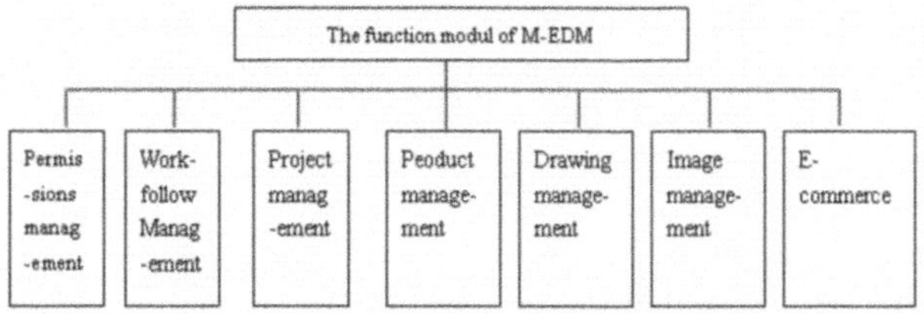

Fig. 2. The function modul of M-EDM

- Authority and organizational management module: Users can only login the system through inspection, unauthorized users can not enter the system, therefore, he can not get any information about the data of engineering drawings. The system users are divided into three levels, namely, the system administrator (super user), project leaders and ordinary users.

- Workflow and process management module: In the long-term production practice, each enterprises has developed its own product development processes, such as the development of a new products typically go through several stages as shown in Figure 3. Preliminary design and type in the product design process also includes a number of workflow. Product design generally includes six stages: design, proofing, standard inspection, audit, validation and approval as shown in Figure 4.Process design includes process planning, proofreading and approval stage.

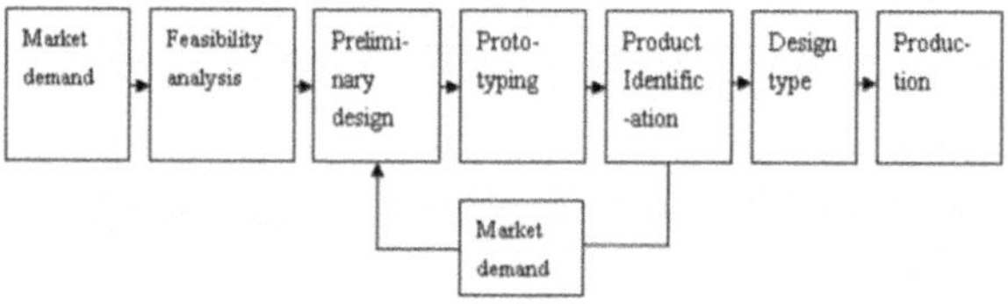

Fig. 3. The several stages of new product developement

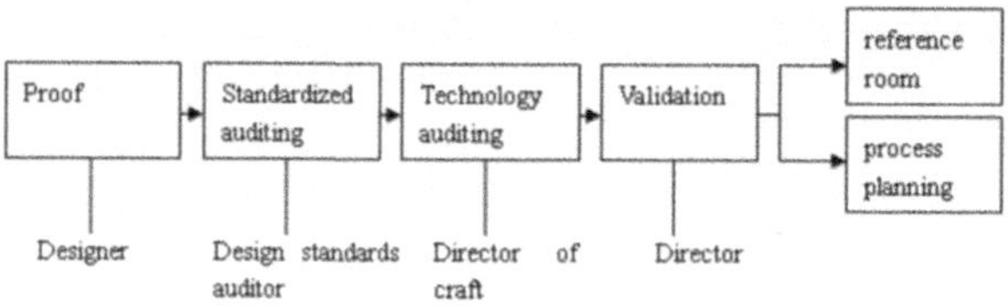

Fig. 4. The diagram of product design approval progress

- Project management: is the project planning, organization, personnel and related data management and configurations in the implementation process, to monitor the operational status of the project and complete the feedback. Project management which based on the workflow management provides a visualia tool,which control the development time and cost, coordinate development activities and ensure the operation of the project.Its main function is beforehand project the all aspects of product development, such as designated officers, provide submitted time of all documents,etc. At the same time, establish a product structure tree based on the project name,the content of the product structure tree must be writed and submitted by the designer and accord with time and other requests.
- Product structure and configuration management modules: In the M-EDM system,we can establisha product structure tree which describe the relationship of product composition,through compartmentalize the product into components and parts,then, components can be divided into biodegradable or non-decomposable, biodegradable components can be further broken down into sub-components and parts. Meanwhile,we should also consider the

concept of replacement parts and interchangeable parts which often appear in the the product structure relation.

- Digital images management: In order to meet the project needs, M-EDM system provides a storage which store the electronic image file. M-EDM system establish a distributed electronic warehouse (Vault) management mechanism through network file services which provided by the operating system to achieve effective management for distributed digital images,as shown in Figure 5. Digital images management includes: electronic image file storage, drawings networking archive, browse ,query, read and comment on scanned image, version management and so on.

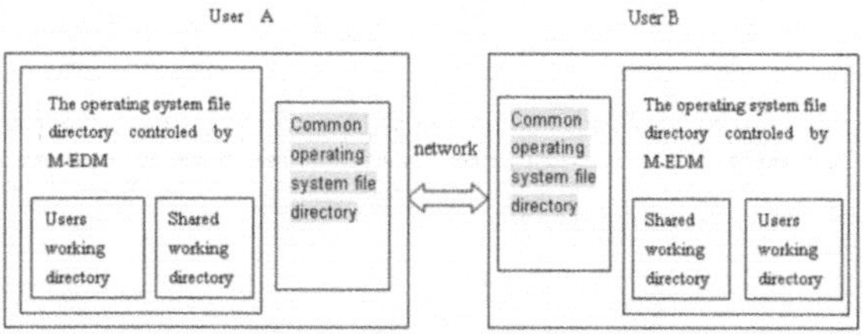

Fig. 5. The diagram of distributed e-warehouse management system

- E-collaboration module: The function of E-collaboration is provide the team working between the product developer in the distributed environment. such as e-mail, whiteboard, video conferencing, online chat and so on. Participants in the design process can be obtained tasks by e-mail and promptly notified when the associated data changes. whiteboard can allow design teams to discuss design issues.
- Image processing module: its function is convert the scanned image into digital image files and store in the system to organize and manage. Including the explain interface of engineering drawings, the interface of physical reconstruction and raster vector hybrid editing.

4 Conclusions

In this paper, we first analysis the principle and the limition of distributed PDM,then we design the machine engineering data management system (M-EDMS) based on the principle of distributed PDM, the whole system is composed of two modules::the architecture design of M-EDM system,which have three layers,the first layer is the M-EDM system support level, the second layer is the core function of M-EDM object layer, the third layer is the user layer;and the function design of M-EDM system,which have the rights and organizational management module, workflow / design process management module, Project management, product structure and configuration management module, electronic collaboration module, image processing module and so on.This proposed system designment scheme is well suited for various applications for medium and small enterprises and has good prospects.

Acknowledgment. We acknowledge the support of High-Tech Research plan of Jiangsu Province (No.BG2007045),we sincerely thank the anonymous reviewers for their constructive comments and suggestions.

References

1. Qiu, Z.M., Wong, Y.S.: Dynamic workflow change in PDM systems. Computers in Industry 58, 453–463 (2007)
2. Chen, Y.M.: Development of a computer aided concurrent net shape product and process development environment. Robotics & ComputerIntegrated Manufacturing 13(4), 337–360 (1997)
3. Czerwinski, A.S., Sivayoganathan, K.: Development of CIM applications fromPDES/STEP informationmodels. Concurrent Engineering: Research and Applications 2, 133–136 (1994)
4. Gum, P., Chan, K.: Productmodelling using STEP. Computer-Aided Design 27, 163–179 (1995)
5. Chan, E., Yu, K.M.: A concurrency control model for PDM systems. Computers in Industry 58, 823–831 (2007)
6. Li, Y.: PDM Technology Strategy Analysis and Practice. Computer Applications (8), 45–47 (2006)
7. Dong, J.: The Study of CAD / PDM Integration. Mechanical Science and Technology (3), 78–79 (2001)

Design Study on Human-Computer Interface in Kpi-System of Enterprises

Zhou Ning, Wang Jian-hai, and Wu Jia-xin

Research Center of Information Resources, Wuhan University, Wuhan 430072, China
zhouning1975@sina.cn

Abstract. This article discusses the methods by which KPI indicator system of enterprises is selected and the indicator database is established. Standardization of indicator description as well as automation of indicator data extracting serves as the basis for intellectualization of human-computer interface. By taking a provincial-level mobile communications company for example, this article focuses on establishment and application of KPI system as well as the design method of human-computer interface.

Keywords: KPI indicators system, Intelligent control, Human-computer interaction, Interface design.

1 Introduction

In the context of international financial crisis, competition among enterprises becomes increasingly intense. Establishment of KPI system serves as an important means to consolidate performance management and enhance competitiveness.

In the establishment of KPI system, the design of human-computer interface counts a critical link, with normalization, standardization and intellectualization of indicator data as its primary content. This article discusses the establishment of KPI system along with the normalization and automation of input interface. Visualization and intellectualization of output interface can be realized by means of data mining[1] and visualization technology[2].

2 Establishment of KPI Indicator System

A. KPI Management

A KPI (Key Performance Indicators) refers to an operable tactic indicator obtained from the layer decomposition of macro strategic objective decision-making, acting as a pointer for monitoring the effect of implementation of macro strategic decision[3]. Generally, KPIs are used to reflect the effect of strategy implementation.

KPIs are the key indicators for measuring the effect of implementation of enterprise strategy. They are used to set up a kind of mechanism to transform enterprise strategy to internal process and activities, so as to constantly enhance core enterprise competitiveness

J. Zhang (Ed.): ICAIC 2011, Part V, CCIS 228, pp. 189–195, 2011.

and continuously acquire high benefits, and to make the assessment system not only an incentive and restrictive means, but a strategy implementation tool.

KPIs are obtained by decomposing the overall strategic objective of an enterprise, reflecting the key driving factors (KDFs) that have the most effective influence on enterprise's value creation. The value of setting up KPIs lies in that, operators and managers may focus on operation that can give the greatest driving force to performance, and may diagnose in time problems in productive and operating activities and take measures to improve performance level.

KPIs conform to an important management principle-- 80/20 Principle[4]. The 80/20 rule exists in an enterprise's value creation process, that is, 20% of backbones create 80% of the enterprise's value; and this rule is also applicable to every employee, namely, 80% of task is finished by 20% of key behaviors. Therefore, we must analyze and measure the 20% key behaviors to grasp the center of performance assessment.

B. KPI System and KPI Warehouse

KPI system is a system formed by layered and classified indicators obtained from layer decomposition of strategic objective. It is a quantitative and behavioral standard system for appraising and managing personal performance. In the process of operation, along with the changes of market environment and internal conditions, operators and managers will set forth different strategic objectives for different periods, and managers will adjust their focuses accordingly. Employees must be guided to pay their attentions to current key points of operation by changing and adjusting performance indicators. The KPI systems centered on by an enterprise in different periods are called strategy-oriented KPI systems, and all the KPI systems make up a KPI warehouse. Enterprises must form dynamic and open KPI resource warehouses by constant perfection and accumulation. They may directly select KIPs from KPI warehouses according to adjusted strategies to carry out assessment and evaluation.

C. Selection of KPIs and Establishment of KPI System

The principle of determination of KPIs is SMART[5]. SMART principle is important for the determination of KPIs. It is an acronym of five words, i.e. Specific (accurately targeted, properly refined, and changeable according to situation); Measurable (quantitative, behavioral, data or information are accessible); Attainable (realizable in an appropriate time limit after efforts are made); Realistic (provable and observable); and T(time-bound, concerned with time unit and efficiency).

Methods for Selection and Determination of KPIs

In general, the following key points should be grasped to determine KPIs:

(1) Individual and departmental objectives should be connected with the overall strategic objective of the company, and the overall situation should be taken into account.

(2) Enterprises should pay attention to the hierarchy of objective while applying KPIs. A very vital task in performance management is to determine the goals for performance management in aspects of organization, departments, flows, and employees. And corresponding performance goals should also be set up from macro to micro and from whole to parts.

(3) The indicators should generally be stable, that is, if the operation flow remains the same, key indicators should not be largely changed.
(4) Key indicators should be simple, clear, and easy to implement, accept and understand.
(5) KPIs should be standardized and defined. A List of KPI Definition and Description should be set for every KPI (as shown in Table 1). Several lists of such kind of an enterprise constitute the concrete content of its KPI system.

An Example: Design and Establishment of KPI System

An enterprise's KPIs are not produced out of some managers' imagination, but the outcome of collective wisdom of specialists, managers and common employers, in which specialists are especially important. Usually, the KPIs of an enterprise compose of the following three levels: enterprise-level KPIs which are evolved from strategic objective; department-level KPIs which are determined according to enterprise-level KPIs and department responsibility; and performance measurement indicators, that is, department-level KIPs which are practically applied to specific positions. Generally, in the KIP system, it is very important to fix the first level because the follow-up KPIs will be determined accordingly. And unreasonable enterprise-level KPIs will result in bad operability of the latter two levels, and further affect the performance of the whole enterprise.

The design of KPI system is very complex, involving determination of work output, establishment of assessment indicators, laying down of standards for assessment, establishment of KPI system, and review of KPIs. In order to make the system fit for the company's actual situation, repeated verification and modification are needed (as shown in Fig. 1).

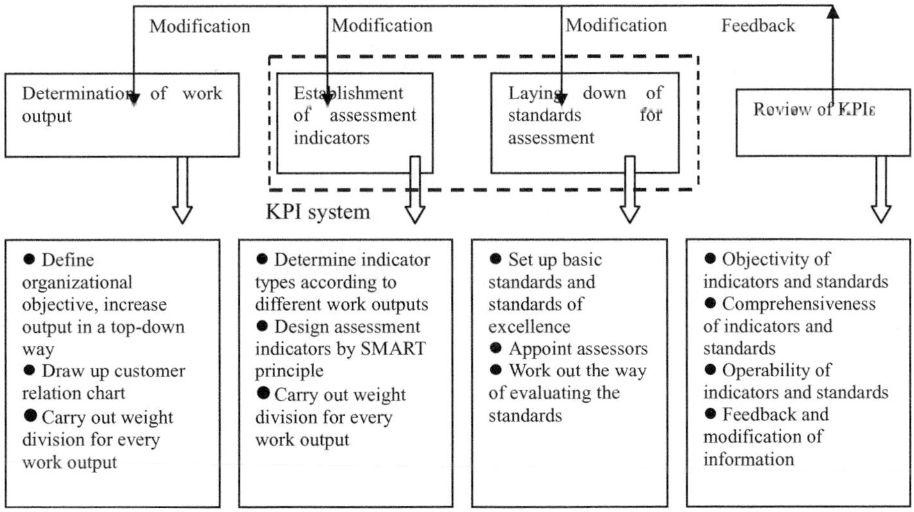

Fig. 1. KPI system design

To construct the KPI system, a certain indicator set and suitable system building approach must be adopted. For the construction of indicator system in the mobile communication company, we utilize indicators which are based on current KPIs and which are worthy to be assessed as well as classification and gradual-layering methods to establish the above system. It is built up on the basis of the following aspects:

(1) Current KPIs and those worthy to be assessed
Since indicators having managerial significance and adhering to the overall corporate strategy should be included in the KPI system, KPIs are the core of the system; Every year, the mobile company carries out performance assessment in its branches by some assessment indicators, in which some overlap with KPIs and others are worthy of being assessed. All of them should be involved into the system.

(2) Establishing KPI system by classification and gradual-layering methods
As the indicators are divided by management level into strategy layer, department layer and position layer, which are laterally classified and longitudinally subdivided, naturally, the KPI system should be established by classification and gradual-layering methods.

We classify indicators by function domain of the company. The grades in classification are dynamic, that is, they are not constant. Indicator grades can be increased or reduced according to the longitudinal depth of indicator management conditional on complete involvement of all the relevant indicators and close connection between every indicator with its father class.

The first grade indicators are divided into seven classes by function domain analysis: finance, network quality, market operation, group customers, capital expenditures, business support, and safety production. Each of them has corresponding sub-indicators.

In consideration of the following aspects, indicators are divided without defining the grades: different department/position functions require different complexity of work, leading to different indicator quantity and complexity, and therefore, more complex function domain calls for more indicator grades, while function domain with fewer indicators demands simpler indicator grades; just like contents of books, chapters that are more inclusive need to be divided into section, subsection, etc., while chapters with less content only need to be divided into sections.. In order to effectively organize indicators and exposit the whole system, we need to conduct indicator encoding. The encoding method should include all the indicators and embody the principle of lateral classification and longitudinal subdivision of indicators at the same time. Based on analysis and weighing, we adopt the method of "class code" plus "indicator number"[6].

The depth of indicator system is grade three. Codes of first grade indicators are two capitals, that is, abbreviation of corresponding Chinese Pinyin. For example, Market Operation class is represented by SC (Shichang); Network Quality class, WL (Wangluo); and Finance class, CW (Caiwu). Classes of second grade and below are expressed by two-digit orderly numbers (range: 01-99).

The whole indicator system is of "tree structure", in which leaf nodes are concrete indicators, and non-leaf nodes may be a kind of subclass nodes or concrete indicators. And there are calculation formula and statistical method for such indicators; they are subclass nodes, too. For example, Node WL03 in the above figure is a concrete indicator, and it also contains several sub-indicators: WL0301, WL0302, WL0303, and WL0304. Therefore, while encoding concrete indicators, we need to cover all the leaf nodes and non-leaf nodes which are concrete indicators. For example, in CW indicators, net income ratio and net profit rate are CW01 and CW0101, respectively. And a standard description sheet is made for every indicator[7] (as shown in Table 1).

Table 1. A case of KPI description

Code	CW0101	Indicator name	Net profit rate
Indicator definition	This indicator mainly assesses the situation of completion of budget of net profit rate of branch office. Assessment and scoring is carried out by objective method.		
Data source	Finance Dept. + Financial statements		
Statistical caliber	Net profit is calculated according to the caliber of financial statement, and operation revenue is calculated in the light of the caliber of finance department.		
Calculation formula	Net profit rate=net profit/operation revenue *100%	Unit of measurement	%
Method of collection	Semi-automatic collection; according to the net profit and operation revenue of monthly financial statement, the indicator is calculated by the formula.		
Relevant department	Finance Dept.	Supported by	Each branch office
Name of indicator	Net profit rate		
Data domain	Jan. 2008 to now		
Statistical frequency	At the end of each month		
Dimensionality	Granularity		
Date	Month/ Year		
Region	Each branch office		
Remarks		Filled out by	xxx

Indicators are divided into seven classes by classification and encoding standards, totally including 107 indicators: market operation (37), network quality (32), safety production (17), financial performance (7), group customer (6), capital expenditures (5), and information and business support (3).

3 Design of Human-Computer Interface

Human-computer interaction design is embodied in the design and realization of system input/output interface. The following passage proceeds in terms of input and output.

A. Standardization of Indicator Data and Automation of Data Acquisition

Automation of data acquisition is based on standardization of indicator data description and storage. In the past, departments within a company form a system that is lacking in coordination. It is not yet scientific and standardized due to lack of unified standards as well as difference in scope of statistics and computing methods.

After 107 description sheets (e.g. Table 1) are made, description codes are produced for every indicator. The indicator data in the indicator survey sheets is first hand information of the company. Then, we will analyze these data and dig out their intrinsic characteristics from different angles.

First, conduct data modeling to produce database of indicators, that is, indicator data sheets that can be stored in database; then, convert the data in indicator survey sheets into indicator database by indicator data extraction program.

Thanks to the establishment of indicator database, indicator values can be directly obtained through the existing system. Approach: open the indicator database;

Then, read the data and select the required indicator values in accordance with database schema; in this way, automatic operation of input interface is achieved.

B. Visualization of Output Interface

On the basis of that, create query views (incl. basic query views and complex statistical views) by SOL. We can conveniently acquire the distribution characteristics of indicators on time, class, and collection method with these views; finally, we convert these characteristics into EXCEL and PAJEK files to conduct data analysis.

EXCEL, a popular software tool developed by Microsoft Corporation, is characteristic of providing result data in the form of histogram, pie and line graph etc. In this way, cut-and-dried data is visualized, which wins popularity among users.

PAJEK[8] is a visualization tool capable of alluding ASCII data into two-dimension color images. This suffices to reveal the data-to-data relation, and output is carried out through the visual result. People can analyze and master its regularity to carry out intelligent control by means of visual result.

4 Conclusions

The establishment and application of KPI system, in spite of spending much manpower and material resources, is an effective measure to enhance competitiveness. First, different departments within enterprises need to formulate various KPI indicatores which not only conform to international standards but also adapt to actual conditions of their own units; the indicator system is then established upon verification by experts, managers and executors with the overwhelming majority of stuff. The mode is defined by relational database technology and the indicator database is created.

For the purpose of successfully establishing and applying indicator database, the design of human-computer interface is the main content, with achieving standardization and automation of input interface as the primary goal plus intellectualization and visualization of output interface as the development orientation.

Acknowledgement. This research was supported by the key project of MOE under Grant 08JJD870225.

References

1. Hand, D., Mannila, H., Smyth, P.: Principles of Data Mining. MIT Press, Cambridge (2001)
2. Chen, C., et al.: Visual analysis of conflicting opinions. In: IEEE VAST 2006, Baltimore MD, USA (2006)
3. Wang, Y., Wei, l.: KPI-Oriented Forecasting model design for Value-added Sercice Operation Quality Prediction. Telecom Engineering Technics and Standardization (2), 87–91 (2009)
4. Jin, R.-l.: Amalgamate 2/8 Principles and Related Regulation. Computer Era (12), 51–52 (2006)
5. Li, Z.-q.: SMART Principle and Its Process Management. Market Modernization (19), 148–149 (2007)
6. Subject Research on Management Information Theme Data Mining of 2008 (08MITDM). Final Project Research Report, November 5 (2008)
7. (08MITDM) China Mobile Enterprise KPI Warehouse, November 5 (2008)
8. Networks/Pajek: Program for Large Network Analysis, `http://vlado.fmf.uni-lj.si/pub/networks/pajek/`

Research on Ductile Mode in High Speed Point Grinding of Brittle Materials*

Yueming Liu, Yadong Gong, Jun Cheng, and Wei Li

School of Mechanical Engineering & Automation
Northeastern University
Shenyang Liaoning, 110819, China
YuemingLiutom@tom.com

Abstract. Through analyze the fracture theory in grinding brittle materials, the critical force and cutting depth are brought in. Consider the affection for cutting thickness from the swivel angle α between the wheel axis and the wheel axis by geometry analysis. The reasonable finite element model has been built to explore the critical conditions from brittle mode to ductile mode according indentation experiment. The conclusions were drawn at last to provide reference basis for grinding experiment of brittle materials.

Keywords: Ductile mode, Critical conditions, Cutting thickness, Simulation, Swivel angle.

1 Introduction

Application of brittle materials in high performance structural applications continues to be elusive despite concerted global research efforts in the last two decades. The economic feasibility of high performance ceramics depends on how efficiently they can be machined by grinding. To get better performance of brittle materials components, the technology of grinding has been applied widely to be used in the manufacturing for them. In the process of grinding brittle materials, the surface quality is affected by the way of removal materials. Unfortunately, the machined components are most likely to contain a deformed layer, surface cracks, residual stresses and other damages, and the brittle materials usually have hard character, which makes that grinding can comprise up to 80% of the total cost. The latest grinding research shows that although the brittle materials still could be machined in ductile mode while adopting appropriate machining parameters, which can improved the performance of brittle materials component increasingly [1-3].

High speed grinding has been used into machining the brittle components, especially for the cylindrical component. The novel cylindrical grindin*g technology is achieved through change the angle between the wheel axis and the workpiece axis, the swivel angle is named α, so is high speed point grinding which make quick-point grinding wide to research the critical conditions in brittle-ductile mode transition. This paper will show

* This work is supported by NSF Grant #50775032 and BSRCU:N100603006.

J. Zhang (Ed.): ICAIC 2011, Part V, CCIS 228, pp. 196–204, 2011.

the basic principle for machining brittle materials in theory, through adding the swivel angle of α to grinding geometry model and carry out the simulation for indentation between the single abrasive and the workpoiece, the new model of cutting depth can be introduced to research the affection for ductile mode[4-5].

2 Theoretical Analysis of Factors for Affecting Critical Conditions

A. Indentation Fracture for Brittle Materials

The traditional method to research breakage of brittle materials is to adopt indentation experiment through changing the normal force to show the mode through measuring the hardness of the brittle materials. Reference [1] introduced the concept of critical normal loading P_c. While the normal load P less than P_c, the deformation of brittle materials belongs to plastic deformation. However, the value of normal force is hard to control, and the shape of the diamond indenter can not stand for grinding process completely. So the indentation experiment could show the facture principle of brittle materials, but not be suit to research the critical conditions into ductile grinding.

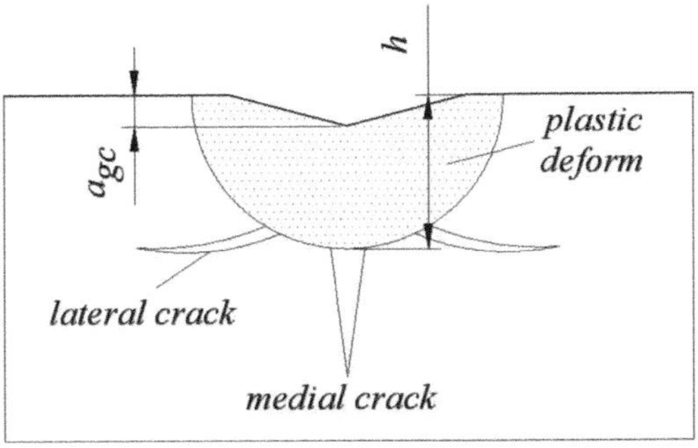

Fig. 1. Indentation phenomenon

The indentation fracture mechanism is show as Fig.1. The shape of the indenter was reflected on the surface of the workpiece, which can be used to research the main factors to affect the plastic deformation and fracture. The deformation zone has been depicted in different fill in Fig.1, and the depth of plastic deformation and the fracture crack are marked in it. There are cracks distributed on the round and bottom of the plastic deformation, which are named lateral crack and medial crack according to their position to research facility. Reference [3] explained the produced phase of medial crack and lateral crack, the former emerges in the loading process while entering the brittle mode from ductile mode, and the later is in the phase of unloading. While assuming the normal load P was implemented on the surface of brittle materials, so the relationship between P and the indenter size could be expressed in Equation 1.

$$P = kHa^2 \tag{1}$$

Where k is the geometry factor of the indenter, H is the hardness of the brittle material, P is the normal load. As P is increasing, the plastic zone will be enlarged until the medial crack emerges. The critical normal load P_c will appear while the micro-crack is intended to be just produced, and P_c could be expressed in Equation 2.

$$P_c = \lambda_0 K_{1c} (\frac{K_{1c}}{H}) \tag{2}$$

where K_{1c} is the fracture toughness of the material, λ_0 is the integrative affected factor. And It could be concluded whether the grinding of brittle materials in ductile mode or not depends on the value of normal load or other related parameters. The basic theory for fracture of brittle materials will provide support for simulation or experiment analysis of critical transition of grinding mode.

B. Analysis of Cutting Depth in High Speed Point Grinding

In high speed point grinding, the diameter of the wheel is much larger than the cylindrical workpiece in normal, and it was usually used in machine all surfaces of cylindrical component by once holding, which can achieve the high grinding productivity and avoid the affection for grinding precision from repeated fixing. So the grinding depth a_p is so important to affect the machining productivity. For researching the affection for critical transition condition of grinding mode from grinding depth, the followings conditions will be assumed to discuss the geometry model of grinding.(i) The abrasives are distributed equidistantly on the round of the wheel; (ii) The cutting edge height of single abrasive is equal; (iii) The contact arc will be seen as a straight line while the wheel diameter is so large. To compare the difference of cutting depth by single abrasive, reference [5] has got the equivalent diameter d_e to facilitate further research while α or not. According to the geometry model got former, the equivalent diameter d_e of the wheel in point grinding can been expressed as Equation 3.

$$d_e = \frac{d_s d_w}{d_w + d_s \cos^2 \alpha} \tag{3}$$

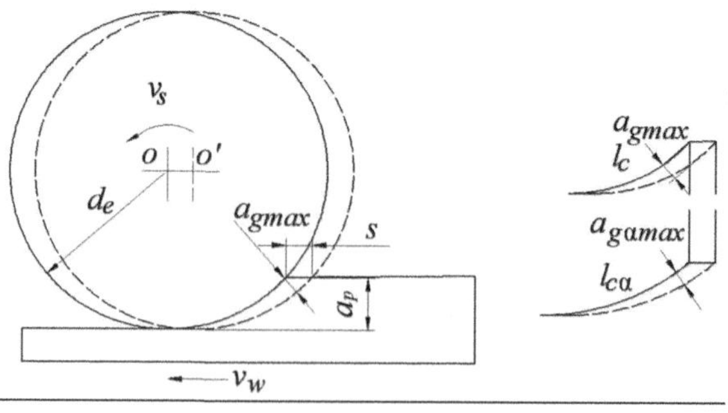

Fig. 2. Diagram of cutting thickness

Where d_s is the diameter of the wheel, d_w is the diameter of the workpiece, α is the swivel angle between the wheel axis and the workpiece axis. While α exists, d_e will be amplified in theory while other parameters wouldn't be altered. Fig.2 shows the diagram of grinding process with the maximum cutting thickness and the affection for it from α, and the maximum cutting thickness a_{gmax} is expressed as Equation 4.

$$a_{g\max} = \left[\frac{4v_w \cos\alpha}{v_s N_d C} \sqrt{\frac{a_p}{d_e}} \right]^{1/2} \quad (4)$$

Where v_w is the speed of the workpiece, v_s is the tangent speed of the wheel, N_d is the number of the dynamic effective cutting edge, C is the coefficient related to the edge density, a_p is the grinding depth. While the other grinding parameters is same except α, a_{gmax} is lager than a_{gamax}, that is the maximum cutting thickness becomes thinner in the affection of α. It can decrease bearing force by single abrasive and reduce the micro crack in surface or subsurface. So the critical cutting depth can be seen as one of the critical factors.

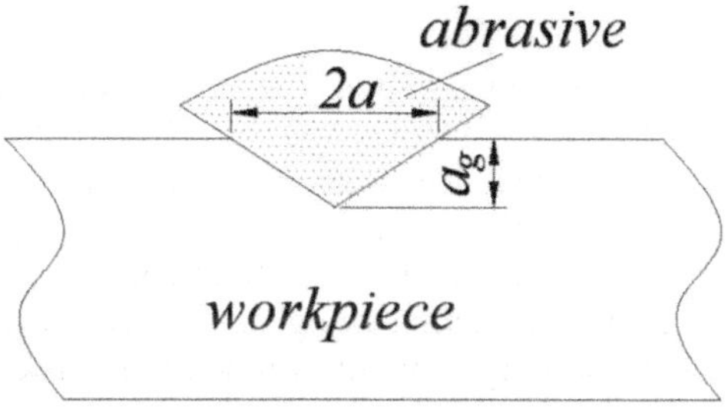

Fig. 3. Actual painting of cutting by single abrasive

$$a_{gc} = \cot(\frac{\alpha_0}{2})\sqrt{\frac{2\lambda 0}{a}} \left(\frac{K_{1d}}{H} \right)^2 \quad (5)$$

The critical cutting depth a_{gc} in ductile grinding is expressed in Equation 4. Where α_0 is the top angle value of the abrasive, a is the size coefficient of the abrasive, K_{1d} is the dynamic fracture toughness of the material.

While only grinding brittle materials belongs to ductile mode, the surface performance will be improved, that is say the maximum cutting thickness must be smaller than the critical cutting thickness a_{gc}. It was concluded that the brittle materials is removed in the ductile mode while $a_{gmax}<a_{gc}$. while the grinding depth a_p is ensured, $a_{gmax}<a_{gamax}$ can be concluded through above analysis. So when the critical cutting depth is ensured according to the special brittle material, lager grinding depth a_p can be achieved in ductile mode but not affect the surface performance in high speed point grinding, which can improve the grinding productivity for brittle materials.

3 Simulation Analysis and Resualt

A. FEM Model of Indentation

Finite Element Method has been used to predict the temperature, the residual stress, the grinding force and the grinding energy distribution induced in the workpiece by the entire wheel during the grinding process[6]. The approach includes many types of grinding such as conventional, creep-feed, cylindrical grinding and high efficiency deep grinding and so on, especially for the value of hard to measure. This paper builds the FEM model to explore the critical conditions in grinding from brittle mode to ductile mode.

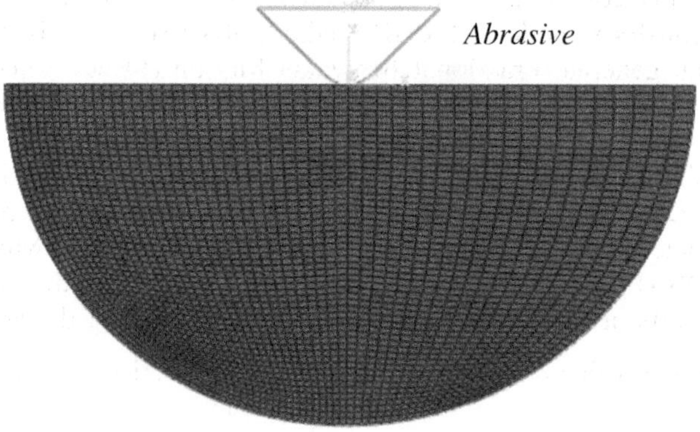

Fig. 4. FEM model of indentation by single abrasive

The built model is shown as Fig.4, which includes the abrasive and the workpiece of brittle materials. The single abrasive was seen as analytic rigid, which couldn't deform while bearing pressure. In real grinding, the diamond will adopted to grinding brittle materials. The property of brittle material used in simulation is shown in Table 1. The fixed velocity along the reversed direction of Y axis is applied on the abrasive. The dynamic explicit analysis was set and the running time would last for 0.3*s* to ensure the enough time until the deformation from initial elastic to plastic, and to the fracture at last. In the process, the semi-circular part of the workpiece was fixed to make the boundary located in the internal part of the workpiece.

Table 1. Material Property

$P(T/mm^3)$	$E(MPa)$	v	$Y(Mpa)$	$E_p(Mpa)$	$\sigma_f(Mpa)$
2.4×10^{-9}	290000	0.18	4500	3.4	0.45

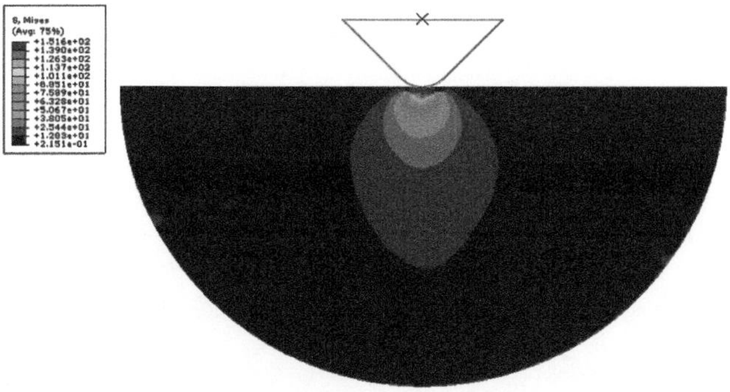

(a)Plastic deformation

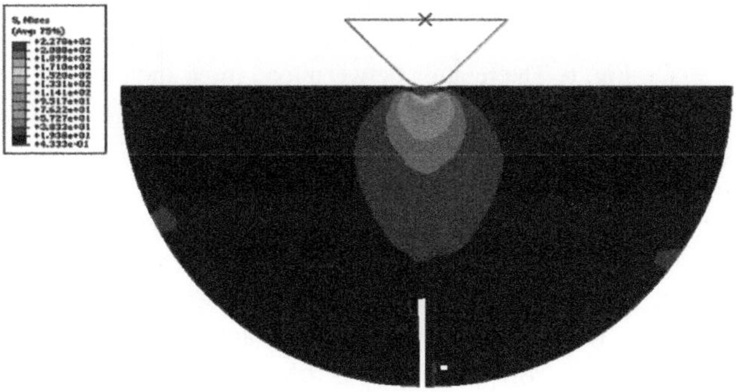

(b)Fracture crack

Fig. 5. Two results in different normal forces

B. The Simulation Result and Discussion

Fig.5 shows two different results while enforce different normal loads, (a) expresses that plastic deformation occurs on the contact area between the abrasive and the brittle workpiece with the normal load. It presents that the load P in (a) is less than the critical load P_c and no fracture happens. On the other hand in (b), the medial crack emerges in the bottom of plastic zone while $P > P_c$. However, lateral crack growth occurs during the unloading phase in above analysis, so but in this simulation, the unloading wasn't added in it to reduce computational time .But it will not affect the conclusion because the purpose of the simulation is to find the critical load or other critical conditions.

To get the critical load in the simulation, the curves presented more information should be extracted in the simulation.

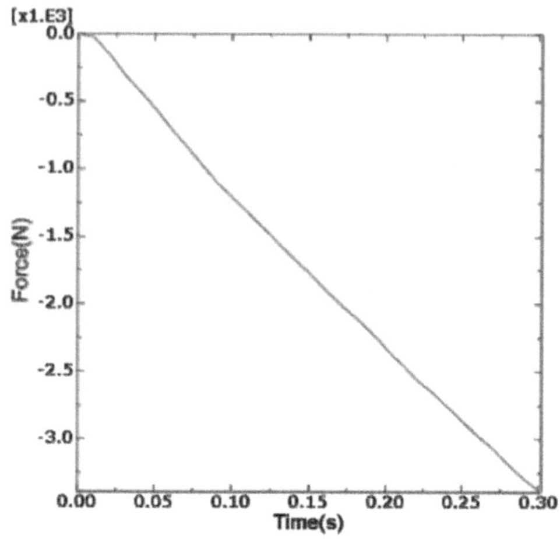

Fig. 6. The relation between load force and time

C. The Critical Condition of Ductile Mode

Fig.6 shows the force extracted from the reference point fixed on the abrasive, which reveals that the against force from the brittle materials is increasing with the simulation running, there is not directly relation between the normal force and the grinding mode either brittle or ductile. But in actual, as the load increases, the grinding mode will transit from ductile to brittle. Because the affection for the force from friction wasn't considered in it, the obvious change hasn't been seen in the curve.

Fig.7 shows the relation between the total energy and the running time, the total energy includes the abrasive and the workpiece. In FEM model, the initial total energy is set to zero, while the crack emerges in the surface, the energy will lose accompanying with the flying cutting scraps. At the time runs on about 0.15s, the energy decreases obviously, which expresses the initial last plastic deformation has changed, that expresses the facture crack occurs in the surface or subsurface. The crack may is very small or in the subsurface, but the phase isn't belongs to ductile mode.

To find the simple parameter to control grinding in ductile mode, the cutting thickness a_g is feasible and controllable. While $a_{gmax}<a_{gc}$, it belongs to ductile mode. In high speed point grinding, the larger grinding depth a_p can be adopted because $a_{gmax.}> a_{gamax}$. Through indentation simulation, the critical grinding depth can be ensured to guide the experiment for machining brittle materials. The method is positively useful to reduce the cost for finding the ductile mode through a lot of experiments, especially for so many kinds of brittle materials. Although it has difference with the real critical value, the method could be adopted to lessen the rang to get the real critical vale through other methods.

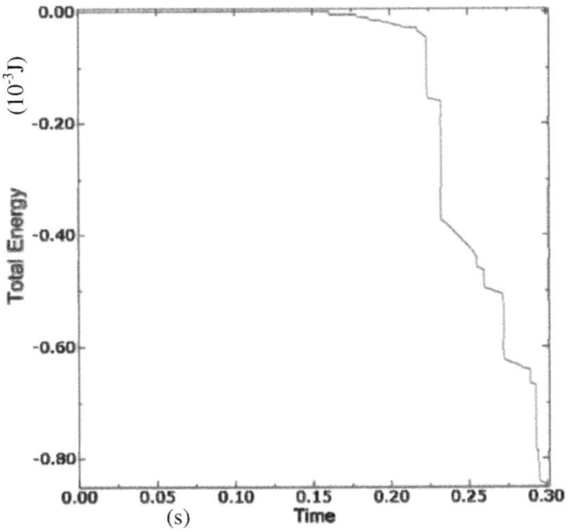

Fig. 7. The relation between total energy and time

4 Conclusions

Through above theoretical analysis and FEM simulation, the conclusions could be drawn as followings:

i. The indentation experiment has been used widely to research the critical conditions from brittle mode to ductile mode while machining brittle materials. The critical condition can be expressed by normal loading, cutting thickness and grinding depth.

ii. In high speed point grinding, the cutting thickness becomes thinner with α is brought in the geometry model at the same other parameters. The larger grinding depth could be set to increase the productivity while grinding cylindrical component of brittle materials.

iii. Through the FEM simulation for indentation by single abrasive, the plastic deformation and facture crack could be shown distinctly, and the total energy will decrease abruptly while the crack emerges in workpiece surface or subsurface. The method can provides theoretical reference for grinding brittle materials in actual.

Acknowledgment. The authors would like to thank the support of No. 50775032 by National Natural Science Foundation of China and the support of N100603006 by Basic Scientific Research of Centre University.

References

1. Chen, M., Zhao, Q., Dong, S., Li, D.: The critical conditions of brittle–ductile transition and the factors influencing the surface quality of brittle materials in ultra-precision grinding. Journal of Materials Processing Technology 168, 75–82 (2005)

2. Malkin, S., Huang, T.W.: Grinding mechanism for ceramics. Annals of the CIRP 45, 569–580 (1996)
3. Sheng, X.M.: Technology of Super-high Speed Grinding, pp. 153–194. Mechanical Industry Press, Beijing (2010)
4. Li, K., Liao, T.W.: Surface/subsurface damage and the fracture strength of ground ceramics. Journal of Materials Processing Technology 57, 207–220 (1996)
5. Xiu, S., Cai, G.: Edge grinding layer model and parameters in quick-point grinding. Chinese Journal of Mechanical Engineering 42, 197–201 (2006)
6. Doman, D.A., Warkentin, A., Bauer, R.: Finite element modeling approaches in grinding. International Journal of Machine Tools & Manufacture 49, 109–116 (2009)
7. Zhang, W., Subhas, G.: An elastic-plastic-cracking model for finite element analysis of indentation crack in brittle materials. International Journal of Solids and Structures 38, 5893–5913 (2001)
8. Cao, J., Shi, Y.: The finite element analysis of example by ABAQUS, pp. 162–174. Mechanical Industry Press, Beijing (2006)
9. Information, `http://forum.simwe.com/`

Research on the Question Answering System in Chinese Based on Knowledge Represent of Conceptual Graphs*

Peiqi Liu[1], Longji Li[1], and Zengzhi Li[2]

[1] School of Information and Control Engineering Xi'an University of Architecture and Technology Xi'an 710055, China
[2] School of Electronics and Information Engineering Xi'an Jiaotong University Xi'an 710049, China
PeiqiLiu2365@yeah.net

Abstract. The core of question answering is a searching based on natural language understanding. Because of the flexibility and complexity of Chinese, the study of the Chinese question answering has become very difficulty. The current Chinese question answering relies on keywords to query and retrieve the answer. The nature of these systems is the keyword matching. It is a kind of simple natural language understanding. Its answering accuracy is low. In this paper, the research methods of the question answering in Chinese syntax and semantic analysis, which based on the knowledge representation of the conceptual graphs, is proposed. According to the methods, the question answering is formal defined, and the system structure is designed. The conceptual graphs of the interrogative sentences in Chinese are also constructed by common statement in Chinese interrogative sentences in this paper. After test, the conceptual graphs of the interrogative sentences reflect the basic situation of existing questions in Chinese, and the solution is feasible. The research methods are valuable for the question answering and natural language understanding.

Keywords: Conceptual graphs, Knowledge representation, Question answering, Natural language understanding.

1 Introduction

The automatic question and answering system (QA for short) is a very stimulate field in the natural language understanding (NLU for short). Its research includes comprehending the question asked by the customer, automatic searching the large-scale information resources, and returning an accurate answer which can satisfy customer demand. The main technique used by QA is NLU. The research about this field is long before abroad, and has already obtained many pleased results. There are some important examples, e.g., Start system which is developed by Massachusetts institute of technology based on the web, and AnswerBus system which is developed by university of Michigan can be used by many languages. In our country, this

* This work is supported by National Natural Science Foundation of China (No.:60673170) and National Technology Support Project of China (No.:2008BAH37B05060).

J. Zhang (Ed.): ICAIC 2011, Part V, CCIS 228, pp. 205–214, 2011.

research is late. Because of the complexity of Chinese syntax, semantics and representation, the successful system about QA is less. In recent years, the well-known system is HKI, which is a QA based on the knowledge developed by Chinese academy of science [1].

In the current Chinese QA, the import of the system is natural language sentences. The QA must split the sentence as keywords, tags words and computes words frequency from the natural language sentences. Actually, it is still belong to keyword match. For example, the sentence "中国首都是哪里(Where is the capital city of China) ?" can be split as keyword sequence "<中国>+<首都>"[2], and expanded as "<中华人民共和国><中国><祖国>+<行政中心><首都><国都>", etc. Obviously, this method is a kind of keyword match, which is the more simple comprehension in NLU. Its effect is also worthy to improve further. In this article, QA based on the knowledge representation of the conceptual graph (QACG for short) is developed by analyzing and researching the existing QA. The system can well comprehend semantics of Chinese sentences and return an accurate answer.

2 Basic Concepts

The main difference of QACG with other QA is that QACG based on the conceptual graphs and comprehend Chinese sentences really. In order to analyze and design QACG, the conceptual graphs must be introduced firstly.

A. Conceptual Graphs

The conceptual graph (CG for short) is a kind of the knowledge representation which gathers linguistics, psychology and philosophy. The foundation of CG is semantic network [3]. Its main advantage is that the language knowledge and deep case relation in sentences can be represented. B.J.Granet etc. has already proved that CG is a kind of excellent knowledge representation which has been widespread accepted by AI scholar.

Definition 1. CG is a knowledge network consisted of concept nodes and the relation nodes. It is a finite, connected and bipartite graph.

$$CG=(C, R, F)$$

Where, C is a set of the concept nodes, R is a set of the relation nodes, $F \subseteq (C \times R) \cup (R \times C)$ is a set of relations between concept nodes and relation nodes.

In CG, the conceptual node can represent any concrete concept, abstract conception, real object and attribute. It is drawn as a box with conceptual type label and conceptual reference which can be individual or a set. The relation node can represent deep layer relation during the concepts, which is drawn as a circle with relational type label. An arc with an arrow that links a box and a circle shows the effect relation during nodes. For example, the sentence "A cat eats meat with paw" can be shown in figure 1.

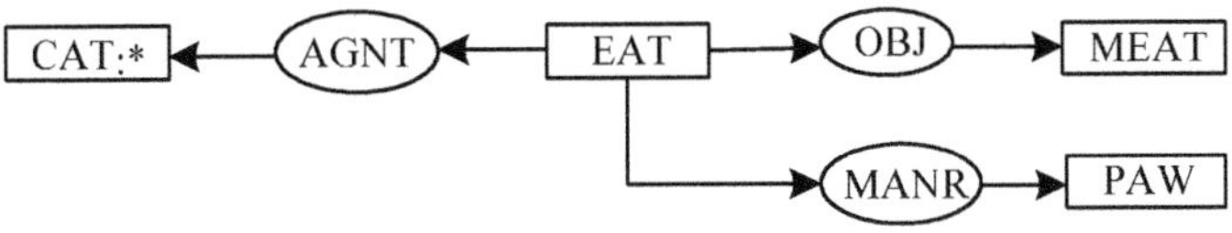

Fig. 1. Conceptual graphs representation

In the figure 1, the relation AGNT is the issuer of EAT; the relation OBJ is the object of EAT. The relation MANR is a method of EAT. The conceptual CAT: * shows that it could be any cat. The figure 1 can't be typed easily at a computer. In order to process easily, the linear form of the CG is used in the computer. In the linear form of the CG, a box bracket is replaced by a square and a circle is replaced by a bracket. In the linear form, the figure 1 can be shown as

[EAT]—(AGNT)→[CAT: *]
(OBJ)→[MEAT]
(MANR)→[PAW]

Obviously, the CG can represent not only the case grammar relation AGNT and OBJ, but also the deep relation MANR. The CG is an effective method to represent natural language.

In the research about QA in this article, the CG is the basic knowledge representation.

B. Formal Definition of QACG

The QACG can generate an accurate answer according to natural language claim by querying database and network. For detailing the structure of QACG, some basic concepts are showed in this section.

Definition 2. A Chinese sentence S is

$$S=q_0q_1q_2\ldots q_n$$

Where, $q_i(i=0,1, \ldots, n)$ is a word in S. For arbitrarily word $q_i(i=0,1, \ldots, n-1)$, the q_i which located in q_{i+1} left is noted as $q_i<q_{i+1}$.

Obviously, Chinese sentence is a kind of order relation of Chinese words. Different word orders represent different sentences. In these words, some are keywords, and others are modifier words. For example, in the sentence of "中国的首都是哪里?", the keyword sequence is "<中国>+<首都>". In QA, the keywords' sequence is very important. According to definition 1, a sentence can be defined as:

Definition 3. After pretreatment ψ, a sentence S is

$$\psi(S)=k_0k_1k_2\ldots k_m$$

Where, $k_i(i=0,1, \ldots, m)$ is a keyword drawing from the S, and $m\leq n$. If the sentence S_1 and S_2 have same keywords sequence after pretreatment, S_1 and S_2 are equivalence and note as $S_1 \triangleq S_2$. For example, "中国的首都是哪里?" and "哪里是中国的首都?" are equivalence sentences.

Obviously, the relation $\triangleq$ is an equivalent relation. By the relation $\triangleq$, the classification of answer sentences can be acquired. The aim of QA is to acquire an answer set through searching the resources and partitioning answer sentences. The perfectly QA can be formal defined as:

Definition 4. QA is a 4-tuples:

$$QA=(S_0,C,F,A)$$

Where, S_0 is a natural language sentence, $\psi(S_0)=k_0k_1k_2...k_m$, $K=\{k_i|i=0, 1,..., m\}$ is keyword, Context C is a resources set drawing the problem, $F\subseteq K\times C$ is the relation between K and C, $A\subseteq F$ is the answering set.

According to QA in definition 4, QACG can be formally defined as:

Definition 5. QACG is also a 4-tuples:

$$QACG= (G_0, C, P, G_A)$$

Where, G_0 is a CG of the S_0, $C=\{G_1, G_2,..., G_k\}$ is resources set. The operation $P=\{\Phi,\Pi\}$,Φ is the biggest conjunction, $G =\Phi(G_{i1}, G_{i2}, ..., G_{ir})$, $\Pi:G_0\rightarrow G$ is G_0 projection on G. The $GA=\Pi\ G_0$ is an answer of QACG about sentence S_A.

Above definitions, the essential difference between QA and QACG is that the answer of QACG is accurate and simple.

3 The Structure of QACG

According to QA structure [2,4], QACG structure is designed in this section. QACG includes the question analysis module, information search module and answering extraction module.

A. The Question Analysis Module

The question analysis module is showed in figure 2. In the figure 2, the preprocessor splits a question sentence into words, and processes synonym. Then the lexical analysis identifies the words and processes vocabulary. If the word is not included in dictionary, the system reports the wrong information. Moreover, the lexical analysis will ignore unimportant words to increase flexible of the system. The parsing program analyzes the structure of sentence based on the vocabulary, dictionary and phrase structure grammar. If structure of sentence is correct, the system will output the parsing tree. The CG maker converts the parsing tree into CG and the match/reasoning infer regular graph which is a CG with positive meaning. The regular graph is a basis of information index and formation answer.

B. The Information Search Module

The information search module (see figure 3) mainly includes the search agent, CG maker and match/reasoning. The search agent starts the searches engine according to the regular graph from the question analysis module and searches the answers in the network, database and knowledge base. The searched results are regular graph sets which are called search set, and the answering extraction module draws the answering by search set.

C. The Answering Extraction Module

This module includes CG clustering, background CG (BCG for short) generating, project matching and natural language builder. First, the answering extraction module computes correlation measure of the regular graph in searching set. According to the correlation measure, CG is clustered and a set of CG is formed in this module. Through the maximum conjunction match in CG which has maximum correlation measure, the BCG is found. In conclusion, the answering CG is formed by the working CG project on the BCG. By means of augmenting phrase structure grammar (APSG for short) database, the language processor converts the answering CG into a natural language. For making the language naturally, the optimizing program turns the natural language more excellently. The answer extraction module is showed in figure 4.

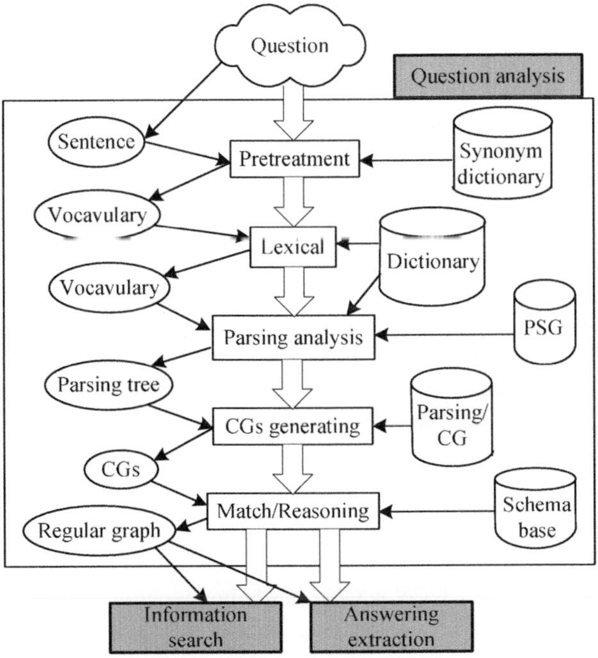

Fig. 2. The question analysis module

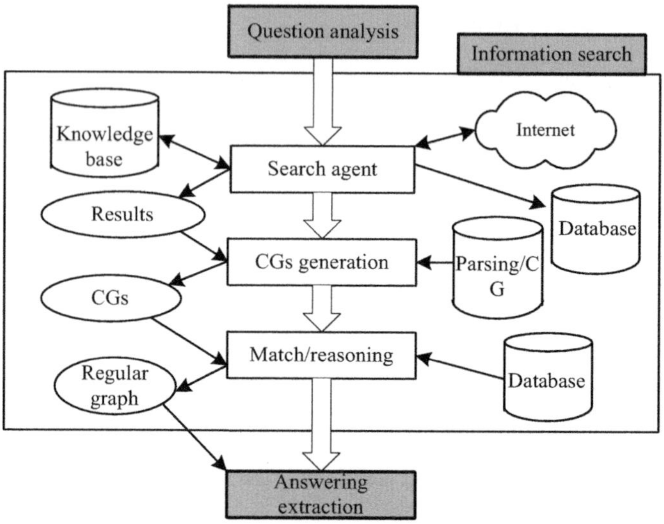

Fig. 3. The information search module

4 The CG of the Question Sentence

The expression of Chinese interrogative sentence is more agility. It can be divided into direct and indirect speech act in the speech act theory [5]. The right and wrong, reiteration and choice belongs to direct speech act. Others as instruction (include claim, suggestion and order) and judgment belongs to indirect speech act. In this section, the interrogative sentences are defined based on the sentences classified and recentness principle [6] between question word and predicate. An interrogative sentence can be defined as:

Definition 6. An interrogative sentences is a 3-tuples:

$$(Q,F_n,S)$$

Where, Q is a question phrase, the F_n is a grammatical functions in the interrogative sentence, S is a generalizing sentence which the question word is extended. E.g., "谁发现了南极大陆(Who discovered New Continent in South Pole)?" is extended into ("谁", "Subject", "某人(Someone)发现了南极大陆") is an example.

Definition 7. CG of an interrogative sentence is:

Q:GS

Where, Q is a question word, GS is CG after being generalized sentence. The function of question word is the reference "*". E.g., the CG of "谁发现了南极大陆?" is:

谁:[发现]—(AGNT)→[PERSON:*]

(OBJ)→[新大陆] →(LOC)→[南极].

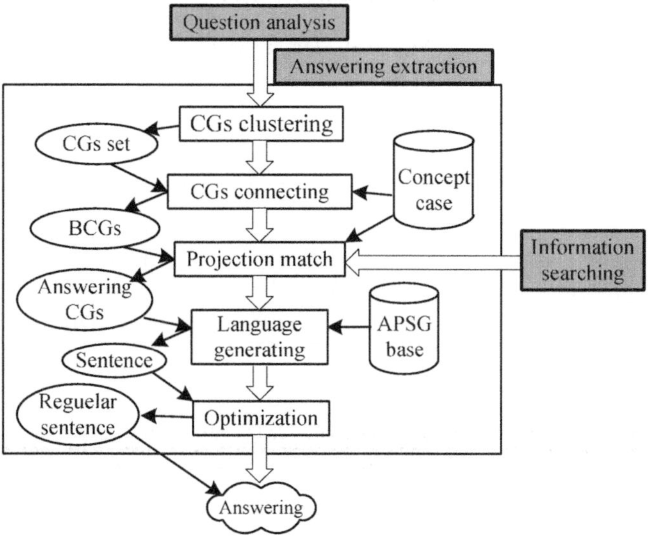

Fig. 4. The answering extraction module

Where, "[PERSON:*]" is a universality of " 谁 " and the reference "*" is indetermination.

In the interrogative sentence, the concept and relation of CG is an abstractive object. In the Object Oriented analysis, an object has many attributes which includes name, size, color and location, etc. In the question sentence, the question words can put forward question about object attribute. There are some familiar concepts in CG in the table 1.

Table 1. Familiar relation

Concept	Function	Concept	Function
PERSON	Asking a person	LOCATION	Asking a place
AGE	Asking an age	NUMBER	Asking an amount
NAME	Asking name	ORG	Asking organization
DATE	Asking date	SATAE	Asking appearance
TIME	Asking time	OTHER	Other characteristics

In generating CG, the relation of concept nodes also need to be defined. The concept's relation is more complicated. It can be divided into including, and dependency etc.[7].This section defines some relations in table 2 based on basic sentence patterns of modern Chinese and case grammar.

Table 2. Familiar relation

Name	Relate	Name	Relate
Include	INCL	Agent	AGNT
Dependency	DEPE	Object	OBJ
Similarity	SIMI	Type	ISA
Same	SAME	Place	LOC
Composition	COMP	Method	WAY
Classification	TYPE	Difference	DIFF

According to table 1 and table 2, the CG of some basic interrogative sentences is designed [5,6]. Following are some CG of common sentences in Chinese.

1) Interrogative pronoun [Subject] || Predicate

In this kind of sentence, an interrogative pronoun acts as subject. Its CG is:

[Predicate]—(AGNT)→[Interrogative pronoun]

(OBJ)→[Object].

The sentence of "谁发现了南极大陆?" is an example.

2) Subject || Verb+ Interrogative pronoun[Object]

In this kind of sentence, an interrogative pronoun acts as Object. Its CG is:

[Predicate]—(AGNT)→[Subject]

(OBJ)→[Interrogative pronoun].

E.g. the CG of "欧盟总部是哪里(Where is the headquarters of EU) ?" is

[是]—(AGNT)→[欧盟总部]

(OBJ)→[位于] →(LOC)→[LOCATATION:*].

Where, the * in "[LOCATATION:*]" is a uncertain place.

3) Subject || Interrogative pronoun [Predicate]

In this sentence, an interrogative pronoun is a state which acts as predicate. Its CGs is:

[Predicate]←(AGNT) ←[Subject]

E.g., the CG of "他怎么了(what's the matter with him) ?" is:

[STATE:*]←(AGNT)←[他]

Where, the * of "[STATE:*]" is an uncertain state.

4) Subject || Interrogative pronoun [Adverbial modifier]+ Verb/Adjective

In this sentence, an interrogative pronoun inquires a method which the subject is formed. Its CG is:

[Predicate]—(AGNT)→[Subject]

(WAY)→[Object].

E.g., the CG of "风是怎么形成的(How is the wind formed) ?" is:

[形成]—(AGNT)→[风]

(WAY)→[METHORD:*].

5) Subject || Verb/ Adjective+ Interrogative pronoun [Complement]

In this sentence, an interrogative pronoun acts as complement which illuminates the method or state of predicate further. Its CG is:

[Predicate]—(AGNT)→[Subject]

(STA)→[Object].

E.g., the CG of "那家公司发展得怎么样了呢(How does the company develop) ?" is:

[发展]—(AGNT)→[那家公司]

(STA)→[STATE:*].

6) Choose interrogative sentences

The selecting part of an interrogative sentence is "W1 or W2" and W1 is similar to W2.In the CG of this sentence, the node can be represented as "[X:{W1|W2}]".E.g., the CG of "你去北京还是南京(Are you going to Beijing or Nanking)?" is:

[去]—(AGNT)→[你]

(OBJ)→[LOCATATION:{北京|南京}].

5 The Experiment Test

The QA in Chinese is a system based on semantic analysis. There are many questions in the research. In this section, we have designed some experiments about the question sentence structures and theirs CG.

In the experiment, we have collected 300 Chinese question sentences from the medium, such as network, newspaper and magazine...etc. These sentences have made experiment tests having representative and extensive. In the test, 300 CG has generated by means of splitting word, marking word and parsing. After the artificial examination, 253 CG are completely correct because the syntax of those sentences is simple and their grammar structure is comparatively clear. CG of 21 sentences is correct basically. In these sentences, the syntax is simple, but there exist some abridge phenomenon. This CG can reflect the meaning of interrogative sentence basically. The correctness of 17 CG' structure can't be judged because of lacking contexts of the sentence. Due to the complexity of 9 sentences, Its CG is entirely wrong.

By the experiment test, 84.3% CG is completely correct, 7% CG is basic correct, 5.7% CG is accuracy indeterminations, and 3% isn't correct. If the structure of the interrogative sentence in the importation is standardized further, the correctness of generated CG can reach to 91.3%. It can satisfy a practical demand basically.

6 Conclusion

The QA in Chinese is a new research field. Because of complexity of Chinese, the research in this field is very difficulty and the succeed cases are less too. In this paper, the method based on CG is presented, and the structure of QACG in Chinese is designed on the Chinese natural language understanding. In order to process the question sentences easily, concepts and relations of CG about the question sentence are summarized, and six kinds of question sentences with their CG are designed. The research of this topic has certain reference values for designing and researching of Chinese question and answering system.

Acknowledgment. In the research of this project, our authors think the help of center of information and network. We also think the help of Lu lin and Jiehan Sun.

References

1. Liu, L., Zeng, Q.-t.: An Overview of Automatic Question and Answering System. Journal of Shandong University of Science and Technology 26(4), 73–76 (2007)
2. Jla, K.-l., et al.: Query expansion based on set theory in Chinese question answering system. Journal of Jiangxi Normal University 32(2), 211–214 (2008)
3. Sowa, J.F.: Conceptual structure. Addison Welslely, UK (1984)
4. Zhang, L., Huang, H., Hu, C.: Research on Chinese Question Answering System Model. Journal of the China Society for Scientific and Technical Information 25(2), 197–201 (2006)
5. Yin, H.-b., et al.: Speech Performance Types of Modern Chinese Interrogatives. Journal of Jianghan University (Humanities Sciences 26(3), 47–51 (2007)
6. Sun, A., Jiang, M.-h., et al.: Chinese Question Answering Based on Syntax Analysis and Answer Classification. Acta Electronica Sinica 36(5), 833–839 (2008)
7. Huang, K., Yuan, C.-f.: Automatic Assessment Technology Based on Domain- specific Concept Network. Application Research of Computers 11, 260–262 (2004)

Research on Optimization Process of Hierarchical Architecture for Virtual Prototype Based on User Experience

Liwen Shi, Qingsen Xie, and Ran Tian

School of Mechanical Engineering Tianjin University
Tianjin, China
LiwenShi1111@126.com

Abstract. The key point of virtual product development in parallel engineering is to build the the model of virtual prototype, which is responsible to supply with various kinds of signal in a whole design cycle of a specified product; The construction method is a main direction of virtual product development and research. In this article, the research condition of virtual prototype is analyzed; and based on the analyzed result, the human-computer interaction theory and design methods ,such as user experience, are applied to the modeling process of virtual prototype. A new flow chart of virtual prototype building process is raised. This flow chart first apply a five-layer analysis method to the key points of building model. In the process of optimization, the effective of building model and the operators' efficiency are highly raised; and the period of the development of product has been shortly, which can satisfy the demand that the quick modeling of complex virtual prototype.

Keywords: virtual prototype, human-computer interaction, user experience, framework architecture.

1 Introduction

With the development of the technology, an increasing number of products have been made; which lead that the period of marketing (from the time the product into the marketing to the time that the product is eliminated by marketing) has been shortened largely. To change according to the users' demands has been as main competitive approaches. The diversity and individuation of users' demands have been the main trend nowadays. How to develop the product which is fit for the marketing has been the most insignificant thing to the manufacturing field. The respond speed of the marketing has been the primary factor of enterprize benefits and survival, which has also been the most important motivation.

The visualization and intergration of product design has been the main trend of development of parallel engineering. Virtual Product Development (VPD) is able to conceive, design, manufacture, test and analysis product to solve that large pressure question that are associated with time, cost, quality and so on in the virtual state. The research of virtual prototype on model theory, method, frame has been hot spot in

J. Zhang (Ed.): ICAIC 2011, Part V, CCIS 228, pp. 215–222, 2011.

recent years. VPD requires product design to have networking, agility and personality features; which is able to create a new manufacture method that connect high-speed and low-cost batch production with the demand of user's personality.

2 Overview of Current Research Status of Virtual Prototype

Virtual prototype is a new technology, which may create an advanced conception that are involved in the design, development and production. It can be applied to the whole life period of product development, the test of product on each behalf, analysis and assessment. Based on the conception that the product development process of virtual prototype is the basis of parallel engineering, it often takes single filed or various kinds of fields simulation technology. So in the initial step of product design can economically and simply analysis and compare varioius kinds of methods to decide the parameters that could affect the function, simulate the behavior characteristic in real environment and optimize the design.

However, there are two defects in virtual prototype now. One is that though the software of virtual prototype(such as ADAMS) is very powerful, the modeling function field is so weak that it is very difficult to build complex components. So the modeling of virtual prototype is often by means of Pro/Engineer, UG these three dimensional CAD softwares. Another defect is that the analysis software of virtual prototype can only build model and make an analysis of its own components. The complex model based on the three dimensional CAD softwares can only be analyzed by single dimension and shape. If, in the design process, the geometry dimension has been changed, such as the geometry modeling, the importing of model, the simulation set, these complex works would influence the whole CAD/CAE/CAM process.That is the disadvantage of the product lifecycle.

3 Construction of Virtual Prototype and User Experience

To model the virtual prototype need to be based on a software platform, whose usability and practicability will affect the speed and effectiveness of building the virtual prototype. Now, the exiting softwares of virtual prototype, including Pro/Engineer, Solidworks and UG, are very complex to model; and the operation sequences of these softwares are difficult to be satisfied. When user want to modify the model, if they are lacking for the guide and help, it is very easy for them to fail to build the right model; which will delay the production process. So, in economic times, how to let users to get the best expressions through human-computer interaction and how to build a virtual prototype frame based on high quality users' experience are instructional for users to build the model quickly. Its essence is a personalized and customized interactive technology tool and method to build fast user-centric model of virtual product development.

The combination of user experience and the process of building the virtual prototype model is based on two parts: one is to optimize user experience with consideration of those fields such as content organization, framework architecture, complex communication support and possible isomorphized design; whose phenomenon performances the achievement of targets for the user, behavior improvement, satisfaction in psychology, emotional dependence and so on. The other one is to measure the feelings of users to the operation difficulty level, the user interface, interactive process, functional design and so on. The usability, intelligibility, aesthetic value and users satisfaction of the software in modeling process is manifested in six aspects:

- To emphasize the development of "user-centric" concept of throughout development
- To encourage individual to attend the interactive building
- To stress the cognitive psychology of user feedback and emotional capacity of their degree of subjective
- To emphasize the interaction of qualitative research and quantity research
- To support to change and reconstruct easily;
- To emphasize the feedback design process which consist of feedback and assessment.

The user experience, in the development process of the whole virtual product, is to make sure that the user experience in modeling process won't be beyond the specific and conscious intent. That is, every action taken by customer in every step should be taken into account and try to know the user's expectation in each step.

Virtual prototype technology will expect to achieve perfect results as follows: to save various kinds of needed machine factors in computer in terms of database; and then input digital drawing sheet, which could be combined automatically to three dimensional model by computer. The next step is to input practical data, index and other possible fatigue into the modeling platform to complete three dimensional simulation, which shows possible productivity, reject ratio and checks that where the bottleneck is and whether each sectors in production line can match product. It is very clear for custom to know the data above and variation curve. It is also able to change the model in accordance with the customers' reviews. With the help of the combination speed of computer, to modify quickly to satisfy customers' demand.

4 Framework for Hierarchical Modeling Process of VPD Based on User Experience Design

A. The Design of Process Framework

It is using a relatively static user research in construction of the traditional process of virtual prototyping, which is essentially a development model divided user and model building into two different pieces. Original developing ideas of information-centric make it difficult to attract users and improve user experience. In view of this, we propose a hierarchical model of VPD based on user experience design, for the purpose of this concept into a structured, operational module design.

The process of virtual prototype building will be five levels of process planning, followed by bottom-up for the strategy level, the scope layer, the structure layer, the skeleton layer and the surface layer, and provide a basic framework. Only buildimg software platform bottom-uply on this infrastructure, we can further discuss the problem of the user experience, and in what ways and means to solve and implement the user experience.

Each level is based on a level below it This dependence means that the decision on the strategy layer will have some upward "chain effect". When we want to select a option outside the "higher layer", we will need to reconsider the decision we made in "lower layer". But it is not to say that each decision making in "lower layer" must be before the design of "higher layer". There are two sides of things, the decision in the "high layer" sometimes leads a re-assessment (or even the first assessment) to the decision in"lower layer". These decisions may have two-way chain reaction. We should plan so that each work in lower layer can not be done before the start of the work in lower layer.A reference model is shown in Fig.1.

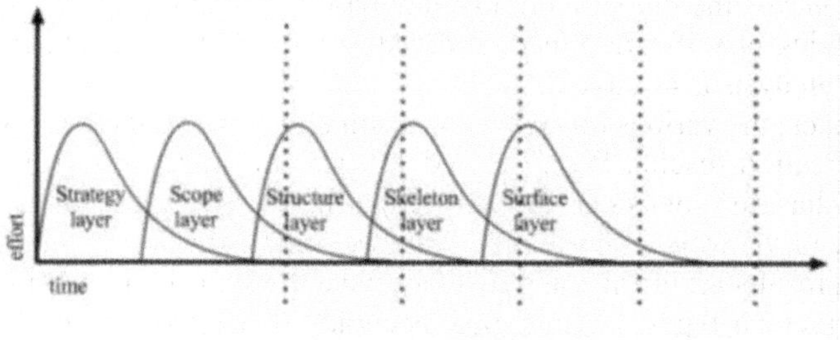

Fig. 1. The working-layer diagram

Virtual prototype is based on integrated products and data management. While virtual product development is the actual product development process to achieve the essence by using computer. It is achieved by using computer simulation and virtual reality technology and to work together in the computer group, through three-dimensional models and animations, to achieve the essence of product design and development process. To solve these basic dual properties of virtual prototype, we separate from the middle of five levels, on the left, which elements are only used to describe the modeling process planning software interface. On the right, we use these elements of data management for virtual prototyping.

- In the side of software interface, our main concern is the task: all operations are incorporated into a process, to think about how people accomplish this task. Here, we consider the planning process of software interface as one or more tasks to complete one or a set of tools.
- In the side of data management, our focus is on information: what information should be provided by the database and what is the meaning of the user form these information. We strive to create an “information space which users can pass through ".

B. Five Hierarchical Construction

Strategy layer: User need and Software target.

- Whether for the software product or information space, the contents of the strategic layer are the same concerns. We must understand what users want from us and we would also like to know how to meet other objectives they expected, when they have achieved these goals.These are the needs of users from outside the enterprise. Corresponding with the user need is our expected Target ,which is about the ultimate realization of functionality, performance, brand, etc based on the software platform.

Scope layer: Functional specification and Content requirement.

- Into the range from the strategy layer to scope layer, it turns to create functional specification in terms of software: It is the detailed description of the "functional group"about products. In the information space, the scope appears the form of content needs: It is the detailed description of requirements of the various elements.

Structure layer: Interaction design and Information architecture.

- On the software side, the structure layer transformed into interactive design from the scope layer.Here we can define how the system responds to user's request. On the side of information space, the structure layer is the Information architecture,that is the distribution of content elements in the information space.

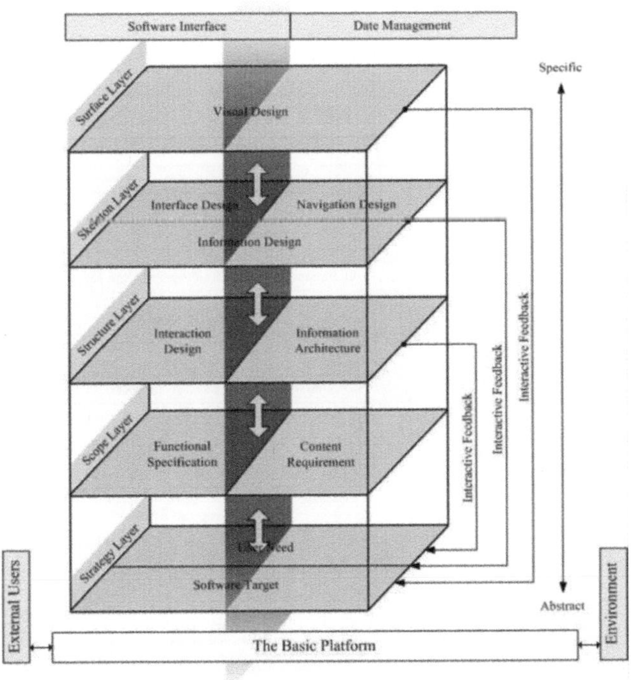

Fig. 2. The five-layer famework

Skeleton layer: Interface design, Navigation design and Information design.

- Skeleton layer is divided into three parts.Whether is software interface or the information space, we must complete the information design. On the software side, the skeleton layer also includes interface design. We should arrange functions which are allowed users to interact with the system interface elements. On the side of information space, this kind of interface is the navigation design. It is by the combination of some elements on the screen that information architecture allow users to pass through.

Surface layer: Visual design.

- In the surface layer, the ultimate concern is the final appearance of the product, the visual design, regardless of software or information space.

A reference model is shown in Fig.2.

C. Construction Method

Specific process is shown in Fig.3.

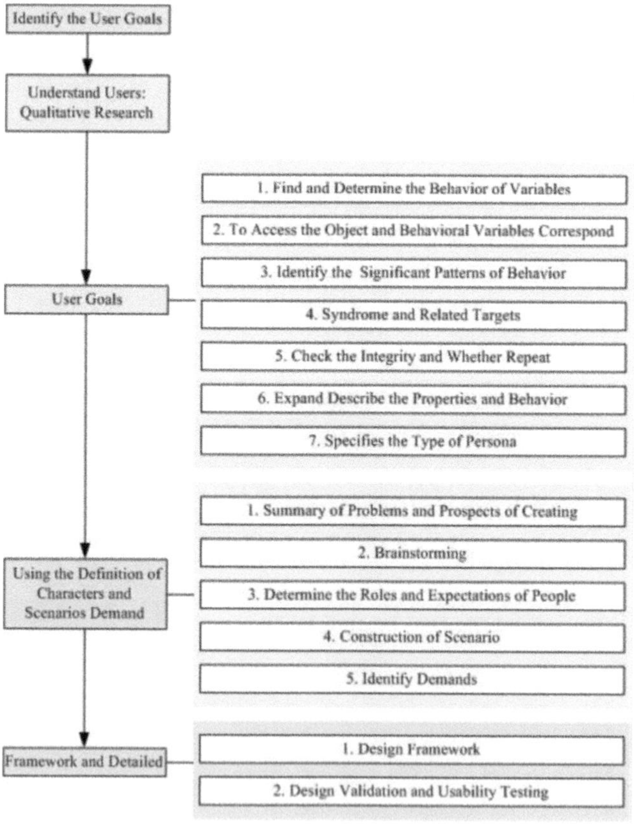

Fig. 3. The construction method

D. Characteristic Analysis

Framework architecture for hierarchical modeling process of VPD based on user experience design has the open, dynamic and interactive development characteristic.

- Open: Through hierarchical structures of VPD based on an open platform for modeling, it will be linked one "information island" to another, laying the foundation for information exchange, open application and data sharing for the complex environment. This is more suitable for the aspirations of the users, which users do not care about the information or services come from, but to expect to obtain the available resources by the highest efficiency in the shortest time.
- Dynamic: We emphasis cycle optimization process on "User needs→ Information architecture→ Personalization construction→Behavior interaction →Interactive feedback→User re-analysis", through the creation of interactive feedback mechanisms to capture the dynamic behavior of the user, thus promoting the improvement of individual users build.
- Development of interaction: it performances of an open software platform-based, five-layer information architecture of interactive feedback. It includes construction of internal information of the individual needs of users, the formation of micro-information architecture as the core to the user experience interactive, as well as groups collaboration-based social information architecture.

5 Conclusion

Now, the user experience research in virtual prototype model building field in China is still in the initial step. There are two reasons: one is because of lack of structural and strong varied gradations in discussed systems. People began to know the importance of user experience, but lack of systematic guide. The other one is that the construction of the virtual prototype model of the user experience in practice has just started, so the related technology is preliminary. Especially for the user to capture emotional and psychological and quantitative research is still in experimental stage, which is still used widely from the practice. These defects reflects the direction of research points and development. In this article, The idea of the framework architecture for hierarchical modeling process of VPD based on user experience design is a new test in the building model of virtual prototype field. Its main idea is to grasp the user experience design and communication intersection, so as to promote the harmonization of hierarchical design model, development model and user mental model. and improve different experience levels of the user experience and psychological feelings.

References

1. Cheng, C.: Interactive Virtual Environment Technology. Chinese Academy of Sciences (2002)
2. Liu, X.: For Virtual Conceptual Design of Mechanical Products and Key Technologies for Behavioral Modeling. Wuhan University of Technology (2003)

3. Li, D., Li, Y.: Virtual Prototyping in Manufacturing Industry and Research. Machinery (06) (2008)
4. Zhong, X.: Personalized Product Design Method for Rapid Response. University of Science and Technology (2008)
5. Cooper, A.: About Face3: the essence of interaction design. Beijing Electronic Industry Press (2008)
6. Hua, M.: Interaction Design Prototyping Method. Popular Literature (11) (2010)

Appraisal to the Competitiveness of Service Industry in Central Districts of Hangzhou and Its Improvement Strategy*

Jianzhuang Zheng[1] and Haixia Jiang[2]

[1] Zhejiang University City College Hangzhou, Zhejiang Province, China
[2] Hangzhou Vocational & Technical College,
Zhejiang University of Technology Hangzhou, Zhejiang Province, China
JianzhuangZheng@tom.com

Abstract. The Competitiveness of the service industry directly reflects the development level and potential ability of service industry in district. This paper studies the current development of the service industry in Hangzhou, constructs an appraisal index system of the service industry in the central district perspective, analyses the advantages and disadvantages in the sustainable development of service industry in each central districts of Hangzhou, and finally, gives suggestions in the strategy to enhance the competitiveness of service industry of Hangzhou.

Keywords: central district, service industry, competitiveness.

1 Introduction

Since the 1980s, the global economics has showed an overall trend changes from 'industry economic' to 'service economic'. The service industry is becoming a pivotal engine in the worldwide economic improvement (Li and Wang 2005). Its crucial role in the aspects of improving country economics, increasing employment rate, adjusting industry structure and upgrading level of living standard is apparent. In the internal view of a district, the development of service industry mainly occurs in the central districts instead of suburbs, due to the fact that central districts, regarded as the 'heart' in economic development of a city, have favorable factors in geographic position, transportation and culture, which not only bring 'magnetic power' towards aggregation of productive factor, population and industry, but also benefit to the market development. Therefore, the central districts are becoming the major zones of service industry (Li 2007). This paper conducts an apprasial based on analysis of the service industry competitiveness in some central districts of Hangzhou, in order to search for an effective strategy to improve the service industry development in cities.

* This work is supported by Zhejiang Provincial philosophy and Social Science planning Subject Foundation of China under Grant 10CGYD93YBX, the Hangzhou philosophy and Social Science planning Subject Foundation of China under Grant C10GL29 and Business Information Analysis Key Laboratory of Hangzhou.

J. Zhang (Ed.): ICAIC 2011, Part V, CCIS 228, pp. 223–229, 2011.

2 The Literature of the Competitiveness of Service Industry

Most academic researches towards the service industry competitiveness are based on their competitive ability in regions. Thereby, the service industry competitiveness can be simply defined as 'the ability and capability of service industry to obtain and utilize resources in the competition'. Researchers in the world are more concentrate on the competitive analysis of a particular service industry, e.g. the commercial service industry (Rubalcaba and Gago 2001), the insurance industry (Hardwick and Dou 1998), and knowledge-intensive service (Windrum and Tomlinson 1999). However in China, emphasis is put on a particular region rather than a particular industry, e.g. 31 main provinces in China (Wu 2003), 6 high-developed provinces in China (Diao and Zhuang 2004), 16 major provinces in China (Su and Zhang 2005). Due to the big amount of factors influencing service industry competitiveness and their complexities, most Chinese scholars hold the proposition to establish a multi-layer integrated appraisal index system such as 'General Objective-Guideline-Sub guideline-Index'. They make reference to the structure of regional competitiveness appraisal system, use quantities parameters as capability of regional economic development, scale of service industry, expanding speed, productive efficiency, science and technology capability and so on, use analytical methods like main element analysis, factor analysis and Data Envelopment Analysis (DEA) appraisal, conduct measurement to the competitiveness of service industry in several provinces.

On the whole, studies on the service industry competitiveness in China mainly target to provinces and cities, while using the integrated data in the scale of provinces and cities. Thus, relevant data about economic in suburbs, especially in rural areas, can not be derived. In fact, the service industry nowadays mainly develops in central districts. Data in these areas can effectively reflect the competition circumstances of service industry in the core region of a city. On the other hand, along with the trend of aggregation development in modern society, even those several central districts in the same city will have different emphasizes on the service industry development. Therefore, only by analyzing favorable and unfavorable factors affected the competitiveness of service industry in each central district and conducting research separately, can the development of the whole city's service industry be improved effectively.

3 The Current Situation of Service Industry in Hangzhou

Hangzhou plays the role of central city in the Yangtze River delta and one of the three biggest integrated transportation hinges. In recent years, aiming to be 'a modern service industry centre in Yangtze River delta' and to construct a modern industry system, Hangzhou made use of the 'Reversed Transmission' in the global financial crisis, speeded the upgrading of industry structure, improved the quality of service industry and encouraged service enterprises to grow up. Significant development in the service industry is obtained. In 2009, the service industry in Hangzhou gained value about 247.352 billion Yuan with the increasing rate about 13.9%, accounted for 48.51%

in GDP. Apparently, the service industry has become a major engine in the economic development of Hangzhou. From the internal structure of the service industry, Hangzhou makes division of ten fields in the development strategy direction, including culture & innovation, tourism, information service, financial service, commercial service, modern logistic, agency service, real estate, community service and technological service. Among them, the amount of culture & innovation, commercial service and financial service industry occupies more than 15% in the overall service industry, while real estate, information service, tourism and modern logistic accounting for nearly 10%. It can be seen from the data above that the industry structure of service industry in Hangzhou is already optimized to some extent, while some superior industries have relatively high adding-value.

However, many problems are confronted by Hangzhou in the service industry development. For instance, the total amount of service industries and the structure modernization degree are far more behind of cities like Beijing, Shanghai, Guangzhou and Shenzhen. What's more, in the scale of Yangtze River delta, Hangzhou falls behind cities of Suzhou in the total amounts of service industries and Nanjing in the proportion occupied in GDP. To Hangzhou, there are still many aspects needed to improve in order to sustain the rapid development of service industry, especially in the central districts. Every central district should be regarded as a basic development unit. Thus, different feasible strategies can be adopted aiming to the advantages and disadvantages of service industry development in each district.

4 Competitiveness Appraisal in Each Central District

A. The Index System

The competitiveness of regional service industry is a function involving several influence factors. Based on former studies, the paper gives a integrated consideration to the comprehensiveness, scientificalness and acquirability, establishes an appraisal index system to the service industry competitiveness, comprising 4 first order indexes of Economic Development Foundation, Development Level, Growing Ability and Technological Capability, with 12 second order indexes under them (Table 1).

B. The Appraisal Process and Outcome Analysis

Concerning to eliminate possible relativity between indexes, this paper adopts the factor analysis method to measure the service industry competitiveness. Based on relevant data of each central district in 2009, the paper selects 7 central districts within the Zhejiang Province and 9 central districts in other provinces, which all have fast-developed service industry. These districts are used as samples in the horizontal comparison to find out the service industry competitiveness situation and the advantages and disadvantages of each central district of Hangzhou.

Table 1. The appraisal Index System to the Service Industry Competitiveness

	F1	F2	F3	F4	F
Xiacheng District, Hangzhou	0.715	-0.546	-0.323	0.257	0.089
Shangcheng District, Hangzhou	-0.289	-1.206	0.075	-2.95	-0.649
Jianggan District, Hangzhou	-0.871	-1.465	-0.274	1.183	-0.484
Gongsu District, Hangzhou	-0.790	-0.595	0.167	-0.251	-0.369
Xihu District, Hangzhou	0.221	-0.028	1.181	0.329	0.302
Binjiang District, Hangzhou	-1.381	-1.690	0.838	1.594	-0.456
Haishu District, Ningbo	0.160	-0.051	-0.662	-0.486	-0.127
Jiangdong District, Ningbo	-0.111	-0.404	-0.623	-0.535	-0.282
Heping District, Shenyang	0.084	0.748	-0.738	1.127	0.176
Dongcheng District, Beijing	1.714	-0.173	-0.517	0.020	0.420
Futian District, Shenzhen	2.479	-0.408	1.590	0.320	1.006
Kuancheng District,Changchun	-1.117	0.906	-0.777	-0.393	-0.346
Zhongshan District, Dalian	0.406	1.420	-1.441	0.733	0.250
Shibei District, Qingdao	-0.494	0.521	-0.411	-0.491	-0.176
Nan'gang District, Haerbin	-1.065	2.120	2.414	-0.208	0.479
Baixia District, Nanjing	0.384	0.379	0.350	-0.446	0.207
Jianghan District, Wuhan	-0.041	0.472	-0.851	0.199	-0.041

According to the eigenvalue greater than 1.0 rule, this paper chooses four common factors, with accumulated variance contribution rate achieving 79.896%. Thereby, the four factors can reflect most information in the original 12 indexes.

The first common factor loads more on the district public revenue, total retail sales of consumer goods, adding value, proportion occupied in GDP, and employment proportion in the whole economy, reflects the regional economic capability and service industry development level in a certain extent. It can be seen as the 'Economic and Industry Capability Factor'. The second common factor loads more on the proportion of training in the general expenditure budget, and proportion of research funding in the total expenditure. Since these two indexes mainly reflect the technological capability of a region, 'Technological Factor' can be the new name of the second factor. The third common factor only loads more on the productivity; it can be explained as the 'Industrial Productive Ability Factor'. The fourth common factor loads more on the

Table 2. Scores of Service Industry Competitiveness in Each Central Districts

First Order Index	Second Order Index
Economical Development Foundation	Per capita disposable income (X1) District public revenue (X2) Total retail sales of consumer goods (X3)
Development Level	Adding value (X4) Proportion occupied in GDP (X5) Employment proportion in the whole economy (X6)
Growing Ability	Growth rate in the amount of enterprises in the last three years (X7) Growth rate in the number of employees in the last three years (X8) Productivity (X9)
Technological Capability	Proportion of training in the general expenditure budget (X10) Proportion of research funding in the total expenditure (X11) Number of granted patents (X12)

two indexes of growth rate in the amount of enterprises in the last three years and growth rate in the number of employees in the last three years. 'Industrial Growth Factor' can stand for it.

Weighted on the contribution rate of each common factor, the appraisal model of service industry competitiveness can be formed:

$$F = 0.31543F1 + 0.20105F2 + 0.17038F3 + 0.11119F4$$

It is apparently that in Table 2, in the six central districts of Hangzhou, only Xihu District ranks in the top 5. Besides, Xiacheng District ranks eighth, while other four districts rank at the last four on the list. Hangzhou does not have a good performance on the service industry competitiveness. Here in detail. On the first common factor — Economic and Industry Capability Factor, Xiacheng District ranks the third position, Xihu District ranks the sixth, while other districts failing more behind. Thus, though Hangzhou is a well-developed city, its central districts do not represent obvious advantages on the economic capability. On the second common factor — Technological Factor, Xihu District ranks the eighth, while all other districts stay at the end. It leads to the conclusion that, Hangzhou's regional technological capability is weaker than other cities'. On the third common factor — Industrial Productive Ability Factor, Xihu District and Binjiang District rank the third and the fourth, other four districts occupy the position from the sixth to the ninth. It infers that the input and output ratio and its production efficiency in the service industry in each district sustain a relatively high level. On the fourth common factor — Industrial Growth Factor, Binjiang District and Jianggan District rank in the top two, while Xihu District and Xiacheng District rank the fifth and the seventh, mainly in the middle section of the list. Therefore, apart from Gongsu District and Shangcheng District, the remaining four central districts can keep a growing tendency.

5 The Improving Strategy of Service Industry in Hangzhou

Through comparison to the competitions of service industry of each central district, it is found that, in despite of the favorable growing trend maintained in Hangzhou and the rational structure of service industry, there is no superiority in the competitiveness as a whole. In order to keep a consistent development in the service industry and achieve the objective of transform from 'Big' to 'Powerful', Hangzhou should lengthen the 'barrelboard' while avoiding the 'short-board effect' according to the specific situation of each central district. Four improving strategies are provided in practice:

A. Increase the Input on the Scientific Research of Service Industry

Government policy should give support to the scientific research for the service industry; help them to invest more on the scientific research section. The third industry should be enriched in its technical content and development level. Information-based infrastructure should be promoted; scientific management methods and principles should be adopted to contribute to the service industry improvement. Special emphases should be put on the Xihu District and Binjiang District to improve their scientific research abilities, for these two central districts involve more high-tech service industries than others.

B. Focus on the Stratification and Multi-layer in the Training System of Human Resource

As the requirements toward employees are quite different in labor-intensive industry and knowledge-intensive industry, multi-layer training system for human resource should be established. Accelerate the cooperation with relevant high schools and colleges, even universities can be taken into consideration. Deliver education based on the position in a forward-looking way, and seek to improve the skill level and qualification of employees in the service industry.

C. Encourage to Implement the Name-Brand Strategy to Enlarge the Profit Space

Focus on using the benefit of a name-brand to obtain profits. Establish several service industry brand based on exploration, innovation and leader effect, while making a greater effort on attracting investment for big-scale and superior enterprises. Build service brands based on distinct features of each district; make it 'shine' over enterprises and industry, for instance, a business service brand of Xiacheng District, a cultural innovation brand of Xihu District and an information service brand of Binjiang District.

D. Continue to Modify the Industrial Structure and Optimize the Spatial Agglomeration

To be objectives-oriented, identify a group of key industries and key fields. Adopt a gradient development strategy while using the minority to bring along the majority, in order to form a distinct featured industry structure with favorable arrangement.

Furthermore, identify some functional agglomeration regions to achieve the objective of improving innovation ability in the service industry, which is, district by district, and field by field.

References

1. Li, W.W., Wang, H.M.: Strategic Consideration on World Industrial Servicization and Development of Shanghai Modern Service Industry. World Economy Study (1), 58 (2005)
2. Li, X.Y.: The Central District's Status and Role in the Development of the Regional Service Industry. Modern Business Trade Industry (7), 26–27 (2007)
3. Luis, R., David, G.: Relationship between Services and Competitiveness: The Case of Spain Trade. Service Industries Journal 21(1), January
4. Philip, H., Wen, D.: The Competitiveness of EU Insurance Industries. Service Industries Journal 18(1), January
5. Paul, W., Mark, T.: Knowledge intensive Services and International Competitiveness: A Four Country Comparison. Technology Analysis & Strategic Management 11(3), September
6. Wu, S.Y.: The Competitiveness of Provincial Evaluation Services. Statistics and Decision (10), 57–58 (2003)
7. Diao, M.R., Zhuang, L.J.: Total Evaluation of the Competitive Power of Service Industry in China's Six Provinces and Cities. Finance and Trade Research (4), 15–19 (2004)
8. Su, G.B., Zhang, X.L.: Evaluating Competitiveness of Tertiary Industry Based on Super-Efficiency DEA Mode. Statistical Research (10), 58–60 (2005)

A Scheduling Method of Maintenance Work Based on Dynamic Priority

Ying-wu Peng, De-jun Mao, Wei-yi Chen, and Rui Wang

Dept. of Weaponry Engineering Naval University of Engineering
WuHan, HuBei Province, China
yxiai23@yahoo.cn

Abstract. The scheduling problem of maintenance work on the equipment support station influences the support efficiency, especially for a station which repairs a large number of failed items from more than one subordinate operating station. The influence of the mean time to repair (MTTR) on system availability was analyzed for maintenance stations, then the factors which determine the maintenance priority of a failed item were modeled, including the demand of the subordinate station, time and maintenance resource to repair the item. Then a scheduling algorithm of maintenance work on equipment support station based on dynamic maintenance priority was proposed. The algorithm was used in an case, and the result shows that comparing to FIFO (first in, first out) method and resource fitting method, the algorithm improves the support efficiency by much while cutting down the demand for maintenance resources.

Keywords: equipment support, dynamic priority, maintenance, scheduling algorithm.

1 Introduction

To the materials that had been deployed and working in the army, improve the efficiency of the support system is a good choice to improve the availability of the materials. To improve the material support efficiency, one can increase the stock number of all the spares that involved when supporting the systems, or by optimizing the structure of the support organization, speed up the transportation, and decrease the repairing time for the failed items. However, when the support expenses is limited and the performing flow of the support system is fixed, which is the typical case we encounter, optimize the scheduling of maintenance work on the equipment support station is another choice to improve the material support efficiency.

The scheduling of maintenance work is somewhat different to the scheduling of producing and transportation, firstly, different type of maintenance work has different requirement to the repairing resources, even to the same kind of failed item, when the fail mode changes, the maintenance work varies in flow and requirement of resources; secondly, the maintenance work initiated by different subordinate station comes from different systems, and with different priorities. This paper takes all the restricting factors into account when scheduling the maintenance work and a scheduling algorithm based on dynamic maintenance priority is proposed.

J. Zhang (Ed.): ICAIC 2011, Part V, CCIS 228, pp. 230–238, 2011.

2 Scheduling Algorithm Based on Dynamic Maintenance Priority

A. How MTTR (Mean Time to Repair) Affects the Availability

For the material system that deployed in an operating station (we use system when referring to the materials if there is no ambiguity), its average availability is calculated like this:

$$A = \frac{MTBF}{MTBF + \tilde{M}_{ct} + WT} \tag{1}$$

Where: MTBF: Mean Time Between Failure;

$\tilde{M}_{ct}$: repairing time given no item shortage.

WT : mean waiting time for the item when there are no spares on the hand.

If take the preventive maintenance and the using ratio of the system into account, we use another one:

$$A_0 = \frac{T_U - T_P}{T_U} * \frac{MTBF}{MTBF + \tilde{M}_{ct} + WT} \tag{2}$$

Where: T_U is the maximum working time during the observed period, T_P is the total time of preventive maintenance in the period. If there are no spares of the required item on the maintenance station, it has to repair the failed and then satisfy the requirement, we get:

$$WT = T_T + TAT \tag{3}$$

Where: T_T is the transporting time for the item go and back between the maintenance station and the station where the system is deployed. TAT is the mean time to repair the failed item.

In case that a maintenance station repairs failed items from more than one subordinate station, if cut down the mean repairing time TAT by optimized scheduling, the availability of the system can be improved.

B. Dynamic Maintenance Priority

When scheduling the maintenance work, the expect time to repair the failed item, the available number of system in each operating station and the available maintenance resource are the most important factors considered. Maintenance resource includes the maintenance devices and installations, tools, spares and staff. If the required maintenance resource is not sufficient, the failed item has to wait.

The scheduling algorithm is to determine which item can be repaired and others should wait, in fact the algorithm assigns the maintenance resource.

For the sake of simplicity, the operating station where the systems are deployed is called station, when the maintenance station which repairs the failed items for several operating stations is called workshop.

1) The Factors that Determine the Dynamic Priority

The number of stations is marked as L, the number of systems that deployed in the station index l is marked as L_l, and there are N kinds of items that need to be sent to the workshop to get repaired, so when a system failed in stations l, after allocating the failed item, the index of the failed item is marked $n(n=1,2,\cdots N)$. However, before sending the item to the workshop, the station check the condition of all the systems, as the number deployed in the station is L_l, the number of available system is L_{lWork}, then the failed system number is:

$$L_{lFail} = L_l - L_{lWork} \quad . (4)$$

The station calculates a recommended priority p for maintenance work, in fact p represents the situation of the station:

$$p = \frac{L_{lFail}}{L_l} = \frac{L_l - L_{lWork}}{L_l} \quad (5)$$

There are chances that the station has to maintain a certain number (L_{lLimit}) of available systems to carry out some kind of missions, so when the number of available is less than L_{lLimit}, the recommended priority is multiplied by a scale factor μ ($\mu > 1$), in order to get the failed item repaired as fast as possible:

$$p = \mu \cdot \frac{L_{lFail}}{L_l} \quad (6)$$

When the workshop receives the failed item, the attribute of the item can be recorded as a vector:

$$FailedItem(n, p, l). \quad (7)$$

Where: n is the index of the failed item, p is the recommended priority from the station, and l is the index of the station who sends out the item.

The workshop checks the failed item and calculates the time to repair it (TTR). For simplicity, only one kind of maintenance resource is considered. The number of required resource is r and the TTR is t, notice that even to the same type of item, r and t may vary. Generally r is a stochastic variable which obeys the Poisson distribution, while t obeys the Normal distribution, the mean value of the distribution is MTTR:

$$t \sim N(T_0, \sigma^2) \quad (8)$$

Where: $T_0 = \text{MTTR}$, σ^2 is the variance of the variable.

A failed item waiting for repair in the workshop can be described as $FailedItem(n, p, l, t, r)$. The waiting queue should be: $(n_1, p_1, l_1, t_1, r_1)$, $(n_1, p_2, l_2, t_2, r_2)$, $(n_2, p_3, l_3, t_3, r_3)$,…, $(n_q, p_k, l_k, t_k, r_k)$, $q = 1,2,\cdots;\ k = 1,2,\cdots$. Notice that the index of the failed item is not unique.

The queue is then sorted by the index of the items, the number of the item indexed n is D_n, and the total number of the queue is D_T. D_n represents the demand for the item indexed n of all the stations, and is an important parameter to the scheduling algorithm. The algorithm also checks t of each item, the maximum is marked as t_{max}, while the minimum is t_{min}; the maximum r is r_{max} while the minimum is r_{min}, the total available resource number in the workshop is R. Since the resources are occupied when the repairing begins and released when repairing is finished, R is not a constant.

a) The influence of t to dynamic maintenance priority

To repair as many items as possible, a higher priority is given to the items with shorter TTR:

$$pt = \frac{t_{\max} + t_{\min} - t}{t_{\max} + t_{\min}} \tag{9}$$

Higher priority is given to the items with greater pt $(0 < pt < 1)$. As t_{max} and t_{min} vary when the queue changes, pt to a certain item is also changeable.

b) The influence of r to dynamic maintenance priority

The more resource needed to repair, the lower priority a failed item gets:

$$pr = \frac{r_{\max} + r_{\min} - r}{R} \tag{10}$$

Higher priority is given to the items with greater pr. pr is also a changeable value to an item.

c) The influence of D_n to dynamic maintenance priority

According to a) and b), if the failed items indexed n always need more time and maintenance resource to repair, these item would be with very low priority, making the D_n noticeably big, the following expression is used to adjust:

$$pd = \frac{D_n}{D_T} \tag{11}$$

The items with greater pd get higher priority.

As the workshop is not so clear with the condition of all the systems deployed to the stations, the recommended priority p is used.

So far, the four factors with greatest influence are all listed out, the dynamic maintenance priority to a failed item is calculated by:

$$\Phi_k = \alpha \cdot p_k + pt_k + pr_k + pd_k \tag{12}$$

Where: $k = 1,2,\cdots$ is the index of item in the waiting queue, α is a weighting factor: if a station has to maintain a certain number of available systems, α with greater value is used, and a lower α is used to keep the total number of available systems of all the stations as bigger as possible. The item with greatest Φ_k gets the highest maintenance priority.

2) An Additional Selecting Rule

Even the item with greatest dynamic maintenance priority is chosen out, there may be an item with the same index but easier to be repaired (less resource needed and shorter TTR):

$$\Phi_{k1} > \Phi_{k2}, t_{k1} \geq t_{k2}, r_{k1} \geq r_{k2} \tag{13}$$

If such an item does exist, the p,l values of the item $k1$ and $k2$ in the queue are interchanged ($k1$ and $k2$ are two items in the waiting queue belonging to the same type). For example, station 1 need an item indexed 3 to be repaired as soon as possible, while station 2 also sent an item indexed 3 to repair but with less urgency and easier to be repaired, then the item from station 2 is immediately repaired but sent back to station 1, when the item from station 1 is repaired, it is sent to station 2.

The additional selecting rule is to make sure that the item be repaired headmost is not only of the type with most urgency, also the easiest one to be repaired of that type (if exists). The item left in the waiting queue is with less importance, or not so easy to be repaired.

3) The Usage Of Idle Maintenance Resource

Generally, the item with greatest dynamic maintenance priority does not make full use of the maintenance resource, there may be some resource left in idle, the number is R_{remain}, but not enough to repair the item with secondary priority. The algorithm now looks up the waiting queue to find out an item which demands less resource and with short TTR. If the minimum remained time to repair (remained TTR) of all the items which are now being repaired is $TTR_{\min}$, then an item is selected out to repair according to the restrains:

$$\begin{cases} t \leq TTR_{\min} \\ r \leq R_{remain} \end{cases} \tag{14}$$

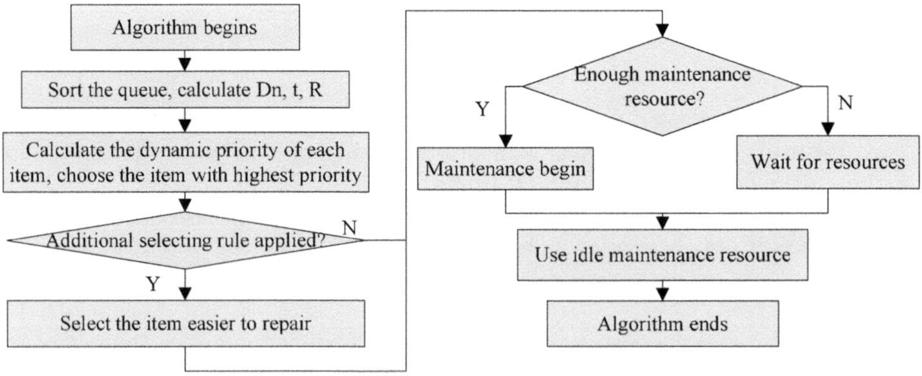

Fig. 1. The flow of the scheduling algorithm

C. The Flow of the Scheduling Algorithm

The scheduling algorithm is performed by the workshop when: (1) a failed item is received; (2) the maintenance of an item is finished; (3) the maintenance of an item is started. However, due to the restriction of the resource, the maintenance of the item with greatest priority may not start until all the demanded resources are available. The flow of the scheduling algorithm is shown in figure 1.

3 An Example of Application

A. The Scenario

A simulate system is constructed to perform the algorithm, and there is a workshop which repairs all the failed items from 6 subordinate stations, 8 systems are deployed to each station, and each system is divided to 10 subsystems (items). The system stops working if one subsystem failed, and the station can carry out mission only if more than 6 systems are available: $L_{lLimit} = 6$. Troubles may hit the systems which are on mission, and the frequency is determined by the reliability parameters, all the failed items are sent to the workshop to repair. There are 45 skilled workers, divided into 3 teams, 8 hours a shift, so there are 15 workers on the shift at any time. The scale factors $\sigma = 2$, $\alpha = 1.5$.

The simulate system runs to simulate the support process for a year, the indicators such as average utilization of resource (workers), total mission time for all the stations, available systems of each station and the average available ratio of all systems are used to estimate the support efficiency of the support organization.

B. Support Efficiency under Different Scheduling Methods

(1) Without maintenance priority, the workshop uses first in first out method (FIFO) to schedule the maintenance work. After performing the scenario for 1,000 times, the average utilization of resource is 78.09%, average available systems of all stations are 39.05, corresponding average available ratio of all systems is 81.35%.

(2) Resource matching method. The workshop arranges the failed item queue according to the arriving time. If the resource is enough to maintenance the first item, the item is immediately put into maintenance process, if not enough the second item is checked. The whole queue is checked until there is not enough resource to maintenance any item.

This method leads to the results: average utilization of resource is 78.09%, average available systems of all stations are 39.05, and corresponding average available ratio of all systems is 81.35%.

(3) Dynamic maintenance priority method. The average utilization of resource is 78.09%, average available systems of all stations is 39.05, average available ratio of all systems is thus 81.35%.

Figure 2 shows the variance of the accumulated mission time for all the stations during 50 times of simulation. Under FIFO, if the item which comes first is not repaired due to the lack of resource, other items would have no chance, the accumulated time for each station to carry out mission is inevitably shorter. Under the resource matching method, the result is better than FIFO. Obviously, the dynamic priority scheduling has the best performance, the accumulated mission time is longer and the variance between different runs is lower.

Figure 3(a) shows the variance of the average utilization of resource, while figure 3(b) showing the variance of average available ratio of all systems.

The total accumulated mission time for all the stations is listed out and compared in table 1.

Table 1. Variance of the total accumulated mission time

Method	Total mission time (Day)	Increment comparing to FIFO
FIFO	1662.5	--
resource matching	1867.3	12.3%
dynamic priority	2106.6	26.7%

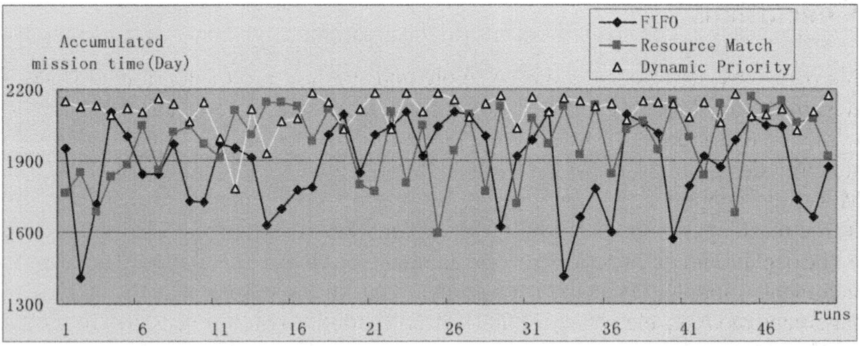

Fig. 2. Variance of the accumulated mission time during 50 runs of simulation under different methods

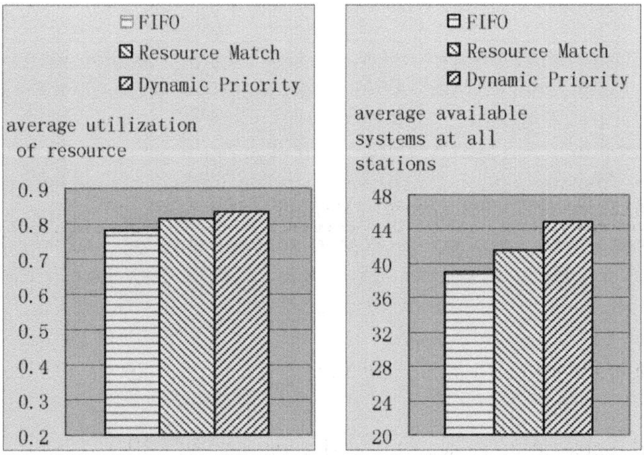

(a) Average utilization of resource (b) Average available systems

Fig. 3. Results of different Scheduling methods

4 Conclusion

The need of the stations, number of demand resources and the maintenance time of the item, the balance of available systems among the stations, are all taken into account by the scheduling algorithm based on dynamic maintenance priority. Comparing to other scheduling methods, this algorithm could make better use of the resources, increase the average available ratio of all systems by much, and the total accumulated mission time for all the stations is also increased.

Better use of the maintenance resources, increment of the average available ratio of all systems, all lead to better economic benefit for the support stations. The scheduling algorithm based on dynamic maintenance priority is not fixed to a certain problem, it could be improved to deal with more support stations, more kinds of maintenance resources, and the weighting factors also could be changed to match the problems encountered while dealing with supporting matters.

References

1. Cohen, M.A., Kamesam, P.V., Kleindorfer, P., Lee, H., Tekerian, A.: Optimizer: IBM's multi-echelon inventory system for managing service logistics. Interfaces (1), 65–82 (1990/2006)
2. Jensen, K.: Tools and algorithms for the construction and analysis of systems. In: 10th International Conference, TEXAS 2004. Springer, New York (2004) ISBN: 354021299X
3. Caglar, D., Li, C.-L., Simchi-Levi, D.: Two-echelon spare parts inventory system subject to a service constraint. IIE Transactions 36, 655–666 (2004/2005)
4. Kang, F.J., Liu, W.D.: Research on Real-time Multitask Scheduling Based on Timed Petri Nets. Journal of system simulation 18(11), 3075–3077 (2006)
5. Sun, J.S., Li, S.J., Lv, Y.M.: Simulation and Research on the Three-Echelon Storage Mode of the Valuable Spare Parts in Weapon Equipment. Acta Armamentarii 29(8), 1314–1316 (2007)
6. Al-Rifai, M.H., Rossetti, M.D.: An efficient heuristic optimization algorithm for a two-echelon (R, Q) inventory system. International Journal of Production Economics 109, 195–213 (2007)
7. Zhang, T., Wu, X.Y., Guo, B.: Operational availability model for repairable system under (m, NG) maintenance policy. Journal of Systems Engineering 22(6), 627–633 (2007)
8. Wang, K., Zhang, H.X., Wang, Z.J.: Study of Aircraft Maintenance and Support Simulation Based on UML. Journal of System Simulation 18(5), 1347–1349 (2006)
9. Frohne, P.T.: Quantitative measurement for logistics, pp. 61–67. Sole Logistics Press, New York (2008)
10. Aharon, B.-T., Boaz, G., Shimrit, S.: Robust multi-echelon multiperiod inventory control. European Journal of Operational Research 199, 922–935 (2009)
11. Cheng, H.L., Kang, R., Xiao, B.P.: Spare model under the constraint of spare sufficiency. Systems Engineering and Electronics 29(8), 1314–1316 (2007)

12. Kochel, P., Nielander, U.: Simulation-based optimization of multi-echelon inventory systems. International Journal of Production Economics 93-94, 505–513 (2005)
13. Chen, C.L., Wang, Y.L., Sun, S.K.: Application of HTCPN in Modeling and Optimization of Equipment Support Work Process. Journal of System Simulation 20(10), 2746–2749 (2008)
14. Lee, L.H., Chew, E.P., Teng, S., et al.: Multi-objective simulation-based evolutionary algorithm for an aircraft spare parts allocation problem. European Journal of Operational Research 189, 476–491 (2008)

New Application of Computational Intelligence Method of EMD*

A Case Study of International Tourist Arrivals to China

Lingling Chen and Zhenshan Lin

College of Humanity
Jingling Institute of Technology
Nanjing, Jiangsu Province, China
linglingchen112@sina.com

Abstract. Empirical Mode Decomposition (EMD), a computational intelligence method dealt with non-linear complex system, was introduced to find out the fluctuation rule of international tourist arrivals in this paper. Main results show that: (1) Through decomposition of long-term and short-term historical data respectively, four IMFs and one residual trend term are both obtained. There are almost the uniform fluctuation periods of the first three IMFs, which are 3, 6, and 12 months separately (Tab.1 and 2). (2) In the long run, the dominant factor to control the change of international tourist arrivals is the residual res whose variance contribution is 83.1%; while in the short term, intense fluctuation of 3 months' period with the biggest variance contribution, 50%, is still the main change characteristic. (3) Intense fluctuation of international tourist arrivals should be paid more careful consideration when establishing recent tourism plan. At the same time, long-term measures to deal with the large tourists flow will also be endeavored in immediately. As one of the best methods of extracting data series, EMD is great beneficial to predict future international tourist arrivals and provide a theoretical guidance for tourism policy.

Keywords: international tourist arrivals, computational intelligence, EMD, multi-scale.

1 Introduction

First 20 years in the 21st century is an important stage for China to achieve the goal of becoming a world-power country in tourism, and a huge demand for tourism consumption will be emerged then. Inbound tourism is always an important part of tourism industry, in which international tourist arrivals is a major factor to measure the strength. Herein the continuing ascent of international tourist arrivals is the basis and guaranty to achieve economic goal of tourism. As a government, we must try our best to establish a scientific strategy for tourism development, which can not only deal with the increasing consumer demand entering China but also promote sustainable development of China's tourism industry simultaneously. Consequently,

* This work is supported by Dr. Start-up Fund in Jingling Institute of Technology (jit-b-201011).

J. Zhang (Ed.): ICAIC 2011, Part V, CCIS 228, pp. 239–245, 2011.

accurate analysis and prediction of international tourist arrivals will directly become the theoretical premise and foundation to establish strategies for tourism development. However, the tourist arrivals change is a very complex non-linear system, leading more difficulties to acquaintance the inherent rule of the historical sequence.

In this paper, a computational intelligence method, Empirical Mode Decomposition (EMD), was firstly introduced to deal with the complex time series in tourism industry. According to historical data of long term and short term respectively, EMD decomposed the international tourist arrivals to China, based on which we detailed analyzed the fluctuation rule. This research will provide a theoretical guidance of China's future tourism policy.

2 Computational Intelligence Method of EMD

In each field of natural science and social science, a great deal of decision problems can not get away from the prediction which is the foundation of the decision. The best method to settle down prediction problems is to detect and find out the law in the dynamic state process or phenomenon. In natural, the needed information is usually insufficient, also the relative theories. Our understanding to the thing is limited by the observed data, namely time series. So we can make use of the existing history data to establish a model to predict the future. Nowadays, there are many intelligence methods applied to study nonlinear systems, such as neural network [1-2], which is a kind of static network that have no processing ability of time so that it cannot identify the time series model. In this paper, we introduced a new computational intelligence method to solve our question, i.e., Empirical Mode Decomposition (EMD).

EMD is a computational intelligence method applied to non-stationary data series which essentially is a steady signal processing [3]. The process is to progressively disaggregate the signal in different scales by different fluctuations or trends, thereby to form a series of different scales of data sequences. Each sequence is called an Intrinsic Mode Function (IMF) component, of which the physical meaning of different scales of data will be retained, and will be well extracted and expressed. The lowest frequency component (IMF, res) is on behalf of the general trend of the original signal or the mean time series. As an application method, EMD decomposition can either extract a series of data trends or remove the mean of the data sequence. Results of various studies show that EMD is one of the best methods of extracting data series trends at present [3-4], and has been successfully applied to nonlinear scientific fields, such as signal management, image management, earthquakes and atmosphere science [5-7].

The detailed process of decomposition is as follows: the method can simply use the envelopes defined by the local maxima and minima separately. Once the extrema are identified, all the local maxima and minima are connected by a cubic spline line as the upper and lower envelopes. Their mean is designated as m_1, and the difference between the data and m_1 is the first component, h_1, i.e.,

$$h_1 = x(t) - m_1 \quad (1)$$

However, h_1 is still not a stationary oscillatory pattern. Then, repeat the above process with h_1 replacing $x(t)$, and m_2 is the mean envelope of h_1,

$$h_2 = h_1 - m_2 \tag{2}$$

By repeating the above process time after time, the sifting process will be stopped by a criterion; standard deviation, SD, and a typical value for SD can be set between 0.2 and 0.3:

$$SD = \sum_{t=0}^{T}\left[\frac{\left|(h_{i-1}(t) - h_i(t))\right|^2}{h_{i-1}^2(t)}\right] \tag{3}$$

c_1 is the first IMF component from the data. We can decompose c_1 from the rest of the data by

$$r_1 = x(t) - c_1 \tag{4}$$

r_1 herein still contains information of longer period components, and it is treated as the new data and subjected to the same sifting process as described above. This process can be repeated on all r_i and the result is

$$r_1 = x(t) - c_1, r_2 = r_1 - c_2, \ldots, r_n = r_{n-1} - c_n \tag{5}$$

$$\text{i.e., } x(t) = \sum_{i=1}^{n} c_i + r_n \tag{6}$$

Thus, we decompose the data into several IMFs, and a residue (r_n), which can be either the mean trend or a constant. Although it is a powerful method, one difficulty encountered when using the EMD method is the influence of the end-point treatment. The envelopes are calculated by using a cubic spline; however, spline is notoriously sensitive to end points. It is important to make sure that the end effects do not propagate into the interior solution. Here, this problem is dealt with by the extrema extending method [8].

3 Decomposition Results Based on EMD

Fig.1 illustrated the historic monthly data change of international tourist arrivals to China from 2000 to 2009, total of which including 120 data. In general, the data went up quickly at first but the ascent speed gradually slower down. At the same time, there were so many fluctuations that we could not get its distribution rule or timescale characteristics. Therefore, EMD was introduced to deal with the historic data from two time scales, i.e., one is long term and the other short term, through which the inherent rule of international tourist arrivals would be deeply acquainted with.

Form the long time scale, EMD decomposed international tourist arrivals in last 10 years, and we get four components of intrinsic mode function (IMF) and its trend res, as showed in Fig.2. Every IMF's signal characteristic was very clear, whose change accorded with the nonlinear characteristic of natural signal. At the same time, Instantaneous Frequencies of every IMF and residual trend res were illustrated in Fig.3. Though the decomposition, we also obtained its fluctuation cycles with a relative steady quasi-period, although it was not a strict functional period. According to the algorithm, MATLAB procedure was then used to get the average period and variance contribution of each IMF, as showed in Tab.1.

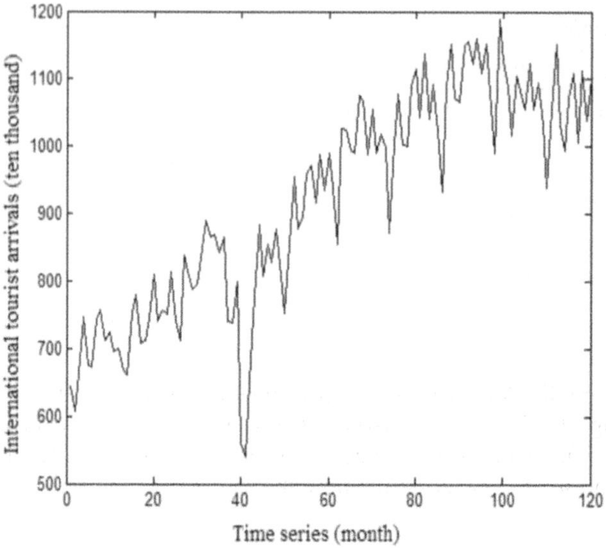

Fig. 1. Monthly change of international tourist arrivals to China in year of 2000 to 2009

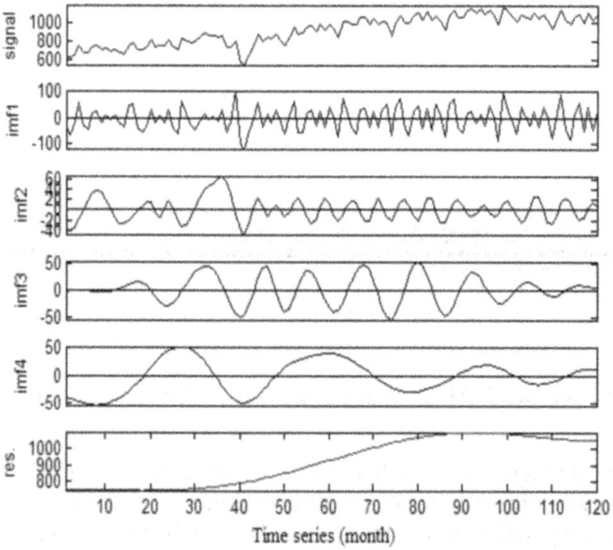

Fig. 2. IMFs and residual trend res of international tourist arrivals to China from 2000 to 2009

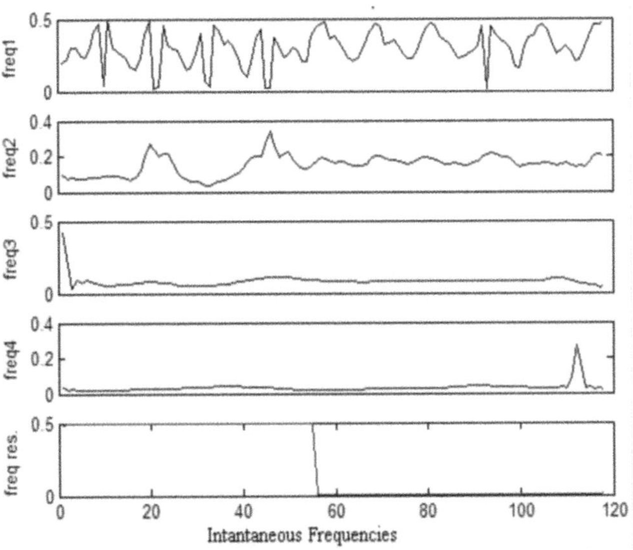

Fig. 3. Instantaneous Frequencies of every IMF and residual trend res

Table 1. The Average Period and Variance Contribution of Each IMF and Trend Res

IMF	IMF1	IMF2	IMF3	IMF4	RES
Average period (month)	3	6	12	32	
Variance contribution	8.5%	1.9%	2.6%	3.6%	83.1%

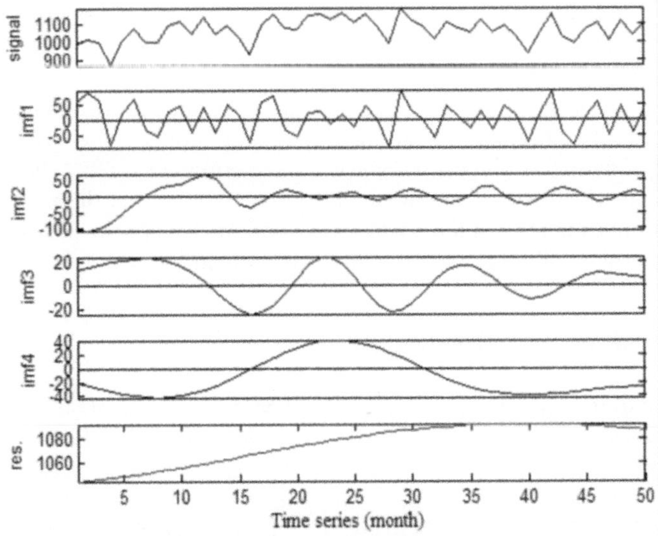

Fig. 4. IMFs and residual trend res of international tourist arrivals to China from 2006 to 2009

Table 2. The Average Period and Variance Contribution of Each IMF and Trend Res.

IMF	IMF1	IMF2	IMF3	IMF4	RES
Average period (month)	3	6	12	28	
Variance contribution	50%	25.2%	4.2%	15.4%	5.2%

Form the short time scale, we decomposed the historic data in about latest 4 years from the current, total of which including 50 data. Through decomposition, Fig.4 clearly illustrated the detailed information of every IMFs and the trend res, and the average periods with its corresponding variance contribution were given in Tab.2.

4 Results Analysis

Through the long-term and short-term time series, fluctuations of the international tourist arrivals growth at multi-time scales were detailed illustrated respectively. Whatever the length of the sequence, the data were decomposed into four IMFs and one residual trend term (Fig.2 and Fig.4). In the long run (Tab.1), IMF1 expresses a period of 3 months' fluctuation, while IMF2, IMF3 and IMF4 denote fluctuations of 6, 12 and 32 months' periods respectively. Herein the main fluctuation periods of the international tourist arrivals are 3 and 32 months according to their corresponding variance contribution, total of which is 13.1%. This result demonstrate that the seasonal factor is the main impact component in tourism industry, while another fluctuation with large time scale often ignored in previous work, about 3 years, must be additionally paid attention to. In spite of obvious period fluctuation clearly shown, the dominant factor to control the change of international tourist arrivals in the long term is the residual res, because its variance contribution reaches up to 83.1%. Above results state clearly that international tourist arrivals to China will rise up continuously in the long run.

However, from the short-term decomposition result, the residual res is no longer the dominant factor because its variance contribution only comes to 5.2%. The period of 3 months with the biggest variance contribution, i.e., 50%, becomes the main fluctuation characteristic of international tourist arrivals, and the period of 6 months' fluctuation comes next with the 25.1% contribution. This result shows that in the short term, the international tourist arrivals to China will still stay in intense fluctuation.

5 Conclusions and Discussion

The computational intelligence method of EMD can be applied to decompose any non-linear system and detailed illustrate the fluctuation rule behind the data. Through our study on international tourist arrivals to China from the long-term and short-term time series respectively, four fluctuation periods, residual res and their corresponding variance contribution are all obtained, which are clearly illustrated in Fig.2, Fig.4, Tab.1 and Tab.2, thus giving us a clear future dynamics of international tourist arrivals. In the long run, international tourist arrivals will rise up continuously, while intense fluctuation with periods of 3 and 6 months are the main change characteristics in recent years. Consequently, intense fluctuation should be paid more careful consideration when establishing recent tourism plan and long-term measures to deal with the large tourists flow will also be endeavored in immediately.

In this paper, we firstly attempt to introduce computational intelligence method of EMD into tourism industry, and time series of international tourist arrivals to China, i.e., an important impact factor for tourism development, is detailed decomposed and analyzed. In fact, many non-linear systems can take advantage of EMD as the first decomposition step to detect the change rule behind complex data series, and this intelligence method will be inevitably applied to more fields. For our research, a computational intelligence system based on EMD method will be established in the next work, and we can use it accurately to predict future international tourist arrivals and provide theoretical guidance for government's tourism policy.

Acknowledgment. L.L. Chen would like to thank Dr. Hui Yang for her great help with this paper and we would like to express our gratitude to the anonymous reviewers.

References

1. Lei, K.W., Chen, Y.: Forecast of inbound tourists to China based on BP neural network and APIMA combined model. Tourism Tribune 22(4), 20–25 (2007)
2. Wu, J.H., Ge, Z.S., Yang, D.Y.: An artificial neural network to forecast the international tourism demand– – Taking the Japanese demand for travel to Hongkong as an example. Tourism Tribune 17(3), 55–59 (2002)
3. Huang, N.E., Shen, Z., Long, S.R.: The empirical mode decomposition and the Hibert spectrum for nonlinear and non-stationary time series analysis. Proceedings of the Royal Society of London 45(4), 903–955 (1998)
4. Huang, N.E., Shen, Z., Long, S.R.: A new view of nonlinear water waves: the Hibert spectrum. Annual Review of Fluid Mechanics 31, 417–457 (1999)
5. Chen, L.L., Lin, Z.S., Guo, J.: Research on Chinese future grain security based on the method of EMD. Scientia Agricultura Sinica 42(1), 180–188 (2009)
6. Zhang, Z.Z., Lin, Z.S., Du, J.L.: Analysis on multi-scale cycles of solar activity with the data of tree-ring (1511-1954). Scientia Geographica Sinica 29(5), 709–714 (2009)
7. Zhang, Y.G., Lin, Z.S., Liang, R.J.: EMD-based prediction on dunamics of land carrying capacity in Shangdong province. Scientia Geographica Sinica 28(2), 219–223 (2008)
8. Huang, D.J., Zhao, J.P., Su, J.L.: Practical implementation of the Hilbert-Huang transform algorithm. Acta Oceanologica Sinica 25(1), 1–11 (2003)

Digital Technology Application on the Teaching of Basic Constitution Professional Design

Li Wang

Department of Art Design,
University of guilin Science and Technology,
Guilin, Guangxi Province, China
LiWang0012@foxmail.com

Abstract. Three constitution design as a professional basic course, since it was introduced in china from bauhause in 1980's have been more than half a century, times changed, especially computer technology has become a basic design instrument, this technology, have give a new fashion and how to adjust such changes should not only in technical, we should with the help of modern technical and the students' actural situation take methods to stimulate students' learning initiative and creativity in teaching.

Keywords: compute technology, art design, three constitution, teching innovation.

1 Introduction

In the early 20th century Bauhaus have set up the basis design program independent from professional design.include plane constitude, color constitude and space constitude courses. Therefore various visuak special design have the same basic identical content. Today the traditional design have entered scientific ear, from early 1980s our country large-scalely introduced bauhaus teaching system, 1990s constitution courses have become design compulsory education, widely applied in design practice. Now, in the 21st century, art design have processed many kinds of changes, also more diversified, especially in recent years the design education becomes a ball of fire. more than ten years ago, the colleges introducted art and design education only few, but today, the colleges and institutions have increased by almost more than one thousand. Professional institutes, colleges, all kinds of integrated vocational schools have trained a large number of specialized students with the development of design education fashion. Which make our domestic aesthetic level up to a new hierarchy, but no matter what the art school was, thinking how to take courses professional, the three constitutions still have its important position, because it is the basic and including catalogue and all kinds art principles, rules and laws, with uncertain external visual form to ensure the permanency of beauty.

For modern art of writing practice, three constitution will increase the ability of thinking, and illuminate the design inspiration, which is the foundation. its ultimate purpose is to extend the design thought, integrate the design method of rational thinking and feelings and make a solid foundation for the future design. "Innovation"

J. Zhang (Ed.): ICAIC 2011, Part V, CCIS 228, pp. 246–251, 2011.

is a hot topic discussed in new era, how to pursue the times change is a big problem we have to face. Since three constitution introducted in our country,its innovation has been the question of educator concerned, especially in recent years with the thought of opening up, and foreign exchange, many universities have conducted a great innovation for performance art design foundation courses. Many schools are to make innovations, reforms to how integrate the three constitution courses and professional courses together more closely. but because of the art and design history of our country is not so long, and the academic system still need to improve, more than to art and design the professional basis system, this is a long way need to explore.

A. New Changes with Modern Technology

The internet is the most notable application in 20th century of human social development from 1969, half a century, the computer network have successfully processed from experimental in the world into commercial operation. tradition education and modern science technology also experiencing from the beginning of the opposition fusion develop coordinated today, they are affected and inflitrated each other. at First gharopiwus and bahaus have the ideal which is treat people as the dimension and they should have balance and in all-round development with the education. especially as the basis leader of the bahause courses, john eton professor, in the spirit of integrated the traditional eastern and western scientific culture progress, in order to overcome the modern crisis. He stressed the importance of inspiring students' personality, and he classify the stydents into three types: spirit expression, reasonable instruct and truth expression, different types should educate in different direction, today the types seems should to add a partition that is the technical performance, because the computer technology give three constituions a lot of changes, and which was just after the computer applied results. It is precisely this new technology has created new problems: traditional education and new technologies how to get along ?

B. Differences between Traditional Education and Modern Technology Applied Education

In our society, traditional chinese design education development from beginning to the present is enormous and profound, but with the spread of new technological evolution, the coming of the sixth phase of the internet era (the first stage : form and sign language phase, the second stage : time phase, the third stage : words phase, the fourth phase : printing phase, the fifth stage: electronic age, the sixth phase: network era). Until now, the function of the art design have reached its extreme development, so the art design education is gradually from the traditional educational methods changed to adapt network era. However at the same time, traditional education still remained its principle which was not inflitrated by the development of modern science and technology, while modern technology and human culture have developed rapidly into a new era. In regard to the relationship between traditional education and computer technology operation, there are two different options.

Some people in our teaching work ignore the computer technology, they think that idea is the most fundamental thing in design, the computer technology is secondary. But we all know undeniably, the computer technology in creating implementation process has been a considerable effect, with the development of science and technology applications we could improve the efficiency of our work, we can rely on computer technology to save more time for training work in mind.

Another point of view is too much attention about computer technology, we oftern find many students of science and engineering, they have a natural fear of the hand technology, they could do nothing without computer. I think this situation should have two reasons, one of them is the popular of computer make many students think without the computer will not be able to design, what is design? They think granted that computer drawing is design. the other reasons is the students graduated from school to seek job. computer technology is the important standards to pass the job examination, many units will ask your ability of operation the photoshop, autocad or 3D, if you say no, you will lost the chance of job. This entails lead the students began to take much time in computer training, even think that hand expressed is nothing. In fact the art design is different with other work, it is one of the visual language, if you want to cast the expression of design feeling, you have to lean to hand ecpression,if not, you will unable to carry out specialized design.

I have the pleasure have worked in my department teaching three constitution courses for five years, from the process of my been taught experience since a students to be a teacher to teach other students. The course: plane, color, and three-dimensional constitutes content have a great change and can not help bringing me a lot of thinking, with my teaching experience, I ventured to talk about under the new technology inviroments how to improve the students creation capacity with our teaching.

2 Combution New Techolonogy and Design Education

Traditional education and modern technology have experienced through decades friction, "innovation" always is the word oftern been referred to when new technology appears in the field of education, the new technology and education about how to combine in constitute teaching, teaching innovation is the key we all need to think of.

From the first stage when three constituions courses just introduced from western, the computer technology just begun, seldom applied in education field. The teaching conditions similar with bahaus school situation, so we costed very large energy on introduction foreign teaching experience in for our teaching content, teaching system with a broad "modern" and public opinion, showing western standard, rational in big position on judge. and disregard truly image expression, the work is mechanical and callous. we all think point, line, plane are abstract, and re-organizing themselves, although this is a rational thinking training phase we must experience. But we see it as the end. In fact the aim of constitution is to learn a way of thinking, its ultimate aim is not the work expression. Now computer is a very popular tools, the internet is an indispensable part, we will have to think of how to combute better this new technical conditions between education and modern technology, how to adjust such changes, which should not only in technical, we should with the help of modern science for better education methods. Because design teaching means how to stimulate students' learning initiative and creativity.

Three constituion course as teaching should take the advantage of the computer and the hand drawing, so that give students a better understanding for form element. My department in the curriculum arrangments on the three constituions and computer skills are closely linked together, in the plane constitute and cad, the plane constitution is 64 programes, coreldraw experiment teaching take 10 Class time, in the period of colours constitutes and cad, 48 class teaching include 10 photoshop experiment teachin, three-dimensional constitution for the period does not link with the computer technology, if the curriculum can include 3d computer technology, will give a great help for the students in the form understanding.

3 Innovation about Constitution Teaching

A. Students Analysis

In the teaching process, the students level are uneven, some of the students did not have good fine art, while part of the students have passed a preparatory training before entering college, therefore some of them have good understanding about teaching content, so that as a professional teacher, teaching should seek methods to adapt to their study ability, for the aim of the plane, color and solid constituions teaching, not only should earnest study and explore in teaching philosophy, pattern, but also should aim at createing ability, teaching people to do some studies. The content of teaching should not be prejudiced, only from narrow education thoughts, regard the bad handdrawing students lack of art feeling, for the reason ignore them. Modern psychology theory prove that logic of science and art imagination is not in contradiction. The art psychologist Arngeim in the <arts and visual consciousness> said : "all thinking involve reasoning, all the visual involve intuition, all the observations involved creation." every student has the potential to be one of the ability on image thinking. the key is how to stimulate their creative thinking and lead them to think with what kind of teaching methods and patterns.

B. Get Rid of Stereotype Training

The stereotypes of training In the teaching include for example the same teaching content and the same homework in training, a lot of heavy hand work wast much time and ignored the computer technology applications.

We all know different design major should have different emphasis on constitution teaching, for example, the environmental art design and industrial design should take more time on the three-dimensions shape while visual communicate design and costumes design should focus on a two-dimensional graphic expression on two-dimensional shape instead of three-dimensional representation, and they all should to understand the materials technical, not all the design major have the same point, line, plane. We should not be a copy teaching instruction.

Therefore homework content should be more target, and the same time we should make good use of computer technology for improving efficiency, such as three-dimensional courses, the students can use a 3d computer software to see the form changes before do it in reality, and then through cut up and fold to strengthen intuitive felling. Because sometime the print machine lost the true color reason, some color constitute job is in the key to understand the difference in color, hence these work should be completed in hand. While the job application of color we can use the computer to do. Almost most of all plane constitution homewok can use the computer to making.

At the last demonstrate work step, the final work should include the draft of driginal ideas at begainig phase and all aspect process untill his last work should be linked. At the same time, students should have the ability to talk about their innovative ideas, so that to exercises their thought logic and oral expression.

C. Strengthen the Ability of Image Thinking Expression

At past, we thought the plane constitution just work about dot, line and plane. Therefore training are stereotypes and mechanical. We neglect the image thinking, the students' paperwork are much like and grim. But the constitution teaching aim is combine sensibiliity and rational ability, so the teaching process should pay more attention on combining different thoughts. The teaching concept of dot, line and plane should be more deep and wide, inspire students to develop creativity, give dot, line and plane new visual form and content. Exercise can do more work for emotional expression, for example do some listening exercise convert to the visual experience, give some smell concept convert to the visual experience and emotiona feeling convert to the visual experience. These special subject training aim to improve the emotional thinking. These subjects also aim to train strong comprehensive ability on the dot, line and plane so that student can learn easily and effectely about the plane constitution knowledge.

Western have strengthen the feeling expression about constitution form since 20 or 30 years of the 20th century, some of the reseachers trying to trace the story and come into practice for a form exploration, to make ideas and form integrate in a more organic method. Constitution design is a curriculum about training the art language and the power of expression, by a large number form training, rich artistic thought and improve the students ability to use the art element freely and the ability to express their feelings.

4 Conclusion

As the basic subjects of design constitution should make new technology link with design closely, this is an inevitable trend, we should make use of computer technology for improving the efficiency of the students work, but as a design work such a special expertise we can't get too much reliance on computer technology.We should with the help of modern science to make better teaching, because education means to stimulate students' learning initiative and creativity.

In the classroom teaching have a considerable amount of homework to strengthen students understanding about emotional expression of constitute form, a constitution design should develop students' ability from three sides : first of all, students can obtain the information from different areas through a variety of techniques and the same time learn to summarize and refine design elements; secondly, take importance on the matter of material feeling and learn to break through the physical limitations;

thirdly, training a all-round control ability and the ability of aesthetic ideas always throughout the design process.

The design times require design education must be good in time, teaching in the classroom instruction cannot meet the needs of students learning, internet age, people work efficiency greatly improved, provide great convenience for our design. How to deal with the matter about technological and artistic design thinking and creativity, I think is the key to education of design innovation, in a few years of design teaching practice, the students' works will cause a lot of educational research. Of course, there are many places that not mature will continue to improve in the future.

References

1. Xue, C.: Indestry history, pp. 35–42. Huazhong University of Science and Technology Press, Wuhan (2005)
2. Wang, Z.: My teach and learn from Product modeling course. Package and Design 106, 46 (2001)
3. Wang, Z.: Plane constitution teaching research in industrial major of science college, pp. 50–53. The National Industrial Design Seminar Collection, Jiangnan University (2003)
4. Wang, S.: Scan and dialysis, pp. 33–35. People's Fine Art Press, Beijing (2001)
5. Li, L.: Humanistic quality, comprehensive ability and personality - Interview. Product Design 3, 53–54 (2003)

Recognition of the OPGW Damper Based on Improved Random Hough

Guoqing Liu[1], Liang Zhao[2], and Yingshi Wu[2]

[1] Department of Aeronautical and Mechanical Engineering Air Force Aviation University Changchun, Jilin Province, China
[2] Changchun Power Supply Company and Department of Information Engineering Northeast Dianli University Changchun and Jilin City, Jilin Province, China
GuoqingLiu116@tom.com

Abstract. Detection and recognition of obstruction is the key technique of OPGW navigational Robot. For the obstruction of OPGW, A method is presented in this paper that takes advantage of image processing technology to detect the damper. After processing the acquiring image, make use of improved random Hough Transform to randomly select one dot and calculate the other tow dot by way of adding the condition ,which improves blindness of randomly selecting three dots.This method can detect the circle part of the obstruction and impose the condition of a fixed restricted shape, then detection of a damper is realized.

Keywords: Navigational Robot, Obstruct, Detection, Random Hough Transform.

1 Introduction

Nowadays with the wide use of OPGW cable in power system, requirements for OPGW cable detection is constantly improved, making detection to be indispensable. At present, there exists several methods for OPGW cable detection.

1) *Ground eye-measurement method*: Making detection for OPGW cable within scope by naked eye or telescope, which is high labor intensity、poor reliability and low efficiency and precision.
2) *Aerial method*: Helicopter navigation. When helicopter flies along OPGW cable, the staff with naked eye or on-board camera observe and record the abnormal. Though such method is close to the cable, which improves the efficiency and precision, the cable quickly pass from the vision of observer or video equipment, causing high technical difficulty and the cost.
3) *Working in the OPGW cable*: Such detection fits for the cable across river、lake and railway whereas it is low-efficiency and high-danger, thus causing accident.

That is why we do research on navigational robot of optical fiber composite overhead ground wire. Because of many obstacles being along the wire, it brings trouble to navigate for robot. In this paper it takes Hough transform to detect and recognize obstacle.

J. Zhang (Ed.): ICAIC 2011, Part V, CCIS 228, pp. 252–259, 2011.

2 Hough Transform

Hough Transform is presented by Hough, it takes image coordinate from one space to another one following voting principle, when the transformed space makes grid then the transformed point falls into sub-grid, the grid number adds up by one. By testing local peak it will recognize the target image. This way is effective for linear detection, but the computation turns to be quite large for circle recognition and it also consumes too much memory, thus presenting random Hough Transform. That is, for circle detection it will randomly selects three points that are not in the same line to make up of circle as candidate and then to make quantity accumulation. For being randomly selected, it calls for massive redundant computation. This paper applies such improved method to damper recognition.

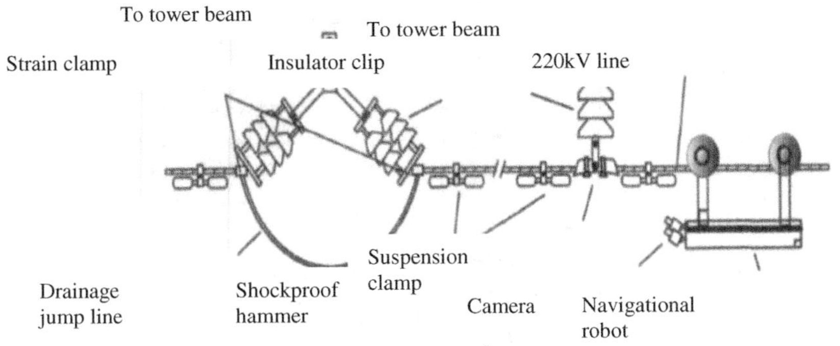

Fig. 1. Structure of OPGW cable

3 Damper Recognition of OPGW Navigational Robot

Figure 1 shows the structure of OPGW cable, in general, the obstacle of OPGW cable includes damper、 strain clamp and suspension clamp. When navigating along the cable, the robot needs recognizing and stepping across the obstracle.

A. Wire Recognition

From all the obstacles wire recognition is the most important. Wire can not be as obstacle while it can restrain to identify the one and wire can be recognized by Hough transform. Above the image the wire tilts from top to bottom, because identification may detect many lines there must be some rules existed. Supposing $\theta_1 \langle \theta \langle \theta_2$ (θ_1 and θ_2 are constraint parameters, θ is angle between wire and horizontal gradient by experiment.) it will make the abscissa of wire endpoints identified to follow childhood sequence, (x_1, x_2, K) subtracts two abscissa nearby like $(x_2 - x_1, x_3 - x_2, K)$, if the difference meets in a certain range ($x_2 - x_1 \le d$, K, d is the certain width made by experiment), that can judge as wire. As follows

$$\begin{cases} \theta_1 \le \theta \le \theta_2 \\ x_2 - x_1 \le d, x_3 - x_2 \le d, \ldots \end{cases} \tag{1}$$

Figures below are the images fit by experiment, θ_1 = 80 °, θ_2 = 90 ° of the pictures figure 3(a) is the outline graph near shockproof hammer acquired and disposed and figure 3(c) is the identified graph of OPGW cable disposed by image process, judging from the figures it can make good recognition for wire by such algorithm.

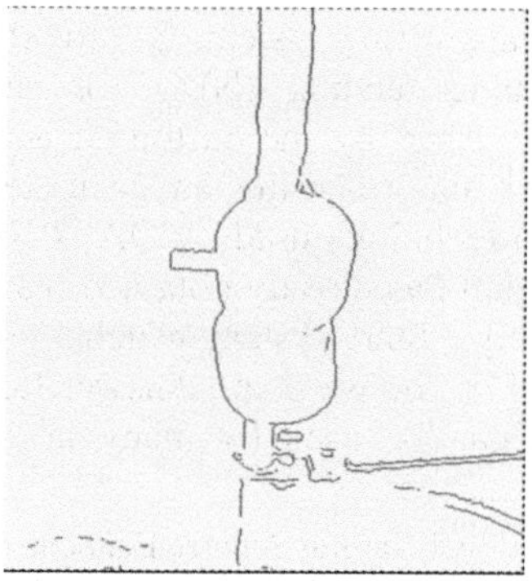

Fig. 2(a). Outline graph

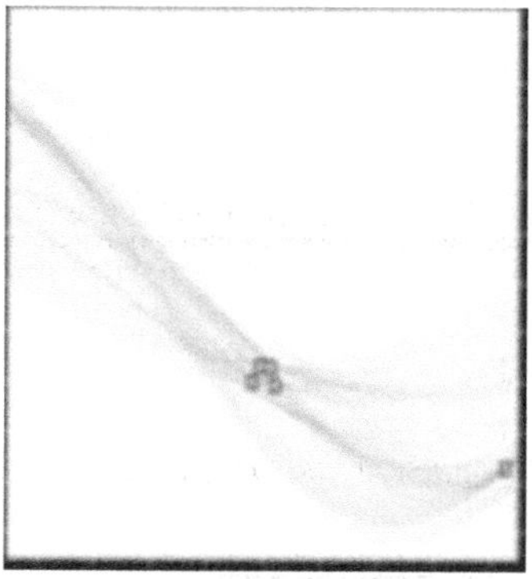

Fig. 2(b). Hough space

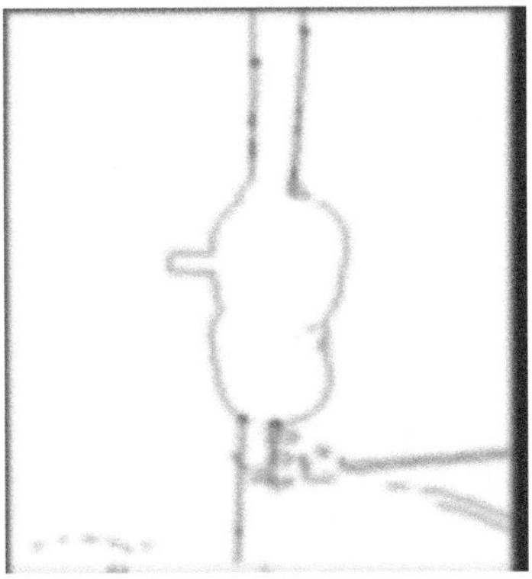

Fig. 2(c). Identified wire

B. Recognition of Shockproof Hammer

Shockproof hammer is installed under the wires, connected by two hollow cylinder. When it projects to camera, its section appears to be arch in the outline graph, checking shockproof hammer being by detecting the arch. The arch detection algorithm is more complex while it can be placed by its circle detection.

1) Selection of random dots

Initialized array stores all the edge points into D, selecting one point randomly from D as Y.

In our opinion if the point is in the circle then the gradient of this point must pass the center of circle, as figure 3 shows, OA and OB can be catenary. Gradient calculation follows reference, gradient of pixel point $f(x, y)$ defines as $\Delta f = [G_x, G_y]^T = \left[\frac{\partial f}{\partial x}, \frac{\partial f}{\partial y}\right]^T$ direction angle of gradient at (x, y) is $\beta(x, y) = \arctan\left(\frac{G_y}{G_x}\right)$,which is the direction of YO based on the axis. Taking Y as datum point, it launches rays to the either side of YO to form angle α. (Angle α is defined by experiment). Due to the continuity of circle, it intersects to the point U and Q ,all the three points mentioned above determine one circle.

We will find two points on the line YU and YQ in the edge point set then compute the distance to the point Y separately in order to decide point U and Q, from which to Y is the equal distance. But there are two problems met them : (1) In the recognition of shockproof hammer to detect arch if angle α is big enough there may calculate that the nearest point to Y is near to Y . (2) The calculated points are not on the circle but on the reverse extension line. For the first question, there solves that the radius of shockproof hammer is between r_1 and r_2 (r_1 and r_2 are the lower limit and upper limit of shockproof hammer radius that are little difference.) For the two rays direction it can calculate length of line YA and YB , $d_a = r_1 \cos\alpha$, supposing $d_\beta = r_2 \cos\alpha$,in the YU and YQ direction, it should find a point from which to Y Y the distance d meets $2d_a < d < 2d_\beta$,if not then try to find another point. For the second question it should calculate the vector from Y to the point we get, and then find out the angle between this vector and the direction of gradient, if it is acute angle, that is the one we want or else if being obtuse angle try to find another one. Above all the three points to form circle is determined.

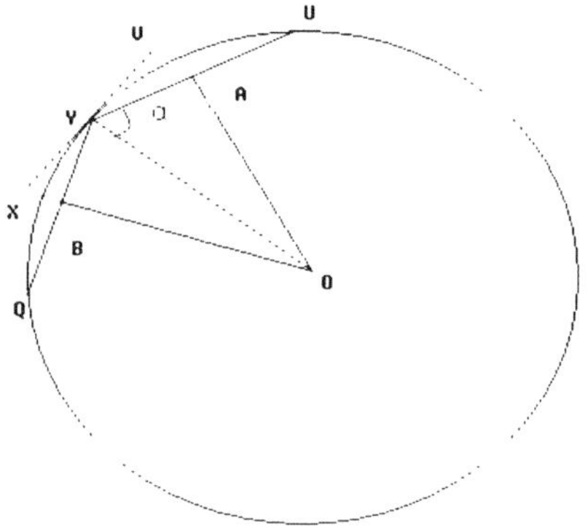

Fig. 3. Moutline graph

2) Circle area determination

After finding the candidate circle center coordinate $O(a,b)$ and radius r , then it compares the distance between the edge point from image space and candidate circle, so as to make sure if it is on the circle. In this paper it excludes all the points not on the circle, avoiding unnecessary computation to improve detection accuracy. Ideally speaking, for the points on the circle O it only addresses in the area between circle circumscribing square ABCD and inner square EFGH (the grey area shows in the figure 4), each square length is $2r$ and $\sqrt{2}r$,the center of which is O. Considering the image is discrete, that needs to be threshold δ $(\delta > 0)$ .Assuming the point processed is $p(p_x, p_y)$,it meets

$$\left|\sqrt{(p_x - a)^2 + (p_y - b)^2} - r\right| < \delta \tag{2}$$

Then point p is the one on the candidate circle.

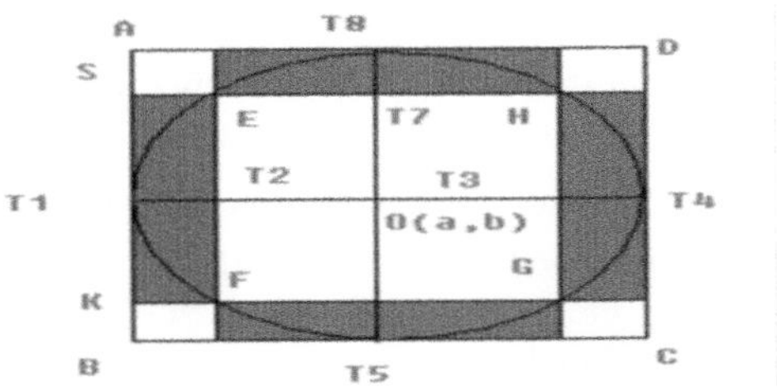

Fig. 4. Candidate circle area

As known above the length of circumscribing square ABCD and inner square EFGH are $2(r+\delta)$ and $\sqrt{2}(r-\delta)$.

Originally

$$T_1 = a - r - \delta, T_2 = a - \frac{\sqrt{2}}{2}(r-\delta), T_3 = a + \frac{\sqrt{2}}{2}(r-\delta),$$

$$T_4 = a + r + \delta, \mathrm{T}_5 = b - r - \delta, T_6 = b - \frac{\sqrt{2}}{2}(r-\delta),$$

$T_7 = b + \frac{\sqrt{2}}{2}(r-\delta)$, $T_8 = b + r + \delta$, for rectangular area the pixel coordinate are

$$\begin{cases} T_1 < P_x < T_2 \\ T_6 < P_y < T_7 \end{cases} \tag{3}$$

The same as the other area as follow

$$\begin{cases} T_3 < P_x < T_4 \\ T_6 < P_y < T_7 \end{cases} \tag{4}$$

$$\begin{cases} T_2 < P_x < T_3 \\ T_5 < P_y < T_6 \end{cases} \tag{5}$$

$$\begin{cases} T_2 < P_x < T_3 \\ T_7 < P_y < T_8 \end{cases} \tag{6}$$

Such method eliminates many points unrelated and improves efficiency. It only judge whether the points from boundary set meet one of the (3)~(6),if so it can calculate the distance to the candidate circle in order to make sure if it is on it.

3) Determination Circle area pixel

Among all the candidate set met the circle edge points $p(p_x, p_y)$,if it meets $\left|\sqrt{(p_x-a)^2+(p_y-b)^2}-r\right| < \delta$

then p can be as the point of candidate circle, counted by one.

By evidence accumulation it counts M1 that is the number of image boundary on the candidate circle, if it is greater than the minimum number of arch, $M_{\min}$ ($M_{\min} = 2\lambda\pi r$, λ is the defined factor) it can consider the candidate as the actual circle, deleted the points on the circle from D, making another circle detection.

4) Additional conditions attached

For this paper confirms the arch being by identifying the circle belonged by arch, there must be some wrong detection, so it should supplement another conditions for the detected circle : the circle radius must be $r_1 < r < r_2$ the location of circle center is constrained by its abscissa, $x_a + d_1 \langle a \langle x_b + d_1$ (x_a, x_b are the endpoint abscissas these are the detected line at its left side and right side, $d_1 = 3|x_b - x_a|$).

5) Experiment

Pictures below are fit by algorithm so as to detect shockproof dammar in the laboratory.

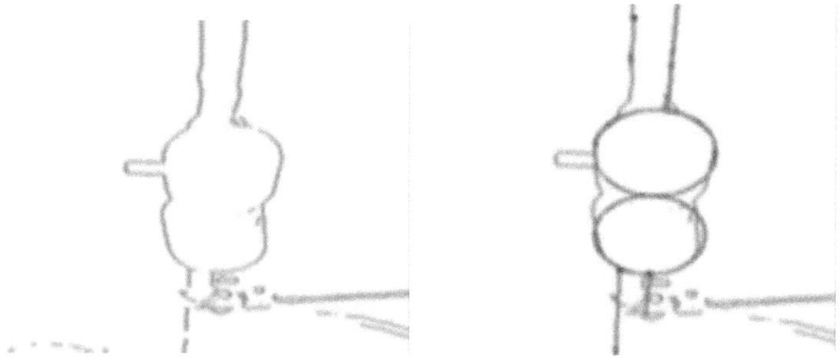

(a) Outline graph (b) Recognized graph

Fig. 5. Simulation graph of shockproof hammer fit in the lab

C. Comparison of Hough and This New Algorithm

Traditional transform is the complex algorithm in three dimension space, which is only detected regular circle. However, the algorithm in this paper not only simplifies the computation spatial dimension but also realizes arch, detection. With the application of this algorithm, detection time consumes 222.48ms for processor frequency of 1G, while time being 1118.974ms for traditional Hough algorithm detection.

4 Conclusion

In this paper it presents a way to detect shockproof hammer based on the visual sight, applying visual sensor to collect images and extracting edge information contained obstacles. It makes typical geometric figure detection with random Hough transform, using operation condition of navigational robot to realize structure restraint and then detects shockproof hammer. But when the light becomes too strong, this method happens to be wrong so it needs further improvement. In addition, based on such method, it can make visual sensor to realize obstacle position、 robot posture detection related to wire and the distance detection between robot and obstacle.

References

1. Sawada, J., Kusumoto, K., Munakata, T.: A mobile robot for inspection of Power transmissions. IEEE Transactions on Power Delivery, 309–315 (1991)
2. Hough, P.V.C.: Method and means for recognizing complex patterns. U.S. Patent 3, 069654, December 18 (1962)
3. Xu, L., Oja, E.: Randomized Hough transform (RHT): basic mechanisms, algorithms and computational complexities. CVGIP: Image Understanding, 131–154 (1993)
4. Hu, C., Wu, G., Jiang, G.: The research for navigational robot obstacle visual inspection and recognition on hihg-voltage power transmission line. Chinese Journal of Sensors and Actuators 21(12) (December 2008)
5. Zhao, G., Huang, S.: The improved alogrithm for circle inspection based on random Hough transform. Computer Technology and Development 18(4) (April 2008)

Pheromone Based Self-routing of Intelligent Products in Dynamic Manufacturing System*

Ke Shi[1], Xinmiao Chen[1], and Xuan Qin[2]

[1] School of Computer Science and Technology Huazhong University of Science and Technology Wuhan, Hubei Province 430074, China
[2] Office of Science and Technology Development Wuhan, Hubei Province 430074, China
keshixin53@163.com

Abstract. This paper illustrates the capacity of a pheromone based self-routing control model for intelligent products to autonomously decide efficient processing operations and routing paths in dynamic manufacturing systems undergoing perturbations. This method is inspired by the behavior of foraging ants that leave a pheromone trail on their way to the food. Following ants use the pheromone trail with the highest concentration of pheromone to find the shortest path to the food. Self-routing of intelligent products in dynamic manufacturing systems imitates this behavior in a way that whenever a product leaves a manufacturing resource, i.e. machine, the product leaves information about the performance at the respective resource. The following products use the data from the past to render the routing decision. The discrete event simulations are analyzed by comparing statistics on throughput time data resulting from the system's behavior in dynamic order arrival and machine breakdown situations.

Keywords: intelligent products, self-routing, pheromone, manufacturing system.

1 Introduction

Nowadays the environment within which manufacturing systems operate is characterized by more rapid change than ever before. The unforeseen disturbances occur frequently and the manufacturing system has to be able to react to these disturbances. One possible solution to deal with this high level of dynamics in manufacturing system is intelligent product driven manufacturing control paradigm. An intelligent product can be understood as today's products enriched with competencies and abilities for decision-making and interaction with its environment. McFarlane [1] describe the intelligent product as a physical and information based representation of an item which:

(1) possesses a unique identification;
(2) is capable of communication effectively with its environment;

* This work is partially supported by the National Natural Science Foundation of China under Grant No.50805058, the Fundamental Research Funds for the Central Universities (HUST 2010MS083), and International S&T Cooperation Program of Hubei Province.

J. Zhang (Ed.): ICAIC 2011, Part V, CCIS 228, pp. 260–268, 2011.

(3) can retain or store data about itself;
(4) deploys a language to display its features, production, requirements, etc.;
(5) is capable of participating in or making decisions relevant to its own destiny.

In intelligent product driven manufacturing control paradigm, the intelligent products manage and control their manufacturing process autonomously. These products are self-routing. Which operation should be the next operation a product undergoes and where this operation will executed is decided by the part itself based on acquired local information.

This leads to a shift from central and static planning in advance to distribute and dynamic manufacturing control in real time. Each intelligent product makes its own decisions by only taking into consideration the necessary information available in its environment. This avoids extensive and time-consuming calculations and offers the possibility to react quickly to sudden changes.

The intelligent products can be realized by novel computing and communication technologies such as wireless sensor networks, Internet of Things, cyber-physical systems, and etc. However, most existing manufacturing systems still consider products to be passive entities: they never communicate, decide or act during the manufacturing process. To promote intelligent product driven manufacturing control paradigm, new product-centric control methods including operation allocating, scheduling, routing should be developed. Furthermore, these methods must be adaptive to resource constrained computing environments.

Given these realities, we have begun to work on the distributed control of dynamic manufacturing systems using the notions of intelligent products and pheromone to provide efficient, real-time, adaptive product routing.

This paper is structured as follows: A short introduction of pheromone based approach applied to manufacturing control is given in Section 2. Section 3 presents the manufacturing context in which our approach is applied. Section 4 explains the proposed pheromone based self-routing approach for intelligent products. Performance evaluation and results are presented in Section 5. Finally, conclusions and perspectives for future research are presented in Section 6.

2 Pheromone Based Approach: Concepts and the State-of-the-Art in Manufacturing Control

Pheromone concepts come from social insect colonies. Social insects leave an evaporating substance called pheromone on their way and the following insects follow the trail with the strongest pheromone concentration.

For intelligent product driven manufacturing, self-routing means the decision on which path to choose through the manufacturing system is not made by a central controller, but by the individual intelligent product itself. Pheromone based approach is similar to the organization of an ant colony, where ants leave chemical messages for the following ants, thereby transmitting the optimal path to the food supply. Whenever an intelligent product leaves a machine, i.e. after a processing step is accomplished, the product leaves information about the performance such as the duration of processing and waiting time at the respective machine like ant leaving pheromone. The following intelligent products use the performance data from the previous products according to certain rules resembling the evaporation of the pheromone to render the decision about the next production step. That is to say, the intelligent products choose their processing routes based on the experience of other products of the same type.

The main advantages of such a self-routing approach are an added robustness against unforeseen changes in production times and machinery breakdown and an increased flexibility of the manufacturing system. Furthermore, our approach is comparatively simple to implement, especially for intelligent products that only possess limited computing and communicating resources, since we consider only simple decision making by using decision rules.

Recently, a few researchers have developed approaches based on Ant Colony [2] inspired scheduling in shop floor control. Most of the applications proposed concern the shop floor routing [3] and permutation flow-shop sequencing problem [4]. The pheromone based approach differs from these previously proposed methods, since no reinforcement of the pheromone trail takes place as there is no equivalent to ants returning their way back to the nest. The intelligent products disappear after completing the manufacturing. This approach is also different from concepts for ant-based routing and load balancing [5], as crowding as a consequence of limited capacity is not addressed.

Furthermore, centralized implementation still dominates in existing control prototypes [6] using Ant Colony Optimization, while distributed implementation is preferred in intelligent product driven manufacturing context. Some implementations [7] claim they adopt agent-based structure that is naturally decentralized and distributed. However, in these implementations, the agents, running on PCs instead of physical objects, depend on some kinds of data collecting and synchronizing mechanism to update their status, simulate the real production processes that the products undergo and make the decision. These kind of implemented systems become more hierarchical than a real distributed one. And the consistency between physical and informational flows is still a problem.

Scholz-Reiter et al. [8] proposed a pheromone based autonomous control method to a shop floor model in matrix format and tested it in different dynamic demand situations. They compared the pheromone approach with a method based on queue length estimator. The performance is the throughput time with a sinusoidal arrival rate of the parts. In contrast to previous work, this paper extends pheromone based approach to self-routing of intelligent products in a generic manufacturing system without imposing any constraints on the system topology or scenario.

3 Manufacturing System Context

The manufacturing system consists of a given number of resources; each resource is able to perform any kind of manufacturing/transporting operation so that the resulting manufacturing system is a pure general-purpose one. In such a system, self-routing of intelligent product decides in what resource the product will perform the next operation; therefore it is a general scheduling/dispatching problem. The manufacturing system has the following characteristics:

(1) Several product types have to be manufactured by the manufacturing system. Each product type has a predefined process plan deciding the operations that this type of product must undertake. Each product is assigned a due date.
(2) Orders for production of different products arrive randomly with an inter-arrival according to a special distribution, i.e. exponential distribution;
(3) The queues are managed by the First In First Out policy in order to investigate only the pheromone approaches policy.
(4) Each resource can breakdown randomly with a special distribution, i.e. exponential distribution.

In this research, the operations of resources include both the transportation operations of the material handling/transporting devices and the manufacturing operations of the machines.

Process plan of a certain product is represented as directed acyclic graphs (DAG) which enumerate all possible operation sequences that can be applied to this product. Graphs are commonly used to represent systems with complex connections between their components. Routing flexibility is well represented by such a graph based model, and efficient routing optimization is enabled.

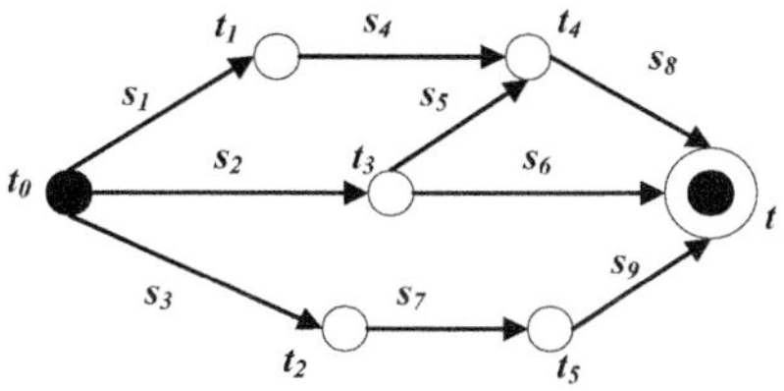

Fig. 1. Routing graph

As shown in Fig.1, each intelligent models its process plan as a DAG G = (S, T) with two distinguished vertices (start and end points). Nodes denote states $T = \{t_j \mid j = 1, ..., n\}$ and edges are labelled to represent certain resources $S = \{s_i \mid i = 1, ...,m\}$ that perform the processing operations. Generally, multiple edges between two nodes are permitted. The processing of an intelligent product begins from the start state t_0. In the end state t the intelligent product finish its processing. A graph G = (T, S) is called a routing graph. Note that branching in the routing graph corresponds to routing choice. For each intelligent product, self-routing problem is transferred into the problem of finding a shortest path from start state to end state in this routing graph.

4 Pheromone Based Approach

In the manufacturing system mentioned above the ants are the intelligent products that flow through the resources; the resources are the edges that ants must go through. When an intelligent product leaves a resource, it deposits a pheromone based on the predefined performance metric in the resource. The pheromone based self-routing approach and its main characteristics can be summarized as follows.

(a) Each intelligent product act concurrently and independently, and communicate in an indirect way, through the information they read and write locally to the resources.
(b) Each intelligent product searches for a minimum cost path joining its start and end states in the routing graph.
(c) Each intelligent product moves step-by-step towards its end state. At each intermediate state the intelligent product chooses the next edge to move according to the amount of pheromone left on this edge.
(d) While moving, pheromone for each edge is modified.
(e) Once the intelligent product arrives at its end state, its routing process ends.

Therefore, the pheromone based self-routing algorithm consists of two parts: routing decision making and pheromone updating. These two parts are described in the next two sections.

A. Pheromone Updating

There are two main problems that pheromone updating algorithm must address: (a) how the pheromone information is formalized, what type of information can be included; (b) the methodology of pheromone evaporation.

There are many measures that can be used to indicate the performance of manufacturing system, including resource utilization, throughput time, Work-in-Process (WIP) level, and etc. Since in most cases the main purpose of intelligent product driven control is to reduce product throughput time as a strategy to meet due dates, the information deposited by the products for the pheromone computation is the throughput time of a product (ant) that is processed on a resource (edge). In particular, the pheromone value deposited by a product on an edge s_j is computed by the following expression:

$$\Delta\tau_j = 1 / \left(t_j^{leave} - t_j^{enter}\right) \tag{1}$$

where t_j^{leave} is the time when the product leaves the resource s_j; t_j^{enter} is the arrival time in the resource s_j of the product.

From the above expression, the difference (t_j^{leave} - t_j^{enter}) concerns the throughput time of the product in the resource s_j. Therefore, if the throughput time is low, then the pheromone deposited by the product is high.

At the start of manufacturing, the initial pheromone value of each resource s_j is the reciprocal of average value of processing time performed by this resource.

Once the product leave a resource, the pheromone value will be updated by the following expression:

$$\tau_j(t+1) = (1-\rho)\cdot\tau_j(t) + \Delta\tau_j \tag{2}$$

ρ is the evaporation coefficient, which can be a real number between 0 and 1.0. Pheromones on all edges evaporate at the rate of ρ so as to diversify the searching procedure into a bigger solution space or out of local optima. In particular, a high value of ρ leads to high evaporation because the value is more significant and vice versa for a low value of ρ.

For each resource s_j, the value of pheromone is updated by all the products that flow through it.

B. Routing Decision Making

The probability to select a resource by the product is directed by both the pheromone amount on the route and the distance from its current state to the end sate. Product in the state t_i chooses the next edge according to the state transition rule in formula (3).

$$p_{ij} = \frac{(\tau_j(t))^{\alpha}\left(\frac{1}{d_j}\right)^{\beta}}{\sum_{j\in allowed-edges}(\tau_j(t))^{\alpha}\left(\frac{1}{d_j}\right)^{\beta}} \tag{3}$$

The parameters α and β tune the relative importance of the pheromone and the heuristic distance in decision making. The heuristic distance d_j in this study is the average sum of the travelling time between the current state to the end state.

5 Performance Evaluation

In this section, a discrete-event simulation model is developed to evaluate the performance of the proposed approach.

A. Simulation Environment

In our simulation, the manufacturing system consists of four general-purpose manufacturing/transporting machines (resources) that are called to manufacture a set of four different products. The minimal processing time per manufacturing step is equally 2 hours. The process plans for these four products are shown in Fig.2.

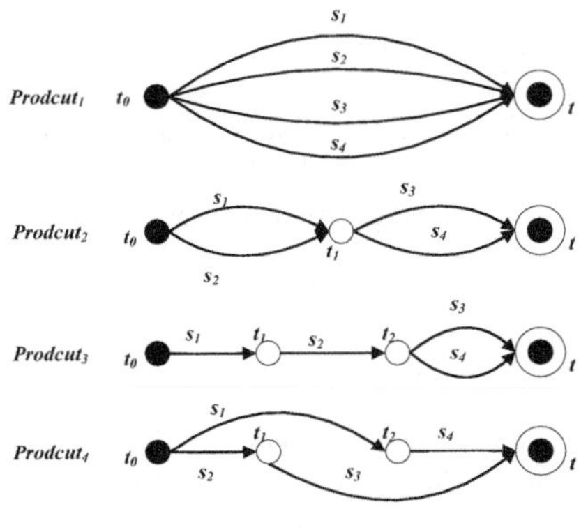

Fig. 2. Process plan for four different products

In order to emulate a dynamic environment the proposed approach have been tested through, the inter-arrival time of the products changes dynamically, and all the resources are subject to faults. The inter-arrival time is oscillating between two values with situation of over and under load. The resource failures occur in accordance with exponentially distributed time between failures, with mean time between failures (MTBF). Repairing times are also exponentially distributed with mean time to repair (MTTR). The simulation runs under different level of dynamic characterized by the changing rate and range of products' inter-arrival time and resource's MTBF and MTTR.

The evaporation parameter of the pheromone approach ρ is fixed to 0.5 in order to evaluate with the same importance between the last value stored and the new pheromone deposited by the product. The parameters α, β and heuristic distance d_j are all set to 1 in order to measure the effects caused by pheromone approach only.

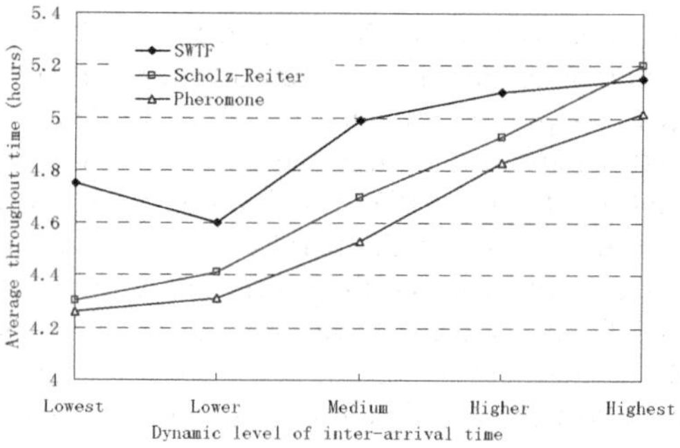

Fig. 3. Average throughout time vs. dynamic level of external demand (inter-arrival time)

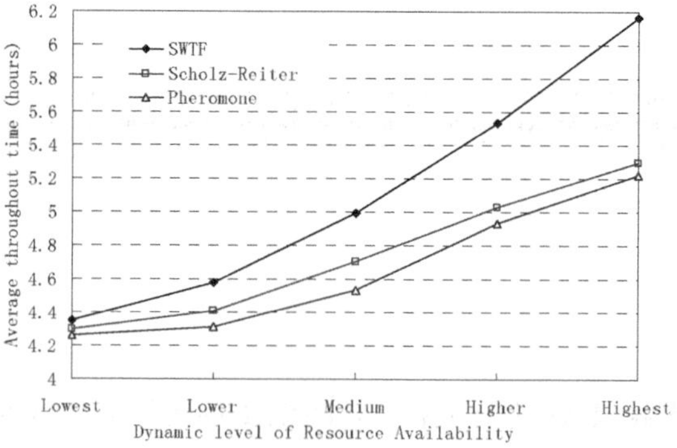

Fig. 4. Average throughout time vs. dynamic level of internal disturbance (resource breakdown)

B. Simulation Results

We mainly concentrate our analysis on the average throughout time of products. We also compare the performance of our approach against heuristic approach (Shortest Waiting Time First, SWTF) and pheromone approach proposed by Scholz-Reiter [8].

As shown in Fig.3 and Fig.4, the proposed pheromone approach leads to best performance as external and internal exceptions occur dynamically in the manufacturing system. When the dynamicity increases, the performance of all these three methods go worsen, and our approach is always the best approach.

Fig.3 illustrates the effects of external exceptions on the performance. Given a medium level of resource dynamicity, pheromone based approaches lead to better performance than SWTF approach. When the demand fluctuation increases, the performance improvement becomes less significant, and Scholz-Reiter's approach is even a little worse than SWTF. This is caused by the fact that the pheromone method needs some time to react on changing conditions and is less affective if the dynamics gets too high. Our approach outperforms Scholz-Reiter's approach because our pheromone updating method reacts faster to the manufacturing system than the moving average method adopted by Scholz-Reiter's approach. And our pheromone updating method, combing the new pheromone value with the old value is simpler than the moving average for the data management in distributed environment.

The effects of the internal exceptions on the performance are shown in Fig.4. Given a level of medium demand dynamicity, pheromone based approaches perform better than SWTF approach. When the resource dynamicity increases, the performance improvement becomes more significant. This is because the performance of SWTF depends on the foresting precision of SWT. When resource availability varies frequently, the processing time can not be predicted accurately. The inaccurate SWT leads to declined performance. For the same reason mention above, our approach still performs better than Scholz-Reiter's approach.

6 Conclusions

The research presented a pheromone based self-routing control model for intelligent products to autonomously decide efficient processing operations and routing paths in dynamic manufacturing systems undergoing perturbations. We also use discrete-event simulation to test the proposed approach in a generic manufacturing system. The simulations results show our proposed approach leads to shorter average throughout time when compared with heuristic approach and Scholz-Reiter's pheromone approach. Further research will include studying influence of the manufacturing system size on the performance measure and integrating with dispatching rules for resource selecting queued products.

References

1. McFarlane, D., et al.: Auto id systems and intelligent manufacturing control. Engineering Applications of Artificial Intelligence 16(4), 365–376 (2003)
2. Bonabeau, E., Dorigo, M., Theraulaz, G.: Swarm Intelligence - From Natural to Artificial Systems. Oxford Press (1999)

3. Sallez, Y., et al.: A stigmergic approach for dynamic routing of active products in FMS. Computers in Industry 60(3), 204–216 (2009)
4. Peeters, P., et al.: Pheromone Based Emergent Shop Floor Control System for Flexible Flow Shops. In: Proc. of IWES, pp. 173–182
5. Dussutour, A., et al.: Optimal traffic organization in ants under crowded conditions. Nature 428, 70–73 (2004)
6. Zhou, R., et al.: Performance of an ant colony optimisation algorithm in dynamic job shop scheduling problems. International Journal of Production Research 47(11), 2903–2920
7. Xiang, W., Lee, H.P.: Ant colony intelligence in multi-agent dynamic manufacturing scheduling. Engineering Applications of Artificial Intelligence 21, 73–85
8. Scholz-Reiter, et al.: Bio-inspired and pheromone-based shop floor control. International Journal of Computer Integrated Manufacturing 21(2), 201–205 (2008)

Parallel Flow Shop Scheduling Problem Using Quantum Algorithm

Yunchen Jiang[1] and Shanshan Wan[2]

[1] Department of Laboratory and Equipment Management Beijing Institute of Technology Beijing, China
[2] Computer Teaching and Network Department Beijing University of Civil Engineering and Architecture
yunchenjiang@sina.cn

Abstract. Quantum algorithm is applied to Parallel Flow shop scheduling problem which has the characters of the parallel machines scheduling and the flow shop scheduling. In this paper, quantum bit encoding strategy of quantum algorithm is designed to a general Parallel Flow shop problem with the objective of minimizing Makespan. All the scheduling problems are represented by the quantum bits. This algorithm is tested on some randomly generated problems with different scale sizes and compared with GA algorithms. Computational results show that the QA and the encoding strategy have shown exciting performance in obtaining the most accurate and optimal solution.

Keywords: Parallel Flow shop, Quantum algorithm, Makespan.

1 Introduction

With the rapid development of our country economics production and manufacturing industries have been the very important part undoubtedly. And the parallel machine and flow shop scheduling problems always attracts hot researches [1-2]. In this paper we draw our attention to the Parallel Flow shop which is the extension of the above two problems. Multiple Parallel Flow processing is a common phenomenon during the production process of electrical appliances and industrial supplies such as TV, cables, steel wires. The corresponding process sequence for each processing consists of the same or different machine model pose, but the processing function is the same. This is a new and special multi-stage multi-machine scheduling problem from the production practice of Parallel Flow shop scheduling problem which has the characteristics of both parallel machines and Flow shop scheduling problem.

Reference [3] studied two-phase Flow shop scheduling problem where the degree of parallelism is 2. It considered the same workpiece's processing time in the working procedure among different Flow shops are proportional and it's on of the special case. In that paper the heuristic algorithm is given and integer programming model and solution is proposed to minimize maximum completion time (Makespan). In [4] the authors introduced a population based on incremental learning algorithm to solve the Parallel Flow shop problem which had not the proportional time constraints.

J. Zhang (Ed.): ICAIC 2011, Part V, CCIS 228, pp. 269–274, 2011.

2 PARALLEL Flow Shop Scheduling Problem

Parallel Flow shop scheduling problem can be described as follows, N workpieces are processed among the K-parallel degree production Flow shop where each flow shop line has M machines. Each job can only be processed in one of the production line, and in this production line the workpiece must have its all procedure to be finished. Assumed that the workpiece is arranged to the permutation order processing mode in the production line, at the same time each machine can only process one workpiece, each workpiece at the same time can be processed by only one machine, and interrupt is not allowed in the processing production. We aims to allocate the workpiece for the flow shop lines and determine the processing sequence for each processing line, so that we can get an optimal Makespan target.

To describe the problem, we give the following definitions.

J- setting of workpieces, J={1,2...N}.
L-parallel degree, setting of parallel processing lines, L={1,2...K}.
i ,j- serial number of workpiece, i , j = 1 , 2 ...N.
k -serial number of parallel processing line, k = 1,2 ...K
m-serial number of processing machine, m = 1,2...M.

P_{ikm} -processing time of workpiece i which is processed by machine m on flow shop k.

C_{ikm} -completed time of workpiece i which is processed by machine m on flow shop k.

S-Large enough positive number.

X_{ik} - if workpiece i is processed on flow shop k then X_{ik} =1,otherwise X_{ik} =0.

Y_{ijkm} -if workpiece i is processed more earlier than workpiece j by machine m on flow shop Y_{ijkm} = 1,otherwise Y_{ijkm} =0.

The scheduling problem math model of flow shop Makespan is given as followings.

$$\underset{i\in J}{\text{minimize}} \quad \max\{\sum_{k=1}^{K} X_{ik} \cdot C_{ikM}\} \tag{1}$$

$$\text{s.t.} \quad \sum_{k=1}^{K}\sum_{i=1}^{N} X_{ik} = N \tag{2}$$

$$\sum_{k=1}^{K} X_{ik} = 1 \qquad i = 1,2 \ldots N \tag{3}$$

$$C_{jkm} - C_{ikm} + (3 - X_{ik} - X_{jk} - Y_{ijkm}) \cdot S \geq P_{jkm}$$

$$i , j = 1,2 \ldots N ;k=1, 2\ldots K ; m = 1, 2 \ldots M \tag{4}$$

$$C_{ikm} - C_{jkm} + (2 - X_{ik} - X_{jk} + Y_{ijkm}) \cdot S \geq P_{ikm}$$

$$i , j = 1,2 \ldots N ;k=1, 2\ldots K ; m = 1, 2 \ldots M \tag{5}$$

$$\sum_{k=1}^{K} X_{ik}(C_{ikm+1} - C_{ikm} - P_{ikm+1}) \geq 0$$

$$i=1,2\ldots N, m = 1,2\ldots M\text{-}1 \tag{6}$$

$$X_{ik} = 1 \text{ or } 0 \tag{7}$$

$$Y_{ijkm} = 1 \text{ or } 0 \tag{8}$$

Where, (1) is the objective function of Makespan. Constraint (2) denotes that all the workpieces must be processed in the processing lines. Constraint (3) shows that each workpiece can only be allocated to one processing line. (4) and (5) represent the sequence constraints for the different workpieces in the same processing line and guarantee that one machine can only process one workpiece as the same time. Constraint (6) describes that each workpiece must be processed according its sequence demand and ensure one workpiece can only be processed by one machine at the same time. (4) and (5) are the decision variable constraints.

3 Quantum Algorithm and Encoding

A. Quantum Algorithm

Quantum algorithm [5,6] is able to represent many quantum states with one quantum bit. A quantum bit can reflect the superposition of two states. $|\alpha_i|^2$ and $|\beta_i|^2$ denote the probabilities that the states | 0>, | 1> are chosen. To meet the normalization condition of constraint, $|\alpha_i|^2 + |\beta_i|^2 =1$.

Problem solutions are encoded with quantum bits. We assume the problem is a n-bit quantum encoding, that's, the n-bit quantum code can identify all possible solution. The formation probability of each state can be measured by the magnitude's product operation. The procedure of QA algorithm is as follows.

Step 1. Initialization. $P_{(t)} = \{p_1^t, p_2^t, \cdots p_i^t, \cdots, p_m^t\}$,m is the number of individuals among each population. p_i^t (i=1,2,…,m) is the ith individual select probability among the tth iterative generation. The length of chromosome is n so the quantum code is shown as (9).

$$p_i^t = \begin{bmatrix} \alpha_1 | \alpha_2 | \cdots | \alpha_j | \cdots | \alpha_n \\ \beta_1 | \beta_2 | \cdots | \beta_j | \cdots | \beta_n \end{bmatrix}. \tag{9}$$

Here, n is the length of quantum chromosome. All the α_j and β_j (j=1,2,…,n)are initialized as $1/\sqrt{2}$.

Step 2. $R(t)$ construction. According to p_i^t, QA constructs the solution population $R(t)$. $R(t)=\{r_1^t, r_2^t, \cdots, r_i^t, \cdots, r_m^t\}$. r_i^t (i=1,2,...,m) is a binary character string with the length n. And each r_i^t is determined by p_i^t.

Step 3. Evaluate each solution and obtain the optimal one or the top optimal more individuals. If termination condition is met output the optimal solution, then stops. Otherwise continue to execute the next step.

Step 4. Update p_i^t with dynamic rotation gate $U(t)$.

$$p_i^{t+1} = U(\Delta\theta) \cdot p_i^t = \begin{bmatrix} \cos(\Delta\theta), -\sin(\Delta\theta) \\ \sin(\Delta\theta), \cos(\Delta\theta) \end{bmatrix} \cdot \begin{bmatrix} \alpha_1 | \alpha_2 | \alpha_3 | \cdots | \alpha_n \\ \beta_1 | \beta_2 | \beta_3 | \cdots | \beta_n \end{bmatrix}. \tag{10}$$

Where, $U(\Delta\theta) = \begin{bmatrix} \cos(\Delta\theta), -\sin(\Delta\theta) \\ \sin(\Delta\theta), \cos(\Delta\theta) \end{bmatrix}$, rotation angle $\Delta\theta_i = k * f(\alpha, \beta)$. $k = 10 * \exp(-t/(5 * pop\_size))$. t is the iterative number and pop_size is the individual quantity of each population.

Step 5. Generate the new solution according to p_i^{t+1}.

Step 6. Output the optimal solution if termination condition is met then stops. Otherwise jump to step2.

This kind of probability amplitude presentation has its predominance in nature that one quantum state of the chromosome carries several of information and a wealth of species can be obtained by the random method of measuring the quantum chromosome in generating new individuals. Thus it can maintain the diversity of population and overcome the premature.

In quantum algorithm the construction of quantum gates is the key and which is directly related to the performance of the algorithm's evolutionary process. Here a dynamic rotation gate strategy is put forward to optimize the quantum algorithm.

B. Algorithm Encoding Design

Parallel Flow shop scheduling problem can be divided into two interrelated processes which are workpiece elicitation and processing sorting .Firstly we use natural integer encoding to figure the encoding strategy. If a n-workpiece and K production lines Flow shop scheduling problem are considered, [$a_{11} \cdots a_{1j} \cdots * a_{21} \cdots a_{2j} \cdots * a_{k1} \cdots a_{kj} \cdots$] is a feasible solution. Where the whole length is N+K-1, different processing lines are separated by symbol *. And a_{kj} means that the jth workpiece in k processing line. So both the workpieces' distribution and processing order can be expressed in the form of the feasible solution.

For QA algorithm binary encoding is used so the code length should be $(N+K-1)\cdot\left|\log_2(N+K-1)\right|$. For example, if N=10, K=3, M=3, the code length for each feasible solution is 48 bits.

4 Computational Experiments

To test the QA performance in Flow shop scheduling problem we take 5 different scale problem as example. The N*M*K is 7*2*2, 10*3*2, 20*4*3, 30*5*4, 40*6*5 respectively. The workpiece processing time P_{ikm} belongs to this range [1, 20]. The program is based on VC++6.0 and the main performance parameter of the computer is Intel(R) Core Duo CPU, P8600, 2.4GHz,2G memory.

To test QA algorithms' performance the algorithm is compared with GA algorithm.

The optimal solutions, corresponding computation time and optimal proportion are shown in tabel1. Optimal proportion is the proportion of feasible solutions which have the optimal values and it illustrates how many optimal solutions are among the pop-sizes. 30*5*4 Flow shop scheduling problem considered evolutional trends comparisons of three algorithms are given in Fig.1 by recording the average objective function value of each iterative population.

Table 1. Results of five problems

Pop Size	Optimal values		Computation Time		Optimal proportion	
	QA	GA	QA	GA	QA	GA
7*2*2	73	73	3.2	3.1	1	1
10*3*2	186	194	5.6	5.8	1	1
20*4*3	118	120	9.1	9.9	0.98	0.95
30*5*4	147	165	13.4	14.7	0.99	0.94
40*6*5	294	319	16.0	18.3	0.95	0.91

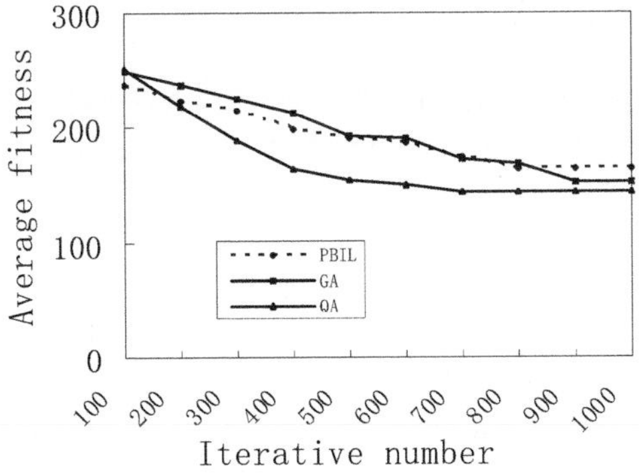

Fig. 1. The evolutional trend of the algorithm

5 Conclusions

According to the general Parallel Flow shop scheduling problem which has the performance of both the parallel machine and flow shop we propose QA to solve the maximum Makespan problem and design the quantum encoding strategy. The workpiece allocation and processing sequence problems are represented by the quantum bit code, where special symbol is used to separate the workpiece. This code strategy is simple and it has amounts of individual's information. The QA is compared with GA algorithm and the results show that these two algorithms has the similar performance in searching the optimal solution when the problem scale is small, and when the scale becomes larger QA show its excellent. QA can obtain the most optimal solution. Long encoding length and various quantum bits considered QA's computation time is slightly longer. Quantum evolutionary algorithm has a better diversity than the traditional evolutionary algorithms and has more extensive application prospects and how to make full use of the quantum bit information and how to shorten the computation time are the future research directions.

Acknowledgment. The authors wish to acknowledge the support of Beijing University of Civil Engineering and Architecture Science Research Foundation of China (No.100903808) and Beijing Technology Plan General Projects (No. KM201010016008).

References

1. Li, C., Cheng, T.: The parallel machine minmax weigthed absolute lateness scheduling problem. Naval Research Logistics (41), 33–46 (1993)
2. Widmer, M., Hertz, A.: A new heuristic method for the flow shop sequencing problem. Eur. J. Opl. Res. (41), 183–193 (1989)
3. Sundararaghavan, P., Kunnathur, A., Viswanathan, I.: Minimizing Makespan in Parallel Flow shop. Opns. Res. Soc. (48), 834–842 (1997)
4. Pang, H.L., Wan, S.S.: Study on Parallel Flow shop Scheduling Problem Using Probability Learning Based Evolutionary Algorithm. Control Theory & Applications (22)-1, 149–152 (2005)
5. Feynman, R.: International Theoretical Physics. (21), 467–488 (1982)
6. Deutsch, D.: Proc. R Soc. London A (400), 97–117 (1985)

Research on the Technology of Virtual Data Private Network Based on 3G Network

Jian Shen, Shoulian Tang, and Yonggang Wang

School of Economics and Management Beijing University of Posts and Telecommunications
Beijing, China
JianShen1112@yeah.net

Abstract. During recent years, with the development of enterprises informatization, telecom operators are demanded to provide a more flexible network mode. With the development of 3G service, the technology of virtual data private network(short for VDPN in this paper) provided by telecom operators based on mobile network and fixed network has been accepted gradually by enterprise users. According to the features of VDPN, this paper has designed the overall system structure of VDPN, illustrated two different modes of access, and provided a new mode of access by comparing the two modes above. The new mode can enhance the operating efficiency and security of the virtual data private network.

Keywords: Virtual Network, APN, VPDN, 3G.

1 Introduction

During recent years, with further development of enterprise informatization, the enterprise users has more and more application demand for remote data transmission and remote login of enterprise network. However, there are some problems for traditional fixed network mode such as wiring difficulties, long implementation cycle and high cost. With the coming out of 3G service, more and more enterprise users expect to realize the remote data transmission by wireless method. But in normal wireless transmission, there are problems such as low security, low reliability and low transmission speed. Therefore, telecom operators need badly a networking technology with features such as nationwide roaming, wireless access at anytime and anywhere, self control of user's internal network access, and total isolation from the internet, and with high security, reliability and transmission speed.

At present, the three telecom operators in China are all all-service telecom operators with business license of fixed telephone, internet and mobile phone. The third generation of mobile communication technology represented by WCDMA extremely enhanced the transmission speed of wireless network, which enabled the operators to provide data value-added service such as video call, mobile internet for the users. At the same time, the operators can provide for the users VDPN products based on the fixed network and the mobile network with the higher data transmission speed of 3G network.

J. Zhang (Ed.): ICAIC 2011, Part V, CCIS 228, pp. 275–281, 2011.

VDPN based on 3G network refers to the process for which the mobile terminal uses 3G network by dialing numbers to access to the enterprise private network, and transmit private data through packet and encryption of net data, reaching private network security class, so as to build a virtual enterprise internal data network through fixed network and mobile network. The enterprise users can enable the wireless terminal and the enterprise private network server to begin data communication at any moment through VDPN access scheme, and can provide authentication and management functions to sub-user access, so that only the wireless terminals which meet the enterprise users' demand can access to the enterprise internal network.

According to the features of VDPN, this paper designs its overall system structure, illustrates two different modes of access and the dialing process, points out the security mechanism of VDPN mode and provides a new mode of access by comparing the two modes above.

2 Virtual Data Private Network

The overall system structure of VDPN is shown in Figure 1. In this paper, the overall structure is divided into 3 parts as A, B and C.

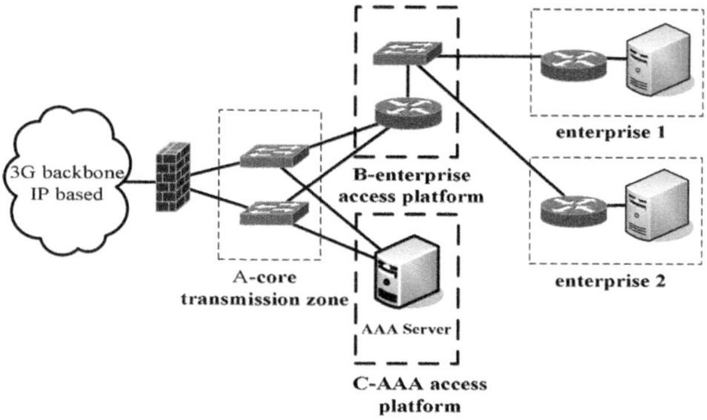

Fig. 1. Overall Structure of Virtual Data Private Network

A zone is the core transmission zone of VDPN access platform, mainly used to provide data routing forward between 3G core packet network and VDPN platform.

B zone is the enterprise access zone, which is used to provide dedicated access for enterprise users, supporting multi-interface types.

C zone is the access management platform for VDPN, which is used to provide enterprise trusteeship authentication service and management service.

The structure isolates 3G packet network from VDPN platform through an active/standby firewalls, so as to guarantee the VDPN will not be attacked from 3G packet network side. For the malicious attack from the enterprise side, the platform security can be guaranteed by configuration of corresponding security rules in the enterprise access router.

Applying this layered system structure, we can accelerate data transmission and rapidly collect data based on the security for the whole system. Based on this system and 3G network, VDPN service can support 3G mobile terminal in APN mode and VDPN mode to access the enterprise user private network.

Below APN mode and VDPN mode will be illustrated separately.

3 Access Point Name Mode

Access Point Name Mode is shortened as APN, which is used to mark the networking type of the wireless terminals, mainly divided into 2 types as visiting internet through 3G network and through enterprise LAN for enterprise users. When the wireless terminal wants to access to the 3G network, APN name is used to mark the access point for every external 3G network user.

The telecom operators provide APN mode access service for the enterprise users, also the enterprise users can agree with the telecom operator for a special APN name. For example: the APN name of the access to enterprise ABC is ABC.BJAPN. After the enterprise starts APN mode access function, a GRE data transmission tunnel will be built between GGSN of telecom operator and the enterprise access router, for which only the legitimate users permitted by the enterprise can pass this tunnel to reach the enterprise internal network, while the other users' visits will be refused. APN access mode provides with dedicated access or through the internet to build private data tunnel. The users usually apply the dedicated access to access the telecom operator's industry application platform, so as to guarantee the security of the physical tunnel. Both the network equipment and the server IP address at the enterprise user side are programmed by the telecom operator, the enterprises are not able to select their IP section for their internal network equipments.

The enterprise user applying APN mode can choose a unified user's password, or different usernames, passwords at different terminals, or empty password, any of these three authentication modes can accomplish the authentication of identity.

The network topological diagram of APN mode is illustrated as in Figure 2.

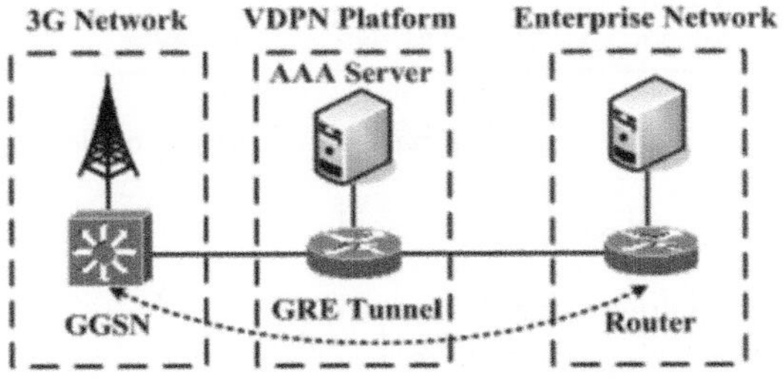

Fig. 2. Network topological of APN mode

The dialing process of APN mode is:

1, GRE tunnel is built.
2, Mobile terminal dialing request reaches GGSN.

3, Visits telecom operator AAA, checks the domain name
4, Terminal builds PPP connection with enterprise LAN
5, Allocates the user with IP address
6, Mobile terminal visits the private network successfully.

4 Virtual Private Dial-Up Network Mode

Virtual Private Dial-up Network is shortened as VPDN, applying the special network encryption and communication protocols, enabling the enterprises to build a secured virtual private network on public network.

VPDN mode refers to the technology realization method of building L2TP tunnel between enterprise router and operator GGSN, which applies a unified access point name and different username to access the enterprise private network. Users in the same industry and same city or province use the same VPDN access name, different enterprise users will no more be differentiated by different APN name, but by the enterprise domain name in the username.

VPDN mode doesn't apply the three-tier GRE tunnel, but a safer two-tier L2TP tunnel to build a private connection between GGSN and the enterprise router. VPDN mode supports the enterprise user to set up his own AAA server, the authentication request of the mobile terminal is sent to AAA server of the enterprise user to make a twice authentication of the user's identification. The enterprise user of VPDN mode can freely set up the network equipment of its internal network and the server IP address, different enterprise users may use the same IP address.

The network topological diagram of VPDN mode is illustrated as in Figure 3.

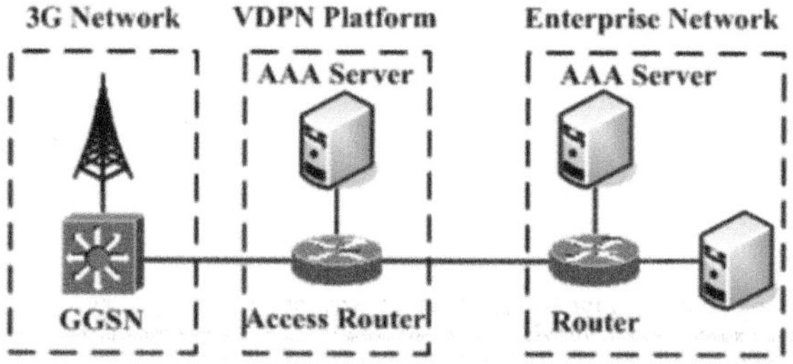

Fig. 3. Network topological of VPDN mode

The dialing process of APN mode is:

1, Mobile terminal dialing request reaches GGSN
2, GGSN visits telecom operator AAA, checks the domain name
3, L2TP tunnel is requested
4, PPP connection is built between mobile terminal and enterprise user router.
5, Visits enterprise user AAA, security verification
6, Private network allocates mobile terminal with IP address
7, Mobile terminal visits the private network successfully

Comparing with APN mode, the biggest advantage of VPDN mode is its security mechanism. This security mechanism is made up by the following 5 parts.

A. Security of 3G Network

Taking WCDMA as example, its code division multiple access technology was used for the military communication with a code length of 128 bit because of its own security feature, that is 2 to the power of 128 times of combinations, which is very difficult to work out. WCDMA network base station applies asynchronous working mode, for which the base station does not need to work synchronously with a high reliability. USIM card upgrades the arithmetic method on security, enhancing the network authentication function of the card, which makes the illegal copy of USIM card almost impossible.

B. Authentication Security at the Telecom Operator Side

First-class access control, all the users who request for accessing enterprise network must pass the AAA server authentication for the enterprise domain name, IMSI, so as to efficiently refuse the illegal access the enterprise network.

C. Private Line and Tunnel Security between GGSN and the Enterprise User Router

Telecom operator GGSN is divided into many sets of equipments, the one applied by enterprise users is called private GGSN. The other GGSN equipments are dealing with the public users' data service. GGSN and enterprise user router are connected by point-to-point dedicated access, isolated from the internet, which reduces the attacked risk. L2TP tunnel is built between GGSN and router, in which the data is transmitted and transparent for the network equipment between them.

D. Authentication Security of AAA Server at the User Side

AAA at the user side operates a twice authentication for the terminal which applies for access the internal network. AAA at the user side manages the usernames and passwords for all the dialing terminals, and is able to modify at any moment.

E. Firewall Set Up at the User Side

For users who request a high security, firewall can be set up behind the brink router. The firewall operates visit control, stops any possible potential attack risk.

5 Mode Combines APN and VPDN

APN and VPDN modes illustrated above have their own features, APN mode has a high access efficiency, but low security, suitable for enterprise users with not high security request, while VPDN mode has a low access efficiency but high security, suitable for enterprise users with high security request. The differences of the two modes are shown in Table 1.

Table 1. Comparison between APN mode and VPDN mode

Comparison	APN	VPDN
Terminal Address	Not support Address Domain function, IP address is only domained by the operators	Support address domain function, can apply enterprise private network IP address.
Transmission protocol	GRE encapsulated transmission through IP	Can be encapsulated through IP, ATM, FR, etc.
Tunnel tier	three-tier tunnel	Two-tier tunnel, more tunnel application can be attached, eg. Twice authentication.
Identity authentication	once authentication	twice authentication
Security	Based on three-tier tunnel, not encrypted, low security	Based on two-tier tunnel, encrypted transmission of control package, high security.
AAA function	Only verify enterprise domain name	Verify enterprise domain name and password

According to the lack of security of above two modes, in this paper a new access mode is designed. The Structure of the new model is same as the VPDN model. The feature of it is that the enterprise user changes domain name regularly, applies management of mobile terminals at different levels and authenticates the mobile terminal number. If the mobile terminal has a high security and pass the number authentication, then it is permitted to access the enterprise private network after the twice authentication. If the mobile terminal does not pass the number authentication, then it is refused to continue the next dialing process.

The dialing process is as follows:

1, Mobile terminal dialing request reaches GGSN.
2, GGSN visits telecom operator AAA, checks domain name and number.
3, If the security is high then GRE tunnel will be built, if low, then L2TP tunnel will be built.
4, If L2TP tunnel is built, then connection will be made according to VPDN process.
5, If GRE tunnel is built, then connection will be made according to APN process.

The above access mode features priority of number authentication, and according to the security level of mobile terminal, the judgement is made by the GGSN of telecom operator to decide whether a twice authentication is necessary, which enhances the access efficiency and strengthens the security management.

6 Conclusion

Virtual data private network refers to the access service for which mobile terminal reaches the network by dailing access, and through packet and encryption of network data to reach security of the private network, so as to apply the 3G network to build the enterprise LAN. VDPN could be a strong complement for the expansion of enterprise private network, applying mobile terminal to access the enterprise private network without fixed network, so as to realize a mixed networking of dedicated access and mobile terminal. This virtual network could guarantee the reliability and stability of enterprise user data stream transmission, and extremely enhance the working efficiency of the enterprise, and reduce the communication cost at the same time.

According to the features, this paper has put forward the concept of the virtual data private network, and designed the overall system structure, which can be divided into 3 parts by function. Then this paper has illustrated the two access modes and dialing process of the network, pointed out the security mechanism of VPDN mode, and provided a new access mode by comparing the two modes. The new access mode can help to enhance the access efficiency of virtual data private network, and strengthen the management by levels for mobile terminal security.

References

1. Erwin, M., Scott, C., Wolfe, P.: Virtual Private Networks, 2nd edn. (December 1998)
2. Carlson, C.J.: Tunnel of secure transmission (virtual private networks in data security) (July 1999)
3. Worldwide Videotex, Globalcrossing Debuts Data-Virtual Private Network Service (March 2000)
4. Richardson, T., Stafford-Fraser, Q., Wood, K.R., Hopper, A.: Virtual Network Computing. IEEE Internet Computing 2 (February 1998)
5. Cisco Press, Access VPDN Solutions Guide (December 2001)
6. Luo, W., Pignataro, C., Chan, A., Bokotey, D.: Layer 2 VPN Architectures (November 2008)
7. Dulworth, M.: The Connect Effect: Building Strong Personal, Professional, and Virtual Networks (January 2008)
8. Yuan, R., Strayer, W.T.: Virtual Private Networks: Technologies and Solutions (May 2001)
9. Meeta Gupta, Building a Virtual Private Network (November 2002)

Research on the Technology of Electric Dispatching Customized Information Service

Yang Jiashi[1], Xiao Xinquan[1], Liu JunYong[2], Huang Yuan[2], and Wu Yan[2]

[1] Telecommunication & Automation Center of SEPC Chengdu, Sichuan Province, China
[2] Department of Electrical Engineering and Information University of Sichuan Chengdu, Sichuan Province, China
yangjiashi009@sina.cn

Abstract. In the context of smart grid development, there are more and more data and information in power system, which does not only make dispatcher' work harder, but also constitute a threat to power grid. On the basis of analyzing three typical dispatching modes, we put forward the technology of Information customization. According to dispatchers' requirement, it can offer them the best needed dispatching information in the best way. It can greatly improve dispatchers' efficiency of getting and understanding information, and make their work easy. For ensuring the grid safe and stable, it is of great importance.

Keywords: Power system, dispatching information, information customization, dispatcher.

1 Introduction

In the context of smart grid development, in order to promote the development of power market and meet the requirement of grid automation, the electric power system is equipped with more and more EMS, DMS, MIS, GIS. They constitute the base of the production and management of power system. How to deal with the massive data they produce has become one of the most important problems which constraint the further development of power grid.

The operation of grid power system on one hand is related to the level of grid automation, on the other hand depends on the skilled dispatcher. Dispatcher need to analyse the massive information. Especially when the grid is abnormal or malfunctioning, a great quantity of alarm information occurs, such as guard signal, tripping signal, SOE and overload information continuously comes to dispatching centre. The massive information greatly hinders dispatchers to master the state of the power grid [1].

The traditional dispatching information system used to deliver the information to every dispatcher in the same way. For this reason, the fact that dispatching terminal and controlling terminal provide is not targeted [2]. This will perplex dispatchers' work further. In fact, for they have different duties and different thinking model, dispatchers needn't see all the information. Moreover, because of the difference in individual habit and preference, dispatchers want the message sent to them in different way.

J. Zhang (Ed.): ICAIC 2011, Part V, CCIS 228, pp. 282–289, 2011.

This article put forward the idea of electric dispatching customized information service. According to their duties, areas, habits and preference, dispatchers can choose their best ways of reading and disposing information. This is called personalized information service.

2 Analysis of Dispatching Modes

Since the grid dispatching technology have been developed for 30~40 years, we have accumulated lots of experience in operation and conclude a range of dispatching modes. Now, we put forward three typical dispatching modes, and introduce them in general.

A. The Dispatching Mode Based in DyLiacco Theory

DyLiacco raised the theory that divides the working condition of electric into four statements (Fig. 1) ,and thus lay the foundation for the frame divide of grid SCADA/EMS real-time working. Not only the actual operation, but also the theory does not beyond this frame. It is the main tool of the electric grid working system.

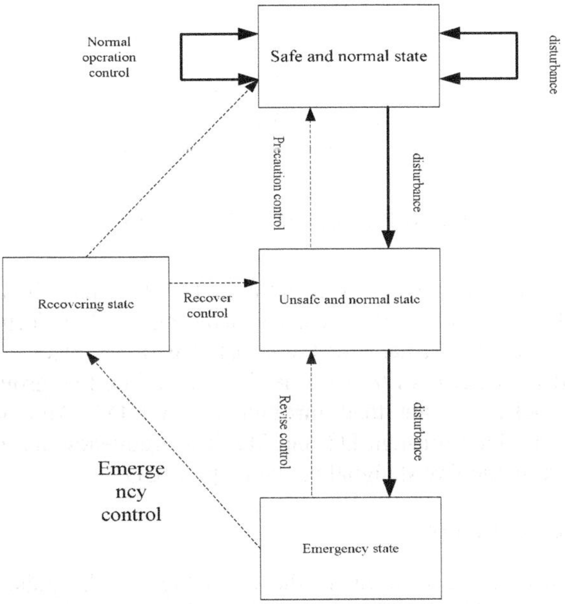

Fig. 1. Four operation statements of power system

The dispatching mode based in DyLiacco theory aims to take action to transit the state of the electric system.

B. The Dispatching Mode Based in Dispatchers' Thinking Model

"Dispatchers thinking model" means putting the dispatcher in the first place, and making the computer and dispatchers constitute the dispatching system together. Based in dispatchers' thinking model, article [3] introduces the perception model of grid state and trend (Fig. 2). The grid state and trend, which are related to the device condition and users' behaviors, refer to the operation state and conversion trend of the grid. The perception of grid state and trend means to get, understand, display, and predict the changing trend of the electric grid. By the means of building a series of index system, which can describe the grid state and trend, dispatchers can have a comprehensive understanding of the grid. According to that dispatchers are able to protect the grid from fault.

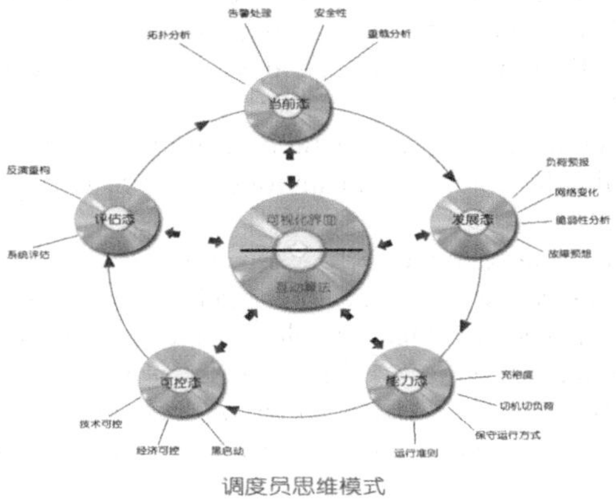

Fig. 2. The perception model of grid state and trend

C. The Dispatching Mode Based in Alarm Signal

The dispatching mode based in alarm signal seeks the origin of the fault from the alarm signal. Although there are varied fault in grid, we can conduct quantitative and qualitative analysis of them from different perspective.

From the perspective of system, alarms can be classified as component alarm and system alarm. From the perspective of the level of fault, faults can be classified as gradient fault, mutation fault and accumulated fault. From the perspective of the number and relation of components, faults can be classified as single fault, combination fault, series fault and simultaneous fault.

In order to make the alarm signal more useful and targeted before sending the messages to dispatchers, screen and classify the messages from the above perspectives (Fig. 3), and then let dispatchers choose which kind of information delivered to them.

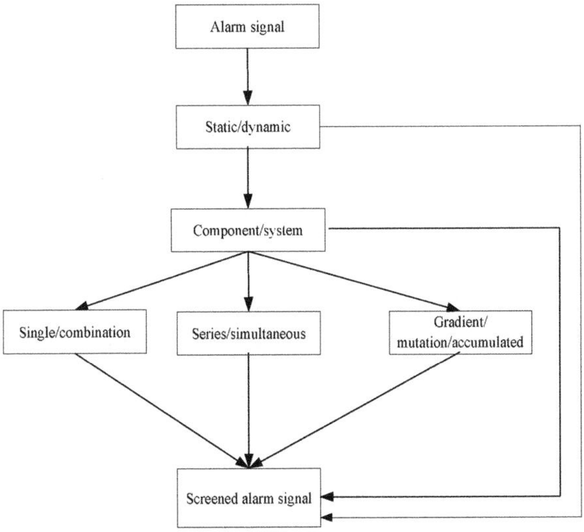

Fig. 3. Process of screening alarm signal

3 Electric Power Dispatching Information Customization

"Customized information establishment" is the main trend of the information service development [4]. It allow users to choose whatever service they want [5], including their categories, display methods etc.

The electric power dispatching establishment belongs to the supply of the power system. to get the service, the dispatcher should first type in their requirement of information, this mainly includes the personal files, the main themes, and then a relevant information collection and combination will be done by the service system based on the database. At last, the system delivers the information to the dispatcher in accord with their requirement. The working process can be divided into the following four steps: getting the information that the dispatcher need, building the database for the dispatcher, checking the information, collection, estimating, screening, processing, combination, delivering information to dispatcher, and feedback [6].

4 The Key Technology of Power Dispatching Information Customization

A. The Organization of Power Dispatching Information

1) Source data integration

The data of the power system database can be mainly obtained from SCADA/ EMS, DMS, MIS, GIS. For different types and storage formats, data in each database cannot be unified managed, so different data sources should be transform into a public data model when constructing dispatching a data platform. Realization of data transformation method can be divided into manual programming and using the special data transformation tool [7]. Due to special data transformation auxiliary tools, it's relatively simple to operate with less people. But the problem is about the poor flexibility. However, the manual programming can be relatively agile. In practice, these two methods are always combined together and embedded handwork weave code while using a special data transformation tool.

2) The organization of information

Using the technology of data warehouse facing the theme of organizational information, based on the three scheduling model, the database information can be divided into three themes: alarm signs, DyLiacco thought, the dispatcher thinking model to organize information. Every theme has several facts attributes, which are in the corresponding with the data sources in dispatching data platform. Topics and data connection relations are realized through dimensional modeling. Dimensional modeling [8] is a logical design technology, which is using object standard frame structure to represent data. Each dimension model is made up of fact tables with compound keys and series of dimension tables. Each dimension table is precisely in corresponding to a certain a component of fact sheets through a primary key. Fact table is made of index entities and index measurement attribute constitute, which treated as all dimension of intersection describe a particular fact, i.e." alarm signals" theme mainly involves dimensions including the temporal dimension, geographical dimension, line dimension, a equipment dimension, second equipment dimension, etc.

B. Intelligent Agent Technology [9]

Intelligent Agent technology is applied to extract and combine information that the dispatcher customized. Agent is one of the basic terms in distributed artificial intelligence (DAI), which is the research focus in the field of artificial intelligence in recent years. Intelligent Agent is made up of containing type software program, undertaking missions with constitute stored in a database contains information [10]. It can effect on itself and environment, and can respond to the environment [11].

Intelligent Agent has two key technical features: intellectualization and acting ability. Intellectualization is to use the system to analyze and explain what is contacted by reasoning, learning, and other techniques, or just submit its various information and knowledge. Acting ability refers to the ability that a Agent can sense its environment and the corresponding action [6].

Intelligent Agent which is applied to dispatch information customization service mainly has the following functions: screening, which is to screen the proper information in large number of information form dispatching data platform according to the dispatcher model; Restructuring information, which is to layout restructuring the selected information according to the dispatcher's favor in different dispatcher model; Intelligent delivery, which is to deliver the information to the dispatcher based on order and mode that dispatcher customized.

C. The Customization Technology of Dispatcher's Hobby and Duty

Dispatcher modeling [12] is the basis of electric dispatching information customization service. The dispatcher model [11] means the computable describe of dispatcher. The dispatcher model ultimately decides the type of the information delivered to the dispatcher. The dispatching information customization system builds a model for every dispatcher. The model can record the dispatcher's personal details. According to the model, system provides the dispatcher with targeted customization service [10].

When a dispatcher first uses the system, he need to type in his personal details, including his hobby and duty. Then the rudiment of the dispatcher model is built. These data is stored in dispatcher database. On the basis of the dispatcher model, Intelligent Agent determines the dimensionality of the needed information, and then collects the information in the dispatching data platform, and at last combines them. For example, the dispatcher who is in charge of load forecasting needs the information dimensions of load, weather, economy development, energy conservation technology.

D. Free Dispatching Information Customization

Although dispatchers can give the system some personal details, in many cases, even the dispatchers themselves can not conclude their hobbies and habit, furthermore, their requirements of information is not invariable. They usually changes in working. Therefore, free dispatching information customization is introduced. It can provide dispatchers with personalized information customization service. Moreover, it can keep modifying and perfecting the dispatcher model in use.

The realization of free dispatching information customization needs the help of machine learning technology [13] [14]. Based on the information dispatchers browse and the way they browse information, the realization is done automatically by a kind of software entity. The software entity has two functions: learning and reasoning [10]. With the learning function, it can trace the dispatcher's using process, and extract the dispatcher's hobby and habit. And modify the dispatcher model in the basis of the information. For example: if a dispatcher always clicks a graph, then we infer he is interested in the graph, if a dispatcher click a page and then close it quickly, consequently we infer he is not interested in the page. With the reasoning function, the software entity can quantize dispatcher's hobby and preference, and make all kinds of hobbies and preference best fit, so as to reflect dispatcher's requirement. In the whole process of building and modifying dispatcher models, massive data mining and pattern recognition technologies are needed. Data mining can draw dispatcher's personal characteristic from dispatcher database. Pattern recognition technology can pick up dispatcher's personal characteristic from broad perspective.

5 The Design of Electric Information Power Dispatching System

The power dispatching system includes four parts, data source, dispatching information platform, information theme agent, and personal information assistant.

Data source: supply the data that the dispatching system need, including SCADA, EMS, DMS, MIS, and GIS.

Dispatching information platform: it is the foundation of smart power grid technique support system. SCADA, EMS and other relevant automated system all work on this platform. It can store and manage various kinds of data for the power system. The feature of the platform can satisfy the demand for short storage cycle, huge data size, excellent continuity, high reliability.

Theme information agent: it is the main part of the whole information dispatching system. There are two kinds of themes it agent, one can be got directly, which depend on the original formation of the information dispatching organization. The other one is got under the guidance of the personal-information-service assistant, by means of choosing the information and combining them into the theme in accordance with dispatcher's requirement, and it can be signed as the dispatcher's name or number.

Personal information assistant (PIA): it is the main part for the dispatcher customization. It has three important functions: the first one is to communicate with the dispatcher, find their needs, and build characteristic library; the second one is to lead the information agent to choose and combine the information; the third one is to deliver the information, according to the dispatcher's request, to the dispatcher timely.

Dispatcher database (DDB): it is mainly used to store the dispatcher's feature information, which includes the customized information the dispatcher set. In this way, it records the information of the dispatcher's requirement.

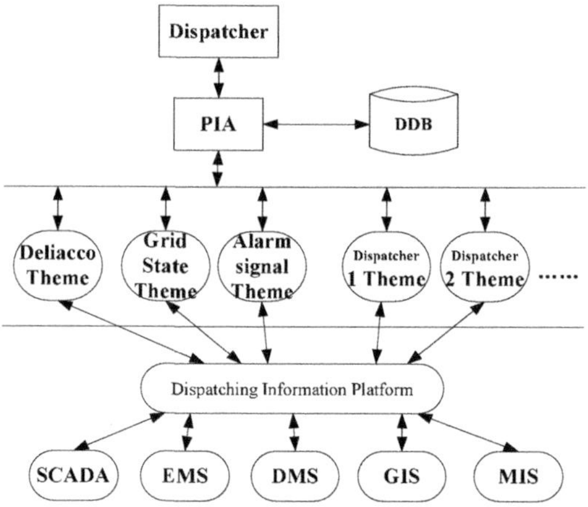

Fig. 4. Building of grid dispatching information customization system

6 Conclusion

The electric grid dispatching information customization technology introduced in this article is the application of information customization technology in electric system. It can greatly improve dispatchers' efficiency of getting and understanding information, thus letting them concentrate their attentions on making decision and judgment. It can emancipate dispatchers from heavy burdens.

References

1. Guan, S., Ma, J., Dai, C., Liu, L., Li, Y.: Research of Grid Dispatching Auxilary Decision Making System. East China Electric Power 37(09), 1450–1453 (2009)

2. Shu, Z.: Design of the Dispatching Information Screening Software. Electric Power Informatization 7(09), 40–42 (2009)
3. Liu, J., Shen, X., Tian, L., Chen, J., Huang, Y., Li, C.: The Develoment of Visual Technology in Smart Grid. Electric Automation Equipment 30(01), 7–13 (2010)
4. Xiao, X., Yang, Y., Zhai, G.: The Design of a Personalized Information Service System Based in Internet. Computer Engineering and Science 24(01), 59–62 (2002)
5. Zhao, W., Hu, H., Xie, D.: Customized Information Service and Retrieval of Biomedicine. Researches in Medicine Education 15(4), 336–339 (2006)
6. Shen, L.: The Organization and Service of Customized Information. Wuhan University, Information Science, Wuhan (2004)
7. Yang, J.: Research on Key Technologies of Data Warehouse-Based Decision Support System. Microcomputer Development 05, 31–33 (2002)
8. Li, S., Yu, G.: Electric Grid Dispatching Database Modelling. The Academic Journal of Shandong Junior College 13(01), 54–60 (2010)
9. http://my.lib.ncsu.edu
10. Geng, Z.: Research on Personalized Service in Digital Archives. Guangxi University for Nationalities, Department of Archival Science, Guangxi (2007)
11. Cai, M.: Design of the internet customized service. Wuhan University, Department of Library Science, Wuhan (2007)
12. Cohen, S., et al.: Mylibrary Personalized Electronic Services in the Cornell University Library. D-Lib Magezine 6(4) (2000)
13. Fu, C.: Push Technology and TargetLink Software Package, http://www.ahetc.gov.cn/cit/199907/005/.htm
14. Ying, X., Dou, W.: Internet Customized Service, http://www2.ccw.com.cn/03/0322/b/0322b51_6.asp

Design and Implementation of Internet of Things for Building Electrical Equipments*

Yan Qiao[1,2,3], Zhang Guiqing[1,2], Wang Ming[1,2], Shen Bin[1,2], and Zhang Lin[1,2]

[1] School of Information & Electrical Engineering, Shandong Jianzhu University, Jinan, 250101, P.R. China
[2] Shandong Provincial Key Laboratory of Intelligent Buildings Technology, Jinan, 250101, P.R. China
[3] School of Electrical Engineering, Shandong University, Jinan, 250061, P.R. China
yanqiao1314@gmail.com

Abstract. To realize the remote control and parameter monitoring of different electrical equipments, such as air conditioning equipments, lights and water pumps, etc., an Internet of Things for building electrical equipments is constructed and its network topology and functional structure are explored in this paper. In this Internet of Things, there exist three main functional blocks ---- the wireless measurement and control block, the room controller and the environment block. The issues on the self-organization and protocol of the wireless sensor network, which is used for information exchange in these functional blocks, are also studied. What is more, an equipment-management system, which can communicate with the room controllers via Ethernet, is provided on the management computer to realize the remote control of building electrical equipments. At last, an example is given to show how the electrical equipments can be connected to the Internet of Things and be controlled remotely.

Keywords: Internet of Things, wireless sensor network, building electrical equipment.

1 Introduction

Along with the increase of building electrical equipments, traditional energy saving control theories and methods for single equipment/system have been unsuitable for the global optimization of all the building electrical equipments. Although the Building Automation System (BAS) can realize the monitor and remote control of all public electromechanical equipments, such as the central air conditioning system, water supplying and sewerage system, power transmission and distribution system, lighting system, etc.. But the BAS often needs to be imported aboard because of its strict demands of hardware and software. Even if some manufacturers have developed their own BASs, their products have not been able to be commercial. Furthermore, the lower openness and extendibility of the BAS make its investment unfavorable in the long run.

* This work is supported by National Natural Science Foundation (61074149).

J. Zhang (Ed.): ICAIC 2011, Part V, CCIS 228, pp. 290–297, 2011.

A new technology -- "Internet of Things", enable the Internet to reach out into the real world of physical objects. Since all the electrical equipments can be connected to the network, it is possible to gather and share the mass data like the equipment states, environmental parameters and so on. It also provides a new way to control and optimize more and more building electrical equipments.

The Internet of Things for Building Electrical Equipments (ITBEE) is a complex network. It highly integrates communication system, calculation system, control system and physical system into a complicated, dynamic and human-oriented system. In the ITBEE, different electrical equipments are connected to the same network through different mediums (wire or wireless) and forms (fixed or mobile). In this unified framework, electrical equipments can get messages (commands, sensor information, etc.) from the network and provide service (information gathering, calculation, equipment control, etc.) for it. Especially when the wireless networks are introduced, it becomes more convenient to construct the ITBEE. Nowadays, in the field of energy-saving control, it becomes very significant to study how to construct the ITBEE to the make building electrical equipments operate with low energy consumption and to provide a safe, comfortable and convenient living or working environment for people.

As a developing technology, the structure design and application of "Internet of Things" is still in the exploring stage in recent years. Being interested in the safety and reliability of the Internet of Things, Kottenstette et al. [1] have presented it a theoretical network architecture based on the concept of passive adaptation. Besides, some scientists built customer-focused network architecture according to the random multi-agent system of Internet of Things [2], [3]. Karsai [4] has designed a totally parallel system using the model-driven approach to realize the communication between the Internet of Things and the models. All these researches provide a useful reference for the architecture design of ITBEE. The ITBEE not only has the common features of network, e.g. complexity, dynamics, reconstruction and reconfiguration, but also has its special characteristic named as "human-centered feature". To the authors' knowledge, no work has studied the topic of ITBEE, so it is a challenging but meaningful thing to study the architecture and application of the ITBEE and to solve the theoretical and technical problems existing in the ITBEE. For this purpose, this paper makes some useful attempt. The network topology and functional structure of the ITBEE are explored in this paper to realize the remote control and parameter monitoring of different electrical equipments.

2 Architecture of Internet of Things for Building Electrical Equipments

Fig. 1 shows the architecture of the ITBEE. In the following, we will introduce it in two aspects ---- network topology and functional structure. Its network topology structure is composed of an industrial Ethernet, a wireless network, WAN and the GPRS/GSM network, etc. In the aspect of the functional structure, as shown in Fig.1, electrical equipments (M_i), e.g. air conditioners, lamps, pumps, have two-way communications with their room controller (C_i) by the wireless networks with the help of the measurement and control block (W_i) embedded in the equipments. Each

M_i has its own W_i, which collects the states and parameters of M_i and sends these information to C_i, at the same time, receives commands from C_i to control M_i. An environment block (E_i) gathers environmental information around the room, such as environment temperature, humidity, lightness and infrared sensor information, carbon dioxide concentration, personnel identity information RFID and the other expanding information if needed. Environmental information in E_i can also be sent to C_i via a wireless network. Each C_i receives data of E_i and W_i, and then send it to a management computer (P) through Ethernet. In other words, the room controller (C_i) not only has the two-way communications with W_i and E_i through a wireless network, but also realize the bi-directional data exchange with P. After receiving the data collected by C_i, the P manages these data in a database, makes decision according to the results from the energy optimization algorithms, and finally sends control commands to C_i to control the electrical equipments through W_i. At the same time, the management computer also provides users a friendly equipment-managing interface in the Browser/Server (B/S) framework, by which the authorized users can observe the interested information (environment parameters, equipment status, etc.) by means of Internet or mobile phone remotely.

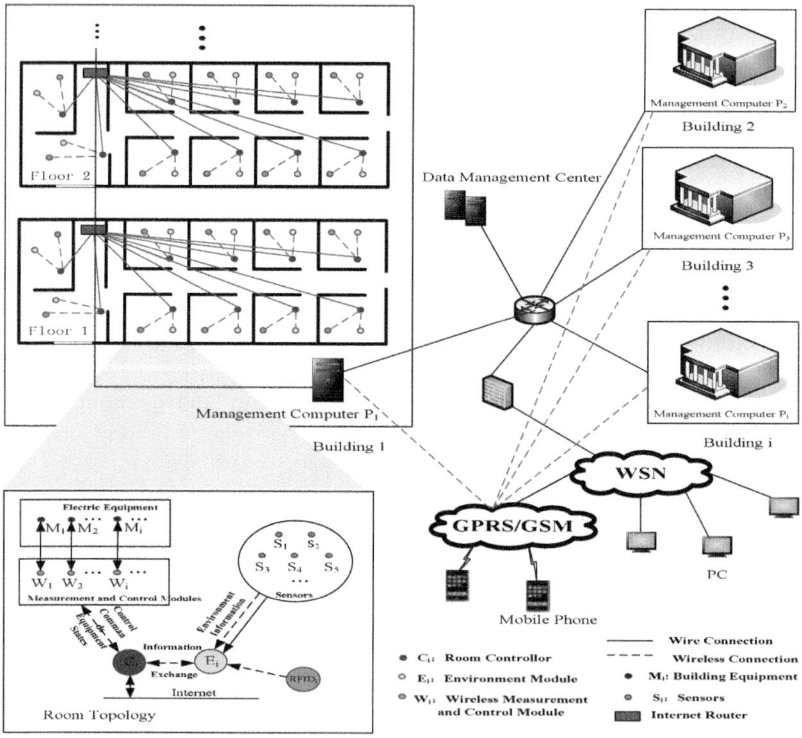

Fig. 1. Architecture of Internet of Things for Building Electrical Equipments

A. Equipments' Connection with Network

If we want to control hundreds of electrical equipments in a building by a global optimization approach, these equipments must be physically connected into our Internet of Things, in which each equipment has a corresponding M_i. The block M_i uploads information to the room controller and at the same time receives commands from it through a wireless network. The structure of the wireless measurement and control block for electrical equipments is shown in Fig. 2.

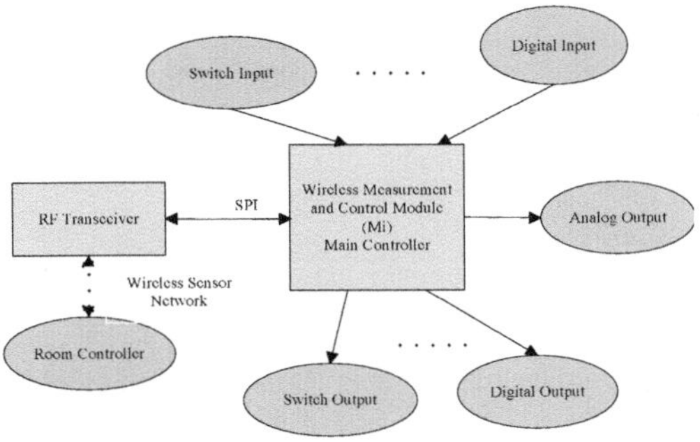

Fig. 2. Structure of Wireless Measurement and Control Block for Electrial Euipments

For such devices that can be controlled by switch signal as lamps, solid state relays (SSR) and buttons are fixed on M_i. So once the block has received the switch commands, on/off control action for lamps could be realized automatically. The main controller in M_i achieves on-off manipulation of the devices by use of the zero-crossing conduction/cutoff characteristics of SSR through the I/O port. Moreover, the port must be read again to confirm the state of device. Manual operation, which has been assumed to own the highest priority, can also be achieved by the buttons on the block. For example if a lamp is operating and at some time the "off" button is pressed down at the interface, M_i will cut off SSR to turn off the lamp. Simultaneously, the block reports the changed state to C_i. And, C_i will transmit it to the P immediately. Next, the "off" state will be shown in the management interface on the management computer. For those equipments controlled by digital signal, e.g. air conditioners, their blocks can control their operation like a remote-control unit. Similarly, the gathered state information will be sent to C_i.

B. Room Controller (Ci)

Room controller (C_i) equipped with RF transceiver, as shown in Fig.3, can obtain environment information of every room from E_i, and get the messages of equipment parameters by two-way communication with W_i. In addition, it is responsible for giving orders to W_i to manipulate the related equipment, and executing instructions given by the P via Ethernet. For instance, when a person walks into a room, C_i in this room will detect the change, and then start up the lighting system automatically and adjust the air-conditioner to the most suitable temperature. On the contrary, when he leaves the room after a while, C_i will ask for the management computer. If the computer answered that the person had left the building, then C_i will shut down the lighting and air-conditioning equipments automatically. The main controller in C_i is MC9S12NE64 or MEGA64. The main controller of RF transceiver is UM2455, which is a kind of the wireless transceiver chips.

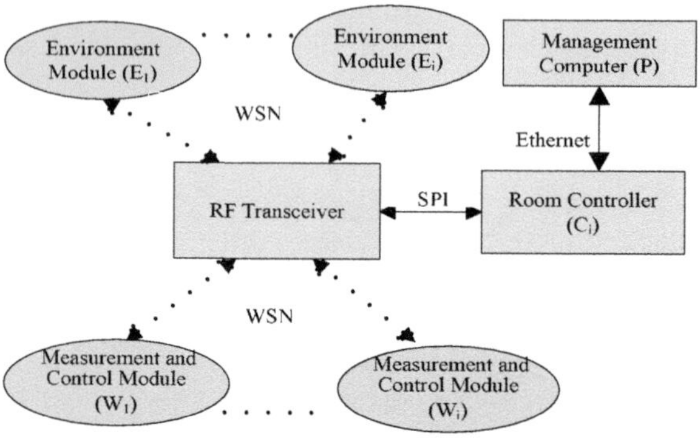

Fig. 3. Relationship of Room Controller and other Modules

C. Environment Block (Ei)

Environment block is used to gather environmental temperature, humidity, illuminance and infrared sensor information, carbon dioxide concentration, personnel information and so on. Thanks to its powerful popularity, the block can be further expanded to collect other information for special requirements. If some parameters become unimportant, some function of this block can also be simplified. Environment block, which has joined the wireless sensor networks by RF transceiver, exchanges with C_i and receives personal identification RFID information. The main controller in E_i is MEGA64. UM2455 is employed as the controller of a RF transceiver. Fig.4 demonstrates the architecture of E_i.

D. Construction of Wireless Sensor Networks (WSN)

In the ITBEE, WSN plays a very important role as a key means of communication between C_i and E_i, C_i and W_i. Since the wireless network is applied, it does not need to run wires and cables in the field. Our Internet of Things can be installed easily due to the powerful self-organizing ability of WSN. Nodes with self-organizing capability in the network can arrange and manage automatically, and finally form a multi-skip network with the help of the topology control mechanism and the internet protocol. For example, once a new W_i or E_i appears, these blocks can distinguish WSNs, and find a node with the strongest signal to enter. When a certain node loses for reasons of power failure or disturbances, the WSN can be rebuilt automatically.

Construction of WSN in a room has two stages. At first, a WSN must be established logically. After initializing, W_i and E_i broadcast an asking frame in the wireless network, and C_i or routers nearby will send messages (including short address of the asking block) to respond. Having received the replied frame, the block will choose a node to join in, and extract its own address from the replied frame. The second stage is the registration of blocks. Due to the dynamics of WSN, the allocated address maybe modified. As a result, E_i and each W_i must register to their C_i. Having stored of these messages in a table, the room controller replies the packet. Once E_i and W_i have received the responded packet, it means that the registration stage is finished successfully.

3 Design of Equipment Management System

In this paper, *B/S mode* and *.Net layered framework* are applied for design of the equipment management system. With access to the Internet, users anywhere in the world can view the environmental parameters and the working status of building electrical equipments, and manipulate the operation of equipment remotely. The equipment management program based on Web server is divided into three parts: receiving data and sending commands, construction of database and Web publishing. Developed with *C#* and *Winsock* programming techniques, the first part program is not only responsible for monitoring the specified ports, distinguishing UDP Socket connection request of each room controller, but also sending orders decided by a host to the corresponding C_i. *SQL Server 2000* is chosen as a database management tool in our scheme. Five tables with different functions are designed as required, which are called *environment information table*, *equipment table*, *log table*, *access information table* and *system configuration information table*. Amounts of the environmental parameters, equipment states and the like are stored in the first two tables. *The log table* is used to record system maintenance information. The access permission of every user can be checked from the fourth table. System configuration is described in the final table. Limited to the space, only data form of *environment information table* is listed in Table 1.

Table 1. Form of Environment Information Table

Name	Type	Null	Default Value	Comments
Rnum	text		0	Room ID
Uname	varchar(20)			User ID
Upass	varchar(21)			Password
Tem	float		0	Temperature
Hum	float		0	Humidity
Light	float	Allowable	0	Illuminance
CO2	float		0	CO_2 Concentration
Person	text	Allowable	0	Person Information
Dflag	char	Allowable	0	Equipment Flag

Having finished the database construction, we assign an IP address to the P and each C_i. The P carries on receiving UDP Socket connection request sent by C_i, and saving information also sent by C_i into database. Similarly, the equipment management system processes the gathered data in real-time, and displays various parameters and charts. When a user opens the Web browser and enters URL address of homepage, it is easy to access the environment and equipment information of each building. Furthermore, it is feasible to control the electrical equipments remotely as long as you are authorized.

4 Application Example

It is a good idea to take the remote control of electrical equipment in a room as an example to illustrate how the ITBEE can be realized. The steps are listed as below:

1) In the room XX308, as shown in Fig.4, each device has its own W_i. all W_is and E_i in XX308 register to the C_i in the same room first. After the registration, the equipment states are uploaded to the C_i via WSN. As well as, E_i transmits environmental information.

2) An IP address is assigned for every C_i and P. After initializing, the equipment management system can realize the two-way communication among them through Ethernet.

3) If one user logs in the system successfully, the management interface of the room XX308 which the controlled equipments lies in can be found with the help of the navigator on the left screen. For example, when the user presses down an on/off button of an air-conditioner, the control procedure begins to run and give an order to the C_i of XX308.

4) Having received the order, the C_i of XX308 finds the W_i of the air-conditioner, and delivers the control command to it. Finally, W_i can complete the "on/off" operation of the air-conditioner by means of SSR successfully.

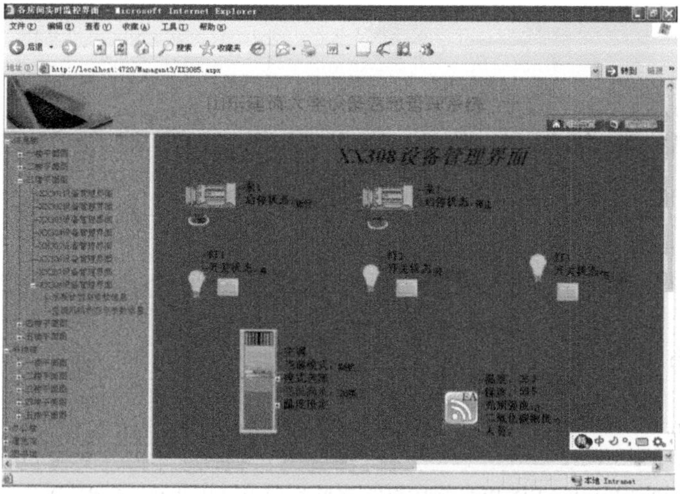

Fig. 4. Interface of Equipment Management System

5 Conclusion

An Internet of Things for building electrical equipments is designed in this paper. Environmental parameters and equipment states are collected by the environment block and the wireless measurement and control blocks embedded on the electrical equipments, such as air-conditioner, lights, water pumps, etc. These data are also sent to the room controller by means of WSN. The self-organization characteristic and protocol problem of WSN are solved ideally. In addition, we provide an equipment management interface in the B/S framework on the management computer to realize the remote control and monitoring of the electrical equipments. By integrating information of each room controller, the management computer can improve the measurement accuracy and achieve fault diagnosis for the electrical equipments.

References

1. Kottenstette, N., Koutsoukos, X., Hall, J., Sztipanovits, J., Antsaklis, P.: Passivity-Based Design of Wireless Networked Control Systems for Robustness to Time-Varying Delays. In: Real-Time Systems Symposium, pp. 15–24 (2008)
2. Kortuem, G., Kawsar, F., Fitton, D., Sundramoorthy, V.: Smart objects as building blocks for the Internet of things. IEEE Internet Computing 14, 44–51 (2010)
3. Shen, B., Zhang, G., Zhang, L., Wang, M.: Muli-Agent System Design for Room Energy Saving. In: The 5th IEEE Conference on Industrial Electronics and Application, vol. 5, pp. 73–76 (2010)
4. Karsai, G., Neema, S., Sharp, D.: Model-driven architecture for embedded software: A synopsis and an example. Science of Computer Programming 73, 26–38 (2008)

Study of Adaptive Cone Taper Measuring Device*

Xiqu Chen[1], Yu Li[2], Jianxiu Su[1], Zhankui Wang[1], Sufang Fu[1], and Lijie Ma[1]

[1] Department of Machine & Electricity,
Henan Institute of Science and Technology
Xinxiang Henan Chna, 453003
[2]Department of Mechanical Engineering,
Henan Polytechnic University
Jiaozuo Henan, China, 454003
XiquChen0090@tom.com

Abstract. A kind of simple practical adaptive quantitative detection taper cone measuring devices is developed in this paper, which can resolve the qualitative measurement which cannot obtain concrete error problem, and avoid to use the measurement of data to calculate in the original quantitative detection method, operates inconveniently, detection device expensive, and not suitable to use in the production site. The measuring device through the relative sliding automatically of main and vice foot adapts to the change of cone diameter, and the change of the cone taper can be obtained by measuring claw around hinge rotation, which is can be measured directly on the machine(measurement accuracy is 0.02 mm, and measuring scope is $0^0 \leq 2\alpha \leq 150^0$), which effectively solved the problems of quantitative detection of the cone taper in production site.

Keywords: Adaptation, Cone, Taper, Measuring Devices, Precision.

1 Introduction

Taper surface inspection has two ways: comprehensive inspection and single inspection. Comprehensive inspection use taper gauges, and single inspection use angle measuring instrument, sine gauge, and taper angle of gauges, angle gauge block and other instrument to measure cone tolerances parameters. Specifically, the detection conical taper cone can use rules to obtain qualitative detection[1] [2], and use sine gauge, single sample great fixed high block, million workers show, ball gauge block set of legal and double balls method, the cone taper apparatus etc to obtain taper cone quantitative detection[3,4]. In addition, qualitative tests cannot measure the specific taper cone error, and measuring instruments are easily damaged, especially when the taper is large, through the scribed line ,the end scribed line, and the end surface are located very closely, and it is difficult to distinguish the direct guidance function of production. But the original quantitative detection method needs certain equipment

* This work is partially supported by education department of henan province naturai science research program (serial number: 2010A460007P).

J. Zhang (Ed.): ICAIC 2011, Part V, CCIS 228, pp. 298–305, 2011.

(some even very expensive), and not directly measured taper cone of values, and not obtain first measure data, then calculated taper cone by value [5], then unloaded the workpiece, so the operation process trouble, low efficiency, and the cost of detection is high and not suitable to use in the production field. This article aims to design an easy and practical adaptive quantitative detection taper cone measuring device to solve above problems.

2 Measuring Device Structure

Measuring device structure is shown in Figure 1, and its structure contains the main foot 1, vice-foot 2, the left (right) Dial 5 (3), left (right) cursor 6 (14), left (right) measured Xenopus 9 (11) , axis 7 (13), left (right) supports 4 and so on. Left (right) Dial is fixed by the main bearing and vice-foot, Vice-foot slide on the main foot to automatically adapt to the change of the diameter of the cone; Left (right) cursor is connected with left (right) measured Xenopus with the hinge 8 (12) by bolt holes as a whole, and rotate around the axis 7 (13) to automatically adapt to the changes of the cone taper; the centerlines of the two axes (7,13) are parallel to each other; The intersect between measuring surface M and positioning surface N is hollow to avoid the impact of glitches. During the measurement, the positioning surface N of left (right) measured is tightly contacted with the cone end surface 10, the measured face M of left (right) measured Xenopus is tangent with conical surface at the symmetry of the generatrix. The angle deflection of measured Xenopus relative to cone centerline can be read out through the dial of the left and right cursor, and reading principles is the same as vernier caliper, and the sum of the two readings is the cone angle.

3 Analysis to the Range of Measurement

A. Taper Measuring Range

When measuring taper gradually increases, the cursor is gradually transferred from the horizontal position close to the vertical position, which is shown in Fig. 2. To avoid interference, the cursor can not go beyond the scope of the dial. Cursor radius R'=63mm, The wheels radius in Axis 7 r=5.5mm, so:

$$\gamma = 2\tan^{-1}\frac{r/2}{R'} = 2\tan^{-1}\frac{5.5/2}{63} = 5.0^{0}$$

Maximum angle measurement:

$$\alpha_{\max} = 90^{0} - 9^{0} - \gamma = 76^{0}$$

The range of measurements taper: $0^{0} \leq 2\alpha \leq 150^{0}$.

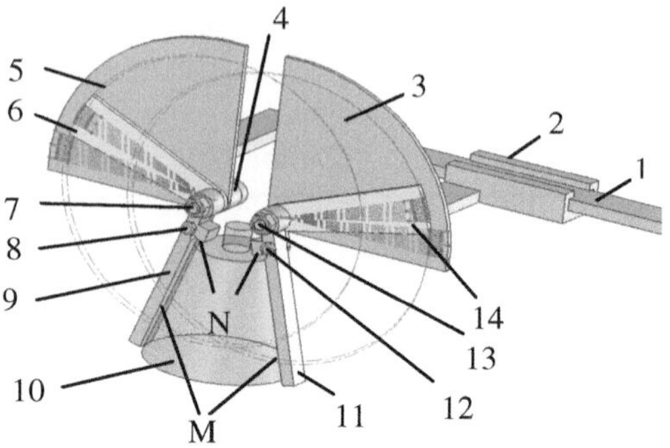

Fig. 1. Measuring device structure

M—The tangent plane of Left (right) paw and measuring surface

N—The positioning surface of Left (right) paw and cone-side surface contacts

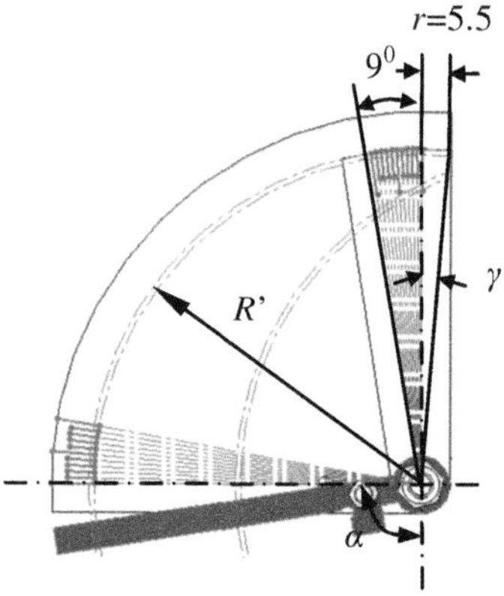

Fig. 2. Analysis to the measurements range of taper

B. Diameter Measurement Range

The maximum measuring diameter is limited by the main foot length, and the smallest measuring diameter is analyzed in the research. The smallest measuring diameter is clearly connected with the taper of the measured cone 2α, hub radius r and the distance of locating surface N on the measured Xenopus to the hole center line L (= 13mm) on, in order to avoid interference with $x \geq 0$,which is shown in Fig. 3.

$$2R = x + 2r + 2L\tan\alpha$$

So

$$2R_{min} \geq 2r + 2L\tan\alpha = 11 + 26\tan\alpha$$

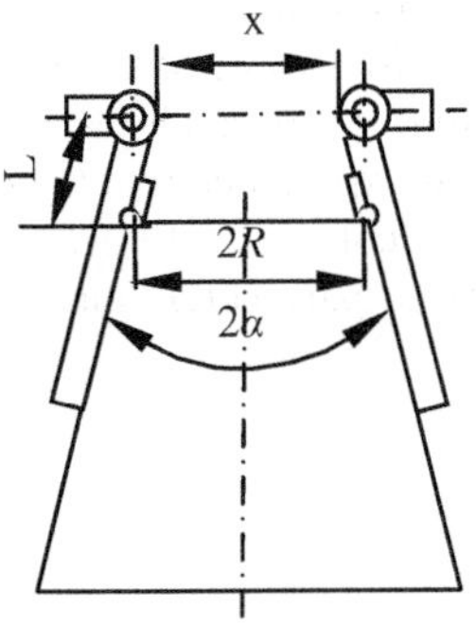

Fig. 3. Analysis to the measuring range of diameter

The smallest measuring diameter is connected with the taper of the measured cone. When the taper decreases, the hub radius r influences it; when the taper increases, L influences it, as is shown in Tab.1.

Table 1. Relationship between Taper and Hub Radius

$2a$	20	50	80	110	140
$2R_{min}$	16	23	33	48	82

4 Accuracy analysis

The influences of measurement accuracy of measuring devices contain the manufacturing errors of measuring devices and the manufacturing errors of the cone:

A. The Influence the Distance error δ1 of the Locating Surface on the Measuring Paw to the Hinge Center Line

Normal measurement, claws locate in the ends of diameter AB, and record the measured angle between the bus AD and BC, so the cone angle of the cone is:

$$(2\alpha=\alpha_1+\alpha_2(\alpha_1=\alpha_2=\alpha)$$

Where α_1 and α_2 respectively stand for the dial reading of the left and right sides.

When the distance of locating surface N on the measured Xenopus to the hole center line is different, the two hinges center connection is not perpendicular to the cone centerline. That is shown in dotted lines in Figure 4, as the two hinges center connection change into mn into mn '. The measurement error $\Delta\alpha_1$ and $\Delta\alpha_2$ respectively appear in the left and right dial, when the dial reading of left side reduced, the right dials reading increases. Obviously, $\Delta\alpha_1=\Delta\alpha_2=\Delta\alpha$, therefore, measurement errorΔ_1 is:

$$\Delta_1=\alpha_1+\alpha_2-\ (\alpha_1-\Delta\alpha_1+\alpha_2+\Delta\alpha_2)\ =0 \quad (1)$$

The distance of locating surface N on the measured Xenopus to the hole center line error δ_1 has no influence to measuring accuracy.

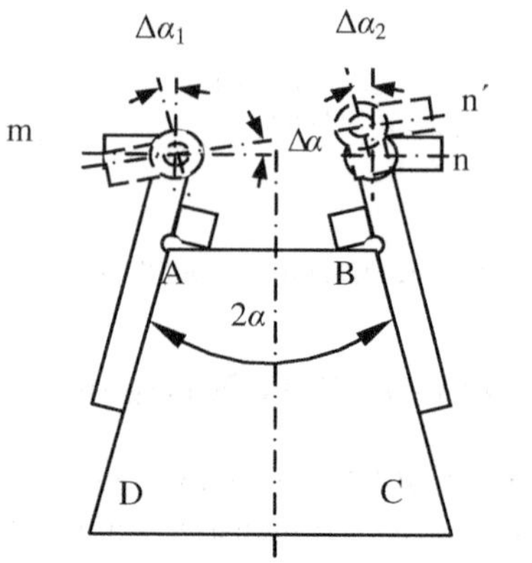

Fig. 4. δ_1 influence to measurement accuracy

B. The Influence of the Two Hinge Centerline Parallelism Error in the Horizontal Plane (the Angle θ1 Indicated)

When the two hinge center line is parallel, the angle between the generatrix AD and BC is measured (the cone angle of cone 2a); when the two hinge center line exists parallelism error θ_1 in the horizontal plane, the angle of the generatrix AD and B 'C'2 ($a+\Delta a$) is measured, which is shown in Figure 5, so it apparently had a measurement error.

Supposed cone height l, the radius of the circle plane of the upper and lower ends is R_1,R_2, (AB = 2 R_1, CD = 2 R_2). Because ΔABB 'is a right triangle (AB' ⊥ BB '), ∠ BAB' = θ_1, so

$$AB' = AB\cos\theta_1 = 2R_1\cos\theta_1$$

Similarly

$$C'D = CD\cos\theta_1 = 2R_2\cos\theta_1$$

Therefore

$$\tan(a+\Delta\alpha) = \left(\frac{C'D}{2} - \frac{AB'}{2}\right)/l = (R_2 - R_1)\cos\theta_1/l$$

And

$$\tan a = \left(\frac{CD}{2} - \frac{AB}{2}\right)/l = (R_2 - R_1)/l$$

So

$$\tan(a+\Delta\alpha) = \tan a\cos\theta_1$$

Arranged that:

$$|\tan\Delta\alpha| = \frac{(1-\cos\theta_1)\tan a}{1+\tan^2 a\cos\theta_1} \le \frac{(1-\cos\theta_1)\tan a}{2\tan a\sqrt{\cos\theta_1}} = \frac{1-\cos\theta_1}{2\sqrt{\cos\theta_1}}$$

$$\Delta_2 = 2\Delta\alpha \approx 2|\tan\Delta\alpha| \le \frac{1-\cos\theta_1}{\sqrt{\cos\theta_1}} \qquad (2)$$

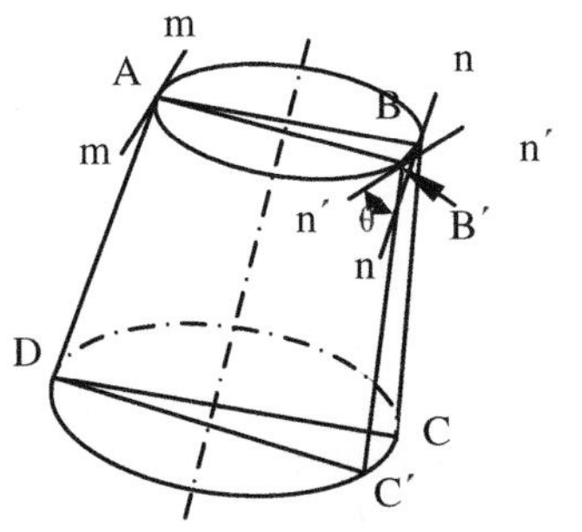

Fig. 5. θ_1 influence to measurement accuracy

C. The Influence of Vertical Error of the End Cone and Axis (Indicated by Angle β)

Fig.6 (a) shows that when the up end of the cone is not perpendicular to the axis, AB largest vertical errors and EF is the direction of the minimum vertical error (0), when measured jaws were stuck in the AB and EF, measurement of both directions is not the same.

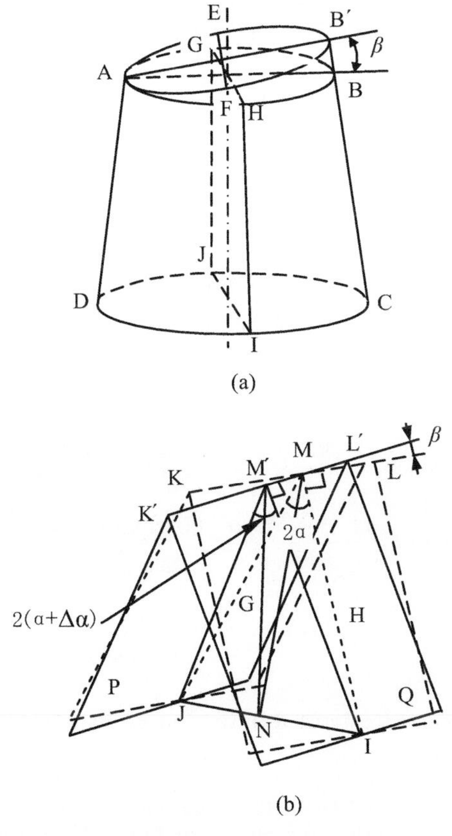

Fig. 6. The influence of β to measuring accuracy

① When the measuring claw measure along the AB direction, it will make the right claw's positioning surface N move above, which cause the two hinge center connection line and the cone axis is not perpendicular, And the situation is similar to (1), therefore, the measurement error:

$$\Delta_3=\Delta_1=0 \tag{3}$$

② When measuring claw survey along the direction of EF, due to the presence of the vertical error, making the measuring device tilt angle β change the dotted line to the solid line position. For minimum value β, two measuring surface P, Q of the jaws and the tangent cone GJ (MJ), HI (MI) can be approximately turn into M'J, M'I, as shown in Fig.6 (b).

Supposed the line of intersection of two jaws measuring surface P, Q is KL, the point of intersection of cone generatrix GJ, HI and KL is M, then GJ (MJ) ⊥ KL, HI (MI) ⊥ KL, surface IMJ ⊥ KL, ∠ IMJ = 2α(the actual cone angle).

For IM '⊥ K'L' in M ', connecting M' with J, we can obtain JM '⊥ K'L' and surface JM'I ⊥ K'L 'by the symmetry of knowledge. Meanwhile, ∠ IM'J =2(α+Δα) (measured cone angle).

Connect J with I, supposed JI midpoint for N, connect M and N, M´ and N´. because ΔMJI and ΔM´JI are an isosceles triangle, Easy to proof: MN⊥JI, M´N⊥JI, MN, M´N were two an isosceles triangle vertex Angle of angular bisectrix.

Because Face IMJ⊥KL,so MN⊥KL; And because Face JM´I⊥K´L´, so M´N⊥K´L´,so ∠M´NM=∠LML´=*β*.

In RTΔM´NI,

$$NI=M'N\tan\alpha'$$

In RTΔMM´N,

$$MN=M'N/\cos\beta$$

In RTΔMN,

$$NI = MN\tan\alpha = (M'N/\cos\beta)\ \tan\alpha$$

Therefore,

$$\tan(a+\Delta a)= \tan\alpha/\cos\beta$$

Sorting,

$$\tan\Delta\alpha=\frac{(1-\cos\beta)\tan a}{\cos\beta+\tan^2 a}\leq\frac{(1-\cos\beta)\tan a}{2\tan a\sqrt{\cos\beta}}=\frac{1-\cos\beta}{2\sqrt{\cos\beta}}$$

$$\Delta_4=2\Delta\alpha\approx 2\tan\Delta\alpha\leq\frac{1-\cos\beta}{\sqrt{\cos\beta}} \tag{4}$$

D. Two Hinge Centerline Restricted in the Parallelism of Error Influence (with Angle θ_2 Represents)

If two hinge centerline restricted in inside parallelism error θ_2, when measuring the think of one side of the state the correct(The dial records on reading correct), the other side of the hinge and card claw happened tilt, this part (2) and (3) the similar, only then produce error is (2) (2) parts of the half. Namely

$$\Delta_5 = \Delta_4/2 = \Delta\alpha \le \frac{1-\cos\theta_2}{2\sqrt{\cos\theta_2}} \tag{5}$$

By measuring device manufacturing precision measuring error caused by the for:

$$\Delta = \sqrt{\Delta_1^2 + \Delta_2^2 + + \Delta_5^2} \le \sqrt{0 + \left(\frac{1-\cos\theta_1}{\sqrt{\cos\theta_1}}\right)^2 + \left(\frac{1-\cos\theta_2}{2\sqrt{\cos\theta_2}}\right)^2}$$
$$= \sqrt{\frac{(1-\cos\theta_1)^2}{\cos\theta_1} + \frac{(1-\cos\theta_2)^2}{4\cos\theta_2}} \tag{6}$$

Obviously for main factors affecting the measurement precision is measuring devices on two hinge parallelism theta, and the hinge on in horizontal parallelism on measuring accuracy influence than vertical plane within the influence of parallelism. Through the analysis of the knowledge, if the measuring device according to IT4 ~ IT5 precision manufacturing, the two hinge parallelism of no greater than for 13/1000, corresponding $\theta \approx 0.75^0$, take $\theta_1 = \theta_1 = \theta$, *can be obtained* $\triangle \le 0.55^0$ ($< 0.02^0/3$) , so the accuracy of measurement equipment for 0.02^0.

5 Conclusions

This paper research the cone taper measurement device minimum measuring diameter and measured by cone taper relevant, taper hour is mainly affected by the influence of r, taper large mainly by L influence. Measuring scope for $0^0 \le 2\alpha \le 150^0$; Measurement accuracy for 0.020.Measuring device simple structure, easy operation, can on a wide range of adaptive cone diameter changed, directly measured data varies cone, can avoid original taper measuring method of defects, and effectively solve the manufacturing taper measurement problems.

References

1. Li, Z., Kang, S.: Conical Degree Test Methods of Taper Socket Gauge. Industrial Measurement 2, 48–49 (2000)
2. Chen, K.: A taper gauge design calculation. Machinist Metal Cutting 02, 21 (1998)
3. Zhou, F., Zhou, P., Zhang, G.: Machinery Manufacturing Measuring Technical Manuals, vol. 1, pp. 137–150. Mechanical Industry Press, Beijing (1999)
4. Zhao, S., Fan, Y.: Roll Cone Taper Detector. Tool Technique 09, 66–68 (2005)
5. Du, M., Zhang, Y.: The Cone Measurement and Data Processing. Measurement And Testing Techniques 04, 10–11 (2002)
6. Fang, K.F.: Appying the Practical Manual, vol. 02, pp. 288–291, 278. Mechanical Industry Press, Beijing (2006)

DR/INS Redundant Integrated Navigation System Based on Multi-model Adaptive Estimation Method*

Kefei Yuan, Gannan Yuan, and Hongwei Zhang

College of Automation,
Harbin Engineering University,
Harbin, 150001, China
KefeiYuan111@foxmail.com

Abstract. The uncertainty of navigation system model will lead to model identification errors. A DR/INS redundant integrated navigation system is proposed basing on multi-model adaptive estimation method to merge models constructed in different considerations and view angles, and formed a practical adaptive estimation algorithm using autoregressive model and hypothesis tests. The redundant integrated navigation system, which acquires information primarily from DR and INS, and TAN information as secondary sources, calculates the fusion information by federal Kalman filter. Simulations are performed and the results show that proposed system can overcome the uncertainty of system modeling, and high precision, fault-tolerance, anti-interference and stability can be obtained.

Keywords: multi-model adaptive estimation, redundant navigation, dead reckoning, federal Kalman filter.

1 Introduction

In multi-sensor integrated navigation system, the accurate sensor error model is the fusion of precision. The uncertainty of navigation system model will lead to model identification errors. But model establishment actually not entirely as desired, because mechanism modeling always cannot describe the complete information, and experimental modeling only can aim at the specific equipment. The multi-model theory is the strong mathematical tool, which approaches the uncertainty error model through many models, reducing the request of precise model in the fusion system and high precision, stability and fault-tolerance can be obtained. In artificial intelligence, process control, biomedical and other areas, multiple model theory has been used perfectly. The paper which uses multi-model adaptive estimation in integrated navigation system, solves the problems that dead reckoning system error model is difficult to describe, improving system performance of redundant navigation system.

* This paper is funded by the International Exchange Program of Harbin Engineering University for Innovation-oriented Talents Cultivation.

J. Zhang (Ed.): ICAIC 2011, Part V, CCIS 228, pp. 306–315, 2011.

2 Multiple Model Adartive Estimation Principle

Any system all has its uncertainty, therefore the model identification is a important system error. The multi-model theory is used to solve the problem. The basic method of multi-model adapted estimation is approaching the uncertainty system by many models. High precision, stability and fault-tolerance can be obtained as result from covering system real condition by many models [1].

Design a group of parallel filters for the different models. Each filter input is the input information for the model, and each filter output is estimation and residual error for the model. The overall estimate can be obtained by detection principle. The multi-model adaptive estimation structure is shown as Figure 1.

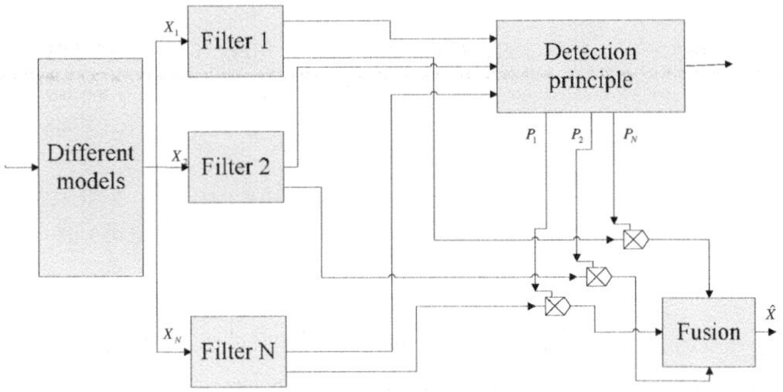

Fig. 1. The multi-model adaptive estimation structure

In the navigation system, a model set for various navigation subsystem should be modeled in a different way to demonstrate different characteristics.

The filter used to estimate current and future state in time. Filter choosing depends on prior knowledge. If we know nothing about prior knowledge, two point extrapolation and least-squares method are used generally. If the measurement error only be known, time series analysis can be used. If the dynamic noise and the measurement error are all known, the minimum variance estimation is the best way, including Wiener filter, etc.

The overall estimate can be obtained by weighted average. the conditional probability can be gained by hypothesis testing algorithm (HTA).

3 Redundant Integrated Navigation System Model

According to the configuration of the navigation sensors in the underwater vehicle, the integrated navigation system is composed of dead-reckoning system (DR), Inertial Navigation System (INS), Global Positioning System (GPS), radio navigation system (loran C), bathymeter and terrain aided navigation (TAN) generally. Usually DR and INS will be the datum navigation system, because of the anti-interference and weatherproof capability. The redundant integrated navigation system is built in the paper, which acquires information primarily from DR and INS, and TAN information as secondary sources. The composing of system is shown in Figure 2.

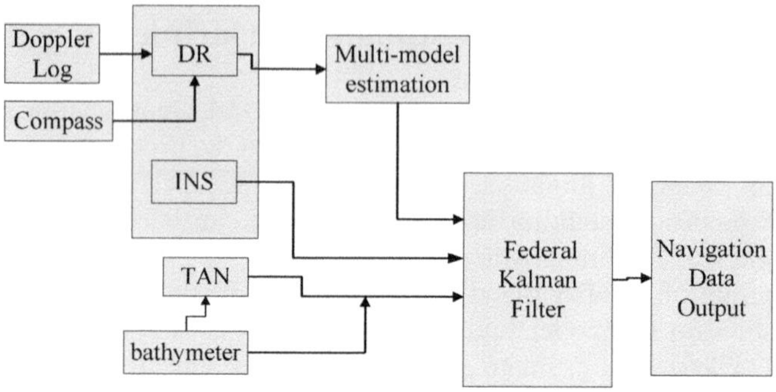

Fig. 2. The composing of redundant integrated navigation system

A. DR Models Set

DR models construct in different considerations and view angles to demonstrate different characteristics.

Singer model is the first model. The dead reckoning process is described as dynamic model which is the time-correlation noise sequence. Acceleration can be considered as zero-mean first-order stationary processes.

$$a(t) = -\alpha a(t) + w_1(t) \tag{1}$$

where α denotes the reciprocal value of time constant, $w_1(t)$ denotes white Gaussian noise.

The autocorrelation function is given by:

$$R(T) = E\left[a(t)a(t+T)\right] = \sigma^2 e^{-\alpha|T|} \tag{2}$$

The state model is:

$$\dot{X}_{D1}(t) = \begin{bmatrix} F_{D1} & & \\ & F_{D1} & \\ & & F_{D1} \end{bmatrix} X_{D1}(t) + \begin{bmatrix} 0 & 0 & 1 & 0 & 0 & 1 & 0 & 0 & 1 \end{bmatrix}^T w_1(t) \tag{3}$$

where, $X_{D1} = \begin{bmatrix} L & V_E & f_E & \lambda & V_N & f_N & h & V_U & f_U \end{bmatrix}^T$ is the error state vector , $F_{D1} = \begin{bmatrix} 0 & 1 & 0 \\ 0 & 0 & 1 \\ 0 & 0 & -\alpha \end{bmatrix}$.

Current statistical model (CS), this model uses the nonzero average value and modified Rayleigh distribution, conforms to the actual situation.

$$a(t) = -\frac{1}{\tau} a(t) + \frac{1}{\tau} \bar{a}(t) + w_2(t) \tag{4}$$

where τ denotes time constant, $\overline{a}(t)$ the nonzero average value of acceleration, $w_2(t)$ denotes white Gaussian noise.

$$\begin{aligned}\overline{a}(k+1) &= E\left\{a(k+1) \mid z^k\right\} \\ &= e^{-\frac{1}{\tau}T} E\left\{a(k+1) \mid z^k\right\} + \left(1 - e^{-\frac{1}{\tau}T}\right)\overline{a}(k) \\ &= e^{-\frac{1}{\tau}T} \hat{a}(k) + \left(1 - e^{-\frac{1}{\tau}T}\right)\overline{a}(k)\end{aligned} \tag{5}$$

The variance is:

$$\sigma_{cs}^2 = \begin{cases} \dfrac{4-\pi}{\pi}\left[a_{\max} - \hat{a}(k-1)\right]^2 & 0 < \hat{a}(k-1) < a_{\max} \\ \dfrac{4-\pi}{\pi}\left[a_{\max} + \hat{a}(k-1)\right]^2 & a_{-\max} < \hat{a}(k-1) < 0 \end{cases} \tag{6}$$

The state model is:

$$\begin{aligned}\dot{X}_{D2}(t) &= \begin{bmatrix} F_{D2} & & \\ & F_{D2} & \\ & & F_{D2} \end{bmatrix} X_{D2}(t) + \begin{bmatrix} 0 & 0 & \frac{1}{\tau} & 0 & 0 & \frac{1}{\tau} & 0 & 0 & \frac{1}{\tau} \end{bmatrix}^T \overline{f}(t) \\ &+ \begin{bmatrix} 0 & 0 & 1 & 0 & 0 & 1 & 0 & 0 & 1 \end{bmatrix}^T w_1(t)\end{aligned} \tag{7}$$

where, $F_{D2} = \begin{bmatrix} 0 & 1 & 0 \\ 0 & 0 & 1 \\ 0 & 0 & -\frac{1}{\tau} \end{bmatrix}$, $\overline{f}(t)$ denotes three channels $\overline{a}(t)$ [2].

Sensors model is composed of the errors of the sensors. The information sources of dead reckoning are log, compass, propeller speed and sea chart. The error of Doppler log is:

$$\delta V = \Omega \times V + \delta K V + w_3(t) \tag{8}$$

where $\Omega = \begin{bmatrix} 0 & -\phi_U & \phi_N \\ \phi_U & 0 & -\phi_E \\ -\phi_N & \phi_E & 0 \end{bmatrix}$ is the anti-symmetric matrix, δK denotes scale factor error, $w_3(t)$ denotes white Gaussian noise.

The error of compass is:

$$\Delta\Psi = A + B\sin\Psi + C\cos\Psi + D\sin 2\Psi + E\cos 2\Psi \tag{9}$$

where A,B,C,D,E denote error pensate scales.

The state model is:

$$\dot{X}_{D3}(t) = F_{D3}(t)X_{D3}(t) + G_{D3}(t)w_3(t) \tag{10}$$

where $X_{D3} = \begin{bmatrix} L & \lambda & h & V_E & V_N & V_U & \phi_E & \phi_N & \phi_U & \delta K & \Delta\Psi \end{bmatrix}^T$, $w_3(t)$ denotes white Gaussian noise [3].

B. INS Model

INS errors include three channels velocity errors, position errors and attitude errors besides the inertial elements errors. Inertial elements errors include the installation error, scale factor error and random error etc, and only the random error is considered here. Assume the gyro drift and accelerometer errors .

Thus, INS state equation is given by

$$\dot{X}_I(t) = F_I(t)X_I(t) + G_I(t)W_I(t) \tag{11}$$

$X_I = \left[\phi_E\ \phi_N\ \phi_U\ \delta V_E\ \delta V_N\ \delta V_U\ \delta L\ \delta\lambda\ \delta h\ \varepsilon_{bx}\ \varepsilon_{by}\ \varepsilon_{bz}\ \varepsilon_{rx}\ \varepsilon_{ry}\ \varepsilon_{rz} \nabla_x\ \nabla_y\ \nabla_z\right]^T$ is the error state vector , $W_I = \left[w_{gx}\ w_{gy}\ w_{gz}\ w_{bx}\ w_{by}\ w_{bz}\ w_{ax}\ w_{ay}\ w_{az}\right]^T$,

$$G_I = \begin{bmatrix} C_b^n & 0_{3\times3} & 0_{3\times3} \\ 0_{9\times3} & 0_{9\times3} & 0_{9\times3} \\ 0_{3\times3} & I_{3\times3} & 0_{3\times3} \\ 0_{3\times3} & 0_{3\times3} & I_{3\times3} \end{bmatrix}^T, F_I = \begin{bmatrix} F_N & F_S \\ 0 & F_M \end{bmatrix}_{18\times18}, F_S = \begin{bmatrix} C_b^n & C_b^n & 0_{3\times3} \\ 0_{3\times3} & 0_{3\times3} & C_b^n \\ 0_{3\times3} & 0_{3\times3} & 0_{3\times3} \end{bmatrix},$$

$$F_M = Diag[0,0,0,-\frac{1}{T_{rx}},-\frac{1}{T_{ry}},-\frac{1}{T_{rz}},-\frac{1}{T_{ax}},-\frac{1}{T_{ay}},-\frac{1}{T_{az}}]\ [4].$$

C. TAN Model

When vehicle voyages to an area, bathymeter measures the s actual depth, and then the actual depth and the position are sent to data calculator, and then use calculating algorithm (the paper uses the iterative closet contour point (ICCP) algorithm) to match, and then send the position and depth to federal Kalman filter.

$$Z_T = \begin{bmatrix} L - L_T \\ \lambda - \lambda_T \\ h - h_T \end{bmatrix} = H_T X + V_T \tag{12}$$

in TAN and DR system, $X=X_D$, in TAN and INS system, $X=X_I$, V_T is zero-mean white Gaussian noise [5].

4 System Algorithm

In this system, the DR and INS are redundant master system, which the input is multi-model adaptive estimation in DR system. Basing on DR/TAN and INS/TAN, the overall estimate can be obtained by federal Kalman filter. System algorithm structure is shown as in Figure 3.

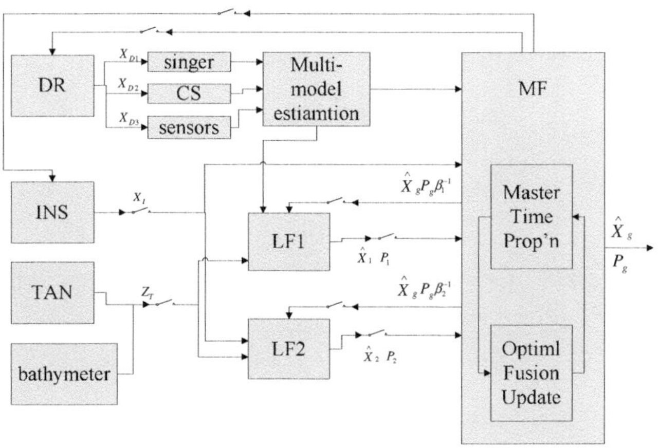

Fig. 3. System algorithm structure

A. Multi-model Adaptive Estimation of DR

According to DR models set, the estimation and residual can be calculated by autoregressive model (AR model). Conditional probability can be calculated by hypothesis testing algorithm (HTA). Then the overall estimate can be obtained.

Assume $x(k)$ is the sequence of model state. The AR model of $x(k)$ is given by:

$$x(k) = c_k + \varphi_N(1)x(k-1) + \varphi_N(2)x(k-2) + \cdots + \varphi_N(N)x(k-N) \tag{13}$$

where N denotes the order of AR model, model error c_k is zero-mean and variance σ^2 white Gaussian noise, φ_N denotes the coefficient of AR model, the power spectral density of $x(k)$ is given by:

$$S_x(\omega) = \sigma^2 |H(\omega)|^2 = \frac{\sigma^2}{\left|1 + \sum_{k=1}^{N} \varphi_k e^{-j\omega k}\right|^2} \tag{14}$$

R_{N+1} denotes the autocorrelation function matrix of $x(k)$.

$$R_{N+1} = \begin{bmatrix} \hat{r}_x(0) & \hat{r}_x(1) & \hat{r}_x(2) & \cdots & \hat{r}_x(N) \\ \hat{r}_x(1) & \hat{r}_x(0) & \hat{r}_x(2) & \cdots & \hat{r}_x(N-1) \\ \hat{r}_x(2) & \hat{r}_x(1) & \hat{r}_x(0) & \cdots & \hat{r}_x(N-2) \\ \vdots & \vdots & \vdots & \vdots & \vdots \\ \hat{r}_x(N) & \hat{r}_x(N-1) & \hat{r}_x(N-2) & \cdots & \hat{r}_x(0) \end{bmatrix} \tag{15}$$

According to Yule-Waker function, model coefficient can be calculated if the autocorrelation function can be gained [6]. Assume $\hat{\varphi}_N(i)$ is the estimation of AR model coefficient $\varphi_N(i)$, $i = 1, 2, \cdots N$, $\hat{\rho}_N^f$ denotes minimum error power of AR model,

$\hat{\rho}_N^f = \sigma^2$. According to Toeplita matrix and Levinson-Durbin recurrence algorithm, model coefficient is given by:

$$\begin{aligned}
\hat{\varphi}_N(i) &= \hat{\varphi}_{N-1}(i) + \hat{q}_N \hat{\varphi}_{N-1}(N-i) \\
\hat{q}_N &= -\left[\sum_{i=1}^{N-1} \hat{\varphi}_{N-1}(i)\hat{r}_y(N-i) + \hat{r}_y(N)\right] / \hat{\rho}_{N-1}^f \\
\hat{\rho}_N^f &= \hat{\rho}_{N-1}^f[1 - \varphi_N^2(i)]
\end{aligned} \tag{16}$$

The order of AR model can be ascertained by FPE principle, AIC principle and CAT principle. FPE principle is:

$$FPE(N) = \hat{\rho}_N \frac{\Delta + (N+1)}{\Delta - (N+1)} \tag{17}$$

where Δ is the length.

Then we could get the prediction.

$$\hat{x}(k) = c_k + \sum_{i=1}^{N} \varphi_N(i)x(k-i) \tag{18}$$

Residual ξ is the prediction error.

$$\xi = x(k) - \hat{x}(k) \tag{19}$$

HTA tests the residuals of filters at the same time. If the model matches to the actual system model, the residual can be described as zero mean white Gaussian noise sequence. Therefore, the conditional probability is:

$$f_{x(k)|\varphi,x^{k-1}}\left(x(k)\middle|\varphi_i, x^{k-1}\right) = \beta_i \exp\{\bullet\} \tag{20}$$

where $\{\bullet\} = \left\{-\frac{1}{2}\xi_i^T(k)A_i^{-1}\xi_i(k)\right\}$, $\beta_i = \frac{1}{(2\pi)^{m/2}|A_i|^{1/2}}$, A denotes variance, m denotes the order of measure vector.

Define the conditional probability of hypothesis model is:

$$p_i(k) = \Pr\left\{\varphi = \varphi_i \middle| x^k\right\} \tag{21}$$

Then the conditional probability can be calculated by:

$$p_i(k) = \frac{f_{x(k)|\varphi_i,x^{k-1}}(x(k)|\varphi_i, x^{k-1}) * p_i(k-1)}{\sum_{j=1}^{K} f_{x(k)|\varphi_j,x^{k-1}}(x(k)|\varphi_j, x^{k-1}) * p_j(k-1)} \tag{22}$$

The overall estimate can be obtained.

$$\hat{X}_D(k) = \sum_{i=1}^{N} p_i(k)\hat{X}_i(k) \tag{23}$$

B. Federal Kalman Filter

1) Information allocation

$$\begin{aligned} P_{i,k-1} &= \beta_i^{-1} P_{k-1} \\ Q_{i,k-1} &= \beta_i^{-1} Q_{k-1} \\ X_{i,k-1} &= X_{k-1} \end{aligned} \tag{24}$$

According to information conservation principle at the same time, $\sum_{i=0}^{n} \beta_i = 1 \cdot$

2) Prediction

$$\begin{aligned} \hat{X}_{i,k/k-1} &= \Phi_{i,k,k-1} \hat{X}_{i,k-1} \\ P_{i,k/k-1} &= \Phi_{i,k,k-1} P_{i,k-1} \Phi^{\mathrm{T}}_{i,k/k-1} + \Gamma_{i,k-1} Q_{i,k-1} \Gamma^{\mathrm{T}}_{i,k-1} \end{aligned} \tag{25}$$

3) Update

$$\begin{aligned} K_{i,k} &= P_{i,k/k-1} H^{\mathrm{T}}_{i,k} (H_{i,k} P_{i,k/k-1} H^{\mathrm{T}}_{i,k} + R_{i,k})^{-1} \\ \hat{X}_{i,k} &= \hat{X}_{i,k/k-1} + K_{i,k} (Z_{i,k} - H_{i,k} \hat{X}_{i,k/k-1}) \\ P_{i,k} &= (I - K_{i,k} H_{i,k}) P_{i,k/k-1} (I - K_{i,k} H_{i,k})^{\mathrm{T}} + K_{i,k} R_{i,k} K^{\mathrm{T}}_{i,k} \end{aligned} \tag{26}$$

4) Overall optimization

$$\begin{aligned} \hat{X}_g &= P_g \sum_{i=1}^{3} P^{-1}_{i,k} \hat{X}_{i,k} \\ P_g &= (\sum_{i=1}^{3} P^{-1}_{i,k})^{-1} \end{aligned} \tag{27}$$

5) Reset the filtering value and the estimation error variance matrix using by global optimal solution.

As a result of using noise variance upper bound technique, every local filter is uncorrelated with each other and amplification state matrix is uncoupled. Local filter results are suboptimum, but the global estimate is optimum after re-synthesizing in the fusion [7].

4 Simulation and Analysis

Simulations of the integrated navigation system are performed with Matlab. Doppler log error is 0.4m/s, compass error is 0.5°. Gyro constant drift is 0.1 °/ h, gyro Markov process time constant T_g is 300s, accelerometer bias is 0.1mg, and accelerometer Markov process time constant T_a is 1000s. The time constant of TAN is 100 ~ 200s.

There is no output of INS in 100 ~ 200s, and no output of TAN in 200 ~ 300s

The system error curves are shown in Figure 4.

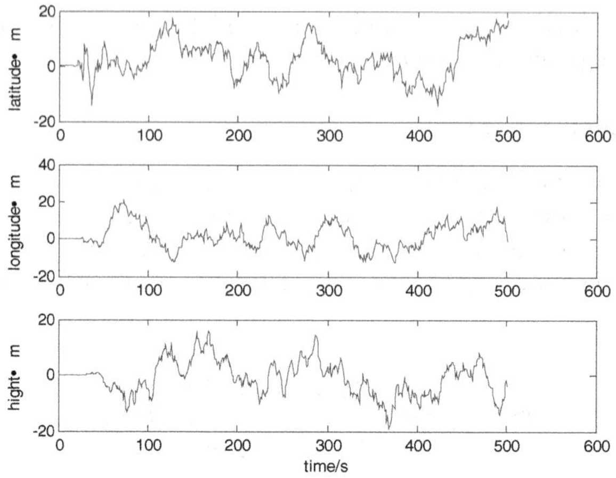

(a) position error curves

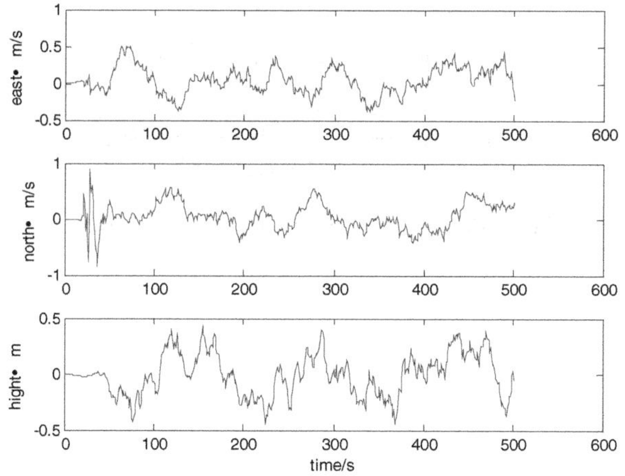

(b) velocity error curves

Fig. 4. The error curves of system

As shown in Figure 4, system can obtain high filtering precision and ensure the navigation precision consequently. When other sensors inactivate, system output is not jump and divergence and the system reliability and stability are improved.

5 Conclusion

By experiment, the proposed fusion system is not divergence and the peak value of position error is 20m, the peak value of velocity error is 1m/s. When INS or TAN system inactivate, redundant system can effectively restrain the interference noise from sensors and performance of proposed system remains stable with different sensor configuration.

In conclusion, modified system can obtain high precision. It has not only better error correction ability, but also is good at anti-interference. Reliability, stability, availability and fault-tolerance of integrated navigation system are improved.

References

1. Ahlers, H.: Mulitisensorikpraxis, Berlin, Heidelberg, German (1997)
2. Li, F., Pan, P.: The Research and Progress of Dynamic Models for Maneuvering Target Tracking. Fire Control and Command Control 32, 17–21 (2007)
3. Zhang, A.: Research of Integrated Navigation and Data Fusion Technology for autonomons underwater Vehicle. Dissertation Submitted to Najin University of Science and Technology, China (2009)
4. Cour-Harbo, A.l., Bisgaard, M.: Inertrial Navigation System. Department of Control Engieering. Aalborg University (2008)
5. Wang, F., Chen, Z.: Information Fusion of Federal Kalman Filtering in INS/GPS/TAN System. Acta Aeronautica et Astronautica Sinica 19, 87–91 (1998)
6. Wang, C., Zhao, W.: An Adaptive Fedrated Filtering Algorithm for Dynamic SINS/Multi-satellite Integrated Navigation System. Information and Control 37, 453–458 (2008)
7. Carlson, N.A.: Federated filter for fault-tolerant integrated navigation systems. In: Proceedings of Position Location and Navigation System 1988, Orlando, pp. 110–119. Winchester, MA (1998)

Automatic Multi-point Mooring Vessel Monitoring System

Gao Shang, Huang Zhen, and Hou PingRen

School of Automation,
Wuhan University of Technology,
Wuhan, Hubei Province, China
gao.shang113@sohu.com

Abstract. In this paper, as a typical engineering ship, Soft-Geotextiles--Laying (abbreviate: SGL) vessel is taken for instance. For the case that the size and the anchor winch number of SGL vessel keep on increasing, the author develops an automatic SGL vessel controlling system with a relatively high rate of integration. The key technologies in this system are listed: the hardware structure, the function and principle of each soft module, the display of graphic interface, the control strategy, etc. This system does provide excellent performance in the actual engineering environment in virtual of its stability, real-time capability, monitoring effect and convenience.

Keywords: Engineering vessel, Automatic monitoring system, Graphic interface, Control strategy.

1 Introduction

Along with the growth of national investment in waterway regulation area, the application of engineering vessels is becoming more and more widespread. As a result, a series of engineering vessels, such as Soft-Geotextiles-Laying (abbreviate: SGL) vessel, dredger vessel, etc, have arisen. As a special kind of ship used in the area of engineering construction only, usually, engineering vessels are non-self-propelled. Thus, engineering vessels mainly realize movement by anchor mooring. Particularly, SGL vessel is designed as an engineering vessel which meets the requirement of implementing construction on soft river bed in the project of waterway regulation. This kind of vessel mainly consists of Jute Geotextiles winches, bridge, turning plate, GPS location system, mooring system, etc. Therefore, the design of positioning system of multiple points anchoring mooring SGL vessel and the realization of its automatic action become an issue worth of research.

The SGL vessel monitoring system designed by the Industrial Network and Intelligent Systems Research Institute has been already used on several SGL vessels such as “YUGONGPAI NO.1”, “YUGONGPAI NO.2”, “CHANGYAN NO.2”, etc. As a result, related studies of SGL vessel automatic working and anchor mooring movement have been under way. Adopting fuzzy control strategy, as it can realize the positioning of the four anchor winches and the automatic laying working, the early designation met the accuracy requirements at that time. However, nowadays, the SGL

J. Zhang (Ed.): ICAIC 2011, Part V, CCIS 228, pp. 316–322, 2011.

vessel working condition has developed from 20 meters laying to 40 meters laying. Meanwhile, the anchor winch number on a SGL vessel has also changed from four to six. Thus, a new kind of control strategy is urgently needed to meet the accuracy requirements in current construction.

In this paper, the author introduces a predictive control strategy based on the practical application of the SGL vessel named "Chang Yan No. 8". Meanwhile, with the utilization of the DXF files which gives rise to the improvement of the graphic interface, the author designs the automatic monitoring system for that vessel as well. The system's hardware components, software functions and control strategy as well as its applied effect will be explained as following.

2 Overview on Automatic Monitoring System

SGL vessel automatic monitoring system adopts "PC+PLC" control scheme. On the one hand, Industrial control computer is chosen as PC for the purpose of guarantee credibility of the system and getting rid of some problems in the industrial environment, such as electromagnetic interference, vibration, dust pollution, high or low temperatures, etc. On the other hand, because of the reason that the size of SGL vessel is relatively huge and the data needed to be collected and controlled is distributed on both sides of the fore, middle, aft of the ship, PLC is chosen to use SIMATIC S7-300 PLC serial which can establish PLC main station and Remote IO. Besides, data transmission is implemented via Profibus, the communication between PLC main station and PC is realized by Ethernet and the GPS data is transmitted to PC to be

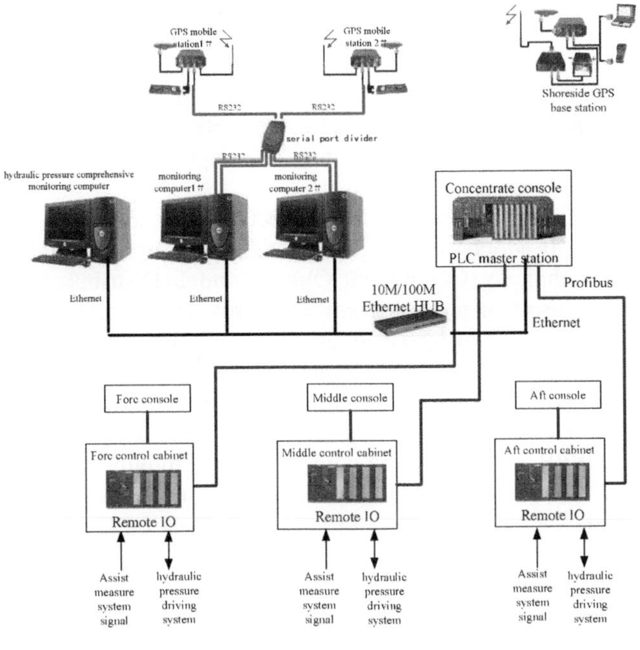

Fig. 1. The structure of the SGL vessel automatic monitoring system

processed through serial communication. The scheme can make full use of the high operation characteristic of PC, implement GPS data processing rapidly and get relevant instructions to control PLC based on the result of data computation which can be finished in accordance with the control Strategy. Then, PLC will lead to the action of hydraulic equipment which can drive the anchor winches. Thus the whole vessel can work according to the predetermined operation route. Figure 1 shows the hardware structure of the automatic monitoring system.

3 Analysis of the Function of the SGL Vessel Automatic Monitoring System

The main functions of the SGL vessel automatic monitoring system are composed of four parts: data collection, integrated information processing, construction control, man-machine interaction. Figure2 shows the main function of the system.

Data collection system consists of the collection of serial data and OPC data. Serial data collection: in order to realize automatic control of the SGL vessel, the system has to gain accurate coordinate via GPS and water depth information via certain sensors. For the reason that the SGL vessel always keeps moving in construction process and the movement of the vessel relates to sailing track and course, DGPS (Dynamic Global Positioning System) technique with the form of 1(base station)+ 2(moving station) RTK(Real-Time Kinematic)is adopted as the positioning system on this vessel to guarantee the real-time capability and the accuracy of the positioning. In this way, the base station should be located at a point whose coordinate has already been known exactly. Thus, the observation modifier in this area can be estimated via the comparison between the coordinate information gained from synchronous observation of this known point and its accurate coordinate. Meanwhile, the two GPS moving station on the vessel accept information from the same GPS satellite and figure out the vessel's coordinate which hold a precision being up to centimeter level based on the modifier transmitted from the base station by radio. Then the PC of the monitoring system can collect the GPS information as well as the water depth data via serial communication to realize vessel positioning[2].

OPC data collection: OPC means "Object linking and embedding techniques in process control". It adopts the OLE (Object Linking and Embedding) technique and COM/DCOM technique (Component Object Model/ Distributed Component Object Model) of Microsoft in process control. It provides process control and industrial automation with a series of interface, attribute and method standards. It serves as the key foundation upon which the whole control system's information interchange between field device level and process management level and openness can be realized. Specifically, in our monitoring system, Simatic.NET is used as the OPC server which implements TCP/IP communication with Ethernet communication module CP243-1 at the PLC main station. In the SGL vessel monitoring system, monitoring computers collect real-time state information of every device via OPC communication protocol and generate control command to drive PLC.

The synthesize information process performs the function of processing the GPS information, water depth information and PLC state information already gained. Then the Man-machine interactive displays the vessel position exactly and the state information of every device on the screen.

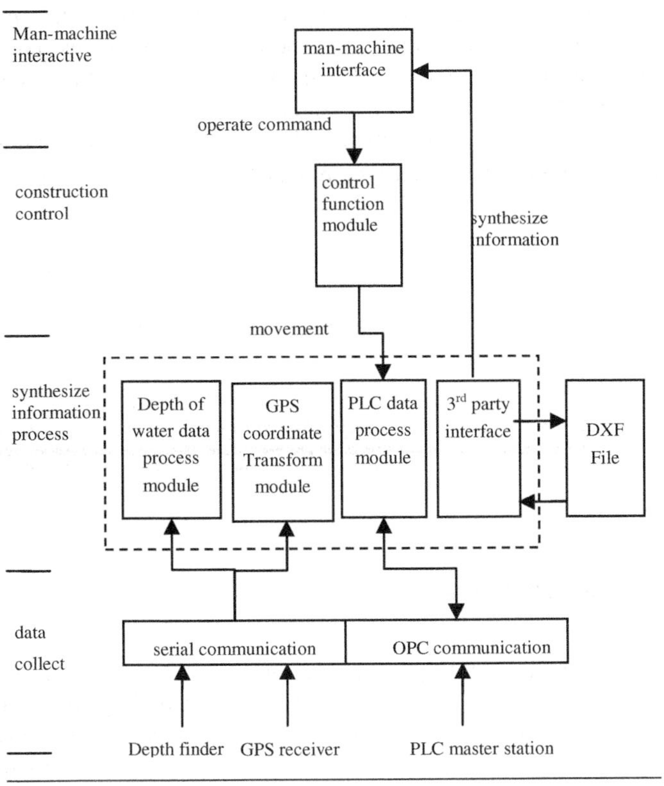

Fig. 2. The diagram of the functions of the SGL vessel automatic monitoring system

4 The Implementation of the Vessel Automatic Monitoring System

Based on the requirement of practical construction which needs the vessel works according to predetermined path, the author introduces DXF graphic files as well as several improvements to control strategy to realize man-computer interaction interface and accurately control winches' movement.

A. SGL Vessel Automatic Monitoring System Interact with the DXF File

In the construction process of the SGL vessel, DXF (Drawing Exchange File) is used to load construction basic diagram. DXF is a kind of drawing exchange file of AutoCAD software. Because of the fact that DXF is standardized ASCII file with regular arrangement, it holds many merits in terms of software developers' writing programs to realize graphic information reading and writing[3]. As a result, in the information processing we need develop 3rd party interface to read DXF files. In SGL vessel working process, the engineers should be familiar to CAD graphic operation and get used to CAD graphic interface. Yet in the system introduced in this paper, CAD graphic can be directly displayed on the interface[4].

In our system, we utilize a kind of 3rd party control called DXFReader to implement the exchanging of DXF files. Specifically speaking: DXFReader control's Filename attribute takes charge of transmitting the DXF files which contain construction basic diagrams into the interface of the monitoring system; its DrawLine method performs the function of plotting the geodetic coordinate information of the vessel on the construction basic diagrams; and its WriteDXF method saves construction information to DXF files in order to facilitate engineers to check out.

Based on the case that DXF files can be loaded directly, the monitoring system's screen is able to display the construction basic diagrams. The fact that the vessel can be plotted on the construction basic diagrams directly through GPS data transformation makes the man-machine interaction more perceivable. Engineers can observe the deviation between the construction state of the vessel and the predetermined working route. Figure3 shows the operational effect of the system.

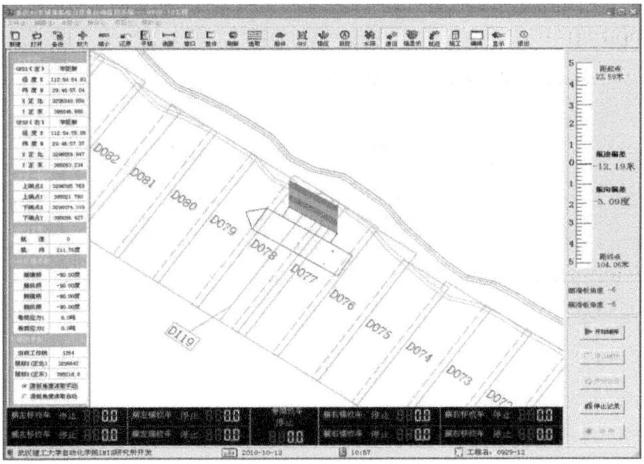

Fig. 3. The diagram of the operational effect of the SGL vessel automatic monitoring system

B. The Control Strategy of the SGL Vessel Monitoring System

In the early monitoring system, as the laying range of the SGL vessel is 20 meters and the laying construction is driven by 4 anchor winches, we may find some kind of fuzzy control strategy to realize appropriate automatic control. However, in the new monitoring system, the laying range enlarges into 40 meters and the number of the anchor winches also becomes 6. As the number of control variable increases, fuzzy control strategy can hardly meet the need of accurate control under current construction situation. Thus, predilection control strategy is introduced in this paper in order to realize the automatic control of the SGL vessel.

As the SGL vessel checks out its coordinate by GPS and the coordinate can never be affected by environment factors, the coordinate of each anchor winch as well as the sailing track and direction deviation can be figured out by GPS coordinate value based on the SGL vessel mathematic model already established. Then based on the coordinate value of every anchor gained above, the length of every anchor chain ($l_1 \sim l_6$) at the current time can also be estimated. For the fact that the predetermined construction route of the SGL vessel is already known, the predilection of the vessel's position coordinate can be got via the combination of path planning method and sailing direction deviation. In the same way, because of in the process of the SGL vessel movement the positions of the anchors keep unchanged, the length of every anchor chain ($l_1^/ \sim l_6^/$) when the vessel arrive the predetermined position can be estimated. Thus, the quantity of the cables that should be released or retracted by the winches ($\Delta d_1 \sim \Delta d_6$) can be checked out. Then, as the time needed for releasing and retracting cables keeps equivalent, the velocity of every anchor winch ($v_1 \sim v_6$) can be estimated. The velocity information should be transmitted to PLC to drive actuators to implement vessel movement. When the vessel arrive the new position, the operation that the deviation can be modified continuously through the comparison between current position and predilection position make sure that the vessel will work according with the construction scheme[5]. Figure3 shows the movement of the SGL vessel.

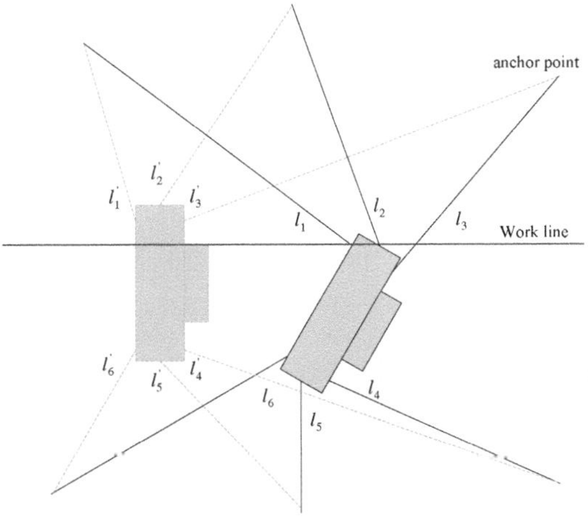

Fig. 4. The diagram of the control strategy of the SGL vessel

5 The Practical Application of the SGL Vessel Monitoring System

In the practical working process of the SGL vessel, for the reason that the DXF construction basic diagrams can be imported to the monitoring system directly, the operation interface becomes more visual, and besides, the position information of the vessel is displayed in the interface more exactly. When the vessel implements construction automatically, the prediction control scheme which coordinates the working motion of the six winches makes the vessel works according to the predetermined construction plan. Thus, it meets the accurate requirement of practical construction. Meanwhile, the state information collected from the whole operation is stored into the construction basic diagrams.

6 Summary

The automatic monitoring system introduced in this paper realizes relatively excellent combination with the construction basic diagrams via the 3rd party control DXFReader. The man-machine of this system is quite perceivable. Meanwhile, the utilization of the predilection control strategy supplies the precision need under current construction situation that the laying range and the number of anchor winches keep increasing. In summation, the system holds many merits in terms of its deviation modification and convenience of operation. However, for the practical capability difference of the winches, in the further studies, the influence of environment should be taken into account in order to perfect the automatic system.

References

1. Huang, Z., Wang, Y.J., Liu, Q.: Research and Development of Engineering ship Integrated Monitoring System. Ship Engineering 30(2), 191–193 (2006)
2. Xue, J., Guo, J.M., Huang, Z.: GPS ship positioning control system design and development. Ship & Ocean Engineering 37(3), 142–145 (2008)
3. Dai, M., Xie, M.: NT control system based on DXF graphics files. Microcomputer Information 22(5), 7–9 (2006)
4. Yu, J., Wang, H.M.: Development and application CAD / CAM interface program based on DXF file. Manufacturing Automation 30(7) (July 2008)
5. Huang, Z., Wang, Y.J., Liu, Q.: development of Engineering ship based on VB dynamic navigation system. Transportation and Computer 23(4), 100–103 (2005)

Optimizing Maintenance Job-Scheduling Based on MAS and Bee Swarm Algorithm

Wang Rui[1], Jin Jia-shan[1], and Huang Xiaodong[2]

[1] College of Naval Architecture and Power Naval University of Engineering,
Wuhan, Hubei Province, China
[2] Electronic Information Engineering Department
Naval Aeronautical and Astronautical University
Yantai, Shandong Province, China
76178669@qq.com

Abstract. Intermit level maintenance unit often need to maintain plenty of navy ships failure. The rule to measure its ability is whether it can get the best maintenance supportability effect with less costs and time or not. Obviously it can improve the abilities of intermit maintenance unit to reduce the maintenance time. In order to reduce the maintenance time, the paper divided the maintenance unit into intelligent Agents. With the transformation of the functions and local optimizing strategy of Honey Bee Swarm algorithm, the spare maintenance materials was dynamically combined into new maintenance unit to optimize the dispatch of maintenance operation based on several agents' cooperation. The experiment indicates that the optimizing dispatch way reduces the fault-maintenance time and improves the maintenance ability of maintenance unit.

Keywords: Maintenance Support, optimizing, job-scheduling, Agent, Bee Swarm algorithm.

1 Introduction

Building maintenance supportability is the key factor to navy ships' equipment battle effectiveness. Its aim is to get the best supporting effect with less costs and time. Obviously it's an effective building measure for reducing maintenance time to optimize the dispatch of maintenance operation.

It can get the optimizing effect to adopt the queue theory to make modeling based on the demand of the maintenance operation personnel. It can also get the optimizing effect to make a reasonable arrange for the maintenance personnel based on the Colored Petri net. But the intermit level maintenance units comprise plenty of the supporting resource, so utilizing the dispatch with the queue theory and Petri net to do the dispatch operation can make some resource in a free state of waiting when dealing with large quantities of operation. And it must adopt the intelligent and dynamic optimizing strategy. The literature adopt Honey Bee Swarm algorithm to make an investigation on the dispatch of maintaining the wartime dilapidated equipment and the optimizing effect is obvious.

J. Zhang (Ed.): ICAIC 2011, Part V, CCIS 228, pp. 323–330, 2011.

It abstracts the participant intermit level maintenance units into Agents to form an Agent MAS. Then the MAS optimizes the dispatch of maintenance operation with the cooperation of Honey Bee Swarm algorithm and Agent.

2 Optimizing Model of the Maintenance Units

The optimization issue is to get the max of target function with the given restriction condition. An example of optimization issue can be expressed as an antithesis <A , f >. "A" is a feasible sets satisfying the restriction. Target function f is a mapping that be defined as $f: A \to R^n$. It refers to solve the max of target function which was called the least optimization. It can be marked:

$$\min(f(x)), x \in A \qquad A = \{x \mid x \in R^n, g_i(x) \geq 0, i = 1 \ldots m\} \tag{1}$$

There in, $g_i(x) \geq 0 \quad i = 1 \ldots m$ is for restriction condition The optimizing restriction condition of supporting maintenance units dispatch is the fixed amount of maintenance resource and its optimizing aim is to do the operation with the least maintenance time. But the optimizing the dispatch of supporting units comes down to the dynamic compound of maintenance resource. It can't be simply described with a function. It must be a dynamic process which is as follows: The equipment department passes them on intermit level maintenance units When it receives a set of repair assignment. Then the intermit level maintenance units will establish maintenance element as much amount as possible.

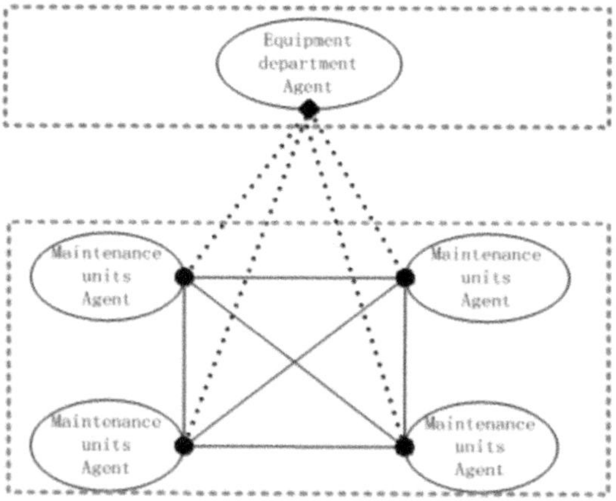

Fig. 1. Communication logic among MAS

After establishing much maintenance element, it is formed a group between equipment department and each maintenance element. The group adopts Agent technology to form an intelligent group called Swarm Intelligence. Individual function among equipment departure is to accept and analysis repair item list and then transforms it into repair task and dispatches it to maintenance element. The equipment department must work out the best way for a repair task which maintenance element can be dispatched. All levels maintenance element Agent can also work out how to make use of its own maintenance resource to get the best maintenance effect after it receives the maintenance tasks. The communication logic relation among MAS is shown as Fig. 1.

At present it's proved that the troop intelligent is a new way which can resolve most of dynamic optimization issue by researching and application of troop intelligent theory. So Honey Bee Swarm algorithm (HBSW) can be used to resolve the dynamic optimization issue of the dispatch of supporting units.

3 Apply HBSW to Optimize the Maintenance Operation Scheduling

The characteristic of Swarm is related with not only inherit ability but also entironment, physic environment and outside condition and their mutual influence. It puts up powerful sociality action, such as cooperation and communion and so on.

The swarm's variety communication, transformation of function and local optimization can be applied to the supporting units' optimization of Scheduling. The strategy, the transformation of function is applied to the dynamic change of the subordinate maintenance unit and the local optimization is applied to equipment department Agent to deal the maintenance item to a unit Agent. Applying the two above-mentioned strategies, the equipment Agent can make the personnel at leisure reform a unit Agent to speed up the maintenance process of intermit level maintenance units.

Both the equipment department Agent and all the maintenance units Agent form an intelligent swarm. The equipment department Agent won't directly apportion all the maintenance operation any more, but exchange information with all maintenance units as the dispatcher of the whole operation and assign the operation to the maintenance unit individually according to the change of condition and environment. The equipment department Agent informs all the maintenance units the information about the rest maintenance item in each simulation step size. All the maintenance units Agent work out the reflection threshold to the rest item and give the reflection threshold and their personnel information that accomplishes the mission go the equipment department Agent in return.

Since a maintenance item needs many professional manpower resource and the item's maintenance working procedure is fixed, so some personnel of a maintenance unit Agent have already finished their mission before the item ends. If these personnel can be transferred to new maintenance item in time, obviously it will quicken the pace to finish the whole maintenance item.

The equipment department Agent adopts the local optimization strategy to choose the assignable maintenance operation for a unit after it receives the reflection threshold from all units Agent and the information about the personnel accomplished one's mission. If the unit Agent needs new personnel, men that have completed the mission can be dispatched to it. From the view of the man being dispatched, his unit's attribute transfers. And from the practical process the staff's transfer will waste some time, so the transfer process must be reduced to the minimum. In other words, for the maintenance operation and unit Agent chosen, the proportion between professional vacant staff and required staff must be the largest, so only few men are needed transferring, even no staff is needed transferring. Based on the several analysis above mentioned, the optimizing process of the schedule to maintenance operation can be worked out, shown as Fig 2:

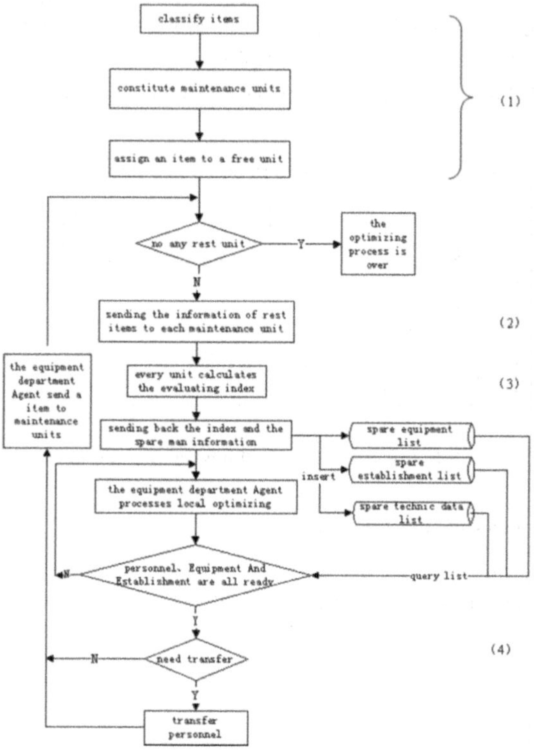

Fig. 2. Optimizing Maintenance Job-Scheduling based Bee Swarm Algorithm

(1) At the classification stage of the elementary maintenance project, its main item is to send the homogeneous fault to the same maintenance unit, so it can avoid reducing the staff transfer. And when establishing the maintenance units, each maintenance unit is in charge of one type defect. Only when failure types are more than the account of building units, the staff transfer can happen.

In fact, most repair mission that the fleet give repair machine is different. That is to say the fault model is seldom repeated. Because intermit level maintenance unit only deals with the best known fault model, it can learn about the all fault model's demand of repair resource by experience. Since the resource most repair project need is different, intermit level maintenance unit don't pay special attention to each repair mission for the repetition rate of repair mission gets lower.

Therefore, the main principles of classification is that the maintenance project abstract and discrete the demand of categories and account of concrete maintenance supporting resource in the process of imitation. As for all maintenance projects of related equipment, as long as their demand of personnel type and account, facility type and account, installation type and account, category, technical data and spare part requested are the same, no matter their frequency is the same or not, they are classified as a category.

All maintenance unit Agent can share many facilities, installation, technical data and storage resources form intermit level maintenance unit in the frame of MAS simulation. When optimization the schedule, the dynamic allocation of these hardware resource costs assign personnel much time and it is less flexible than assign personnel. Therefore, in this paper, the realization of the dispatch optimizing algorithmic is discussed.

In the process of simulation the experience of maintaining a fault i is described with one quad {M, L, N, T} array AR Rb[k]., of which the number of element denotes that maintaining the fault needs k sets of working procedure. They are the first one, the second one, ... the k-th one. For each working procedure, M denotes specialty type of the repairing personnel. L denotes their technical grade. N denotes the number of repairing personnel. T denotes the time when the professional personnel finish the working procedure.

(2) Unit Agent identifies the working procedure it adopts and works out time of all working procedure according to the fault information they receive. When the simulation time advances to Ti,, it can find out the personnel who have accomplished their mission according to the following formula:

$$\sum_{i=1}^{I} t_i < T_i < \sum_{j=1}^{I+1} t_j \tag{2}$$

If there is no congener one after the first working procedure, the maintenance staff of the first one have accomplished the mission and the maintenance operation is carrying on the (I+1) working procedure.

(3) After the equipment department Agent sends the information about the rest project, unit Agent works out its reflection threshold η to the rest project.

① When one rest project is the same type as the unit's current project, the thresholdηis equal to 1 if finished, or it is equal to 0.

② If one rest project isn't the same type as the unit's current project, then:

Therein, M is the specialty type number needed by the rest project; N is the same specialty type number between the unit's current project and the rest project; Li is the personnel number needed by the i-th same specialty of the rest project; Hi is the personnel number needed by the i-th same specialty of the unit's current project. When Hi is big then Li, , the value of Hi is set to the value of Li, so the value of Hi equal to min(Hi/ Li,1).

The local optimizing process of an equipment department Agent is to find an optimization tuple A(the i-th rest project, the j-th unit Agent). First the equipment department Agent selects the corresponding A (i, j) of the maximum η according to the responding threshold sent by every unit Agent, then see if the tuple A satisfies the following conditions

① the facilities, the installation and technical data that A's rest item i needs are free or not, if not, the equipment department chooses the next the biggest tuple over again according to the reflection thresholdη. Or

② if the vacant staff the j-th unit Agent is dealing with just satisfy the personnel scheme of the rest item i, the equipment department will transfer the rest item i to the j-th unit Agent. Or

③ the equipment department Agent examines the free personnel from other unit Agent and judges if the personnel scheme of the rest item can be satisfied after other free personnel are transferred to the Unit J Agent. If satisfied, the transferring process of personnel will start. Otherwise, the equipment department Agent chooses the next the biggest tuple over again according to the reflection thresholdη.

④ If there is no tuple A satisfying the condition, the equipment department will continue to send the message and repeat above-mentioned operation.

4 The Effect Test of Optimizing Maintenance Operation

In order to test the above capability of Honey Bee Swarm algorithm, a set of maintenance projects first are chosen as the test example. Taking test example for the Fig 3 and Fig 4, the supporting resource required contain fault name, the name of resource required, the account and resource type required and so on. As for each equipment, the triple in the first line in the chart denotes the number of persons with high professional title, intermediate title and low-level titles. The second line is about working procedure. The third line is the time the corresponding working procedure costs. The first line in the list "workshop" in the chart is denoted with the number of workshop required by the fault. The third line is about the time that occupying corresponding workshop costs. It's marked in the line matching the last list.

When the method of dynamic compounding the resource isn't adopted, the equipment department first establishes three units, matching three maintenance task: a, c and d. The third unit continues to maintain Task g after it accomplishes Task c. When the second maintenance unit accomplishes Task c, the third unit also accomplishes Task g. At that time Task e and f are assigned to the two units. One hour later, the first unit accomplishes Task a. Then Task b is assigned to it. As shown in Fig 5, the whole course lasts 30 hours.

When the method of dynamic compounding the resource is adopted, the equipment Agent also establishes three units which respectively undertake three task: Task a, c and d. Because it takes only one hour to accomplish' the first working procedure of Task a, one engineering and two technical personnel with intermediate title attending the working procedure turn to be free. It makes Task b satisfied with the demand of manual labor resource, but the workshop is not enough, so the Task b can't start. After 4 hours, the third unit with Task d releases one workshop. Therefore there are two free workshops, but Task b needs three workshops. Now Task b can't still start. Only one hour later, the second unit with Task c releases a workshop. At that time there are three free workshops. The manual labor resource and material resources are satisfied with maintaining Task b. The fourth unit can be established to repair b. the whole course lasts 23 hours as shown in Fig 6.

Fau-lt	techni cian	lathe-man	repair-man	electr-ician	pipe-man	erector	welder	work-shop	la-the	data
a	(1,2,0)	(1,2,0)	(0,2,3)	(1,1,2)	(0,0,0)	(0,0,1)	(1,2,0)	2	1	2
	1	3	2	4		5	6			proc-ess
	1	2	3	2		4	3	(14,15)		spend-time
b	(1,2,3)	(0,1,1)	(1,0,1)	(0,2,0)	(0,1,0)	(0,1,0)	(0,0,2)	3	1	1
	3	4	2	1	6	5	7			proc-ess
	1	2	2	3	2	2	4	(6,9,15)		spend time
c	(1,1,2)	(0,2,3)	(1,2,0)	(0,0,2)	(0,0,0)	(0,0,2)	(0,2,0)	2	1	2
	2	6	5	1		4	3			proc-ess
	1	2	2	3		4	2	(5,14)		spend time
d	(0,1,2)	(1,0,1)	(0,0,2)	(0,1,1)	(0,0,0)	(0,0,0)	(1,0,1)	2	1	0
	3	4	5	2			1			proc-ess
	1	1	2	3			1	(4,8)		spend time
e	(0,2,3)	(1,2,0)	(0,0,0)	(0,2,3)	(0,1,1)	(1,1,2)	(0,0,0)	3	1	0
	1	3		4	5	2				proc-ess
	2	1		2	2	1		(2,8,8)		spend time
f	(0,2,2)	(0,2,0)	(1,2,0)	(0,0,0)	(1,2,0)	(0,0,0)	(0,1,1)	2	1	0
	2	1	4		3		5			proc-ess
	1	1	2		2		2	(5,8)		spend time
g	(0,1,3)	(0,0,0)	(1,2,0)	(0,1,1)	(0,0,0)	(0,2,3)	(0,0,0)	2	1	1
	4		1	2		3				proc-ess
	1		1	2		2		(6,6)		spend time
h	(1,1,3)	(0,2,3)	(0,0,0)	(0,0,0)	(1,1,2)	(0,0,0)	(0,2,0)	3	1	1
	2	1			4		3			proc-ess
	1	2			2		2	(7,7,7)		spend time

Fig. 3. List of supporting resource in the process of optimization

technic-ian	latheman	repair-man	electr-ician	pipe-man	erector	welder	work shop	la-the	data
(2,4,6)	(2,4,6)	(2,4,6)	(2,4,6)	(2,4,6)	(2,4,6)	(2,4,6)	7	4	10

Fig. 4. Restriction on resource in the process of optimization

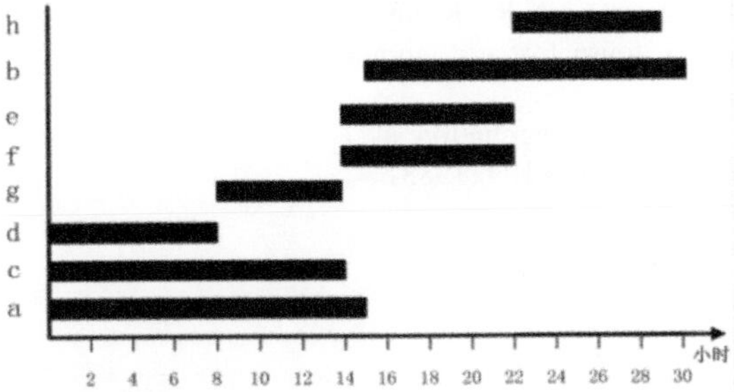

Fig. 5. Spending time in non-dynamic compounding model

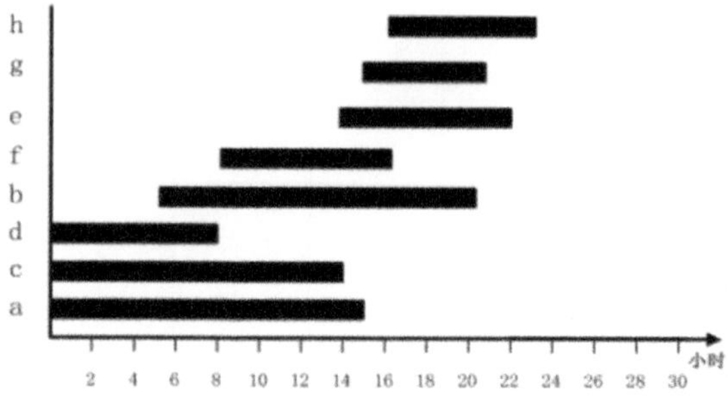

Fig. 6. Spending time in dynamic compounding model

5 Conclusion

In the mode of pier supporting, intermit level maintenance unit needs to schedule a big number of operation to improve maintenance effect. When it is researched as an intelligent troop MAS all Agents inside can adopt Honey Bee Swarm algorithm to dynamically combine free maintenance resource into new maintenance unit by means of cooperation and assign a new job to it in time. Experiments show that the item's maintenance time can be reduced and the maintenance capability of intermit level maintenance unit can be improved after it adopts MAS and intelligent Honey Bee Swarm algorithm.

References

1. Yang, G.-h., Song, J.-s., Qu, X.-r.: A Model of Personnel Need for Equipment Maintenance Based on Queue Theory. Command Control & Simulation 29(2), 116–120 (2007)
2. Dong, C.-x., Wu, D.-w., Tai, N.-j., He, J.: Application of Colored Petri Nets to the Maintenance Support Effectiveness for Navigation Equipment. Fire Control and Command Control 32(11), 102–104 (2007)
3. Li, Y., Wu, C., Cao, H.-Q.: Application of CPN—TOOLS in Making Weapon Service Schema. Journal of System Simulation 18(1), 200–203 (2006)
4. Wang, H.: Job-Scheduling for Damaged Equipments in Wartime based on Wasp Colony Algorithm. Fire Control and Command Control 34(34), 141–144 (2009)
5. Wooldridge, M., Jennings, N.R., Kinny, D.: The Gaia methodology for Agent-oriented analysis and design. Journal of Autonomous Agents and Multi-Agent Systems 3(3), 285–312 (2000)
6. Davis, R., Smith, R.G.: Negotiation as a Metaphor for Distributed Problem-Solving. Artificial Intelligence 20, 63–109 (1983)

Design of the Double Polar Pulse Power for Plating

Yu Yadong, Ruan Xieyong, and Pan Zhangxin

Department of Physics and Electronic Information
Shaoxin University,
Shaoxin, Zhejiang Province, China
yuyadong1979@sina.com

Abstract. The double polar pulse digital power source was designed for plating. dsPIC30F3011 acted as MCU for producing PWM pulse in the design. M57962L was used to drive IGBT. So as to isolate controller from main loop of power by opticalcoupler, LM331 was used to process conversion of A/D. Arithmetic of PI was adopted to control output voltage and current. The tryout proved the controller has good stability and it can output double pole pulse whose frequency is adjustable from 10HZ to 10KHZ.

Keywords: double pole pulse, digital power source, PWM.

1 Introduction

Plating power is one of key device in the field of plating industry. DC power and pulse power are two kings of power widely used at present. DC power output continuous smooth current. And the direction of current doesn't change along with the change of time in the course of plating. Although the value of current density is big while DC power is adopted, current efficiency is low and quality of plating layer is bad. Because crystal of plating layer is small under pulse power condition, more subtle and symmetrical plating layer can be obtained[1]. And raw material can be saved, so pulse power has been developed rapidly in plating domain[2].

At present, researches on pulse power concentrate on changeable polar pulse and high power[3]. Double polar pulse power whose pulse width and frequency is adjustable designed by self is introduced in this paper. The main technologic parameters are listed as follow. Amplitude of pulse is range from 0 to 24V. Frequency of pulse is range from 10 to 10KHZ. And the duty ratio of pulse can be changed from 0.3 to 0.9.

2 Design of Hardware

A. The Main Loop of Pulse Power

Because positive voltage and negative voltage are not balance when power works, namely duty ratio and work time of positive and negative pulse is commonly different, two set of parameters which can be adjust individually are adopted. Fig. 1 is the block diagram of the main loop of pulse power.

J. Zhang (Ed.): ICAIC 2011, Part V, CCIS 228, pp. 331–336, 2011.

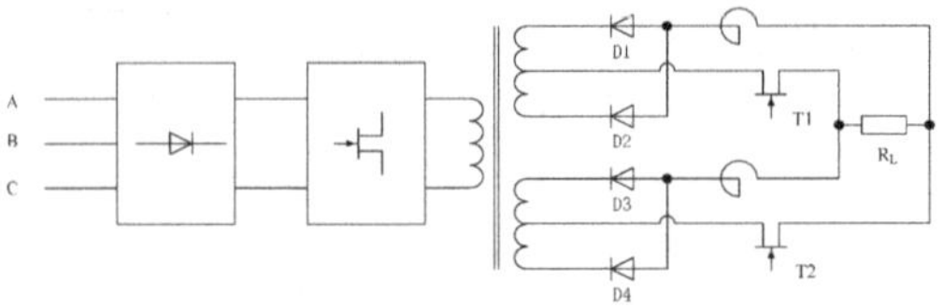

Fig. 1. The main loop of pulse power

At first, three phase AC power is commutated in the main loop. Then the DC power chopped by is sent to the primary winding of transformer. Voltage of secondary winding of transformer will be adjusted with the change of duty ratio of IGBT. The secondary winding of transformer is divided into two groups. One of them provides positive pulse for the load through D1 and D2. And another provides negative pulse for the load through D3 and D4. The frequency and duty ratio of the pulse are controlled by the PWM signal provided by MCU.

B. Control Circuit of Pulse

The block diagram of control circuit is shown in Fig. 2. The pulse is produced by MCU which is dsPIC30F3011 in this design. There are three channels PWM produced by dsPIC30F3011. The first PWM controls the duty ratio of IGBT connecting with primary winding of transformer to adjust the voltage of secondary winding of transformer. The others PWM pulses control the IGBT connecting with secondary winding of transformer to produce positive and negative pulse. The duty ratio of IGBT is obtained by a closed loop system which is realized by software. Voltage of the load which processed by voltage detection circuit and A/D converter is an input of the closed loop system. The keyboard is used to input parameters such as frequency of pulse and output voltage etc. displayer displays voltage and current when the system works.

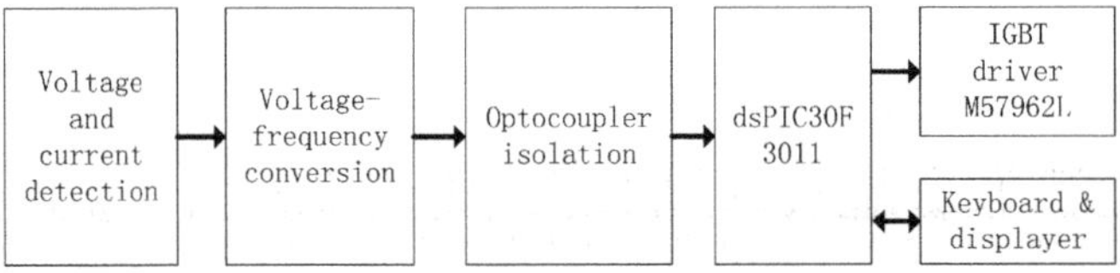

Fig. 2. Block diagram of control circuit

C. A/D Converter

Voltage-frequency converter is used to process A/D conversion for isolating the control circuit from the main loop of power. A/D conversion circuit is shown in Fig. 3. After filtered and limited amplitude the sample voltage is sent to LM331 which performs Voltage-frequency conversion. LM331 converts the sample voltage which is analog signal to pulse whose frequency changes with sample voltage. The pulse will be isolated by optocoupler and sent to I/O port of MCU. MCU counts the number of pulse and compute the value of voltage according to the number of pulse in one second.

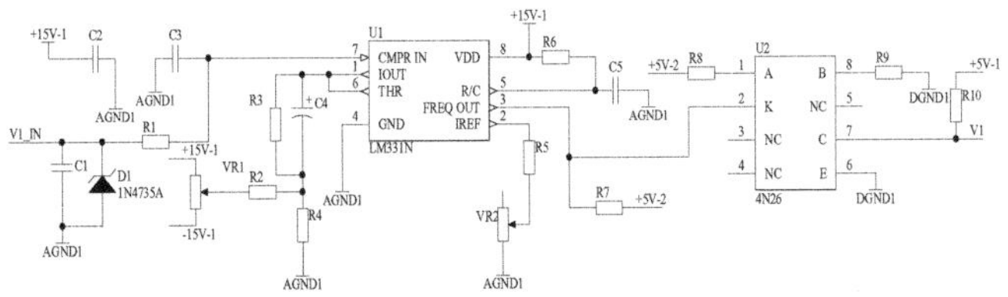

Fig. 3. A/D conversion circuit

D. Driver of IGBT

IGBT is drived by mixed integrated circuit M57962L whose work frequency is up to 40KHZ. There are two sources (+15v and -10v) in the circuit which is shown in Fig. 4.

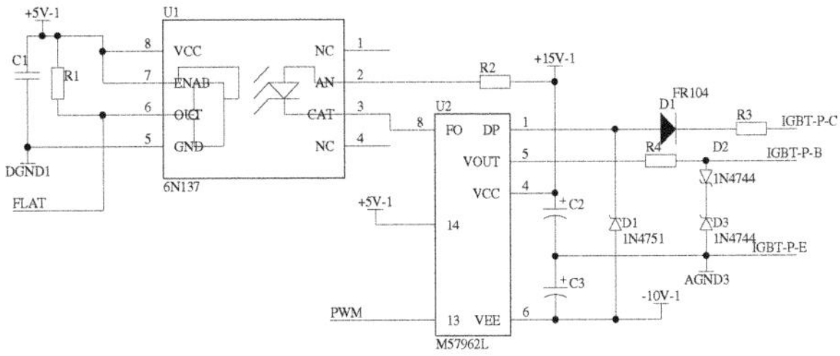

Fig. 4. Driver of IGBT

There is an optocoupler which can isolate voltage up to 2500v, over current protection circuit and output terminal of protection signal in the interior of M57962L. M57962L is a high speed driver can drive the IGBT module which can work under the voltage at 600v.

When the circuit works, M57962L begins to self-test at first. It detects whether there is short and overload or not. If there is short or overload, potential of collector of IGBT will rise, and current from fast recovery diode (D1) to detection circuit will rise too. Then grid of IGBT will be cut off and a direction signal will be output through the eighth pin of M57962L. The signal will be isolated by optocoupler and be sent to fast interrupt pin of MCU. The thirteen and fourteen pins of M57962L are interface of external circuit and M57962L.They are connected with a optocoupler in M57962L's interior. The PWM signal output by MCU will enter into the thirteen pin of M57962L and be amplified by output driver stage.

3 Design of Software

A. Structure of Program

Main program and interrupt service program of timer executed all function of the system. Timer T1 serves as a work timer. T1 will overflow once per second. The function of its interrupt service program is calculating the total running time of the system. Timer T2, T3 and T4 serve as timer for PWM pulse. Timer T5 counters the number of input pulse in unit time, then value of voltage or current can be computed. Main program scans keyboard and processes the user's instruction. It will also work as a PID controller while the system is running. And main program controls output signal. Flow chart of main program is shown in Fig. 5.

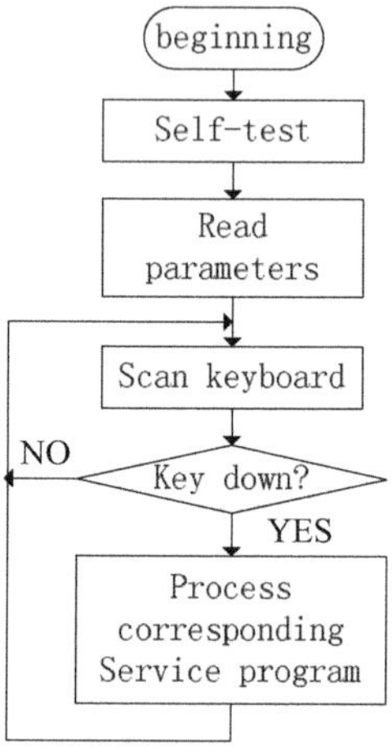

Fig. 5. Float chart of main program

B. Algorithm of Adjustment of Voltage and Current

Values of voltage and current will change along with the duty ratio of PWM pulse produced by MCU. The more the ratio of PWM pulse, the longer on-time of IGBT. The duty ratio of PWM pulse is adjusted by using digital PI controller. There are voltage and current feedback in the design. The relation between last output of controller and new output of controller is as follows:

$$u(n) = u(n-1) + K_p\left[e(n) - e(n-1)\right] + K_p \frac{T}{T_i} e(n) \tag{1}$$

$u(n)$ is new output of controller while $u(n-1)$ is last output of controller in Equation 1. And K_p is Proportional (P) term. T is sampling interval while T_i is time of integral. $e(n)$ is the error signal which is formed by subtracting the desired setting of the parameter to be controlled from the actual measured value of that parameter while $e(n-1)$ is last of it. The PI controller responds to an error signal in the closed control loop and attempts to adjust the controlled quantity in order to achieve the desired system response. The effect of the P term will tend to reduce the overall error as time elapses. The Integral (I) term of the controller is used to fix small steady state errors.

4 Experimental and Results

The double polar pulse digital power source for plating designed in this paper satisfied the needs of design. It can output double polar pulse which can be adjusted with frequency and duty ratio. We simulated the output of the system in Matlab7.1. The experimental and simulated potential responses with different pulse periods are shown in Fig. 6 and Fig. 7. The load was set as a resistor and duty ratio of positive pulse was set as 0.5 while negative pulse was 0.3. Fig. 6 and Fig. 7 shows the simulated responses accord with the parameters of setting.

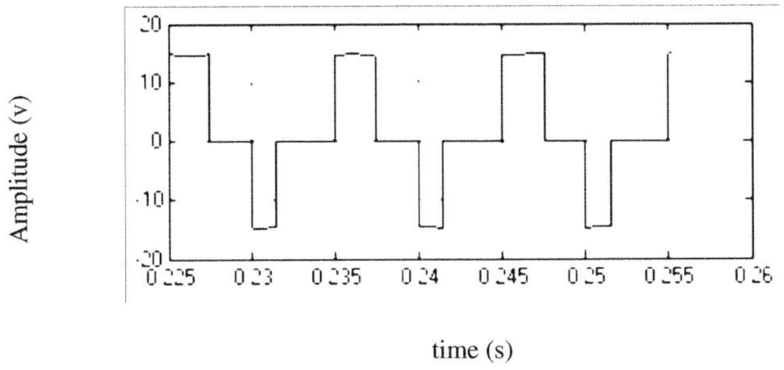

Fig. 6. Potential responses while pulse period is 10ms.

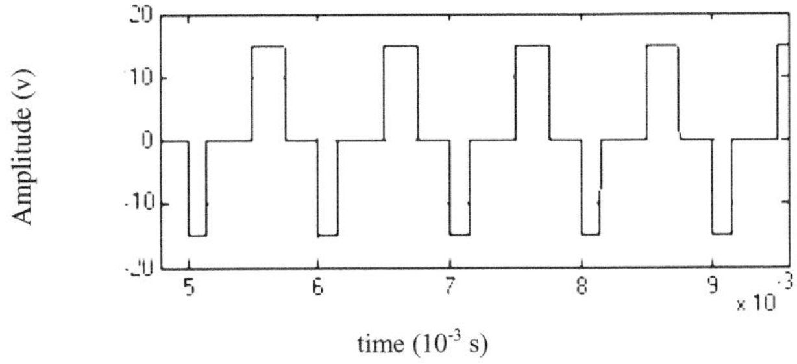

Fig. 7. Potential responses while pulse period is 1ms

The interference between main loop circuit and control circuit was decreased because they were isolated by optocoupler. And probability of damage of IGBT decreased because of short and overload protection circuit. The control board has been tested in corporation. Results show performance parameters are according with design demand.

Acknowledgment. This research was supported by Department of Education of Zhejiang Province, China, under grant number Y201018515 in 2010.

References

1. Tsai, W.-C., Wan, C.-C., Wang, Y.-Y.: Mechanism of copper electrodeposition by pulse current and its relation to current efficiency. Journal of Applied Electrochemistry 33(12), 1143–1153 (2003)
2. Yung, K.C., Chan, K.C., Yue, T.M., Yeung, K.F.: The effect of waveform for pulse plating on copper plating distribution of microvia in PCB manufacture. The International Journal of Advanced Manufacturing Technology 23(3-4), 245–248 (2004)
3. Zhang, H., Liu, Z., Han, J., Wang, J.: Development of Plating Power Supply on Adjustable-Voltage and Frequency with Bipolarity Pulse. Journal of Northern Jiaotong University 27(4), 101–104 (2003)

A Cloud Model Method for Reliability Evaluation of Distribution System

Chunwen Yang[1,2], Lixin Liu[1], Zhigang Li[1], and Lingling Li[1,*]

[1] School of Electrical Engineering,
Hebei University of Technology,
Tianjin, China
[2] Kailuan Group,
Research and Development Center,
Tangshan, China
cwyang1598@163.com

Abstract. This article is first proposed a cloud model method for reliability evaluation of distribution system based on the fault traversal algorithm .the theory of cloud-based distribution system reliability model, to solve the uncertainty between distribution system reliability parameters and load parameters. The definition of the cloud model, its features and normal cloud generation algorithm are described in detail in this article. Processing the original parameters of distribution grid equipment with the cloud model, cloud model representation of reliability index is derived, then combined with traversal algorithm , to obtain the algorithm flowchart. The proposed algorithms is applied to the RBTS-Bus6 subsystem model, and discuss the difference between the exact value and this model, it analyzed the affect degree to reliability index under different conditions, and verified the usefulness and correctness of the algorithm.

Keywords: distribution system, reliability evaluation, cloud model, fault traversal algorithm, RBTS –Bus6.

1 Introduction

The distribution system is directly contacted with the users, and the power supply capacity and quality to user from power system must be reflected by power distribution system. The reliability indices of power distribution are a concentrated reflection of whole power system structure and operation characteristic. To modern society, more users rely on high quality electric power, and expecting reliable power protection. The distribution power reliability assessment just met the urgent needs of enterprises and users [1].

The original data of reliability from distribution network equipment is the basis of the reliability assessment, and the precision degree has a great influence to assessment

* Corresponding author.

J. Zhang (Ed.): ICAIC 2011, Part V, CCIS 228, pp. 337–345, 2011.

results. The reliability data of equipment from Reliability Management Centre is an average data across the country. The technology levels of distribution network equipment manufacturers are different, which will directly led the equipment failure outages have random. In addition, some statistical data of equipment is insufficient or existing statistical error. And the assessments to specific distribution network model will introduce large error if direct use these data. In conventional reliability evaluation methods, device parameters are treated as exact numbers, then, the obtained reliability indicators are the exact numbers. This situation is clearly inconsistent with the actual engineering, it is necessary to consider the uncertainty of statistical results from components reliability data.

Cloud is an uncertainty transformation model between a qualitative concept and its quantitative expression expressed with the natural language value, in other word, cloud model is an uncertainty transformation model to realize the transform from qualitative to quantitative. [2].

This paper introduced the cloud theory to the reliability evaluation of distribution system, and proposed reliability evaluation cloud model algorithm of distribution system combined with fault traversal method. Cloud model represents the original parameters and the reliability indices that can be better taken into account the influence of random events and fuzzy events, and provide more information. Then the validity and practicality of the algorithm is further verified by using it in RBTS-Bus6 system.

2 Cloud Model

A. The Definition of Cloud

Cloud is a quantitative and qualitative uncertainty conversion model, which was proposed by Professor Deyi Li based on the traditional fuzzy set theory and probability statistics.

Cloud is an uncertainty conversion model between a qualitative concept and its quantitative representation, which is expressed by natural language values. That is to say, the cloud model is an uncertainty model used to achieve the conversion between qualitative and quantitative.

Suppose U is a quantitative domain expressed by precise values, and C is a qualitative concept on the domain. If the quantitative value $x \in U$, and x is a random realization to the qualitative concept C , whose membership $\mu(x) \in [0,1]$ for C is a random number with stable tendency:

$$\mu : U \to [0,1], \forall x \in U, x \to \mu(x)$$

Then, the distribution of x on the domain is called as cloud, and each x is called as droplet [3].

B. The Number Features of Normal Cloud

Cloud is made up of many cloud droplets, and a single cloud droplet is a specific realization of the qualitative value in number. Its abscissa value represents the quantitative value corresponding to qualitative concept, and the ordinate value expresses the membership degree of the quantitative value on behalf of the qualitative concepts. The three number features of cloud are expectation E_x, entropy E_n and hyper entropy H_e [4 ~ 7].

1) E_x

Expectation best representatives the value of the qualitative concept, and it is usually the x value corresponding to the gravity of cloud, reflected the center value of corresponded qualitative concept.

2) E_n

Entropy represents the measure to the fuzzy degree of the qualitative concept, the size of which directly determines the number of elements that can be accepted by the qualitative concept on the domain, and also reflects the margin of qualitative value based on both this and that.

3) H_e

Super entropy expresses the uncertainty measurement of entropy, and that is the entropy of entropy. The size of super entropy indirectly reflects the cloud's thickness.

The $3E_n$ rules of cloud refers to that the total contribution of all elements on the domain U to the qualitative concept C is 1.That is, 99.7% of the cloud droplets will fall into the range ($E_x - 3E_n$, $E_x + 3E_n$). Thus, cloud fell on the outside of this scope are small probability events for a qualitative linguistic values concept, and it can be ignored.

The generation algorithm of normal cloud is as follows:

1) $x_i = G(E_x, E_n)$

Generating the normal random number x_i that the expectation is E_x and the variance is E_n.

2) $E_{ni}' = G(E_n, H_e)$

Generating the random number E_n' that the expectation is E_n' and the variance is H_e.

3) $\mu_i = \exp[-(x_i - E_x)^2 / 2E_{ni}'^2]$

Generating the cloud droplets (x_i, μ_i).

Normal cloud is the most common to express the linguistic values.

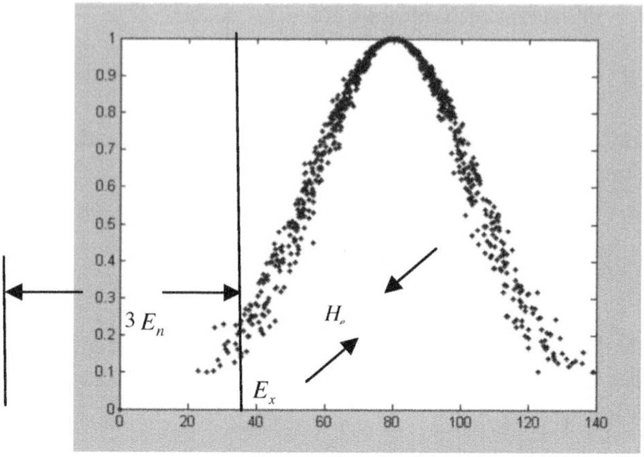

Fig. 1. Simulation result of Cloud（80,24,2）

Figure 1 shows the simulation results of cloud (80,24,2) when the number of cloud droplets is $n = 1000$.

3 Cloud Model Algorithm of Distribution System Reliability Assessment

A. Fault Traversal Algorithm

Literature [8] proposes a reliability evaluation algorithm based on fault traversal. Fault enumerate is the basic idea of the algorithm, traversal technology of the data structure is a mean. The algorithm can be effectively used to the complex distribution system reliability evaluation of having the sub feeder and effectively takes capacity constraints of system into account. It unites the flow calculation with the reliability calculation taking full advantage of distribution network own characteristics.

This paper will do the reliability calculation with the algorithm as the basis. When a component in the network (mainly consider the transformer, line, circuit breaker etc. components failure and the combination of components) fails, fault traversal algorithm determine the scope of the switch according to forward search breaker and backward search disconnect switches, and the load point are be divided into the following 4 types.

1) Normal node, namely the node is that the correct action of switch is not affected by the failure after failure event;

2) The node is that isolation operating hours;

3) The node is that fault time is equal to isolation time plus switch operating hours;

4) The node is that fault time is equal to repair times of components.

Then form block subsystems, determine the load point type and calculate reliability indices of each load point.

Figure 2 is reliability calculation procedure based on the fault traversal algorithm and cloud model.

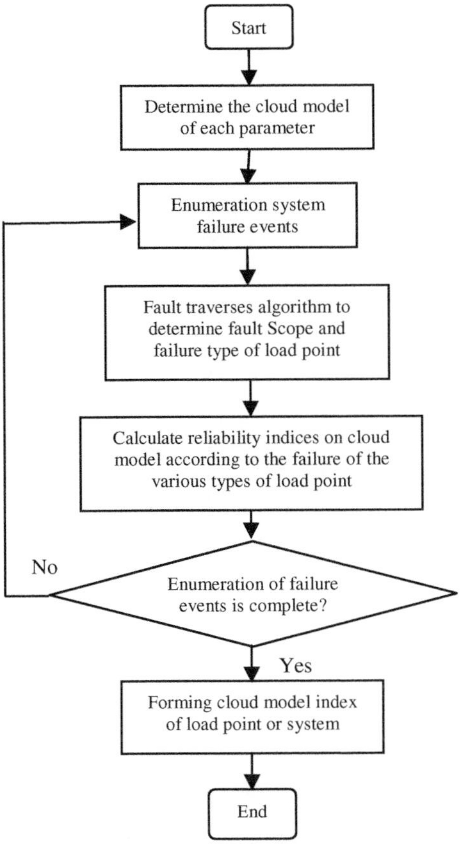

Fig. 2. Algorithm flow chart based on cloud model and fault traverses

B. System Reliability Index

Load point i 's failure rate, fault duration λ_i, and average annual outage time U_i is the classic indicators of reliability engineering, because they can not fully characterize distribution system characteristics and response, so need to calculate a specific set of distribution system reliability indices.

Common distribution system reliability indices are six:

1) System average interruption frequency index SAIFI
2) System average interruption duration index SAIDI
3) Customer average interruption frequency index CAIFI
4) Customer average interruption duration index CAIDI
5) Average service availability rate index ASAI
6) Energy not supplied ENS

The expectation of cloud model of reliability parameters is still taken for the original certainty parameters in this paper, while the entropy and super entropy will be determined according to expert experience or many testing.

4 Example Application

This paper select subsystem composed of main feeder F4, F5, F6 and F7 of the IEEE-BUS6, shown in Figure 3.

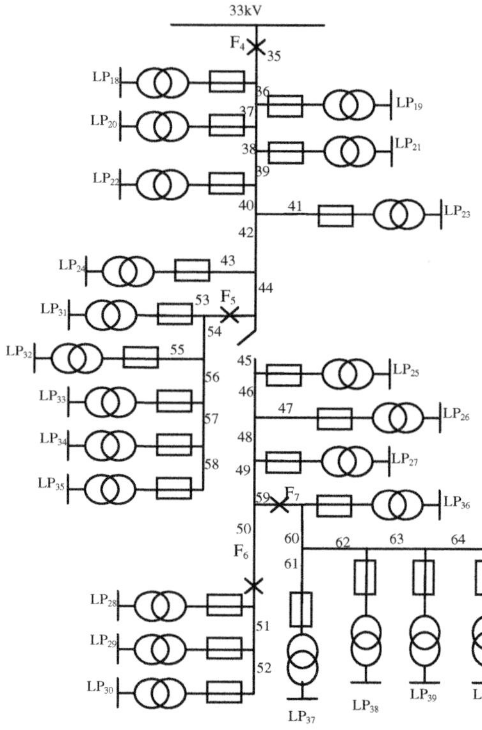

Fig. 3. IEEE-Bus6 subsystem wiring diagram

The system has thirty lines, 23 fuses, twenty-three distribution transformers, twenty-three load points, four circuit breakers and a disconnect switch [9-10]. The subsystem has fifty-three static components and five operational components, the number of users of each load point, load size, line length etc. original data see [9], and failure parameters of each component see [10]. This article assumes that the fuse (or branch circuit breaker) refuses to move, branch circuit breaker (or main circuit breaker) operates reliably, that is not consider double miss trip. 6 situations are being considered as follows:

Case A: considering the contact line, and without considering protection equipment reliable action rate and transformers standby;

Case B: consider reliable action rate of fuses and circuit breakers, without considering the contact line and transformer spare;

Table 1. Load point reliability parameters comparison between case A and the reference value

load point	λ	
	the reference value	this paper
23	1.7115	(1.7115,0.0907,0.0171)
32	2.5890	(2.5890,0.1372,0.0258)
40	2.5110	(2.5110,0.1331,0.0251)
	r	
	the reference value	this paper
23	5.02	(5.0228,0.4934,0.0855)
32	5.02	(5.0151,0.4965,0.0859)
40	6.16	(6.1649,0.6179,0.1067)
	U	
	the reference value	this paper
23	8.60	(8.5965,0.71090.1185)
32	12.98	(12.9840, 1.0858,0.1806)
40	15.48	(15.4800,1.3224, 0.2189)

Case C: consider reliable operation rate of fuses and circuit breakers and contact line, and without considering transformer standby;

Case D: consider reliable action rate of fuses and contact line, without considering reliable action rate of circuit breaker and transformer spare;

Case E: consider transformer standby and contact line, without considering reliable operating rate of fuses and circuit breakers.

Case F: consider transformer standby, contact lines, reliable action rate of fuses and circuit breakers.

Table 2. System reliability parameters comparison between six cases and the references value

	SAIFI	SAIDI
	1.0067	6.6688
Case A	(1.0067,0.0534,0.0101)	(6.6688,0.5593,0.0930)
Case B	(1.2281,0.0650,0.0122)	(10.72004,0.9046,0.1501)
Case C	(1.2281,0.0650,0.0122)	(9.2929,0.7798,0.1296)
Case D	(1.0706,0.0567,0.0107)	(8.9166,0.7498,0.1245)
Case E	(1.0066, 0.0533,0.0100)	(3.6714, 0.3037,0.0512)
Case F	(1.2281, 0.0650,0.0122)	(4.2776, 0.3513,0.0587)
	ASAI	ENS
	0.999239	72.81531
Case A	(0.999239,0.000064,0.000011)	(72.6414,6.1181,1.0161)
Case B	(0.99877,0.000103,0.000017)	(115.8614,9.7916,1.6249)
Case C	(0.998939,0.000089,0.000015)	(109.6162,9.2452,1.5349)
Case D	(0.998982,0.000086,0.000014)	(104.2471,8.8096,1.4620)
Case E	(0.999581, .000035,0.000006)	(48.1218,4.02603,0.672049)
Case F	(0.999512, .000040,0.000007)	(55.0593,4.5845,0.763441)

Comparison load point reliability parameters in Table 1's and Case A's in Table 2 with the reference value, when entropy and hyper entropy is taken to be 0, only consider the expectations, the cloud model reduces to the exact value model, indicators and the reference value of reference [9] [10] are fully consistent.

In addition, comparison the system reliability parameters between six cases and the reference value in Table 2, whether considering the protection of the equipment reliable action rate affect SAIFI greatly. Whether the spare transformer is considered has greater impact on the ENS. Figure 4 is the cloud model of ENS.

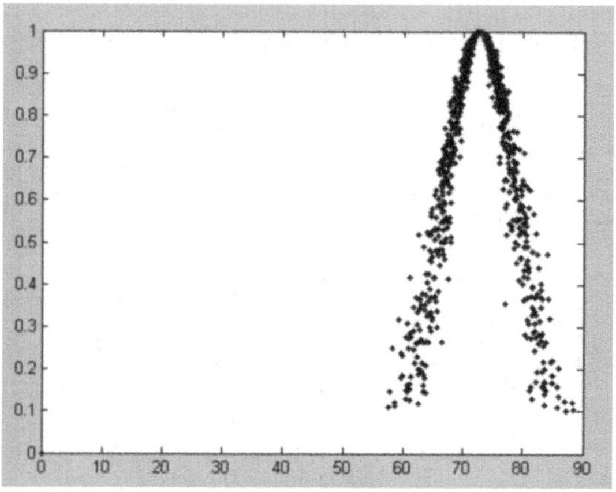

Fig. 4. Cloud model of ENS

It can be seen from the figure 4, the cloud model reflects the changes and distribution information of the reliability index, more fully reflect the uncertainty of the reliability index .This is more in line with the actual situation, contains more information, electrical workers may take into account changes of indicators flexibility based on the actual situation.

5 Conclusion

Cloud theory is first introduced to the distribution system reliability evaluation, the establishment of a distribution system reliability evaluation algorithm based on cloud model; this will take into account the randomness and fuzziness, more in line with the actual situation.

It verified the validity of the algorithm through the calculation of IEEE RBTS-Bus6 subsystem. Application of the cloud model simulation also gets more information as reference for electricity staff.

However, the reliability evaluation result of the model depends on the selection of a series of factors, such as the entropy of each cloud model. These parameter values need assessment engineer's actual experience and through repeated research to obtain. Therefore, the comprehensive evaluation for future smart grid distribution intelligent level, how to determine the relevant parameters more reasonably and accurately is the focus of the next step.

Acknowledgment. This work was supported by the National Natural Science Foundation (60771069), the Hebei Province Natural Science Foundation (F2010000151) and China Postdoctoral Science Foundation (20100470778).

References

1. Zhang, P., Wang, S.: The interval algorithm for reliability evaluation of large-scale distribution system. In: Proceedings of the CSEE, vol. 24(3), pp. 77–83 (2004)
2. Yu, T.: Distribution network reliability and economy of comprehensive assessment methods. North China Electric Power University, a master's degree thesis, 16–22 (December 2008)
3. Wen, X.: Power system based on cloud theory of operation risk assessment.North China Electric Power University, a master's degree thesis, 29–31 (December 2008)
4. Pan, R., Liu, J., Guo, X.: Smart grid power systems for cloud computing. Sichuan Electric Power Technology, 6–10 (December 2009)
5. Shi, Y., Gao, X., Zhang, A.: Cloud theory in aviation electronic warfare system performance evaluation application. Computer Engineering and Applications 45(22), 241–243 (2009)
6. Pu, T., Zong, R., Hu, J., Shi, J., Chang, J., Wang, P.: Under the cloud-based reliability assessment of electric power communication network. In: China Institute of Communications Sixth Annual Conference, vol. (26), pp. 493–497 (November 2009)
7. Wang, J., Xiao, W., Zhang, J.L., Deng, B.: An improved performance assessment method based on cloud model. Fire Power and Command Control, 97–99 (May 2010)
8. Li, Z., Li, W., Liu, Y.: Radial distribution system reliability evaluation of fault traversal algorithm. Power Systems and Automation 1(25), 53–56 (2002)
9. Billinton, R., Jonnavithula, S.: A test system for teaching overall power system reliability assessment. IEEE Trans. on Power Systems 11(4), 1670–1676 (1996)
10. Allan, R.N., Billinton, R., Sjarief, I., et al.: A reliability test system for education purposes-basic distribution system data and results. IEEE Trans. on Power System, 813–820 (1991)

Electric Field Simulation of Electrospinning with Auxiliary Electrode*

Wenbing Zhu, Jingjing Shi, Zhongming Huang,
Pei Yu, and Enlong Yang**

Department of Polymer Material and Textile Engineer,
Jiaxing University,
Jiaxing, Zhejiang Province, China
WenbingZhu2011@tom.com

Abstract. The low production rate of electrospinning process may limit the industrial use of single needle system. Multiple needles can be used to improve the throughput. But the jets bend outward because adjacent electric fields interfere with one another. In this paper, auxiliary electrode was used to release electric field interference. Effect of auxiliary electrode on electric field interference was analyzed by electromagnetic analysis software. The influence of needle distance on electric field force of jet's stability section was also investigated. The results indicate that jet's y and z direction electric field force decreases along with the increase of distance between needles. When distance between needles is 30 mm, the influence of electric field is smaller and jet's elongation greater.

Keywords: Electrospinning, Nanofiber, Simulation, Electric Field.

1 Introduction

Electrospinning is a straightforward and cost effective method for fabsubmicron fibers. [1, 2]. Due to unique properties such as high surface area to volume ratio, small pore sizes, high porosity and so on, the ultrafine fiber membranes prepared by electrospinning process have been extensively studied and widely used for its potential applications in filter media, composite materials, biomedical applications, etc [3-7]. The major technical barrier for manufacturing electrospun fibers for applications is the production rate. To achieve high throughput industrial production, several researchers investigated multiple jets electrospinning. Yarin [8] provided a new approach employing a ferromagnetic liquid sub layer. With the external electric field, the perturbations of the free surface became sites of jetting directed upwards.

* This work was financially supported by the Education of Zhejiang Province Foundation (Y200909763), Jiaxing University Foundation (70110X11BL) and Zhejiang Province University Student Science and Technology Innovation Program (851909120).

** Corresponding author.

J. Zhang (Ed.): ICAIC 2011, Part V, CCIS 228, pp. 346–351, 2011.

Dosunmu and Varabhas [9, 10] employed a porous hollow tube. The mass production rate from the porous tube is 250 times greater than from a typical single jet. Liu and He [11, 12] employed bubble electrospinning. Multiple jets are observed during the electrospinning process as predicted for it is easy to many bubble-induced cones on the solution surface.

In addition to the increase in production rate, multiple nozzles have been used in electrospinning. The technical problem for mass production of electrospun fibers is the assembly of nozzles during electrospinning. A straightforward multi-jet arrangement as in high-speed melt spinning cannot be used because adjacent electrical fields often interfere with one another, making the mass production scheme by this approach very impractical [13]. Chu [14, 15] guided the design of multi-jet electrode arrays by a two-dimensional finite element analysis (FEA) for electromagnetic, through the use of software by Field Precision. Yang [16] calculated and analyzed electrical field distribution of a special aligned multiple jets setup. The results demonstrate that the special setup with different heights of the 7 needles could produce uniform fibers according to SEM images. Theron [17] describes the results of the experimental investigation and modeling of multiple jets during the electrospinning of polymer solutions. The results demonstrate how the external electric fields and mutual electric interaction of multiple charged jets influence their path and evolution during electrospinning. Kim [18] fabricated nanosize fibers using an extra-cylindrical electrode connected with single and multiple nozzles of an electrospinning process to stabilize the initial spun jets. The result indicates that the modified electrospinning technique shows a possibility as a useful method for increasing the production rate of nanofiber manufacturing. Tomaszewski [19] tested three types of multi-jet electrospinning heads, series, elliptic and concentric. The concentric electrospinning head was selected as the best type with respect to both the efficiency and quality of the process. Kim[20] invented a bottom-up multiple nozzles electrospinning device to improve the productivity and quality of nonwoven webs. Though lots of literature was on multiple needles electrospinning, fewer works were about electric field simulation of electrospinning with auxiliary electrode. In the present work, effect of auxiliary electrode on electric field interference was investigated, and electric field force of jet's stability section was also analyzed by a three-dimensional finite element analysis for electromagnetic.

2 Experimental

Polyvinyl alcohol (PVA) powder (Mw=88 000 g mol^{-1}, 88% hydrolyzed), purchased from J & K Chemical®, was used to prepare a solution that was employed as the working fluid. The polymer powder was dissolved in distilled water at 10% weight concentration with the aid of mechanical stirring. The electrospinning apparatus is schematically shown in Fig. 1. Electrospun voltage was 15 kV. Tip-to-collector distance was 200 mm.

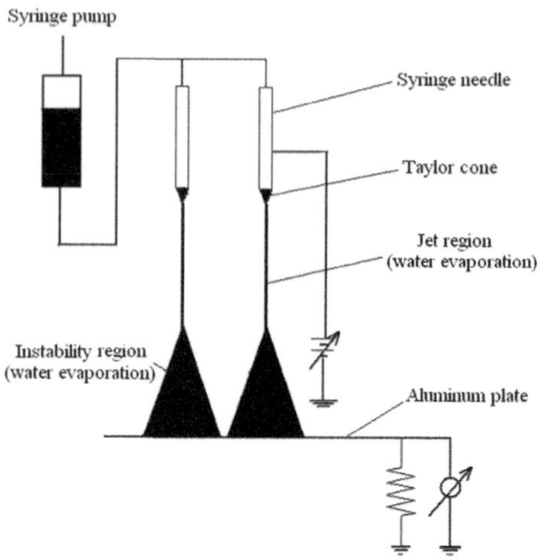

Fig. 1. Schematic diagram of electrospinning apparatus

3 Modeling

The electric field of electrospinning was analyzed by a three-dimensional finite element analysis (Ansoft Maxwell) for electromagnetic. The model of double needles electrospinning with auxiliary is shown in Fig. 2. The applied voltage of needle and collector is 15000 and 1440 V separately. The length of needle and jet's stability section is all 20 mm. The tip to collector distance is 40 mm. The top and down radius of auxiliary electrode are 6.4 and 7.7 mm separately, and the height is 16 mm. The applied charge on jet's surface is 5.94×10^{-8} C [21]. The radius of jet is only 0.2 mm, so mesh operation of model is difficult if tip to collector distance is 200 mm and applied voltage of collector is 0 kV. In this paper, we used a smaller tip to collector distance and a proper collector applied voltage to simulate electric field.

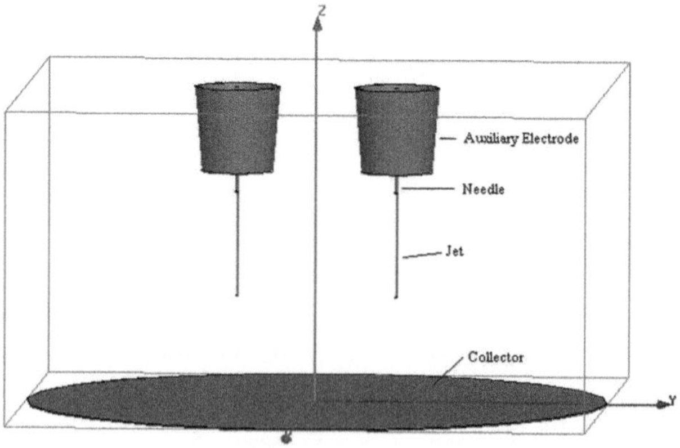

Fig. 2. The electromagnetic analysis model of double needles electrospinning with auxiliary electrode

4 Results and Discussion

Fig. 3 compares the electric field from needle tip to collector with different needle distance and tip to collector distance. 10L200 represents needle distance 10 mm and tip to collector distance 200 mm in the figure. As is shown in Fig. 3 that electric field strength on the line is almost the same when the distance between needles changes from 200 to 40 mm. The results indicate that model shown in Fig.3 can simulate electric field near needle tip. Electric field is much stronger near needle tip, so we predict that electric field force of jet's stability section may be the main reason for jet incline.

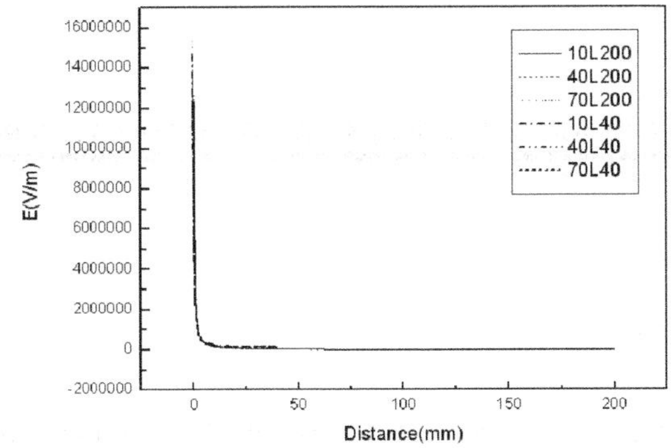

Fig. 3. Influence of tip to collector distance on electric field strength

Fig.4 shows electric field strength ratio of xy and z direction from needle tip to collector with and without auxiliary electrode. Exy and Ez ratio without auxiliary is greater than that with auxiliary electrode, which indicates that auxiliary electrode can decrease electric field interference.

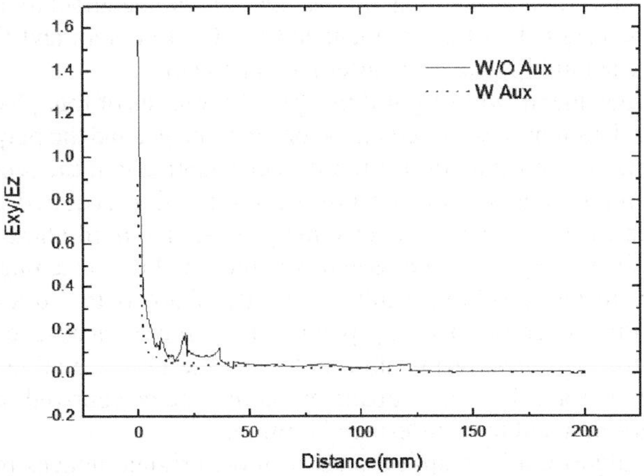

Fig. 4. Electric field strength ratio of xy and z direction from needle tip to collector

The jet's y direction electric field force of double needles electrospinning with auxiliary electrode is shown in Fig. 5. It can be seen from Fig.5 that jet's y direction electric field force of double needle electrospinning decreases along with the increase of distance between needles. When needle distance is above 30 mm, jet's y direction electric field force has little change. The reason for this is that the electric field interference decreases along with the increase of needle distance. When needle distance is above 30 mm, electric field interference is weak enough.

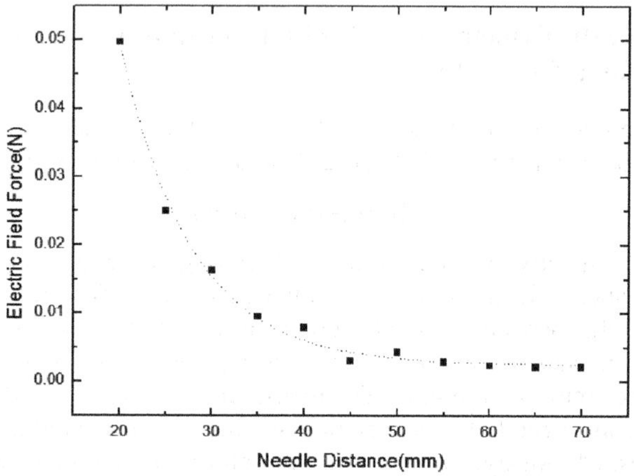

Fig. 5. Jet's y direction electric field force of double needles electrospinning

Jet's z direction electric field force of double needles electrospinning with auxiliary electrode is shown in Fig. 6. It can be seen from Fig. 6 that jet's z direction electric field force of double needle electrospinning decreases along with the increase of distance between needles. The jet elongates due to z direction electric force. So the electric field interference is smaller and z direction electric force greater when needle distance is 30 mm.

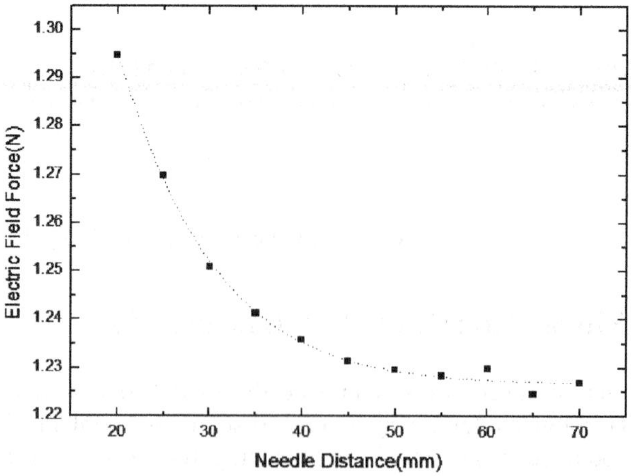

Fig. 6. Jet's z direction electric field force of double needles electrospinning

5 Conclusions

The electric field force of jet's stability section was analyzed by electromagnetic analysis software. The influence of needle distance on electric field interference was investigated in this paper. Electric field strength on the line is almost the same when the distance between needles changes from 200 mm to 40 mm and applied voltage of collector changes from 0 to 1440 V, which indicates that the model in this paper can simulate electric field near needle tip. Jet's y direction force of double needle electrospinning with auxiliary electrode decreases along with the increase of distance between needles. Jet's z direction force of double needles electrospinning increases along with the increase of distance between needles. When the distance between needles is 30 mm, the influence of electric field is smaller and jet's elongation greater.

References

1. Baumgarten, P.K.: J. Colloid. Interf. Sci. 36, 71 (1971)
2. Reneker, D.H., Chun, I.: Nanotechnology 7, 216 (1996)
3. Yang, E.L., Qin, X.H., Wang, S.Y.: Mater. Lett. 62, 35–55 (2008)
4. Park, S.W., Bae, H.S., Xing, Z.C., Kwon, O.H., Huh, M.W., Kang, I.K.: J. Appl. Polym. Sci. 112, 112 (2009)
5. Doshi, J., Reneker, D.H.: J. Electrostat. 35, 151 (1995)
6. Dzenis, Y.: Science 304, 17–19 (2004)
7. Wang, S.D., Zhang, Y.Z., Yin, G.B., Wang, H.W., Dong, Z.H.: J. Appl. Polym. Sci. 113, 26–75 (2009)
8. Yarin, A.L., Zussman, E.: Polymer 15, 29–77 (2004)
9. Dosunmu, O.O., Chase, G.G., Kataphinan, W., Reneker, D.H.: Nanotechnology 17, 11–23 (2006)
10. Varabhas, J.S., Chase, G.G., Reneker, D.H.: Polymer, 42–26 (2008)
11. Liu, Y., He, J.H.: Int. J. Nonlin. Sci. Num. 8, 393 (2007)
12. He, J.H., Liu, Y., Xu, L., Yu, J.Y., Sun, G.: Chaos Soliton Fract. 37, 643 (2008)
13. Fang, D., Hsiao, B.S., Chu, B.: Polym. Prepr. 44, 59 (2003)
14. Chu, B.: U S Pat 6713011 (2004)
15. Burger, C., Hsiao, B.S., Chu, B.: Annu. Rev. Mater. Res. 36, 333 (2006)
16. Yang, Y., Jia, Z.D., Li, Q., Hou, L., Gao, H.F., Wang, L.M., Guan, Z.C.: IEEE1-4244-0189-5, 940 (2006)
17. Theron, S.A., Yarin, A.L., Zussman, E., Kroll, E.: Polymer 46, 28–89 (2005)
18. Kim, G., Cho, Y.S., Kim, W.D.: Eur. Polym. J. 42, 20–31 (2006)
19. Tomaszewski, W., Szadkowski, M.: Fibres. Text East Eur. 13, 22 (2005)
20. Kim, H.Y., Park, J.C.: US Pat WO/2005/073441 (2005)
21. Theron, S.A., Zussman, E., Yarin, A.L.: Polym. 45, 2017–2030 (2004)

Research on Scale-Free Network Model of C2 Organizational Structure

Hongxia Wang and Xiuli Ma

School of Computer Science and Engineering,
Shenyang Ligong University,
Shenyang, China
hongxiawang1969@sohu.com

Abstract. In order to better optimize and design the C2 organizational structure, from the perspective of complex network theory, the C2 organizational structure network topology model is constructed and the C2 organizational structure network nodes growth characteristics is analyzed, then network scale-free model algorithm is established. Simulation result shows that degree distribution of the scale-free model which was constructed by the algorithm followed power-law distribution, it prove that the proposed algorithm could efficiently construct the C2 organizational structure scale-free network model, meanwhile it lay the foundation for optimal design of the C2 organizational structure.

Keywords: complex networks, command and control, organizational structure, network model.

1 Introduction

Command and control system performance mainly depends on advanced technology and equipment, and the C2 organizational structure rationality, while in the C2 organizational structure, rational allocation of resources and effective management largely affect whole operational effectiveness of the command and control system. In recent years, complexity science, especially complex network, has become the focus of research. It gives us a new view, to help us grasp network war complexity from the whole, to better understand their network topology and corresponding kinetics characteristics. Therefore, in this article, we use complex network theory to research the C2 organizational structure, we have studied the C2 organizational structure scale-free network model with emphasis. First, military command and control systems would be abstracted into a complex network, which makes up a typical complex network which includes reconnaissance nodes, firepower nodes, all kinds of command and control nodes, and among corresponding relationships. Then we propose to establish network scale-free model algorithm. Structure determines function is the basic point of system science[1]. this article which researches the C2 organizational structure scale-free network model must lay a good foundation to optimize the C2 organizational structure.

J. Zhang (Ed.): ICAIC 2011, Part V, CCIS 228, pp. 352–358, 2011.

2 Analysis of Scale-Free Network Model

In order to explain power-law distribution mechanism, Barabasi and Albert proposed a scale-free network model, now known as the BA model[2-5]. They believed that many previous network models didn't take into account the following two important characteristics s in real network:

(1) Growth characteristics: The size of network is growing. For example, a large number of new research articles are published every month, a large number of new pages emerge everyday, the increase of transit network sites during a certain period.

(2) Priority connection characteristics: New node tends to connect with those "big" nodes with higher connection degree. This phenomenon is also known as "the rich get richer" or "Matthew". For example, new published articles are more inclined to cite some important documents which are widely quoted; new personal homepage hypertext links are more likely to point to the Sohu, Yahoo, and other famous sites.

Based on characteristics of web growth and priority connection, BA scale-free network model algorithm is constructed as follows:

Growth: Begin with a network with m0 nodes, it brings in a new node once, and it connects to m-existing nodes, where $m_0 \leq m$ ..

Priority connection: Πi is the probability of a new node connected to an existing node i , k_i is the degree of node i and k_j is the degree of node *j*, satisfy the following relationship:

$$\Pi i = \frac{k_i}{\sum_j k_j} \tag{1}$$

The BA model mainly constructs the network structure of power-law distribution. We will use master equation law to analysis the network structure of power-law distribution.

$p(k,t_i,t)$ is defined that the degree of the node *i* which joins at the moment of t_i is exactly the probability of *k* at the moment of *t*. In the BA model, when a new node is brought into the system, the degree increase probability of node *i* is $m\prod_i = k/2t$, Otherwise, the node's degree is unchanged. From above, we can obtain the following recursion relationships.

$$p(k,t_i,t) = \frac{k-1}{2t} p(k-1,t_i,t) + (1-\frac{k}{2t}) p(k,t_i,t) \tag{2}$$

The network degree distribution is

$$p(k) = \lim_{t->\infty} (\frac{1}{t} \sum_{t_i} p(k,t_i,t)) \tag{3}$$

Finally, the BA network degree distribution function is

$$p(k) = \frac{2m(m+1)}{k(k+1)(k+2)} \propto 2m^2 k^{-3} \tag{4}$$

3 The C2 Organizational Structure Network Topology Analysis

In order to understand the C2 organizational structure network nodes growth characteristics, and then build the C2 organizational structure scale-free network evolution model. This article analyzes the C2 organizational structure network topology.

The C2 organization consists of a kind of whole orderly conduct which is formed by fight resources entity because of fight mission driven under combat environment, and coordinated with the C2 structure relationships, and its orderly conduct is the C2 organization completes its mission action or task process, coordinated C2 structural relationships matching with the orderly conduct entity among the C2 structural relationships. The C2 organizational structure consists of the C2 organization decision-making entity, the platform entity and the C2 relation expression of tasks. The organizational structure decides cooperation among each entity in the organization interior, while in organization interior, among each entity, the cooperation relations is the key of procedure execution [6-8]. In the policy-making entity, the platform entity and the tasks relations, the C2 organization includes: (1) The cooperation relationships of platform (or task) and policy-making entity; (2)command control relationships of policy-making entity and platform; (3) During policy-making entity's policy-making level relationships. The C2 organizational structure entity structure drawing is shown in the Figure 1.

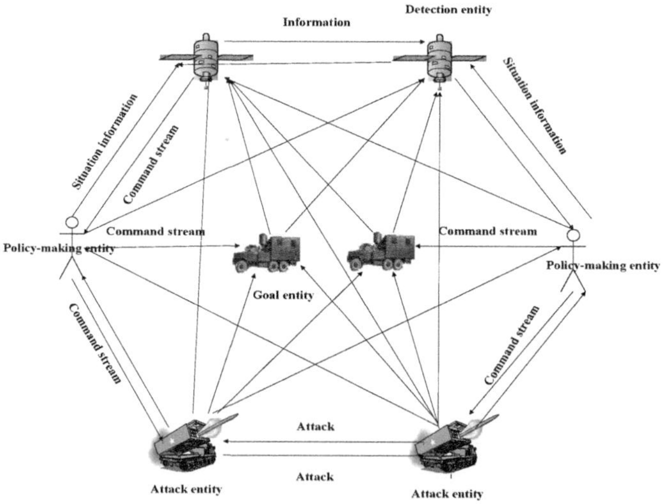

Fig. 1. The C2 organizational structure entity network topology

In the figure 1, detection entity refers to some sensors, its function is to receive possible observation information from other nodes, and send these messages to policy-maker. The policy-making entity usually refer to combat's command staffs, its function is to receive messages from the detection entity, and make the decision to deploy current and future other nodes by current situation messages. Attack entity refers to the executor who carries out order in the combat, its function is to receive policy-maker's order, with the other nodes interaction, and affect these nodes state. The goal entity refers to all nodes which have military value, except detection entity.

4 C2 Organizational Structure Network Scale-Free Model Establishment

The BA model's evolution mechanism is the growth and priority connection. In the network war, all kinds of nodes request intercommunication, interconnection, the size of network is growing, which can cause some military network to conform to the two kinds of production mechanisms which are the growth and the priority connection, so we can use the BA model to research these networks. But the military network still has some nature which BA model doesn't refer to during the military network expansion and evolution, therefore when construct these network topology model, we need to make certain expansion to the BA model.

In order to make the C2 organizational structure network better to accord with reality, we hope to construct such dynamic evolution (growth) network model:

(1) Permits new nodes join, namely permits new entity to join in C2 organization. For example, to join a command control node or firepower attack unit node and so on.

(2) In networking war's network, the importance of node relate with its function, for instance in the network, all levels of control center are obviously different with some front firepower attack unit in importance, when they join in the network, they have the number of connections naturally has widely differences. When we construct the C2 organizational structure network model should consider this point.

(3) Connection rules. In BA model, when we establish a connection only to consider the priority connection, in other words, the possibility of node connected with new joined node only relates with the node degree. Under the network war condition, the C2 organizational network is divided into many operational units which carry out different tasks, each operational unit function is different, and its connection possibility is also different. Expression lies in command control node has priority connection.

Based on above consider, we establish the following network model, its definite algorithm is:

(1) Growth: Begin with a network with m0 nodes, to bring in a new node once and connect it to the m-existing nodes, where $m_0 \leq m$. Importance of each node selects with probability distribution $\rho(\gamma)$.

(2) Priority connection: $\prod_i$ is the probability of a new node connected to an existing node i, with k_i is the degree of node i, k_j is the degree of node j and γ_i is the importance, which satisfies the following relationship.

$$\prod_i = \frac{\gamma_i k_i}{\sum_j \gamma_i k_j} \tag{5}$$

5 Simulation

We can establish algorithm according to the C2 organizational structure network model which is proposed in the previous section, use value simulation language MATLAB to carry on compilation simulation, the test platform is Windows XP. Because sparse() which is a function of MATLAB uses sparse matrix to memory data, to effectively enhance operating efficiency. Its concrete program flow diagram is shown in Figure 2. According to the above algorithm, use MATLAB to simulate, we can get two diagrams about the C2 organizational structure which are model diagram and degree distribution diagram, as shown in Figure 3 and Figure 4.

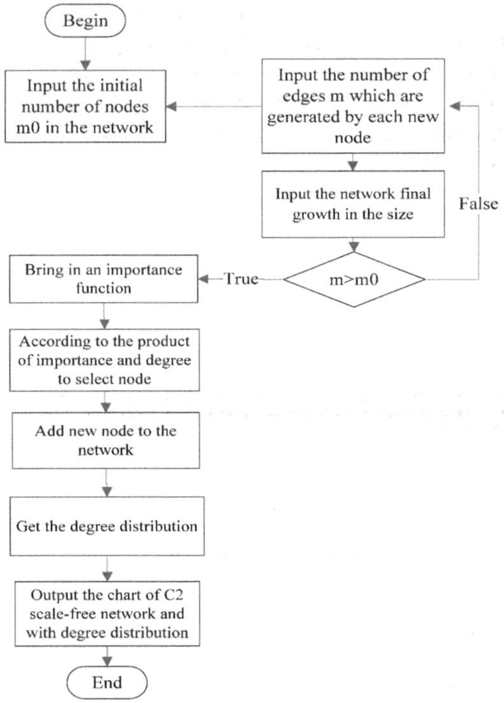

Fig. 2. The C2 organizational structure scale-free model program flow chart

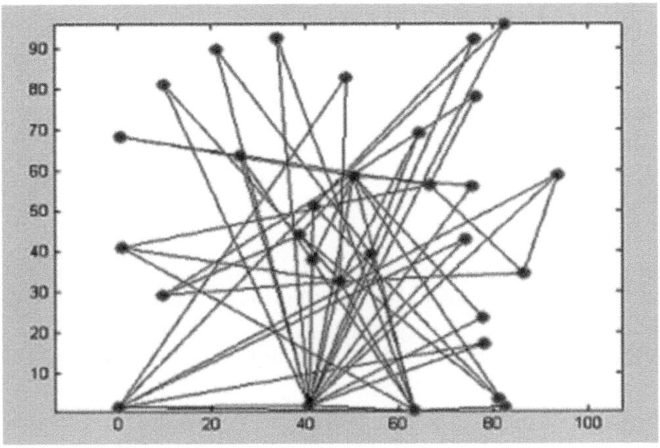

Fig. 3. 20 nodes in the C2 organizational structure network model chart

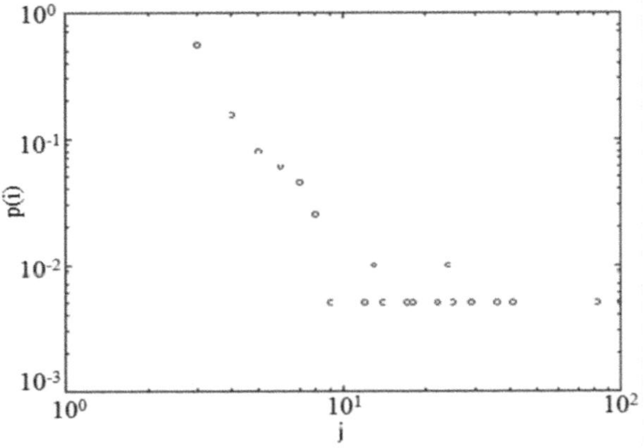

Fig. 4. The C2 organizational structure distribution chart evolution along with time

It can be seen from Figure 4, the scale-free model degree distribution which is constructed by the algorithm follows power-law distribution, proves that the proposed algorithm can efficiently construct the C2 organizational structure scale-free network model.

6 Conclusion

Based on the analysis of the C2 organizational structure network topology model and network nodes growth characteristics, we established the network scale-free model algorithm. The simulation had confirmed that the scale-free model degree distribution which was constructed by the algorithm followed power-law distribution, in other words, the algorithm could effectively construct the C2 organizational structure network scale-free model, and it laid a good basis for the C2 organizational structure optimal design.

Acknowledgment. This project was supported by school of computer science and engineering, Shenyang Ligong University.

References

1. Xu, G.: Systems Science. Shanghai Scientific and Technological Education Publishing House (2000)
2. Barabasi, A.L., Albert, R.: Emergence of scaling in random networks. Science 286(5439), 509–512 (1999)
3. Liu, H., Cai, S., Zhang, Y.: Research of scale-free network model. University Physics 27(4), 41–45 (2008)
4. Bollobas, B., Riordan, O.: Mathematical results on scale-free random graphs. In: Handbook of Graphs and Networks, pp. 1–34 (2003)
5. Cohen, R., Havlin, S.: Scale-free networks are ultrasmall. Phys. Rev. Lett. 86, 3682–3685 (2003)
6. Xiu, B.: Design Methodology of C2 Organizational Structure and Its Analysis of Robustness and Adaptivity. Doctor Dissertation Research Paper of Graduate School of National University of DefenseTechnology, pp. 24–28 (2006)
7. Cares, J.: Distributed networked operations: The foundations of network centric warfare. Alidade (2005)
8. Carely, K.M., Krackhardt, D.: A PCANS Model of Structure in Organization. In: International Symposium on Command and Control Research and Technology, Monterray, CA, pp. 15–20 (1998)

Blind Signal Separation Algorithm Based on Bacterial Foraging Optimization*

Lei Chen[1,2], Liyi Zhang[1,2,**], Ting Liu[1,2], and Qiang Li[1]

[1] School of Electronic Information Engineering,
Tianjin University, Tianjin 300072, China
[2] School of Information Engineering,
Tianjin University of Commerce, Tianjin 300134, China
LeiChen1980@yeah.net,
zhangliyi@tjcu.edu.cn

Abstract. A novel blind signal separation algorithm based on bacterial foraging optimization was proposed. Fourth-order cumulant of signal was used as objective function for separation and Givens transform method was used for reducing the number of variables in objective function. The bacterial foraging optimization algorithm was used for optimizing the objective function transformed and source signal could be separated from mixed signal. Simulation results show that the separation algorithm based on bacterial foraging optimization can achieve the blind separation successfully and the separation precision is high.

Keywords: bacterial foraging optimization, blind signal separation, fourth-order cumulant, Givens transform.

1 Introduction

Blind signal separation is the technology for separating out the independensignal from the mixed signal[1,2]. After nearly twenty years of development, blind signal separation technology has been widely used in the fields of voice, images, communication and biomedicine. Bacterial foraging optimization(BFO) algorithm [3] is a new intelligence optimization algorithm proposed by Passino according to the bacterial foraging behavior of E. coli bacterium. The BFO algorithm has been successfully used in the field of controller optimization design[3,4], machine learning[5], predict of stock index[6] and image compression[7].

In this paper, BFO is used for blind signal separation for the first time. Fourth-order cumulant of signal is used as objective function for separation and Givens transform method is used for reducing the number of variables in objective function. The BFO algorithm is used for optimizing the objective function transformed and source signal was separated from the mixed signal.

---

* The research grants National Natural Science Foundation of China 60802049.
** Corresponding author.

J. Zhang (Ed.): ICAIC 2011, Part V, CCIS 228, pp. 359–366, 2011.

2 Description of Blind Signal Separation

Blind signal separation problems can be described as follows

$$\boldsymbol{x}(t) = \boldsymbol{A}\boldsymbol{s}(t) \tag{1}$$

$\boldsymbol{s}(t) = [s_1(t), s_2(t), \cdots, s_N(t)]^T$ is N independent source signal vector and $\boldsymbol{x}(t) = [x_1(t), x_2(t), \cdots, x_K(t)]^T$ is K mixed signal vector. Under normal circumstances, source signal is instantaneous linear mixed and $N = K$.

$\boldsymbol{A}$ is an $N \times N$ constant matrix. The problem need to be solved for blind signal separation is to get the separation matrix $\boldsymbol{W}$ and the estimate $\boldsymbol{y}(t)$ of source signal can be got from $\boldsymbol{x}(t)$.

$$\boldsymbol{y}(t) = \boldsymbol{W}\boldsymbol{x}(t) \tag{2}$$

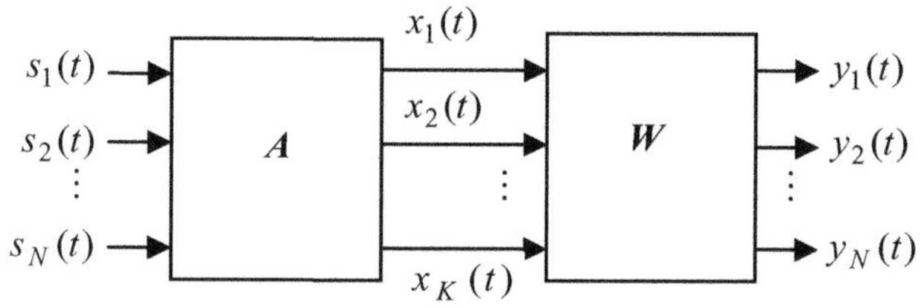

Fig. 1. Diagram of blind signal separation

3 Bacterial Foraging Optimization Algorithm

BFO algorithm is a new intelligence optimization algorithm proposed by Passino[3]. In the BFO algorithm, chemotaxis, swarming, reproduction, elimination-dispersal behavior of bacterial is used for optimization of complex problem. The algorithm can be described as follows

1) Chemotaxis: Chemotaxis behavior is composed of swimming and tumbling. The bacteria will do some judgement according to the environment in the process of foraging. Bacteria will do swimming using their flagella according to attractant responses and do tumbling according to repellant responses.

$$\boldsymbol{\theta}^i(j+1,k,l) = \boldsymbol{\theta}^i(j,k,l) + C(i)\phi(j) \tag{3}$$

Where $\boldsymbol{\theta}^i(j,k,l)$ denotes the location of the i th bacterium. Let j be the index for the chemotactic step. Let k be the index for the reproduction step. Let l be the index of the elimination-dispersal event. Let $C(i)$ denote a basic chemotactic step size and $\phi(j)$ be the unit length direction vector.

2) Swarming: In the process of bacterial foraging behavior, each bacterium is not just individual and information transmission behavior exists in the bacteria group.

Bacteria achieve the swarming phenomenon according to the attraction and repellent between each other.

$$J_{cc}(\theta, P(j,k,l)) = \sum_{i=1}^{S} J_{cc}^{i}(\theta, \theta^{i}(j,k,l))$$
$$= \sum_{i=1}^{S} \left[-d_{attract} \exp\left(-\omega_{attract} \sum_{m=1}^{p} (\theta_m - \theta_m^i)^2 \right) \right]$$
$$+ \sum_{i=1}^{S} \left[h_{repellant} \exp\left(-\omega_{repellant} \sum_{m=1}^{p} (\theta_m - \theta_m^i)^2 \right) \right] \quad (4)$$

Where $d_{attract}$ is the depth of the attractant released by the bacteria and $\omega_{attract}$ is a measure of the width of the attractant signal. Let $h_{repellant}$ be the height of the repellant effect and $\omega_{repellant}$ is a measure of the width of the repellant. $\theta = [\theta_1, \theta_2, \cdots, \theta_p]^T$ is a point on the optimization domain and. θ_m^i is the m th component of the i th bacterium position. This formula shows the resultant force to the ith bacterium from all other bacteria.

3) Reproduction: After N_c chemotactic steps, a reproduction step is taken. For reproduction, the population is sorted in order of ascending accumulated cost. The healthy bacteria are reserved and reproduced, the unhealthy bacteria die. For the given j, k and l, the health of ith bacterium is represented as follows

$$J_{health}^{i} = \sum_{j=1}^{N_c+1} J(i,j,k,l) \quad (5)$$

Where $J(i,j,k,l)$ is the objective function value.

4) Elimination-dispersal: After N_{re} reproduction steps, elimination-dispersal is used for preventing the bacteria falling into local extremum. Let N_{ed} be the number of elimination-dispersal events, and for each elimination-dispersal event, each bacterium in the population is subjected to elimination-dispersal with probability p_{ed}, and then, the ability for bacteria group to find out the global extremum is improved.

4 Blind Separation Algorithm Based on BFO

The key problem for resolving the blind signal separation problem using bacterial foraging optimization is as follows

(1) Selection of objective function

(2) Bacteria coding according to the variable need to be optimized in objective function

(3) Optimizing the objective function using bacterial foraging optimization algorithm for finding out the separation matrix.

A. Criterion for Blind Signal Separation

In the actual calculation of blind signal separation problem, fourth-order cumulant are generally used as the criterion for independence[8].

Fourth-order cumulant of y_i can be expressed as follows

$$k_4(y_i) = E\left[y_i^4\right] - 3\left(E\left[y_i^2\right]\right)^2 \tag{6}$$

The objective function of separation suggested in [9] can be expressed as follows

$$J(W) = \sum_{i=1}^{N} \left|k_4(y_i)\right| = \sum_{i=1}^{N} \left|E\left[y_i^4\right] - 3\left(E\left[y_i^2\right]\right)^2\right| \tag{7}$$

Set the source signal is zero mean, unit variance stationary ergodic random signal and mixing matrix is non-singular matrix. Whitening matrix $\boldsymbol{M}$ need to be constructed.

Let $\boldsymbol{C}_x = E\{\boldsymbol{x}\boldsymbol{x}^T\}$ be the covariance matrix of $\boldsymbol{x}$. $\boldsymbol{D} = diag(d_1, d_2, \cdots, d_n)$ is the diagonal matrix with the element comes from eigenvalue of $\boldsymbol{C}_x$ and $\boldsymbol{E}$ is the matrix composed of eigenvector of $\boldsymbol{C}_x$. Observation signal $\boldsymbol{x}(t)$ can meet the demand $E[zz^T] = \boldsymbol{I}$ using linear transformation $z(t) = \boldsymbol{M}\boldsymbol{x}(t)$.

When we use fourth-order cumulant as the criterion for signal separation, some algorithm need to be used for finding out the separation matrix $\boldsymbol{W}$ who can maximizing $J(\boldsymbol{W})$.

B. Bacteria Coding and Optimization Process

According to the description of blind signal separation, the substance of blind signal separation algorithm is to find the separation matrix $\boldsymbol{W}$.

$$\boldsymbol{W} = \begin{bmatrix} w_{11} & w_{12} & \cdots & w_{1N} \\ w_{21} & w_{22} & \cdots & w_{2N} \\ \vdots & \vdots & \vdots & \vdots \\ w_{N1} & w_{N2} & \cdots & w_{NN} \end{bmatrix} \tag{8}$$

The number of variables need to be identified in separation matrix $\boldsymbol{W}$ is N^2 and the bacteria coding in the bacterial foraging optimization is $[w_{11}, w_{12}, \cdots, w_{1N}, w_{21}, w_{22}, \cdots, w_{2N}, \cdots, w_{N1}, w_{N2}, \cdots, w_{NN}]$. To optimization problem, the difficulty for optimization will increase if the dimension of objective function increases. Therefore, we should lower the difficulty of optimization by reducing the dimension of objective function. We use the Givens transform method proposed in [10], and then, $\boldsymbol{W}$ is transformed to product of some Givens matrix and the number of variables need to be identified decreases. Therefore, the calculated amount and optimization difficulty can both be reduced.

When $N = 3$, separation matrix $\boldsymbol{W}$ is 3×3 square matrix in blind signal separation problem. The bacteria coding is $[w_{11}, w_{12}, w_{13}, w_{21}, w_{22}, w_{23}, w_{31}, w_{32}, w_{33}]$

and the number of variables need to be analyzed are 9. Using the Givens transform method, separation matrix $\boldsymbol{W} = \boldsymbol{T}_{2,3} \cdot \boldsymbol{T}_{1,3} \cdot \boldsymbol{T}_{1,2}$. Where

$$\boldsymbol{T}_{2,3} = \begin{bmatrix} 1 & 0 & 0 \\ 0 & \cos\alpha_1 & \sin\alpha_1 \\ 0 & -\sin\alpha_1 & \cos\alpha_1 \end{bmatrix} \tag{9}$$

$$\boldsymbol{T}_{1,3} = \begin{bmatrix} \cos\alpha_2 & 0 & \sin\alpha_2 \\ 0 & 1 & 0 \\ -\sin\alpha_2 & 0 & \cos\alpha_2 \end{bmatrix} \tag{10}$$

$$\boldsymbol{T}_{1,2} = \begin{bmatrix} \cos\alpha_3 & \sin\alpha_3 & 0 \\ -\sin\alpha_3 & \cos\alpha_3 & 0 \\ 0 & 0 & 1 \end{bmatrix} \tag{11}$$

Then

$$\boldsymbol{W} = \begin{bmatrix} 1 & 0 & 0 \\ 0 & \cos\alpha_1 & \sin\alpha_1 \\ 0 & -\sin\alpha_1 & \cos\alpha_1 \end{bmatrix} \cdot \begin{bmatrix} \cos\alpha_2 & 0 & \sin\alpha_2 \\ 0 & 1 & 0 \\ -\sin\alpha_2 & 0 & \cos\alpha_2 \end{bmatrix} \cdot \begin{bmatrix} \cos\alpha_3 & \sin\alpha_3 & 0 \\ -\sin\alpha_3 & \cos\alpha_3 & 0 \\ 0 & 0 & 1 \end{bmatrix} = \begin{bmatrix} w_{11} & w_{12} & w_{13} \\ w_{21} & w_{22} & w_{23} \\ w_{31} & w_{32} & w_{33} \end{bmatrix} \tag{12}$$

Therefore, the number of unknown variables need to be identified reduces to 3 and the bacteria coding is transformed to $[\alpha_1, \alpha_2, \alpha_3]$. Then, the objective function of separation is transformed to $J(\boldsymbol{\alpha})$ and separated signal can be got when we use the BFO algorithm to optimize the objective function.

5 Simulation and Analysis

In order to verify the validity of this algorithm we proposed, experiment for blind signal separation on three linear mixed voice signal is conducted. The number of sampling point for source voice is 5000 and they are mixed by randomly generated mixing matrix $\boldsymbol{A}$.

$$\boldsymbol{A} = \begin{bmatrix} 0.6500069 & 0.0525514 & 0.6277587 \\ 0.4240006 & 0.4095178 & 0.1720583 \\ 0.0437448 & 0.3347499 & 0.7183708 \end{bmatrix}$$

The parameter of bacterial foraging optimization algorithm is as follows. We choose $S = 30$, $N_{ed} = 2$, $N_{re} = 4$ and $N_c = 50$.

Three source voice signal are shown in Figure2. Three mixed signal are got under the action of matrix $\boldsymbol{A}$ and are shown in Figure3. The source voice signal are separated using the algorithm we proposed in this paper and the results are shown in Figure4. From the experimental results, we can see that source voice signal can be correctly recovered using the blind signal separation algorithm based on BFO we proposed.

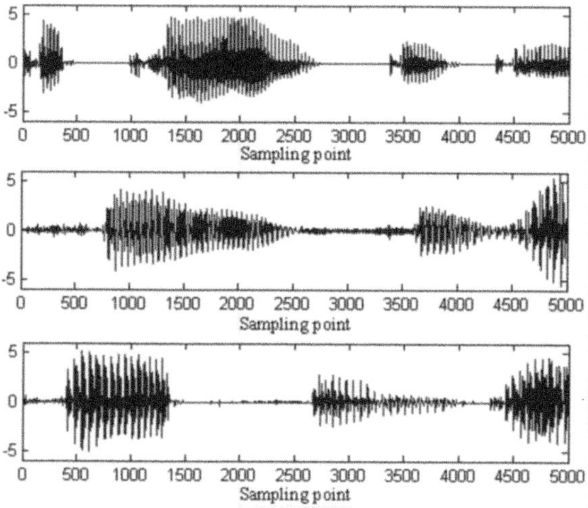

Fig. 2. Source signal

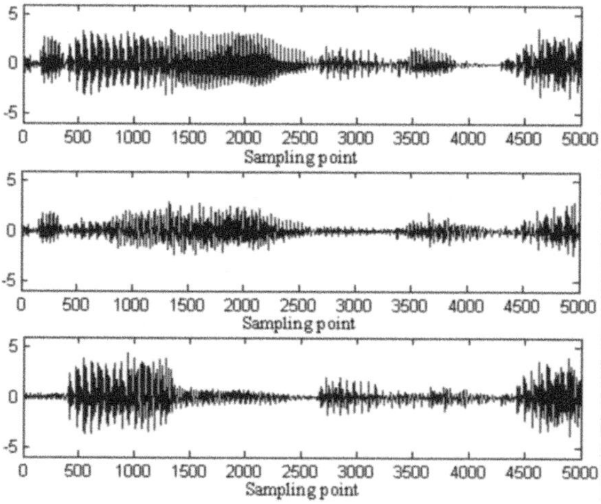

Fig. 3. Mixed signal

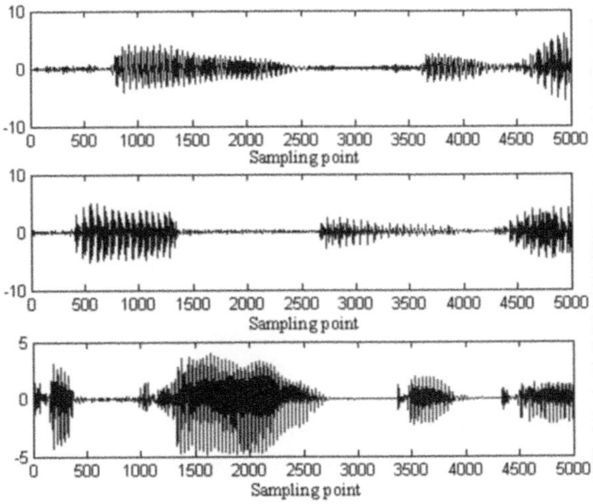

Fig. 4. Separated signal

Next, we will evaluate the similarity between separated signal and source signal using absolute value of correlative coefficient. From table 1, we can draw an conclusion that the source voice signal are recovered efficiently using the algorithm we proposed and the average value of correlative coefficient absolute value in ten simulation exceeds 0.99.

Table 1. Correlative coefficient of separated signal and source signal

	Source signal 1	Source signal 2	Source signal 3
1	0.99874	0.99875	0.99632
2	0.99316	0.98515	0.97440
3	0.99990	0.99797	0.99990
4	0.99871	0.99340	0.98534
5	0.99740	0.99169	0.98480
6	0.99515	0.99367	0.99197
7	0.99702	0.99886	0.99666
8	0.99567	0.99345	0.99989
9	0.99245	0.98878	0.99409
10	0.99626	0.99022	0.99144
Average value	0.99645	0.99319	0.99148

6 Conclusion

A new blind signal separation algorithm based on bacterial foraging optimization was proposed. Fourth-order cumulant was used as objective function for separation and Givens transform method was used for reducing the difficulty of optimization. The BFO is used for optimizing the objective function to recover the source signal. Simulation results show the validity of the blind signal separation algorithm based on BFO.

References

1. Jutten, C., Herault, J.: Blind separation of sources, Part I: An adaptive algorithm based on neuromimetic. Signal Processing 24(1), 1–10 (1991)
2. Comon, P.: Independent component analysis, a new concept? Signal Processing 36(3), 287–314 (1994)
3. Passino, K.M.: Biomimicry of bacterial foraging for distributed optimization and control. IEEE Control Systems Magazine 22(3), 52–67 (2002)
4. Liu, Y., Passino, K.M.: Biomimicry of social foraging bacteria for distributed optimization models, principles, and emergent behaviours. Journal of Optimization Theory and Applications 115(3), 603–628 (2002)
5. Mishra, S.: A hybrid least square-fuzzy bacterial foraging strategy for harmonic estimation. IEEE Trans. on Evolutionary Computation 9(1), 61–73 (2005)
6. Majhi, R., Panda, G., Majhi, B., et al.: Efficient prediction of stock market indices using adaptive bacterial foraging optimization (ABFO) and BFO based techniques. Expert Systems with Applications 36(6), 10097–10104 (2009)
7. Chu, Y., Shao, Z.B., Mi, H., et al.: An application of bacteria foraging algorithm in image compression. Journal of Shenzhen University Science and Engineering 25(2), 153–157 (2008)
8. Hyvarinen, A., Karhunen, J., et al.: Independent Component Analysis. Wiley, New York (2001)
9. Yi, Y.Q., Lin, Y.P., Lin, M., et al.: Blind source separation based on genetic algorithm. Journal of Computer Research and Development 43(2), 244–252 (2006)
10. Qin, H.R., Xie, S.L.: Blind separation algorithm based on QR decomposition and penalty function. Computer Engineering 29(17), 55–57 (2003)

A Multi-agent Simulation Framework for Organization Behavior of Large Scale Engineering Project*

Hua Meng[1], Qiang Mai[2], and Shi An[2]

[1] School of Management Harbin Institute of Technology,
N0.92 West Da-Zhi Street, Harbin, China
[2]School of Transportation Science and Engineering Harbin Institute of Technology
73 Huanghe Road, Harbin, China
HuaMeng1982@126.com

Abstract. This paper proposed a Multi-Agent simulation framework for such organizations in order to analyze how organization behavior impacts the project performance. Based on Information processing view of organizations, the paper grouped the interdependent activities among the different sub-teams or actors in organization and explicitly models these activities. Then a Multi-Agent simulation framework was established for mapping from the real organization into the simulated one. Basic elements of the model were agents, which could perform interdependent activities, and they were organized as Multi-Agent framework through information communication. The information processing selection-mechanism and operation-mechanism were developed to simulate those organizations by the Multi-Agent models. The Multi-Agent simulation framework has been tested internally, and evaluated externally.

Keywords: Large scale engineering project, Organization behavior, Multi-Agent simulation, Interdependent activities, Information processing.

1 Introduction

Large scale engineering project faces the challenge of managing and coordinating distributed organizations developing novel products, such as space project. This causes complex interdependent activities among different sub-team or actors in organizations. And in some large scale engineering projects, the schedule and the cost are always reduced. In order to meet the new plan, it is needed that the organization design must change which will cause different organization behaviors.

The computational organization models provide a new way to analysis the interdependent activities in organizations. Cyert and March (1963) are the pioneers of

* This work is supported by CNSF for Young Scholars Grant #70903017, Research Fund for the Doctoral Program of Higher Education of China Grant#20092302120012, China Special Postdoctoral Science Foundation Grant#201003431, China Postdoctoral Science Foundation Grant#20090460898, Heilongjiang Provincial Natural Science Foundation of China Grant# QC2010087 to Qiang Mai.

J. Zhang (Ed.): ICAIC 2011, Part V, CCIS 228, pp. 367–375, 2011.

simulation of manufacturing organizations. They provided early examples of the theoretical insights that could be gained from simulating organizational decision[1]. Masuch and LaPotin (1989) proposed an AAISS system which using non-numerical computing paradigms to model organizational decision making in clerical tasks[2]. Carley and Prietula (1994) developed a organization modeling for detailed task[3]. In practice, some organizations uses simulation model to develop new organization capable of performing complex modeling and analysis task, such as Systems Analysis Integrated Discipline Team (Carroll etc.), Design Structure Matrix (Steward 1981), OrgConTM (Burton and Obel, 2004), SimVisionTM used by NASA[4][5].

Agent technique can be used to simulate the active behavior. Besides its object-orientation, an Agent can actively perform tasks. Kraus(1997) described Agent as automata[6]. [7] proposed communication languages for communicating between multiple Agents. Kraus etc (1995) investigated the interdependent activities and communications in the Multi-Agent environments[8]. Agents can also be used to model qualitative decision-makers[9].

The main intent of this paper is to propose a Multi-Agent simulation framework for organizations of large scale engineering projects in order to analyze how activity interdependencies impact the project performance. The first phase of this paper groups the interdependent activities among organizations based on Information processing view of organizations. The second phase proposes the Multi-Agent simulation framework in which agents perform interdependent activities. And the mechanism used to constrain the simulation is developed in this phase. The final phase discusses the mapping from the real organization into the simulated one.

2 Organization Behavior of Large Scale Engineering Project

Large scale engineering project are often have complex organization behaviors which have many interdependent activities. Based on Information processing view of organizations, the paper groups the interdependent activities among the different sub-teams or actors in organization and explicitly models these activities.

A. Basic Organization Behavior

Based on the Information Processing theory, we group the organization behaviors into two sets: liability activity and cooperation interdependent activity. Because all activities are viewed as information processing, so we can model the in terms of work volume which is the product of people needed and the days needed. Then, if we assume the total work volume of the project be TW, liability work volume be BW, and cooperation work volume be CW, we have

$$TW = BW + CW \tag{1}$$

If the liability activity cause defect in the product, then a rework activity is needed. Then, the liability can be divided into two subsets: originally planed liability activity, whose work volume is BW_O, and the rework activity arising due to the failure of original liability activity, whose work volume is BW_r. From equation (1), we have:

$$TW = BW_O + BW_r + CW \quad (2)$$

B. Cooperation Interdependent Activities

Besides the liability activity, organizations of engineering project must spend time on cooperate each other which can be defined as coordinate interdependent activities. According to different characters, this kind of activity can be divided into two subsets.

One is the information communication work volume caused by interdependency. The interdependency has three kinds. The first is the pooled interdependency which means one activity is a part of the work of a sub-organization, and can't exist by its own. The second is the sequential interdependency which means the activities have successor relationship used in CPM. The third is the reciprocal interdependency which means the change in one activity will cause the change in the other activity.

The second is the decision activity made by information processing actors. For some upper level sub-organizations, they must make decisions for the requests from lower level sub-organizations.

If we assume CW_{cm} representing the information communication work volume, CW_{cd} representing the decision-making work volume, then the equation (1) and (2) can be expressed as

$$TW = BW_O + BW_r + CW_{cm} + CW_{cd} \quad (3)$$

C. Information Communication

The information communication consists of the dynamic procession of the project organization. The communication item, CW_{cm}, have the characters such as the sender, receiver, PRI and the arrival time.

The information communication also needs communication media to realize the communication. According to Nass and Mason (1990), these media can be described as some functional attributes: Synchronicity, cost, recordability, capacity and bandwidth[10]. Different media has different functional attributes.

D. Restraint Factors of Organization Behavior

In order to forecast the work volume, it is needed to know the restraint factors of organization behavior which will determine the liability and coordinate work volume. Here, we figure out 4 kinds of restraint factors.

The first kind of restraint factor is the organization attribute. The restraint factors influencing liability activities are the factors influencing the work volume and skill requirement.

The second restraint factor is the organization structure. The organization structure is the management subject relationship, and can be defined as Supervise/Report-to relationship according to information processing view.

The third restraint factor is the organization culture. Organization culture which realized through long term time is the value view all the number of the project organization identify with. In different industry, the organization culture has little difference. This paper uses the failure rate in analogous projects which the organization take part in to describe the organization culture.

The forth restraint factor is the communication structure. Besides the organization structure, the communication structure is influenced by communication mode. A more formalized organization relies on scheduled formal meetings for cooperation, but the more time and cost will be spent, and the effect is better.

According to the above analysis, the restraint factors influencing the organization behavior can be described as table 1.

Table 1. Restraint Factors Of Organization Behavior

Restraint Factor	Element	Description
Organization attributes	Agents	Set of sub-organization involved in the project
	Skill of faculty	Amount of technical ability each agent represents
	Experience	Amount of similar projects each agent has worked
Organization Structure	Highly centralized organization	High level sub-organization makes the decision
	Decentralized organization	Low level sub-organization makes the decision
Organization Culture	Project error rate	Normal probability of a project exception after project completion
Information Communication Structure	Information communication rate	The volume of information communication that agents use
	Information communication media	Determine agent's preference for meeting vs. informal communications

2 Multi-agent Simulation Framework for Large Scale Engineering Project Organization

A. Agent in the Large Scale Engineering Project

An Agent is a computing entity who has autonomy, reactive, proactive, social, evolutional attributes in some environment. We can think about the large scale project organization by this definition. The sub-organization can be viewed as a Agent which receives information about task in the project, and execute different activities to realize the project objective as well as apperceiving the development of the project, and obey the rules of activities proposed in advance.

The Structure of an Agent in the large scale project organization is shown in figure 1.

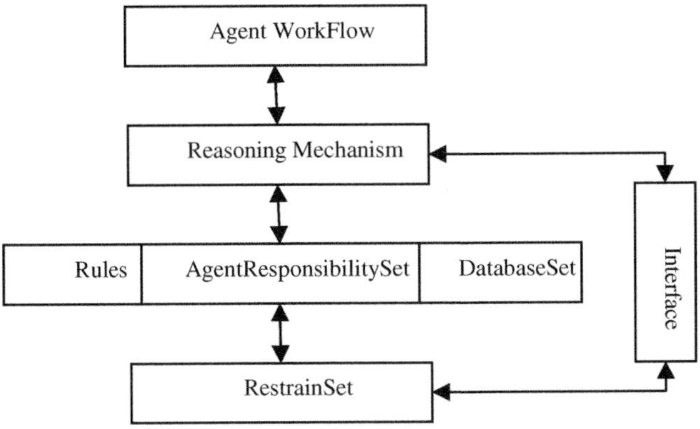

Fig. 1. Agent Structure

As showed in figure 1, the interdependency between activities and restraint factors in the "in-tray". In the "out-trey", the relative behaviors are the rework, cooperation and decision.

B. Multi-agent Simulation Framework for the Large Scale Engineering Project

Using the above Agent structure, we can model the organization of large scale engineering project.

1) Multi-Agent Simulation Framework: The Multi-Agent simulation framework has three levels, which carries out the Monitor Agent level who is on the top, the President Agent level who is on the middle, and the Execution Agent level who is on the bottom.

The liabilities of the Monitor Agents is to communicate with other Monitor Agent, plan the task and constitute the object, manage the President Agent, and Monitor the activities of President Agent.

The President Agent which contains a set of Agent who simulates some organization behaviors is the core of the Multi-Agent simulation framework.

The Execution Agent performs the defined task. Because this kind of Agent is doing the passive objects, the inheritance relationships form the inheritance hierarchy between the classes[11]. This can be represented as $(A_1(t_b,t_e),\ldots,A_n(t_b,t_e))$ to reflect the time factor of activities, where t_b and t_e is the beginning and the ending of an activities respectively.

The inter-related information communication is controlling the Multi-Agent simulation framework. Then the Multi-Agent simulation framework can be described as:

$$\text{Multi-AgentOrganisation} ::= \langle(\text{MonitorAgentSet},\ \text{Inter-MonitorInforFlow}),$$
$$(\text{PresideAgentSet},\ \text{Inter-PresideInforFlow}),$$
$$(\text{ExecuteAgentSet},\ \text{Inter-ExecuteInforFlow})\rangle$$

$$\text{Agent} ::= (\text{Interface, ReasoningMechanism, Rules, AgentResponsibilitySet, DatabaseSet IntraAgentWorkFlow, RestrainSet})$$

2) Information in the Multi-Agent Simulation Framework: This paper assumes that an information communicated between Agents in the same level of the Multi-Agent framework and between different levels can be described by a form with the following feature[12]:

$$\begin{aligned}&\text{InForName}:[\text{Signature}:\text{InputType}\rightarrow\text{OutputType};\ \text{Behaviour}:(\text{TextDescription},\text{InForSet})\\&\text{Condition}:\text{Condition}_1\wedge\cdots\wedge\text{Condition}_n;\\&\text{Restraint}:\text{I/O}\quad\text{and}\quad\text{time}\quad\text{restraint}]\end{aligned}$$

3) Information Communication in the Multi-Agent Simulation Framework: The information communication between Agents is realized through "publish" and "receive" behavior. The information structure can be described as:

$$\begin{aligned}&\text{InForTool} ::= (\text{Version}, \text{Chapter})\\&\text{Chapter} ::= \{\text{Information}\}^+\\&\text{Information} ::= (\text{Title}, \text{Author}, \text{Task}, \text{TimeStamp}, \text{Status})\end{aligned}$$

4) Nest-Information processing in the Multi-Agent Simulation Framework: A nest-information processing should guarantee three kinds of consistency. The first is temporal consistency. The second is consistency between the liability and activity. The third is consistency between "in-tray" and "out-tray". The initial input flow and the final out flow of low level Agent should be the same as the input and output of its direct up level Agent.

C. Activity Rules of Multi-agent Simulation Framework

The input information to Agent will be processed by the rules set, then the output will be got. Here, we will define two kinds of rules set, one is the information selecting rules, the other is the information processing rules.

1*) Information Selecting Rules*: Under the condition that more information come into the "in-tray" of the Agent at the same, there needs rules to determine the sequential of the information processing. We think that there are four kinds of rules to keep the consistency among the information processing.

(1) Priority Rules: In the simulation framework, priority are measured on a scale from one to nine, with nine being the highest, liability activity has priority 5, a decision about how to handle an exception has priority 8.

(2) Time Rules: In the situation that the information about activity has the same priority, we must selecting information based on the length of time in the in-tray, i.e. FIFO.

(3) Random Rules: Although the information selecting is always based on sense of the Agent, but the organization theory, "bounded rationality", assumes that the organization behavior sometimes is random.

2*) Information Processing Rules*: In the simulation framework, Agent process both liability activities and cooperation activities through various information processing cycles.

(1) Information Processing Rules for Liability: When an Agent picks up a liability activity A, the Agent allocates ΔT_A to complete the liability activity. When it is finished, the Agent will check it whether a communication is needed and whether there is an error in it.

(2) Exception Processing Rules: When upper Agent receives the reported exception information, the decision made by this level Agent has three choices: complete rework, partially correct and ignore the error. The choice is determined stochastically, depending on the organization structure and organization culture.

(3) Information Communication Rules: This paper assumes that the information communication can be divided into formal meeting and informal communication. Regular meetings are set up before the simulation. The Agent will receive the meeting notice before each meeting during the simulation.

3 The Development of the Multi-agent Simulation Framework

There are two basic requirements for the development of the Multi-Agent simulation framework. Firstly, the model must capture enough detail of both activities contents and dependencies. Secondly, the development must be able to map the real project data to model attributes. In order to realize these requirement, some researches proposed many methods, such as Quality Function Deployment (Hauser and Clausing, 1988) and Design Structure Matrix (Gebala and Eppinger, 1991)[13][14]. Because this paper uses Agent technology to describe the real organization of project, the development of the simulation framework will adopt the mapping from organization into Multi-Agent simulation framework.

The rules for establishing the mapping can help to describe the simulation framework. Four basic elements of the rule set are as follows:

Rule 1: Map the low level sub-organization into the Execution Agent.

Rule 2: Map the middle level sub-organization into the President Agent.

Rule 3: Map the up level sub-organization into the Monitor Agent.

Rule 4: Mapping the interdependencies between the sub-organization into the information communication.

A proper mapping must satisfy the mapping restraint set. The four basic elements of the set are as follows.

Restraint1: Agent can only interact with agents within the same level. Every Agent receive the task through its upper Agent.

Restraint2: A Monitor Agent can only manage its President Agent.

Restrain3: Every sub-organization should have a corresponding Agent in the Multi-Agent simulation framework.

Restrain4: The mapping should be carried out top-down.

Based on this instinct, the Multi-Agent simulation framework is developed on Windows using JSP, JAVA, SQI Server 2000 and so on.

4 Conclusion

This paper analyzes the organization behavior of large scale engineering project by the information processing view which looking project organization as an information processing and communication system consisted of information processing actor with limited capacity under restraint factors. According to this theory, the organization behavior is divided into liability activities (original activities and rework activities) and cooperate activities (information communication activities and decision activities). These activities can be quantified by work volume. In order to realize this object, this paper proposes the restraint factors, including organization attributes, organization structure, organization culture and communication structure.

Based on the above organization behavior analysis, this paper proposes a three level Multi-Agent simulation frame for large scale engineering project by viewing the sub-organization as an Agent. The simulation framework consists of three level Agents, which are Monitor Agent, President Agent and Execution Agent. In order to connect these Agents, we design the information model and information communication model. We also establish the rules set for information processing in order to realize the simulation process. At last, this paper proposes the development of the Multi-Agent simulation framework.

Acknowledgment. The authors thank the editor and the anonymous referees for their helpful comments on the earlier version of this paper.

References

1. Cyert, R.M., March, J.G.: A Behavioral Theory of the Firm. Prentice-Hall, Englewood Cliffs (1965)
2. Masuch, M., LaPotin, P.: Beyond Garbage Cans: An AI Model of Organizational Choice. Administrative Science Quarterly 34, 38–67 (1989)
3. Carley, K., Prietula, M.: ACTS Theory: Extending the Model of Bounded Rationality. In: Carley, K.M., Prietula, M.J. (eds.) Computational Organization Theory. Lawrence Erlbaum, Associates, Publishers, Hillsdale (1994)
4. Steward, D.V.: The design structure system: A method for managing the design of complex systems. IEEE Trans. Engrg. Management 78(3), 71–74 (1981)
5. Burton, R.M., Obel, B.: Strategic Organization Diagnosis and Design: The Dynamics of Fit, 3rd edn. Kluwer Academic Publishers, Norwell (2004)
6. Kraus, S.: Negotiation and cooperation in multi-agent environments. Artificial Intelligence 94, 79–97 (1997)
7. Covington, M.A.: Speech acts, electronic commerce, and KQML. Decision Support Systems 22, 203–211 (1998)
8. Kraus, S., Wilkenfeld, J., Zlotkin, G.: Multiagent negotiation under time constraints. Artificial Intelligence 75, 297–345 (1995)
9. Brafman, R.I., Tennenholtz, M.: Modelling agents as qualitative decision makers. Artificial Intelligence 94, 217–268 (1997)
10. Nass, C., Mason, L.: On the Study of Technology and Task: A Variable-Based Approach. In: Faulk, J., Steinfield, C. (eds.) Organizations and Communication Technology, pp. 46–67. Sage Publications, Beverly Hills (1990)

11. Zhuge, H.: Inheritance rules for flexible model retrieval. Decision Support Systems 4(22), 379–390 (1998)
12. Batory, D., O'Malley, S.: The design and implementation of hierarchical software systems with reusable components. ACM Transaction Software Engineering and Methodology 1(4), 355–398 (1992)
13. Hauser, J., Clausing, D.: The house of quality. Harvard Business Review (May 1988)
14. Gebala, D., Eppinger, S.D.: Methods for analyzing design procedures. In: Third Intnl. ASME Conference on Design Theory and Methodology, Miami, Florida (1991)

The Parameters Selection for SVM Based on Improved Chaos Optimization Algorithm*

Yong Wang, Yong Liu, Ning Ye, and Gang Yao

Institute of Intelligent Vision and Image Information,
China Three Gorges University, Yichang Hubei 443002, China
yongwang1971@sina.cn

Abstract. The parameters selection of support vector machine decides its study performance and generalization ability. The SVM model is greatly influenced by penalty factor C and the kernel function parameter such as σ for the radial basis function (RBF) kernel. To searching the best compound of parameters, a new algorithm is proposed based on improved chaos optimization strategy to realized automatic parameters selection for SVM. Chaos optimization algorithm is a global searching method in which the complexity and dimension of variables need not to be considered. Compared with the algorithms based on GA and PSO, the classification efficiency is improved greatly.

Keywords: Support vector machine, Chaos series, Chaos optimization, Parameter selection.

1 Introduction

Support vector machines (SVM) pioneered by Vapnik in 1992 is a new machine learning method which is systematic and properly motivated by statistical learning theory. Compared to the others learning algorithms such as neural network, SVM has characteristics of small sample machine learning and powerful generalization, which can effectively avoid overfitting, local minimum and curse of dimensionality [1]. The idea of SVM classifier is to use nonlinear transformation to translate nonlinear problems in original sample space to linear problems in high space. The nonlinear transformation is realized by defining a kernel function which satisfies Mercer theorem, so the performance of classifier is directly influenced by parameter selection of kernel. At the same time, the penalty factor C of SVM model which is a compromise between structural risk and sample error also influences performance of SVM. When using SVM, two crucial problems are confronted: how to choose the optimal input feature subset for SVM, and how to set the best kernel parameters σ and penalty factor C .At present, there is lack of mature theory to instruct the parameters selection which are acquired by experience, experiments and large space searching. The essence of parameters selection of SVM model is a process to solve combinatorial optimization problem. Its object is to choose optimal parameters to

* This work is partially supported by NSF Grant # 50805087 and # 60972162 to Yong Liu and Research Foundation for Talented Scholars of China Three Gorges University Grant #2010PY047 to Gang Yao.

J. Zhang (Ed.): ICAIC 2011, Part V, CCIS 228, pp. 376–383, 2011.

minimize the classification error rate of the SVM model [2]. In reference [3], an algorithm which is called Bilinear Grid Search Method (BGSM) to search optimal parameters compound in the parameters set. At first, the parameter C corresponding to high generalization ability for linear SVM is computed, then the parameter C is fixed, and the parameter σ of RBF kernel is searched corresponding to the highest generalization ability to obtain the compound (C , σ). Although the number of SVMs is decreased, the second linear searching stage relies heavily on the first stage parameter. The optimal parameters are computed by steepest descend method in reference [4], but the shortage is that this algorithm is sensitive to initial values. If the initialization is far from the optimal value, it is easy to fall into local optimization. An algorithm based on Genetic Algorithm(GA) to optimize the SVM parameters is presented in reference [5], and a more optimal parameters is acquired to use this method, but it is also easy to fall into local optimization and the iteration is large and it is time-consuming. A another algorithm based on particle swarm algorithm(PSO) is showed in literature[6] which is more efficient than GA and a global optimization is calculated, but the weakness is that early maturity appears.

Chaos optimization algorithm [7] is a global searching algorithm which is new and intelligent optimization strategy. The complexity and dimensions of variables is not considered when selecting parameters of SVM model. A method based on chaos optimization algorithm is showed to optimize the parameters of SVM model and the object function is the average minimum error rate which is computed by cross-validation. The improved chaos optimization algorithm is used to search the optimal combined parameters.

2 Model of SVM Parameter Optimization

A. SVM Model Analysis

SVM model transforms the learning process to an optimization issues according to the principle of minimize struck risk. To obtain smallest wrong divided sample and maximal margin ,that is optimal hyperplane, for training samples (x_i , y_i)(i= 1, 2,...,l, $x_i \in R^n$, $y_i \in \{1,-1\}^l$),SVM model needs to solve the problem that minimizes the object function(1) which meets the constraints of (2)

$$\phi(\omega,\xi) = \frac{1}{2}\|\omega\|^2 + C\sum_{i=1}^{l}\xi_i \tag{1}$$

$$\text{s. t.} \quad y_i[(w^T \bullet z_i) + b] \geq 1 - \xi_i \,;\, \xi_i \geq 0 \,,\, i=1,2,\ldots,l; \tag{2}$$

where training samples x_i is mapped to higher dimension space through function $Z_i = \phi(x_i)$, $\omega \in R^n$ is coefficient vector of hyperplane, $b \in R$ is threshold, ξ_i is slack variable, C ($C > 0$) is penalty factor for samples which is divided wrongly.

Mentioned problem is transformed its dual problem through using Lagrange polynomial, which is searching Lagrange multipliers to maximize the object function as following:

$$Q(\alpha)=\sum_{i=1}^{l}\alpha^{i}-\frac{1}{2}\sum_{i=1}^{l}\sum_{j=1}^{l}a_{i}a_{j}y_{i}y_{k}K(x_{i},x_{j}) \tag{3}$$

subject to:

$$\sum_{i=1}^{l}\alpha_{i}y_{i}=0\ ;0\leq\alpha^{i}\leq C\ ,\text{i=1,2,\ldots,}l \tag{4}$$

So the corresponding classification decision function is

$$f(x)=\text{sign}\{\ \sum_{i=1}^{m}\alpha_{i}^{*}y_{i}K(x_{i},y_{i})+b^{*}\ \} \tag{5}$$

where non-negative α_i^* is support vector . m(m<l) is the number of support vectors, b^* is threshold corresponding to α_i^*, $K(x_i,y_i)=\phi(x_i)\bullet\phi(x_j)$ is kernel function which satisfies Mercer theorem. RBF kernel function is used in this paper which is $K(x_i,y_i)=\exp(-\|x_i-y_i\|)/\delta^2)$,where δ is the parameter of kernel.

B. Object Function of SVM Parameter Optimization

The generalization of SVM classifier is greatly affected by the parameters of C and δ .When C is very small, underfitting appears, but when C very large, overfitting appears that makes generalization worsened. When δ is very small, overfitting appears and δ is very large, underfitting appears. So the parameter δ is crucial to the generalization. In this paper, the error rate E_ζ of classifier is used to estimate the performance of SVM model. So the object is to minimize the E_ζ, which is selecting a optimized compound parameters to minimize the E_ζ . Let domain of penalty factor C and δ is respectively [a_1 , b_1] and [a_2 , b_2], so the optimization issues is

$$\begin{aligned}&\text{Min}\quad E_{\xi}(M)\\&\text{s. t.}\ M=\{C,\sigma\}\ \ a_1\leq C\leq b_1\ ,a_2\leq\sigma\leq b_2\end{aligned} \tag{6}$$

C. The Analysis of Shortage of Presented Parameters Optimization of SVM Model

SVM model has good learning ability and generalization, but its performance depends on the selection of the parameters. Some scholars have studied and discussed from different angles for the parameter selection problem, which are divided three main methods. The first one is that users select the model parameters based on experience; the second one, which is the commonly used method currently, is cross-validation method; the third one is using modern intelligent optimization algorithm, such as the parameters selection based on GA, PSO and SA. The above three mentioned methods also have some disadvantages, which included too many human factors, requirement of the function which is continuously differentiable, long computing time and local minimum. Those shortcomings impact on the practical application of SVM seriously. Therefore, an improved chaos optimization algorithm which can be used to optimize the parameters of SVM model is presented in this paper.

3 The Parameter Optimization Based on Chaotic Series

A. The Analysis on Characteristics of Chaotic System

By taking advantage of ergodicity and random of chaotic motion, that is chaotic motion can traverse all states, Li Bing et al.[8]first proposed a chaos optimization algorithm (COA).The basic idea is that chaotic states is introduced to optimization variables like carrier wave with Logistic mapping ($x_{n+1} = \mu x_n (1 - x_n)$).The traversing range of chaotic motion is extended to value range of optimization variables and then all fields is searched by chaos variables, that is used to solve optimization problem which is continuous and complicated. The essence of parameter optimization is a process of optimization search and this problem is always multi-peak that exists a plurality of local extremum. The ergodicity of chaos can avoid falling into local minimum in searching process, so chaos optimization algorithm is more efficient to obtain optimization parameters by means of its powerful global searching ability.

B. The Improved Chaos Optimization Algorithm Oriented SVM Parameter Optimization

Chaotic motion can traverse all state space. The larger the space is, the longer the traversal time. It is necessary to consider the reduction of search space of variable. Before narrowing the search space, traversal search needs to conduct many times in order to ensure that the optimal solution is in the narrowed search space. The search times are difficult to determine. Generally, it relates to the complexity of optimization function and the size of search space. So the search times just can be determined according to specific application. This reduces the versatility of algorithm. At the same time, because the termination condition is not given, the search is full of blindness and increases the amount of computation [9]. In order to reduce the blindness of search and improve the efficiency of search, the proposed improved and accelerated chaos optimization algorithm has the following advantages:

1) By introducing two-chaotic optimization mechanism, enhance the adequacy of the search, reduce the times of blind search and promote the efficiency of search. Chaos mapping is selecting as following:

$$x_{k+1} = \mu x_k (1 - x_k) \tag{7}$$

and

$$y_{k+1} = \alpha y_k^3 - \alpha y_k + y_k \tag{8}$$

where (7) is logistic (μ =4); (8) is cube mapping which is chaos mapping when $\alpha \in$ [3.3, 4]. However, the possibility of the random value obtained throughout the entire search space is larger when α is a little bigger. So 3.9 is assigned to α .

2) The second improve for chaotic optimization search formula. The original chaos optimization algorithm searches only in the unilateral neighbourhood of Logistic map. When μ is assigned to 4, according to inequality 0< x^i <1, the product of adjustment coefficient α^i and chaotic variable x_i is greater than zero (α^i x_i >0) or less than zero (α^i x_i <0). It results in that the optimal solution x^* is searched in the unilateral neighbourhood. In this paper, a modified formula is proposed. It avoids the disadvantage of unilateral search. The modified formula is expressed as

$$x_i^r = x_i^* + \alpha_i (2x_i^r - 1)R \tag{9}$$

where x_i^r , r are chaotic variable and iteration times of the second chaos optimization search respectively, $r = 1, 2, \cdots$; x_i^* is the optimal variable for the current; Here the introduction of pseudo-random number R can be further reduced the search range and avoid the further deviation of optimal solution caused by the artificial introduction of adjustment factor, and speeds up the convergence.

C. The Steps of Improved Chaos Optimization Algorithm

Step 1: Initialize. According to (7) and (8), generate n initial chaotic variables with different trajectories.

Step 2: Map x_k^i , y_k^i to the range of the optimization variables respectively, have $mx_i^k = a_i^r + x_i^k (b_i^r - a_i^r)$, $my_i^k = a_i^r + (1/2)(y_i^k + 1)(b_i^r - a_i^r)$.

Step 3: Search using chaos variables. If $f(mx_i^k) < f_x^*$, have $f_x^* = f(mx_i^k)$, $mx_i^* = mx_i^k$; If $f(my_i^k) < f_y^*$, have $f_y^* = f(my_i^k)$, $my_i^* = my_i^k$; Else continue the search.

Step 4: Assign k+1 to k : $k = k + 1$,have $x_i^k = \mu x_i^k (1 - x_i^k)$; $y_i^k = \alpha (y_i^k)^3 - \alpha y_i^k + y_i^k$

Step 5: Repeat step 2 and 3. When the repeating times is more than h, judge whether the inequality $\|mx^* - my^*\| < \gamma \|a^r - b^r\|$ sets up, where $\gamma \in (0, 0.25)$ and h can be assigned to experienced 2000. If the inequality does not set up, repeat step 2 and 4 until this inequality meets. Then continue the following steps.

Step 6: Narrow the search range of each variable according to the following equations

$$a_i^{r+1} = \min(mx_i^*, my_i^*) - \xi\gamma \|mx_i^* - my_i^*\|;$$

$$b_i^{r+1} = \max(mx_i^*, my_i^*) + \xi\gamma \|mx_i^* - my_i^*\|.$$

where $\xi \in [1,2]$. If $a_i^{r+1} < a_i^r$,then $a_i^{r+1} = a_i^r$;if $b_i^{r+1} > b_i^r$,then $b_i^{r+1} = b_i^r$ .The two conditional statements ensure the new range is not out of bound.

Step 7: $a_i^r = a_i^{r+1}$, $b_i^{r+1} = b_i^r$, $r = r + 1$;

Step 8: Return to step 2 until the best solution is found.

Step 9: Do the second carrier wave according to the equation $x_{i,n+1}^n = x^* + \alpha_i x_{i,n+1}^{'}$,where $x_{i,n+1}^{'}$ is chaos variables within the interval [-1,1]; α_i is adjustment coefficient; x^* is optimal solution for the current.

Step 10: Continue the iterative search using chaos variables after the second carrier wave. If $f(mx_i^k) < f_x^*$, then $f_x^* = f(mx_i^k)$, $mx_i^* = mx_i^k$.

Step 11: If the termination condition is met, terminate the search and output the optimal solution x^* , f^* ; otherwise return to step 9.

The flow chart of above algorithm is showed in figure 1.

4 Simulation Experiment

In order to validate the effectiveness of the proposed algorithm, experiments are conducted on four classical classification data sets in UCI machine learning database. The parameter selection methods based on GA and PSO are used to compare with the proposed algorithm. The experimental results are displayed as table 1.

As shown in table 1, compared with the mentioned two methods, the computational time of the proposed method is greatly reduced under the same classification accuracy. It shows that the proposed improved chaos optimization algorithm for parameter selection not only improves the classification accuracy, but also reduces the computational time.

Table 1. Comparison of three methods

UCI data set		wine	Breast cancer	Heart disease	Australia credit card
The proposed algorithm	Time(s)	8.204	29.2291	37.5301	113.4846
	(c,g)	(195.155,4.37)	(33.512),2.162)	(144.145,0.033)	(174.8745,0752)
	Classification accuracy	99.8%	100%	99.7%	100%
The method based on PSO	Time(s)	10.8765	63.2585	47.0720	142.5834
	(c,g)	(126.47,3.9)	(46.177,3.871)	(136.068,0.291)	(177.392,0.504)
	Classification accuracy	98.8%	99.4%	97.9%	99.5%
The method based on GA	Time(s)	31.7380	188.4100	177.8151	608.1626
	(c,g)	(58.785,4.601)	(12.177,1.094)	(67.079,0.342)	(49.025,0.537)
	Classification accuracy	99.2%	98.3%	97.3%	98.5%

(c,g) is the optimization of compound (C , σ).

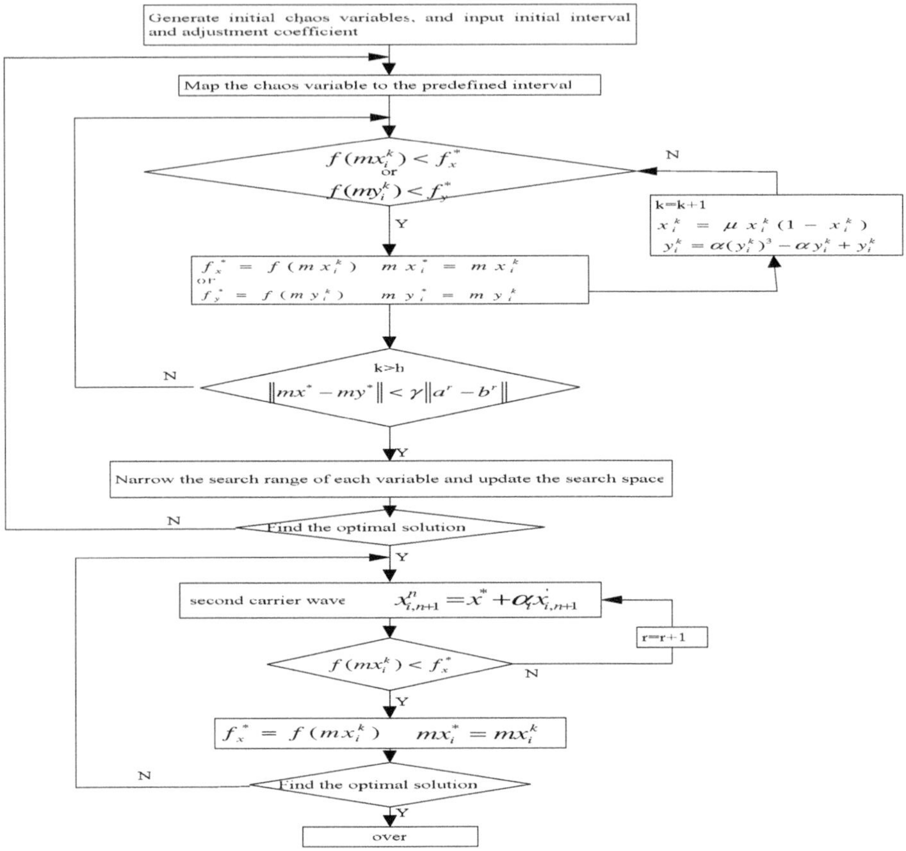

Fig. 1. The flow chart of chaos optimization algorithm

5 Conclusions

The performance of the SVM based on the structural risk minimization principle greatly depends on the parameters selection. Usually, the parameters are selected empirically and it limits the application of the SVM. In this paper, the improved chaos optimization algorithm is introduced into the parameter selection for SVM. The results of experiments demonstrate that the proposed improved chaos optimization algorithm for parameter selection method improves the generalization ability of the SVM, and make that the parameters selection does not depend on experience. It points out a new way for the application of the SVM.

Acknowledgment. This research has been supported by the Natural Sciences Foundation of China (NSFC) (Grant No. 50805087 and 60972162) and the Graduate Innovation Foundation of China Three Gorges University (Grant No.200914) and Research Foundation for Talented Scholars of China Three Gorges University (Grant No.2010PY047). It also has been done under the help of the team of the Institute of Intelligent Vision and Image Information.

References

1. Vapnik, V.N.: An overview of statistical learning theory. IEEE Trans. Neural Networks 10, 988–999 (1999)
2. Cherkassky, V., Ma, Y.Q.: Practical selection of SVM parameters and noise estimation for SVM regression. Neural Networks 17(1), 113–126 (2004)
3. Keerthi, S.S., Lin, C.J.: Asymptotic behaviour of support vector machines with Gaussian kernel. Neural Computation 15(7), 1667–1689 (2003)
4. Keerthi, S.S.: Efficient tuning of SVM hyperparameters using radius/margin bound and iterative algorithms. IEEE Transactions on Neural Network 13(5), 1225–1229 (2002)
5. Frohlich, H., Chapelle, O., Scholkopf, B.: Feature selection for support vector machines by means of genetic algorithms. In: Proceedings of the 15th IEEE International Conference on Tools with Artificial Intelligence (ICTA 2003), November 3-5, pp. 142–148. IEEE Computer Society, Los Alamitos (2003)
6. Ma, C., Ruan, Q.Q.: Parameter Selection for SVM Based on Discrete PSO. Computer Technology and Development 12, 20–23 (2007)
7. Zhang, H., He, Y.: Comparative study of chaotic neural networks with different models of chaotic noise. In: Wang, L., Chen, K., S. Ong, Y. (eds.) ICNC 2005. LNCS, vol. 3610, pp. 273–282. Springer, Heidelberg (2005)
8. Li, B., Jiang, W.: Chaos Optimization Method and Its Application. Control Theory & Applications 14(4), 613–615 (1997)
9. Xiu, C., Liu, X.: Optimization algorithm using two kinds of chaos and its application. Control Theory & Applications 18(6), 724–726 (2003)

Extraction and Sharing the Water Surface Coverage of Poyang Lake Based on the ASAR Data

Chaoyang Fang[1], Jingyuan Yang[2], and Yuanlin Yuan[2]

[1] Key Laboratory of Poyang Lake Wetland and Watershed Research, Ministry of Education, Nanchang, 330022, Jiangxi, China
[2] Geography and Environment College of Jiangxi Normal University, Ziyang Road, Nanchang 330022, Jiangxi, China
ChaoyangFang001@tom.com

Abstract. ENVISAT ASAR data's ability of penetrating through cloud and rain belt is stronger. It is an important data source to extract water coverage during cloudy rainy weather conditions on the flood reason of Poyang Lake. This paper used ENVISAT ASAR data, monitored the variations of Poyang Lake water area in 2009 by the partition threshold method. Finally get Poyang Lake water coverage's area and spatial distribution. Through the construction of sharing service platform achieve real-time information sharing of Poyang Lake water coverage thematic maps, make users at all levels can obtain real-time, rapid, low cost monitoring information of Poyang Lake, and provide the flood disaster quasi real-time monitoring and warning service, which provide a reliable and important data support for Poyang Lake water allocation, flood and drought disaster reduction, disaster assessment and post-disaster compensation.

Keywords: Poyang Lake, ENVISAT ASAR data, water area coverage, information share.

1 Introduction

Poyang Lake (longitude 115° 50′-116 ° 44′, latitude 28° 25′-29 ° 45′) is the largest freshwater lake, which is located in northern Jiangxi Province, south of Yangtze River. It takes in Gan Jiang, Fu River, Xin Jiang, Rao River and Xiushui Rivers such as runoff, after regulation and storage then flows into the Yangtze River through Hukou. It is a water carrying, throughput, seasonal lake. Poyang lake area generally enters into the flood season from late march. The water level began to rise and reaches at the highest level in July. Since then, under the influence of the Yangtze River's flood periods during 7 ~ 9 months, the level of the lake maintains to October start steady decline, to the following January ~ February the water level declines to bottom out. Because water levels change greatly, lakes area changes also greatly. In the flood season, the lake water level rises sharply, the water surface is vast; dry period level falls, lake beach is bared into trough, the lake surface with only a few winding waterway, presents "high water is a lake, low water like river" natural landscape. More than six months' flood season makes flood disaster is one of the biggest natural disasters in Poyang Lake basin. In order to maximize the losses caused by flood, it is necessary to monitor and

J. Zhang (Ed.): ICAIC 2011, Part V, CCIS 228, pp. 384–389, 2011.

evaluate flood timely. Remote sensing has characteristic of wide coverage and short revisit interval, which provides the technical guarantee for quick acquiring flooding information. However, when flooding occurs, it is difficult to get clear images of visible remote sensing. Radar remote sensing is not limited by day or night and clouds, thus it has become the main means of flood water information acquisition, particularly space-borne radar remote sensing [1]. The ASAR (Advanced Synthetic Aperture Radar) instrument on board the ENVISAT-1 satellite is one of the most powerful synthetic aperture radar sensors [2]. This paper using 2009 a complete year multi-date ASAR image extracted water coverage information of Poyang Lake, and through sharing service platform realized real-time Poyang Lake water coverage information sharing, provide reference data for Poyang Lake water allocation control, Poyang Lake flood and drought disaster reduction and building national strategy of Poyang Lake eco-economic region.

2 Experimental Data and Basic Principle

This study used a complete year (Feb. 2 2009 until Dec.14 2009) of 10 pieces ASAR image to extract water area, which received by the Chinese university of Hong Kong Joint Laboratory for Geo-information Science satellite ground stations. ASAR workes in C-band, wavelength 5.6 cm, revisit interval 35 days, each orbit can continuously obtain 30min images. This study used ASAR data for Wide Swath model Level 1B products, image width about 405km, spatial resolution 150m x 150m, polarization modes for VV, angle of 17 ° ~ 43 °. Compared to ASAR's wavelength, land surface roughness is the rough surface, reflects radar beam in all directions, the echo strength is higher, shows grey or dark grey tonal in radar image. While water surface roughness is the smooth surface, because of the radar beam's side scan, specular reflection make water's backscatter value is always in low value area, water appears dark tonal in radar image. Microwave attenuates less in atmosphere, its ability of penetrating through cloud and rain belt is stronger, basically is not affected by smoke, clouds, rain and fog [3]. This characteristic is significant for monitoring Poyang Lake's water coverage, where rainfall days account for more than 1/3.

3 The Method and Procedure

More than six months' flood reason makes flood disaster in Lake Region frequently. In order to resist flood, there are hundreds of dams in Lake Region. These dams divide Poyang Lake's water into two parts: the main lake body connected to Yangtze River and the lake water within dams. The main lake body connected to Yangtze River is divided into the river channel and the main lake body by Songmen Mountain which seated between Duchang and Wucheng. The lake in the northwest of Songmen Mountain is narrow and it's actually a lathy channel. The lake in the southeast of Songmen Mountain is vast and it's the main body [4]. As shown in figure 1, the sum of the three parts water area is considered as the water area of Poyang Lake described in this study.

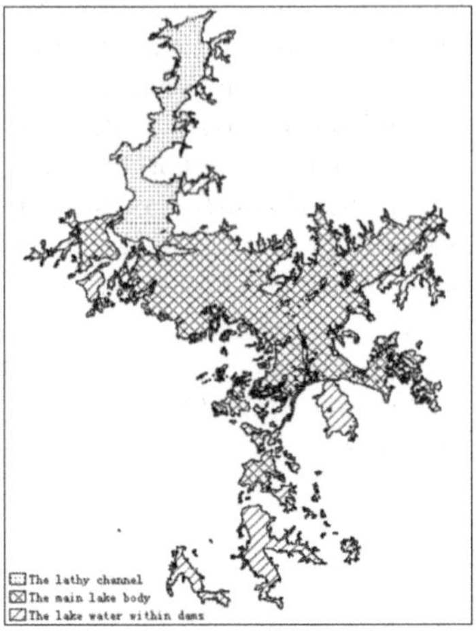

Fig. 1. The water zoning map of Poyang Lake

For single-band radar images, generally use threshold value method for water recognition [5]. When extract water in ASAR image with threshold value method, the man-machine interaction way to determine water gray threshold is often used, above this threshold will be considered as non-water, below this threshold will be regarded as water. In fact, when the image area is large, due to the topography and geomorphology, hydrological and meteorological aspects are often in great differences, couple with the characteristics of ASAR's side scan, the water recognition threshold suitable for different regions is not the same[6]. In view of using one threshold to deal with the entire Poyang Lake area often causes misjudgment, this article extracts water with different threshold according to different parts of Poyang Lake.

Poyang Lake water coverage information extraction procedure is as follows: firstly calculate radar image's backscatter coefficient, convert the geographic coordinates of ASAR's backscatter coefficient data into equal area projection coordinate system via calibration. Then carry on filtering processing in order to reduce the speckle noise of ASAR image. Secondly, select the suitable threshold to extract each part water scope of Poyang Lake:

If $\sigma^0 \langle T$ Then this pixel is the water pixel

If $\sigma^0 \rangle T$ Then this pixel is non-water pixel

σ^0 is the backscattering coefficient of ASAR image, T is the threshold for distinguishing water.

Through the above threshold for image segmentation each part of Poyang Lake water bodies can be extracted, then combine with Poyang Lake DEM data to remove mountain shadow, each part water area can be calculated out (Table 1), finally get Poyang Lake water area' s spatial distribution as figure 2 showed.

Table 1. Poyang Lake Water Area and Water Level Table

Date	The main lake body connected Yangtze River (km^2)	The lake water within dams (km^2)	Total area(km^2)	The water level of Xingzi Station (m)
09-2-2	301.84	453.86	755.70	7.68
09-3-9	1710.30	459.45	2169.75	12.52
09-4-13	1299.70	505.13	1804.83	10.74
09-5-18	1363.32	503.10	1866.42	13.72
09-6-22	1685.25	527.47	2212.72	15.39
09-7-27	2032.12	520.36	2552.48	15.7
09-8-31	2205.83	530.59	2736.22	16.19
09-10-5	893.61	510.24	1403.85	11.4
09-11-9	569.69	487.75	1057.44	8.91
09-12-14	606.74	462.32	1069.06	7.95

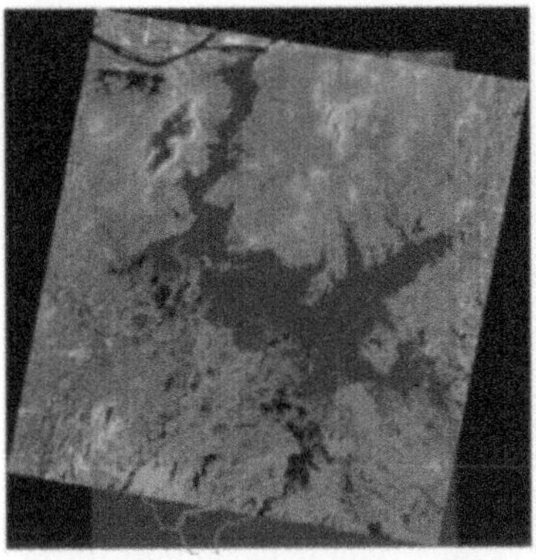

Fig. 2. Poyang Lake water coverage extracted by ASAR

4 The Water Coverage Information Sharing Service

The multi-source spatial information network share and interoperability platform software Geo-Globe4.0 (hereinafter referred to as spatial information sharing service platform) is self-developed by Wuhan University Geo-information technology Co. Ltd. The platform can publish Vector data, satellite remote sensing data, terrain data and information extraction model on line and realize real-time information sharing of Poyang Lake water coverage thematic maps. GeoGlobe products can carry out a series of work such as loading data sources, design data sets scheme, preview data symbolic display effect, making tiles data set, management data set and data issued etc.. Then GeoGlobe client can display the sharing data sets; using the Internet makes users at all levels can obtain real-time, rapid, low cost monitoring information of Poyang Lake, providing the flood disaster quasi real-time monitoring and warning service. Figure 3 shows the users inquire interface of water coverage information.

Fig. 3. Water coverage information inquiring interface

5 Conclusion

ASAR data has the advantage of unlimited by day and night or clouds and rain weather conditions; it is an important data source for quick acquiring flooding scope during the flood. This paper used ENVISAT ASAR data monitored the variations of Poyang Lake water area in 2009 by the partition threshold method, and through the sharing service platform achieved real-time information sharing of Poyang Lake water coverage thematic maps. Thus governments and scientific research institutions at all level can obtain real-time, rapid, low cost monitoring information of Poyang Lake, which provide a reliable and important data support for Poyang Lake water allocation, flood and drought disaster reduction, disaster assessment and post-disaster compensation.

Acknowledgment. This study supported by the Opening Foundation of Key Laboratory of Poyang Lake Wetland and Watershed Research, Ministry of Education (No. PK2008007) and Science and Technology Program of Jiangxi Provincial Department of Education (No. GJJ10391).

References

1. Yang, C.-j., Zhou, C.-h.: Application of complementary information of RADARSAT SWA SAR and LANDSAT TM in deciding the flood extent. Journal of Nature Disasters 10(2), 79–83 (2001)
2. Huang, Q.-n., Tang, L.-L., Dai, C.-d.: Envisat-1 ASAR Data Products' Characteristics and the Potential Application. Remote Sensing Information, 56–59 (March 2004)
3. Mei, A., Peng, W., et al.: An Introduction Remote Sensing. Higher Education Press, Beijing (2001)

4. Poyang Lake, http://baike.baidu.com/view/4173.htm
5. Yang, C., et al.: Investigation on Extracting. The Flood Inundated Area From JERS-1 SAR Data. Journal of Nature Disasters 7(3) (August 1998)
6. Tan, Q.-l., et al.: Measuring Lake Water Level Using Multi-Source Remote Sensing Images Combined with Hydrological Statistical Data. Journal of Beijing Jiao Tong 30(4) (August 2006)

Axial Antenna Current Distribution in Rectangular Mine Tunnel

Meifeng Gao

State Key Laboratory of Coal Resources and Safe Mining
China University of Mining and Technology (Beijing)
Research Institute of Communication Engineering
Jiangnan University
Wuxi, Jiangsu Province, China
meifeng_gao@sina.com

Abstract. Using dyadic green's functions, the axial line antenna integral equation was deduced in the rectangular mine tunnel, and the line antenna current distribution was obtained by MoM. It is discussed that the current distribution is affected by the feed point position, location of the line antenna in mine tunnel, and excitation frequency. The simulation results show that: when the axial dipole antenna is placed in the middle of the rectangular mine tunnel, exciting current distribution of it is similar to in free space. The antenna current magnitude increases when the antenna is near to the tunnel wall, and the current magnitude is decreases as the frequency increases.

Keywords: line antenna, antenna current distribution, rectangular mine tunnel.

1 Introduction

The electromagnetic wave propagation in the mine tunnel has being studied[1-2], and that the antenna exciting in the mine tunnel is one of the key science and technology problem of the tunnel wireless communication[3]. [4] and [5] respectively studied line source and ring source excitation field characteristics in the circular tunnels. [6] analyzed radiation characteristics of the electric dipole in the rectangular tunnel.

In order to predict and analyze electromagnetic environment in mine tunnel, the antenna exciting in the mine tunnel need to be studied. Using dyadic green's functions (DGF), the axial line antenna integral equation is established in the rectangular mine tunnel. The line antenna current distribution is got by MoM, and that the current is affected by the feed point position, location of the line antenna in tunnel, and excitation frequency is discussed.

2 Theory Analysis

A. System Model

The mine tunnel surrounding rock is generally made of high conductivity rock, so that the coal mine tunnel can be approximated as a waveguide[7]. The coordinate of

J. Zhang (Ed.): ICAIC 2011, Part V, CCIS 228, pp. 390–396, 2011.

rectangular mine tunnel is showed in Fig.1. The Cartesian coordinates with z axial, x horizontal, and y vertical. On the assumption that the rectangular mine tunnel horizontal length is a and vertical height is b.

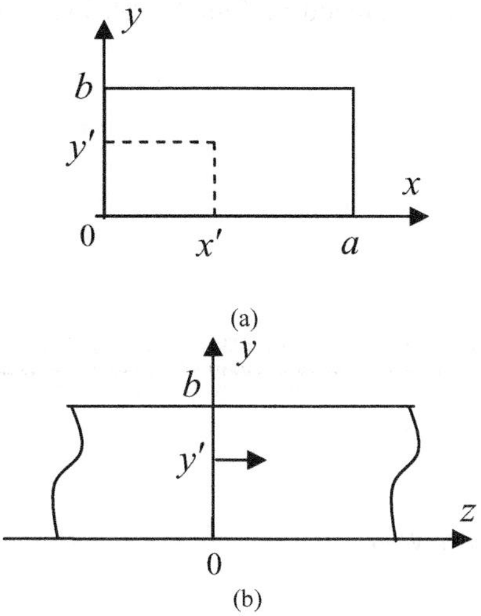

Fig. 1. Rectangular mine tunnel with an axial line antenna: (a) Cross- section of the mine tunnel. (b) Longitudinal- section of the mine tunnel.

Known a point current source along z is located in $\vec{r}'(x', y', 0)$, viz.

$$\mathbf{J}(\mathbf{r}') = \hat{z} I_z dz' \delta(\mathbf{r} - \mathbf{r}') \tag{1}$$

From the electromagnetic theory, the electric field strength generated by the point current source is

$$\mathbf{E}_0(\mathbf{r}) = -j\omega\mu \overline{\overline{\mathbf{G}}}_{e1}(\mathbf{r}, \mathbf{r}') \cdot \mathbf{J}(\mathbf{r}') \tag{2}$$

Where, $\overline{\overline{\mathbf{G}}}_{e1}(\mathbf{r}, \mathbf{r}')$ is DGF in a rectangular waveguide[8].

Substituting $\overline{\overline{\mathbf{G}}}_{e1}(\mathbf{r}, \mathbf{r}')$ expression to (2), the electric field strength along z, which is generated by the point current source along z, is gained

$$\mathbf{E}_{0z}(\mathbf{r}) = -\hat{z}\left[\frac{j\omega\mu}{abk^2}\sum_{m,n}\frac{(2-\delta_0)k_c^2}{\gamma_{mn}}e^{-\gamma_{mn}|z-z'|}S_m S_m' S_n S_n' - \frac{j\omega\mu}{k^2}\delta(\mathbf{r}-\mathbf{r}')\right] I_z dz' \tag{3}$$

Where, $S_m = \sin\left(\frac{m\pi}{a}x\right)$, $S'_m = \sin\left(\frac{m\pi}{a}x'\right)$, $S_n = \sin\left(\frac{n\pi}{b}y\right)$, $S'_n = \sin\left(\frac{n\pi}{b}y'\right)$

$$\gamma_{mn} = \sqrt{\left(\frac{m\pi}{a}\right)^2 + \left(\frac{n\pi}{b}\right)^2 - k^2}, \quad k = \omega\sqrt{\mu\varepsilon}$$

B. Current Distribution of Axial Line Antenna in Rectangular Mine Tunnel

When the axial line current source $\bar{i}_s(\bar{r}) = \hat{z}i_{sz}(\bar{r})$ is in the rectangular mine tunnel, the electric field strength generated by the line source is a linear superposition of point source because of the homogeneous boundary conditions, so the electric field strength along z can be evaluated from (4).

$$E_z = -\hat{z}\int_{-L}^{+L}\left[\frac{j\omega\mu}{abk^2}\sum_{m,n}\frac{2k_c^2}{\gamma_{mn}}e^{-\gamma_{mn}|z-z'|}S_m S'_m S_n S' - \frac{j\omega\mu}{k^2}\delta(\mathbf{r}-\mathbf{r}')\right]i_{sz}(z')dz' \tag{4}$$

Using the boundary conditions of the antenna exterior total electric field along z equaling to zero, we obtain the line antenna integral equation in rectangular mine tunnel, viz.

$$Ai_{sz}(z') = E_z^{inc} \tag{5}$$

Making $A_{mn} = \frac{j2\omega\mu}{abk^2}S_m S'_m S_n S'_n$, then the operator of the equation (5) is

$$A = \sum_{m,n} A_{mn}\frac{k_c^2}{\gamma_{mn}}\int_{-L}^{+L}e^{-\gamma_{mn}|z-z'|}dz' - 2A_{mn}\int_{-L}^{+L}\delta(z-z')dz' \tag{6}$$

Equation (5) can be solved by MoM. Taking the rectangular-pulse function as basis function $u_i(z')$ and using Galerkin method, to do the inner product on the both sides of equation (5), then (5) can be changed into linear equations, viz.

$$\left[z_{ji}\right]\mathbf{I} = \mathbf{e} \tag{7}$$

Where, $\mathbf{I}$ is the antenna current spread coefficient, $z_{ji} = \int_{-L}^{+L} Au_i(z')\cdot w_j^*(z)dz'$, $w_j^*(z)$ is a conjugate of the weight function $w_j(z)$, $\mathbf{e} = \left[e_j\right] = \left[\left\langle E_z^{inc}, w_j(z)\right\rangle\right]$.

Solving linear equations (7), the line antenna current numerical values are obtained.

3 Numerical Results

In this paper, the system model parameters are: the mine tunnel horizontal width $a = 6\,\text{m}$, and vertical height $b = 4.5\,\text{m}$. Because the dipole antenna is one of basic line antennas, the dipole antenna model is used. The feed voltage of the antenna is unit voltage. In Fig.2 and Fig.3, the frequency $f = 900\,\text{MHz}$.

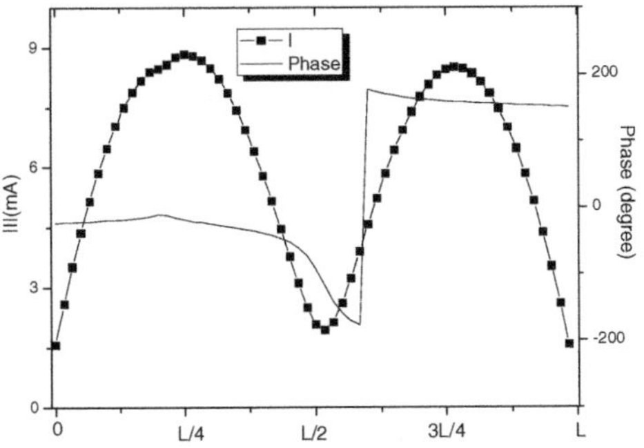

(a)

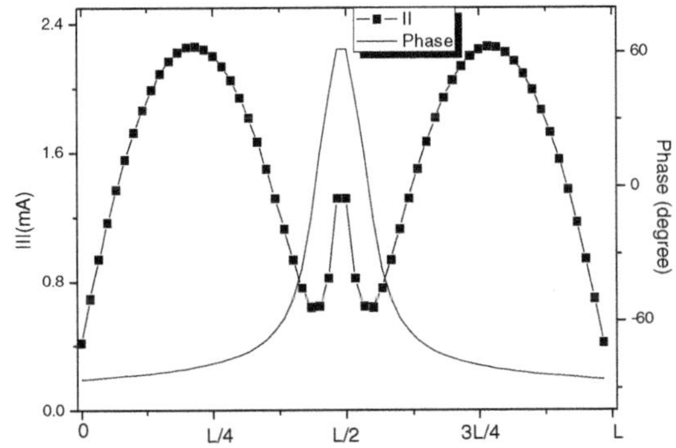

(b)

Fig. 2. The antenna current amplitude and phase when feed point is changed: (a) Feed point in L/4. (b) Feed point in L/2.

Fig.2 shows that the current amplitude and phase of the line antenna at the rectangular mine tunnel center are changing along with feed point changing. It can be seen from the simulation results that the current distribution generated by the axial full-wave dipole antenna in the rectangular tunnel is similar to in free space, so it is reasonable that the line antenna current distribution is supposed sinusoidal, when the electromagnetic field distributions are discussed in the mine tunnel with a line antenna.

Fig.3 shows that when the antenna is set near to corner of the tunnel, the current amplitude increases and the phase also greatly changes, due to the antenna is affected by the tunnel wall.

When the frequency is 100MHz, 900MHz, 1.2GHz and 1.9GHz, the half-wave dipole antenna current amplitude and phase distributions are showed as Fig.4. The current amplitude is decreasing with the bigger of frequency, but the phase shape changes a little when the antenna is located in middle of the mine tunnel.

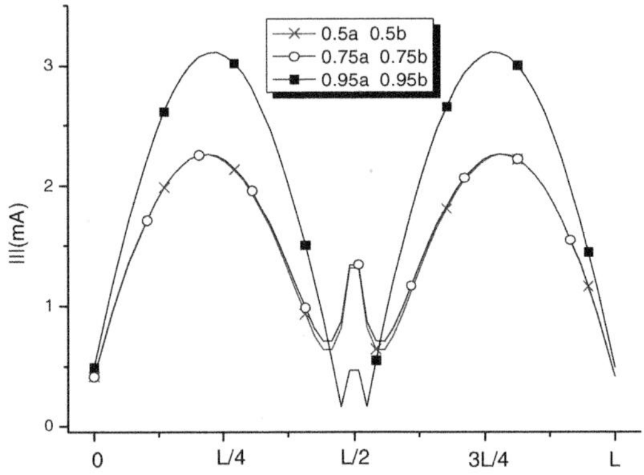

(a)

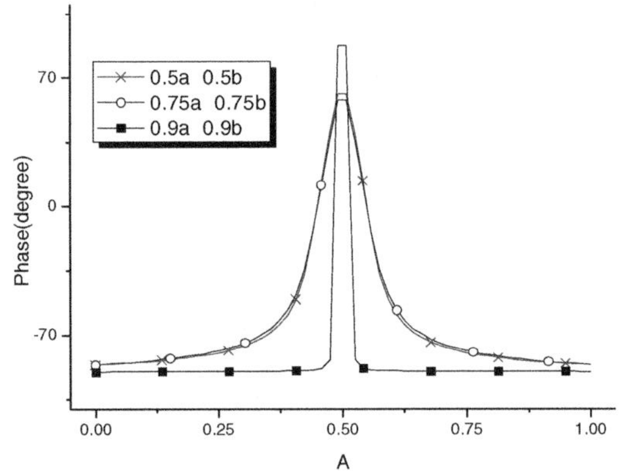

(b)

Fig. 3. The current amplitude and phase when the antenna is diagonal moved: (a) Current amplitude. (b) Phase.

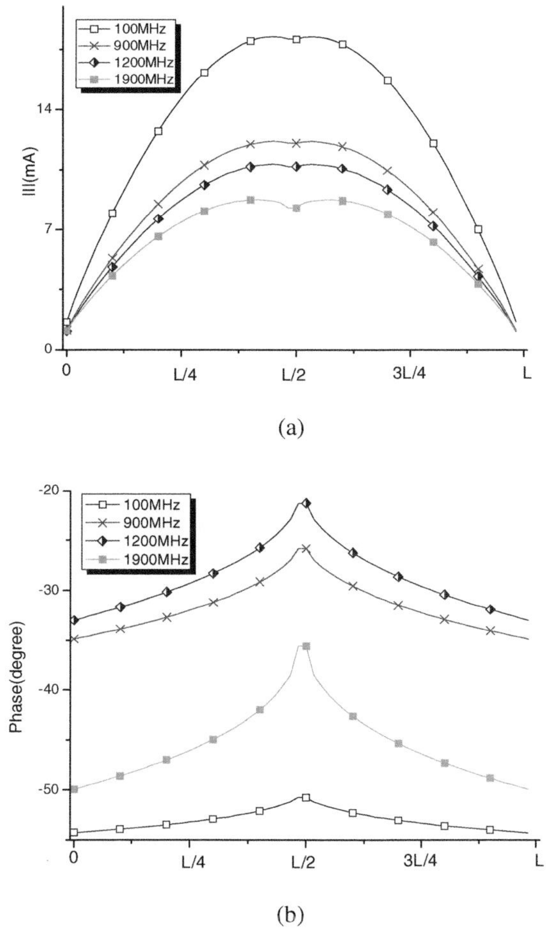

Fig. 4. The antenna current amplitude and phase with different frequency: (a) Current amplitude. (b) Phase.

4 Conclusion

With dyadic green's functions, the integral equation of an axial line antenna, which is in the rectangular mine tunnel, is established, and the antenna current distributions are obtained by MoM. The simulation results show that: it is reasonable to assume the line antenna current to be sinusoidal distribution when the electromagnetic field in the mine tunnel with an axial antenna is analyzed, because the tunnel has large electrical measurement. The effect of tunnel wall is not neglected when a line antenna is near to the wall. The current amplitude of the antenna in the middle of tunnel is decreasing with the bigger of frequency, but the phase of the antenna change is not obvious. Furthermore, studying on the axial line antenna current distribution in the rectangular coal mine tunnel provides an analytical means for the coal mine tunnel complex electromagnetic environment.

References

1. Zhang, Y.P., Zheng, G.X., Sheng, J.J.: Radio Propagation at 900 MHz in Underground Coal Mines. IEEE Transactions on Antennas and Propagation 49(5), 757–762 (2001)
2. Dudley, D.G., Lienard, M., Mahnmoud, S.F., Degauque, P.: Wireless Propagation in Tunnels. IEEE Antennas and Propagation Magazine 49(2), 11–26 (2007)
3. Sun, J.-p.: Present Situation and Key Problems of Science and Technology of Mine Mobile communication. Industry and Mine Automation (7), 110–114 (2009)
4. Dudley, D.G., Mahmoud, S.F.: Linear Source in a Circular Tunnel. IEEE Transactions on Antennas and Propagation 54(7), 2034–2047 (2006)
5. Dudley, D.G.: Propagation in Circular Tunnels: Ring Source Excitation. In: IEEE Antennas and Propagation Society International Symposium, vol. 3, pp. 2987–2990 (June 2004)
6. Sun, J., Li, J., Lei, S.: Analysis of Radiation Characteristics of the Electric Dipole in the Rectangular Tunnel. In: Proc. Wireless Communications, Networking and Mobile Computing (WiCom 2007), pp. 1124–1126. IEEE Press, Los Alamitos (2007)
7. Delogne, P.: Lesky Feeders and Subsurface Radio Communications. Peter Peregrinus Ltd. (1982)
8. Zhou, X.: Complete Eigenfunction Expansion of thr Electrical Dyadic Green's Functions for Rectangular Waveguides and Cavities. Journal of Microwaves 12(3), 108–204 (1996)

A Color Image Quantization Method Used for Filling Color Picture Production*

Yuzheng Lu and Feng Wang

School of Digital media, Jiangnan University,
Wuxi, China, 214122
yuzhenglu0091@sohu.com

Abstract. A simple color image quantization method is described in this paper to simplify the filling color picture production process, which will give much more choice and save much more time for the designers. Firstly, the image prepared for the production should be filtered to eliminate the noise and some details not suitable for painting in the image, and the colors appearing in the images are calculated for the further analysis. Then the K-means method is carried to cluster the colors obtained from the former calculation, and the number of the cluster centers, which is very important to the K-means process, is equal to the color categories required by the manufacture. The color value is transformed from RGB color space to HSB color space to calculate the distance parameters during the K-means process and color matching process, and the final result proved that this process can be proposed to the designer to accelerate the development process of the filling color image products.

Keywords: image quantization, K-means cluster, color space, filling color picture.

1 Introduction

Filling color picture is a very popular product in the market, which provides the customs to complete an artistic work after filling the picture with corresponding colors set in the pictures. At first, the filling color picture was only designed for the children to train their color vision, but now, it has been provided for adults to satisfy their aesthetic appreciation.

The manufacturer should prepare the pigments or watercolor etc. for the customers to fill the picture, and the number of the color categories should be limited about thirties to simplify the filling process and the manufacture process. And the purpose of the manufacturer is to reduce the color categories under the condition that preserve the largest information about the original picture. The first step for the designer is to block the color in the image. It will spend designers several days bordering the color blocks and modifying some details, especially the details that are not important to the picture but need some other color categories. The designers' task can be partly substituted by the computer calculation, and the image quantization process can be carried to group

* Supported by" the Fundamental Research Funds for the Central Universities" JUSRP10923.

J. Zhang (Ed.): ICAIC 2011, Part V, CCIS 228, pp. 397–404, 2011.

the colors in the image in a certain categories, which can provide the designers some useful information during their rectifying about the image.

Many researches are focused on the image compression process based on the image quantization [1-5] .The image quantization is a very important process of representing true color images using a small number of colors in a colormap. True color images typically use 24bits/pixel, each byte for red, green and blue channels, and the possible colors of each pixel can be reached to $2^{24}\approx 17$ million colors, which give the true color image larger file size. However, with a good color quantization algorithm, the same image quality (visually) can mostly be restored from a much smaller file after the true color is quantized to smaller color categories. During the manufacture about the filling color picture, the original picture is quantized to 30~ 40 colors and the visual effect about the picture should be preserved after the quantization.

The color image quantization algorithm typically consists of four phases [6]. the first phase is the calculation process based on the original image, and the purpose of this step is to find some useful information for the further quantization; the second phase is to chose the best representative colors from the information acquired above, a lot of method has been carried in the color selection process; the third phase is to match the pixel in the original images to a representative color, and the matching principle is an important factor to the matching result; the fourth phase is to redraw the image by replacing the original color in every pixel with a representative color. The most important task for the color quantization is to find the representative colors, and the existing algorithms for color image quantization may be categorized into two classes: the splitting algorithm and the clustering algorithm.

The splitting algorithm is to split the color in the original images into two disjoint cells according to some criteria, and the splitting procedure is iterated until the iteration condition is reached. Finally, the centers of the cells are set as the representative colors, and all the pixels in the original image are grouped to the representative colors. The popularity algorithm is provided based on the histogram of the images to find K colors with the highest occurrence rate as the representative colors [7], but some colors in the image with smaller occurrence rate will be eliminated under this algorithm. The octree algorithm [8] is another typical splitting algorithm, it use the terminal nodes of the octree as the representative colors for the image quantization.

The clustering algorithm performs clustering the color space into K-desired clusters, and there are a lot method provided to complete the cluster process, such as K-means, C-means, Maximin-distance and so on [9-12]. All the clustering process involves an initial selection of cluster centers followed by repeatedly updating them (corresponding representative colors). The final representative colors are related to the original images' color distribution as well as the clustering methods. Of course, the clustering algorithm can give suboptimal solutions at the cost of long execution times.

The next step for the image quantization is to match the pixels in the original image with the representative color, and the nearest-neighbor search finding the nearest representative. The distance between the color in each pixel and the representative color is related to the color space and the calculation process.

All the image quantization methods are not suitable for the filling color picture product, and most of the quantization is focused on global color distribution, which the details information with lower occurrence can not be easily set out. So a new image quantization method with more efficiency is provided in this paper to shorten the filling color picture production.

2 Proposed Color Quantization Algorithm

Since the clustering method based on the global image is time consuming and will eliminate some minor parts of the image, a new algorithm is provided to solve this problem. As the image prepared for the filling color product should have some objects in it, such as the natural scenery, people, animal, even cartoon characters and so on, and the colors are provided to express them. Generally, every object can not be expressed by only one or two colors in a clear image, and a color group should be carried to finish this tasks. The color delegates are separated from the image based on its occurrence rate, and then each object will has its own color delegates. After this process, the delegates are grouped with some clustering process, and the size of the object won't be an influential factor to the cluster result. The colors in higher occurrence rate will have the same status with those in the lower occurrence rate, and the following clustering method can preserve all the information from the objects efficiently.

The structure about the new color quantization algorithm is composed with the followings: filtration step to eliminate absolute pixels, making the statistic about the every color occurrence rate in the filtered images and setting a threshold on the occurring times to select the prepared color information, using clustering method to find the representative colors, matching the pixels in the filtered images with the representative colors. The detail about each step is discussed next.

A. Median Filtering Process

The first step is to eliminate the noisy pixel in the original picture. The production about the filling color picture is originated from the picture selection, and there should be have some absolute pixels colors on the picture. This kind of pixels should not appear in the final product because it is impossible to filling colors in the absolute pixels, and it also will add burden to the following statistic work and clustering work, so the media filtering is carried to smooth this kind of pixels [13]. The media filtering formula is shown as followings:

An $M \times M$ template is chosen for the median filtering, where M should be odd. The pixels occupied by the template are ordered according to its color values, and then median value in the order is set as the value of the pixel in the center of the template. After the template scanned the whole image, the median filtering process is completed and the filtered image can be obtained.

Since the original picture is true color image, so the median filtering should be carried with the three channels R G & B separately. After this process, the absolute pixels will be smoothed by their neighbor pixels.

B. Threshold Selection Based on the Statistic about the Colors

This step is carried to select some color delegates for the following analysis. The threshold is set to border all the color categories in the filtered images, and the occurrence rate of the colors bigger than the value of the threshold can be selected as the delegates for the following calculation. The threshold can be adjusted according to the image characteristic and the requirements from the designers. Generally speaking, the bigger threshold value will eliminate some details in the images and will make the image with bigger color blocks. If the details in the image should be preserved clearly, the value of threshold can be set as 0, and all the colors left in the filtered image can be carried to the following clustering process.

As the picture prepared for the filling color process should express some topic, such as people, scenery, activities and so on, the colors distributed in the image should be multiplicating. The clustering process based on the color delegates will be much faster than that based on the global images, and can acquire the cluster centers frequently even all the colors are set as delegates (threshold value is set as 0).

C. K-means Clustering Process

In order to fasten the clustering speed, the simplest clustering method K-means method is carried to cluster the color delegates. The K-means algorithm can be expressed as followings:

Given a set of n data patterns, $X = x_1, \ldots, x_n$, the K-means algorithm is to minimum the target E, which is the distance sum of all the data to the K-cluster centers.

$$E = \sum_{i=1}^{k} \sum_{x \in Q_j^i} D(x, u_j^{(i+1)}) \tag{1}$$

where K is the number of cluster centers, and D is the distance between the data and the cluster center $u_j^{(i+1)}$ after i+1 times iteration, and Q_j^i is the corresponding groups based on the cluster center.

A solution of the objective function E can be obtained via an iterative process as followings:

Step 1: $\forall u_1^{(1)} \cdots u_k^{(1)}$

Step 2: $x \in Q_l^{(i)}$ if $D(x, u_l^{(i)}) < D(x, u_j^{(i)})$ $(j = 1, \cdots K, l = 1, \cdots K, j \neq l)$

Step 3: $u_i^{k+1} = \frac{1}{N_j} \sum_{x \in Q_j^{(i)}} x$, where N_j is the number of x in the $Q_j^{(i)}$ cluster.

Step 4: if $u_j^{(i+1)} = u_j^{(i)}$, the iterative process will be ended, otherwise go to step 2 to continue the iteration.

During the K-means process, there are two important factors, one is the cluster center number K, and the other is the distance D. The value of k is equal to the final product requirements about the color number, but the distance D should be discussed in detail.

The distance between colors is influenced greatly by the color space, and the distance under different color space can be ranged very much, which will greatly influence the final clustering result. The systematical color space is RGB color space, but it is not a very good color space to match the people vision experience. The HSB color space with Hue, Saturation and Brightness is proposed to calculate the distance between colors.

In the HSB model, Hue is a color attribute that describes what a pure color (pure red, yellow, or blue) is. As hue varies from 0 to 360°, the corresponding colors vary from red, through orange, yellow, green, blue, and magenta, back to red. Saturation is a measure of the degree to which a pure color is diluted by white light. Greater values in the saturation channel make the color appear stronger. Lower values (tend to gray) make the color appear much washed out. As saturation varies from 0 to 1.0, the corresponding colors vary from unsaturated (shade of gray) to fully saturated. Brightness varies from 0 to 1.0, the corresponding colors become increasingly brighter.

The transformation from RGB to HSV color space is accomplished through the following equations:

$$\max = \max(r, g, b)$$
$$\min = \min(r, g, b)$$
$$\text{if } \max = \min$$
$$B = \max$$
$$S = 0$$
$$H = \text{unidentified}$$
$$\text{if } \max \neq \min$$
$$B = \max$$
$$S = 1 - \frac{\min}{\max}$$
$$H = \begin{cases} \cos^{-1}\left[\dfrac{(r-g)+(r-b)}{2\sqrt{(r-g)^2+(r-b)(g-b)}}\right] & b \le g \\ 360^\circ - \cos^{-1}\left[\dfrac{(r-g)+(r-b)}{2\sqrt{(r-g)^2+(r-b)(g-b)}}\right] & b > g \end{cases} \quad (2)$$

In the color area, the calculation about the distance in the HSB model should be modified since the color space is a circular cone. The pixels' coordinate in HSB color space (h, s, b) would be transformed to (c_1, c_2, c_3) to cal the Euclid distance with the following equations [14]:

$$\begin{aligned} c_1 &= s \cdot \cos(h) \\ c_2 &- s \cdot \sin(h) \\ c_3 &= b \end{aligned} \quad (3)$$

$$d(x_k, x_i) = \sqrt{\sum_{j=1}^{3} (c_{kj} - c_{ij})^2} \quad (4)$$

The cluster centers acquired from *K*-means clustering method based on (c_1, c_2, c_3) should be easily inverse transformed to HSB color space according to equation (3), and the representative colors is expressed in HSB color space.

D. Color Matching Process

In this process, the pixel's color should be expressed in HSB color space prepared for the matching with the representative colors. Of course, the color expressed in HSB should be transformed to (c_1, c_2, c_3) according to equation (3), and the distance between the pixel's color and representative color should be calculated as equation (4). The pixel's color is grouped as the step 2 in *K*-means algorithm, and all the pixel's color is replaced by the corresponding representative color in its group.

Then the filtered image can be redrawn and can be expressed in the setting number of colors. Generally speaking, the redrawn image can not satisfy the final production as some details or similar colors are combined, but the redrawn image can provide the designers some useful information to modify the original image in the designing software such as Photoshop. The algorithm can be carried on the modified image repeatedly until the final redrawn image can reach the products requirements. The quantization algorithm is efficient and only takes several minutes to acquire the redrawn images. Compared with the several days development cycle in manual designing, this process is more efficient.

3 Experiment Result

A complicated true color images is prepared for the experiments, and the original image is shown in figure 1.

Fig. 1. The original image

After median filtering process with M = 5, the figure is smoothed as figure 2.

Fig. 2. The image after median filtering with R,G & B three channels

All the colors in the filtered images occurrence over 20 times is selected as the delegates colors, and the *K*-means is carried based on the delegates colors with *K*=36. The image can be quantized in 36 colors, and the quantized image is shown in figure 3.

Fig. 3. The quantized image

In figure 3, the sky part as well as the mountain part is quantized perfectly, but it is still not suitable for as the final color filling product. Some details in the image should be modified, such as the roof and bottom of the car, the shades of sky in mountains, the shade of the car on guard bar (lower left part of the image) and so on, because the corresponding parts in original image and the filtered image are also with smaller colors difference. The designers can modify the corresponding parts exposed in the quantized image, and the quantization process can be provided on the modified image repeatedly until the image satisfies the final product requirements.

During the whole quantization process, there three adjustable parameters: the median filtering template size, the threshold for the delegate colors and the number of cluster centers in K-means algorithm, which can be adjusted by the designers according to requirements of the final products.

4 Conclusion

A fast quantization algorithm was provided in this paper to shorten the filling color picture product development cycle. In this algorithm, the original image is filtered with median filtering method to eliminate the isolated pixels' colors, and then the color of the picture is separated according to its occurrence rate and is prepared to make clustering analysis with *K*-means algorithm. After the clustering process, the representative colors can be obtained, and every pixel's color is substituted by the nearest color in the representative colors. The distance parameter, important to the color matching process and the K-means clustering process, is calculated in HSB color space to simulate the people's vision experience. The redrawn image can give adequate information to the designer to rectify the image and will accelerate the development cycle about the filling color picture products.

References

1. Li, R.Y., Kim, J., Al-Shamakhi, N.: Image compression using transformed vector quantization. Image and Vision Computing (20), 37–45 (2002)
2. Wu, Y.-G., Wu, C.-H.: Image vector quantization codec indices recovery using Lagrange interpolation. Image and Vision Computing (26), 1171–1177 (2008)
3. Rovetta, S., Masulli, F.: Vector quantization and fuzzy ranks for image reconstruction. Image and Vision Computing (25), 204–213 (2007)
4. Shelley, P., Li, X., Han, B.: A hybrid quantization scheme for image compression. Image and Vision Computing (22), 203–213 (2004)
5. Ce, Z.: A new subsampling-based predictive vector quantization for image coding. Signal Processing: Image Communication (17), 477–484 (2002)
6. Sirisathitkul, Y., Auwatanamongkol, S., Uyyanonovara, B.: Color image quantization using distances between adjacent colors along the color axis with highest color variance. Pattern Recongnition Letters (25), 1025–1043 (2004)
7. Heckbert, P.: Color image quantization for frame buffer display. ACM Trans. Computer Graphics (SIGGRAPH) (16), 297–307 (1982)
8. Gerrautz, M., Purgathofer, W.: A simple method for color quantization: octree quantization. Graphics Gems, 287–293 (1990)
9. Celeng, M.: A color clustering technique for image segmentation. Computer Vision Graphics Image Process (52), 145–170 (1990)
10. Goldberg, N.: Color image quantization for high resolution graphics display. Image Vision Comput. (9), 303–312 (1991)
11. Hsieh, I.S., Fan, K.C.: An adaptive clustering algorithm for color quantization. Pattern Recognition Letter (21), 337–346 (2000)
12. Lim, Y.W., Lee, S.U.: On the color image segmentation algorithm based on the thresholding and the fuzzy C-means techniques. Pattern Recognition (23), 935–952 (1990)
13. Sun, T., Neuvo, Y.: Detail-preserving median based filters in image processing. Pattern Recognition Letters (15), 341–347 (1994)
14. Wang, T., Hu, S., Sun, J.: Image retrieval based on color-spatial feature. Journal of Software (13), 2031–2036 (2002) (in Chinese)

Production Administration Management System in Network Environment*

Yadong Fang[1] and Lei Zhang[2]

[1] The Institute of Mechanical and Electrical Engineer,
Xi'an Technological University Xi'an,
ShaanXi Province, China
[2] International Business College Qingdao University Qingdao,
ShanDong Province, China
Yadong_Fang@yeah.net

Abstract. With more and more construction of manufacturing enterprise information, manufacturing execution system in plant will become increasingly important. In the paper, business process of plant production management is introduced firstly, and then production administration management system architecture is also discussed. More ever, system analysis and design process is illustrated by orient object method. Lastly, key technology of production administration management system are discussed, including generation technology of dynamic structural tree, generation technology of report based on Web and gantt generating method based on JFreeChart.

Keywords: Production management, Dynamic Structural Tree, Report, Gantt.

1 Introduction

In the 21st century, Computer networked technology developing make it possible to realize information transmit and exchange for modern manufacturing enterprise. The paper researches manufacturing execution system technology, and develops production administration management system based on network for metal cutting machine shop. Sectors related to workshop production management mainly include production department, material supplying department, and enterprise technology center, etc. The relation of departments is shown as Fig.1, and main business process of plant production management is illustrated as Fig.2.

* This work is partially supported by ShaanXi Province Education Department Natural Science Project Grant # 09JK479 and ShaanXi science technology research & development planning project Grant # 2010K01-076 to Yadong Fang and ShanDong Province Natural Science Project Grant # Q2008H01 to LeiZhang.

J. Zhang (Ed.): ICAIC 2011, Part V, CCIS 228, pp. 405–412, 2011.

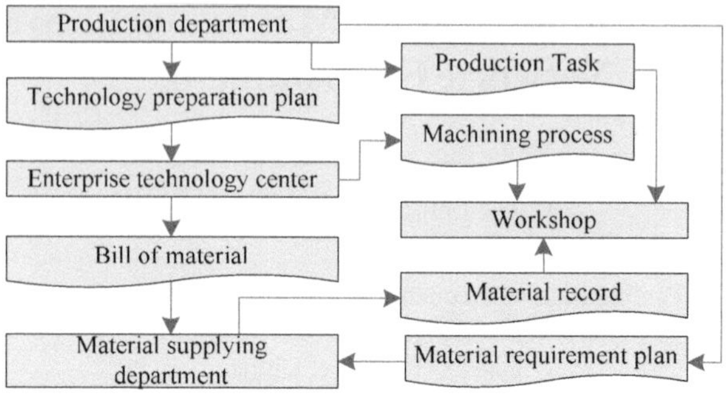

Fig. 1. Relation diagram of related departments

2 Production Administration Management System Architecture

The framework of production administration management system should be open and extensible in terms of dynamic and scalable feature for production management and it is shown as Fig.3. From analysis of above diagram, system can be divided into four layers, i.e., client layer, request receiving layer, business logic layer and data storage layer. The system general function is as follows:

(1) Enterprise personnel or user info management and configuration of system role and operation are provided.

(2) Product data information and enterprise organization information management is realized by tree structure component.

(3) Manufacturing task in network environment new, edit, delete and allocation is supported, and part process information import and interactive management are provided.

(4) Allocated manufacturing task and processing route are selected, sorted, scheduled and dispatched by dispatcher.

(5) The status information of equipment resource and tool resource can be tracked in the system.

(6) Equipment load and dispatching info can be shown in the form of Gantt chart and report, and man-hour analysis, resource cycle counting and processing card report are output.

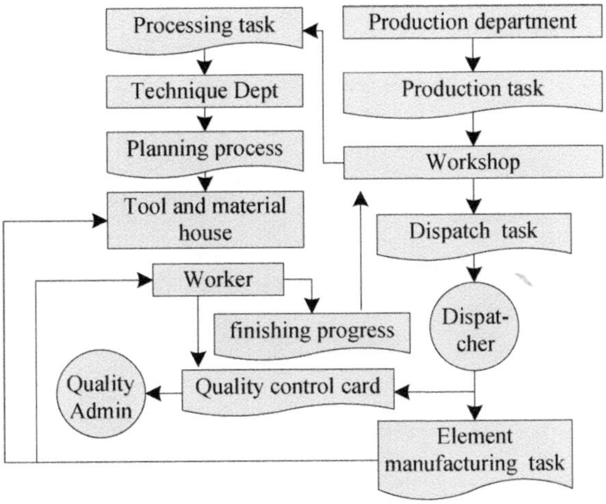

Fig. 2. Business process of plant production management

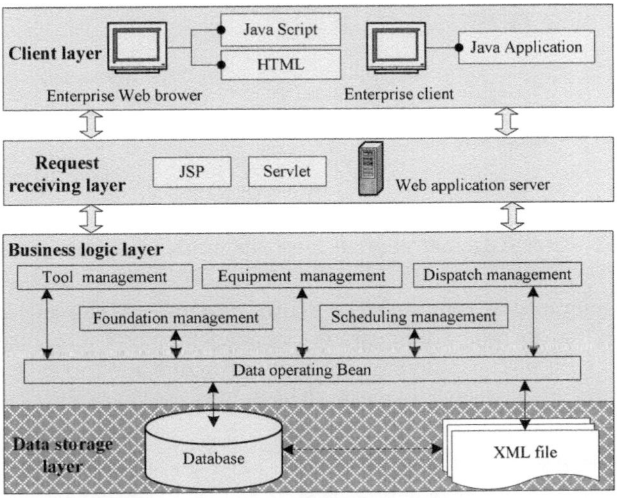

Fig. 3. Framework of production administration management system

3 System Analysis and Design

The function analysis for system is described by Rational Rose software in UML method. The actor of system includes system administrator, dispatcher, technologist, workshop directors, group leaders, quality inspector, equipment administrator and tools stock controller. The use case diagram in UML is shown as Fig.4, and function of system is made up of dispatch management, foundation data management, query and stats, resource management, process management, quality treatment and so on. Different function is operated by different user according to their operation privilege.

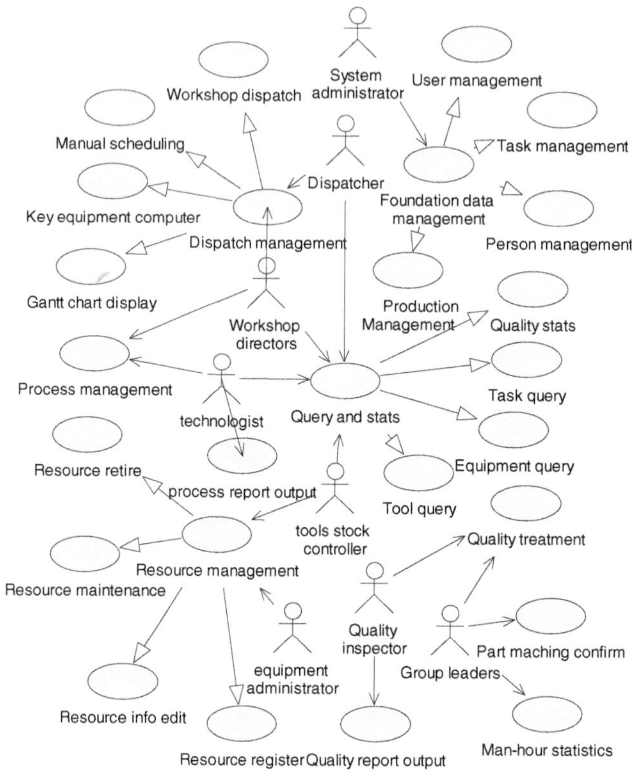

Fig. 4. Use case diagram of production administration management system

Sequence diagram is represented as two dimension diagram according to its interaction. In production administration management system, user just goes to system online login screen and enters correct user name and pin to operate authorized function. The general sequence diagram of system is shown as Fig.5.

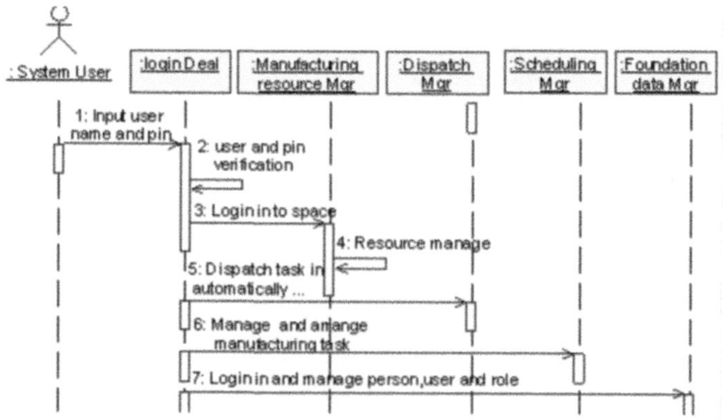

Fig. 5. General sequence diagram of production administration management system

4 Key Technology of Production Administration Management System

A. Generation Technology of Dynamic Structural Tree

The information of tree node can be acquired from back-end database, and then structural tree is constructed dynamically. In database, node information table for structural tree involves identification number, child node name, parent node name, node statue and so on. The generation process of dynamic structural tree is shown as Fig.6.

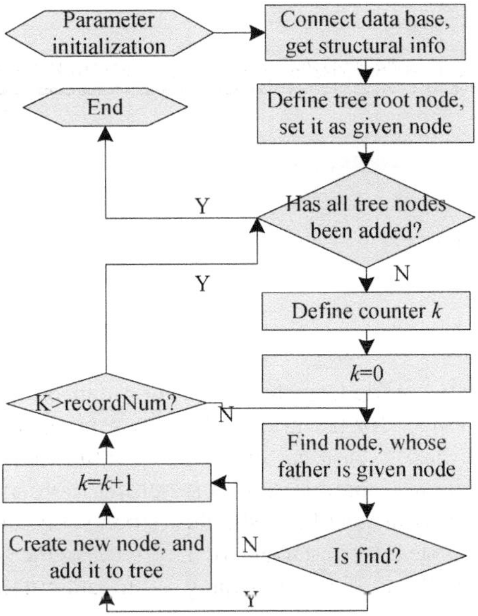

Fig. 6. The generation process of dynamic structural tree

B. Generation Technology of Report Based on Web

JSP can resolve dynamic structural tree generation problem for its combining Java and HTML language. Therefore, generation process of report style and dynamic data in JSP is introduced by taking technological process card for instance. Report can be divided into table header, body and footer in descriptions of report template. Three constituent part of report is described respectively by HTML tag <*table*>, <*tr*> and <*td*>. In defining table or its cell, the border and color of report is described by attribute *border*, *cellpadding*, *cellspacing* and *style*, such as <table width="100% border="0" cellpadding="0" cellspacing= "0" style="border-left:3px #000000 solid" >. Report template of technological process card is shown as Fig.7. The dynamic data of report is acquired from data base by Java data operating class. In JSP file, Java relevant package or class, such as *java.sql* is imported in terms of <%@

page import="java.sql. %>*. In addition, command *page* means package *java.sql* is imported for whole JSP file page. Tag *<jsp:useBean>* tells Web server how to identify and call one Java Bean. Taking *<jsp:useBean id="DataOpeBean" scope="session" class=" capdsBean. DataOpeBean"/>* for instance, *DataOpeBean* is one object of class *capdsBean. DataOpeBean*, and conventional data base operating methods by SQL Statement are encapsulated in the class. The class diagram of data operation class is shown as Fig.8. Further more, dynamic data from database is displayed in page by Java method *out.println(String "displayContent")*.

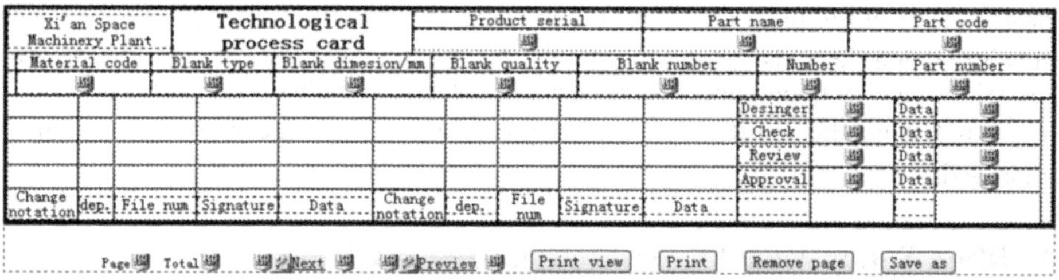

Fig. 7. Technological process card report template

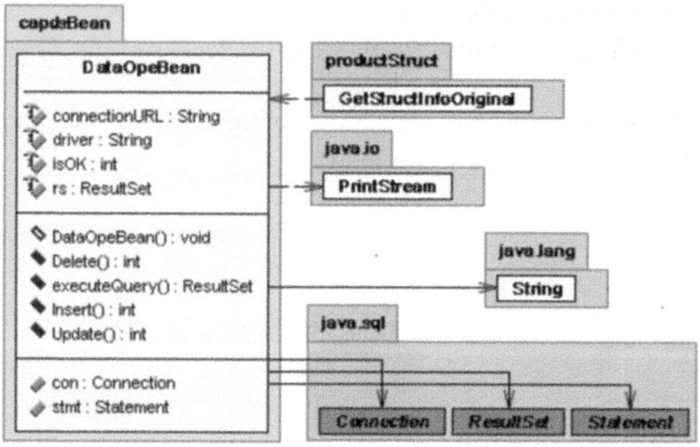

Fig. 8. Class diagram of data operation class

C. Gantt Generating Method Based on JFreeChart

JFreeChart is Java chart engine, which is open source project of *SourceForge.net* website. Package *JFreeChart.jar* and *Jcommon.jar* are deployed in J2EE application server, and gantt chart based on web can be realized by their API. Various charts, such as pie chart, column chart, line graph, area map and gantt chart are developed by the component. There are *ChartFactory*, *ChartUtilities*, *JFreeChart*, *renderer*, *plot*, *axis* class in package *org.jfree.chart*, and package *org.jfree.data* include class *gantt*, *category* and so on. The accession of gantt data is realized by class *ChartDataSet*, and its class diagram is shown as Fig.9.

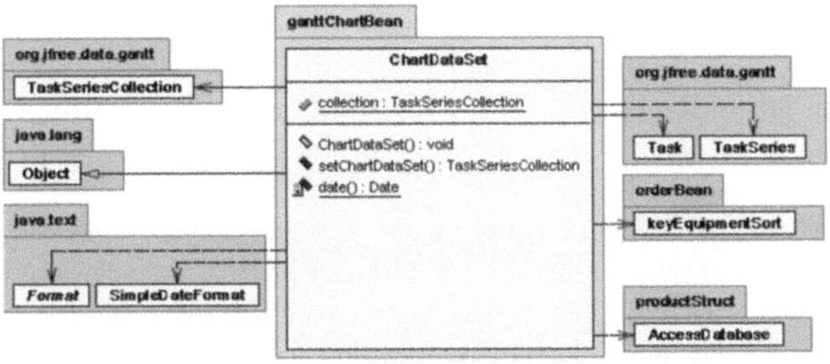

Fig. 9. Class diagram of gantt data accession

Class *AccessDatabase* is used to operate production administration management system database to get dynamic information, such as manufacturing task, equipment and group. In addition, class *keyEquipmentSort* is used to determine key equipment in task dispatch process, and gantt chart for key equipment of loading conditions is illustrated as Fig.10. Class *ChartDataSet* is described in detail as following:

(1) Key equipment code is computed by class *keyEquipmentSort*, and dispatched manufacturing task information, such as task code, begin time and end time is queried by class *AccessDatabase*.

(2) Desired data for gantt diagram display is constructed by class *Task*, *TaskSeries* and *TaskSeriesCollection*. Task instance is created by class *Task*, such as *t*1, *t*2 and *t*3, and operating time of equipment for each task is also put into the instance. Secondly, task series instance s1 is created by class *TaskSeries*, task instance *t*1, *t*2 and *t*3 are added to s1.

(3) Task series collection object is created by class *TaskSeriesCollection*, and it provides realization of gantt series collection interface. In class *TaskSeriesCollection,* method setChartDataSet returns gantt series collection.

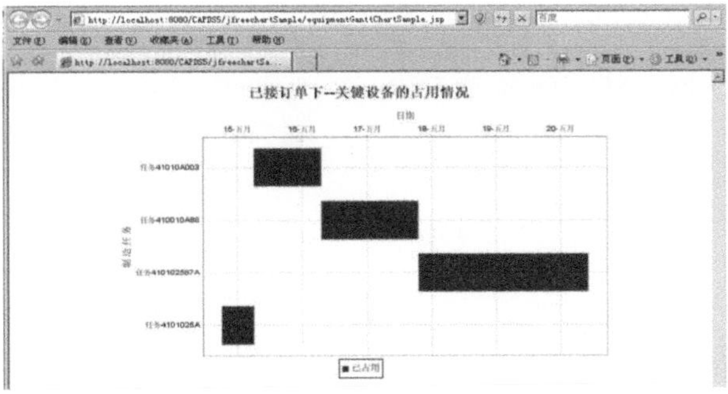

Fig. 10. Gantt diagram of key equipment operating state

1) Paper Size: Prepare your paper in full-size format on US letter size paper (8.5 by 11 inches).

2) Type Sizes and Typefaces: Follow the font type sizes specified in Table I. The font type sizes are given in points, same as in the MS Word font size points. Times New Roman is the preferred font.

3) Paper Margins: Paper margins on the US letter size paper are set as follows: top = 0.75 inches, bottom = 1 inch, side = 0.625 inches. Each column measures 3.5 inches wide, with a 0.25-inch gap between the two columns.

4) Paper Styles: Left- and right-justify the columns. On the last page of your paper, adjust the lengths of the columns so that they are equal. Use automatic hyphenation and check spelling and grammar. Use high resolution (300dpi or above) figures, plots, drawings and photos for best printing result.

5 Conclusions

In order to improve manufacturing plant information condition, the paper discusses manufacturing execution system technology for discrete manufacturing in detail, and production administration management system development process is also introduced. The theory and method discussed in the paper provides reference for machine manufacturing enterprise to realize plant information.

References

1. Fang, Y.D., Kang, P., Zhang, L.: Research and application of machine tools life-cycle management system based on web. In: 2009 Second ISECS International Colloquium on Computing, Communication, Control, and Management (CCCM 2009), pp. 117–120. IEEE Press, Los Alamitos (2009)
2. Lee, B.-E.: An architecture for ubiquitous product life cycle support system and its extension to machine tools with product data model. International Journal of Advanced Manufacturing Technology 42, 606–620 (2009)
3. Song, H.: Manufacturing Execution System based on Web in Networked Collaboration Eenterprise. Manufacturing Automation 23(2), 20–23 (in chinese)
4. Wang, Z.: Package Manufacturing Execution System and its Application. Press of China Power, Beijing (2006) (in Chinese)

Research and Design of a MSD Adder of Ternary Optical Computer*

Hui He, Junjie Peng, Yanping Liu, Xianchao Wang, and Kai Song

School of Computer Engineering and Science Shanghai University,
Shanghai 200072, China
huihe0009@sina.cn

Abstract. The further improvement of basic theory of ternary optical computer (TOC) and the implementation of thousand-bit ternary optical experimental system have promoted the rapid development of TOC. However, lack of adder has confined the applications of TOC to logic operations. The research and design of the adder which is the core component of TOC has become an urgent research topic. Based on the research of feedback strategy, the paper presents the design of a TOC adder based on Modified Signed-Digit (MSD) number system. Combined with the characteristics of thousands bits of the experimental system, the adder is implemented based on the arithmetic unit which is constructed following MSD regulations based on decrease-radix design together with data feedback strategy. Experiment results show that the structure and design of the adder is correct and it has many advantages over the traditional adders with carry.

Keywords: Ternary Optical Computer, Adder, MSD Number System, Data Feedback.

1 Introduction

With the increase of the integration of electronic circuits and the restrictions of the physical limits of semiconductor materials, the improvement of the performance of electronic computer has become harder and harder. However, more and more complex scientific problems have much challenge on the capability of electronic computers. This make many researchers and engineers have to shift their attention to the computers in all possible new areas, in which optical computer is one of the most popular focus. Generally, optical computing has at least the one obvious advantage. The spatial parallelism of the light makes the optical computer have a large number of data bit, so the optical computer can own a higher data width than the electronic computer.

---

* This work is partially supported by the innovation project of Shanghai Municipal Education Commission (No. 11YZ09), Shanghai Leading Academic Discipline Project (No. J50103), the Fundation of Key Laboratory of Computer System and Architecture, Institute of Computing Technology, Chinese Academy of Sciences. the innovation project of Shanghai University (No.SHUCX102174).

J. Zhang (Ed.): ICAIC 2011, Part V, CCIS 228, pp. 413–420, 2011.

The progress of thc research and development of TOC [1] has been obtained since it was put forward in 2000, especially, the proposing of the decrease-radix design theory [2] has much contributed and accelerated the process of the progress. Finish building the thousand-databit ternary optical logic processor has set the research of the TOC into a new era. The adder is still the core component of the ternary optical computer, but the adder of the TOC is not implemented. So the research and realization of the adder of TOC is to be an increasingly urgent and important topic. In 2010, Jin Yi et al put forward the design theory [3] of the adder, basing on the predecessor's research of MSD (Modified signed - digit) digital system [4]. In the study, they used the symbol substitution rules of MSD notation system to avoid the addition of serial carry without need consideration of carry delay. However, the design and implementation of the MSD adder is still to be sovled. This is very important to the research and applications of TOC into the numerical computation field.

Based on the research of the data feedback strategy MSD notation system and combined with the design principle of the adder of TOC, this paper puts forward a design scheme of the adder. The second section presents the adder hardware construction principles, section three mainly introduces the implementation scheme of the design of adder. Section four is the experiment and the results which demonstrates the correctness and feasibility of the scheme. Section 5 is the conclusion remarks.

2 The Design Principle of the Adder

A. The Principle of the MSD Adder

In 1961, Algirdas Vizienis etc. first proposed SD digital system [5], then got in-depth research on it. However, the MSD digital system has never been used in digital computer system because of the wide use of the binary digital system and limit bits of numbers in digital computer system, B.L Drake et al brought it into the field of light calculations in 1986 [6], and studied its optical implementation and logical structure. Later, various transformations and improvements were brought into MSD digital system, which also were used to study optical computing, optical calculators and computers by researchers.

The MSD digital system implements carry-free addition operations by alternating sequence of the five MSD's transform rules in turn, which can effectively avoid the transmission of addition carry. That is, at the first step in an MSD addition, using T, W transform to the addend and augend, then the results of the T and W are used as inputs to execute T ', W' transform, and ultimately results of addition operations is got by the T rule transform with T', W' transform results as input. To implement the addition process, we can design four rules by decrease-radix design theory and control the addition in pipeline manner if there are multiple addition applications. Since the MSD number crunching process, the T transform will be used twice in each addition, the paper only demonstrates the optical design rule and architecture of T transform, all the design of other transforms is quite the same as the T transform does.

Based on decrease-radix design principle, we can design the optical architecture of the T transform as figure 1 shows. For more details about the transforms, we would to recommend the interested readers refer the literature.

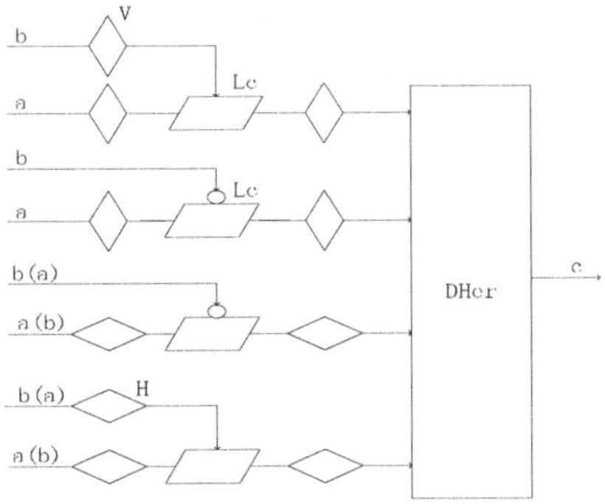

Fig. 1. The optical of the T rules' transform

In the fig.1, a and b represent two bit data which will be added, the vertical diamond which indicated a symbol V represents the vertical polarizer which only can pass the vertical polarized light. The horizontal diamond expressed by symbol H represents the horizontal polarizer which only can pass the horizontal polarized light, the Parallelogram represents constant rotating LCD, which can change the polarization of light when the control signal is 1. The circle represents non-logical, the recessed control can make up the output light, c represents one bit output data, there is only one output in the all of the operators T converter, which can output the results of T transform by the recessed control.

B. The Architecture of the Addition Units

Ternary optical computer is a photoelectric hybrid computer, using horizontal polarized light, vertical polarized light, dark state three independent states to express values in the TOC and use for optical processing, optical transmission, electric control. From decrease-radix design principle we know that optical processor be composed of Liquid crystal affixed by polarizer on both sides. As the polarizer on both sides is different, we can divide the optical processor into four partitions such as VV, VH, HH, HV. The four partitions of optical processor are shown in fig.2. Black rectangle stripe represents the horizontal polarizer, gray rectangle stripe represents the vertical polarizer, white rectangle represents the liquid crystal, the arrows indicate the control light signal. The hardware design of the adder is mentioned in another article.

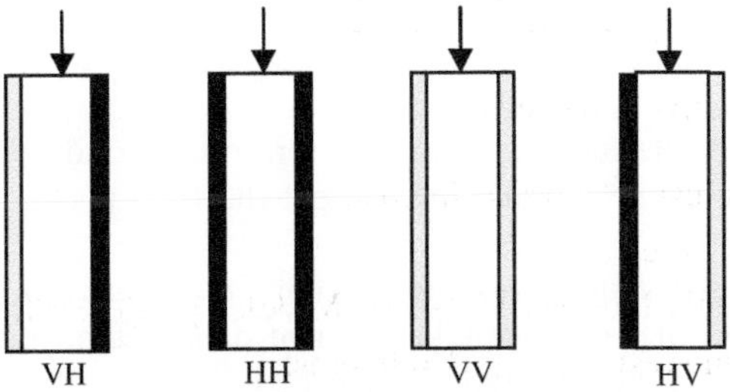

Fig. 2. Four partitions of optical processor

3 The Design and Implement of the Adder

A. The Working Process of the Adder

In view of MSD addition operation can solve the serial carry problems and has the completeness, it is the advantage that results can be involved in the next operation as input without any change expressed by numbers. Ternary optical computer adder was constructed by using MSD number. To take full advantage of LCD pixels resources, optical computing adder can be divided into five fixed logic parts, reconstructed five calculators with using reconstruction circuit technology, each part were used to complete the T, W, T ', W', T2 transformation rules separately, but the order of rules in the computing process is fixed and can't be changed. The Optical processor division is showed in Figure 3, and the division of optical processor is the same in the four partitions.

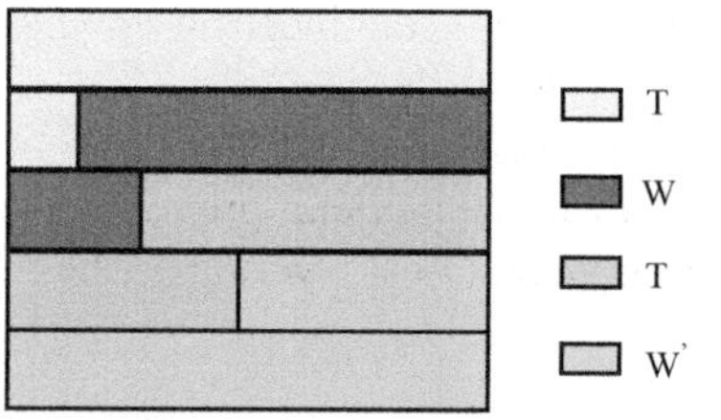

Fig. 3. The division of optical processor

Form figure 3 we can get the computing strategy of ternary optical computer' adder based on MSD which is shown as fellow:

1) Take a set of data, get the data bits aligned and then sent them into T and W rules operations, last, gain the intermediate results u and v;

2) Take out another set of data, and get the data bits aligned again. Then fill zero on the intermediate results which gain from the first set, u on low bit and v on high bit, then sent them into adder at the same time. The second group data sent into adder on T and W rules operations, gain the intermediate results m and n. The intermediate results u and v which filled bits with zero, and then sent into adder with T 'and W' rule logic operations, gain the intermediate results u' and v';

3) Fill zero on the intermediate results u' on low bit and v' on high bit, then sent them into T rules adder to operate, last we get the final results and give them to the users; then we fill zero on the intermediate results m on low bit and n on high bit, then sent them into T' and w' rules adder to operate, gain the intermediate results m' and n';

4) Fill zero on the intermediate results m' on low bit and n' on high bit, then sent them into T rules adder to operate, last we get the final results and give them to the users;

5) Judge whether the data processing is complete, if not, according to the principle of the first come and first out, we take a set of data to the first step; Otherwise, the data processing is complete, terminate data scheduling.

B. Data Feedback Strategy of the Adder

Ternary optical computer adder is through five kinds of transformation T, W, T', W' T to realize the addition with free-carry. The adder adopted the line stream structure to the adder operation, please see Figure 4. However, due to the pipelining process, T, W transformation and T', W' transformation will produce intermediate results, such as Figure 4 shows. So the problem how to conduct data feedback must to be solved in the adder. The stand or fall of data feedback strategy will directly influence the denominator MSD's speed and adder line stream working's efficiency, this paper designs the feedback strategies as follows:

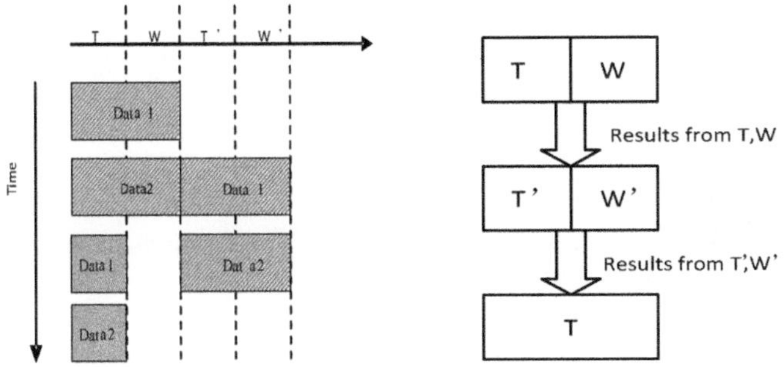

Fig. 4. The addition process of MSD

1) According to the mapping relationship between three light configuration and coded information, we convert decoding results to the corresponding coding information, and then contrapuntal combine into information encoded string;

2) According to T, W, T', W' operation's base address and offset address, we respectively store their encoded string information in memory;

3) Fill the intermediate results with 0 in the data storage space of the T, W, T', W 'operation, the strategy is seen in Table 1;

Table 1. The Strategy of Filling 0

Rule transformation	The strategy of fill 0
T	Fill 00 in lower bits
W	Fill 00 in higher bits
T'	Fill 11 in lower bits
W'	Fill 11 in higher bits

4) Then get the data communication, and sent the encoded message strings in the storage space of T operation through communication to the PC network client, after clear the information stored in storage space after completed rule transformation;

5) Respectively store the encoded information string which filled with 0 in the specified storage space, to generate the next operation control information in main optical path and controlling optical path, storage way to be seen in Table 2;

Table 2. The change address of storage space

The encoding string	The storage space
T	The storage space of the main optical from T' base address and the storage space of the control optical from W' base address
W	The storage space of the main optical from W' base address and the storage space of the control optical from T' base address
T'	The storage space of the main optical from T base address
W'	The storage space of the control optical from T base address

6) Remove the storage space of T, W, T', W' operation, wait for getting the decoding results of the next optical computing.

C. Data Synchronization in the Adder Operation

The design scheme of ternary optical computer adder is combined with the thousands welterweight characteristics of ternary optical computer, through the data feedback process will intermediate results back to restart the code segment encoder participate in next time operations. However, if there are many sets of data needing to process, the data synchronization strategies become the obstacles of the optical processor's speed. Data synchronization strategy is a synchronization method basing on optical processor free degree to real-time choice data into unit complete data processing. The synchronization mainly set a sync signal bits, when signal bit is 0, it says that the optical adder is occupied. We cannot send data into the adder; when signal bit is 1, it says that the adder is at leisure, optical data can be sent into the adder. When two sets of data occupy the optical adder on addition operation, we set synchronized signal bits 0; after the second group data complete T operation, the synchronized signal bit changes to 1. Then send synchronous signals to the PC, PC synchronized signal obtained from the queue which will have to deal with the data according to the priority and advanced first out the principle of combining remove two sets of data into optical adder on the denominator.

4 Experiment and Results

A. Experiment

The eight-bit MSD adder is constructed with pipe line manner in experiment. It is two sets of data to do experiment: 100 and 001, 010 and 001, these sets of data are sent to the optical processor to do the transformation rules for the results of addition operation in turn.

At first, the first set of data is sent to optical processor for addition, the experimental results shown in the four images at the first row of the figure 5; next the second set of data is sent to optical processor for addition, the first set of data will do T', W' rules operation, the experimental results shown in the four images at the second row of the figure 5; next the first set of data is sent to optical processor for doing T rules operation for the final result, the second set of data will do T', W' rules operation, the experimental results shown in the four images at the third row of the figure 5; until the finally addition results of the second set of data is obtained, the experimental results shown in the four images at the fourth row of the figure 6. The experimental results as follows:

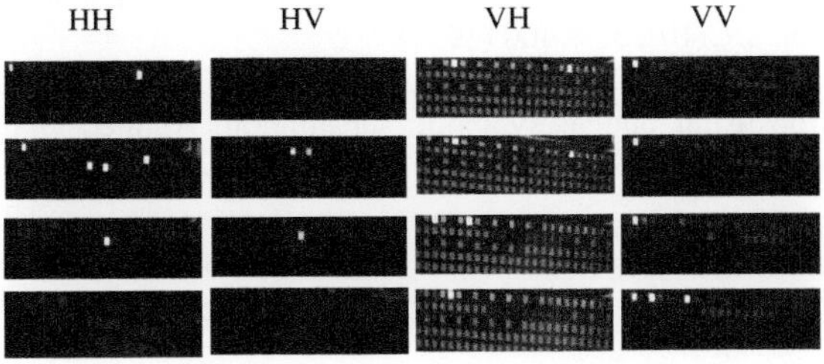

Fig. 5. The experimental results

B. Results and Analysis

The fig.5 shows that the adder can operate correctly and the structure of the adder is reasonable. We divide optical process into four partitions for constructing the rules' change, which can expand the data-bit of addition operation and play better advantage of high data bits; in addition, two sets of data can be addition operations at the same time, compared with the computer, the ternary optical computer adder have a certain parallelism, nor only improves the ability of parallel processing of data, but also improves the data capacity of fan-in-fan-out at the same clock cycle, and has a large numerical advantage.

5 Conclusion

Based on MSD ternary optical computer's adder giving full play to the light of "spatial calculation giant parallelism" and "space irrelevant sex" advantage, it made ternary optical computer to get the result of the add calculation by processing the transform of the rules, which has certain parallelism, in different extent increased the efficiency of ternary optical computer adder, and improved the ability of numerical computation of ternary optical processor, expanded the application of ternary optical computer in the areas of the numerical value. The realization of the ternary optical computer will simplify the architecture of computer system, reduce the energy consumption of the large scale of computer, so as to promote the development of the computer science.

References

1. Jin, Y., He, H., Lu, Y.: Ternary Optical Computer Principle. Science in China (Series F) 46(2), 145–150 (2003)
2. Yan, J., Jin, Y., Zuo, K.: Decrease-radix design principle for carrying/borrowing free multi-valued and application in ternary optical computer. Science in China (Series F) 51(10), 1415–1426 (2008)
3. Jin, Y., Shen, Y., Peng, J., Ding, G.: Principle and Construction of Ternary Optical Computer's MSD Adder. Science in China (Seriers F) 53(11), 2159–2168 (2010)
4. Cherri, A.K., Karim, M.A.: Modified-signed digit arithmetic using an efficient symbolic substitution. Applied Optics 27(18), 3824–3827 (1988)
5. Avizienis, A.: Signed-digit number representations for fast parallel arithmetic. IRE Trans. Electron. Comp. EC(10), 389–400 (1961)
6. Bocker, R.P., Drake, B.L., Lasher, M.E., Henderson, T.B.: Modified signed-digit addition and subtraction using optical symbolic substitution. Applied Optics 25(15), 2456–2457 (1986)

An Adaptive Digital Watermarking Scheme Based on HVS and SVD

Jun Yang[1,2,*], Jinwen Tian[1], Fang Wei[1,2], and Chuan Cheng[3]

[1] Institute for Pattern Recognition and Artificial Intelligence, Huazhong University of Science and Technology, Wuhan, Hubei, China 430074
junyang1280@126.com
[2] Wuhan Polytechnic, Wuhan, Hubei, China 430074
[3] Wuhan Maritime Communication Research Institute, Wuhan, China 430074

Abstract. Digital watermarking technique can be the basis of many applications, including media asset management and security. How to improve the practicality of the watermarking scheme in different applications is an urgent problem. In this paper, we proposed a new watermarking scheme for images based on Human Visual System and Singular Value Decomposition in the wavelet domain. The secret information that can be seen as a watermarking is hidden into a host image, which can be publicly accessed, so the transportation of the secret information will not attract the attention of illegal receiver. The experimental results show that the method is invisible and robust against some image processing.

Keywords: Digital watermarking, human visual system, singular value decomposition.

1 Introduction

With the development of communications, signal processing and computer networks, information hiding technique has attracted much attention in recent years， it becomes important in a number of application areas. For example, in military application, it is very important to send critical data in a safe method. The adversary should not be aware of transmission, because he will attempt to extract the massage. If he couldn't extract the massage, he would attempt to suppress the transmission or at least to destroy it. Data hiding is a form of steganography, digital watermarking is a specific branch of steganography, the most important properties of watermarking include imperceptibility, robustness, and security. [1-4] In this paper, we proposed a kind of wavelet domain adaptive image digital watermarking scheme using human eye visual property and Singular Value Decomposition. The secret information that can be seen as a watermarking is hidden into a host image, which can be publicly accessed, With our approach, a large number of uncorrelated, random-like, yet deterministic chaotic sequenced can be exploited to encrypt the original watermark signal and the secret

---

* Corresponding author.

J. Zhang (Ed.): ICAIC 2011, Part V, CCIS 228, pp. 421–428, 2011.

information is embedded in the wavelet domain. The wavelet coefficients of carrier image is divided into different classes by the background luminance and the texture mask characters of human eye visual property, the watermark image is embedded in the third level sub-band of frequency domain by SVD for the sake of resisting geometrical attacks. The secret information can be reliably extracted without resorting to the original carrier image. The experimental results show that the scheme is invisible and robust against some image processing such as JPEG lossy compression, median filtering, additive Gaussian noise and cropping attacks and so on.

The paper is organized as follows: In section 2, Human visual system and Singular Value Decomposition are introduced. Section 3 presents the adaptive watermarking scheme based on HVS and SVD. The experimental results, which show this method has robustness to some attacks, and discussions are given in section 4. Finally, there is a conclusion in section 5.

2 Human Visual System and Singular Value Decomposition

A. Human Visual System and Partition Based on the Background Luminance and the Texture Mask Characters of HVS

Try Human visual system (HVS) has been deeply studied recently. Similarly, it is today widely accepted that robust image watermarking techniques should largely exploit the characteristics of the HVS, the human visual masking effect, so that watermarking imperceptibility and robustness can be improved. [5, 6]

Some characteristic about human eyes have been found as follows: the eye is less sensitive to noise in high-resolution bands than low-resolution bands; the eye is less sensitive to noise in the areas of the image where background brightness is high or low; the eye is less sensitive to noise in textured areas than smooth areas; the characteristics above illuminate that a small disturb which is acceptable can be easily hidden and can't be seen by human eyes. In image watermarking algorithm, imperceptibility is one of the most important properties, which is pivotal to watermarking quality.

We know from the luminance mask characters, the brighter the background are, the higher the contrast sensitivity threshold (CST) is, the less sensitive the HVS is. According to the luminance value, we divided coefficients of noise sensitivity into sensitivity, less sensitivity and non-sensitivity.

Similarly, we know from visual mask effect, the more complex the texture is, the higher the contrast sensitivity threshold is, and the less sensitive the HVS is. We can use texture energy to measure texture complication. Refer to [8], in which Safranek classified texture energy is variance of a region and square sum of all the coefficients in this region. The calculating formula is as follows:

$$texture(\lambda,x,y)=\sum_{k=1}^{N-\lambda}16^{-k}\sum_{\theta}^{HH,HL,LH}\sum_{i=0}^{1}\sum_{i=0}^{1}\left(c(k+\lambda,\theta,i+\frac{x}{2^k},j+\frac{y}{2^k})\right)^2 \quad (1)$$
$$+16^{N-\lambda}\,\delta(c(3,LL,\{1,2\}+i+\frac{x}{2^{N-\lambda}},\{1,2\}+i+\frac{y}{2^{N-\lambda}}))$$

According to texture energy values, we classify the region of the texture into 4 categories as complex, less complex, simple and smooth.

B. Singular Value Decomposition

Number In the numerical analysis based on linear algebra, SVD is a numerical technique for diagonalizing matrices, and is a powerful tool for many applications including the solution of least squares problem, computing pseudo-inverse of a matrix and multivariate. In addition SVD has been used in image processing applications such as image coding, noise estimation and more recently in image watermarking. For watermarking applications, SVD has the following desirable properties [7, 8]:

SVD is able to efficiently represent the intrinsic algebraic properties of an image, where singular values correspond to the brightness of the image and singular vectors reflect geometry characteristics of the image.

Singular values have good stability, which means a small perturbation added to an image will not significantly change the corresponding singular values.

An image matrix has many small singular values compared with the first singular value. Even ignoring these small singular values in the reconstruction of the image, the quality of the reconstructed image will not degrade a lot.

A digital image A can be decomposed into a product of three matrices $A = U\Lambda V^T$, where U and V are orthogonal matrices, $U^TU = I$, $V^TV = I$, and $\Lambda = diag(\lambda_1, \lambda_2 ...)$. The diagonal entries of Λ are called the singular values, the columns of U are called the left singular vectors and the columns of V are called the right singular vectors. This decomposition is known as the Singular Value Decomposition of A, and can be written as:

$$A = \sum_{i=1}^{r} \lambda_i U_i V_i^T \tag{2}$$

Where r is the rank of matrix A.

3 The Proposed Adaptive Watermarking Scheme

A. Pretreatment of Watermarking

Use Let H be $M \times N$ binary image watermarking. Before embedding it into host image, we will do some previous work.

Watermarking scrambling can eliminate correlation of pixels and improve robustness of algorithm. In this paper, in order to mask meaningful content of the watermark, we use two kinds of chaotic sequence to encrypt.

①According to Logistic mapping, using secret key μ to generate chaotic sequence x_k ,and then generate gray level scrambling Matrix G through x_k.

②According to Tent mapping, using secret key α to generate chaotic sequence y_k, and generate magic cube Matrix through y_k.

③$I = H(i, j) \oplus G(i, j)$, where I is the transformed image.

④Scrambling the transformed image according to transformation of magic cube, we will get the encrypted image W.

③We use pseudorandom sequence and divide watermarking W into two parts that where represented by $W1$, $W2$. The adjusting method is: Generating a positive integer pseudorandom sequence S which has the same length with watermarking, the adjusting formula as follows:

$$W1(i) = W(S(i)) \quad W2(i) = W(S(i+L)) \qquad i = 1,2,\cdots,L \tag{3}$$

B. Watermarking Embedding

①Wavelet decomposition of original image and selection of embedding sub-bands

We use wavelet transform because it matches with multi-channel model of human visual system, With a N-level wavelet transform, the host image is decomposed into a low frequency sub-band LL_i and vertical and horizontal components HL_i, LH_i, HH_i (i = 1, 2, ……, N) according to each level. LL_i sub-bands are not suitable to embed watermarking because they are approximation of original image, if watermarking embed into these bands then the image will distort. Although embedded watermarking at detailed images HL_1, LH_1, HH_1 is imperceptible, there are lots of compress algorithms which also select these sub-bands to compress images. Perform the third discrete wavelet decomposition of the original image to transform an image into wavelet domain. To enhance the robustness and transparent of the embedded watermarking image, we embed the watermarking into the third level sub-band HL_3, and LH_3, half of watermarking at sub-band HL_3 and LH_3 respectively.

②embedding watermarking

Watermaking will be embedded into host image by SVD, in which some modifications are made:

$$A = U_1 \Lambda_1 V_1 \tag{4}$$

$$B = \Lambda_1 + amW \tag{5}$$

$$B = U_2 \Lambda_2 V_2^T \tag{6}$$

$$A_w = U_1 \Lambda_2 V_1^T \tag{7}$$

③After the inverse discrete wavelet transform, we can get the watermarked image.

C. Watermarking Extraction

Watermarking extraction algorithm is the inverse process of the watermark embedding algorithm.

①Three-level wavelet decomposition of the embedded watermark image.

②Classification calculation of background luminance and texture complication of HL_3 and LH_3 sub-bands embedding points.

③The watermark can be extracted using the inverse procedure of embedding:

$$A_w^* = U_2^* \Lambda_2^* V_2^* \tag{8}$$

$$W' = \frac{U_2 \Lambda_2^* V_2 - \Lambda_1}{am} \tag{9}$$

④Combination watermarking $W'1$ and $W'2$.

Pseudorandom number S will generate using the pseudorandom seeds in embedding algorithm. Watermarking sequence W' can be combined as following formula:

$$W'(S(i)) = W1(i)\,,\quad W'(S(i+L)) = W2(i)\,,\quad i = 1,2,\cdots,L \tag{10}$$

After inverse encrypted process W', we'll get the decrypted H'.

4 Experimental Results and Discussion

The The experiment has been performed on a binary watermarking image of size (64×64), into an original image of size (512×512).The similarity measurement [9] is introduced to provide objective judgment of the extraction fidelity, the similarity measurement between original watermarking H and extracted watermarking H' is defined as follows:

$$Sim(H,H') = \frac{\sum_{i=1}^{M}\sum_{j=1}^{N} H(i,j)\cdot H'(i,j)}{\sqrt{\frac{W'}{\sum_{i-1}^{M}\sum_{j-1}^{N} H'^2(i,j)\cdot \sum_{i-1}^{M} H^2(i,j)}}} \tag{11}$$

Given a adjusting threshold ΔT, referring to document of Cox [10]. It is known that the probability of false judgment is the same as the one that Gauss distributed random variable exceeding its mean by more than $\Delta T \times \sqrt{\sum_{i=1}^{M}\sum_{j=1}^{N} W0^2(i,j)}$ times variance. The probability of Gauss distributed random variable exceeding its mean by more than six times variance is $2\int_6^{\infty} \frac{1}{\sqrt{2\pi}} \exp(-\frac{x^2}{2})dx = 1.973\times 10^{-9}$, so when $\Delta T \times \sqrt{\sum_{i=1}^{M}\sum_{j=1}^{N} W0(i,j)^2} \ge 6$, the probability of false judgment is approximate to zero.

Fig.3 shows the watermarked image. The quality of the watermarked image approximates to the original carrier image. The watermarking which is extracted directly from the watermarked image is same as the original watermark and the Sim between them is 1.

The robustness test of the watermarked image against common attacks has been performed. Fig.5 illustrates relation of the correlation coefficient Sim to the quality factor of compression when the watermarked image is attacked by JPEG lossy compression. The extracted watermark image is shown in Fig.4 while the quality factor equals to 20 and compression proportion is 22 to 1.

Fig. 1. Original image

Fig. 2. Binary watermarking image

Fig. 3. Watermarked image

Fig. 4. Extracted watermarking

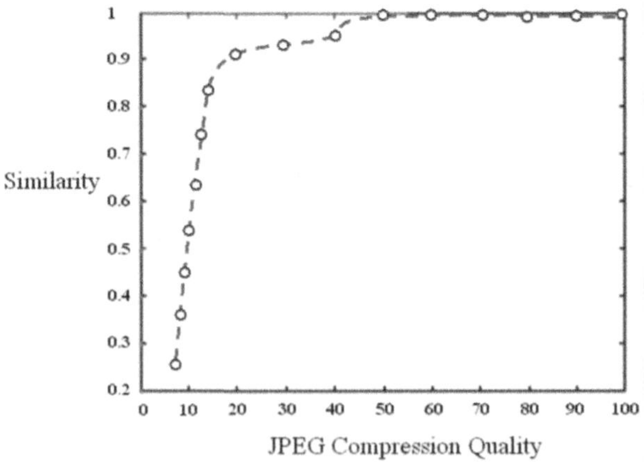

Fig. 5. Robustness testing of JPEG lossy compression

Table 1 shows the extracted watermark image and the correlation coefficient *Sim* between the original watermark and the retrieved one when the watermarked image is attacked by additive Gaussian noise, median filtering, cropping, Gaussian smoothing and grads sharpening.

The figures and the tables of the experimental results show that the embedded watermark can be retrieved under the attacks of common image processing, and the

Table 1. Experimental results with various types of attacks

Acctack types	Extracted watermark	Sim value
3×3 median filtering		0.9103
5×5 median filtering		0.6812
cropping 1/16		0.9046
cropping 1/4		0.7539
Magnify 1024×1024		1.0000
Gaussian smoothing		0.9208
Gaussian white noise (0,0.002)	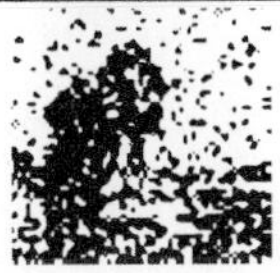	0.7476

correlation coefficient between the original watermark and the retrieved one is greatly larger than the correlation threshold ΔT . Hence this watermarking scheme has strong robustness.

5 Conclusion

In this paper, a wavelet domain adaptive image watermarking scheme based on HVS and SVD is proposed. We use the chaotic sequence to encrypt the watermarking in order to improve the security, use the HVS characters to improve imperceptibility, und SVD technique to resist the geometrical attack, experiments show that the method has strong robust against some actual attacks and can be applied.

References

1. Bender, W., Gruhl, D., Morimoto, N., Lu, A.: Techniques for data hiding. IBM System Journal 38(4), 313–346 (1996)
2. Katzenbeisser, S., Petitcolas, F.A.P.: Information hiding techniques for steganography and digital watermarking, pp. 57–112. Artech House, Norwood (2002)
3. Oppliger, R.: Security Technologies for the World Wide Web. In: Oppliger, R. (ed.) Intellectual Property Protection, 2nd edn., pp. 347–357. Artech House, Norwood (2003)
4. Petitcolas, F.A.P., Anderson, R.J., Kubn, M.G.: Information hiding-a survey. Proceedings of the IEEE 87(7), 1062–1077 (1999)
5. Lewis, A.S., Knowles, G.: Image compression using the 2-D wavelet transform. IEEE Transactions on Image Processing 1(2), 244–250 (1992)
6. Podilchuk, C.I., Zeng, W.J.: Image-adaptive watermarking using visual models. IEEE Journal on Selected Areas in Communications 16(4), 525–539 (1998)
7. Andrews, H., Patterson, C.: Singular Value Decomposition (SVD) Image Coding. Communications. IEEE Transactions on (legacy, pre 1988) 24(4), 425–432 (1976)
8. Andrews, H., Patterson, C.: Singular value decompositions and digital image processing. IEEE Transactions on Acoustics, Speech, and Signal Processing (see also IEEE Transactions on Signal Processing) 24(1), 26–53 (1976)
9. Hsu, C.-T., Wu, J.-L.: Hidden digital watermarks in images. IEEE Trans. on Image Processing 8(1), 58–68 (1999)
10. Cox, I.J., Kilian, J., Leighton, F.T., Shamoon, T.: Secure spectrum watermarking for multimedia. IEEE Transactions 6(12), 1673–1687 (1997)

An Intelligent System for Emergency Traffic Routing*

Yafei Zhou and Mao Liu

Center for Urban Public Safety Research,
Nankai University,
Tianjin, 300071, China
YafeiZhou111@tom.com

Abstract. Under natural disasters such as typhoons, hurricanes, floods and so on, it's necessary to evacuate the affected population to safe shelters quickly, and the large-scale emergency traffic routing becomes more and more important. An intelligent system for emergency traffic routing was proposed in this paper. The evacuation road network was extracted from GIS, the cell transmission model (CTM) was used for the allocation of evacuation routes. Then the numerical solution was tested by the traffic evacuation software CORSIM. After that the determined evacuation routes were shown in GIS, so that the traffic evacuation can be guided better and more intuitively. With a case study, it shows that the system can provide helpful guidance and assistance for emergency evacuation.

Keywords: GIS, emergency traffic routing, cell transmission model, CORSIM.

1 Introduction

Natural and man-made emergencies such as hurricanes, floods, fires or chemical spills, may have serious risks and threats on the human health and safety. In this case, large-scale evacuation or shelter in place (shelter-in-place) has been used as a means of protecting the population from potential hazards (Sorensen et al, 2004). If the population is exposed to chemicals or hazardous materials, or the shelter can provide adequate protection, shelter in place is preferred. However, evacuation is usually used in that the infrastructure or communities were damaged by the hurricane, flood or fire (Perry and Lindell, 2003). As there will be a large number of evacuees and the unpredictable load on the transport infrastructure, the scientific or formal analysis of evacuation process is very complex. So it's important to have an accurate traffic routing in the large-scale traffic routing.

Southworth has conducted a comprehensive review of region evacuation model, summed up that the information of transport structure, the spatial distribution of population, the use of vehicles in emergency situations, the individual response time to emergency circumstances, the choice behavior of evacuation routes and destination, etc. were needed in the simulation of large-scale evacuation of the

* This work is partially supported by the National Natural Science Foundation Project (Nos. 70833003).

J. Zhang (Ed.): ICAIC 2011, Part V, CCIS 228, pp. 429–436, 2011.

population. Lee D.Han simulated emergency traffic evacuation around a nuclear power station in Tennessee with VISSIM which is a microscopic traffic simulation system based on time and behaviors. Cova et al. proposed an approach of evacuation paths based on driveways. Church and Sexton studied the influence mode of traffic conditions on evacuation time. Zhou et al. proposed an emergency traffic evacuation model under toxic gas leakage. Sherali et al., Kongsomsaksakul and Yang took into account of the location of shelters in the study of evacuation time for the hurricane / flood evacuation plan. Feng and Wen studied the various traffic control strategies in the raid zones during the last earthquakes of and propose some models to reduce the time of disaster relief to a minimum. Yi and Özdamar proposed a dynamic logistics coordination model support for disaster response and evacuation. Purasls and Garzon took into account the building evacuation problems and proposed a model for selecting right transfer routes in an emergency. Urbanik (2000) described the mechanism of traffic routing as load balancing, with evacuation traffic being diverted from routes of excess demand to those of excess capacity. Most of the evacuation models are used in the plan formulation stage, some models provide traffic information as a time of function, while others are based on simulation models initially designed for traffic problems.

Note that traffic routing is different from those route selection models widely used in most simulation-based software packages, which are for simulating the route selection behavior of drivers, based on the prevailing network conditions. Examples of such studies include NETVAC1 (Sheffi et al., 1982), which allows dynamic route selection in each interval at each intersection, based on traffic conditions directly ahead; MASSVAC (Hobeika, et al., 1994; 1998), that determines routes for evacuees departing from their origins with static traffic assignment algorithms; and CEMPS (Pidd et al., 1996; de Silva and Eglese, 2000), whose route selection mechanism has evolved from an immediate-congestion-based mechanism in its earlier versions to a shortest-path-based mechanism. Most of such route selection models are myopic in nature.

In this paper, an intelligent system for emergency traffic routing was proposed. The incident impact prediction was input to GIS and the impact scope can be determined, the evacuation road network was extracted from GIS. Based on cell transmission model (CTM), the road network was converted. Then the traffic routing solution of the CTM model was compared with the CORSIM simulation results to have a calibration. Finally, the routing program can be shown in GIS to have an intuitive display. Section 2 describes the system framework for emergency traffic routing. Section 3 presents a case study. Finally, the section 4 offers a summary and the conclusion of the work.

2 Description of Emergency Traffic Routing System

A. System Framework

The system framework can be seen in the following Fig. 1.

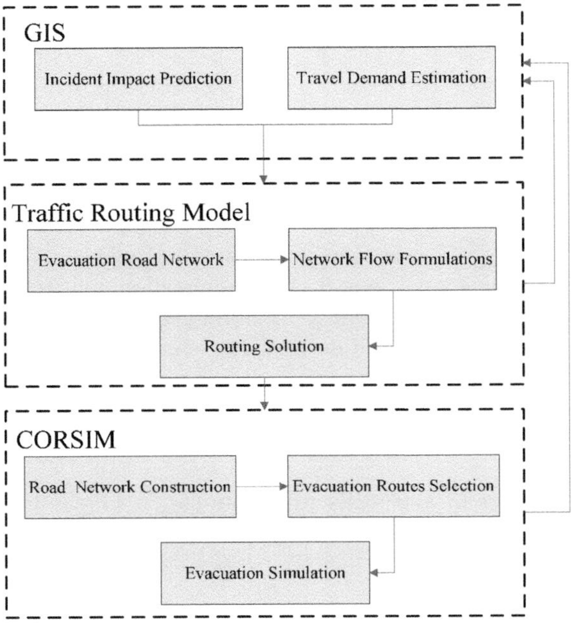

Fig. 1. Framework of intelligent emergency traffic routing system

B. Network Conversion

The cell transmission model was proposed by Daganzo (1994; 1995) to convert roadway links into equal-sized segments, or called cells, that could be traversed in a unit time interval at the free-flow speed. Then, the movements of vehicles among these cells are defined with two types of relations, namely, flow propagation relations to decide flows between two cells based on upstream/downstream traffic conditions and flow conservation equations to depict the evolution of the cell status (i.e., the number of vehicles in each cell) over time.

To successfully apply the cell transmission model for emergency traffic routing, the road network should be converted into a set of connected cells, based on the following steps:

1) Identifying homogenous road segments: homogeneity is defined by the same free flow speed, same number of lanes, same jam density, same saturation flow rate, and no ramps within a segment.

2) Defining unit time interval: the maximal unit interval τ is constrained by the shortest time to traverse a homogenous segment, as (1).

$$\tau = \min\left\{\frac{\text{length of a segment}}{\text{corresponding free flow speed}}\right\} \tag{1}$$

3) Converting road segments to cells: every homogenous segment is converted to a cell, and the cell size l is defined by (2).

$$l = INT\left\{\frac{\text{length of segment}}{\text{corresponding free flow speed} \times \text{unit interval length}} + 0.5\right\} \tag{2}$$

4) Defining connectors between cells: A connector is defined to indicate the potential traffic flows between two connected segments.

C. Emergency Traffic Routing Model

1) Objective function

The evacuation duration T was supposed to be enough for evacuating all the population out of the dangerous area. So the optimization aims to maximize the total throughput within the specified T. Since the throughput can be represented with the total number of vehicles entering all destinations over the study period, the objective function can be formulated as follows:

$$\max \sum_{i\in S_S}\sum_{k\in\Gamma(i)}\sum_{t=1}^{T} y_{ki}^{t} = \sum_{i\in S_S} x_i^{T+1} \tag{3}$$

where S_S is the set of sink cells (destinations), $\Gamma(i)$ is the set of upstream cells to cell i.

2) Constraints

$$x_i^{t+1} = x_i^t + \sum_{k\in\Gamma(i)} y_{ki}^t - \sum_{j\in\Gamma^{-1}(i)} y_{ij}^t, \quad i\in S\cup S_S \tag{4}$$

$$x_r^{t+1} = x_r^t + d_r^t - \sum_{j\in\Gamma^{-1}(r)} y_{rj}^t, \quad r\in S_r \tag{5}$$

$$\sum_{k\in\Gamma(i)} y_{ki}^t \le Q_i^t, \quad i\in S\cup S_S \tag{6}$$

$$\sum_{k\in\Gamma(i)} y_{ki}^t \le N_i^t / l_i, \quad i\in S\cup S_S \tag{7}$$

$$\sum_{k\in\Gamma(i)} y_{ki}^t \le N_i^t - x_i^t, \quad i\in S\cup S_S \tag{8}$$

$$\sum_{j\in\Gamma^{-1}(i)} y_{ij}^t \le Q_i^t, \quad i\in S\cup S_r \tag{9}$$

$$\sum_{j\in\Gamma^{-1}(i)} y_{ij}^t \le N_i^t / l_i, \quad i\in S\cup S_r \tag{10}$$

$$\sum_{j\in\Gamma^{-1}(i)} y_{ij}^t \le x_i^{t-l_i+1} - \sum_{j\in\Gamma^{-1}(i)}\sum_{m=t-l_i+1}^{t-1} y_{ij}^m, \quad i\in S\cup S_r \tag{11}$$

$$y_{ij}^t \le Q_{ij}^t \tag{12}$$

$$\sum_{i\in S_S} x_i^{T+1} = \sum_{r\in S_r} D_r \tag{13}$$

where,

x_i^t is the number of vehicles in cell i at the beginning of interval t ;

y_{ij}^t is connector flows from cell i to cell j during interval t;

d_r^t is evacuation demand from origin r during interval t;

Q_i^t is the number of vehicles that can flow into/out of cell i during interval t;

N_i^t is the number of vehicles that can be accommodated in cell i during interval t;

l_i is the size of cell i;

$\Gamma^{-1}(i)$ is the set of downstream cells to cell i;

$\Gamma(i)$ is the set of upstream cells to cell i;

$x_i^{T+1}, i \in S_S$ is the number of vehicles that has arrived at the destination i after the evacuation time window T;

$D_r, r \in S_r$ is the total evacuation demand generated at origin r;

The subscript r is the index of source cells;

i, j, k is the index of other cells.

Among the above constraints,

(4) is the flow conservation equation for both general cells and sink cells;

(5) is the flow conservation equation for source cells;

(6)-(8) describe the relaxed flow propagation constraints related to the receiving capacity of any downstream cells;

(9)-(11) describe the relaxed flow propagation constraints elated to the sending capacity of any upstream cells;

(12) describes the flow capacity constraints for connectors, which can model the reduced capacity of ramps or the right/left turning movements at the intersections;

(13) is the demand related constraints.

3) Model solution

The emergency traffic routing model was solved using LINGO on a 200 MHz Pentium PC. LINGO is a comprehensive tool designed by LINDO system company that was established by Professor Linus Schrage of Chicago University, it has been widely used to make building and solving linear, nonlinear and integer optimization models faster, easier and more efficient.

D. Model Test

The solution of the emergency traffic routing model was tested with a micro simulation software packages for traffic flow modeling TSIS/CORSIM, which was developed by the U.S. Federal Highway Administration. It's capable of simulating surface streets, freeways, and integrated networks with various traffic control options.

The road network of the study area extracted from GIS was input into the software TSIS/CORSIM, with the obtained routes solution, the evacuation progress can be simulated and analyzed.

3 Case Study

The intelligent emergency traffic routing system was tested with a case study in Tianjin, China. The evacuation origin, road network and shelters can be seen Fig. 2, about 10, 000 people need be evacuated from a stadium to 5 shelters around with traffic under emergency.

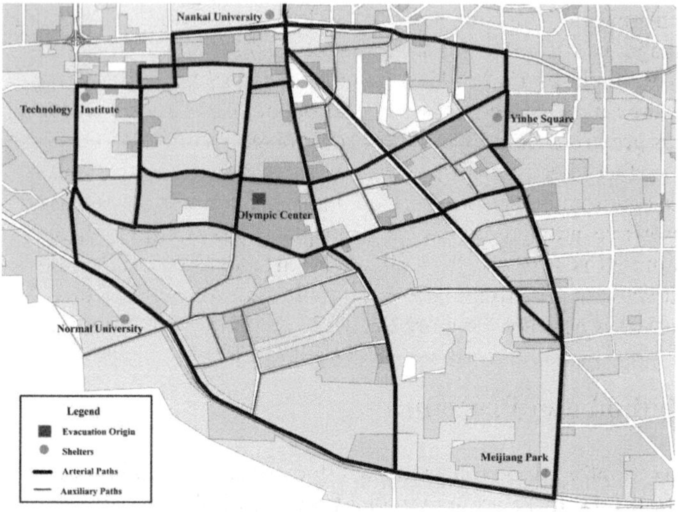

Fig. 2. Evacuation scenario description

Following the network conversion procedures, the evacuation network was converted to a cell-connection diagram as illustrated in Fig. 3. Note that the number in each parenthesis indicates the size of the cell.

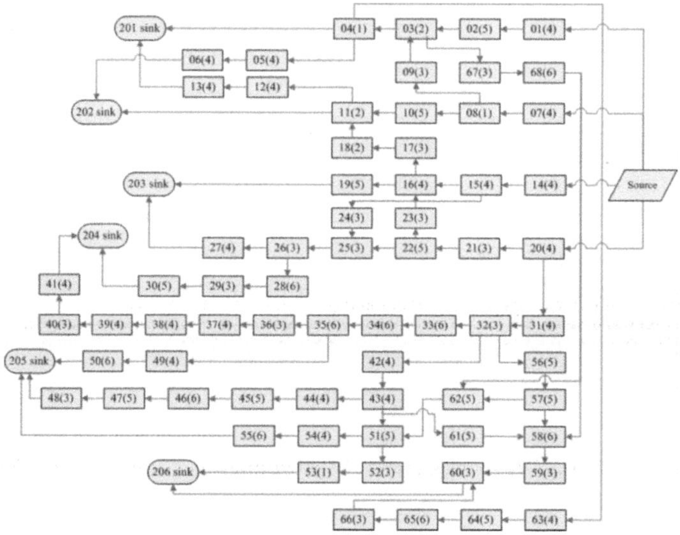

Fig. 3. Cell connection diagram for evacuation network

Then, with the road network in GIS and the emergency traffic routing solution, the evacuation progress can be simulated in TSIS/CORSIM. The road network is shown in Fig. 4.

Fig. 4. Evacuation network in CORSIM

The simulation results were quite similar with the time-varying network traffic conditions of the emergency traffic routing model solution, which thus demonstrates the potential of the proposed model in accurately formulating traffic flows for large-scale networks and in efficiently generating the optimal set of evacuation strategies.

4 Conclusions

In order to increase the efficiency of large-scale traffic evacuation, an intelligent system for emergency traffic routing was proposed in this paper. The incident impact prediction was input into GIS to determine the impact scope, the population need to be evacuated and the traffic demand. The road network was converted to cell-connection diagram for based on the traffic routing model. Then the model solution was compared with the CORSIM simulation results to have a calibration. The case study indicated that the proposed emergency traffic routing model can accurately formulate traffic flows for large-scale networks and generate optimal evacuation routes to increase the evacuation efficiency.

In the future study, the proposed intelligent system should be combined with the other evacuation optimization model to improve its performance such as contraflow operation model, staged evacuation model and signal control model.

References

1. Sorensen, J.H., Shumpert, B.L., Vogt, B.M.: Planning for protective action decision making: Evacuate or shelter-in-place. Journal of Hazardous Materials 109(1), 1–11 (2004)
2. Perry, R.W., Lindell, M.K.: Preparedness for emergency response: Guidelines for the emergency planning process. Disasters 27(4), 336–350 (2003)
3. Southworth, F.: Regional evacuation modeling: a state of the art review. Oak Ridge National Laboratory, Washington DC (1991)

4. Han, D.L., Yuan, F.: Evacuation modeling and operations using dynamic traffic assignment and most assignment and most desirable destination approaches. In: The 84th Transportation Research Board Annual Meeting, pp. 964–969. TRB, Washington DC (2005)
5. Cova, T.J., Johnson, J.P.: A network flow model for lane-based evacuation routing. Transportation Research Part A: Policy and Practice 37(7), 579–604 (2003)
6. Church, R.L., Sexton, R.M.: Modeling small area evacuation: Can existing transportation infrastructure impede public safety?. University of California, Santa Barbara (2002)
7. Zhou, Y., Liu, M., Zhang, D.: Area-wide emergency evacuation under toxic gas leakage. In: Proceedings of the 2008 International Symposium on Safety Science and Technology, China Occupational Safety and Health Association, pp. 465–471. Science Press USA Inc., Beijing (2008)
8. Busvine, J.R., et al.: The toxicity of ethylene oxide to Calandra oryzae, C. granaria, Tribolium castaneum, and Cimex lectularius. Annals of Applied Biology 25(3), 605–632 (1938)
9. Sherali, H.D., Carter, T.B., Hobeika, A.G.: A location-allocation model and algorithm for evacuation planning under hurricane/flood conditions. Transportation Research Part B 25(6), 439–452 (1991)
10. Kongsomsaksakul, S., Yang, C.: Shelter location-allocation model for flood evacuation planning. Journal of the Eastern Asia Society for Transportation Studies (6), 4237–4252 (2005)
11. Feng, C.M., Wen, C.C.: Traffic control management for earthquake-raided area. Journal of the Eastern Asia Society for Transportation Studies (5), 3261–3275 (2003)
12. Yi, W., Özdamar, L.: A dynamic logistics coordination model for evacuation and support in disaster response activities. European Journal of Operational Research (179), 1177–1193 (2007)
13. Pursals, S.C., Garzon, F.G.: Optimal building evacuation time considering evacuation routes. European Journal of Operational Research 192(2), 692–699 (2009)
14. Urbanik II., T.: Evacuation time estimates for nuclear power plants. Journal of Hazardous Materials 5, 165–180 (2000)
15. Sheffi, Y., Mahmassani, H., Powell, W.: A transportation network evacuation model. Transportation Research A 16(3), 209–218 (1982)
16. Hobeika, A.G., Kim, S., Beckwith, R.E.: A decision support system for developing evacuation plans around nuclear power stations. Interfaces 24(5), 22–35 (1994)
17. Hobeika, A., Kim, C.: Comparison of traffic assignments in evacuation modeling. IEEE Transactions on Engineering Management 45(2), 192–198 (1998)
18. Pidd, M., de Silva, F., Eglese, R.: A simulation model for emergency evacuation. European Journal of Operational Research 90(3), 413–419 (1996)
19. de Silva, F., Eglese, R.: Integrating simulation modeling and GIS: spatial decision support systems for evacuation planning. Journal of the Operational Research Society 51, 423–430 (2000)
20. Daganzo, F.C.: The cell transmission model: A dynamic representation of highway traffic consistent with the hydrodynamic theory. Transportation Research Part B 28(4), 269–287 (1994)
21. Daganzo, F.C.: The cell transmission model part II: Network traffic. Transportation Research Part B 29(2), 79–93 (1995)
22. Linus Schrage, LINDO System Inc., `http://www.lindo.com`
23. Federal Highway Administration (FHWA), Freeway Management and Operations Handbook, FHWA Report No.: FHWA-OP-04-003, Office of Transportation Management, Federal Highway Administration (2005a)

Study on Friction Characteristics between Two Bodies under Vibration Working Case

He Li, Wenbo Fan, and Bangchun Wen

School of Mechanical Engineering and Automation,
Northeastern University,
P.O. Box 319, No. 3-11 Wenhua Road,
Shenyang, China
heli_121@sina.cn

Abstract. This paper mainly focuses on vibration friction and discusses the dynamic characteristics of the friction under vibration working case. Many practical projects have already proved that under vibration working case, the friction coefficient and the friction force among work pieces and materials can be reduced obviously, and the wear and tear of the parts of machine can be obviously lightened, and the work efficiency can be improved greatly. In this paper, the influence on sliding friction of vibration both parallel and perpendicular to the sliding direction has been studied. Models for friction characteristics between two bodies under longitudinal and transverse vibration working case are set up. On the basis of these models, longitudinal vibration produces greater reduction in friction than transverse vibration at the same amplitude and frequency.

Keywords: vibration friction, modal, longitudinal vibration, transverse vibration.

1 Introduction

With the rapidly development of engineering techniques it is discovered that with the vibration induced in friction, its characteristic can be efficiently changed. Many practical projects also proved that under vibration working case, the friction coefficient and the friction force among work pieces can be reduced obviously, the wear and tear of the parts of machine can be obviously lightened and the work efficiency can be improved greatly [1,2,3]. Which creates obviously economic and society value for the enterprise and the nation, also leads out the vibration friction theory. Vibration friction is the friction between or inside of solid parts, within dry-loose materials and also in wet-loose material under vibration situation. It is not so broadly in vibration friction studies, there are lots of problems need to solve. But reality applications need theory to direct; it is necessary to set up systemic theory for vibration frictions.

Depending upon the direction of vibration, the friction reduction can be explained either in terms of a change in the magnitude and direction of the resultant sliding

J. Zhang (Ed.): ICAIC 2011, Part V, CCIS 228, pp. 437–444, 2011.

velocity vector, or in terms of a change in magnitude of the normal load [4,5]. The former effect occurs in the case of vibration acting parallel (longitudinal vibration) or perpendicular (transverse vibration) to the sliding direction in a plane containing the interacting surfaces, and the latter applies in the case of vibration normal to the plane of the surface. In this paper, the influence on sliding friction of vibration both parallel and perpendicular to the sliding direction has been studied. Models for friction characteristics between two bodies under longitudinal and transverse vibration working case are set up. On the basis of these models, longitudinal vibration produces greater reduction in friction than transverse vibration at the same amplitude and frequency.

2 Models for Friction Characteristics between Two Bodies under Longitudinal Vibration Working Case

In the case of longitudinal vibration, the directions of vibration and sliding are collinear. So during each cycle of vibration, the friction force on the interacting surfaces is in the same direction.

As shown in Fig. 1 (a), body A, to which a time harmonic vibration motion with amplitude a and angular frequency ω along the same line of action as that of V_c is applied, is assumed to slide with a constant velocity V_c over body B. And Suppose Body B is motionless. An actual velocity V of body A comprising of a vibration velocity $V(t)$ and a constant velocity V_c is expressed as

$$V = V(t) + V_c \tag{1}$$

where $V(t) = a\omega\sin(\omega t)$.

An time averaged velocity of body A in a cycle of vibration is expressed as

$$V_a = \frac{1}{T}\int_0^T V\,\mathrm{d}t = V_c \tag{2}$$

where T is the period of vibration velocity $V(t)$, $T = 2\pi/\omega$.

From Eq. (2), for a whole motion of body A, a vibration velocity $V(t)$ appling to body A may be negligible.

As shown in Fig. 1 (b), when $V_c \geq a\omega$, the direction of friction force F is against the direction of a constant velocity V_c of body A.

As shown in Fig. 1 (c), when the absolute value of instantaneous vibration velocity $V(t)$ of body A during the interval CF is greater than V_c, the friction force F will reverse its direction and acts in the same direction as V_c. The average effects in the whole vibration cycle can be analyzed as follows.

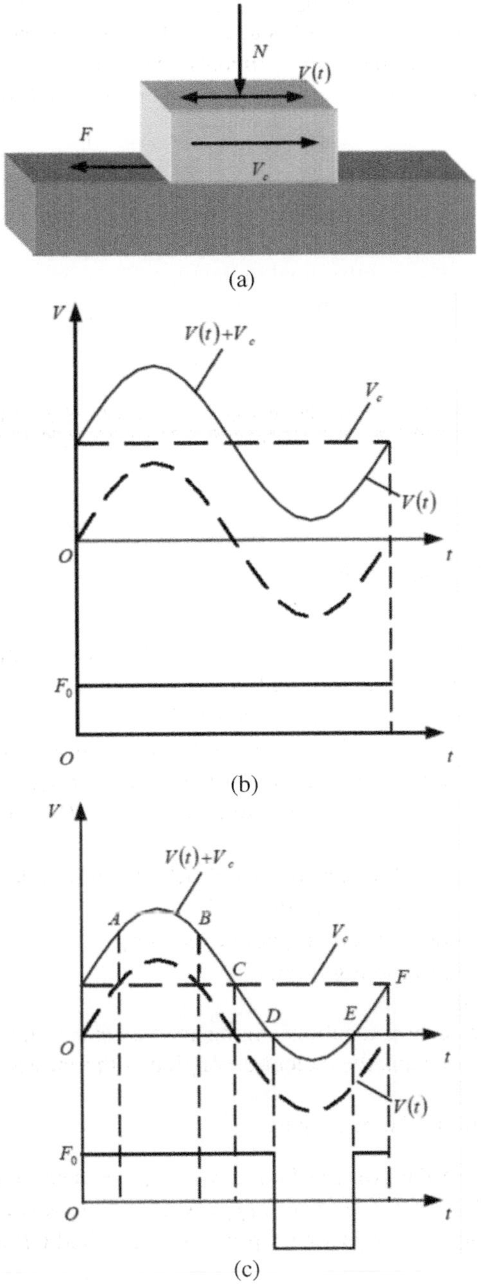

Fig. 1. Friction effected longitudinal vibration (a) Vibrating body A sliding over a support B. (b) Variation of vibration velocity with time and corresponding change in direction of frictional force $V_c \geq a\omega$ (c) Variation of vibration velocity with time and corresponding change in direction of frictional force $V_c < a\omega$.

It is assumed that there is no dependence of coefficient of friction on velocity.

The time t_s taken for the vibration velocity $V(t)$ to reach the sliding velocity V_c is given by:

$$t_s = \frac{1}{\omega}\arcsin\left(\frac{V_c}{a\omega}\right) \tag{3}$$

During the interval OA, $V_c > V(t)$ and hence the friction force on body A is positive. For the interval AB, $V(t) > V_c$ and the friction force is negative. Over the remainder of the cycle BE, the friction is again positive. By symmetry, the time AB for which the friction force is negative is exactly equal to the time CD over which it is positive and in the time average; these two will cancel each other out. The resultant average frictional force over the whole cycle F_a is thus given by

$$F_a = \frac{F_0}{T}(4t_s) = \frac{2F_0}{\pi}\arcsin\left(\frac{V_c}{a\omega}\right) \tag{4}$$

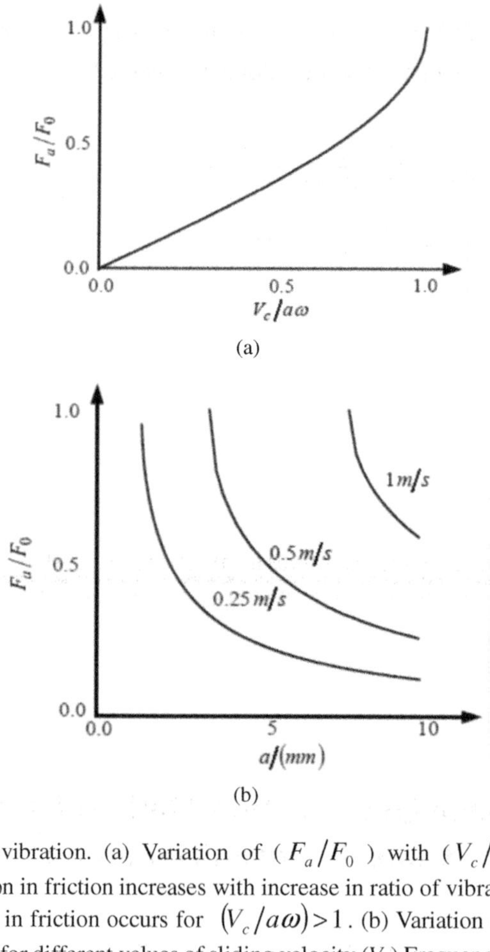

Fig. 2. Longitudinal vibration. (a) Variation of (F_a/F_0) with ($V_c/a\omega$) for longitudinal vibration. The reduction in friction increases with increase in ratio of vibration velocity to sliding velocity. No reduction in friction occurs for $(V_c/a\omega)>1$. (b) Variation of frictional force with amplitude of vibration for different values of sliding velocity (V_c) Frequency of vibration is 20 Hz.

where F_0 is the frictional force in the absence of vibration and T is the period of the oscillation. This equation is valid only for $V_c < a\omega$. The variation of (F_a/F_0) with $(V_c/a\omega)$ predicted by Eq. (4) is shown in Fig. 2(a).

At high sliding velocities and/or low values of V_c, Eq. (2) ceases to apply and no reduction in friction is predicted, because the peak vibration velocity becomes less than the sliding velocity and the friction force then opposes the direction of macroscopic sliding for the whole of each cycle of vibration. At low sliding velocities, even small amplitude of vibration is sufficient to reduce the average friction to a very low value. The variation of frictional force as a function of vibration amplitude for various sliding velocities is given in Fig. 2 (b) for vibration at 20 Hz.

3 Models for Friction Characteristics between Two Bodies under TransverseVibration Working Case

In this case, the direction of vibration is orthogonal to the direction of sliding, as shown in Fig. 3 (a).

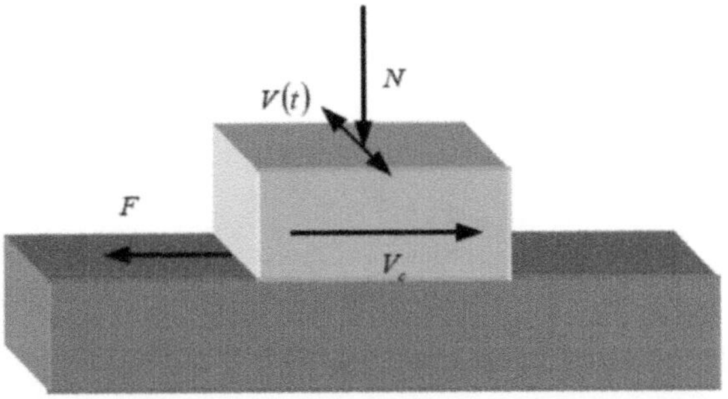

(a)

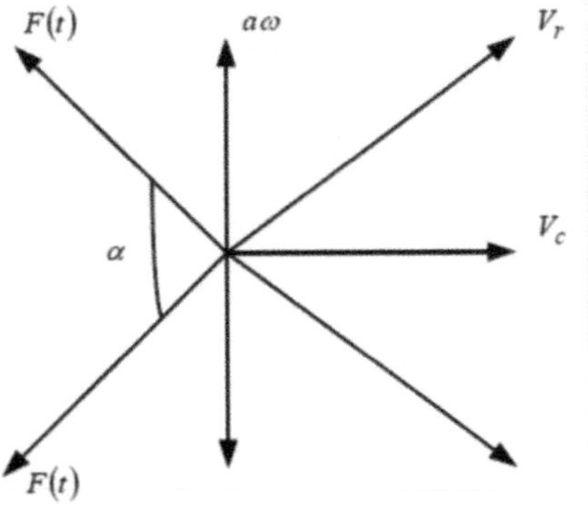

(b)

Fig. 3. Friction effected transverse vibration. (a) Body A sliding over a support B. (b) When slider moves at right angles to the direction of vibration, effective direction of sliding deviates from the bulk sliding direction along the resultant velocity. The broken lines represent the corresponding friction force vector at the two peaks of vibration velocity.

Like the case of longitudinal vibration, it is assumed that $a\omega$ is the amplitude of the vibration velocity and V_c is the sliding velocity. The resultant velocity (V_r) at maximum transverse velocity will be at an angle α to the sliding direction, as shown in Fig. 3 (b).

At any instant, the direction of the frictional force will be in the opposite direction to the resultant sliding velocity. Therefore, during one cycle of vibration, the direction of the frictional force will oscillate between angles (-α) and (+α) to the bulk sliding direction V_C. The angle between the two extreme directions of the frictional force vector will be 2α, as shown in Fig. 3 (b).

The resultant time-averaged frictional force in the bulk sliding direction will be the time average of the component of the frictional force vector acting along the direction of the relative sliding velocity vector. If his the angle between these two vectors at any instant, the instantaneous component of the frictional force $F(t)$ in the macroscopic sliding direction is given by

$$F(t) = F_0 \cos\theta \tag{5}$$

where $\theta = \arctan((a\omega/V_c)\sin\omega t)$.

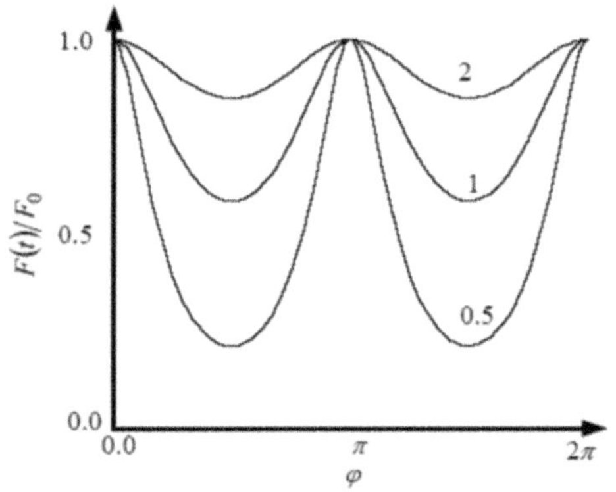

(a)

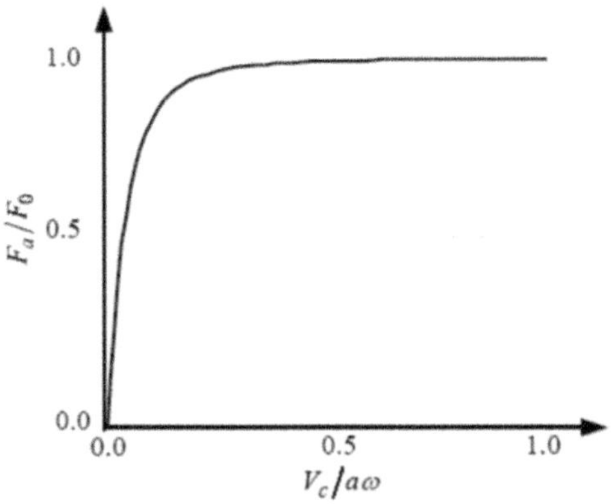

(b)

Fig. 4. Transverse vibration. (a) Expected variation in frictional force during one cycle of transverse vibration predicted by Eq. (5) for different values of $(V_c/a\omega)$. (b) Variations of $(F(t)/F_0)$ with velocity ratio $(V_c/a\omega)$.

The time-averaged frictional force over one cycle F_a is given by

$$\begin{aligned} F_a &= \frac{1}{T}\int_0^T F_0 \cos\theta \mathrm{d}t \\ &= \frac{1}{T}\int_0^T F_0 \cos\left(\arcsin\left(\frac{a\omega}{V_c}\sin\omega t\right)\right)\mathrm{d}t \\ &= \frac{1}{2\pi}\int_0^{2\pi} F_0 \cos\left(\arcsin\left(\frac{a\omega}{V_c}\sin\varphi\right)\right)\mathrm{d}\varphi \end{aligned} \tag{6}$$

where $\varphi = \omega t$, the phase angle.

The integral in Eq. (6) can be evaluated numerically to obtain the time-averaged resultant frictional force in the direction of sliding.

The variation of $F(t)/F_0$ with phase angle φ for different ratios of peak vibration velocity to sliding velocity $a\omega/V_c$ as predicted by Eq. (5) is shown in Fig. 4 (a). It can be seen that the vibration effect is not present at phase angles φ, π, and π. At these phase angles, the instantaneous vibration velocity is zero. Maximum effect is obtained at phase angles of $\varphi = \pi/2$ and $\varphi = 3\pi/2$. For these values, the vibration velocity is max and the effect on the resultant instantaneous frictional force depends strongly on the value of $V_c/a\omega$ as shown in Fig. 4 (a).

The variation of F_a/F_0 with $V_c/a\omega$ according to Eq. (6) is shown in Fig. 4 (b). There is a gradual increase in time-averaged resultant frictional force with increasing velocity ratio. As the velocity ratio reduces, the resultant velocity vector lies close to the vibration direction for an increasing proportion of the vibration cycle, resulting in a greater reduction in the measured frictional force.

4 Conclusions

In this paper, the influence on sliding friction of vibration both parallel and perpendicular to the sliding direction has been studied. Models for friction characteristics between two bodies under longitudinal and transverse vibration working case are set up. The reduction in friction by longitudinal vibration was greater than by transverse vibration. The ratio between the friction forces with and without vibration is a function of the ratio between the peak vibration velocity and the mean sliding velocity. The reduction in friction can be predicted by simple analytical models.

Acknowledgements. This research was supported by National Natural Science Foundation of China (Grant No. 10702014), NCET of MOE (Grant No. NCET-10-0271), PCSIRT of MOE, the Fundamental Research Funds for the Central Universities (Grant No. N100503002, N100703001), the Research Fund for the Doctoral Program of Higher Education (Grant No. 20070145076).

References

1. Wen, B.C., Li, Y.N., Zhang, Y.M., Song, J.W.: Vibration Utilization Engineering. Science Press, Beijing (2005) (in Chinese)
2. Wen, B.C., Liu, F.Q.: Theory of Vibrating Machines, and Its Application. Mechanical Industry Press, Beijing (1982) (in Chinese)
3. Wen, B.C., Liu, S.Y., He, Q.: Theory and Dynamic Design Methods of Vibrating Machinery. Mechanical Industry Press, Beijing (2001) (in Chinese)
4. Littmann, W., Storck, H., Wallaschek, J.: Sliding friction in the presence of ultrasonic oscillations: superposition of longitudinal oscillations. Archive of Applied Mechanics 71, 549–555 (2001)
5. Hess, D.P., Soom, A.: Normal vibration and friction under harmonic loads: Hertzian contacts. Transactions of ASME: Journal of Tribology 113, 80–86 (1991)

A Method of Regional Recognition Based on Boundary Nodes

Yanlan Yang, Hua Ye, and Shumin Fei

School of Automation,
Southeast University,
Nanjing, Jiangsu, China
yanlan1288@126.com

Abstract. This method is to get the regional information of the target place based on boundary nodes of the regions. Firstly, the structure of regional data in the original map file was studied and each region can consists of a set of boundary nodes. The regional scope for searching nodes was designed into a belt shaped area in order to ensure valid searching work. Then, rules which indicate the implicit association within a region and its boundary nodes were made for further regional recognition. Finally, the threshold settings for the size of the searching regional scope were discussed to make the recognition efficient. Experimental results showed that the method is proved to be useful and effective in recognizing regions and acquiring regional information. It makes great sense to expand function of spatial querying in the GIS applications.

Keywords: Spatial Querying, Regional Recognition, Boundary Nodes, Regional Scope.

1 Introduction

In normal life, people's cognizance of a target place is usually not by its position in coordinates but depends on the geographical information with a description on where the target is located based on spatial relationships [1]. According to different forms of the associated spatial objects, there are several kinds of spatial relationships, including the distance between points, the ownership between a point and a line or a region. For example, if people want to find the location which they are interested in, they can get such information to help them, like the interests around, the roads nearby, or the region that it belongs to. And the proceeding for information refers to a series of queries on spatial relationships for point-to-point, point-to-line, and point-to-region.

Actually, the existing SDK packages for GIS secondary development have already included some tools for professional spatial management. And it becomes a popular solution for spatial querying. However, for some special use of GIS, there are drawbacks in feasibility, occupation and performance. Here, they are summarized into following three points:

Firstly, commercial GIS (as MapInfo, ArcInfo), as well as their SDK packages, manages the electronic map in a mixed way both with spatial data files and the

J. Zhang (Ed.): ICAIC 2011, Part V, CCIS 228, pp. 445–451, 2011.

relational database [2]. The files store all spatial data and their topological relations, so any spatial querying operation must base on the file access, which costs memory and restricts the executing speed.

Secondly, The SDK package has shielded the working details on the map managing and spatial querying to the developers. Thus, developers can do nothing to improve the performance on treating map or querying no matter how hard they work.

Thirdly, one of the advantages for commercial GIS is the map visualization. However, there are cases where only the spatial querying is required but no need to show the map. If the SDK package is implanted forcibly for its incidental tools on spatial management, not only the system resources are wasted unnecessarily, some problems on instability may be caused when the SDK package was used improperly.

Therefore, it is necessary to find new methods to make the spatial querying available based on original map data with no dependence on any SDK package. The idea has meanings in improving the performance of spatial querying and expanding the scope of GIS applications. The method discussed in this paper is to get regional information in which the target place is located based on boundary nodes of the region.

2 Structure of the Regional Data

By studying the structures of spatial data with the format of MapInfo, including points, lines and regions, it has been found that spatial data in different forms is composed of a set of nodes. For example, a line consists of the nodes sampled from its central axis, while a region with the nodes from its boundary [3]. In fact, the structure discussed above is exactly the storage format of spatial data in the MIF file, one type of MapInfo files which is used to exchange data with other formats. As shown in Fig. 1, each region consists of one polygon at least, where "numpolygons" denotes the number of polygons contained in this region. And "numpts1" indicates the count of nodes sampled from boundaries of the first polygon, followed by a group of coordinates denoted by x and y in their neighbor order [4]. So boundary nodes of the region are listed regularly, which makes it possible to do the regional recognition.

With acknowledge of the regional boundary, if the target place is surrounded by boundaries, their corresponding region is the exact one that contains the target. In order to make such a judgment based on boundary nodes, rules indicating the implicit association within the region and its boundary nodes should be made. And the main idea of regional recognition is to search the boundary nodes around the target first, and then use rules to find the region which matches the target.

Considering the regional style which makes the rules different, there are mainly two styles referring to the regional distribution [5]. The one is that regions are adjacent closely and share one boundary or more, referring to administrative region division. And the other is that regions separate from each other with a certain distance and no shared boundary exists, such as the distribution of resident areas. The method discussed in this paper is applied for the regional recognition with the first style.

```
REGION numpolygons
numpts1
x1 y1
x2 y2
:
[numpts2
x1 y1
x2 y2]
:
[PEN (width, pattern, color)]
[BRUSH (pattern, forecolor, backcolor)]
[CENTER x y]
```

Fig. 1. Structure of the regional data

3 Method for Regional Recognition

In this method, three questions are to be discussed:

1) Design a regional scope which ensures effective searching for boundary nodes around the target.

2) Make several rules for regional recognition based on their boundary nodes.

3) Set thresholds for the size of the searching region which make it cover enough valid boundary nodes with less useless nodes that may lead to misjudgments.

A. Designing of Searching Regional Scope

Because the region boundary is quite irregular in most cases, the distribution of the boundary nodes is accordingly complex. Meanwhile, considering the possible location that the target anchors within a region, it may be close to the boundary or around the center precisely. There is no exact distance relationship between the target and boundary nodes. Therefore, it is suggested to design a regional scope shaped like a belt to make sure that it contains some boundary nodes.

With the regional distribution shown in Fig. 2, there are shared boundaries for the region marked as R1 with the others denoted by R2-R6. Suppose the target is located in R1, set the meridian passing through it as an axis and then mark out a belt zone with a width of ΔL in degrees. The regional scope must cover part of boundaries owned by R1. Divide the belt zone into halves by the latitude line through the target. And the north half is named as North Belt in this paper, while the south half named as South Belt. The two half-belts compose the final scope of the searching area.

The searching job is designed as follows: Search the North Belt first to find one boundary node with the minimum latitude, which is denoted by Pn. Then search the south part for another boundary node denoted by Ps with the maximum latitude. The Pn and Ps make a pair of feature nodes and are main judging basis for region recognizing.

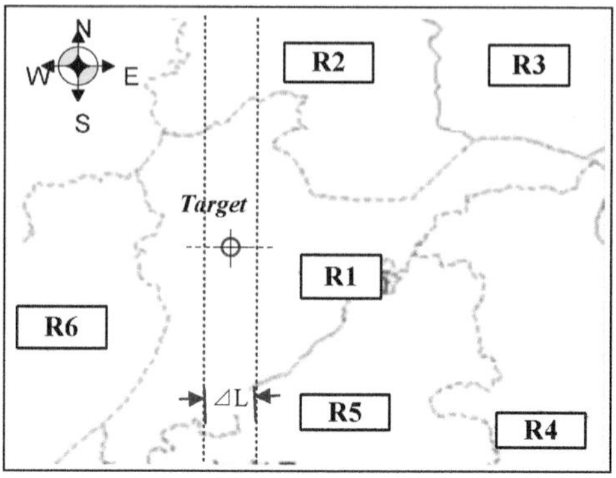

Fig. 2. Searching regional scope

B. Rules for Region Recognition Based on Boundary Nodes

There are two aspects which may influent the result of the regional recognition. One of them is the distribution of the target and the feature nodes, and the other is the boundary characteristic that if it is a shared one or isolated. According to these above, four regional associated rules are made to help make regional recognitions.

Rule 1: If the boundaries where Pn and Ps are located satisfy that at least one of them is not shared by others, and there is only one region which marked as R1 contains Pn and Ps, then the target should belong to R1.

As shown in Fig. 3(a), the boundary containing Pn belongs to R1 only, and another boundary which Ps is located in belongs to R1 and R6 at the same time. So the target is decided to be located in R1 by rule 1.

Rule 2: If both of the boundaries which Pn or Ps is located in are shard by different regions, while R1 is the only region contains both Pn and Ps, then the target belongs to R1.

Take the case shown in Fig. 3(b) for example, the node Pn belongs to regions marked as R1 and R3, and Ps is associated with R1 and R4. So that R1 is the selected region that contains both Pn and Ps. Concluded by rule 2, R1 is the exact region where the target is located.

Rule 3: If the node Pn and Ps are located in the same boundary which is shared by two neighbor regions, then the region matches the target should be either of them.

As shown in Fig. 3(c), Pn and Ps are in the same boundary which is shared by R1 and R6, so the target either belongs to R1 or R6 by rule 3. This is a quite special case and more work should be done to help make a choice. A good suggestion is to find another pair of feature nodes by changing the searching regional scope. Here, the center of the region is used as an important basis to set a new searching regional scope. If the new found pair of boundary nodes meets the condition stated in rule 1 or 2, the right region containing the target can be easily confirmed. There re-searching work includes following three steps:

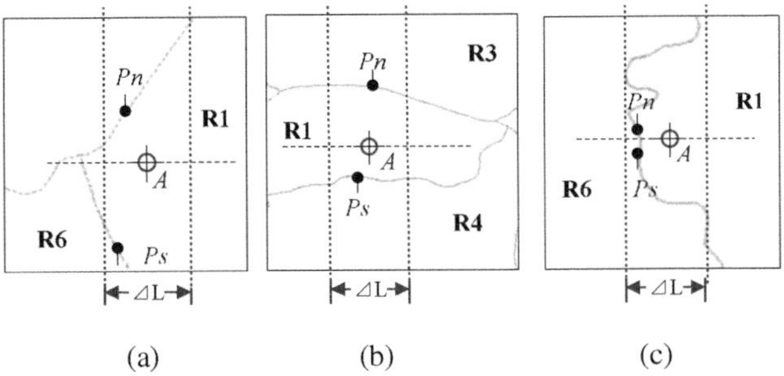

Fig. 3. Three typical regional recognition situations

Step 1: Get two candidate regions R1 and R6 by rule 3. Choose one of them, taking R6 for example, and calculate the difference of latitude denoted by ⊿lat for the target and the center of R6, as well as the difference of longitude which is denoted by ⊿lon.

Step 2: According to the comparative result of the two differences, make a new division of the searching regional scope and find another pair of boundary nodes.

If |⊿lat| ⩾ |⊿lon|, keep the original scope unchanged. Then make the division by two latitude lines, one of which passes through the target and the other passes through the center of R6. That is to say, the latitude line with larger degrees will be the new border of North Belt, while the other one be the new border of South Belt. And another pair of nodes to be searched is marked as Ps' and Pn'.

If |⊿lat| < |⊿lon|, the searching regional scope should be rebuilt by setting the latitude line passing through the target as its axis with the same width. The new scope can be vividly viewed as that the original one rolls around the target for 90 degrees (shown in Fig. 5). Try to divide it into halves with the meridian passing through the target, and they are named as West Belt and East Belt. Correspondingly, the boundary node with the smallest longitude is marked as Pe and the one with the largest longitude is marked as Pw. The Pe and Pw are the second pair of feature nodes.

Step 3: If rule 1 or 2 is applicable to the second pair of feature nodes, it indicates that the target belongs to the region chosen in Step 1. But for rule 3, please repeat Step 1-2 and choose the center of R1 for another calculation. If there are no rules applicable, it can be confirmed that the first selected region (R6) would not contain the target. In other words, the other region (R1) is the exact one. For example, Pw and Pe in Fig.5 do not apply any rules, so that R1 is determined to be the final choice.

When it is still hard to find the correct region after all of three steps above, the target is considered to be located near the regional boundary.

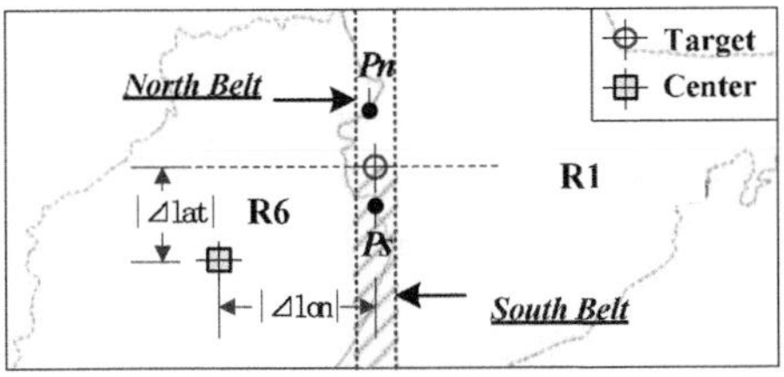

Fig. 4. First division of searching regional scope

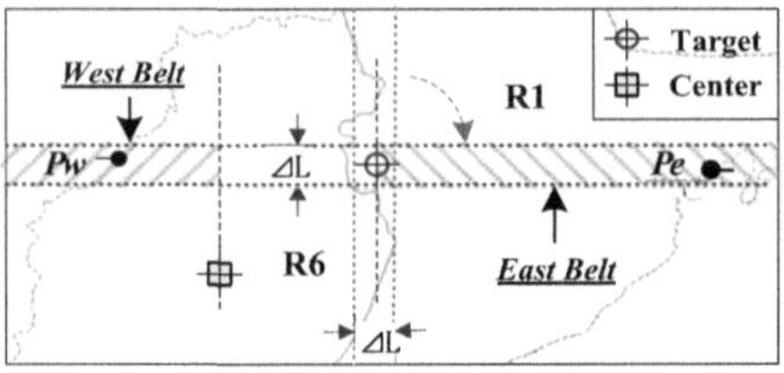

Fig. 5. Second division of searching regional scope

Rule 4: If the location of the node Pn or Ps is completely the same as the target, then the target should be located in the same region as Pn or Ps.

C. Threshold Setting for the Width of the Searching Area

According to these rules, this method depends on the distribution of boundary nodes. So the searching for boundary nodes will be quite important here. To make sure that each half of the searching area contains one boundary node at least, the width marked as ⊿L should be set within a appropriate range.

First of all, the value for ⊿L should not be too small. Because all the boundary nods are distributed separately (See in Fig.6), there must exist differences between neighbor nodes either in longitude or latitude [6], which may causes a problem that no boundary node was found when ⊿L is less than the difference although the area has covered part of the boundaries of some region.

On the other hand, the width of the regional scope should not be too large. As a matter of fact, the feature node is defined as the closest one to the target either in longitude or latitude. If the value for ⊿L is very large, there may be two predictable bad results when searching for nodes. Firstly, the searching efficiency will be reduced because the number of candidate nodes increases dramatically. Secondly, rules are no longer applicable when the searching area is wide enough to cover an entire region.

Thus, the value for ⊿L must refer to the prerequisite and the density of boundary nodes. In this paper, the density is defined as the larger one between two values which are the maximum value of the latitude interval within each neighbor nodes in all the boundary nodes and the maximum value of the longitude interval.

Experimental data has shown that the density of boundary nodes is much smaller than the length of the short side of the minimum bounding rectangle (MBR) of the region. It means that the prerequisite will always be guaranteed as long as the value for ⊿L keeps a bit larger than the density. And the especially unmoral case that the searching regional scope covers a whole region can be avoided as well. Results indicate that the value for ⊿L is set to 1.5 times of the density would be appropriate. And this reasonable value makes the work of regional recognition go on smoothly and efficiently.

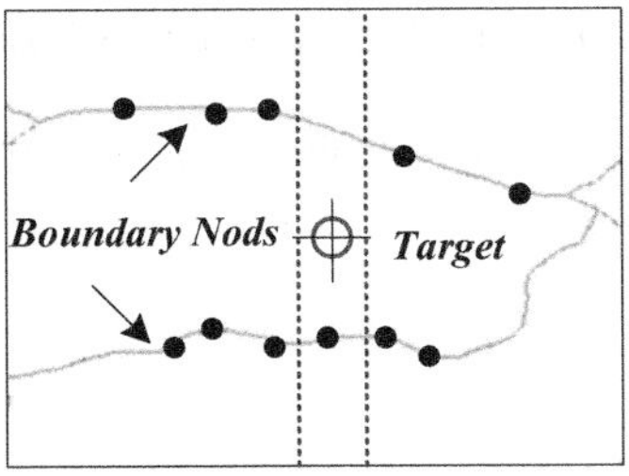

Fig. 6. The case when ⊿L is too small

4 Conclusion

Experimental results show that the method of regional recognition based on boundary nodes is proved to be useful and efficient when the width of the searching regional scope is set to an appropriate value. In most cases, the region can always be recognized successfully after searching for three times at most. Take the target in Fig. 2 for example, the pair of feature nodes was found only by searching once, and region (R1) was recognized as the final one that contains the target by rule 2. And a message that the target was located in Nanjing City was gotten from the regional information of R1. This method can be applied to various types of algorithm for geographic information querying and provides some regional information to the target for further inquiries. Besides, because the implementations of the method is no long relying on commercial GIS software, its application is with advantages of less resources occupation, less processing time and the wider applicability.

References

1. Shi, Y., Ma, X., Chen, J.: Analysis of Spatial Data Model. Geomatics & Spatial Information Technology 32(6), 96–98 (2009)
2. Gao, J.: Evolvement of GIS spatial Data Research. Science and Technology Information of Surveying and Mapping 1, 18–21 (2006)
3. Wang, C., Zhou, C.: MAPINFO Format Map Analysis Programming. Measurement & Control Technology 27(9), 47–50 (2008)
4. Zhang, Y., Zhao, W.: Analysis and Application of MapInfo Common data exchange format. Computer Programming Skills & Maintenance 21, 19–21 (2009)
5. Li, D., Wang, S., Li, D.: Spatial data mining theories and applications, pp. 1–40. Science Press, Beijing (2006)
6. Li, W., Li, H., Liang, R., Liu, J.: Research on Feature-based Spatial Data model. Geomatics & Spatial Information Technology 33(1), 42–45 (2010)

Research on Stability Analysis Methods for Intelligent Pedestrian Evacuation in Unconventional Emergency*

Rongyong Zhao, JianWang, and Weiqing Ling

CIMS Research Center Tongji University Shanghai, 20092, China
rongyongzhao111@sohu.com

Abstract. To evaluate the functional and mathematical performances of the current established and intelligent pedestrian evacuation models in unconventional emergency, the stability analysis is becoming an effective method to evaluate the various models methods for forecasting the time required for evacuation from various places under a variety of conditions especially. This paper introduces the basic stability theory mathematically, then studies on the current analysis methods used for the established pedestrian evacuation system. The mathematics principle, features and computation performance of the main stability analysis method or model are analyzed and compared completely. The schematic of stability analysis for evacuation models is presented in the different engineering context. The conclusions are drawn to show a research guide in the real pedestrian evacuation system or other crowded situation for some especially important public and collective activities in the future to prevent possible damage to personal safety and property in unconventional emergency.

Keywords: Stability analysis, Pedestrian evacuation, Model, Unconventional emergency.

1 Introduction

To date, the safe and high-effective pedestrian evacuation in some unconventional emergency is becoming the vital issue in the modern society. Researchers have developed various models to provide designers with methods for forecasting the time required for evacuation from different places under a variety of disasters. And the other important roles played by the crowd evacuation models are to support numerical simulation and provide the emergency managers to make the reasonable decision in some unconventional emergency respectively.

Different methods from a variety of disciplines, such as computer graphics, robotics, evacuation dynamics, etc. have been successfully used to study the pedestrian evacuation model. A few of these are the swarm intelligence model[1],the Lattice Gas Model [2,3], Social Force Model [4],Cellular automata [5,6], Discrete Choice Model [7] and Ant Trail Model [8].While the there still are obvious

* This work is supported by Key Program of National Natural Science Foundation of China(Grant #91024023).

J. Zhang (Ed.): ICAIC 2011, Part V, CCIS 228, pp. 452–458, 2011.

limitations of modeling objects and the computational complexity of the pedestrian evacuation model.

In order to evaluate the functional and mathematical performances of the current established pedestrian evacuation models in unconventional emergency, some typical stability analysis theory are used such as the stability analysis method presented by Chen-xi Shao [9].In [9], the interaction and feedback were observed between variables to analyze the consistency between qualitative prediction matrix and quantitative prediction matrix of the system by exerting disturbance on the variable. Analyzing the system stability can judge whether the system structure is right. The higher stability the better, while it is necessary to adjust the system structure to reduce the collision in the course of evacuation in the case of lower stability.

In this paper, we introduce the main stability theory concisely, then analyze the current stability analysis methods used in evaluating the mathematical performances under small perturbations of initial pedestrian, building and disaster conditions. The many features are extracted completely. Finally the suggestions are presented for the solution to find a appropriate stability theory to analyze the pedestrian evacuation model, and to guarantee support from a evacuation model to a real evacuation in a unconventional emergency.

2 Stability Theory

In mathematics, stability theory addresses the stability of solutions of differential equations and of trajectories of dynamical systems under small perturbations of initial conditions. An orbit is called Lyapunov stable if the forward orbit of any point in a small enough neighborhood of it stays in a small (but perhaps, larger) neighborhood. Various criteria have been developed to prove stability or instability of an orbit. Under favorable circumstances, the question may be reduced to a well-studied problem involving eigenvalues of matrices. A more general method involves Lyapunov functions [10-12].

Stability means that the trajectories do not change too much under small perturbations. Many parts of the qualitative theory of differential equations and dynamical systems deal with asymptotic properties of solutions and the trajectories: what happens with the system after a large time? The simplest kind of behavior is exhibited by equilibrium points, or fixed points, and by periodic orbits. If a particular orbit is well understood, it is natural to ask next whether a small change in the initial condition will lead to similar behavior [10]. The opposite situation, where a nearby orbit is getting repelled from the given orbit, is also of interest. In general, perturbing the initial state in some directions results in the trajectory asymptotically approaching the given one and in other directions to the trajectory getting away from it. There may also be directions for which the behavior of the perturbed orbit is more complicated (neither converging nor escaping completely), and then stability theory does not give sufficient information about the dynamics.

Considering a pedestrian evacuation network model as a non-linear autonomous systems based on the obvious reasons that the psychological model (such as panic and hesitation) the physiological model (the gender-speed, age-speed relation) and behavior model are non-linear, and additionally all these three models interact with each other with a complex principle but all the pedestrian network can be autonomous based on the human being self-organization. Therefore we can use the general analysis method for non-linear autonomous systems as follows.

Asymptotic stability of fixed points of a non-linear system can often be established using the Hartman–Grobman theorem. Assume that v is a C1-vector field in Rn which vanishes at a point p, V(p)=0. Then the corresponding autonomous system S(x),the derivative of S(x) is

$$S(x)' = V(x) \tag{1}$$

S(x)' has a constant solution as

$$S(x) = p \tag{2}$$

Consider that J = Jp(V) is the nxn Jacobian matrix of the vector field V at the point p. If all eigenvalues of J have strictly negative real part then the solution is asymptotically stable.

We can look Lyapunov functions as functions which can be used to prove the stability of a certain fixed point in a dynamical system or autonomous differential equation. Named after the Russian mathematician Aleksandr Mikhailovich Lyapunov, Lyapunov functions are important to stability theory and control theory. A similar concept appears in the theory of general state space Markov Chains, usually under the name Lyapunov-Foster functions[13].Functions which might prove the stability of some equilibrium are called Lyapunov-candidate-functions. There is no general method to construct or find a Lyapunov-candidate-function which proves the stability of an equilibrium, and the inability to find a Lyapunov function is inconclusive with respect to stability, which means, that not finding a Lyapunov function doesn't mean that the system is unstable. For dynamical systems (e.g. physical systems), conservation laws can often be used to construct a Lyapunov-candidate-function.

According to [13], this paper introduce the basic definitions used to evaluate the established pedestrian evacuation models as follows

Definition 1. Lyapunov candidate function

Assume that $V: R^n \to R$ is a scalar function.V is a Lyapunov-candidate-function if it is a locally positive-definite function, i.e.

$$\begin{aligned} &V(0)=0 \\ &V(x)>0 \quad \forall x\in U\setminus\{0\} \end{aligned} \tag{3}$$

Here U is a neighborhood region around x=0.

Definition 2. The equilibrium point of a system

Assume that

$$\begin{aligned} &g: R^n \to R^n \\ &\dot{y}(x) = g(y) \end{aligned} \tag{4}$$

is an arbitrary autonomous dynamical system with equilibrium point y^*:

$$g(y^*) = 0 \tag{5}$$

There always exists a coordinate transformation , such that:

$$\dot{x} = g(x + y^*) = f(x) \tag{6}$$

$$\dot{f}(0) = 0$$

So the new system $f(x)$ has an equilibrium point at the origin.

3 Main Stability Analysis Methods

In typical unconventional emergency, including un-predictable natural disasters(such as Wenchuan earthquake in China, May,2008, earthquake in Haiti,Feb.2010) and social disasters(such as the hostage of Hangkang tourists in the Philippines,Aug.2010), the safe evacuation for pedestrian or crowd shows the necessity to minimize Casualties of the human being.

The stability analysis aims to evaluate the whole dynamic action of the pedestrian moving. In general the pedestrian swarm consists of several kinds of members, such as pedestrians, rescuers, building workers, policemen and decision makers. The decision makers judge the emergency kind and state, then call on the whole pedestrian crowd to move immediately. The pedestrian accept most evacuation disciplines, and follows the group behavior. The rescuer gets into the emergency area and picks up the harmed, disabled, the old people to help them leave as soon as possible. The policeman clear the blocks in escaping way. The building worker direct the pedestrian swarm to exits because of his/her close knowledge of the inner structure and local space conditions. All kinds of the evacuation members come into being of the whole group behavior based on a collective principle.

Until now researchers studied several stability analysis methods to evaluate the established evacuation models, typically as Qualitative and Quantitative Comparison method and Lyapunov method.

3.1 Qualitative and Quantitative Comparison Method

According to [9] and [14], it is a feasible stability analysis method based on qualitative results and quantitative results comparison. The judgement principle is that if the qualitative analysis conclusions accord with qualitative conclusions, then we can think the system is stabile, and vice versa.

In qualitative analysis, the interactions between all kinds of evacuation member s(pedestrians, rescuers, building workers, policemen and decision makers) are modelled into negative effect, non effect and positive effect, finally an effect matrix is established for the interaction analysis and perturb diffusion in the whole pedestrian crowd.

In quantitative analysis, the interactions between all kinds of evacuation members are further converted into real number, and the initial matrix is converted into real number too. With quantitative matrix, we can not only analyse the effect nature between the evacuation members but also calculate the effect degree between the members. Finally we can judge the stability of the evacuation crowd in the whole moving from emergency site to safe area. Moreover, by the ratio of qualitative result to quantitative result, we can classify the evacuation system into three stability degrees, such as stable, asymptotic stable and unstable. Correspondingly, the evacuation system/model is stable when the ratio is equal to 1, asymptotic stable when the ratio is between 0.5 to 1, unstable when the ratio is less than 0.5 referring the definitions in [9].

3.2 Lyapunov Method

In this paper, we introduce the typical Lyapunov functions to analyse the stability of pedestrian/people trajectories, i.e. to judge whether all the escaping trajectories figured out by evacuation models can link to the exits to the safe areas.

Typically the PSO (Particle Swarm Optimize, PSO) model is appropriate and practical to describe the pedestrian evacuation process[1]. As an evolutionary computation technique, Particle Swarm Optimization was first developed by Kennedy and Eberhart [15]. The original idea of PSO was to simulate the social behavior of a flock of birds trying to reach an unknown destination (fitness function), e.g., the location of food resources when flying through the field (search space). For the problem considered here, the destination would represent any of the exits from an emergency site to safe areas. In PSO, each bird in the flock is looked as a "particle", which, for the purposes of the present study, also refers to each person fleeing from a public space in unconventional emergency. Initially, a number of particles (people) are generated and distributed randomly around the study area. Different kinds of particles can be generated according to classification of evacuation members (pedestrians, rescuers, building workers, policemen and decision makers), exhibiting different physical and/or psychological attributes. Then, particles evolve in terms of their individual and social behaviors and mutually coordinate their movement towards their destination [16].

According to equation (2), we define point p as a stable point of the evacuation system when we introduce Lyapunov theory to analyse the stability of an evacuation model.

Based on differential equation (1), we can introduce and apply the basic Lyapunov theorems to judge the stability of the evacuation system based on the current models as follows.

Assume that an equilibrium of the evacuation system as

$$x^* = 0 \tag{7}$$

Assume the one-order differential state variable x to time t of the Lyapunov-candidate-function V has the relation as

$$\frac{dV}{dt} = \frac{\partial V}{\partial x} \cdot \frac{dx}{dt} = \frac{\partial V}{\partial x} \cdot f(x) \tag{8}$$

1) Stable condition

For some small neighbourhood Γ of x, if the Lyapunov-candidate-function V is locally positive definite and the time dericative of the Lyapunov-candidate-function is locally negative semi-definite as

$$\frac{dV}{dt} \leq 0 \quad \forall x \in \Gamma \tag{9}$$

Then the evacuation system is stable.

2) Locally asymptotically stable condition

For some small neighbourhood Γ of x, if the Lyapunov-candidate-function V is locally positive definite and the time dericative of the Lyapunov-candidate-function is locally negative definite as

$$\frac{dV}{dt} < 0 \quad \forall x \in \Gamma \tag{10}$$

Then the evacuation system is locally asymptotically stable.

3) Globally asymptotically stable condition

If the Lyapunov-candidate-function V is globally positive definite, radially unbounded and the time dericative of the Lyapunov-candidate-function is globally negative definite as

$$\frac{dV}{dt} < 0 \quad \forall x \in \mathbb{R}^n \tag{11}$$

Then the evacuation system is globally asymptotically stable.

4 Application of Stability Analysis Methods

This paper put forward a general application schematic of stability analysis for evacuation models shown in Fig.1.

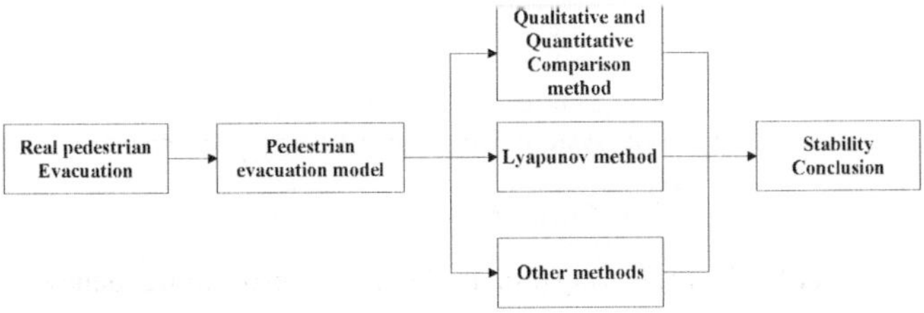

Fig. 1. Application schematic of stability analysis for evacuation models

In Fig.1, the pedestrian evacuation model can be established from the real pedestrian evacuation process and phenomenon happened, with the introduction of main evacuation model described in paragraph 2. Then the main stability analysis method including qualitative and quantitative comparison method and Lyapunov method can be employed to evaluate the stability of pedestrian evacuation system based on the models, stable, locally asymptotically stable, Globally asymptotically stable, or unstable.

5 Conclusions

There are many powerful modelling theories which can be used in establishing the pedestrian evacuation model for the preparedness for unconventional emergency and the decision support in unconventional emergency. While the evacuation of stability of the modeled pedestrian evacuation system is getting more and more attention. The appropriate stability analysis method should depend on the analysis object-the pedestrian evacuation models from various disciplines. So the further research should be addressed on the method selection strategy, the quantitative analysis and new stability analysis methods.

References

1. Izquierdo, J., Montalvo, I., Pérez, R., Fuertes, V.S.: Forecasting pedestrian evacuation times by using swarm intelligence. Physica A: Statistical Mechanics and its Applications 388(7), 1213–1220 (2009)
2. Muramatsu, M., Irie, T., Nagatani, T.: Jamming transition in pedestrian counterflow. Physica A: Statistical Mechanics and its Applications 267, 487–498 (1999)
3. Helbing, D., Isobe, M., Nagatani, T., Takimoto, K.: Lattice gas simulation of experimentally studied evacuation dynamics. Physical Review E 67 (2003)
4. Lakoba, T.I., Finkelstein, N.M.: Modifications of the Helbing Molnár Farkas Vicsek social force model for pedestrian evolution. Simulation 81(5), 339–352 (2005)
5. Kirchner, A., Schadschneider, A.: Simulation of evacuation processes using a bionics-inspired cellular automaton model for pedestrian dynamics (2008), http://arxiv.org/PS_cache/cond-mat/pdf/0203/0203461v1.pdf
6. Pelechano, N., Malkawi, A.: Evacuation simulation models: Challenges in modeling high rise building evacuation with cellular automata approaches. Automation in Construction 17(4), 377–385 (2008)
7. Antonini, G., Bierlaire, M., Weber, M.: Discrete choice models of pedestrian walking behavior. Transportation Research Part B: Methodological 40(8), 667–687 (2006)
8. Nishinari, K., Sugawara, K., Kazama, T., Schadschneider, A., Chowdhury, D.: Modelling of self-driven particles: Foraging ants and pedestrians (2008), http://arxiv.org/PS_cache/cond-mat/pdf/0702/0702584v1.pdf
9. Shao, C.-x., Zhang, Y., Zhang, J.-t., et al.: Qualitative Prediction on Group Evacuation via Feedback Analysis. Journal of System Simulation 17(4), 978–981 (2005) (Chinese)
10. http://en.wikipedia.org/wiki/Stability_theory
11. García-Cerezo, A., Ollero, A., Aracil, J.: Stability of fuzzy control systems by using nonlinear system theory. Annual Review in Automatic Programming 17, 121–126 (1992)
12. Sun, H., Wang, Y., Zhao, W.: Comparison of theories for stability of truss structures. Part 1: Computation of critical load. Communications in Nonlinear Science and Numerical Simulation 14(4), 1700–1710 (2009)
13. http://en.wikipedia.org/wiki/Lyapunov_function#Stable_equilibrium
14. Jeffrey, M.: Dambacher Qualitative predictions in model ecosystems (2003)
15. Kennedy, J., Eberhart, R.C.: Particle swarm optimization. In: IEEE International Conference on Neural Networks, Perth, Australia. IEEE Service Center, Piscataway (1995)

A New Information Retrieval System Based on Fuzzy Set Theory and T-Conorm

Songxin Wang

Department of Computer Science and Technology,
Shanghai University of Finance and Economics,
Shanghai, China
songxinwang@sina.cn

Abstract. The main activity of an Information Retrieval System (IRS) is the gathering of pertinent archived documents that better satisfy the user queries. How to retrieval required information is the key problem in this kind of information retrieval system. In this paper we present a new information retrieval system based on prioritized fuzzy information and T-Conorm, for dealing with such a problem. The advantage of this system with respect to others is that the use of the prioritized weighting information facilitates the expression of information needs and improves the latter issue. Central to our approach of modelling priority weighting information is that the importance of lower priority criteria will be based on its satisfaction to the higher priority criteria, and that a T-Conorm operator is used to calculate the priority in the process.

Keywords: information retrieval, fuzz y set,T-Conorm.

1 Introduction

An Information Retrieval System is a information system whose aim function is to evaluate users queries for information based on a content-analysis of the archived documents. The main activity of an Information Retrieval System (IRS) is the gathering of the pertinent led documents that best satisfy user information requirements (queries). An interest question in the IRSs is how to facilitate the IRS-user interaction. In many applications, linguistic variables are used to represent the input and output information in the retrieval process of IRSs to improve the IRS-user interaction.

Basically, IRSs present three components to carry out their activity according to [1]:

1. A Database: which stores the documents and the representation of their information contents (index terms).

2. A Query Subsystem: which allows use s to formulate their queries by means of a query language.

3. An Evaluation Subsystem: which evaluates the relevance of each document for a user query by means of a retrieval status value(RSV).

A promising direction to improve the effectiveness of IRSs consists of representing in the queries the users' concept of relevance. This is a very complex task because it

J. Zhang (Ed.): ICAIC 2011, Part V, CCIS 228, pp. 459–466, 2011.

presents subjectivity and uncertainty. To do so, a possible solution consists in the use of weighting tools in the formulation of queries. By attaching weights in a query, a user can increase his expressiveness and provide a more precise description of his desired documents. Usually, the most linguistic IRSs assume that users provide their weighting information needs by means of linguistic values represented by the linguistic variable "Importance" assessed on a label set S. Then, the activity of the IRS involves evaluating the linguistic weighted queries and providing the linguistic RSVs of documents represented by the linguistic variable "Relevance", which is also assessed on S. There are many works along this line[2-5].

The drawback is that the use of the numerical or linguistic weighting information. In many applications, elicitation of precisely specified attribute weights may be difficult, we cannot give a numerical or even a fuzzy number to the weight, instead, we only have a partitioning of the set of criteria into M disjoin class, M_k ,k=1 to m, such that all the criteria in the same class are tied and for a i<j,all criteria in Mi are said to have a higher priority than the criteria in M_k. We effectively have ordered equivalence classes. Although this is rather common in realistic applications, none of the works mentioned above can deal with these situations.

This paper presents a new information retrieval system based on prioritized fuzzy information and T-Conorm, for dealing with such a problem. The advantage of this method with respect to others is that the use of the prioritized weighting information facilitates the expression of information needs and improves the latter issue. Central to our approach of modeling priority weighting information is that the importance of a lower priority criteria will be based on its satisfaction to the higher priority criteria, and that a T-Conorm operator is used to calculate the priority in the process. The method presented in this paper is an improvement of our works given in [6].

This paper is organized as follows. In Section 2, we firstly give some preliminary knowledge of fuzzy number and then give a new a prioritized aggregation operator, In Section 3, we presents our method based on linguistic information. And finally some concluding remarks are pointed out.

2 A Prioritized Aggregation Operator

In this section, we firstly give some preliminary knowledge of fuzzy number and T-norm then give a new a prioritized aggregation operator.

A fuzzy number is a convex fuzzy subset of the real line R and is completely defined by its membership function. A popular fuzzy number is the normal triangular fuzzy number.

Definition 1: Normal triangular fuzzy number

$$\mu_{\tilde{A}}(x) = \begin{cases} \frac{x-a}{b-a}, & \text{if } a \le x \le b \\ \frac{c-x}{c-b}, & \text{if } b \le x \le c \\ 0, & \text{otherwise.} \end{cases}$$

Let $A_1 = (a_1, b_1, c_1)$ and $A_2 = (a_2, b_2, c_2)$ be two triangular fuzzy number. The following fuzzy arithmetic operations are defined.

Definition 2: Addition

$$A_1 \oplus A_2 = [a_1 + a_{2,} b_1 + b_{2,} c_1 + c_2]$$

Definition 3: Scalar multiplication

$$k \odot A = (k \times a, k \times b, k \times c)$$

As discussed in [7],the fuzzy numbers resulting from these fuzzy operations retain their original triangular forms.

In many situations fuzzy numbers must be first mapped to real values, this assignment of a real value to a fuzzy number is called defuzzfication. It can take many forms, but the most standard defuzzifiction is through computing the centroid which is defined as the center of gravity of the curve describing a given fuzzy quantity.

Definition 4: Let $\tilde{B} = (a, b, c)$ be a normal triangular fuzzy number, the centroid of the $\tilde{B}$ is

$$C_{\tilde{A}} = \frac{\int_a^d x\mu_{\tilde{B}}(x)\mathrm{d}x}{\int_a^d \mu_{\tilde{B}}(x)\mathrm{d}x} = \frac{a+b+c}{3}$$

Next we give the definition of T-norm and T-Conorm according to [8].

Definition 5: A binary operator R:H×H→H is called an uniform if it satisfies:

1 Symmerty:R(x,y)= R(y,x)
2 Associativity:R(x,y,z)=R(x,R(y,z))= R(R(x,y),z))
3 Monotonicity: If a⩾c and b⩾d then R(a,b)⩾R(c,d)
4 h is an identity element. For any a∈H, R(a, h_q)=R(h,a)=a

If we choose H =[0, 1], then an uniform R is called T-norm if h=1,and R is called T-Conorm if h=0. In situations where H is not[0, 1], the T-norm and T-Conorm can be given similarly so long as identity element is suitable choose.

Note that although a uniform R is binary, this operator is easily to extended to the situation where the arguments is many due to the associatively nature.

Now we give the new a prioritized aggregation operator.

Formally we have following problem settings:

We have n alternatives to evaluate, $X_1,\ldots X_n$, assume that we have a collection of criteria partitioned into Q distinct categories $H_1,\ldots H_q$ such that $H_i=\{C_{i1},\ldots C_{in}\}$, here C_{ij} are the criteria in category H_i.

We assume a prioritization between these categories $H_1 \geq H_2 \geq \ldots\ldots \geq H_j$,

The criteria in the class H_i have a higher priority than those in H_k if i≥k. We assume that the total number of criteria is

$$n=\sum_{i=1}^{q} n_i ,$$

We assume that for any alternative $x \in X_i$ we have for each criteria C_{ij} a linguistic values, that is a normal triangular fuzzy number V_{ij} indicting its value.

The aggregation operator is given as follows:

For each priority category H_i we calculate S_i, where

$$S_i = \underset{j}{Min}\{COA(V_{ij})\},$$

Here S_i is the value of the least satisfied criteria in category H_i under alternative x.

Using this we will associate with each criteria C_{ij} a value u_{ij}. In particular for those criteria in category H_1 we have u_{1j}=1. For those criteria in category H_2 we have $u_{2j}=S_1$. For those criteria in category H_2 we have $u_{3j}=S_1 * S_2$, here $*$ is a T-Conorm. For those criteria in category H_4 we have $u_{4j}=S_1 * S_2 * S_3$. More generally u_{ij} is the product of the least satisfied criteria in all categories with higher priority than H_i.

We can more succinctly and more generally express u_{ij}= Ti where Ti is defined as follows:

Definition 6:

$$T_i = \prod_{k^i-1}^{i} S_{k-1}$$

Note that although the usual $\prod$notation is used here for concise the operation between S_i is in fact a T-Co-norm.

With the understanding that S_0= 1 by default, we note that we can also express T_i as

$$T_i = S_{i-1} * T_{i-1} ,$$

Here $*$ is a T-Conorm.

This equation along with the fact that $T_1=S_0=1$ gives a recursive definition at T_i.

We now see that for all $C_{ij} \in H_i$ we have $u_{ij}= T_i$. Using this we obtain for each C_{ij} a weight wij with respect to alternative x such that $w_{ij}= u_{ij}$. We further observe that $T_i \geq T_k$ for $i < k$. From this it follows that if $i \leq j$ then $w_{ij} \geq w_{ke}$ for all j and e.

Using these weights we then can get an aggregated score x under these prioritized criteria as follows:

Definition 7:

$$F(x) = \frac{1}{n} \sum_{\oplus} u_{ij} \otimes V_{ij}(x)$$

3 The IRS with Linguistic Information

In this section we present a linguistic IRS which uses linguistic term set to express the linguistic assessments in the retrieval process, this IRS accepts prioritized queries and provides linguistic retrieval status values (RSVs). The components of this IRS are presented in the following subsections.

3.1 Database

The database stores the finite set of documents

$$D=\{d_1,d_2,......,d_m\},$$

and the finite set of index terms

$$T=\{t_1,t_2,......,t_l\}.$$

Documents are represented by means of index terms, which describe the subject content of the documents. A indexing function F: D×T→A exists where A is a fuzzy number. F weighs index terms according to their significance in describing the content of a document in order to improve the retrieval of documents.

3.2 The Query Subsystem

The query subsystem accepts prioritized weighted queries whose terms can be weighted in forms of a partitioning of the set of criteria . By associating this kind of weights to terms in a query, the user is asking to see all documents whose content represents the concept that is more associated with the most interest terms than with the less interest ones.

Then, each prioritized query is expressed as a combination of the index terms and a a partitioning of the set of criteria H, that is

$$p=\langle A_1, A_2,......A_n : H\rangle.$$

3.3 The Evaluation Subsystem

The subsystem evaluates prioritized weighted queries by means of a constructive process. Documents are evaluated according to their relevance to user's interest.

In this step, the prioritized aggregation operator proposed in the section2 is applied to evaluate the prioritized weighted query. At the end a satisfy degree is assigned to each document with respect to the whole query. Then, the evaluation subsystem presents the retrieved documents arranged in ascending ordering.

Let pi be a query and dj be a document, Evaluation of a prioritized weighted query is defined as

$$E(p_i,d_j)=COA(F(d_j)),$$

Note that each indexing function is treated as a criteria when prioritized aggregation operator is used.

Once the evaluation has obtained, they may be used to obtain the set of alternatives with maximal evaluation, in this way, the document that the user is most interested in is selected.

4 A Numerical Example

This section presents a numerical example to illustrate the method proposed in this paper.

It is important that in the example a multiply operation is used as a concrete T-Conorm, however, other operation may be used according to different situations.

We will assume there are eight attributes of interest to the users in the process of information retrieval, they are divided into $H_1= \{C_{11,} C_{12}\}, H_2= \{C_{21}\}, H_3= \{C_{31,} C_{32},$ $C_{33}, H_4= \{C_41, C_{42}\}$. Assume that we have prioritized weighted query and a document x as candidate.

The linguistic term sets and associated semantics of labels, which is a triangular fuzzy number, used here are given in following table.

1: Definitely uninterest	(0:2; 0:3; 0:4)
2: Very uninterest	(0:3; 0:4; 0:5)
3: Uninterest	(0:4; 0:5; 0:6)
4: Middle	(0:5; 0:6; 0:7)
5: Interest	(0:6; 0:7; 0:8)
6: Very interest	(0:7; 0:8; 0:9)
7: Definitely interest	(0:8; 0:9; 1:0)

Assume for an alternative x we have

$V_{11}(x)$ = Uninterest, $V_{12}(x)$ = Interest ,
$V_{21}(x)$ = Very Interest,
$V_{31}(x)$ = Definitely Interest, $V_{32}(x)$ = Middle, $V_{33}(x)$ = Interest,
$V_{41}(x)$ = Very interest , $V_{42}(x)$ = Interest.

We shall use the model proposed in this paper to solve this problem.
We first calculate

S_1= Min[COA(V_{11}(x))), COA(V_{12}(x))] = 0.5,
S_2= Min[COA(V_{21}(x))] = 0.8,
S_3= Min[COA(V_{31}(x)), COA(V_{32}(x)), COA(V_{33}(x))] = 0.6,
S_4= Min[COA(V_{41}(x)), COA(V_{42}(x))] = 0.7.

Using this we get

$T_1= 1,$
$T_2= S_1T_1= 0.5,$
$T_3= S_2T_2= 0.4,$
$T_4= S_3T_3= 0.24.$

From this we obtain

$u_{11}= u_{12}= T_1= 1,$
$u_{21}= T_2= 0.5,$
$u_{31}= u_{32}= u_{33}= T_3= 0.4,$
$u_{41}= u_{42}= T_4= 0.24.$

In this case then we have

$w_{11}= w_{12}= 1$,
$w_{21}= 0.5$,
$w_{31}= w_{32}= w_{33}= 0.4$,
$w_{41}= w_{42}= 0.24$.

We now calculate

$$F(x) = \frac{1}{n} \sum_{\oplus} u_{ij} \otimes V_{ij}(x)$$
$$= (0.30, 0.41, 0.25).$$

5 Conclusion

Information retrieval involves the development of computer systems for the storage and retrieval of textual information. The main activity of an Information Retrieval System is the gathering of the pertinent led documents that best satisfy user information requirements (queries).

In this contribution we have presented an information retrieval system (IRS) based on linguistic information for dealing with such a problem. The advantage of this linguistic IRS with respect to others is that the use of the prioritized weighting information and T-Conorm to facilitates the expression of information needs and improves the latter issue. Central to our approach of modelling priority weighting information is that the importance of lower priority criteria will be based on its satisfaction to the higher priority criteria. This methodology allows us to improve the performance of IRS by increasing the classification levels of the documents.

Acknowledgment. The paper is supported by 211 project of Shanghai University of Finance and Economics and Shanghai Innovation Project (08dz112500) .

References

1. Salton, G., McGill, M.H.: Introduction to Modern Information Retrieval. McGraw-Hill, New York (1983)
2. Herrera-Viedma, E., López-Herrera, A.G.: A model of an information retrieval system with unbalanced fuzzy linguistic information. Int. J. Intell. Syst. 22(11), 1197–1214 (2007)
3. Bordogna, G., Pasi, G.: An Ordinal Information Retrieval Model. Int. J. of Uncertainty, Fuzziness and Knowledge Based Systems 9, 63–76 (2001)
4. Herrera-Viedma, E.: Modeling the Retrieval Process for an Information Retrieval System Using an Ordinal Fuzzy Linguistic Approach. Journal of the American Society for Information Science and Technology 52(6), 460–475 (2001)
5. Herrera-Viedma, E.: An IR model with Ordinal Linguistic Weighted Queries Based on Two Weighting Elements. Int. J. of Uncertainty, Fuzziness and Knowledge-Based Systems 9, 77–88 (2005)

6. Wang, S., Huang, H.: An Information Retrieval System with weighted Querying Based on Linguistic information. In: Proceedings of 3rd International Coference on Multimedia and Ubiquitours Engineering, Qingdao, China, pp. 14–18 (2009)
7. Chang, P.T.: Fuzzy regression analysis, Ph.d.dissertation, Department of industrial and manufacturing Engineering. Kansas State University (1994)
8. Yager, R., Rabalov, A.: Uniform aggregation operators. Fuzzy set and Systems 80, 111–120 (1996)

Combining Different Classifiers in Educational Data Mining

He Chuan, Li Ruifan, and Zhong Yixin

School of Computer Science,
Beijing University of Posts and Telecommunications,
Beijing, China
hc1258@yeah.net

Abstract. Educational data mining is a crucial application of machine learning.The KDD Cup 2010 Challenge is a supervised learning problem on educational data from computer-aided tutoring. The task is to learn a model from students' historical behavior and then predict their future performance. This paper describes our solution to this problem. We use different classification algorithms, such as KNN, SVD and logistic regression for all the data to generate different results, and then combine these to obtainthe final result. It is shown that our resultsarecomparable to the top-ranked ones in leader board of KDD Cup 2010.

Keywords: data mining, logistic regression, k-nearest neighbor, singular value decomposition, classifiers combination.

1 Introduction

In KDD Cup 2010, the task is to predictstudent algebraic problem performance giveninformation regarding pasperformance. This prediction task presents not only technical challenges for researchers,but is also of practical importance, as accurate predictions can be used, for instance, tobetter understand and ultimately optimize the student learning process.Specifically, participants were provided with summaries of the logs of student interactionwith intelligent tutoring systems. Two data sets are available: algebra 2008-2009 andbridge to algebra 2008-2009. In the rest of this paper, we refer to them as A89 and B89, respectively. Each data set contains logs for a large number of interaction steps. Some interaction log fields areincluded in both training and testing sets, such as student ID, problem hierarchy includingstep name, problem name, unit name, section name, as well as knowledge components (KC)used in the problem and the number of times a problem has been viewed. However, somelog fields are only available in the training set: whether the student was correct on the firstattempt for this step (CFA), number of hints requested (hint) and step duration information. The details are list in Table 1.

J. Zhang (Ed.): ICAIC 2011, Part V, CCIS 228, pp. 467–473, 2011.

Table 1. Dataset statistics

Datasets	Algebra 2008-2009	Bridge to Algebra 2008-2009
Lines (train)	8,918,054	20,012,498
Students (train)	3,310	6,043
Steps (train)	1,357,180	603,176
Problems (train)	211,529	63,200
Section (train)	165	186
Units (train)	42	50
KC (train)	2,097	1,699
Steps (new on test)	4,390	9,807

The competition regards CFA, which could be 0 (i.e., incorrect on the first attempt) or 1, as the label in the classification task. For each data set, a training set with knownCFA is available to participants, but a testing set of unknown CFA is left for evaluation. The evaluationcriterion used is the root mean squared error (RMSE). In the competition, participantssubmitted prediction results on the testing set to a web server, where the RMSE generatedbased on a small subset of the testing data is publicly shown. This web page of displayingparticipants' results is called the "leader board."

Facing such a complicated problem, we use different classification algorithms such as KNN, SVD [2, 3] and logistic regression [2] for all the data to generate different results, and then combine these to get final result. In particular, logistic regression needs many proper features to work well, so feature engineering is a necessary step. KNN and SVD, which are transferred from collaborative filtering community, will exploit the basic information in the given data. In the following sections, we make the arrangement: Section 2 shows our method in details, including some preprocessing, data grouping, feature engineering, logistic regression and trust region optimization, and combining methods. Section 3 gives the final result and the discussion for our method.

2 Our Method

This section is the main part of our paper. It explains the whole procedure of our method.

A. Validation Set Generation

Because we do not have all the ground truth labels for test data, we have to generate validation sets by ourselves. Table 2 shows the number of samples in validation set of Algebra 2008-2009 and Bridge to algebra 2008-2009.

Table 2. The number of samples in validation and training set of Algebra2008-2009 and Bridge to algebra 2008-2009

Datasets	algebra 2008-2009	bridge to algebra 2008-2009
Training(V)	8,407,752	19,264,097
Validation()	510,303	748,402
Training(T)	8,918,055	20,012,499
Training()	508,913	756,387

B. Feature Engineering

Basic features: Some basic yet important features considered in our early experiments can be categorizedinto two types: student name, unit name, section name, problem name, step name and KC arecategorical features, while problem view and opportunity are numerical features.

Combining features: Because all feature meanings are available, we are able to manuallyidentify some useful pairs offeatures. For example, hierarchical information can be modeledby indicating the occurrences of the following pairs: (student name, unit name), (unit name,section name), (section name,problem name) and (problem name, step name). In addition to two-feature combinations, we have also explored combinations of higher-order features.

Temporal features: Because learning is a process of skill-improving over time, temporal information shouldbe taken into consideration. There are some well-established techniques, utilizing a quitedifferent data model than traditional classification problems, to model student latent attributessuch as knowledge tracing and performance factor analysis. We considered a simple and commonapproach to embed temporal information into featurevectors. For each step, step name and KC values from the previous few steps were added asfeatures.

Other features: KCs are obtained by splitting the KC string with"~~"following the suggestion at the "Data Format" page on the competition website.Then binaryfeatures are used to indicate the KCs associated with each step. However, thissetting results in many similar KCs. To remedy this problem, we tokenized KC strings and used each token as a binary feature.Our experiences indicated thatthis method for generating KC features is very useful for the data set A89. We also usedtechniques such as regular expression matching to parse knowledge components, but didnot reach a setting clearly better than the one used for the baseline.For problem name and step name, we tried to group similar names together via clusteringtechniques. This approach effectivelyreduced the number of features without deteriorating the performance. We have also tried to model the learning experience ofstudents. A student's performance may depend on whether that student had previouslyencountered the same step or problem. We added features to indicate such information.Results showed that thisinformation slightly improves the performance.Earlier problem view was considered as a numerical feature. In some situations, wetreated it as a categorical one and expanded it to several binary features. This setting ispossible because there are not too many problem view values in training and testing sets.One student sub-team reported that this modification slightly improved RMSE.

C. Logistic Regression

After feature generation, we are building a classifier with feature vectors as input. We here use logistic regression as the classification algorithm, which is used for prediction of the probability of occurrence of an event by fitting data to a logit function logistic curve. In particular, the logistic regression model in [2] is:

$$P(y|x) = \frac{1}{1+\exp(-yw^T x)} \tag{1}$$

where x is feature vector of a sample and y is the label of the sample (here y is CFA). Then a regularized logistic regression model is:

$$\min_{w} \frac{1}{2} w^T w + C\sum_{i=1}^{l} \log(1+e-y_i w^T x_i) \tag{2}$$

where w is the weight vector to be obtained through training, and the first term is the l2-regularization and C is a penalty parameter. In our experiments, we use liblinear toolkit [1] to build the models and test the data. Liblinear is a linear classifier which can deal with data with millions of instances and features. It supports to solve seven optimization problems which of course includes *l2*-regularized logistic regression, and is very efficient for training large-scale problems even much faster than libsvm. Liblinear adapts trust region method to optimize *l2*-regularized logistic regression problem.

D. KNN and SVD

In this section, we explain two classification methods which have been working in collaborative filtering (CF) for a long time. In the educational data mining problem, we consider the students are users in CF and the steps are items such as books or movies in CF. Except for these two fields in given data, we basically ignore other fields which can provide a bunch of information though.

In machine learning, the k-nearest neighbor algorithm (k-NN) is a method for classifying samples based on closest training examples in the feature space. *K*-NN is a type of instance-based learning, or lazy learning where the function is only approximated locally and all computation is deferred until classification. The *k*-nearest neighbor algorithm is amongst the simplest of all machine learning algorithms: an object is classified by a majority vote of its neighbors, with the object being assigned to the class most common amongst its k nearest neighbors (*k* is a positive integer, typically small). If $k = 1$, then the object is simply assigned to the class of its nearest neighbor.

The singular value decomposition is well known in linear algebra and states that an arbitrarymatrix *C* with real or complex entries can be decomposed into a product of three matrices. In the collaborative filtering context the matrix*C* is sparse, meaning that most of the elements are unknown. The basic idea of SVD todecompose the matrix *C* can be still applied. During the Netflix competition, this methodbecame very popular because of two reasons. First it delivers good results out of the box andsecond it is easy to tune and customize. Over the last years there were lots of publicationsdescribing extensions to the basic SVD. In collaborative filtering, SVD in [3] is equivalent to non-negative matrix factorization, that is

$$\min_{A,B} \left\| C - AB^T \right\|_C^2 \tag{3}$$

To keep it from over-fitting, the formula can be written as:

$$E(A,B) = \|C - AB^T\|_C^2 + \lambda(\|A\|_F^2 + \|B\|_F^2) \tag{4}$$

With stochastic gradient descent algorithm, it can be solved efficiently.

E. Combination

Combination is the last not the least step in our method.The individual predictions stemmingfrom the methods described above are combined. This is a typical approach to improve the prediction accuracy, which wasshown in [5][6].It is widely known that an ensemble of predictions performsbetter than the best individual result.

3 Experiment Results and Discussion

A. Results

To compare the performance of all the classifiers, we experiment on KNN, SVD, logistic regression and their combination. Table 3 to 4 show the results from all the classifiers over the given two datasets.

Table 3 is the results from KNN algorithm, where *K* is the number of nearest neighbors.

Table 3. Results from KNN algorithm

Dataset	RMSE	Meta parameters
Algebra 2008-2009	0.3257	$K=41, \alpha=12.9, \beta=1.5, \delta=6.2, \gamma=-1.9$
Bridge to Algebra 2008-2009	0.3049	$K=41, \alpha=12.9, \beta=1.5, \delta=6.2, \gamma=-1.9$

The following table is the results from SVD, where *N* is the latent factor number, it is the learning rate, and lambda is the L_2 regularization coefficient.

Table 4. Results from SVD

Dataset	RMSE	Meta parameters
Algebra 2008-2009	0.446277	$N=10, \eta=0.002, \lambda=0.02$
Bridge to Algebra 2008-2009	0.3049	$N=10, \eta=0.002, \lambda=0.02$
Bridge to Algebra 2008-2009	0.3159	$N=20, \eta=0.002, \lambda=0.01$
Bridge to Algebra 2008-2009	0.3178	$N=20, \eta=0.002, \lambda=0.03$

Table 5 is the result from logistic regression:

Table 5. Result from logistic regression

Dataset	RMSE
Algebra 2008-2009	0.2895
Bridge to Algebra 2008-2009	0.2985

Table 6 is the result after combination

Table 6. Result after combination

Dataset	RMSE
Algebra 2008-2009	0.2820
Bridge to Algebra 2008-2009	0.2850

B. Discussion

From the four tables above, we can see the combination method works better than any of the single classifier. Logistic regression has the best performance among the three methods; due to not only it exploits almost all the information in the given data but also it uses very fine feature vectors as input. Even so, the combined classifier can reach new lower RMSE value than it over the two dataset.

As for CF-like methods, we can also try to make use of other information such as KC components, problems, unit, which is different from the features used in logistic regression, but can provide more discriminative information. Besides the combination method, there are some other ways which are considered to be interesting such as dynamic Bayesian network. HMM in [4] is one of DBNs considering the changing math ability of students. In a word, there would be a great number of ways to be explored.

Acknowledgement. This paper is supported by Project 60873001 of National Natural Science Foundation of China and by Project 2009RC0212 of Chinese Universities Scientific Fund.

References

1. Fan, R., Chang, K., Hsieh, C., Wang, X., Lin, C.: LIBLINEAR: A libraryfor large linear classification. Journal of Machine Learning Research 9, 1871–1874 (2008)
2. Yu, H.-F.: Feature engineering and classifier ensembling for KDD Cup 2010. Journal of Machine Learning Research, Workshop and Conference Proceedings (2010)
3. Töscher, A.: BigChaos, Collaborative Filtering Applied to Educational Data Mining. Journal of Machine Learning Research, Workshop and Conference Proceedings (2010)

4. Pardos, Z.A.: Using HMMs and bagged decision trees to leverage rich features of user and skill. Journal of Machine Learning Research, Workshop and Conference Proceedings (2010)
5. Freund, Y., Schapire, R.: A decision-theoretic generalization of on-line learning and anapplication to boosting. Journal of Computer and System Sciences 55(1), 119–139 (1997)
6. Jahrer, M., Töscher, A., Legenstein, R.: Combining predictions for accuraterecommender systems. In: KDD: Proceedings of the 16th ACM SIGKDD International Conference on Knowledge Discovery and Data Mining (2010)

A Model for Looper-Like Structure in Web-Fed Printing Press

Yi Wang[1], Kai Liu[2], and Haiyan Zhang[1]

[1] School of Printing and Packaging Engineering,
X'an University of Technology,
Xi'an 710048, China
ywk1298@163.com

[2] School of Mechanical and Precision Instrument Engineering,
X'an University of Technology,
Xi'an 710048, China

Abstract. In a multi-unit web-fed system, tension control is important because it affects the strip quality. Moreover, the most difficult challenge in the controller design and control performance comes from the interaction between looper angle and strip tension. In this paper, the structure of looper in a web span with idlers is presented and the model of looper-like structure is proposed to effectively eliminate the 'tension transfer' phenomenon among web spans in a multi-unit web-fed system.

Keywords: Index Terms, looper-like structure, model, web span.

1 Introduction

In a multi-unit web-fed system, the web material supply and accumulation is finished by the unwinding and wingding roll. A view of the simplified schematic of web-fed printing units is shown in Fig. 1. Tension estimate and control is a very important part of process control in web-fed printing machine. The dynamic characteristics of the mechanical components and the physical properties of the web material affect the steady-state and dynamic behavior of the web in the longitudinal direction. Precision tension control for high quality product is requested in web-fed printing system. In general, it is difficult for us to effectively reduce the impact of speed variations on tension control, especially when the plant model contains uncertain parameters.

Early work by Shin [1] derived a mathematical tension model considering tension. Whereas Song and Sul [2] proposed a feed forward control with a tension observer to improve the transient performance. Dou [3] provided a reference of winding process control for achieving robust stability in a multistage printing system. Although in the winding process, the desired web speed and tension can be estimated and controlled through actuators, the distinct phenomenon that a tension change in a certain web span will affect the tension in the following span cannot be eliminated. Based on the relation between the tension variation and web length change of a web span, dancer rolls and loopers discussed in the following details are usually be used to be tension control actuators.

J. Zhang (Ed.): ICAIC 2011, Part V, CCIS 228, pp. 474–478, 2011.

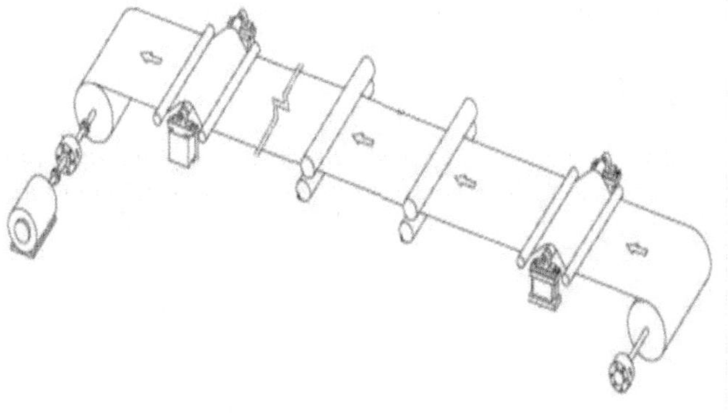

Fig. 1. Simplified schematic of web-fed printing units

2 Mathematical Model of Web Span System

Figure 2 shows a single intermediate web span, the most fundamental primitive element in web processing systems.

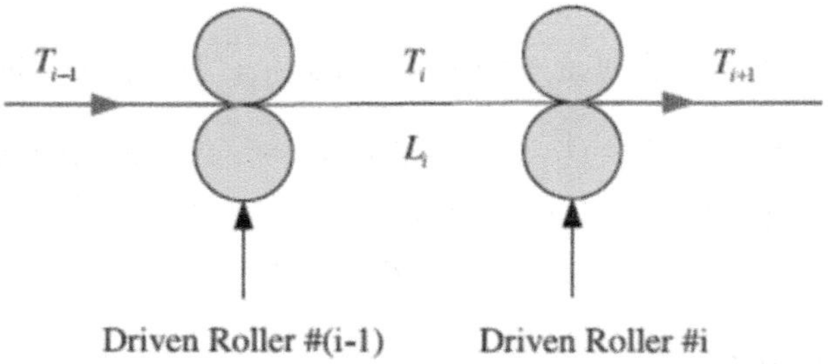

Fig. 2. A single web span

If the dynamics of load cell and idle rollers are neglected, a linearized dynamic model of the *i*th web span can be derived as:[1]

$$T_i = -\frac{v_{i0}}{L_i}T_i - \frac{aE}{L_i}V_{i-1} + \frac{aE}{L_i}V_i + \frac{v_{(i-1)0}}{L_i}T \tag{1}$$

$$V_i = -\frac{R_i^2}{J_i}T_i - \frac{B_{fi}}{J_i}V_i + \frac{R_i^2}{J_i}T_{i+1} + \frac{R_i^2}{J_i}K_{mi}I_i \tag{2}$$

Where:

T_i; V_i; I_i: Changes in tension, velocity, and motor current input in the *i*th web span;

V_{i0}: Steady-state web velocity of the *i*th web span;

L_i:The *i*th web length between two driven rollers;

$a\,;E$: Cross sectional area and Young's modulus of the web material;

$R_i\,;J_i\,;\;B_{fi}\,;K_{mi}$: Radius, inertia, friction coefficient, and motor torque constant of the *i*th driven roller.

From equation (1), a tension change in a certain web span (the *i*th) will affect the tension in the following span---the change is propagated backward.

3 Mathematical Model of Looper Structure

A looper installed at inter–stand positions reduces tension variations by changing its angle, so it can contribute to the quality of products. Ideally, the looper angle needs to keep a desired value during operation to reduce the tension variation and to have the flexibility to achieve large changes in loop length during an abnormal rolling condition.

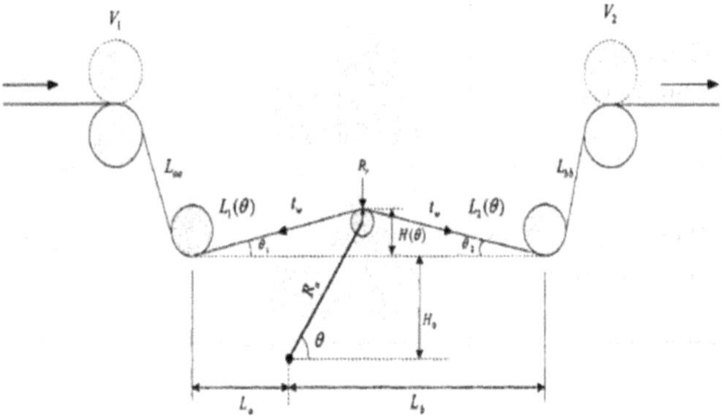

Fig. 3. The structure of looper in a web span with idler rolls

Figure 3 shows that there are usually idler rolls between the two end drive motors of a web span. This structure will be adopted as actuators to attenuate tension fluctuations resulted from driven roller velocity variations or tension disturbances from adjacent web spans. Tension change in a looper-controlled web span will not bring tension transfer to the following web spans, since no driven roller velocity has to be changed in order to achieve the tension control target.

An important issue is that the looper cannot be moved to any angle. The rotation should be restricted within a meaningful range. Looper angle constraints are defined as:

$$\theta_{max} = \frac{\pi}{2} \tag{3}$$

$$\theta_{min} = \sin^{-1}\left(\frac{H_0 - R_r}{R}\right) \quad \text{when} \quad H(\theta) = 0 \tag{4}$$

If there is no web delay between any nipped driven rollers, then the mathematical model of looper in a web span with idlers can be described by the following equations:

$$t_w(\theta) = t_{w0} + T_w = t_{w0} + T_{w1} + T_{w2} \tag{5}$$

$$T_{w1} = aE\frac{L(\theta) - L_0}{L_0} \tag{6}$$

$$T_{w2} = \frac{1}{L_0}[-v_{20}T_{w2} + aE(V_2 - V_1) + v_{10}T_u] \tag{7}$$

Where

$$L(\theta)=L_{aa}+L_1(\theta)+L_2(\theta)+L_{bb}=L_{aa}+\sqrt{H^2(\theta)+(L_a+R_a\cos\theta)^2}+\sqrt{H^2(\theta)+(L_b-R_a\cos\theta)^2}+L_{bb} \tag{8}$$

$$L_0 = L(\theta_0) = L_{aa} + L_1(\theta_0) + L_2(\theta_0) + L_{bb} \tag{9}$$

$$H(\theta) = R_a \sin(\theta) + R_r - H_0 \tag{10}$$

The moment balance description is derived by applying Newton's second law.

$$J_L\theta = \gamma\tau_m - \tau_t - \tau_f = \gamma K_m i_m - \tau_t - B_f\theta \tag{11}$$

Where, τ_m is the toque generated by the looper driving motor; τ_t is the toque generated by web tension; i_m is the current input of the driving motor.

$$J_L = J_r + J_a = \frac{1}{2}M_r R^2_r + M_r R_a^2 + \frac{1}{3}M_a R_a^2 \tag{12}$$

$$\begin{aligned}\tau_t &= t_w R_a \sin(\theta+\theta_2) - t_w R_a \sin(\theta-\theta_1) \\ &= t_w R_a[\sin(\theta+\theta_2) - \sin(\theta-\theta_1)]\end{aligned} \tag{13}$$

$$\theta_1 = \tan^{-1}\frac{H(\theta)}{L_a + R_a\cos\theta} \tag{14}$$

$$\theta_2 = \tan^{-1}\frac{H(\theta)}{L_b + R_a\cos\theta} \tag{15}$$

4 Summary

In this paper, we introduced the tension variation in web span system and presented the structure of looper in a web span with idlers. The model of looper-like structure is proposed and with the designed inactive actuator, tension control is no longer performed by controlling web driving motors and is separated from line speed control, so tension variation within a single web span will not be propagated to following spans.

References

1. Shin, K.H.: Tension Control. Tappi Press, Technology Park/ Atlanta (2000)
2. Song, S.H., Sul, S.K.: A new tension controller for continuous strip processing line. IEEE Transactions on Industry Applications 36(2), 633–639 (2000)
3. Dou, X., Wang, W.: Robust control of multistage printing systems. Control Engineering Practice 18, 219–229 (2010)
4. Chen, C.L., Chang, K.M., Chen, C.M.: Modeling and control of a web-fed machine. Applied Mathematical Modeling 28(10), 863–876 (2004)
5. Asano, K., Yamamoto, K., Kawase, T., Nomura, N.: Hot strip mill tension–looper control based on decentralization and coordination. Control Engineering Practice 8, 337–344 (2000)

Semantic Association Based Image Retrieve and Image Index

Haiying Zhou[1,2] and Zhichun Mu[3]

[1] School of Information Engineering, University of Science and Technology Beijing, Beijing, 100083, China
Huaguoshan, Jileshijie Province, China
[2] School of Electronics and Computer Science and Technology, North University of China, Taiyuan, Shanxi, 030051, China
[3] School of Information Engineering, University of Science and Technology Beijing, Beijing, 100083, China
HaiyingZhou1980@tom.com

Abstract. The challenges to content based image retrieve (CBIR) are various image retrieval requirements as well as the complex and hard described image content, and the gap between the digital array of image expression and the conceptual information universally accepted by human being. In this paper, a semantic association based image retrieve is proposed. Based on semantic association, a semantic representation vector for a scenic category is formed so that users can organize the similar images in perception together to form perceptional context and from which users can explain and mark images without need to give the connotation description for images. In addition, a kind of image neuron index method is proposed, which can speed up and guide the image or image block detection.

Keywords: Semantic Association, Image Retrieve, Image Index.

1 Introduction

With the evolution of Internet and the application increasing of image acquisition devices, the digital instrument of image search and browse find wide use in various application fields. Many image retrieval systems emerge because of demand. The image retrieval system applied in present basically is divided into two kinds of frame: text based retrieval and the content based retrieval. The text based image retrieval is carried out on manual annotations which are the key words describing the basic connotations in an image. This kind of system is managed through database, the shortcoming is time consuming and arduously, and due to the individual subjective factors the annotation often is inexact. For overcoming the above shortcoming, content based image retrieve (CBIR) appeared in1980's, this kind of retrieval system is make use of visual content as the index of image, such as color, texture, and shape and so on. As we know, the images in an image base have certain latent internal relations between them. By setting up the image indices the relations will be revealed so as to form the image clustering and image classification [1].

J. Zhang (Ed.): ICAIC 2011, Part V, CCIS 228, pp. 479–485, 2011.

In recent years, many kind of commercial system and some testing production have appeared, such as QBIC, Photobook, Virage, Visual SEEK, Netra, Simplicity etc. In CBIR, relevant feedback technology would improve results of retrieval, through the process of relevant feedback, each image is assign a relevant or irrelevant measurement, so that make the retrieval proceeding in a vector space and the retrieval result is a set of neighboring points in the space while the optimal result retrieved is consist of several images close with the query point which consist the k-nearest eigenvectors. In the relevant feedback retrieval system, through learning it would produce an approach of locating and adjust the distance matrix between neighboring points.

2 Semantic Based Image Associations

The challenges to the CBIR consist in the various image retrieval requirements as well as the complex and hard describe image content, with the gap between the digital array of image expression and the conceptual information universally accepted by human being, which is called the semantic gap, referring to the separation of the low level information in image and the conceptual semantic description.

In the early time of image retrieving, most system are based on texts, the connection and classification are mainly depend on the key words, headline textual clues and through algorithm assist the query criterion. The image retrieval system applied in present basically be divided into two kinds of frame: text based retrieval and the content based retrieval. The text based image retrieval is carried out on manual annotations which are the key words describing the basic connotations in an image. As we know, the images in an image base have certain latent internal relations between them. By setting up the image indices the relations will be revealed so as to form the image clustering and image classification. In CBIR, relevant feedback technology would improve results of retrieval, through the process of relevant feedback, each image is assign a relevant or irrelevant measurement, so that make the retrieval proceeding in a vector space and the retrieval result is a set of neighboring points in the space while the optimal result retrieved is consist of several images close with the query point which consist the k-nearest eigenvectors. In the relevant feedback retrieval system, through learning it would produce an approach of locating and adjust the distance matrix between neighboring points [2].

From the viewpoint of human vision, image has three level of features: (1) the low level features: which include the color quantization, color histogram features in color aspect and the texture aspect features, such as the texture co-occurrence matrices, random texture features, for instance Gauss-Markov random field texture features, texture spectrum features and fractal feature and son on; (2) The middle level features, which refer to the content of shape expression aspect, including shape histogram, shape feature point distance and angle, radius of gravity center of shape, chain vector code and so on. In general, by means of the inflection point near the image edge region to express the middle level features; (3)The high level features (semantic): these features actually refer to some interesting regions through image segmentation which are combined together to express the semantics of image[3].

Generally, in multimedia content expression, the image features are separated into several levels: on the bottom level, they are consist of some statistics feature and on the top level they are composed of literary description, vision expression, and feedback mechanism and the contents concerned with learning and image organization. In vision perception, the similar or not in the aspect of color, texture, shape and configuration to a large extent depends on the image features introduced which provide necessary information to the image database when carry out the image retrieval and knowledge exploring.

The deep concordance of image database exploring system is to help users to interpret database on the whole and to find the latent relation and inherent law. The building of this system consists of three stages: the first stage includes feature extraction of images as well as building efficient form for the image expression, organization and storage so that in the image retrieval process the result would approach to vision perception of human being; the second stage is to building hierarchical clustering, the prototype images of clustering be regarded as the classification identifications which are aim of in the appropriate form constructing the abstract for the clustering so as to make convenient to understand their component elements. The classification identification, in fact, is an image whose sum of square distance from the other images in a cluster maximize. Besides the effect of description in abstract for an image database, classification identification can provide an access to another dataset of classification identification; the third stage is to establish multidimensional scaled marks to provide visualization in various detail grades for image database. In addition, they can organize the similar images in perception together to form perceptional context by which users can explain and mark images without need to give the connotation description for images, accordingly by this form reducing the annotation time for an image dataset.

The image semantic associations accomplish through association linklist (see Table 1). The association link list pointer would link to some scenery semantics groups, for instance, from "boat", "sand beach", "people", and "water", it could associate to "beach scenery", "forest", "city" etc. classes.

Table 1. SCenery Semantic Association

Scenery label	Association image set	Semantic group	Association linklist (pointer)	Database association (pointer)
ID01	Set1	{...}	AL_pointer	DB_pointer
ID02	Set2	{...}	AL _pointer	DB_pointer
ID03	Set3	{...}	AL _pointer	DB_pointer
...	...	...	...	...

The structure attributes of scenery expression include space frequency distribution and orientation, power spectrum: for a pixel of image $p(x,y)$, the power spectrum $\Gamma(f_x,f_y)=|FT(\{p(x,y)\})|^2$, where FT is Fourier Transform, f_x and f_y are the space frequencies of $p(x,y)$ in the directions of x and y, $\Gamma(f_x,f_y)$ is energy density characterized by frequencies and (f_x,f_y) is the orientation eigenvector, normalization form is (f_x, f_y) , which become the orientation information, where $D = \sqrt{f_x^2 + f_y^2}$.

3 Neuron Index for Image

Neuron index for image is an association between image categories and image blocks, realizing through a neural relevant feedback network which is 3-layer feed forward structure. In this network, the nodes in input layer represent some of image blocks which be regarded as query terms and the nodes in the output layer represent conceivable image classification labels which be regarded as document terms. During the training stage, select a set of typical images as a category (represented as a label) image samples, to set association between image blocks and category label [4]. On the base of introducing relevant feedback, all the term image blocks present in a relevant category are ranked according to following weighting function: for a query image block term i , $W_i = \log(\frac{R_{i1} \cdot IR_{i0}}{R_{i0} \cdot IR_{i1}})$ where R_{i1} , R_{i0} , IR_{i0} , IR_{i1} represent separately the numbers in a training samples for a term image block i being relevant or irrelevant to a category of image: R_{i1} (block i appears and relevant); R_{i0} (block i no appears and relevant); IR_{i0} (block i no appears and irrelevant); IR_{i1} (block i appears and irrelevant).

The process of neuron indexing for image includes the image meaningful region detection and the image indexing stage, there are five stages in whole process. We introduce different granular grade of perception clues, such as pixel grade clues, region grade clues and the high level semantics clues. Through neural self-organization training, the classification be formed and memory and stored in local patterns. By means of frequent pattern mining a stable connection between perception clues forms, playing the speeding up and guiding effect for the later detection. If introducing relevant feedback in this process, it also has neural learning function. In the first stage, in order to establish image index, we have a “moving eye” to move on an image. During the stage, when the moving eye to somewhere, the information there (belong to low grade information) such as texture, orientation, color be compared with the semantics base and carried out similarity judgment. In the second stage, carry out the semantics judgment and regional semantic fusion, in which the relevant feedback would introduce. In the third stage, form conceptual classification and correlation to establish semantic fusion [5]. The fourth stage is learning stage, which forms association and strengthening memory and the result be stored in semantic network database. In the fifth stage, establish image base index and carry out the fast classification for the input images. During this stage, we adopt the combination method which the rough classification combines with elaborative clustering.

4 Image Block Associations and Match

In relevant association and query, an opening problem is how to measure semantics, in other words, the homogeneity of the semantic image to certain class of images. Establish a semantic association path to a class of images and set up a semantic group according to some demands is of an important significance for speeding up image retrieve and association recognition. Presently, there are many methods establishing associations between pixels, however, these association can merely reflect the local low level visual features which could provide help to the classification and retrieve in the aspect of visual features association, however, to the understand and recognition of image content, it Still has certain distance. Therefore, establish the association between low level visual features and the high level semantics is always a work being full of challenge, even called semantic gap. Presently, the image segmentation could make homogeneous pixels classified into the same area, which is in fact homogeneity in visual feature [6] [7] [8]. One of method for pixel level homogeneity judgment is to measure the homogeneity of connection path. We definite the homogeneity of pixels as the similarity of low level features of pixels. During the process of image semantics association, the semantic association is essentially the matching of special image block to stand for semantics according to already existed semantic lexicon. For two matched blocks the similarity is definite as follows:

$$S(q_i,t_i)=\frac{1}{1+2D_h(q_i,t_i)+D_s(q_i,t_i)+D_v(q_i,t_i)}$$

Where q_i and t_i are represent two matched blocks, $(h_{q_i},s_{q_i},v_{q_i})$ and $(h_{t_i},s_{t_i},v_{t_i})$ are their corresponding hue, saturation and value of pixel for the two image blocks, and D_h,D_s,D_v use to measure the similarity of hue, saturation and value of image blocks :

$$D_h(q_i,t_i)=1-\cos^2((\frac{1}{256})\|Hq_i-H_{t_i}\|\frac{\pi}{2}) , \quad D_s(q_i,t_i)=\frac{\|S_{q_i}-S_{t_i}\|}{256} ,$$

$$D_v(q_i,t_i)=\frac{\|V_{q_i}-V_{t_i}\|}{256} .$$

5 Experiments

In this section we give some experiment examples in out image match and retrieve system (Fig.1). Fig.2 is a horse input image and showed horse head region according to the retrieve criterion "horse head" and the source input image. Fig.3 is a detected cavity region of an ear according to an outline of an ear block picture.

Fig. 1. Interface of our image retrieve system

Fig. 2. Matched the horse head region

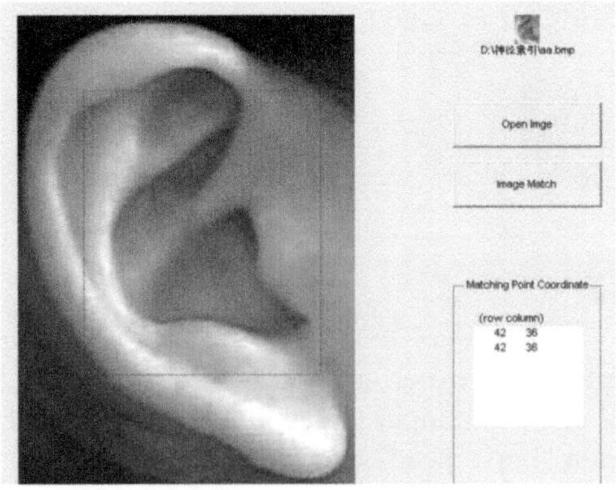

Fig. 3. A matched cavity region of an ear

6 Conclusions

Semantic based image associations and retrieve can overcome the inherent shortcoming in CBIR and from the view of cognition, the results of retrieve are more approach to human being interpretation, which will become the future field of investigation in image retrieve technology, and meanwhile, the neural indexing technology of image data would also contribute to accelerate the process of image database browsing and matching. All above will be our future investigation work.

References

1. Gosselin, P.H., Cord, M.: Combining visual dictionary, kernel-based similarity and learning strategy for image category retrieval. Computer Vision and Image Understanding 110(3), 403–417 (2008)
2. Lim, J.H., Tian, Q., Mulhem, P.: Home photo content modeling for personalized event-based retrieval. Journal of IEEE Transactions on Multimedia (4), 24–37 (2003)
3. Fuh, C., Cho, S., Essig, K.: Hierarchical color image region segmentation for content-based image retrieval system. Journal of IEEE Transactions on Image Processing (9), 156–162 (2000)
4. Palmer, S.E.: Hierarchical structure in perceptual representation. Journal of Cognitive Psychology (9), 441–474 (1977)
5. He, X.F., King, O.: Learning a semantic space from user's relevance feedback for image retrieval. IEEE Transactions on Circuits and Systems for Video Technology 13(1), 39–48 (2003)
6. Zhou, W., Yinglin, Y., Zhang, D.: Best neighborhood matching: an information loss restoration technique for block-based image coding systems. IEEE Transactions on Image Processing 7(7), 1056–1061 (1998)
7. Hayakawaa, H., Shibata, T.: Block-matching-based motion field generation utilizing directional edge displacement. Signal Processing and Communication Systems 36(4), 617–625 (2010)
8. Anastasia, D., Andreopoulos, Y.: Software designs of image processing tasks with incremental refinement of computation. IEEE Transactions on Image Processing 19(8), 2099–2114 (2003)

A Service Mining Approach for Time Estimation of Composite Web Services

Xizhe Zhang, Ying Yin, and Bin Zhang

College of Information Science and Engineering,
Northeastern University,
Shenyang, China
xizhezhang111@sina.com

Abstract. Response time of web service is one of the most important features of SOA-based application. This paper proposes a time estimation method for composite web services based on mining historical execution information of web service in the dynamic environment. Firstly, it gives frame and format for the service log collection, and then it analyzes the extracting service execution time and network transmission time, and constructs service network based on the service behavior interaction. Secondly, the paper presents the method of forecasting composite service execution time and also provides the measure of environmental availability and time availability.

Keywords: Web services, execution time, composite service, data mining.

1 Introduction

Along with the maturation of web service technology, a large number of web services appear on internet, and service calculation is going to a large-scale application stage. Subsequently mass service execution historical data contain abundant information, and the knowledge supporting service intelligence extracted from the information is an important method to solve application problems.

Web service execution time is the leading indicator of quality of service (QoS), and execution time of service prediction is significant to the users and service suppliers. Users can see the information of service environment and performance based on the execution time. It guarantees the quality of service choice and service composition.

Due to the independence and self-contained performance of web service, the users can know about the service only by the description from service suppliers. Mean while, estimated execution time of composite service is significant as the composite service is the main application methods of web services. However, the execution time of composite service is not the simple accumulation of the composed services, as the service compatibility and the environment must be considered.

The prediction of web service execution time can be come down to the prediction of the application program execution time. Smith [2] estimated the execution time by historical information, and sought for the similar program running records to predict the execution time base on the procedures similarity template defined by users.

J. Zhang (Ed.): ICAIC 2011, Part V, CCIS 228, pp. 486–492, 2011.

Similarly Krishnaswamy [3] used algorithm of rough sets to solve the problem of selecting the best similarity template. The main ideas of above methods are to define the similarity of the current application and the historical data, and to predict execution time by average or regression analysis based on these similar programs. But duo to the self-contained feature of web service, it's difficult to classify similar web services. So it is not easy to use above methods on prediction of web service time.

With respect to the prediction for composite web service execution time, a new solution needs to be proposed, because the different services might be exist in different network environment, and the actual execution time is not equal to the execution time of independence service resulted from the unparallel of each web service.

This paper proposes a new forecast method for composite service performance based on service execution log. The method can analyze the environment feature of service according to service's historical log, and seek for similar service so as to estimate the execution time. Firstly it introduces the collection structure and format of the log. Secondly it proposes the definition and construction method of service interactive network. Lastly it gives the algorithm of forecasting composite service execution time based on web interactive network.

The paper is organized as follows, section 2 introduces related concepts and construction methods of service network; section 3presents the prediction algorithm of composite service response time; conclusions in section 4.

2 Basic Concepts of Web Services

As the current service logs are recorded by the web server, which is lack of high level service execution information, so it cannot meet the requirement of service mining. So we propose a web service log collecting architecture, which records service execution information through service composition execution engine.

Fig.1 shows a web service composition framework which supporting web service logging. The framework logs web service execution information by service composition execution engine. The execution engine is response for execute business process and exchange message between the web service and record log simultaneity.

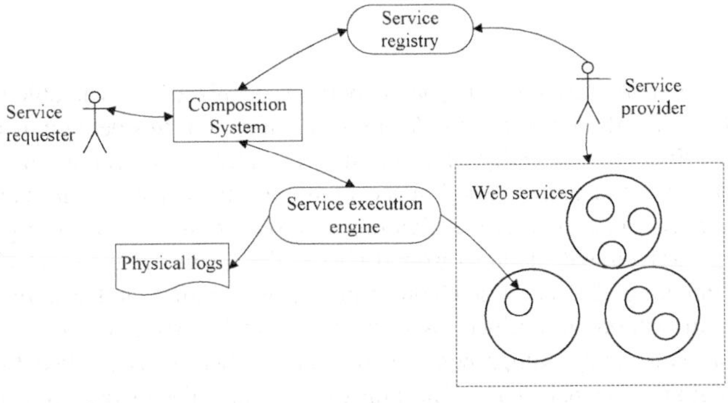

Fig. 1. A composite service log architecture

The log records the interaction information between execution engine and web service. SOAP message is the physical basic unit of interaction information and information transferring between engine and service is the minimum logic unit.

The service usage log can be defined as message record during a certain period of service execution. The service execution log is denote as $L=\{l_1, l_2, \ldots, l_n, \ldots\}$, among which l_i is called as log item. It can be described as $l=\{PID, op, time, size, category, para\}$, among which:

PID represents the ID of business model which current log item belong; it is generated by composite service execution engine.

op is the operation which current log item executed;

time is the record time of the log item;

size is the message size;

Cata indicate the role of execution engine in the service interaction; its value is {*Response*, *Require*}, it means the execution engine is the message sender or message receiver.

Para is the parameter in the messages;

Definition 1. Composite Pattern. Let $S=\{s_1, s_2, \ldots, s_n\}$ be a set of n Web services. Let $CP=\{cp_1, cp_2, \ldots, cp_n\}$ $(i=1,2\ldots,n)$, where $cp_i=\{s_1\ s_2\ s_3 \ldots s_l\} \subseteq S$ $(l=1,2,\ldots,n)$ be a subset of services to complete a given business process. We called CP a **C**omposite **P**attern set.

Definition 2. Service Behavior. Given a Web service s_i, its behavior can be denoted by 6-tuples $B(s_i)=<f(s_i), q_{ij}(s_i), i(s_i), o(s_i), i_{.order}(s_i), o_{.order}(s_i)>$, where $f(s_i)$ is the function of Web service s_i, $q_{ij}(s_i)$ is m qualities of Web service s_i where $q_{ij}=\{q_{i1}, q_{i2}, \ldots, q_{im}\}$ $(j=1,2,\ldots,m)$, $i(s_i)$, $o(s_i)$, $i_{.order}(s_i)$ and $o_{.order}(s_i)$ are input parameters, output parameters of Web service and their parameters order respectively. we say B is the abbreviation of Behavior.

Definition 3. Composite Behavior. Given a composite pattern $cp_i=\{s_1 s_2 s_3 s_l\}$ $(l=1,2,\ldots,n)$. The behavior of a cp_i can be also denoted by 6-tuples $CB(cp_i)=<F(cp_i), Q_{ij}(cp_i), I(cp_i), O(cp_i), I_{.order}(cp_i), O_{.order}(cp_i)>$, where $F(cp_i)$ is the function of composite pattern(cp_i), $Q_{ij}(cp_i)$ is m qualities of composite pattern cp_i where $Q_{ij}=\{Q_{i1}, Q_{i2}, \ldots, Q_{im}\}$ $(j=1,2,\ldots, m)$, $I(cp_i)$, $O(cp_i)$, $I_{.order}(cp_i)$ and $O_{.order}(cp_i)$ are input parameters, output parameters of composite pattern and their parameters order respectively.

Definition 4. Web service interaction network. Given Web service interactive log, we can construct Web service Interaction network, short by *WSIN*. *WSIN* is 5-tuple: $WSIN = <S_b, S, E, TW, S_f>$, where S_b is an unique beginning state, S_f is an unique final state, S is a set of services, $s_i s_j \in E$, TW is a set of weights, $tw_{ij} \in TW$, that are assigned to the corresponding edge of s_i and s_j.

Definition 5. Service Community. Let $C=\{C_1, C_2, \ldots, C_n\}$. For each Web Service C_i and C_j $(1 \leq i, j \leq n)$, C_i and C_j have the frequently interaction relationship, we say C is a Web Service Community.

3 Construction of Service Interaction Network

This section discusses the construction of service interaction graph. As the execution engine is responsible for service execution and message exchange, web service doesn't know the object which it interact. The service usage log is only the service message record by execution engine, so match for message and reconstruction of service interaction are needed.

The web service interaction relationship can be defined as service connection based on certain business semantic. It exhibit in the service log as the operation pair that in the same business process and have time sequence relationship with matched parameter. The operation op_i and op_j have interaction relationship if and only if the following condition is satisfied:

1. There exists log sequence $l_1,l_2,\ldots,l_n$, $i,j \in [1,n]$, in which $l_i.op=op_i$ and $l_j.op=op_j$;

2. $l_i.PID= l_j.PID$; $\forall k \in [i+1, j-1], l_k.PID \neq l_i.PID$

3. $l_i.para = l_j.para \wedge l_i.cata = Response \wedge l_j.cata = Require$

The web service interaction relationship denote as $op_i \rightarrow op_j$.

Condition 1 represent there is time sequence relationship between operation op_i and op_j in the service usage log; Condition 2 represent the two operations belongs same composite service model, and there is no other operations which in the same composite service model between them; Condition 3 represent that the two operation have parameter exchange. If the operation pair satisfies the above condition, we say that they have an interaction behavior,

We can represent the interaction behavior of service operation as a directed multigraph, called operation interaction graph, denote as $G(S,E)$. It figures the interaction relationship between operations of the service space in a certain time horizon. The operation set is denoted as S and the edge set E represent the operation interaction relationship, the direction of edge represent the time sequence in the interaction.

The operation interaction graph is directed multigraph for operation is repeatedly executed commonly. So we can classify the operation set by the density of interaction and we have the following operation group concept:

The operation group is a community in which the operations have tightly interaction. From the business view, the operation group often relative to the same business goal. So we can get the service composition model by finding operation group.

Fig.2 shows the construction method for operation interaction graph based the definition of operation interaction.

```
Input : Service usage log L
Output : web service interaction graph G(S,E)
ConstrctionOperationGraph(L)
{   let n be the number of log item in  L;
      for( i=0; i++; i<n)
      {   read log item l_i;
            If l_i.op ∉ S ,   add(op, S);
            for( j=i; j++; j<n)
            {
                read log item l_j;
                if
  l_i.op = l_j.op ∧ l_i.para = l_j.para
  ∧l_i.cata = response ∧ l_j.cata = require ,
                          add( op_i → op_j ,E)
            }
      }
}
```

Fig. 2. Construction method for operation interaction graph

4 Predict Response Time of Composite Web Service

The response time for web service in dynamic network environment depends on below factors:

(1) The calculation scale of requested issue;

(2) The Algorithm internal mechanism: We can obtain different execution efficiencies for the same issue if we adopt different methods;

(3) The performance and load of the service;

(4) If programs stress data calculation, they rely on execution speed of web service hardware environment;

(5) If programs stress the exchange of data, they rely on the quality of network environment and load of transmission, such as bandwidth, delay and data bulk of transmission.

We can sort above aspects into two sections. The first one is time of network transmission, including occupation ratio of bandwidth, time used to transfer and so on. The second one is the execution time of service which mainly rely the running task. Now we will talk about how to predict time of transmission and time of execution.

A. Definition of Service Similarity

Similar applications often have similar execution time, and this is the base of prediction of execution time of composite service. So we can predict the response time based on the log if they are similar application. The realization of this method focuses on two issues:

1. Definition of similarity. For example, one user submits two applications, and the applications have same time, using same device, same parameter, same point and so on.

2. Generation of prediction. According to the definition of similarity, we can classify previously performed task, and the similar tasks have similarity. Then we can calculate the result of new task using similar tasks. For example, we can regard the middle value as the prediction value.

At first, we provide the definition similarity of web service. We define the interactive similarity of web service as the contact ratio between the sets of interior point and the exterior point.

Definition 6. The adjacent point set of point i in the web service interactive network G(S,E) is :

$$\alpha(i) = \{ j \in S \mid (i, j) \in E \} \ ; \ \beta(i) = \{ j \in S \mid (j,i) \in E \}$$

We call $\alpha(i)$ as exterior point of i , and $\beta(i)$ is interior point of i;

We can define the set of adjacent point of C as follows:

$$\alpha(C) = \bigcup \alpha(i) - C \bigcap \left(\bigcup \alpha(i) \right), i \in C$$

$$\beta(C) = \bigcup \beta(i) - C \bigcap \left(\bigcup \beta(i) \right), i \in C$$

Above we define a collection which obtains all adjacent point of set C excluding the collection C.

For the web service interactive network, if service have similar interactive relationship (for example. A point has a similar collection of adjacent point comparing with B's, then we think that A and B are similar.).

The similarity of i and j can be defined as follows:

$$sim(i, j) = \frac{\alpha(i) \bigcap \alpha(j) + \beta(i) \bigcap \beta(j)}{\alpha(i) \bigcup \alpha(j) + \beta(i) \bigcup \beta(j)}$$

B. Prediction of Composite Servicer Execution Time

The time of network transmission have a relationship with bandwidth, the current network load, the transmission quantity of data. Because network load is always changing as time going on, different time to call leads different transmission time. We can get a set of transmission time and call time.

We can get different results even though we use the same service, the same business, the same call of different service. The reason is that they maybe exist in different network environment which leads the increasing of transmission time.

As for the time of transmission, we can evaluate it with the interactive ws_i/ws_j and predict it with average execution time of historical log.

As for composite service CW={ws_1, ws_2, ..., ws_n} , we can detach it into some sections such as ws_i/ws_j based on execution sequence.:

The execution time of CW is denoted as following:

$$ExecTime(CW) = \sum_{i=1}^{n} time(ws_i, ws_{i+1})$$

As for composite service, we can search the similar service with ws_i/ws_j based on below tactics.

1. Detaching composite service into a pair such as ws_i/ws_j.

2. Considering the collection of *WSL*, which is similar with ws_i, we can operate it as follows:

a) Searching for the service collection of *WSL*, which is similar with ws_j

b) Searching for the interactive log record of *WSI*/*WSJ*.

3. Predict the transmission time of ws_i/ws_j using the historical log of *WSL* and *WSI*.

5 Conclusion

This paper proposes a response time prediction method for composite web service based on web service interactive log. By analysing the historical log of web service this paper formulate a new causality and timing relationships, and then we construct an interactive web service network. We can find a similar collection by the method of similarity of web service, which we use to predict the services execution time. In this way we can predict web service execution time.

Acknowledgment. This work is sponsored by the Natural Science Foundation of China under grant number 60903009 and the Fundamental Research Funds for the Central Universities under grant number N090104001.

References

1. Lelli, F., Maron, G., Orlando, S.: Towards Response Time Estimation in Web Services. In: IEEE International Conference on Web Services, July 9-13, pp. 1138–1139 (2007)
2. Smith, W., Foster, I., Taylor, V.: Predicting Application Run Times Using Historical Information. In: IPPS/SPDP 1998 Workshop on Job Scheduling Strategies for Parallel Processing (1998)
3. Krishnaswamy, S., Zaslavsky, A., Loke, S.W.: Estimating computation times to support scheduling of data intensive applications. IEEE Distributed Systems Online (Special issue on Data Management, I/O Techniques and Storage Systems for Large-scale Data Intensive Applications) 5(4) (2004)
4. Da Cruz, S.M.S., Campos, M.L.M., Pires, P.F., Campos, L.M.: Monitoring e-business Web services usage through a log based architecture. In: Proceedings. IEEE International Conference on Web Services, pp. 61–69 (2004)
5. Xu, M.W., Hu, C.M., Liu, X.D., Ma, D.F.: Research and Implementation of Web Service Differentiated QoS. Journal of Computer Research and Development, 669–675 (2005)

An Item-Based Efficient Algorithm of Learning from Examples

Yuan Lin and Ruiping Geng

Department of Computer Science and Technology,
College of Information Engineering
Minzu University of China
Beijing, 100081, China
yuan_lin@sohu.com

Abstract. This paper presents a novel algorithm of learning from examples based on items, while the traditional algorithms could be considered as example-based. The deference is that, though both use frequency matrixes as main heuristic information , the traditional algorithm obtain the matrixes through the line-by-line scanning of examples, while the presented algorithm through scanning of items and then carrying out some calculations according to some laws found by the author in examples space. Since a item may contains a lot of examples and the cost of scanning of a example is as the same as of scanning of a item, the presented algorithm possesses great ability to be able to handle those learning tasks with great number of examples , many variables and assignments, while the traditional algorithms could not treats with those tasks.

Keywords: machine learning, learning from example, frequency matrix, item based, new laws in examples space.

1 Introduction

Many algorithms of learning from examples, such as AQs, use frequency matrixes as main or important heuristic information [1]. Those algorithms cannot deal with learning tasks with a large number of examples or variables, since they obtain those matrixes through the scanning of every examples and the example space will grow exponentially with the linear increase of the number of variables or assignments. So, the traditional algorithms could be called as *example-based* ones. The presented algorithm obtains exactly those same matrixes through scanning of some conjunctions of examples (called as *item*) and then carrying out some reasoning and calculations based on some laws found by the author in examples space [2]. So, the presented algorithms could be called as *item-based algorithm*. As one conjunction (item) may contain a lot of examples, the presented algorithm gets rid of the limitations in the number of examples, variables, or/and assignments, and possesses a great ability to handle learning tasks with great number of examples variables, or/and assignments, while the traditional algorithms could not be able to treat with those tasks.

J. Zhang (Ed.): ICAIC 2011, Part V, CCIS 228, pp. 493–502, 2011.

2 Basic Concepts

Suppose that $E= D_1 \times D_2 \times \cdots \times D_n$ as n-dimensional space of finite examples, $j \in N$; $N=\{1, 2, \cdots, n\}$ as *attributes' (variables')subscript set*; E's element, $e= < v_1 , v_2 ,\ldots, v_n>$, called *example*; $v_j \in D_j$ called the *assignment* of E's j-column; subsets PE, NE and $DE \in E$, called *positive-example set, negative-example set* and *independent set*, and $E= PE \cup NE \cup DE$.

Selector is like the statement, $[x_j=A_j]$, in which x_j is the *jth attribute* and $A_j \subseteq D_j$. *Formula* or *item* is the *conjunction* of selectors. *Rule* is the *disjunction* of formulas or items. Obviously, e is also a special formula as $[x_1=A_1][x_2=A_2]\cdots[x_n=A_n]$.

It is agreed that there is no common element between any two of PE, NE and DE.

For formula (item) $L=[x_1=A_1]\cdots[x_n=A_n]$ and formula $L'=[x_1=A_1']\cdots[x_n=A_n']$, it is defined that

1) L contains L' ($L \supseteq L'$), iff $\forall\ j \in J,\ A_j \supseteq A_j'$;
2) L and L' can be combined (merged) into one new formula, $[x_1=A_1]\cdots[x_j=A_j \cup A_j']\cdots[x_n=A_n]$, iff $\exists j \in J$, $A_j \neq A_j'$ and $\forall\ i \neq j$, $A_i=A_i'$, and the new formula is marked as *comb(L , L')*.
3) L intersects with L', iff $\forall\ j \in J$, $A_j \cap A_j' \neq \Phi$, and the new intersection formula (item) is marked as *inter(L , L')* , and constructed as $[x_1=A_1 \cap A_1]\cdots[x_n=A_n \cap A_n]$.
4) If L intersects with L' and there are totally m selectors in *inter(L , L')*, $m \le n$, as $[x_{j1}=B_{j1}],\cdots,[x_{jm}=B_{jm}]$ and $A_{j1}' \supset B_{j1},\cdots, A_{jm}' \supset B_{jm}$, and $\forall\ j \in J \backslash \{j_1,\ldots,j_m\}$ other selectors in *inter(L , L')* are $[x_j=B_j']$ and $A_j'=B_j$. Then, m new *rules*, $L_1',\cdots, L_m'$, called *Remaining Disjoint Formulas* for *L' against L* , can be constructed according to the following method:

   $\forall\ i \in \{1, \cdots, m\}$, L_i' is obtained from L' by changing L's selectors, from j_1th to j_{i-1}th attributes, into $[x_{j1}=B_{j1}],\cdots,[x_{j\ i-1}=B_{j\ i-1}]$, changing L's j_ith selector into $[x_{ji}=A_{ji}' \backslash B_{ji}]$, and remaining L's other selectors unchanged.
5) For any formula L (item), the all examples covered by L in E, PE and NE form three sets, called E_L, PE_L and NE_L, separately. Each of E_L, PE_L and NE_L has a related *Frequency Matrix*, marked as FM_E_L, FM_PE_L and FM_NE_L separately. It is defined that, for any formula S (item), its *Frequency Matrix*, FM_S, is constructed in the way that FM_S 's element of ith row and jth column takes the value equal to the occurrences of the x_j's assignment, $i \in D_j$, in all examples of the set S.

For example, $D_1=\{0,1,2\}$, $D_2=\{0,1\}$ and $D_3=\{0,1\}$ construct the example space, E, with total 12 examples in three sets, PE, NE and DE, separately, as shown in Figure 1.

Figure 1 is also used to illustrate roughly the traditional algorithms' running steps.

(0)Input: only the examples in PE and NE (without DE)

PE: $x_1x_2x_3$	NE: $x_1x_2x_3$	DE: $x_1x_2x_3$
e_1=<0 1 0>	e_5 =<0 0 0 >	e_9 =<0 1 1>
e_2=<1 0 1>	e_6 =<0 0 1 >	e_{10} =<1 1 1>
e_3 =<1 1 0>	e_7 =<1 0 0 >	e_{11}=<2 0 0>
e_4 =<2 1 0>	e_8 =<2 1 1 >	e_{12}=<2 0 1>

(1)First, set L=empty formula, then PE_L=PE , NE_L=NE.
Scan e_1~e_4 in PE_L to get Scan e_5~e_8 in NE_L to get

FM_PE_L=

	x_1	x_2	x_3
0	1	1	3
1	2	3	1
2	1		

FM_NE_L =

	x_1	x_2	x_3
0	2	3	2
1	1	1	2
2	1		

(2)By some strategy, choice the selector [x_2=1] added into L.
Then L=[x_2=1], and PE_L ={e_1, e_3, e_4 } , NE_L ={e_8}.
Scan e_1 e_3 e_4 to in PE_L to get Scan e_8 in NE_L to get

FM_PE_L=

	x_1	x_2	x_3
0	1	0	3
1	2	3	0
2	1		

FM_NE_L=

	x_1	x_2	x_3
0	0	0	0
1	0	1	1
2	1		

(3) By some strategy, add one more selector into L , then
L=[x_2=1][x_3=0]. So, PE_L ={e_1, e_3, e_4 } , NE_L =Φ, and
Scan e_1 e_3 e_4 in PE_L to get Scan Φ NE_L to get

FM_PE_L=

	x_1	x_2	x_3
0	1	0	3
1	1	3	0
2	1		

FM_NE_L=

	x_1	x_2	x_3
0	0	0	0
1	0	0	0
2	0		

(4)Remove, from PE, the positive examples, e_1, e_3 and e_4, which are covered by L. Then, PE={e_2}, left to be covered.

Fig. 1. PE, NE, DE, frequency matrixes and the running steps in a traditional algorithm based on examples

As Figure 1, for the given *PE* and *NE*, a temporary formula(item) *L* keeps changing, from the empty content into *[x_2=1]* and next step into *[x_2=1][x3=0]*, trying to cover as many *positive examples* in *PE* as possible, and as less *negative examples* in *NE* as possible. As soon as *L* no longer covers any negative example in *NE*, output the content of *L* as *one element (formula)* of the *result rule for some class*. Then, remove the positive examples covered by *L* from *PE*. Again, *L* is reset empty to repeat the next process, until *PE* becoming empty set and the final *rule for some class* having been obtained.

The best result rule is considered to have the least numbers of both formulas and selectors. That is however known as a NPC problem. So, the algorithm uses the *Frequency Matrix*es and the following heuristic strategy for the approximate best result:

1) Select the assignment of a variable to construct the selector to be added into *L*, which has the greatest number of corresponding occurrences in *FM_PE_L*, at the time. That will make the result rule have fewer formulas. If there are more than one such assignments, the one is first selected which has less value of corresponding occurrences in NM_PE_L, at the time. That will reduce the number of the selectors in the result formula.

2) Select the assignment of a variable to construct the selector to be added into L, which has the non-zero number of corresponding occurrences in FM_PE_L, and has the zero number of corresponding occurrences in FM_NE_L. That will immediately make L becoming a result formula, so as to reduce the number of the selectors in the formula. If there are more than one such assignments, the one is first selected which helps to cover more positive examples in PE.

3) 2) is more prior than 1).

It should be noted that the great increase of the efficiency or ability of the presented algorithm is the focus of this paper, rather than the heuristic strategy to choice the selectors.

The presented algorithm is described in chapter 4. Before that, some theories are first introduced in chapter 3, which lays down the basic for the presented algorithm to greatly increase its efficiency or ability.

3 Theorems

In order to get the heuristic information, the *Frequency Matrix*es, such as FM_PE_L, and NM_PE_L, etc., the traditional algorithms need to scan every example in PE and NE (or PE_L and PE_L) .So , as the numbers of the variables and assignments increasing, the efficiency of the algorithms will decrease greatly, even fail to complete the tasks.

Differently with the traditional algorithms , the presented algorithm combines the examples in PE or NE into some items firstly, scans the items secondly, and then carries out some reasoning or calculations to finally get the same heuristic information, FM_PE_L, and NM_PE_L. Since one item may contains a lot of examples and the cost to scan a item is the same as to a example, the presented algorithm processes a great efficiency or ability, suitable to deal with the learning tasks with a large number of variables, assignments and examples, with which the traditional algorithms may not be able to deal.

The following theorems present the basic of the reasoning or calculations applied in the presented algorithm, which are summaries of some laws founded by the author in the example space.

In order to save space, the proofs of those theorems have been omitted.

Theorem1: For any two items (formulas) L and L', and any examples set E , if L and L' can be combined (merged) into one new formula, then $E_L \cup E_{L'} = E_{comb(L\ ,\ L')}$.

Theorem2: For any two items L and L', if L and L' can be intersected with each other, then $E_L \cap E_{L'} = E_{inter(L\ ,\ L')}$.

Theorem3: For any item (formula) $L = [x_1=A_1]\cdots[x_n=A_n]$, $A_j \subseteq D_j$, $j \in N$, and any examples set E , the FM_E_L can be calculated (reasoned) according only to the L's literal structure and the following simple arithmetic formula:

$$\forall\ j \in J,\ i \in D_j\ ,$$

$$FM\_E_L[i, j] = |A_1| \times |A_2| \times \cdots \times |A_{j-1}| \times |A_{j+1}| \times \cdots \times |A_n| .$$

Theorem4: If there are m items (formulas), L_1 , L_2 , $\cdots L_m$, and any two of them do not intersected with each other, and any examples set E , then remark $S = E_{L1} \cup E_{L2} \cup \cdots \cup E_{Lm}$, and

$$FM\_S = FM\_E_{L1} + FM\_E_{L2} + \cdots + FM\_E_{Lm} .$$

Theorem5: If item (formula) L contains item (formula) L' ($L \supseteq L'$), then $E_L \supseteq E_{L'}$.

Theorem6: If item (formula) L intersects with item (formula) L' and L does not contain L' , then among L' 's *Remaining Disjoint Formulas(against L)*, L'_1 , $\cdots$, L'_m , there is such fact that any two of them do not intersect with each other, ant that $E_{L'} \setminus E_L = E_{L'1} \cup E_{L'2} \cup \cdots \cup E_{L'm}$.

In the presented algorithm, *Theorem1* is repeatedly applied to combine (merge) any two examples or items (formula) into one new item (formula), in *PE* and *NE* separately, until there is no longer such combination to be done. Then, PE and NE become the sets of items, their volumes are reduced accordingly, the scanning cost is reduced greatly, and the efficiency or ability of the algorithm is increased greatly too.

It is obviously, that the more the examples are there in *PE* and *NE*, the more the combinations will take place among them, and finally the greater the volumes of *PE* and *NE* are reduced. Therefore, the presented algorithm is very suitable to deal with the kind of tasks with great deal of the data to be treated.

After the combination and the temporary formula (item) L has been written with some content, and then the algorithm will find out all the intersecting items of L with every item in *PE*. According to *Theorem2*, those intersecting items are the equivalent alternatives to PE_L. As agreed that there is no duplication of examples in *PE*, any example does not intersects with any others in *PE*, and each item, after the combination, does not intersects with any other in *PE* also. Therefore, those intersecting items do not intersect with each other. So, for any item p, as one of those intersecting items, and the examples set E consisted of all the examples covered by p, EM_E_p can be calculated out according to *Theorem3*. Then, according to *Theorem4*, cumulatively add those EM_E_p together to obtain the EM_PE_L. The EM_NE_L can be obtained in the same way.

As soon as L becoming a result formula, the positive examples covered by L should be removed from *PE*. So, the algorithm checks every item p in *PE*. If $L \supseteq p$, p will remove from *PE* into *pkeep* link to save; if L intersects with p, and L does not contains p, p will also remove from *PE* into *pkeep* link to save , but some special items, called as *Remaining Disjoint Formulas* for p against L, are constructed according to *Definition (4)* in Chapter 2 and attached with some mark , then added into *PE*. When those *Remaining Disjoint Formulas* (items) with attached mark be removed latterly from *PE*, they should be released to system space rather than into *pkeep* link to save. The *Theorem5* and *Theorem6* guarantee the correctness to do that algorithm.

4 The Algorithm: IBA-I

The algorithm is given a name, IBA-I, to mean as the Item-Based-Algorithm I. A different algorithm, called IBA-II, will be given in another paper, which is also a item based algorithm.

Set *k* as the number of classification of cases. *L* is the temporary formula. Pointers *plink* and *nlink* direct two lists which stand for *PE* and *NE* separately. Pointer *pkeep* directs the list to save temporarily the items removed from *PE*. *R* is the result rule obtained in the learning process for the *i*th class examples.

The algorithm is as following:

```
0).for(i=1;i<=k;i++)
{
    Input the examples of the ith class;
    Combine the examples (or items) with each other in the ith class until such
      combination does not exist;
    Using pointer Ci to direct the list consisted of all items in the ith class;
}
1).for(i=1; i<=k; i++)execute every steps in following: (2~10)
                /*the ith loop is for the ith class's learning*/
2) plink=Ci; nlink = the list consisted of all Cj(j≠i);
pkeep=null;  R=Φ ;
3). firstpass=true ;  L= empty formula;
4).if (pkeep == null && firstpass==true)
{
  calculate1(FM_PE_L, plink); calculate1(FM_NE_L , nlink );
  FM_PE_0= FM_PE_L; FM_NE_0= FM_NE_L;  firstpass=false;
}
else if (firstpass==true)
{
  FM_PE_L= FM_PE_0; FM_NE_L= FM_NE_0;  firstpass=false;
}
5).Compare the values, with each other, of the FM_PE_L and FM_NE_L, apply the
heuristic strategy, described in Chapter 2, to choice some i and j, form the selector
[x_i=v_j], and add the selector into L.
6) calculate2(L, FM_PE_L , plink);
   calculate2(L, FM_NE_L , nlink );
7).if(FM_NE_L== 0 matrix) /*L no longer covers any e∈NE*/
      goto  step 8) ; /*go to add L into R as a disjunctive */
else   goto  step 4) ; /*go to add more selector into L */
8).if (R==Φ)  R=L;   else   R =  R +"∨" + L;
9).remove_and_update(L, plink, pkeep, FM_PE_0);
10).if ( plink≠null ) goto  step 3);
else
{
  Output R as the rule for the ith class;
```

```
  Ci=pkeep;
}
The subroutines called in the algorithm are given as following:
calculate1( FM, link)
{   /*all items in list link consist a example set, marked as E*/
 FM= 0 matrix;
 From first to last, for each item p in link
 {                                            /* This is a loop */
     Calculate out FM_Ep, according to Theorem3;
     FM= FM+ FM_Ep;   /* Accumulate FM_Ep into FM */
 }
}
calculate2(L, FM, link)
{   /*all items in list link consist a example set, marked as E*/
 FM= 0 matrix;
 From first to last, for each item p in link
 {                                            /* This is a loop */
     if ( L intersects with p)
     {
        q=inter(L, p);
        Calculate out FM_Eq, according to Theorem3;
        FM= FM+ FM_Eq;/*Accumulate FM_Eq into FM*/
     }
 }
}
remove_and_update(L, plink, pkeep, FM_PE0)
{
 From first to last, for each item p in plink
 {                                            /* This is a loop */
     if ( L intersects with p)
     {
        q=inter(L, p);
        Calculate out FM_Eq, according to Theorem3;
        FM_PE0= FM_PE0 — FM_Eq;
                         /* Regress FM_Eq from FM_PE0 */
     }
 }
}
```

Now, the presented algorithms' running steps is illustrated as following, taking the same *PE*, and *NE* as in Figure 1 above, as example.

(I). Step(0), input the examples in *PE* and *NE* as *C1* and *C2* classes:

PE (*C1* class) :

e1=<0 1 0> e2=<1 0 1> e3 =<1 1 0> e4 =<2 1 0> ;

NE (*C2* class) :

e5=<0 0 0 > e6 =<0 0 1> e7 =<1 0 0 > e8 =<2 1 1 >.

After combinations (Theorem1), examples turned to be items, named as some *ps* and *qs*, separately as following:

$$
\begin{array}{llll}
PE\ (plink): & p1 = <-\ \ 1\ \ 0> & (\text{contain e1, e3, e4}), \\
 & p2 = <1\ \ 0\ \ 1> & (\text{contain e2}). \\
NE\ (nlink): & q1 = <0\ \ 0\ \ 0/1> & (\text{contain e5, e6}), \\
 & q2 = <1\ \ 0\ \ 0> & (\text{contain e7}), \\
 & q3 = <2\ \ 1\ \ 1> & (\text{contain e8}).
\end{array}
$$

Meanwhile, '–' stands for that the related attribute could take any of the all its assignments, such as, in *p1*, x_1=0/1/2.

(II).Step (4) L=empty formula, then PE_L=PE , NE_L=NE.

Scan p1, p2 in PE_L to calculate (according to *Theorem 3*):

$$
FM\_E_{P1} = \begin{array}{c|ccc} & x1 & x2 & x3 \\ \hline 0 & 1 & 0 & 3 \\ 1 & 1 & 3 & 0 \\ 2 & 1 & & \end{array}, \quad FM\_E_{P2} = \begin{array}{c|ccc} & x1 & x2 & x3 \\ \hline 0 & 0 & 1 & 0 \\ 1 & 1 & 0 & 1 \\ 2 & 0 & & \end{array}.
$$

Then, according to *Theorem 4*:

$$
FM\_PE_L = FM\_E_{P1} + FM\_E_{P2} = \begin{array}{c|ccc} & x1 & x2 & x3 \\ \hline 0 & 1 & 1 & 3 \\ 1 & 2 & \textcircled{3} & 1 \\ 2 & 1 & & \end{array}.
$$

In the same way, scan q1, q2, q3 in NE_L to calculate (according to *Theorem 3)*:

$$
FM\_E_{q1} = \begin{array}{c|ccc} & x1 & x2 & x3 \\ \hline 0 & 2 & 2 & 1 \\ 1 & 0 & 0 & 1 \\ 2 & 0 & & \end{array}, \quad FM\_E_{q2} = \begin{array}{c|ccc} & x1 & x2 & x3 \\ \hline 0 & 0 & 1 & 1 \\ 1 & 1 & 0 & 0 \\ 2 & 0 & & \end{array},
$$

and

$$
FM\_E_{q3} = \begin{array}{c|ccc} & x1 & x2 & x3 \\ \hline 0 & 0 & 0 & 0 \\ 1 & 0 & 1 & 1, \\ 2 & 1 & & \end{array}
$$

So, according to *Theorem 4*:

$$
FM\_NE_L = FM\_E_{q1} + FM\_E_{q2} + FM\_E_{q2} = \begin{array}{c|ccc} & x1 & x2 & x3 \\ \hline 0 & 2 & 3 & 2 \\ 1 & 1 & 1 & 2 \\ 2 & 1 & & \end{array}.
$$

(Please check that the FM_PE_L and FM_NE_L here have the same values as the ones in Fig.1(1), separately. However, we see in Fig.1 that the traditional algorithms obtain the *Matrix*es by scanning every example, while the presented algorithm obtains them by scanning less items and doing some simple calculations).

(III) Step (5). Check the two matrixes, FM_PE_L and FM_NE_L. Since $FM\_PE_L[2,1]=3$>other value, choice the selector [x2=1] added into L. Then L=[x2=1].

Step(6), for the updated L, calculate out its FM_PE_L and FM_NE_L. Firstly, L intersects with only p1 in PE, and L intersects with only q3 in NE. The two intersectional items are

p=[x2=1][x3=0] and q=[x2=1][x3=0] separately.

So, according to *Theorem 3* and *Theorem 4*, here are:

$$FM\_PE_L=FM\_E_p=\begin{array}{c|ccc} & x1 & x2 & x3 \\ \hline 0 & 1 & 0 & \textcircled{3} \\ 1 & 2 & \underline{3} & 0 \\ 2 & 1 & & \end{array}, \quad FM\_NE_L=FM\_E_q=\begin{array}{c|ccc} & x1 & x2 & x3 \\ \hline 0 & 0 & 0 & 0 \\ 1 & 0 & 1 & 1 \\ 2 & 1 & & \end{array},$$

(note: the mark '$\underline{3}$' means the assignment '3' already selected)

Step(7), check $FM\_NE_L \neq$ 0matrix. It means that *L* still covers some negative example in NE, still not becomes result formula (see *L* still covers $e_8 \in$NE).So, goto step (4).

(IV) Again, step (4), do nothing. And step (5), $FM\_PE_L[3,0]=3$ > other value unselected, choice the selector [x3=0] added into L. Then

L=[x2=1] [x3=0].

Step(6), for the updated L, calculate out its FM_PE_L and FM_NE_L. Firstly, L intersects with only p1 in PE, and L does not intersect with any items in NE. So, according to *Theorem 3* and *Theorem 4*, here are:

$$FM\_PE_L=FM\_E_p=\begin{array}{c|ccc} & x1 & x2 & x3 \\ \hline 0 & 1 & 0 & \underline{3} \\ 1 & 2 & \underline{3} & 0 \\ 2 & 1 & & \end{array}, \qquad FM\_NE_L=\begin{array}{c|ccc} & x1 & x2 & x3 \\ \hline 0 & 0 & 0 & 0 \\ 1 & 0 & 0 & 0 \\ 2 & 0 & & \end{array},$$

Step(7), check $FM\_NE_L=$ 0matrix. It means that *L* no longer covers any negative example in NE, has already become a result formula. So, goto step (8).

(V) Step 8, make R=*L* . So, R= $[x_2=1]$ $[x_3=0]$.

Step 9, p1 is removed from *plink* into *pkeep*, FM_E_{p1} is calculated , and FM_PE_0 is updated as following:

$$FM\_PE_0 = FM\_PE_0 - FM\_E_{p1} .$$

and,

PE(*plink*): { p2 }; NE(*nlink*): {q1,q2,q3}.

(VI) In next loop, the second result formula $L = [x_1=1]\ [x_3=1]$ will be obtained, and *plink* becomes empty. Then, the *C1* class learning task is completed with the C1 class's result rule:

$$R1=[x_2=1]\ [x_3=0] \vee [x_1=1]\ [x_3=1] .$$

In the same way, the *C2* class's result rule will be obtained as

$$R2=[x_2=0]\ [x_2=0] \vee [x_1=2]\ [x_3=1] \vee [x_2=0]\ [x_3=0].$$

The algorithm is finished finally.

Note that R1 and R2 both cover some examples in *DE*.

5 The Complexity of the Algorithm

The algorithm's running time is mainly spent on step 4, in which every item in *plink* and *nlink* is scanned to calculate out the *Frequency Matrix*es taken as the heuristic information needed.

Set $m=\max|Ci|$, $i=1 \cdot\cdot k$. So, *km* is the total number of all the items in *plink* and *nlink*. In the worst cases, the temporary formula *L* intersects with each of all the above items. Therefore, to check all those intersection items and calculated out their *Frequency Matrix*es need totally *2kmnl* compares to construct one selector, meanwhile $l = \max|Dj|$, $j \in N$. There are *n* selectors in a result formula at most, and there are *m* formulas, at most, in the result rule of one class, and there are k classes to be learnt. Therefore, the presented algorithm has the *Complexity* of about $O(k^2m^2n^2l)$.

6 The Experiment and Conclusion

Experiment:
For comparison of the efficiencies between a traditional algorithm with the presented algorithm, a Boolean function minimization task, with 15 variables, 30 items, and 96 literals, has been carried out between a traditional algorithm and the presented algorithm.

The traditional algorithm spent about *90 minuets (5400 seconds)* to yell a result of 18 items, and 36 literals.

While the presented algorithm spent less than *1 second* to yell a better result of 16 terms and 31 literals.

Conclusion:
This paper reveals some laws among the examples' subsets, items, and *Frequency Matrix*es, and has given the methods to calculate the *Frequency Matrix*es according to the structures of the literals in items. Based on those, the algorithm, IBA-I, has shown great efficiency and ability in experiments. So, IBA-I is suitable to deal with the learning tasks with a large numbers of attributes and variables.

References

1. Michalski, R.S., Mozetic, I., Hong, J.R.: The Multipurpose Incremental Learning SystemAQ15 and Its Testing Application to Three Medical Domains. In: Proc. AAAI (August 1986)
2. Lin, Y.: A Deductive Reasoning Algorithm for Learning from Positive Examples. Journal of the Central University for Nationalities 5(2) (1996)

New Approach to Measure Tuning-Off Degree and Automatic Tune Arc Suppression Coil in Compensated Network

Lingling Li[1], Lixin Liu[1], Chuntao Zhao[1,*], and Chunwen Yang[1,2]

[1] School of Electrical Engineering, Hebei University of Technology, Tianjin, China
[2] Kailuan Group Research and Development Center Tangshan, China
Lingling_1Li@yeah.net

Abstract. To measure the tuning-off degree of compensated power network in running condition, a new approach named as resonating frequency measuring method was proposed. In this method, by the way of injecting frequency scanning signal with frequency range from 30Hz to 70Hz to the low voltage side of potential transformer (PT), the resonance frequency of power network was measured and tuning-off degree V and capacitance current IC can be worked out according to correlation theory. Then, arc suppression coil can be tuned by taking the tuning-off degree ν as controlled variable. Designed was not only measurement circuit but also fuzzy controller to implement fuzzy tuning based on fuzzy theory.

Keywords: compensated power network, tuning-off degree ν, automatic tuning, fuzzy control, resonance frequency.

1 Introduction

Compensated power network, which is also called resonant grounding network, is the network with neutral grounding through arc suppression coil. It belongs to non-efficiently grounding way just like neutral non-grounding and grounding through high resistance or high impedance. Compensated network is widely used in middle voltage power network. In China it is especially used in the network with the voltage of 10kV and 35kV. In France, the resonating grounding method has been used in all mid-voltage power networks. In Japan among the distribution power network of 11~33kV, about 28% of neutral point is grounded through arc suppression coil, about 30% is grounded by resistor. From this point of view, the measuring and tuning method for tuning-off degree of power network is of much more practical importance for ensuring the quality of power supply and human safety.

In recent year the study in this field drives more attentions, especially in China. Literature [1] introduced a new method to capacitor current by combining resonance method and curve-fitting method. Some factors to influence the measurement precision of capacitor current and a revision equation were given in [2]. A new type

[*] Corresponding author.

J. Zhang (Ed.): ICAIC 2011, Part V, CCIS 228, pp. 503–512, 2011.

of arc suppression coil based on symmetry and phase tuning principle and the method to trace the measurement of capacitor current was presented in [3]. In [4] the tuning-off degree of arc suppression coil and some parameters such as grounded current of capacitor have been calculated in 66kV earthing system. A new method to compensate the power network was proposed in [5]. Some new auto-tracing and compensating equipment are coming out.

2 New Theory and Realizing Way in Measuring Tuning-Off Degree ν and Capacitor Current IC

The arc suppression coil in fact is an induction coil which inductance can be adjusted. It has two advantages in compensated power network (simply called network below):

1) Lessen the short-circuit current of grounding fault point so that the arc is easily self-quenching.

2) Lessen the rising speed of arc gap recovery voltage so that the arc is not easily re-burned.

There exists distributing capacitance C which is the sum of that between each of the three-phase transmission line of network and ground. If anyone phase grounding fault happens, quite heavy capacitance current I_C will flow into the grounding fault point.

If the fault current exceeds 30A, steady and persistent grounding arc will be formed and intense arc will lead to short current between phase lines and equipment damages. If the grounding current is more than 5A but less than 30A, unsteady and intermittent arc will be formed and the sequent arc grounding over-voltage will be a threat to the lower insulation equipment within network and the transmission line with lower insulation property. The existence of arc suppression coil will generate induction current I_L which is opposite to the direction of I_C and compensates to I_C to lessen the fault current of grounding point. So I_L is called compensation current. The fault current after the compensation is defined as grounding residual current I_d, and smaller grounding residual current not only guarantees the safety of the equipment within network but also can avoid electromagnetic disturbance to the communication system nearby.

$$I_d = |I_C - I_L| \tag{1}$$

A. The Theory on Measuring ν

The key parameters of network are tuning-off degree ν, capacitance current I_C and grounding residual current I_d. And ν has some functional relation with I_C and I_d. When ν changes, I_d changes correspondingly. So the turning of the arc suppression coil always refers to the tuning of ν. Tuning-off degree ν denotes the departure degree from the network resonance state. It is defined as:

$$\nu = 1 - \frac{I_L}{I_C} \tag{2}$$

It is called over-compensation when ν<0; full-compensation when $\nu = 0$; under-compensation when ν>0.

Industry developed countries usually controls v in the range of [-5%, 5%], the maximal variation will not exceed [-10%, 10%], otherwise Id cannot be lessened effectively. According to the formula (2), the traditional way of measuring v is translated into measuring I_L and I_C. However, the lack of on-line measurement equipment on spot brings difficulties to debugging work. For this reason, the paper brings out a new way, resonance frequency measurement method to measure v and I_C. The theory is based on the below:

$$v = 1 - \frac{I_L}{I_C} = 1 - \frac{X_C}{X_L} = 1 - \frac{\omega_0^2}{\omega^2} = 1 - \frac{f_0^2}{f^2} \quad (3)$$

where, C is the total distributing capacitance between each of three-phase transmission lines of network and ground; L is inductance value of arc suppression coil; X_c is equivalent capacitance of C; X_L is inductance of arc suppression coil; f0 and ω_0 is resonance frequency and resonance angle frequency respectively; f and ω is power frequency and power angle frequency respectively and f is usually 50Hz, ω is usually 314rad/s. From formula (3), we know that v can be worked out as long as the f_0 is measured out, so direct measurement on I_C can be avoided. It is the principle of the resonant frequency measurement method in this paper.

B. Method of Measurement

The measuring circuit of network resonant frequency f_0 can be seen from figure 1.

In figure 1, C', L', r', U_0' is respectively the value converted to low voltage side of arc suppression coil potential transformer (PT) of the total distributing capacitance C , inductance L of arc suppression coil, damp resistance r in series with L, displacement voltage value U_0 of network neutral point. The measurement device designed in the paper of tuning-off degree v is shown in the dashed line frame in figure 1, $\dot{U}s$ is frequency -scanning voltage source controlled by microcomputer which can produce a series square-wave signals with virtual value 300V, frequency variety from 30Hz to 70Hz with frequency increase ±0.1Hz.

The frequency of $\dot{U}s$ is decided by the duration of high level from pin of CPU, which can be controlled by software. R is current-limiting resistance of 300Ω/600W; RQ is sampling resistance of 1Ω/5W. When the measurement device is used on the spot, it should be connected to the low voltage side of PT and can take a monitor on v ranging from –96%~64%. Of course, the value of composition components of the device can be changed but have some limit. For example, to eliminate the influence of neutral point displacement voltage U_0, $\dot{U}s$ cannot be too small but should be within 200-300 volts. The value and power of R should matches $\dot{U}s$ so that the current $\dot{I}s$ can be about 1A in the end.

Under the control of microcomputer, $\dot{U}s$ takes a frequency-scan to the right circuit from point M, point N and the voltage $\dot{U}_{RQ}$ of R_Q and voltage $\dot{U}_{MN}$ between point M and point N are taken back to the microcomputer system for phases analysis. Because R_Q is a pure resistance component, its phase of voltage and current is same. So $\dot{U}_{RQ}$ has the same phase with the current $\dot{I}s$ in the circuit.

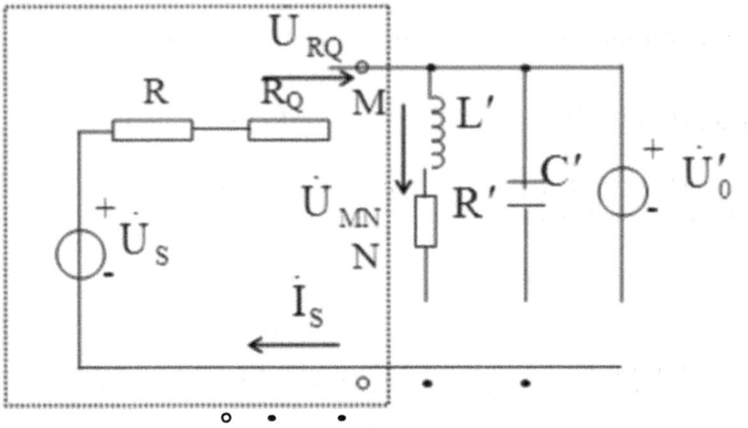

Fig. 1. The measuring circuit of network resonant frequency

When $\dot{U}s$ is scanning at certain frequency, the whole power network appears inductive if the phase of $\dot{U}_{MN}$ is forward to $\dot{U}_{RQ}$ (voltage forward to current) which analyzed by the computer, resonant frequency f_0 must be larger than the current scanning-frequency; otherwise, f_0 must be less than current scanning-frequency if $\dot{U}_{MN}$ lags behind $\dot{U}_{RQ}$; if the phase difference between $\dot{U}_{MN}$ and $\dot{U}_{RQ}$ is zero, it shows that parallel resonance takes place in the right side circuit from M, N, and the current frequency of $\dot{U}s$ is the resonance frequency f_0 of power network and ν can be calculated by formula (3).

C. Characteristic of the Measurement Devices

1) New measurement theory is adopted to measure directly the tuning-off degree ν, need no other parameters like I_C, I_L etc.

2) The method of measuring ν in the paper can be used in any patterns of arc suppression coil.

3) If there are several arc suppression coils installed in the power network, only one of them instead of all of them needs to be measured and the result of measurement is the value of ν of the whole power network.

4) In theory, the scanning-frequency of $\dot{U}s$ is not restricted and can be in any range. In the view of the real-time measurement and the probability of ν when power network is operating, it is ranged from 30Hz to 70Hz and the measurement time is less than 20s under parameters designed in the paper.

5) At the low voltage side of PT does the whole measurement process proceeds, which has changed traditional operation mode at high voltage side and has good security.

As we known, when power network is running in good condition, its neutral-point displacement voltage U_0 usually reaches a high level of hundreds of volt and can cause more influence to measurement accuracy of f_0.

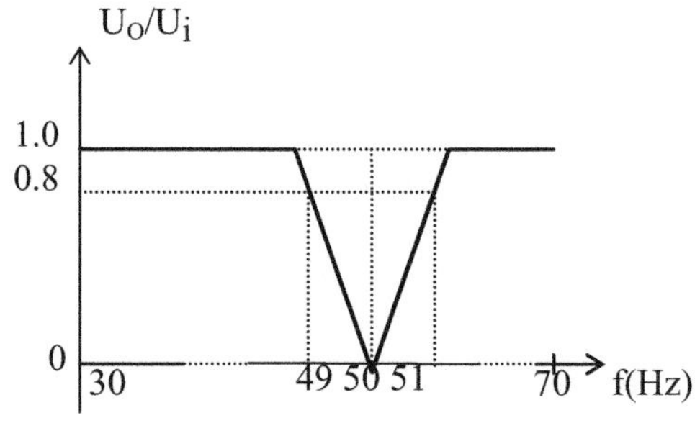

Fig. 2. Input-output characteristic of the filter

For this reason, an active band-stop filter is design to leach power voltage of 50Hz. The two voltage signals, $\dot{U}_{MN}$ and $\dot{U}_{RQ}$ (in figure 1) are treated by the filter before they are put into microcomputer for phase analysis so that measurement accuracy of f_0 can keep the same value even with the disturbance of power network operation. Input-output characteristic of the filter is shown in figure 2, in which U_i is input signal and Uo is output signal.

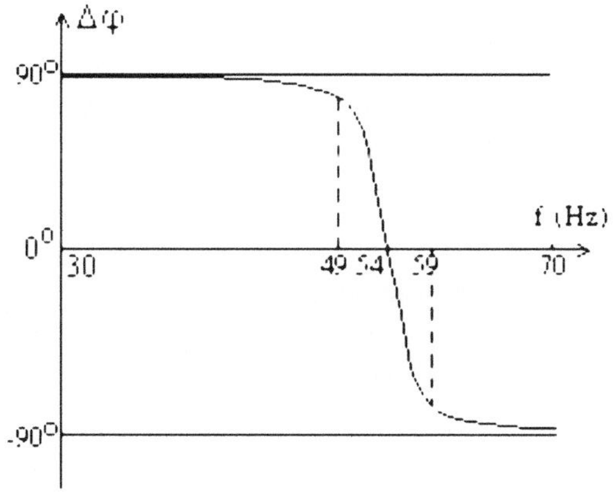

Fig. 3. The corresponding relation of $\Delta\varphi$ and f

It should be noted that when scanning -frequency is around 50Hz, $\dot{U}_{MN}$ and $\dot{U}_{RQ}$ of 50 Hz are also leached and can't be put into microcomputer because of the existence of filter. Once phase analysis of the measurement device testifies that resonance frequency f_0 of power network ranges from 49Hz to 51Hz, the value of f_0 can be calculated according to the rule that per 0.1 Hz increase of scanning-frequency f' of $\dot{U}s$ leads to 1.8^o decrease of phase difference $\Delta\varphi$ of $\dot{U}_{MN}$ and $\dot{U}_{RQ}$. The figure 3 shows the corresponding relation of $\Delta\varphi$ and f' in the case of C=22μf, L=0.5H, r=5Ω.

3 New Approach of Realizing Power Network Auto-Tunning

A. The Deficiency of Existing Tuning Methods

The tuning of power network is actually to tune the key parameter ν. Its step is to change compensation current I_L of arc suppression coil and to make I_L track the variation of the power network capacitance current I_C in time. At present, a largely practical method is to tune according to the max of neutral point displacement voltage U_0 in power network. And U_0, turning-off degree ν, asymmetric degree ρ and damping ratio d of power network has the relation as follow:

$$U_0 = \frac{|\rho| U_\varphi}{\sqrt{\nu^2 + d^2}} \tag{4}$$

When $\nu = 0$, U_0 reaches the maximum and power network is running in resonance state. This tuning method seems to be reasonable but is not so in fact. Because when ν is near to 0, the variation of U_0 is slight but is badly influenced by the voltage fluctuation of power network, so the resonance point can be determined always through many times comparisons and will not be determined under the condition of frequent fluctuation. Otherwise, there are many factors to influence U_0.

Power network needn't to be tuned under many conditions such as breakage, insulator leakage, and single phase arc grounding and so on. However, tuning mechanism has difficulty in identifying these disturbance phenomena so that the error action of tuning mechanism arises frequently, which does harm to safe running of power network. Moreover, other methods have individual deficiencies that bring adverse influences to tuning precision or response speed.

B. New Approach of Tuning Strategy

ν is measured directly according to the design mentioned in this paper. Power network can be tuned by using accurately measured ν as manipulated variable rather than by indirect tuning strategy like many present tuning methods. According to formula (1) and (3), if residual current I_d at single phase grounding faults point and tuning-off degree ν are off-limit, compensation current I_L of arc suppression coil is immediately adjusted till Id and ν are controlled in allowable range. On the premise of known f_0, the total distribution capacitance C can be obtained by the following formula:

Where, L refers to the sum of both inductance of arc suppression coil and zero-sequence equivalent inductance of Z-type grounding transformer which should be measured before system is debugged.

After distribution capacitance C is gained by formula (5),

$$f_0 = \frac{1}{2\pi\sqrt{LC}} \tag{5}$$

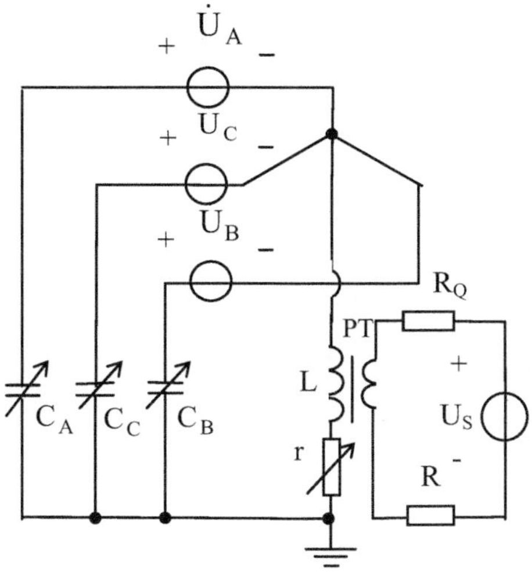

Fig. 4. The circuit diagram of te simulated test

capacitance current IC can be calculated by the formula:

$$I_C = \frac{U_\varphi}{X_C} = \omega C U_\varphi = \frac{U_\varphi \omega}{(2\pi f_0)^2 L} \tag{6}$$

where, ω is equal to 314 rad/s. At last, the ideal value of IL can be calculated according to v and formula (3). So, the power network can be tuned and the tuning process is simple and direct, has a short response time and eliminates radically tuning malfunction.

4 Fuzzy Control Scheme of Power Network Auto-Tunning

A set of fuzzy control scheme to realize power network auto-tuning using fuzzy modus reasoning is designed in this paper:

If A then B

Now A_1

Conclusion B_1=A1o$R_{A\to B}$ (7)

where, "o" is synthesizing operation of A to $R_{A\to B}$ and B_1=A_1o$R_{A\to B}$ is the token of synthesizing consequence law.

The type of fuzzy controller is single input-output; input language variable is the error E between measured value and true value of v; output language variable is the variation ΔI_L of I_L.

In this paper, a standard discussion field {-6,-5,-4,-3,-2,-1,0,1,2,3,4,5,6} is chosen for E and ΔI_L; and 5 language values of NB,NS,0,PS,PB are adopted individually. In every fuzzy subset, the factor with maximal subordination degree is defined as:$\mu_{PB}(x)=1$—x=+6; $\mu_{PS}(x)=1$—x=+3;$\mu_0(x)=1$—x=0;$\mu_{NS}(x)=1$—x=-3; $\mu_{NB}(x)=1$—x=-6; linear function (triangle) is taken as subordination function and then evaluation list of E and ΔI_L comes into being. Fuzzy control rule database is obtained by summing-up manual tuning strategy, and fuzzy judge from maximal subordination degree rule and weighted average rule is used as final judge. The founded state query form is table 1.

Table 1. State query table of the auto-tuning fuzzy controller

$e(x_i)$	-6	-5	-4	-3	-2	-1	0
$\Delta I_L(y_i)$	+6	+5	+4	+3	+2	+1	0
$e(x_i)$	+1	+2	+3	+4	+5	+6	
$\Delta I_L(y_i)$	-1	-2	-3	-4	-5	-6	

If E=NB then $\Delta I_L = PB$ or If E=NS then $\Delta I_L = PS$ or If E=0 then $\Delta I_L = 0$ or If E=PS then $\Delta I_L = NS$ or If E=PB then $\Delta I_L = NB$.

5 Measurement Error and Simulated Results

1) The differential coefficient of formula (3) is as follow:

$$dv = e_v = \frac{\partial v}{\partial f_0} df_0 = -\frac{2f_0}{50^2} df_0 = \pm 0.008\% f_0 \tag{8}$$

Where, e_v is the basic error of measurement value of v and df_0(±0.1Hz) is the increase of scanning-frequency. In this paper, the formula (8) proves the new method about v whose maximal measurement error e_v is not more than ±0.56% when f_0 is equal to 70Hz, with high accuracy and scanning process lasts not more than 20s with good real-time characteristic.

2) The error of I_C calculated from the equation (6) is derived from f_0. According to the error transferring theory,

$$e_{IC} = 2\times\frac{df_0}{f_0} = \pm 2\times\frac{0.1}{f_0} = \pm\frac{0.2}{f_0}\times 100\% \tag{9}$$

where, e_{IC} is the basic error of measurement value of I_C.

From the equation (9), the maximum value of e_{IC} is only ±0.67% under the condition f_0 =30.0Hz when v =64%. In practical power network, the tuning-off degree v is usually less than 64%, so e_{IC} should be also less than 0.67%.

3) The simulation of influence of damping ratio d and asymmetry ρ

The simulation of zero-phase equivalence loop of network as follow was carried out to test the influence of ρ and d on the measurement value of v. The test circuit is shown in figure 4. Where, power source $\dot{U}_s$, current-limiting resistor R and sampling resistor R_Q are connected to the secondary side of PT, r was connected with L in series and contains the resistance of the first side of PT. The measurement values of the tuning-off degree under different values of ρ and d is listed in table 2 and table 3. Where $C = C_A + C_B + C_C$. The data in the two tables show that asymmetry ρ has no influence on the measurement result of v and the damping ratio does only negligible influence on it.

Table 2. The measured result of v under different value of d

$C(\mu f)$	$L(H)$	$\|\rho\|(\%)$	$r(\Omega)$	$d(\%)$	$f_0(Hz)$	$v(\%)$
27.002	0.32	≈ 0	1.628	1.901	54.2	-17.51
27.002	0.32	≈ 0	4.637	5.405	54.1	-17.07
27.002	0.32	≈ 0	8.213	9.531	54.0	-16.64

Table 3. The measured result of v under different value of $|\rho|$

$C(\mu f)$	$L(H)$	$\|\rho\|(\%)$	$r(\Omega)$	$f_0(Hz)$	$v(\%)$
27.002	0.320	≈ 0	1.628	54.2	-17.51
27.002	0.320	19.228	1.628	54.1	-17.51
27.002	0.320	38.479	1.628	54.0	-17.51

It needs to say that v in table 2 varies slightly when d is different. It is because that the test circuit simulates serial ground mode of arc suppression coil L and damping resistor r. If shunt ground mode or ground with no resistor mode is simulated, d has no influences on v.

6 Conclusion

The resonant frequency measuring method is used for real-time measuring capacitance current I_C and tuning-off degree v of network, and features in the following aspects:

1) High accuracy of measurement. When resonant frequency of network is changing from 30~70Hz (namely v belongs to[-96%, 64%]), the system errors of the measured values of I_C and v are not more than ±0.56% and ±0.67% respectively.

2) No requirement for starting tuning mechanism of arc suppression coil again and again, so as to prolong the work life of the tuning mechanism and provide the measurement system with favorable real-time quality.

3) Used widely mainly at the two aspects as follow:

The arc suppression coil can be any type; the inductance of it can be adjusted continuously or not. There is no special requirement for the network parameters, just as stated above, the asymmetry Kc and the damping ratio d has no distinct influence on the measurement result of ν.

4) Auto-tuning scheme directly takes ν as a controlled variable and has good state identification function and anti-disturbance ability.

5) This fuzzy control scheme carries tuning out by seeking state query table and requests several devices only, with the advantages such as good real-time characteristic, strong generality of control rules and quite high applied value.

Acknowledgment. This work was supported by the National Natural Science Foundation (60771069), the Hebei Province Natural Science Foundation (F2010000151) and the Hebei Province Construction Science and Technology Research Project (No. 2009-250).

References

1. Chen, Z., Wu, W., Zhang, Q., et al.: A newautomatic tuning method for multi-tap arc-suppression coil. High Voltage Apparatus 41(5), 370–372 (2005)
2. Chen, Z., Wu, W., Chen, J., et al.: Study of measuring and tracing of arc suppression coil with automatic tuning. Relay 32(18), 24–28
3. Ji, F., Wang, C., Mou, L., et al.: New type of arc suppression coil based on symmetry and phase tuning principle. Automation of Electric Power Systems 28(22), 73–77 (2004)
4. Yang, T., Li, S., Du, T., et al.: The fault waveform analysis of arc-suppression coil earthing system and calculation of tuning parameters. Heilongjiang Electric Power 26(4), 268–270 (2004)
5. Lian, H., Tan, W., Pei, S., et al.: Principle and Realization of presetting and following-setting compensation mode in resonant grounded system. Automation of Electric Power System 29(21), 79–84 (2005)

Knowledge-Based Decision Support Model for Supply Chain Disruption Management*

Lihua Wu and Yujie Zang

School of Information Management,
Beijing Information Science & Technology University,
Beijing, China
lihuawu111@sina.cn

Abstract. Frequent risk evens disrupt normal supply chain operations in recent years. It becomes a strategic issue to determine which action to undertake for managing disruption risks. This paper proposed a decision support model which integrates a case-based reasoning mechanism to help supply chain members cope with disruption risks. The model retrieves and reuses past experience knowledge to solve current disruption cases. It reduces the decision efforts as well as improves the decision performance and acceptance.

Keywords: Supply chain disruption; Decision support; Case-based reasoning.

1 Introduction

Supply chain disruption is an unplanned event that interrupts the normal flow of materials, goods and information within a supply chain[1-3]. In recent years, threats from operational contingencies, man-made catastrophes and natural calamities [4-5] increase the disruption risks that break down the different supply chain components. Managing and mitigating such risks becomes a strategic issue for organizations who are dedicated to improve their supply chain performance.

Ref. [5] identified four kinds of strategies to manage and mitigate supply chain disruption risks: proactive strategy, advance warning strategy, coping strategy and survival strategy. The first three groups are prevention strategies which increase the robustness of supply chain against disruptions. The fourth, however, pertains to remediation actions for supply chain resilience once a disruption occurred. Since the total prevention and protection is impossible, the ability to mitigate the consequences after a disruption is seen as a critical capability that every company should develop [6-7].

There are a variety of approaches to help a company select appropriate countermeasures to mitigate the consequences brought by a disruption. One of them is to use past examples of survival to evaluate and identify possible strategies suitable

* This work is supported by Social Science Research Common Program of Beijing Municipal Commission of Education (No. SM200910772005); Beijing Natural Science Foundation(No. 9072005); Funding Project for Academic Human Resources Development in Institutions of Higher Learning Under the Jurisdiction of Beijing Municipality(No. PHR201008437).

J. Zhang (Ed.): ICAIC 2011, Part V, CCIS 228, pp. 513–521, 2011.

for the current breakdowns. Case-based reasoning (CBR) is often chosen as the primary tool to conduct this task. CBR is a problem-solving and continuous learning technology in artificial intelligence field. It learns from past experiences incrementally and then uses learnt knowledge to explore feasible solutions to a new situation by analogism. This paper proposed a decision support model based on CBR technique to help supply chain members determine which actions to undertake in order to mitigate disruption risks.

This paper is structured as follows. Firstly, the framework of the decision support model is introduced. Then the decision reasoning mechanism employing CBR is established. Finally the proposed model is applied to an example test set and a conclusion about the model is made.

2 Framework of Decision Support Model

The proposed model integrates CBR as the core component. A decision process supported by the model is decomposed into five phases: case representation, case retrieve, case reuse, case revise and case retain, as shown by Fig.1.

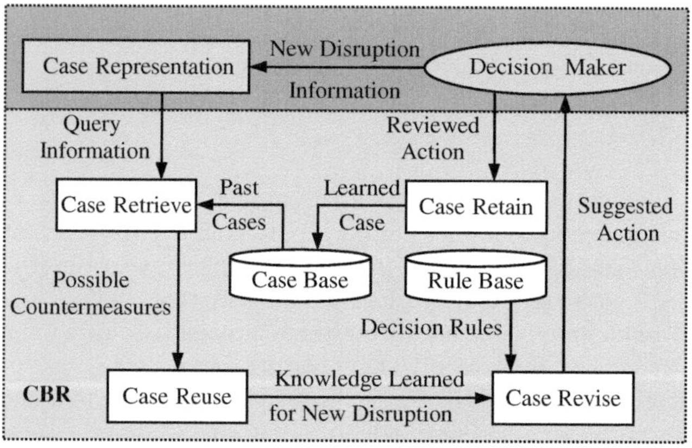

Fig. 1. Framework of decision support model based on CBR

The rationale of the model is as follows:

1) When a new disruption occurs in supply chain, it will be defined as a new pending problem and represented with an appropriate structure so that it can enter into the CBR cycle for problem-solving.

2) The CBR cycle is triggered once a new enquiry arrives. It starts to retrieve the past disruption cases stored in the case base and finds out some candidate according to the features of new disruption risk.

3) If the candidate case matches the new disruption case well, then the countermeasures of the candidate case will be reused as the solution to the new disruption case; otherwise, a revision is needed to make the retrieved countermeasures suitable for the current situation.

4) The revised countermeasure is suggested to the decision maker for reviewing to see whether the solution is acceptable or need further revision. After verified by the decision maker, the revised countermeasure is exported as a feasible solution for the new disruption and then retained into the case base for the future usage.

The model makes use of the experiential knowledge on how to deal with the disruption risks of supply chain fully, which prompts the decision with less efforts and higher acceptance. At the same time, the model accumulates new knowledge sustainedly in virtue of the incrementally learning capability of CBR. This ability permits the model to adjust the decisions to adapt to changes in the environment and makes the decision support process more robust and intelligent.

3 Enabling the Model

CBR provides a universal framework for problem solving. However, a particular CBR reasoning mechanism is needed to deal with some given situation. A detailed introduction of the mechanism, including case representation form, case retrieval technique, and case revision rule, is described in this section.

A. Case Representation

Case representation is about what features should be used to describe a case and how to organize the case base for effective retrieval and reuse. After a survey of supply chain disruption literature, six attributes are chosen to define a supply chain disruption risk. Table I summarizes the case attributes.

1) Case Name: Indicates the abstracted content information of a certain disruption case.

2) Disruption Source: Illustrates the causes of disruptions in supply chain. Three categories of disruption source are identified: operational contingency, man-made catastrophe and natural calamity [4-5]. The operational contingency involves the interior risks related to supply chain operations. The man-made catastrophe and natural calamity, by contrast, reveal the exterior roots of supply chain disruptions. Generally speaking, a certain disruption mainly roots from one kind of these sources, so the single-choice data type is appointed to the attribute.

3) Disruption Element: Describes the objects which may fail during a disruption. According to the research of [8], we distinguish the disruption elements into four aspects: value stream, fixed & mobile assets, organizations & inter-organizational network and environment. The value stream checks the supply chain vulnerability from the perspective of the product and process. The process involves a series of activities and workflows realizing the value-addition of supply chain. The fixed & mobile assets depict the physical resources available in supply chain. They consist of the infrastructure, mobile devices and the links between infrastructures. The organizations & inter-organizational network include all roles who are concerned with the supply chain operations. The contractual relationships between these participators are also considered in this category. The last factor is the environment within which organizations do business. It is further divided into the political, economic, social, technological, and legal elements of the operating environment as well as the natural geological, meteorological and pathological phenomena.

Table 1. Case Attributes

Attributes	Range	Data type
Case name	N/A	Text
Disruption source	Operational contingency Man-made catastrophe Natural calamity	Single-choice

Disruption element	Value stream
	  Product
	  Process
	Fixed & mobile assets
	  Infrastructure
	  Mobile assets
	  Links of infrastructures
	Organizations & inter-organizational network
	  Member
	  Customer
	  Partner
	  Employee
	  Contractual relationship
	Environment
	  Political environment
	  Economic environment
	  Social environment
	  Technological environment
	  Legal environment
	  Geological phenomenon
	  Meteorological phenomenon
	  Pathological phenomenon
		Multi-choice

Disruption mode	Supply disruption Production disruption Demand disruption Logistics disruption Information disruption	Multi-choice
Disruption severity	High Medium Low	Single-choice
Solution	N/A	Text

4) Disruption Mode: Considers the effects of the disruption on supply chain. The basic disruption modes include supply disruption, production disruption, demand disruption, logistics disruption and information disruption[7, 9]. A direct focus on the disruption mode can help companies evaluate the consequences and gain the necessary commitment to increase the security and resilience[7]. For a particular disruption situation, more than one mode can appear at the same time. Hence companies need take mixed effects into account.

5) Disruption Severity: Corresponds to "how much" a disruption might affect a supply chain. It is differentiated into three levels: high, medium and low[5]. Different severity requires totally different survival strategies to mitigate the disruption risk.

6) Solution: Provides the feasible mitigation strategies for the given disruption risk. It stores the domain knowledge and experience knowledge, and may be used to conquer other similar disruption risks under certain conditions.

Cases can be represented in a variety of forms such as script, frame, rule, object, predicate logic, semantic net and so on. According to case attributes confirmed by Table 1, the frame structure is used to represent a disruption case.

B. Case Retrieve

Case retrieve is to find the most similar case from the case base to match with the new problem. This is the core step of the CBR cycle. There are two sides to carry out the case retrieval: one is deciding the importance of the case attributes; the other is comparing cases by a similarity measure.

1) Weight Calculation: To assess the importance of each case attribute, a weight is needed to be calculated. The analytical hierarchy process (AHP) method is widely used to assess the attribute weights and has been applied in many studies. The basic steps of AHP are as follows:

The first step involves defining the problem and construct-ing a hierarchical structure for weight assessment.

Next a questionnaire survey should be carried out to establish the pair-wise comparison matrix *A*.

To validate the reliability or consistency of the importance judgment of attributes, the consistency ratio (CR) is used as the ultimate judgment criterion. It can be calculated from (1):

$$CR = \frac{CI}{RI} \tag{1}$$

where *CI* is the consistency index and *RI* is the random index. *CI* could be obtained according to (2):

$$CI = \frac{\lambda_{\max} - n}{n-1} \tag{2}$$

where $\lambda_{\max}$ is the largest eigenvalue of matrix *A*, *n* is the number of attributes.

Under most cases, when $CR \leq 0.1$, matrix *A* is supposed to have a satisfied consistency; otherwise, the matrix *A* should be adjusted until the consistency is achieved.

2) Similarity Measure: This process is used to appraise the matching degree between the target case and source cases in case base. The nearest neighbor algorithm is employed here to conduct the task. For the nearest neighbor algorithm, the similarity of two cases is calculated as followings:

$$similarity(N,R) = \frac{\sum_{i=1}^{n} w_i \times sim(N_i, R_i)}{\sum_{i=1}^{n} w_i}. \tag{3}$$

In (3), N is the new case; R is the retrieved case; N_i and R_i are the feature i of N and R respectively; w_i is the weight of feature i; and $sim(N_i, R_i)$ is the similarity function of the feature i of N and R. The value of $sim(N_i, R_i)$is calculated as in

$$sim(N_i, R_i) = 1 - \left| \frac{N_i - R_i}{|N_i| + |R_i|} \right|. \tag{4}$$

C. Case Reuse and Case Revise

Case reuse is a process where solution of the retrieved case is directly applied to solving the new problem. In this work, two predefined thresholds, θ_1 and θ_2, are used to control the reuse process. The control rule can be expressed as:

If

$$similarity(N,R) \geq \theta_1 \tag{5}$$

and

$$sim(N_i, R_i) \geq \theta_2 \quad \forall i \in n \tag{6}$$

then the retrieved case is reused directly; otherwise, case revision should be conducted.

Case Revision is about adjusting the retrieved cases to fit the current case. The decision-makers generally need modify the suggested solution according to their expertise and experience. According to the control rules (5) and (6), the case revision process of this work is further differentiated into two kinds of situation:

1) Both (5) and (6) are not satisfied. It means the target case is a totally new problem and there is not ready knowledge for it. Then the decision-makers should combine the results of similarity measure and some rules from rule base to acquire available solution.

2) Only (6) is not satisfied. For example, the similarity of attribute i of the target case and retrieved case is very small. In this case, the decision-maker should first find out a case set where the similarities of all cases to target case are bigger than θ_1. Then in the case set, the decision-maker selects top-n cases that have a similarity bigger than θ_2 based on the attribute i. Finally, the decision-maker can tune the retrieved solution according to the top-n cases.

4 Model Application

The model is applied to an example test set consisting of ten disruption cases. Among these cases, one is used as the target case, and the others belong to the source cases.

A. Quantifying Case Attributes

For six attributes described in Table 1, disruption source, disruption element, disruption mode and disruption severity are the main features characterizing a disruption risk. They will take part in the similarity measure during the model application. They are all choice data type and the quantified rule for them is simple. When they are true, they are assigned with 1; otherwise, they are assigned with 0.

B. Weight Calculation

Before applying AHP to determine attribute weights, the preference scale for pair-wise comparison should be ascertained. Table 2 shows the scale adopted by this work.

Table 2. Preference Scale for Pair-wise Comparison

Important level	Numerical rating
Extremely not important	1/7
Distinctly not important	1/5
Not Important	1/3
Equally important	1
Important	3
Distinctly important	5
Extremely important	7

According to the attributes of Table 1, the assessment on weights is divided into three tiers. Due to the space limitation, we only present the relative importance weights of the first-level attributes in Table 3.

Table 3. Attribute Weights by Using AHP

Attribute	Disruption source	Disruption element	Disruption mode	Disruption severity	Attribute weight
Disruption source	1	1/5	1/7	1/5	0.0519
Disruption element	5	1	1/3	1	0.2100
Disruption mode	7	3	1	3	0.5281
Disruption severity	5	1	1/3	1	0.2100

λ_{max} =4.0735, CI=0.0245, and CR=0.0272.

C. Similarity Measure

By using (3) and (4), the attribute similarities in each tier and the total similarities between target case and source cases can be calculated. Table 4 lists the total similarities between target case and other nine source cases.

Table 4. Similarity Measure

Case	1	2	3	4	5	6	7	8	9
Sim.	0.499	0.836	0.965	0.790	0.653	0.702	0.796	0.356	0.665
Rank		2	1				3		

From Table 4, case 3 is the most similar to the target case. By examination, the attribute similarities of each tier between target case and case 3 are bigger than θ_2=80%. So the solution of case 3 is reused directly to cope with the target risk.

5 Conclusion

Managing and mitigating supply chain disruption risks involves the development of survival options after a disruption event occurs. This paper provides an intelligent model based on the case-based reasoning technique to help supply chain members make contingency decisions. The model applies historical knowledge on how to deal with disruption risks to a new risk case by a knowledge reuse mechanism, avoiding the decision-making process from scratch.

Since the attributes describing the disruption risk is very important for knowledge retrieval and reuse, future research should be laid on the further development of disruption representation features. At the same time, the model should be applied to more practical situations to validate its effect and improve its performance continuously.

References

1. Xu, J.-X.: Managing the risk of supply chain disruption: towards a resilient approach of supply chain management. In: Proceedings of 2008 ISECS International Colloquium on Computing, Communication, Control, and Management, pp. 3–7. IEEE Computer Society, Piscataway (2008)
2. Skipper, J.B., Hanna, J.B.: Minimizing supply chain disruption tisk through enhanced flexibility. International Journal of Physical Distribution & Logistics Management 39(5), 404–427 (2009)
3. Li, J., Wang, S.-Y., Cheng, T.C.E.: Competition and cooperation in a single-retailer two-supplier supply chain with supply disruption. International Journal of Production Economics 124, 137–150 (2010)
4. Kleindorfer, P.R., Saad, G.H.: Managing disruption risks in supply chains. Production and Operations Management 14(1), 53–68 (2005)
5. Stecke, K.E., Kumar, S.: Sources of supply chain disruptions, factors that breed vulnerability, and mitigating strategies. Journal of Marketing Channels 16(3), 193–226 (2009)
6. Pochard, S.: Managing risks of supply-chain disruptions: dual sourcing as a real option. PHD Thesis, Massachusetts Institute of Technology (2003)

7. Sheffi, Y., Rice, J.B., Fleck, J.M., et al.: Supply chain response to global terrorism: a situation scan. In: EurOMA POMS Joint International Conference (2003)
8. Helen, P.: Drivers of supply chain vulnerability: an integrated framework. International Journal of Physical Distribution & Logistics Management 35(4), 210–232 (2005)
9. Chen, T., Ji, G.-J.: Study on supply chain disruption risk. In: Proceedings of the 2009 6th International Conference on Service Systems and Service Management, pp. 404–409. IEEE Computer Society, Piscataway (2009)

Improvement of Long-Term Run-Off Forecasts Approach Using a Multi-model Based on Wavelet Analysis*

Xue Wang, Jianzhong Zhou*, Juan Deng, and Jun Guo

School of Hydropower and Information Engineering
Huazhong University of Science and Technology
Wuhan, China
xzpycy@gmail.com

Abstract. This paper promotes a new multi-model method of long-term run-off forecasting based on wavelet analysis to produce an combined estimated output to be used an alternative to that obtained from a single individual long-term run-off forecasts model. Three methods of long-term run-off forecasting are considered, namely Nearest neighbor bootstrapping regressive model (NNBR), Support vector machine model (SVM), Wavelet analysis (WA). Firstly, the original annual runoff series is decomposed into different component sequences with WA, and then the NNBR and SVM models are used to get the forecast results of different component sequences, at last they are reconstructed with WA again. The observed discharges of a 42-year period getting from Yichang station, Yangtze River basin, are used to test the performance of the proposed method. The results confirm that better discharge estimates can be obtained by the multi-model approach based on wavelet analysis compared with the single NNBR and SVM models.

Keywords: long-term run-off forecasts, WA, NNBR, SVM.

1 Introduction

Long-term hydrological forecast is scientific forecast for a period of hydrological elements based on the objective laws of hydrological phenomena, using pre-hydrological and meteorological data through certain mathematical methods. Since the time series analysis model is proposed, it has been widely applied to the long-term hydrological forecast [1]. A non-parametric time series analysis model, namely, nearest neighbor bootstrapping regressive model (NNBR) had been presented and was applied to the hydrology and waterresources with single variable time series and multivariable time series [2]. At present, some new forecast methods, such as ANN, grey system analysis, fuzzy logic model, SVM develop quickly. SVM, developed by Vapnik and his coworkers, have become one of the most popular methods in Machine Learning in the past few years and have been successfully applied to many fields for classifying, regressing and predicting, etc [3]. Ouyang Yongbao applied wavelet transform methods to the Yangtze River to discuss its applying prospect to hydrological trend

* Corresponding author.

J. Zhang (Ed.): ICAIC 2011, Part V, CCIS 228, pp. 522–528, 2011.

forecast [4]. Instead of relying on one individual model, this paper proposes a new forecast method which uses wavelet analysis methods to identify the various frequency components and local information of the original hydrological series to get several runoff series of different frequency components. Then NNBR model is used to forecast decomposed low-frequency approximate series, while SVM model with global optimization and good generalization performance is used to forecast decomposed high-frequency details series. Finally, results predicted respectively are coupled by the principle of wavelet reconstruction to get final predicted series. This new method fully exploits the effective components of the data in order to improve the forecasting precision, and it can overcome the shortcomings in the forecasting process using a single model. Example shows the results of the new long-term run-off forecasts approach using a multi-model based on wavelet analysis is more accurate and practical, and has good generalization performances which can response to the actual process of runoff.

2 Description of Selected Models

A. Nearest Neighbor Bootstrapping Regressive Model (NNBR)[2]

Time series variable $X_t(t=1,2,3,\ldots\ldots n)$are given depending on the adjacent historical value which are X_{t-1}, X_{t-2} ...X_{t-p}. According to the sampling weights of the nearest neighbor which are $W_t^{(i)}$, we can predict the follow-up value by follow equation:

$$X_t = \sum_{i=1}^{K} W_t^{(i)} X_i \tag{1}$$

Where $W_t^{(i)} = \dfrac{\sum_{i=1}^{K} \frac{1}{r_t^{(i)}}}{r_t^{(i)}}$, $r_t^{(i)}$ is the Euclidean distance between any two feature vectors, K is the number of nearest neighbor, $\sum_{i=1}^{K} W_t^{(i)} = 1$.

B. Support Vector Machine Model (SVM)[3]

SVM model uses a kernel function mapping. Given the time series x_1, x_2... x_n, $x_i \in R^m$, y_i is the associated output value of x_i, $y_i \in R$ and N is the number of samples for training. The nonlinear mapping between x_i and y_i is described by a nonlinear regression function $f(\bullet)$ after training. Given a known x_i, we can get the associated y_i which is the runoff to be forecast.

Vapnik shows the mapping form of the f [5]:

$$f(x,\alpha,\alpha^*) = \sum_{i=1}^{N} (\alpha_i^* - \alpha_i) K(x, x_i) + b \tag{2}$$

Where $\sum_{i=1}^{N}(\alpha_i^* - \alpha_i) = 0$, λ is an adjustable parameter. This paper chooses RBF kernel function $K(x,y) = \exp(-\|x-y\|^2 / 2\sigma^2)$ which is more common and more conducive to the selection of parameters. Then SVM model can be abstracted as:

$$y = f(x \mid C, \varepsilon, \sigma) \tag{3}$$

C. Wavelet Analysis

In this paper, our discrete approach to the wavelet transform is implemented with a version of the so-called *A Trous* algorithm.

Wavelet decomposition: It's supposed that scaling function $\phi(t)$ and wavelet function $\varphi(t)$ are given:

$$\frac{1}{2}\phi\left(\frac{t}{2}\right) = \sum_{n=-\infty}^{+\infty} h(n)\phi(t-n) \tag{4}$$

Where *h (n)* is a discrete low-pass filter associated with the scaling function $\phi(t)$.The smoothed sequences $A_j(t)$ at *j* scale and at a position *t* is the scalar product result:

$$A_j(t) = \sum_{n=-\infty}^{+\infty} h(n)A_{j-1}(t+2^j n) \tag{5}$$

The signal difference sequences $W_j(t)$ at *j* scale between two consecutive resolutions are:

$$W_j(t) = A_{j-1}(t) - A_j(t) \tag{6}$$

$\{W_1(t), W_2(t) \ldots W_J(t), C_J(t)\}$ represents the wavelet transform of the sequences at *J* scale. *J* is the maximum number of decomposition scales.

Wavelet reconstruction: A series expansion of the original image $A_0(t)$ in terms of the wavelet coefficients is now given as follows. The final smoothed sequences $A_j(t)$ is added to all the differences $W_j(t)$:

$$A_0(t) = A_J(t) + \sum_{j=1}^{J} W_j(t) \tag{7}$$

3 New Forecast Approach

A. Forecast Process

Figure 1 illustrates the whole flowchart of multi-model coupled forecast method based on wavelet analysis. The detail steps are described in the following.

Step1:The original runoff sequences are decomposed into two different component sequences using *A Trous* wavelet algorithm including high-frequency sequences $W_1(t)$ and $W_2(t)$ for each layer and low-frequency sequences $C_2(t)$ for the final layer.

Step2: NNBR and SVM models are built and then SVM model is trained using training samples.

Step3: Decomposed low-frequency sequences $C_2\ (t)$ are simulated and predicted by NNBR model, getting predication sequences $C_2(t)'$. Meanwhile, decomposed high-frequency sequences $W_1\ (t)$ and $W_2\ (t)$ are trained and tested with SVM model, getting predication sequences $W_1\ (t)'$ and $W_2(t)'$.

Step4: The different individual predication sequences are reconstructed with *A Trous* wavelet algorithm and then the final forecast sequences can be got.

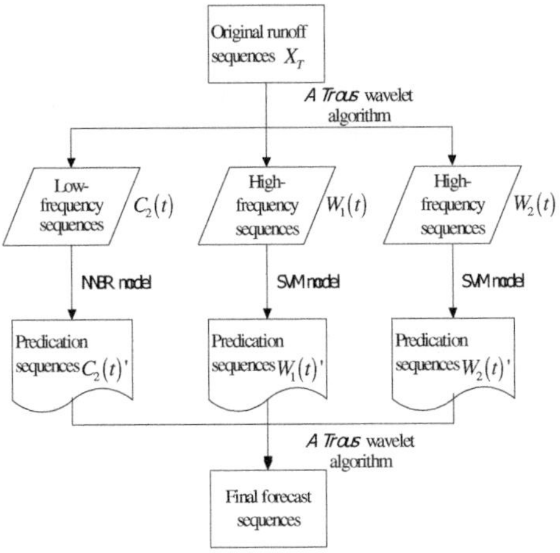

Fig. 1. Flow chart of multi-model coupled forecast method based on wavelet analysis

B. Model Parameters

In this paper, autocorrelation analysis, partial correlation analysis and trial and error methods had been used to determine the dimension of feature vector p which is the main parameter of NNBR model, using p=4. The number of nearest neighbor is given by $K = \text{int}\left(\sqrt{n-p}\right)$, where n is the sample size.Sample data with a 6-year period was used to determine the model structure of SVM model. The optimal values, identified with the trial method, are 1, 0.01 and 0.5 for C,ε,σ. Parameters for low-pass filter of *A Trous* wavelet algorithm are given by

$$h(7)=\left\{-\frac{\sqrt{2}}{32},0,\frac{9\sqrt{2}}{32},\frac{\sqrt{2}}{2},\frac{9\sqrt{2}}{32},0,-\frac{\sqrt{2}}{32}\right\}.$$

4 Case Study

A. Study Area and Basic Data

Annual mean runoff of totally 127 years from 1882 to 2008 from Yichang station was selected as X_T. It was decomposed with *A Trous* wavelet algorithm to get corresponding approximate sequences $\{W_1(t),W_2(t),C_2(t)\}$ for low-frequency and detailed sequences for high frequency. High frequency detailed sequences from 1987 to 1966 were chosen for the training sample of SVM model, and data from 1967 to 2008 to testing the model. Meanwhile, low frequency approximate sequences from 1967 to 2008 were used to simulate and predict with NNBR model. At last, the final forecast sequences can be obtained from the prediction results reconstructed by each component series using *A Trous* algorithm. The results show in Figure 2.

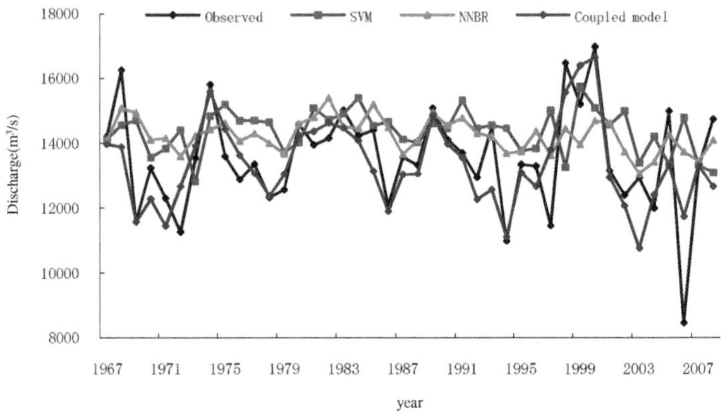

Fig. 2. Comparison chart of forecast results use different models

B. Discussion of Results

In this research, we have demonstrated the application of proposed multi-model coupled forecasting model. Statistical analysis for the forecast results of these three models in different testing periods are shown in Table 1 and Table 2

Table 1. The proportion falling into different intervals of different models in testing period(%)

Models	Testing periods				
	[0,10]	(10,20]	(20,30]	(30,40]	(40,100]
NNBR model	66.7	21.4	9.5	—	2.4
SVM model	50.0	33.3	9.5	4.8	2.4
Multi-model coupled method	81.0	14.3	2.4	2.4	—

Table 2. Statistical results of different models evaluation index

Index	NNBR	SVM	Multi-model coupled method
Maximum error(%)	62.54	75.02	38.82
Minimum error(%)	0.07	0.31	0.04
Average error(%)	9.85	11.87	5.92
Average error variance	114.18	172.88	51.62
Qualified rate(%)	88.1	83.3	95.2

We can see from the above testing and statistical results of different models:

(1) The NNBR model may fall into the high error interval when the future trajectory can not coincide with its own laws obtained from historical data. The prediction result of SVM model which can solve the problems of nonlinear, pattern recognition with high dimensional and local minimum efficiently is not satisfactory, because its performance will be affected by the model parameters and samples.

(2) A higher accuracy of prediction has been got by the multi-model coupled forecast method based on wavelet analysis. The proposed method can fully mine the regularity and randomness of hydrological data. It's not only can fit out of the trend of original runoff series, but also keep the qualification rate with which relative error within 20% is above 95%.

(3) The simulation accuracy between the natural and calculated runoff decreases after 1997, as the successfully completed damming of the Three Gorges Reservoir, which has some impact on the natural runoff. The forecasting results of all the three models are not satisfactory in 2006, because this year is an extraordinary dry year which makes that its laws can not be accurately simulated by historical data. However, we still can see that the results of long-term run-off forecasts using multi-model coupled forecast method is better than that of the single prediction model.

5 Conclusions

Multi-model coupled forecasts method based on wavelet analysis simulates the future runoff based on the hydrological laws extracted from original hydrological runoff data. NNRR model is used to predict the approximate sequences to avoid the application of the "assumption - fitting - forecast" mode such as the SVM model, which well coincides with the hydrological rules and also reflects its own superiority in the simulation. To get the global optimal solution, SVM model which has excellent generalization performance is used to predict the details of the sequences. The results show that the qualification rate of forecasting is 95.2 %(relative errors are less than 20%), so the new method in this paper has high accuracy and reliability. It can be used for long-term forecast in Yangtze River basin.

Acknowledgments. This research was partially supported by the National Scientific and Technological Support Projects of China under Grant No.2008BAB29B08; the Special Research Foundation for Public Welfare Industry of the Ministry of Science and Technology and the Ministry of Water Resources under Grant No. 200701008.

References

1. Mohammadi, K., Eslami, H.R., Kahawita, R.: Parameter estimation of an ARMA model for river flow forecasting using goal programming. Journal of Hydrology 331(1-2), 293–299 (2006)
2. Wang, W., Xiang, H., Ding, J.: Predication of Hydrology and Water Resources with Nearest Neighbor Bootstrapping Regressive Model. International Journal Hydroelectric Energy 19(2), 8–14 (2001)

3. Cao, L.J., Tay, F.E.H.: Support vector machine with adaptive parameters in financial time series forecasting. IEEE Transactions on Neural Networks. IEEE Computational Intelligence Society, 1506–1518 (2003)
4. Ouyang, Y., Ding, H.: Wavelet theory and its application to hydrological forcast and prediction. Haihe Water Resources (6), 44–46 (2006)
5. Vapnik, V.N.: The Nature of Statistical Learning Theory. Springer, New York (1995)

Application Research of Improved Dynamic Programming for Economic Operation of Gezhouba Hydropower Station

Zhenzhen Wang, Jianzhong Zhou*, Yongqiang Wang, and Wen Xiao

School of Hydropower and Information Engineering
HuaZhong University of Science and Technology
Wuhan, China
ZhenzhenWang009@tom.com

Abstract. This paper presents a algorithm based on priority list coupled with successive approximation dynamic programming for solving unit commitment problems of GeZhouba hydropower station to minimize water consumption as the optimization criterion, which can resolve effectively the "dimension disaster" problem in traditional dynamic programming. It is confirmed that the optimization speed of the algorithm improve significantly under the premise of the same optimal water consumption by comparing the water consumption and the optimization time with the traditional dynamic programming, successive approximation dynamic programming, priority list dynamic programming. The algorithm meets the daily operational requirements of Gezhouba hydropower station.

Keywords: dynamic programming, successive approximation, priority list, economic operation, global optimization.

1 Introduction

As a important work of hydropower station management, the main purpose of ecnomic operation is to determine the operating mode in order to get the maximum economic benefits under the premise of safety engineering and production, equipments in good condition, reliable operation. Which is of great significance for guiding the station determine the operating mode to get the maximum economic benefits. Currently researchers have proposed a variety of optimization algorithms, such as equal incremental principle, dynamic programming algorithm, genetic algorithm, in which dynamic in which dynamic programming algorithm has been most widely used, Wang Yide proposed to make a list of optimum unit commitment and counts of units in operation by using fast dynamic programming algorithm[1], which can be queried when load is distributed. But for a giant hydropower station with variety of units, traditional dynamic programming algorithm can easily lead to the curse of dimensionality problem. There are a variety of methods to improve the dynamic programming algorithm, such as successive approximation[2] and priority list[3], but

* Corresponding author.

J. Zhang (Ed.): ICAIC 2011, Part V, CCIS 228, pp. 529–535, 2011.

the effect is not significant. In tjis paper, the economic operation of giant hydropower station according to the operation situation of Gezhouba hydropower station is investigated by using dynamic programming algorithm based on priority list coupled with successive approximation.

2 Mathematical Modelling

Ecnomic operation of hydropower station is to arrange and organize the operation of power plant equipments reasonably under the premise of safety, reliable and quality generation of electrical energy, to obtain the greatest ecnomic benefits. Its optimization criterion is the minimum water consumption under certain power load conditions[4].

A. Space Optimization Model

The objective of space optimization is to choose units and the optimal load distribution reasonably under the certain power load conditions, the mathematical model is shown in the following formula:

$$\begin{cases} Q = \min \sum_{i=1}^{m} Q_i(H, N_i) \\ s.t \quad N = \sum_{i=1}^{m} N_i \\ N_i \in D_i \end{cases} \tag{1}$$

Where

Q: Minimum total water consumption under the certain total load condition;

$Q_i(H, N_i)$: Reference flow of No.i when the water head is H, the load is N_i;

N: Total load given by dispatching centre;

D_i: Load feasible region of No.i (including load limit, cavitation pitting zone, vibration limit);

m: Number of unit put into operation;

i: Unit No.($i = 1, \ldots, m$).

B. Time Optimization Model

Time optimization is to investigat the optimization problem within a period Ton the basic of space optimization. Not only the optimization of each period, but also the effects on optimization according to the unit commitment caused by the changes in load and the water consumption due to the start-stop of units are considered. Space optimization is a foundation, while time optimization is global optimization. The mathematical model is shown in the following formula:

$$\begin{cases} Q^*(T) = \min\sum_{t=1}^{T}\sum_{i=1}^{m}[Q_i(H(t),N_i(t)\cdot T_i + S_i(t)(1-S_i(t-1))\cdot Q_{si} \\ +S_i(t-1)(1-S_i(t)\cdot Q_{si}) \\ st.N(t) = \sum_{i=1}^{m} N_i(t) \\ N_i(t)\in D_i \end{cases} \tag{2}$$

Where

$Q^*(T)$: Minimum total water consumption under the certain total load condition within a period T;

$Q_i(H(t), N_i(t))$: Reference flow of unit i when the water head is $H(t)$, the load is $N_i(t)$ within a period t;

T_i : Time span of period t;

$N(t)$: Total load given by dispatching centre within a period t;

Q_{si}, Q_{ci} : Water consumption due to the start-stop of units;

S_i : State variable of the start-stop of units ('0' is outage; '1' is power on).

3 Model Solution

A. Dynamic Programming(DP)

There are generally two steps in solving the problem of economic operation by using DP algorithm [5]. First step, create a economic operation summary list of the power station based on the optimal number of the units, unit commitment and optimal allocation of units according to the possible load determined by economic operation model and DP algorithm in the rang of the power output. Second step, according to the economic operation summary list, station load $N(t)$ and the upstream water level H at the beginning of a day etc. which are given by the cascade dispatching system. The algorithm uses k as session variable, uses $\overline{N}_k$ as state variable, uses N_k as decision variable, the recursion relations is shown in the following formula:

$$\begin{cases} Q_k(\overline{N}_k) = \min\limits_{N_k\in R_k}[Q_k(N_k) + Q_{k-1}(\overline{N}_{k-1})] \\ \overline{N}_{k-1} = \overline{N}_k - N_k (k=1,2,...,n) \\ Q_0^*(\overline{N}_0) = 0, \forall \overline{N}_0 \end{cases} \tag{3}$$

Where

$\overline{N}_k$: Total load of unit k~1;

$Q_k^*(\overline{N}_k)$: Total water consumption according to the economic operation of unit k~1;

$Q_k(N_k)$: Water consumption of unit k when the load is N_k;

$Q_0^*(\overline{N}_0)$: Boundary conditions, means the water consumption is 0 before the the initial period.

There are $2^N \times T$ status when the system of N units is considered in the period T by using DP algorithm, if N and T increase, the number of status increase sharply, which lead to curse of dimensionality. It takes large space to store and long time to optimize.

B. Dynamic Programming Based on Priority List (PLDP)

RanYao[6], R.M.Burns and C.A.Gibson[7] proposed to determine the optimal unit commitment by using improved DP algorithm with priorities. The key due is to determine the order of unit commitment according to minimum water consumption per unit and unit characteristic curve before the implementation of DP algorithm. Minimum water consumption per unit means minimum water consumption with the change of load in unit time. The mathematical model is shown in the following formula:

$$\mu = \frac{q}{P} \tag{4}$$

Where

q : minimum water consumption in a unit time.

Those unit status which don't go with the priority list are abandoned during the DP algorithm optimization, which reduce the number of status to a certain extent[8], The PLDP algorithm is flexible[9], it takes less memory, the optimization results is close to optimum combination for small system of unit commitment, but for giant hydropower station which has many units, the amount of calculation is too large, the speed can not be improved significantly, the improvement effects is not fully up to expectations.

C. Successive Approximation Dynamic Programming

The basic thought of SADP algorithm[10] is solving the start-stop plan of each unit on condition that the load of other units is fixed, and repeat the process until the start-stop plan won't change[11]. For giant hydropower station, it takes too many iterations by using the SADP algorithm, the velocity of convergence is not fast, the improvement effects is not satisfactory either.

D. Dynamic Programming Based on Priority List Coupled with Successive Approximation(PL-SADP)

PL-SADP algorithm is formed by priority list and SADP algorithm. The main idea of PL-SADP algorithm is solving unit commitment problem with SADP algorithm according to the minimum water consumption per unit obtained by formula (4). The procedure is as follows:

Step1: Set the initial state, that is, all generating units shut down;
Step2: Determine the priority list according to the minimum water consumption per unit;
Step3: Set up the loads of unit 2~N according to the priority list, determine the start-stop plan of unit 1 every period by using DP algorithm;
Step4: Set up the loads of unit 1, 3~N according to the priority list, determine the start-stop plan of unit 2 every period by using DP algorithm;
Step5: Repeat step3 until the start-stop plan won't change.

Those unit status which don't go with to the priority list are abandoned during SADP algorithm by using DP algorithm optimization, which lead to a substantial increase in convergence rate. The optimization speed is improved significantly under the premise of required optimization accuracy in ecnomic operation of giant hydropower station.

4 Example

To illustrate the effectiveness of the proposed method, application of ecnomic operation of Gezhouba hydropower station is considered. The highest water head of the station is 27 meters, the lowest water head is 8.3 meters. The system has two stations named Dajiang station and Erjiang station, Gezhouba hydropower station, with a total installed capacity of 2757 MW, consists of 21 units, in which 14 units of Dajiang station are Xiaoji, with installed capacity of 125 MW each unit, including unit 14 named Dongdian, with installed capacity of 146 MW; the other 7 units belong to Erjiang station, in which unit 1~2 are Daji, with installed capacity of 170 MW each unit, unit 3 is Hadian, with installed capacity of 146 MW, unit 4~7 are Xiaoji, with installed capacity of 125 MW each unit, in addition, there is a 20MW unit of Zhi-fa. Now set the water head $H=21m$, the daily load curve of the station is set as the following Fig.1.

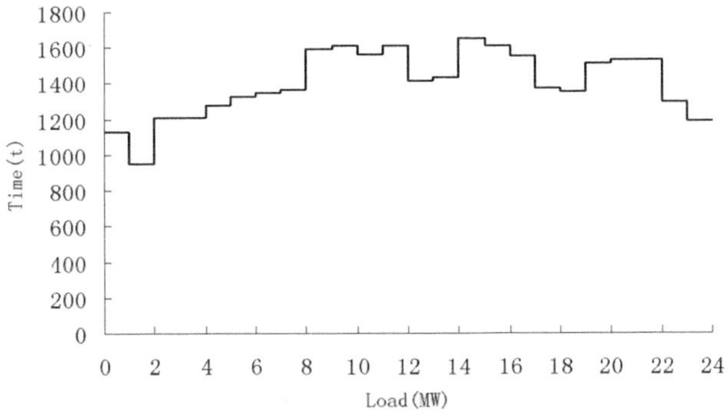

Fig. 1. Daily load curve

The traditional DP algorithm, the PLDP algorithm, the SADP algorithm and PL-SADP algorithm are used to determine the start-stop plan of Gezhouba hydropower station. As shown in Table 1, by using traditional DP algorithm, the total consumption of water is 0.649 billion, but the optimization time is 54 minutes, while the optimization times are 40 minutes, 20 minutes and 6 seconds by using PLDP algorithm, SADP algorithm and PL-SADP algorithm, by contrast, PL-SADP algorithm can better meet the requirement of daily allocation of burden. The optimal load distribution of each unit of Gezhouba station obtained by PL-SADP algorithm are shown as Fig. 2.

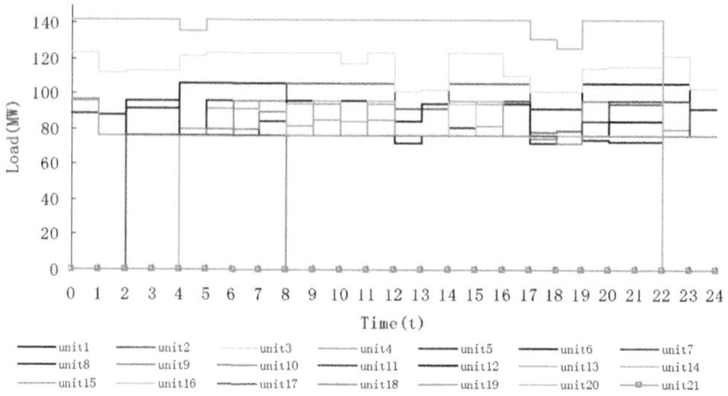

Fig. 2. Start-stop plan of GeZhouba station

Table 1. Comparison with DP,PLDP,SADP AND PL-SADP

Index	DP	PLDP	SADP	PL-SADP
Water Consumption (billion m3)	0.649	0.649	0.649	0.649
Optimization Time	54 min	40 min	20 min	6 s

5 Conclusions

The unit commitment is an important nonliner optimization split in ecnomic operation of hydropower station, the paper proposes PL-SADP algorithm for solving unit commitment problem of giant hydropower station, algorithm can effectively alleviate the “Curse of Dimensionality” problem because of the large number of the units of giant hydropower station. At last, by application example of Gezhouba hydropower station, the feasibility and effectiveness are verified, the algorithm provide a simple, effective way to solving the unit commitment of giant hydropower station.

Acknowledgment. This work is partially supported by the National Scientific and Technological Support Projects of China under Grant No.2008BAB29B08; the Special Research Foundation for Public Welfare Industry of the Ministry of Science and Technology and the Ministry of Water Resources under Grant No. 200701008.

References

1. Wang, Y., Nan, H., Liu, D.: Application of Fast Dynamic Programming for Operated Unit Commitment in Load Distribution. Water Resourse and Hydropower Engineering, 32(6), 28–30 (2001)
2. Hu, M., Li, Y.: Application of Successive Algorithm for Optimal Opaeratiom of Reservoir. Computer and Modernization, 9-10 (June 2008)

3. Zhang, L., Lin, Z., Liu, J.: Improved Dynamic Programming for Solving Unit Commitment Problems. Power and Electrical of Fujian, 7–8 (December 2006)
4. Zhang, Y.: Economic Operation theory of Hydropower Station, pp. 124–131. China WaterPower Press (1998)
5. Snyder Jr., W.L., Powell Jr., H.D., Rayburn, J.C.: Dynamic Programming Approach to Unit Commitment. IEEE Trans. on Power Systems 2, 339–350 (1987)
6. Ren, Y., Zhang, X.: Research of Improved Dynamic Programming for Unit Commitment of Power System. Techniques of Automation and Applications 2(5), 6–8 (2010)
7. Burns, R.M., Gibson, C.A.: Optimization of Priority Lists for A Unitcommitment Program. IEEE/PES, Summer Meeting, Paper A 75 453-1 (1975)
8. Tanabe: A practical algorithm for Unit Commitment based on Unit Decommitment Ranking. Electrical Engineering in Japan (English translation of Denki Gakkai Ronbunshi) (January 30, 2002)
9. Sheble, G.B.: Solution of the Unit Commitment Problem by the Method of Unit Periods. IEEE Trans. on Power Systems 5(1), 257–260 (1990)
10. Zürn, H.H., Quintana, V.H.: Generator maintenance scheduling via successive approximations dynamic programming. IEEE Transactions on Power Apparatus and Systems 94(2), 665–671 (1975)
11. Yu, B., Wang, J.: Successive Approximation for Solving Hydropower Unit Commitment Problem. Central China Electric Power, pp. 2-3 (June2004)

Fuzzy Synthetic Assessment of Geo-engineering Suitability for Urban Underground Space*

Wenling Xuan

School of Earth and Space Sciences,
Peking University.
3403, YiFuEr Bldg., No.5 YiHeYuan Rd.
HaiDian District, BeiJing 100871
WenlingXUAN111@yeah.net

Abstract. Urban underground space is being recognized as a valuable resource for sustainable city development. A fundamental step in this development is the understanding of the quality of urban underground living conditions and the capacity for urban underground space planning and construction. This paper adopts fuzzy comprehensive evaluation methods to focus on an engineering suitability assessment of underground space. Through the National Code for Urban Planning Engineering Geotechnical Investigation and Surveying, the assessment criteria are defined, as well as the sub-factors associated with each criterion. This fuzzy comprehensive assessment method is applied to Binhai New Region of TianJin City. The aim of obtaining the distribution of urban underground space suitable for geo-engineering, both in quantity and quality, is realized.

Keywords: Urban underground evaluation, Analytic Hierarchy Process, fuzzy set, Fuzzy Synthesis Evaluation, Geo-Engineering Suitability.

1 Introduction

Urban development takes place most frequently on the ground and in the air space above the ground. The underground normally is an area and space which is not visible, not easily accessible, with no detailed nor dense measurements of its exact spatial and geological composition. Due to the accelerating process of urbanization, some big cities in China are confronted with strong and critical pressures of development on the ground surface such as serious pollution, traffic congestion, lack of space, risk of natural disasters, deteriorating infrastructure etc. These detrimental effects of string and extensive area development all point to a desirable and immediate change in current over-ground planning. Underground space can therefore be considered as a valuable resource which has the capacity to hold lots of urban

* This work is partially supported by Key Lab of Earth Exploration & Information Techniques of Ministry of Education in Chengdu University of Technology.

J. Zhang (Ed.): ICAIC 2011, Part V, CCIS 228, pp. 536–543, 2011.

functioning construction and optimize the utilization of existing surface infrastructure. However, methods of scientific and geological investigations and evaluations have to be developed first to reduce the risk and uncertainties of building in underground space as well as ensuring their social success. Thus, this paper studies the engineering suitability assessment issue for urban underground space.

Underground space is a typical grey area from the viewpoint of information theory, as the geological phenomena is unseen to human beings, and much knowledge of a specific space is derived from sparse geological drilling data and professional analysis. More over, the geological environment is a complex system. Geological structure, engineering geology condition, hydrological geology status, rock/soil body physical characteristics, adverse geological factors such as land subsidence, sand liquefaction etc., all contribute to the parameters defining Geo-engineering suitability. We have to admit that these multiple sources information do not possess characteristics of being precise, certain, exhaustive and unambiguous. Actually the fuzzy nature of geological science is pervasively present in the geological factors. Therefore, these geological factors ought to be treated as fuzzy nominal information, e.g. one type of factor case comprises a fuzzy set, and a degree of membership of an element, between 0 (non membership) and 1 (complete membership), in a proper mathematical model. For all the related factors, correspondingly fuzzy sets can be formed and valued. Fuzzy comprehensive analysis with the aid of fuzzy weight matrix can be the solution to the whole matter.

2 Fuzzy Comprehensive Assessment Method

A. Fuzzy Set and Fuzzy Numbers

Fuzzy set theory, first introduced by Zadeh in 1965, deals with ambiguous or ill-defined situations. Fuzzy theory includes elements such as fuzzy set, membership function, and the fuzzy numbers used to efficiently change vague information into useful data.

Fuzzy set theory uses groups of data with boundaries that feature lower, median, and upper values that are not sharply defined. Because most of the decision-making in real world takes place amidst situations where pertinent data and the sequences of possible actions are not precisely known, the merit of using the fuzzy approach is that it expresses the relative importance of the alternatives and the criteria with fuzzy numbers rather than crisp ones. A fuzzy set is characterized by a membership function, which assigns a membership range value between 0 and 1 to each criterion and alternative.

Triangular Fuzzy Numbers (TFN) are usually employed to capture the vagueness of the parameters related to the selection of the alternatives. In order to reflect the fuzziness which surrounds the decision makers when they select alternatives or conduct a pair-wise comparison judgment matrix, TFN is expressed with boundaries instead of crisp numbers. In this research, we use TFN to prioritize underground space suitability in the fuzzy hydrogen energy sector. TFN is designated as $M_{ij}=\left(l_{ij},m_{ij},u_{ij}\right)$. m_{ij} is the median value of fuzzy number M_{ij}, l_{ij} and u_{ij} is the left and right side of fuzzy number M_{ij} respectively. Consider two TFN M_1 and M_2, $M_1 = (l_1, m_1, u_1)$ and $M_2 = (l_2, m_2, u_2)$.

Their operations laws are as follows:

$$\left(l_1,m_1,u_1\right)\oplus\left(l_2,m_2,u_2\right)=\left(l_1+l_2,m_1+m_2,u_1+u_2\right) \tag{1}$$

$$\left(l_1,m_1,u_1\right)\otimes\left(l_2,m_2,u_2\right)=\left(l_1\times l_2,m_1\times m_2,u_1\times u_2\right) \tag{2}$$

$$\left(l_1,m_1,u_1\right)^{-1}=\left(l_1^{-1}{}_2,m_1^{-1},u_1^{-1}\right), \tag{3}$$

B. Fuzzy Analytic Hierarchy Process (AHP)

Under the AHP, the decision-making process is modified into a hierarchical structure. At each level of the hierarchy, AHP uses pair-wise comparison judgments and matrix algebra to identify and estimate the relative priorities of criteria and alternatives. This in turn is carried out by breaking down a problem into its smaller constituent parts. The AHP thus leads from simple pair-wise comparison judgments to priorities arranged within a hierarchy. However, the AHP cannot take into account uncertainty when assessing and tackling a problem effectively. However, the fuzzy AHP can tackle fuzziness or the problem of vague decision-making more efficiently by using fuzzy scales with lower, median, and upper values. This can be contrasted with the AHP's crisp 9-point scale and synthesis of the relative weights using fuzzy sets, membership functions, and fuzzy numbers.

Although the AHP is employed herein to capture experts' knowledge acquired through perceptions or preferences, the AHP still cannot effectively reflect human thoughts with its crisp numbers. Therefore, fuzzy AHP, an extension of the AHP model, is applied to resolve hierarchical fuzzy decision-making problems.

C. Fuzzy Synthesis Evaluation

Fuzzy synthesis evaluation process consists of six fundamental elements: (1) U - Evaluating factor domain, which is the set of a series of evaluation criteria. (2) V - Ranking domain, the sub-set of changing scope with single evaluating criterion. (3) R - Fuzzy relationship matrix, the 2D matrix of membership number for each ranking index. (4) W - Fuzzy weight matrix, used for weighted analysis of evaluating factors. (5) *W*R* - Synthesizing operator. (6) B - Evaluating result.

The six steps in Fuzzy synthesis evaluation is drawn as Fig.1

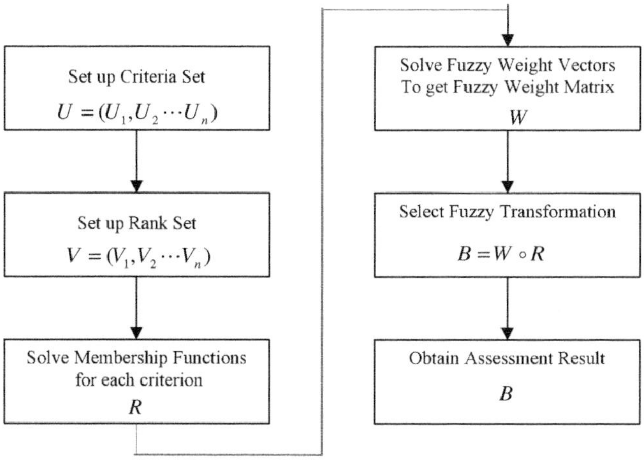

Fig. 1. Fuzzy synthesis evaluation workflow

3 Assessment of Engineering Suitability for Underground Space Based on Fuzzy Synthesis Method

A. Identify Evaluating Index

The assessment index for underground space engineering application is based on guidelines in "Code for urban planning engineering geotechnical investigation and surveying (GJJ 57-94)" considering the following criteria and sub-factors:

_ Engineering Geology Criterion
Influence of active fault
Landslide and carse
Earthquake intensity

_ Hydrological Geology (based on adjacent non-farm activities) Criterion
Existence type of groundwater
Groundwater lever and replenishment
Unit emerged water quantity
Permeability coefficient
Groundwater corrosion

_ Soil body criterion (based on terrain characteristics and infrastructure availability)
Cohesive strength
Internal friction angle
Normal value of bearing capacity
Compression modulus

B. Hierarchy of the Criteria

The hierarchy structure of the criteria is shown in Fig.2

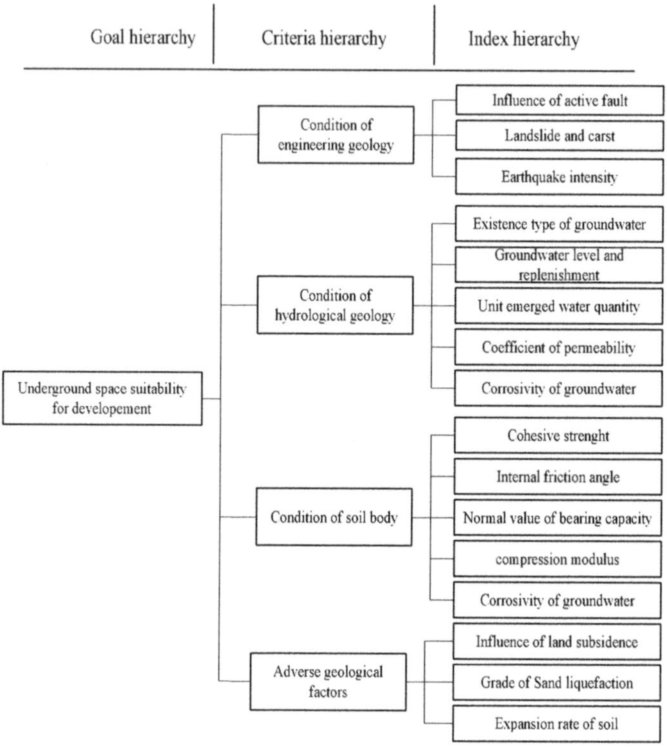

Fig. 2. Assessing criteria hierarchy model for underground space suitablity

At the top of the control hierarchy, there exists the goal of this problem. The goal is to assess underground space suitability in a relative fashion. At Level 1, there exist four criteria: condition of engineering geology, condition of hydrological geology, condition of soil body, and adverse geological factors. Each criterion in Level 1 consists of some sub-criteria, as demonstrated in Fig 2.

C. Ranking Rules of Engineering Suitablity for Underground Space

There are different ranking rules for underground space suitability. In our study, we give five grades for the suitability of underground space taking into account of assessment and practical requirement. They are very bad, bad, normal, good and very good, and represent by I, II, III, IV and V,

D. Normalization of the Individual Evaluating Factor

Due to the dimensional differences among different evaluating factors, as well as the scale effect, we need to normalize indivial evaluating factor.

Assume C_i represents the observed value for ith factor,

S_i represents the avaerage value for ith factor,

Then, the weight for the ith factor W_i :

$$W_i = \frac{C_i}{S_i} \tag{4}$$

We normalize individual weight for the sake of fuzzy operation, then, the normalized weight for ith factor:

$$V_i = \frac{C_i / S_i}{\sum_{i=1}^{m} C_i / S_i} \tag{5}$$

4 A Case Study in Binhai New Region

Study scenario

In the port city of Tianjin BinHai new region, with a CBA (Central Business Area) of say 20 km^2, the average height of all buildings above ground level is 10 stories and the average structure depth below ground level is 2 stories of basement levels mainly used for services such as car parks, transport routes etc. The space under the ground, for large scale development, is largely ignored and certainly under utilised. This leads to a practical procedure of planning both new over ground and new underground development together and complimentarily. Conversely, the redevelopment of over ground areas can be made incorporating new underground development within the same project. Pilot projects must be executed to test methods of underground space development.

Geophysical and geochemical experments as well as geoengineering exploration work has been finished. Discret data, multi-sources, geological statistics method for the assessment of the given area is also implemented.

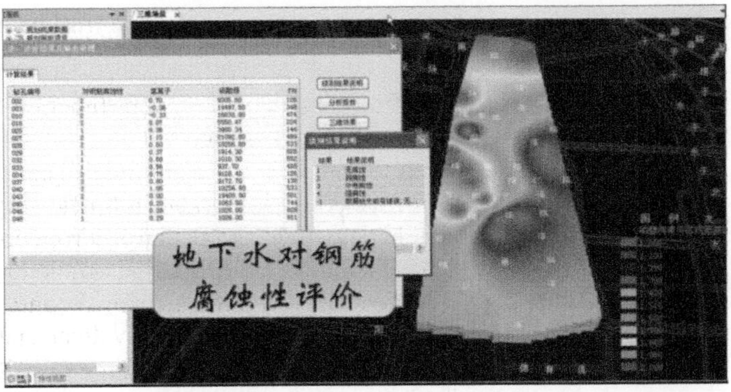

Fig. 3. Ground Water Effects Assessment

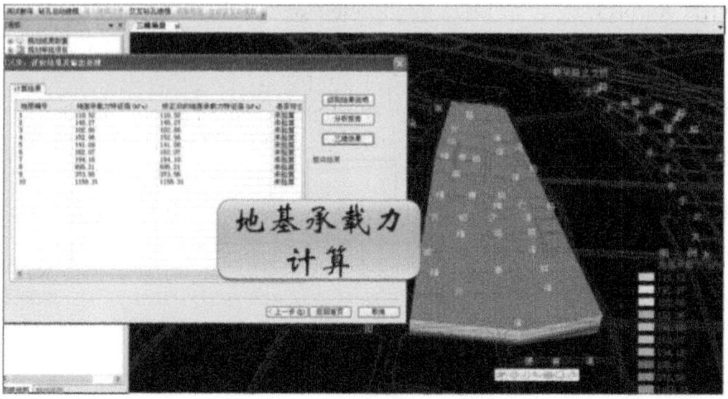

Fig. 4. Bearing Capacity of Underground Soil Assessment

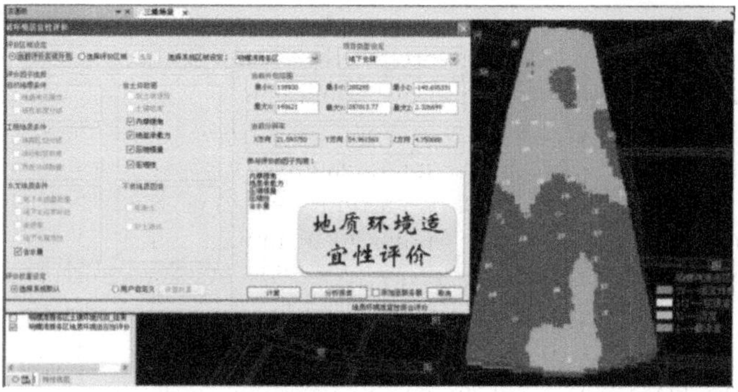

Fig. 5. Geo-engineering Suitablity Assessment

According to the assessing criteria hierarchy model for underground space, conditions of Engineering Geology, Hydrological Geology,Bearing Capacity of Underground Soil etc. are evaluated. The results are illustrated in Fig 3, 4, 5 respectively.

Acknowledgment. The study has attained the support from Ministry of Housing & Urban - Rural Development of China, geological information laboratory of Peking University, Planning Bureau of BinHai New Region of TianJin city.

Special thanks go to Beijing ChaoWeiChuangXiang company for their software platform.

References

1. Tong, L., Zhu, W.: The evaluation and develop planning of urban underground space resources. China construction industry press (2009)
2. School of Earth & Space Sciences of Peking University, Technical design of 3D Underground Information System for TianJin Tanggu Planning Bureau (2008)

3. National Standard of People's Republic of China, Underground Water Quality Standard GB/T 14848-93
4. Code for urban planning engineering geotechnical investigation and surveying (GJJ 57-94)
5. Joerin, F., The'Riault, M., Andre'Musy, et al.: Using GIS and outranking multicriteria analysis for land-use suitability aassessment. Geographical Information Science 15(2) 153-174 (2001)

Large Signal Model of AlGaN/GaN High Electron Mobility Transistor

Jiang Xia[1], Yang Ruixia[1], Zhao Zhengping[2], Zhang Zhiguo[2], and Feng Zhihong[2]

[1] College of Information Engineering Hebei University of Technology
Tianjin 300130, China
[2] The National Key Lab of ASIC The 13 Research Institute,
CETC Shijiazhuang 050051, China
JiangXia1969@tom.com

Abstract. A accurate nonlinear large signal model was extracted in this paper. The model is capable of correctly modeling the DC characteristics of AlGaN/GaN HEMT. The parameters of model are obtained from a series of pulsed I-V characteristics and S parameters adopting the on-wafer testing technique and narrow pulse testing technology. The model is implemented in ADS software, simulations and measurements data show its availability and can be used for GaN MMIC's designing.

Keywords: AlGaN/GaN HEMT large signal model ADS MMIC.

1 Introduction

AlGaN/GaN high electron mobility transistors (HEMT) are emerging as a frequently employed technology in high power RF circuits. This is primarily due to desirable attributes such as high drain-source current, high drain-source breakdown voltage and wideband RF operation[1] [2]. As a result, these devices show great promise in the development of compact, high power microwave circuits[3]. An accurate nonlinear large signal model is essential in designing applications for GaN HEMTs such as high efficiency and broadband power amplifiers.

In this paper, we described the technique and method of modeling, extracted a large signal nonlinear model adopting on-wafer testing technique and IC-CAP software, validated the model's availability in ADS software.

2 Divice Structure

AlGaN/GaN HEMT investigated in this work is primarily developed for power applications. The device is fabricated on a sapphire substrate, gate wide is 1mm. Fig.1 shows the schematic diagram of the AlGaN/GaN HEMT.

J. Zhang (Ed.): ICAIC 2011, Part V, CCIS 228, pp. 544–548, 2011.

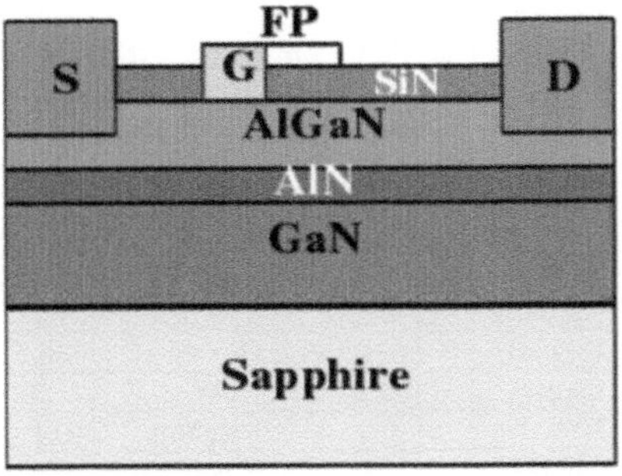

Fig. 1. Schematic diagram of the AlGaN/GaN HEMT structure

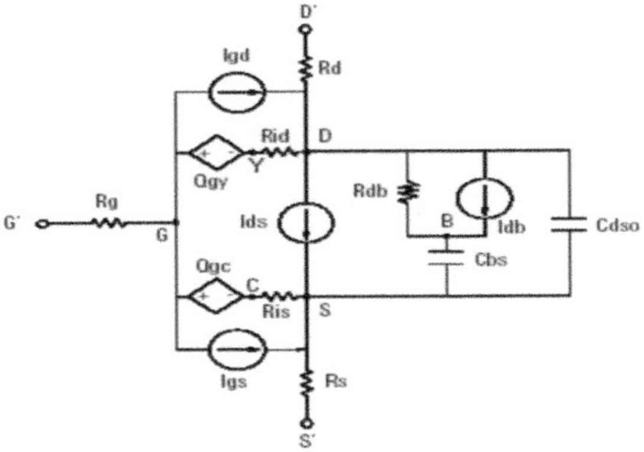

Fig. 2. The equivalent circuit of EEHETM

3 Extraction of Model

The technology of modeling for AlGaN/GaN HEMT is not mature, there is no professional software can be used to model AlGaN/GaN HEMT. So, this paper adopts EEHEMT model based on GaAs power device[4], the model's equivalent circuit is show in Fig.2.

We get model parameters adopting the method of part measure parameter. The process comprize five parts. (1) get gate current and voltage parameters from the measure data of pulse gate current, (2) extract extrinsic elements including resistance and inductance of source, drain and gate from measure data using improved cold-FETS[5], (3) get drain-source current parameters from measure data of pulse drain-source current, (4)get intrinsic model parameters from measure data of S parameter using bias measurement, (5)extract gate capacitance model parameters.

Based on above method, we extracted the model parameters of AlGaN/GaN HEMT. The extracted parameters value show in Table 1.

Table 1. The Value Of Extracted Parameters

Extrinsic Parameters	Value	Intrinsic Parameters	Value
Rd(ohm)	0.65	Cgd(pF)	0.0864
Rs(ohm)	0.55	Cds(pF)	0.16
Rg(ohm)	0.324	Cgs(pF)	1.39
Lg(nH)	0.05	Gm(mS)	36.5
Ls(nH)	0.1	τ(ps)	1.5
Ld(nH)	0.1	Ri(ohm)	0.1
Cpg(pF)	0.027	Rds(ohm)	9.6
Cpd(pF)	0.068		

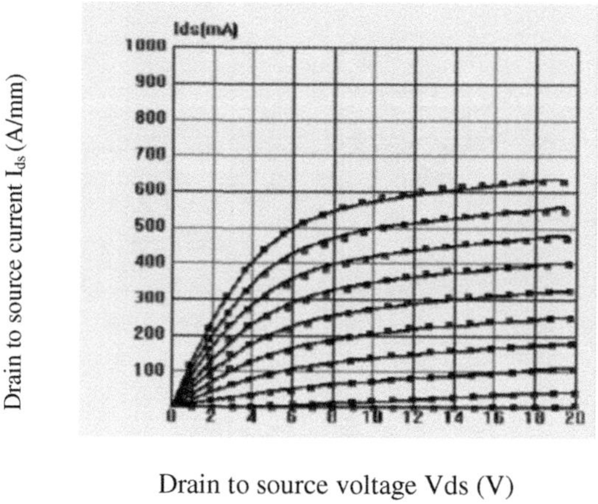

Fig. 3. DC output characteristic of AlGaN/GaN HEMT

The device model was embedded from two-port S parameters by a HP8510 vector network analyzer. We get Fig.3 from modeling software, which shows the DC output characteristic of AlGaN/GaN HEMT. In this figure, solid lines denote model's simulation results, dots denote measurement data. As seen in Fig.3, the model's simulation result is good associated with measurement data.

4 Validation

In order to validate the accuracy of model, which was imported into the ADS software[6], the simulation circuit and simulation result are show in Fig.4 and Fig.5 respectively. Compare Fig.5 and Fig.3, we can see that, there is a good agreement between simulations and measurements. This results show that model extracted is accurate.

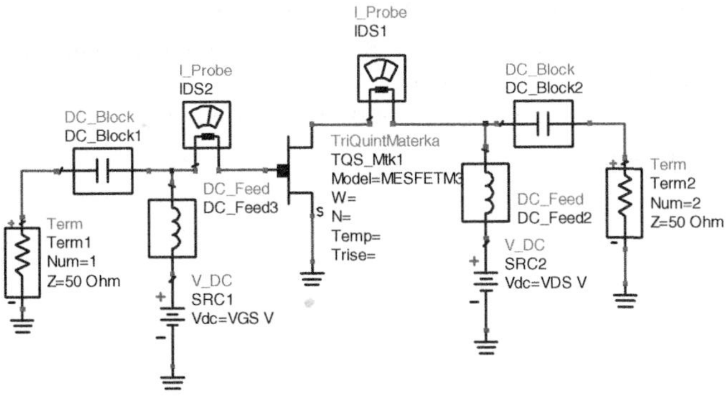

Fig. 4. The simulation circuit of AlGaN/GaN HEMT model in ADS

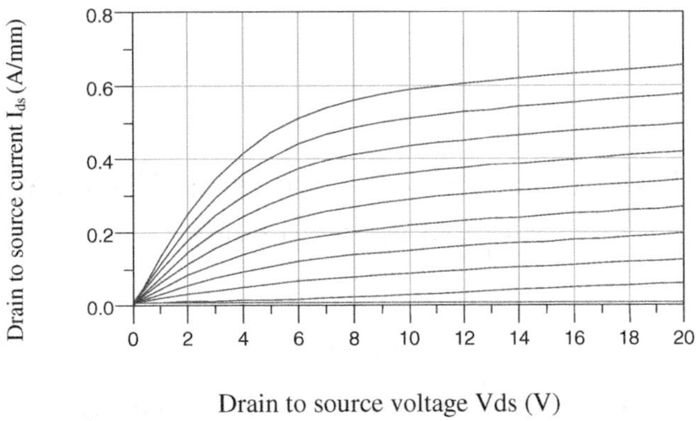

Fig. 5. Variation of drain current Ids with voltage Vds in ADS

5 Conclusions

By microwave on-wafer test and IC-CAP software, we have extracted a accurate large signal nonlinear model for AlGaN/GaN HEMT with total gate width 1mm. This model has been implemented ADS software and proved available, can be used for the design of GaN MMIC directly, this is our emphasis in latter work.

Acknowledgment. Jiang Xia thanks Zhao Zhengping, Zhang Zhiguo and Feng Zhihong for their contribution in the experimental part of this work.

References

1. Barker, A.E., Skellern, D.J.: A realistic large-signal MESFET model for SPICE. IEEE Trans Microwave Theory and Techniques 45(9), 1563–1568 (1997)
2. Islam, S.S., Anwar, A.F., Webster, R.T.: A physics based frequency dispersion model of GaN MESFETs. IEEE Transactions on electron devices 51(6), 846–853 (2004)

3. Behtash, R., Tobler, H., Berlec, F.J.: AlGaN/GaN HEMT power amplifier MMIC at X-band. IEEE MTTS Int Microwave symp digest 3(11), 1657–1659 (2004)
4. Zhang, S., Yang, R., Gao, X.: The large signal modeling of GaAs HFET/ PHEMT. Chinese Journal of Semiconductors 28(3), 439–441 (2006)
5. Dambrine, G., Gappy, A., Heliodore, F.: A new method for determining the FET small signal equivalent circuit. IEEE Trans. Microwave Theory Tech. 36(7), 1151–1159 (1988)
6. Chen, Y., Li, Z., Xia, W.: Detailed application of ADS-Design and simulation for Radio Freqency circuit. Poster and Telecom Press, China (2008)

Analysis of Performance of a System with Reservation Mechanism under Repair Support

Ye Hu[1], Tao Hu[2,*], and BaoHong Li[3]

[1] Information Center Jiangsu Standardization Institute Nanjing 210029, China
[2] Tourism College Hainan University Haikou 570228, China
[3] Library Hainan University Haikou 570228, China
hewu1978@sina.com

Abstract. Reservation mechanism for workflow system has been arisen much more considerations for many researchers. But the repairing of them has not been given reasonable discussions. However, it may be an important issue about how to assure the system to be ready when some of them are out service. If it is repaired in a short time and the system will be in order soon. In this paper, we plan to fill the gap of this. First, we present the study of a system with repair mechanism. Secondly, a workflow system with different repairing mechanism is given. Then, it is discussed that algorithm of them for the repairing. Fourthly, we show a case study about it. The result of the simulation shows that it will be effective to provide the repairing mechanism especially for a system without redundant support.

Keywords: Reservation, Workflow, Repair.

1 Introduction

Workflows have received considerable attention recently, and more and more researchers are concerning the mechanism of advance resources reservations about workflows system. Some of them are complete production-ready systems like VIEW [1LL2008], Kepler [2],Taverna [3], and so on. While they provide sophisticated support for the execution of Scientific Workflows or for analyzing the provenance of workflow results, none of them address the predictability of the execution and repairing [4].

Langguth et al. describe resource reservation system which includes the AR component to be used to find a combination of resource reservations at the providers of the individual operations that can satisfy the client's needs in terms of QoS, and to set up the required reservations, using negotiations based on WS-Agreement, so that the following execution of the workflow is guaranteed to meet these requirements [5]. Furthermore, Christoph Langguth et al. presented the DWARFS approach that employs Advance Resource Reservations to enable the predictable execution of Scientific Workflows with user-definable QoS criteria. Based on the providers' specifications of the resources needed for executing the operations they offer, DWARFS supports the definition and usage of reservations. To that end, they have

* Corresponding author.

J. Zhang (Ed.): ICAIC 2011, Part V, CCIS 228, pp. 549–557, 2011.

devised a Genetic Algorithm that can be used to find near-optimal combinations of the required resource allocations to meet the requested QoS, and provided an evaluation of its qualitative and quantitative performance [4].

In research works about reservations resources for running an system, many systems for allocating tasks on grids, (e.g. DAGMan [6]), currently allocate each task individually at the time when it is ready to run, without aiming to globally optimize the workflow schedule. [7] introduces a novel approach for workflow scheduling based on the HEFT algorithm and resource brokerage for a heterogeneous set of computers. It is demonstrated the effectiveness of our approach with two real-world applications and compare our techniques against the widely known DAGMan Condor scheduler, and show that it has a better performance even using less resources than others.

In [8], several techniques are explored for predicting the runtimes of applications and deriving the queuing time of a new compute job. In contrast, [9] [10] propose mechanisms for deriving the future state of resources not only providing a single value for a new job, but rather a set of probes over the range of start times and service levels. Each probe may contain multiple metrics used in the selection of co-reservations.

These papers providing advance resource reservations for the workflow system. On the other hand, it is also important to do some work about repairing reservations for the same system as reservation support is important to provide the time for repairing, and this is mostly ignored by many researchers.

This paper gives a quantitative analysis of a web service based system under different reservation and repair mechanism. We show that by a case study the redundant system's performance with the consideration of repair.

2 Availability of Single and Redundant System

In order to get a better performance of a workflow system, it has been studied that the availability of web service in protections system [11-16].

A workflow system may be very complicated, it is not clear that a web service some time may be breakdown by software bug or software clash by an exception. As web activities will go error in some time, it is important to present some repairing support service in this situation. It is necessary to discuss the reliability of a system under different reservation mechanisms.

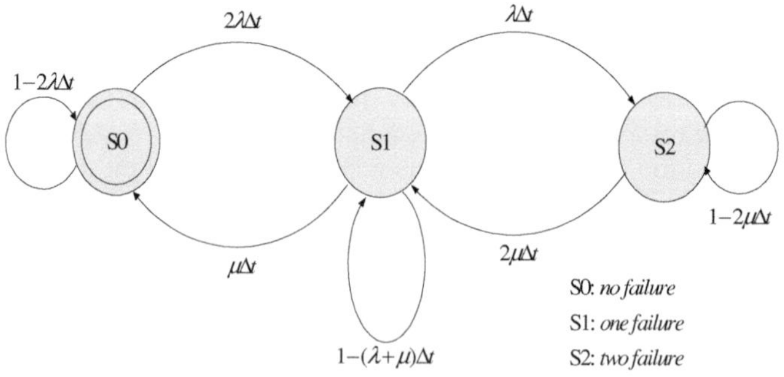

Fig. 1. State Transition Diagram of 1:1 Redundant System

Definition 1. Category-type of protection system ::={P1,P2,P3,P4,P5}. Where:

P1: single;
P2: 1+1 protection system;
P3: 1:N protection system;
P4: M:N protection system;
P5: N+1 protection system;

We will give a first look about the availability of a web activity in protection system. It is given in Table 1.

The availability of single system without protection is $\frac{\mu}{\mu+\lambda}$. The 1+1, 1:N, M:N, N+1 system are also presented.

Table 1. Availability in redundant system

System Category	Reliability Block Diagram	Availability (A)
single		$A_1 = \frac{\mu}{\mu+\lambda}$
1+1	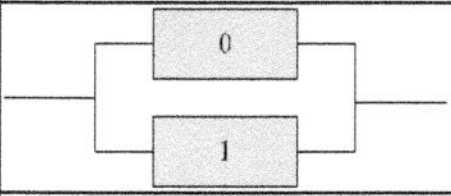	$A_2 = \frac{2\lambda\mu+\mu^2}{(\mu+\lambda)^2}$
1:N	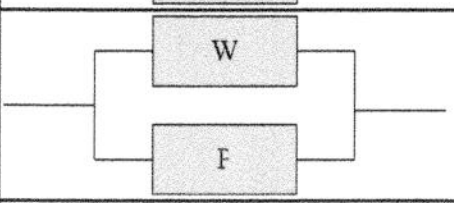	$A_3 = \frac{2\mu(\mu+(N+1)\lambda)}{2\mu^2+\lambda(N+1)(2\mu+\lambda)}$
M:N	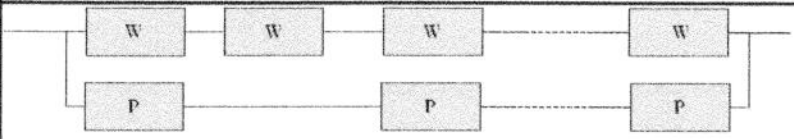	$A_4 = \frac{\sum_{q=1}^{M}\frac{(M+N)!}{q\,(M+N-q)!}\left(\frac{\lambda}{\mu}\right)^q+\sum_{u=1}^{N-1}\frac{(N-r)(M+N)!}{N(M+r)!(N-r)!}\left(\frac{\lambda}{\mu}\right)^{M+r}}{1+\sum_{u=1}^{M+N}\frac{(M+N)!}{u\,(M+N-u)!}\left(\frac{\lambda}{\mu}\right)^u}$
N+1	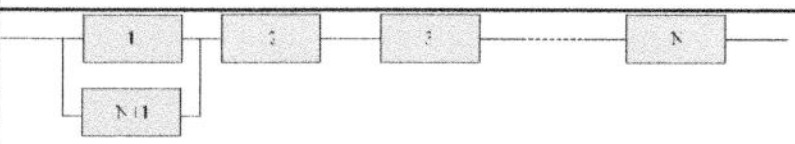	$A_5 = \frac{\mu^{N+1}+(N+1)\lambda^N\mu}{(\mu+\lambda)^{N+1}}$

Remarks: 1. λ : failure rate μ : repair rate.

3 Mechanism of Resources Reservation Model for Workflow System Based on Web Activity

Definition 2. Reservation type is a 4-tuple: RM-type ::=(NP|PC|PR|FR, (UR|RR), λ, μ). PC is the type of prepare the conditions (including resources, set the running context of the activity.) of running an activity. NP it doesn't prepare any resources. PC means that only the precondition of the activity is prepared, and it is not beginning to run. PR means the activity finishes some jobs of it, and it is not fully finished. FR means the activity is fully finished. It may be similar the parallel connection in workflow. UR means the repair resources unreserved, RR means the repair resources (R, P1, H, S, P2, D) reserved. λ is failure rate, and μ is repair rate.

Where:
R: Role;
P1: Person;
H: Hardware;

S: Software;
P2: Privilege;
D: Data;
At the beginning, some rules of the mechanism should be given:

Rule 1: the resources of an activity should be given small granusity. If the activity is detailed, it is much easy to get the resources to be reserved in advance.

Rule 2: the activities should be arranged in parallel at most. If the activities are executed in same time, it will save time and the resources can be reserved in the meantime.

Rule 3: the activities should be given mostly in or connection. If the activities are in or connection, the time cost will be less and the resources will be saved.

Rule 4: Put the resources or activities which often have information change together. If the information for activities' running is changed often, it could be in large probability to change in parallel. Gathering the resources and those activities in advance, and let them run in the meantime, the time cost will be reduced mostly.

4 The Algorithm of Repair Resources Reservation Model

The algorithm of repair resources reservation model can be described as follows:

```
Begin
    Algorithm for UR;
    Algorithm for RR:
    Begin
      Case NP:
          Preparing the repair resources,
          Select the first activity in the workflow to run;
          While tasks not completed
              When errors, using the repair resources to fix the activity;
              Preparing the preconditions of running the next activity;
              Run the next activity;
      Case PC:
          Preparing the repair resources,
          Preparing the conditions of running the first activity;
          While tasks not completed
              When errors, using the repair resources to fix the activity;
              Run the next activity;
```

Case PR:
    Preparing the repair resources,
    Preparing the conditions of running the first activity;
    Running other activities in the meantime;
    While tasks not completed
        When errors, using the repair resources to fix the activity;
        Continuing the next activity to complete the remain processes;
Case FR:
    Preparing the repair resources;
    Preparing the conditions of all activities;
    Running the first activity and run other activities in the meantime;
    While tasks not completed
        When error occurs, using the repair resources to fix activity;
        Running the remaining activity;

End of repair resource reservation module
....
End

5 Case Study

Figure 2 shows the booking diagram of a travel for a guest. The guest will first login the booking system. Then he (she) shall browser the website in detail. After that, he will book flight and hotel. These activities may be executed for several times. Booking car may be necessary for some people. When finished, he (she) may select online payment. The corporate may send the travel order of e-tickets to him (her). The traveler may begin his (her) journey with that. If he (she) chooses offline payment, the company may send the tickets by UPS. At last, the booker logouts.

We suppose that the protection system is 1+1 redundant system. It is shown clear that the 1+1 system has a better performance than single system. If the system is fully reserved, the time cost is nearly one-fifth of NP system.

Therefore, it is necessary to provide redundant system under reservation mechanism as many as possible. On the other hand, the system will perform better if it is provided repairing reservation no matter NP, PC, PP or FP supported system.

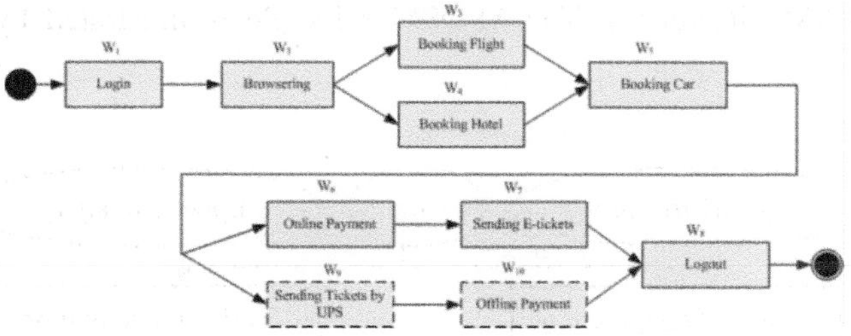

Fig. 2. The Booking Diagram for a Travel

Table 2. Time Cost under UR reservation mechanisms

	NP		PC		PP		FP	
	single	1+1	single	1+1	single	1+1	single	1+1
W_1	2.01	2.0001	1.01	1.0001	1.01	1.0001	1.01	1.0001
W_2	4.02	4.0002	2.02	2.0002	2.02	2.0002	2.02	2.0002
W_3	6.03	6.0003	3.03	3.0003	2.53	2.5003	2.02	2.0002
W_4	6.03	6.0003	3.03	3.0003	2.53	2.5003	2.02	2.0002
W_5	8.04	8.0004	4.04	4.0004	3.04	3.0004	2.02	2.0002
W_6	10.05	10.0005	5.05	5.0005	3.55	3.5005	2.02	2.0002
W_7	12.06	12.0006	6.06	6.0006	4.06	4.0006	2.02	2.0002
W_8	14.07	14.0007	7.07	7.0007	4.57	4.5007	3.03	3.0003
W_9								
W_{10}								
TS	14.07	14.0007	7.07	7.0007	4.57	4.5007	3.03	3.0003

Remarks:

1. λ: failure rate, suppose its value is 0.001
μ: repair rate. Suppose its value is 0.100;
2. Suppose that the system is 1+1 redundant system.
3. TS: the Total System.

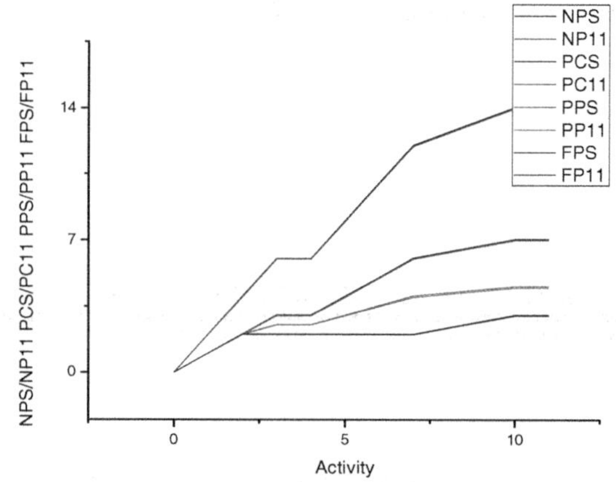

Fig. 3. Time Cost for single system and 1+1 redundant system without repairing reservation (NP/PC/PP/FP)

Table 3. Time Cost under RR reservation mechanisms

	NP		PC		PP		FP	
	Single	1+1	single	1+1	single	1+1	single	1+1
W_1	2.001	2.00001	1.001	1.00001	1.001	1.00001	1.001	1.00001
W_2	4.002	4.00002	2.002	2.00002	2.002	2.00002	2.002	2.00002
W_3	6.003	6.00003	3.003	3.00003	2.503	2.50003	2.002	2.00002
W_4	6.003	6.00003	3.003	3.00003	2.503	2.50003	2.002	2.00002
W_5	8.004	8.00004	4.004	4.00004	3.004	3.00004	2.002	2.00002
W_6	10.005	10.00005	5.005	5.00005	3.505	3.50005	2.002	2.00002
W_7	12.006	12.00006	6.006	6.00006	4.006	4.00006	2.002	2.00002
W_8	14.007	14.00007	7.007	7.00007	4.507	4.50007	3.003	3.00003
W_9								
W_{10}								
TS	14.007	14.00007	7.007	7.00007	4.507	4.50007	3.003	3.00003

Remarks:

1. λ: failure rate, suppose its value is 0.001
μ: repair rate. Suppose its value is 0.100;
2. Suppose that the system is 1+1 redundant system.
3. TS: the Total System.

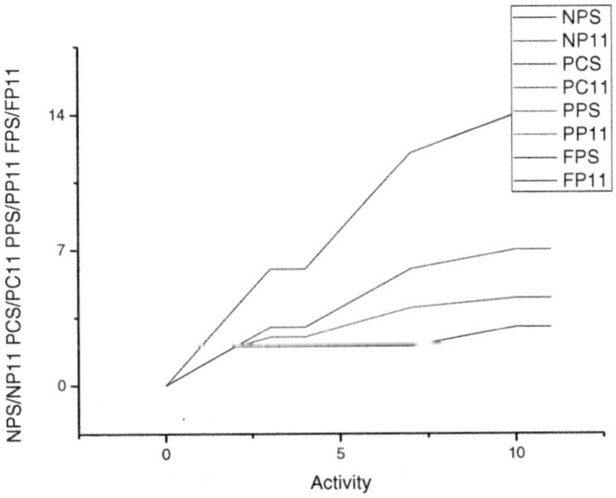

Fig. 4. Time Cost for single system and 1+1 redundant system with repairing reservation (NP/PC/PP/FP)

6 Conclusions

In this paper, it is discussed the time cost for single system and redundant system under different reservation mechanism such as NP, PC, PP, and FP. It is gotten that the redundant system has a better than single system. Especially, if the system is provided repair reservation, the time cost is small or the reliability will be perfect. It is only given the simulation of simple system of booking service of a traveller. And there are much more works to do such as quantitative analysis of some real system and the behavior of them under different reservation mechanism for running and repairing. However, maybe there are many other factors to be considered for the true system such as down time, lost, error rate and so on.

Acknowledgment. This work was supported by National Science Fund of P. R. China (under No. 70561001& 70761002), Natural Science Fund of Hainan Province (under No. 808139), Research on the Application of Tourism Information Service Based on XML/RDF(s) and Web Service(s) (under No. Hjkj2008-10) funded by Hainan Provincial Department of Education.

References

1. Lin, C., Lu, S., Lai, Z., et al.: Service-Oriented Architecture for VIEW: A Visual Scientific Workflow Management System. In: IEEE SCC, pp. 335–342 (2008)
2. Ludäscher, B., Altintas, I., Berkley, C., et al.: Scientific workflow management and the Kepler system. In: Concurrency and Computation: Practice and Experience (2006)
3. Oinn, T., Greenwood, R., Addis, M., et al.: Taverna: Lessons in Creating a Workflow Environment for the Life Sciences. Concurrency and Computation: Practice and Experience 18(10), 1067–1100 (2006)
4. Langguth, C., Schuldt, H.: Optimizing Resource Allocation for Scientific Workflows Using Advance Reservations. In: Gertz, M., Ludäscher, B. (eds.) SSDBM 2010. LNCS, vol. 6187, pp. 434–451. Springer, Heidelberg (2010)
5. Langguth, C., Ranaldi, P., Schuldt, H.: Towards Quality of Service in Scientific Workflows by using Advance Resource Reservations. In: IEEE 2009 Third International Workshop on Scientific Workflows, SWF 2009 (2009)
6. The Condor Team. Dagman (Directed Acyclic Graph Manager), `http://www.cs.wisc.edu/condor/dagman/`
7. Jugravu, A., Fahringer, T.: Advanced Resource Management and Scheduling of Workflow Applications in JavaSymphony. In: Bader, D.A., Parashar, M., Sridhar, V., Prasanna, V.K. (eds.) HiPC 2005. LNCS, vol. 3769, pp. 235–246. Springer, Heidelberg (2005)
8. Li, H., Groep, D., Templon, J., Wolters, L.: Predicting job start times on clusters. In: CCGRID 2004: Proceedings of the 2004 IEEE International Symposium on ClusterComputing and the Grid, pp. 301–308. IEEE Computer Society, Washington, DC (2004)
9. Röblitz, T., Schintke, F., Reinefeld, A.: Resource Reservations withFuzzy Requests. Concurrency and Computation: Practice and Experience 18(13), 1681–1703 (2006)
10. Röblitz, T.: Global Optimization For Scheduling Multiple Co-Reservations In The Grid. In: From Grids to Service and Pervasive Computing, II, pp. 93–109 (2008)
11. Ma, J., Chen, H.-p.: A Reliability Evaluation Framework on Composite Web Service. In: 2008 IEEE International Symposium on Service-Oriented System Engineering (2008)
12. [JH2008-2] Huang, J., Sun, H.: A Reliable Web Service Implementation Approach for Large Commercial Applications. In: International Symposium on Electronic Commerce and Security 2008 (2008)

13. Hwang, S.-Y., Lim, E.-P., Lee, C.-H., Chen, C.-H.: On Composing a Reliable Composite Web Service: A Study of Dynamic Web Service Selection. In: IEEE International Conference on Web Services, ICWS 2007 (2007)
14. Zo, H., Nazareth, D.L., Jain, H.K.: Measuring Reliability of Applications Composed of Web Services. In: Proceedings of the 40th Hawaii International Conference on System Sciences (2007)
15. Hu, T., Guo, S., Guo, M., Tang, F., Dong, M.: Analysis of the Availability of Composite Web Services. In: IEEE FCST 2009 (December 2009)
16. Hu, T., Guo, M., Guo, S., Ozaki, H., Zheng, L., Ota, K., Dong, M.: MTTF of Composite Web Services. In: IEEE ISPA10 2010 (September 2010)

Dynamic Evaluation and Implementation of Flood Loss Based on GIS Grid Data

Tian Xie, Jianzhong Zhou, Lixiang Song, Yuchun Wang, and Qiang Zou

School of Hydropower and Information Engineering
Huazhong University of Science and Technology,
Wuhan, P.R. China
1278887919@qq.com

Abstract. Based on GIS grid data, this paper present a design of flood loss dynamic evaluation model and its application in Jingjiang flood diversion district. In this model the basic information of the demonstration areas and the flood evolution simulation results are managed and analyzed by using traditional evaluating method combined with GIS spatial analysis, raster computation, zonal operation and data spatial-distribution technology.

Keywords: GIS grid data, Submerging analyst, Dynamic flood loss evaluation, Hydrodynamics, Jingjiang flood diversion district.

1 Introduction

The flood loss assessment plays an important role in flood control planning, flood risk management, flood insurance, benefit Evaluation of flood control and disaster mitigation and formulation of related laws and regulations. The traditional method of assessment which Based on Statistical Data do not use the new GIS methods which has a powerful analyst capability as well as display function. Meanwhile, the submerging simulation performed by GIS network algorithm analysis can't accurately reflect the situation of submergence for the limit of the theory itself, which leads to the inaccurate of the Flood loss assessment.

In this paper, the mode of dynamic flood loss assessment which based on GIS grid data and 2-dimensional flood evolution results is provided. And then use it to evaluate the loss of Jingjiang flood diversion zone. The evaluation model has been implemented in the C# & Arc Engine platform. The flood submerge process of flood diversion zone is simulated by unstructured finite volume model 1. Different with static flood loss evaluation nowadays, the model developed by this paper can directly display the situation of submerge and evaluate the loss caused by flood in each time units. Therefore, in every moment of the flood evolution, the submerge area, the depth of water, the velocity of flow and flood loss values can be saved and displayed in the format of grid, which provide useful information to decision-maker.

J. Zhang (Ed.): ICAIC 2011, Part V, CCIS 228, pp. 558–565, 2011.

2 Overview of Grid Based Flood Loss Dynamic Evaluation Model

The model using traditional evaluating Method combined with GIS methods is divided into two slave modules: submerging analysis module and Flood Loss evaluation module.

A. Submerging Analyst Module Based on Grid Data

This module precedes submerging analysis by using spatial analysis functions in ESRI Arc Engine development package. Arc GIS software offered a powerful GIS framework and Arc GIS Engine provides a lower-cost, highly customizable option that can access to GIS component 2 3. The Arc Engine spatial analyst module works closely with the Raster Data Objects. The Raster Data Objects model is used to access, analyze, manage, convert, and visualize raster data. It can take analysis on single cells in the raster or between a set of raster and even the Whole raster layers. The spatial analysis process in GIS can be summed up as four types: Point-by-point operations, Neighborhood Operations, Zonal Operations and Global Operations. Compared with Vector Data Objects, the Raster Data Objects is more suitable for presenting the continuously distributed elements. Consequently the water depth and the flow velocity data simulated by the flood evaluation model can be well presented, analyzed and computed in GIS platform as raster data. GIS-based spatial analyst functions allow users to perform an efficiency and flexible query from both spatial and attribute information. Moreover, it provides a visual environment and offered enough relative information for the use of decision-making aided.

Grid-based flood analysis model and the processing steps are as follows:

1) Using the result of the Two-dimensional hydrodynamic model, the water depth and the flow velocity are recorded in the nodes, then use GIS spatial analyst component, IDW Inverse Distance Weighted interpolation method to obtain the Flood submerge raster.

2) In between the different raster layers, a series of map algebra process is performed to generate new ready to evaluate raster: Each flood evolution raster is reclassified by six levels of water depth and then mixed up with land utilization raster converted from land use data through certain map algebra algorithm. After taking this process each cell of the raster carry the information of two types: one is the land utilization type and the other is the water depth of this cell.

3) Regional submerge analyst methods based on GIS raster computation and related algorithm is used to determine the specific submerge area in the flood region and for the further use, to evaluate and analyze the degree of submerge in varies conditions and environment. Methods for zonal submerging areas statistics based on integrated raster and vector dataset. As the ready to evaluate raster has been generated, embedded with the administrative boundary which has a vector data structure, the area submerging data in each administrative division can be worked out categorized by different type of land utilization and various classifications of water depth.

4) A designed batch-processing approach to analyze the time sequence flood evolution raster and based on which to perform the time sequence submerge area statistics.

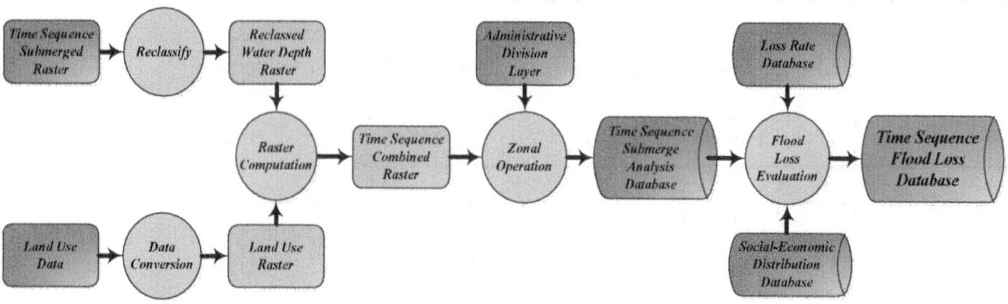

Fig. 1. The process of flood submerging analyst and flood hazard loss evaluation

B. Submerging Analyst Module Based on Grid Data

The model has a former part which is formed by loss rate database, social-economic spatial distribution database and submerging analyst database. Considering of three factors, the equation of the flood disaster evaluation model is as follow:

$$W = \sum_i \sum_j \sum_k \alpha_i A_{ij} \eta_{jk} \tag{1}$$

Where *i, j, k* is respectively the cell code of calculation grid, the land utilization type, the level of the water depth. A_{ij} represent for the *i* calculation unit the j type of disaster-affected-body original related social-economic cell value. η_{jk} stand for the loss rate of j type of disaster-affected-body combined with k level of water depth. α_i is representative for the state of the calculation grid. When *i=1*, means the land utilization type is we concerned to calculate the loss; when *i=0*, means the cell land utilization type is others.

3 Methods of Parameter Decision and Preprocessing of Flood Routing Data

A. GIS Representation of Two-Dimensional Flood Routing Results

In this paper, the two-dimensional shallow water equations are used for governing the floods

$$\frac{\partial \mathbf{U}}{\partial t} + \frac{\partial \mathbf{E}}{\partial x} + \frac{\partial \mathbf{G}}{\partial y} = \mathbf{S} \tag{2}$$

The governing equations are solved on triangular grids by cell-centered finite volume method. Integrating (2) over arbitrary cell and discretizing the following equations, then

$$\Omega_i \frac{\mathrm{d}\mathbf{U}_i}{\mathrm{d}t} = -\sum_{k=1}^{3} \mathbf{F}_{i,k} \cdot \mathbf{n}_{i,k} L_{i,k} + \mathbf{S}_i \tag{3}$$

where $\mathbf{F}=[\mathbf{E},\mathbf{G}]^T$ is the flux; Ω is the area of the cell; $\mathbf{n}=(n_x,n_y)^T$ is the unit outward vector; L is the length. The numerical fluxes are computed by HLLC solver. The second order accuracy in space and time is achieved by the MUSCL and two-step Runge-Kutta scheme. Details on the unstructured finite-volume model for numerical resolution of two-dimensional shallow water equations are given in Song et al 1, in which model validation and verification are also given.

This article analyzes and display the information based on grid data under the platform of GIS, because of which the two-dimensional flood routing data need to be transferred to GIS platform compatible format. The result of hydrodynamic calculation is triangular mesh text data. The points which carry information are the point of a triangle, and their Information format is the position of the point and its water depth and flow velocity. In order to using GIS system to display and analyze, the text date during the time of all the flood routing procedure have to be transferred into gird data in GIS platform, so batch data transformation is needed.

The transmission process simplified the actual conditions:

1) The water body of the submerging area is processed as lots of discrete water body unit.

2) In each unit, assuming the submerge depth are of the same value, which equals the water depth of the position at the center of the unit.

B. Determination of Flood Disaster Loss Rate

Flood disaster loss rates of the hazard-bearing bodies are generally defined as the values that the original worth in the normal year divided by the loss when encountering with the flood. It is an important indicator of the flood damage evaluation. Factors that affect the rate of flood damage mainly contain regional types, hazard-bearing bodies, inundated water depth, flooding duration and so on 4.

This paper considered the characteristics of factors that influence the loss of flood, which mainly contain the relation between inundated water depth, the hazard-bearing bodies and damage condition. The inundated water depth is the most important index of flood damage estimate, which can be divided into six levels: <0.5m, 0.5 ~ 1m, 1 ~ 2m, 2 ~ 3m, 3 ~ 4m and > 4m. Through the establishment of flood loss rate, the inundated water depth and hazard-bearing body relationship table, the different hazard-bearing bodies and different submergence levels corresponding to different loss rates.

After analyzing the actual situation of Jingjiang flood diversion area and the available statistics on the previous year flood disaster, and according to the loss rate investigation data in adjacent areas and similar region. This article using disaster rate of rural housing, urban housing disaster rate, disaster rate of industrial, forestry, animal husbandry and fishery disaster rate to carry out dynamic evaluation of flood damage.

C. Spatial Distribution of Social-Economic Statistic Data

Flood submerge area do not fit well with administrative boundary in most case for its irregularities. Only parts of administrative areas are submerged by water, but ordinary statistical analysis of social-economic data is measured in administrative district, and the social-economic information is unevenly distributed in each administrative district, which leads to the difficulties in flood hazard loss evaluation and the inaccurate of evaluation result.

By the technical means of RS and GIS, more accurate land use type information of demonstration area can be obtain. The spatial distributions of social-economic data can be acquired through relating administerial unit code with spatial distribution social-economic database 5 6.

According to the type of land utilization in administerial unit, the land use type can be divided into seven groups. The agricultural population and rural tenement, urban population and buddings, industrial output value, value of agricultural output, forestry, animal husbandry and fishing can be distributed to country land, city, industrial land, cultivated land, wooded land, grassland and lake respectively. The formulas of distribution are the followings:

$$Dij = \frac{Vij}{Bik} \tag{4}$$

Where D_{ij} Represent for the j type of social-economic statistic data distribution density in i administrative unit. V_{ij} Represent for the total social-economic statistic data of j type in number i administrative unit. B_{ik} is the total area of the k land utilization type in i administrative unit.

4 Application in Jingjiang Flood Diversion District

The Jingjiang flood diversion project is settled to control the flood in Yangtze River by active flood diversion 7. In this paper, the first flood diversion process measured in 1954 was selected as the input of the flood routing procedure.

The time sequence submerging raster of 24.2hour, 48.3hour, 72.1hour and 130.9hour are showed as follows: The total submerging areas are respectively: 246.7km^2, 511.1 km^2, 768.8 km^2, 927.8 km^2.

The detailed time sequence submerging analyst data and the loss evaluation data of the entire administrative unit is calculated by the designed submerging analyst and loss evaluation model. Taking BuHe Town as an example, this paper gives the submerging curves through the whole flood diversion duration and the submerging areas of the different land utilization type.

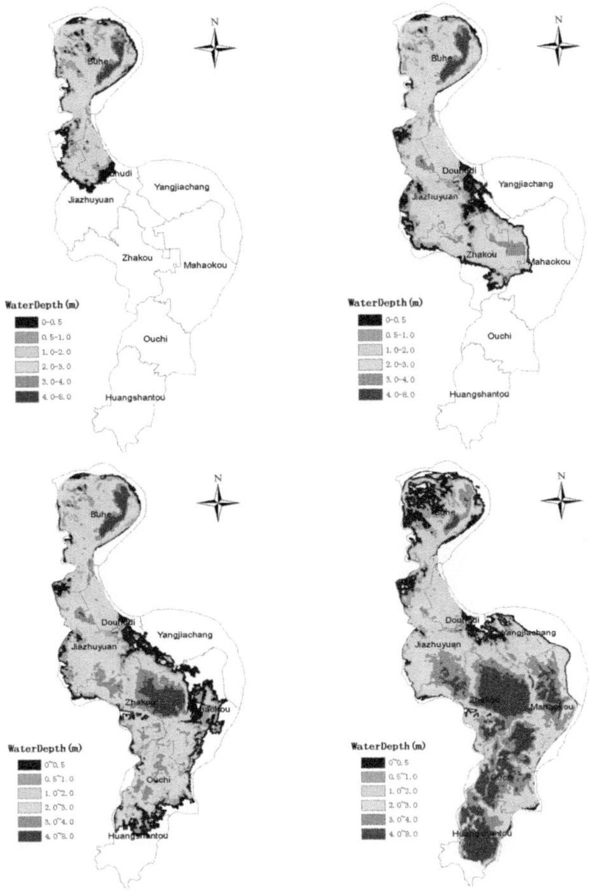

Fig. 2. Submerging raster in different flood routing period

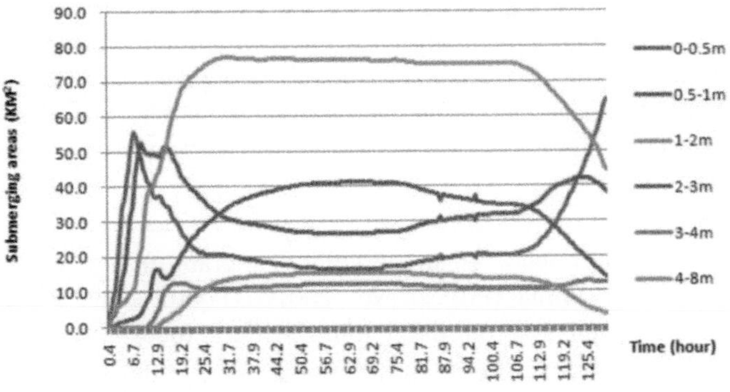

Fig. 3. Time sequence Flood submerging curves for BuHe Town

Table 1. Detailed Description of Submerging Situation in Buhe Town at 130.9 Hour

Submerging Analyst table of BuHe Town at 130.9 hour (km^2)					
Water Depth(m)	Residential land	Industrial land	cultivated land	forestry land	Water body Area
0~0.5	6.03	6.98	47.05	4.31	0.01
0.5~1.0	4.63	3.45	29.00	0.87	0.16
1.0~2.0	4.37	1.76	36.02	1.28	0.73
2.0~3.0	1.12	0.09	12.16	0.04	0.65
3.0~4.0	0.21	0	9.84	0	2.30
4.0~8.0	0	0	2.55	0	0.96

Table 2. Loss Evaluation of all the Administrative Unit in Jingjiang Flood Diversion District at 130.9 Hour

Loss value / Town	Total submerging Area(km^2)	House number	Industry	Planting	Forestry	Animal Husbandry	Fishery	Direct Economic loss
			(Ten thousand Yuan)					
BuHe	186.21	24614	7441.73	22260.49	186.68	32.12	10.66	29931.68
DouHuDi	83.10	37315	9032.57	9617.81	201.60	36.15	15.19	18903.33
YangJiaChang	60.48	4287	1464.52	6683.63	128.99	11.82	4.83	8293.80
MaHaoKou	167.17	27484	5929.71	20300.79	144.67	21.30	21.80	26418.28
OuChi	92.09	17279	5737.39	9701.07	483.82	40.72	11.36	15974.35
HuangShanTou	88.31	11768	7.67	6330.81	174.77	20.45	11.63	6545.33
ZhaKou	123.66	17574	2355.12	8649.18	347.86	44.13	20.33	11416.63
JiaZhuYuan	126.81	14388	2424.80	14270.33	262.21	38.69	25.69	17021.73

5 Conclusion

The model of flood loss evaluation based on GIS grid data is proposed and implement in this paper. Firstly, this paper did researches on spatial distribution method to solve the problem that Social-economic data is unevenly distributed in each administrative unit which lead to inaccurate input data of the loss evaluation model. Then combined with designed dynamic submerging analyst model, a Grid based flood disaster dynamic evaluation model has been proposed. In this model functions of GIS management and spatial data analyst are well used and developed, and batch processing method of the large amount of flood routing data is discussed and used at processing of the Jingjiang flood diversion routing data. This model combined with mathematic ways and GIS analyst methods offers an accurate and efficient method of flood loss evaluation.

Acknowledgment. This work is supported by a grant from the National Basic Research Program of China (Project No. 2007CB714107) and a grant from the Special Research Foundation for the Public Welfare Industry of the Ministry of Science and Technology and the Ministry of Water Resources (Project No.200901010).

References

1. Song, L., Zhou, J., Li, Q., Yang, X., Zhang, Y.: An unstructured finite volume model for dam-break floods with wet/dry fronts over complex topography. Int. J. Numer. Methods Fluids (2010), doi:10.1002/fld.2397
2. Jonge, T.D., et al.: Modelling floods and damage assessment using GIS. In: Application of Geographic Information Systems in Hydrology and Water Resources Management, HydroGIS 1996, vol. 235, pp. 299–306. IAHS publ. (1996)
3. Zhu, X., Pan, S., Zhang, J.-I.: Review of the Integration of Watershed Hydrological Model and Geographic Infonation System. Geography and Geo-Infonation Science 9(3), 10–13 (2003)
4. Wang, L., Jiang, N., Zhou, Y.: The evaluated loss model of fooding and waterlogging in Taihu Basin. Science of Surveying and Mapping 28(2), 35–38 (2003) (in Chinese)
5. Hang, J., Wu, L., Xu, J.: Digital earth based food routing model visualization. Computer Engineering and Applications 45(36), 1–4 (2009)
6. Zhao, X., Wang, P., Gong, Q., Huang, S.: A GIS-based approach to food risk zonation area. Acta Geographica Sinica 55(1), 15–24 (2000) (in Chinese)
7. Li, X., Guo, K.: 100 Questions in Great Flood of 1998, p. 22, III. China Water Power Press, Beijing (1999)

Character Extraction of Vehicle Plate Directly from Compressed Video Stream*

Zhou Qiya[1], Yu Xunquan[1], He Lian[1], and Yang Gaobo[2]

[1] Department of Information Technology Hunan Vocational College of Railway Technology Zhuzhou, Hunan Province, China
[2] School of Information Science and Engineering Hunan University Changsha, Hunan Province, China
zhouqiya@tom.com

Abstract. Character extraction of vehicle plate is conducted directly in MPEG compressed domain. Firstly, vehicle plate region is detected by thresholding the weighted horizontal texture energy, in which adaptive threshold is used, followed by a series of post processing, e. g. morphological operations, connected component analysis, filtering with aspect ratio constrains. Secondly, block's DC+2AC texture intensity projection profile is used to implement the character of vehicle plate localization accurately. Then after minimal decompression processing and contrast adjustment on the located region, OTSU thresholding technique is used for character extraction. Experiment results show the effectiveness of the proposed algorithm.

Keywords: character extraction, compressed domain, vehicle plate.

1 Introduction

Character extraction of vehicle plate can be used in many applications, including automatic traffic violation control, automatic parking lot billing, etc. Cui etc. presented a method for character extraction of license plates from raw video, which is modeled as a Markov random field. The randomness is used to describe the uncertainly in the label assignment of pixels. With this MRF modeling, extraction can be formulated as a problem of maximizing the posteriori probability given prior and observations [1]. However, the main disadvantage of this method is intensively time-consuming.

It's well known that text is a very compact and accurate clue for image and video indexing and summarization. Yet, it is not an easy problem to reliably detect and localize text embedded in images and videos. As an increasing proportion of images and video are being stored in the compressed form, there is an emerging trend to detect text directly in the compressed domain transferred from pixel domain. By manipulating features directly in the compressed domains, we can save the resources (computation time and storage) needed for decompressing. Moreover, many

* This work is supported by Research Fund of Hunan Vocational College of Railway Technology (Project No. HNTKY-KT2010-3).

J. Zhang (Ed.): ICAIC 2011, Part V, CCIS 228, pp. 566–573, 2011.

compressed domain features such as DCT coefficients and motion vectors are actually very effective in several applications, including text detection. The DCT coefficients capture the directionality and periodicity of local image blocks, which are usually used as texture measures to identify text region in compressed domain.

There are already a few text detection algorithms conducted in compressed domain proposed. For example, caption text regions are segmented from background images using their distinguishing texture characteristics, i.e. the intensity variation information encoded in the DCT domain [2]. An efficient method for extraction texts in MPEG compressed videos for content-based indexing is presented in [3]. They make the best use of 2-level DCT coefficients and macroblock type information in MPEG compressed video. In [4], the average AC energy is used as its edge-ness measure for each block in an I-frame. These AC energy values are then thresholded to obtain the blocks of sharp edges. A block is declared as a candidate text block if its AC energy value exceeds some threshold. This simple thresholding procedure detects all blocks with a high AC energy value and thus also picks up some non-text blocks of sharp edges.

In fact, the vehicle plate is composed of text complying with certain constrains. The characters on the vehicle plate belong to typical scene text. For the character extraction of vehicle plate directly from compressed stream is not reported until now. So in this paper, we aim to implement this type of scene text extraction directly from compressed domain. Generally speaking, the process of text extraction includes several steps, i.e. text detection, text localization, text tracking, text segmentation and binarization, character recognition [5][6]. We focus on three steps: text detection, text localization, text segmentation and binarization. The rest of the paper is organized as follows. The theory basis is presented in Section 2. Vehicle plate region detection method and the character of vehicle plate localization method are given in Section 3 and Section 4 respectively. Section 5 present segmentation and binarization operations. Experiment results are presented in section 6 and conclusions are given in section 7.

2 Theory Basis

Discrete cosine transform (DCT) is widely used in image and video coding to reduce spatial redundancies. The general equation for a N×N (N is usually a power of two) 2D DCT is defined by the following equation [2]:

$$F[u,v]=\frac{1}{N}A(u)A(v)\sum_{i=0}^{N-1}\sum_{j=0}^{N-1}\cos\frac{(2i+1)u\pi}{2N}\cos\frac{(2j+1)v\pi}{2N}f[i,j] \tag{1}$$

where u and v denote the vertical and horizontal frequencies (u , v =0,1,…,N-1) and

$$A(u),\ A(v)=\begin{cases}\frac{1}{\sqrt{2}} & u,\ v=0\\ 1 & u,\ v\neq 0\end{cases}$$

. For most image blocks, much of the signal energy lies at low frequencies; these appear in the upper left corner of the DCT matrix. Compression is achieved since the lower right values represent higher frequencies, and are often small - small enough to be neglected with little visible distortion.

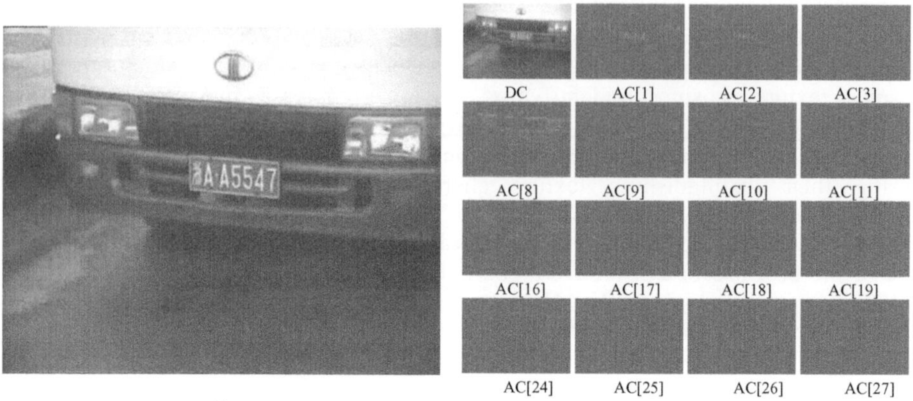

(a) (b)

Fig. 1. Feature extraction from DCT coefficients. (a)768×512 input image (b)DCT feature from the intensity image.

The values of DCT coefficients F[u,v], which are amplitudes of harmonic waves, denote the relative amount of various 2D spatial frequencies contained in N×N blocks. They can be used as measures of spatial periodicity and directionality, when frequencies in the u and v dimensions are properly tuned. Fig. 1a shows an 768×512 input image and Fig 1b shows the absolute values of the DCT coefficients (only luma component considered). Each subimage in Fig. 1b represents one DCT channel of the input image. Each pixel in the subimage is the energy in this channel for the corresponding DCT block of the input image. This figure demonstrates the effectiveness of following rules:

1)The top left channels, which represent the low frequency components, contain most of the energy, while the high frequency channels, which are located at the bottom right corner of Fig. 2b, are mostly blank. For space limitations, not all the frequency channels are plotted in this figure.

2)The channel spectrums capture the directionality and coarseness of the spatial image; for all the vertical edges in the input image, there is a corresponding high frequency component in the horizontal frequencies, and vice versa. That is, the DCT domain features characterize the texture attributes of an image.

3 Vehicle Plate Region Detection Based on Text Detection

The vehicle plate region comprises text with special constraints, which must fit the national standards. For example, the vehicle plate in Fig. 1a belongs to a widely used type in China, called "92" style, which is composed of seven characters and one space mark in horizontal arrangement. Among the seven characters, the first one is Chinese character, the second and the third one is English or numeric character, others are numeric characters. Each character is middle-located in a rectangle area with size 90 mm×45mm.

Due to the horizontal alignment of characters the vehicle plate region exhibits a periodic horizontal intensity variation, which motivated us to detect vehicle plate region based on text detection using texture features in compressed domain. For each DCT block (i,j), the weighted horizontal texture energy is computed as:

$$E_{hor}(i,j) = \sum_{v_1 \le v \le v_2} (0+1)*(v+1)*|F_{0v}(i,j)| \quad (2)$$

From formulation (2), it can be seen that the horizontal texture energy is magnified according to the frequency component position, which is different from Zhong's algorithm [2]. In our experiments, v_1 and v_2 is set 1 and 3 respectively. These weighted horizontal text energy values are then thresholded to obtain the blocks of large horizontal intensity variations. A block is a text candidate if its horizontal text energy is above some threshold. An adaptive threshold is used here, which is computed according to the following formulation:

$$T_{threshold} = AVG_{E_{hor}} + k*(MAX_{E_{hor}} - MIN_{E_{hor}}) \quad (3)$$

Where $AVG_{E_{hor}}$ is the average weighted horizontal texture energy of the whole image, $MAX_{E_{hor}}$ and $MIN_{E_{hor}}$ is maximal and minimal weighted horizontal texture energy of the whole image, k is an empirical value.

(a) original image (b) horizontal direction (c) vertical direction

Fig. 2. Thresholded results of the weighted horizontal or vertical texture energy

Thresholded results of the weighted horizontal texture energy of Fig. 2a are illustrated in Fig. 2b. From which we can see that the vehicle plate region consist of mainly vertical strokes, which result in a high frequency component in the horizontal frequencies, thus leading to high horizontal texture energy in this area, marked as black blocks. The horizontal edges in the input image will result in corresponding high vertical texture energy. See Fig. 2c. Except the desired vehicle plate region is detected in Fig. 2b, there are some noise areas. The following post processing is needed.

Morphological operations are applied to remove the noise blocks and also to merge the nearby disconnected blocks into coherent regions. We apply a closing operation to fill the holes between detached blocks. An opening operation is used afterward to smooth the image and remove the spot-like noise blocks. In our experiments, the structured element is with size 1×3 pixels. The element allows the text blocks to be merged into horizontal lines. Then dilating is conducted on the output of closing and opening operations. Finally, the connected component analysis (CCA) algorithm is performed to obtain the connected regions. Size and shape filters will be utilized to eliminate the regions with too small or too large sizes. Note that the aspect ratio of the vehicle plate is a constant value, which is used as an important constraint condition to filter the detected regions in our algorithm.

4 The Character of Vehicle Plate Localization

In general, the more accurate the text lines are located the less the influence of the background region to the text segmentation and recognition. Usually, the accurate text box of each text line can be obtained in pixel domain using the edgy density or pixel intensity information. A text localization method in compressed domain by using the block's DC+2AC texture intensity projection profile is proposed in this paper.

In our algorithm, the DC and AC[1], AC[8] coefficients are selected to present the block's texture intensity. The method of construction of DC+2AC image refers to our previous work [7]. Let Verpro denote the vertical texture projection of the detected vehicle plate region, which is expressed as follows:

$$\text{Verpro}\ (i) = \frac{\sum_{j=0}^{S_{Height}-1} \text{abs}(\text{P}_{\text{DC}+2\text{AC}}[i,j])}{\text{S}_{\text{Height}}}, \quad 0 < \text{i} < S_{Width} - 1 \tag{4}$$

Where $\text{P}_{\text{DC+2AC}}[i, j]$ is DC+2AC texture energy at position (i,j). S_{Height} and S_{Width} is the height and weight of the candidate vehicle plate region.

a. Detected vehicle plate regions

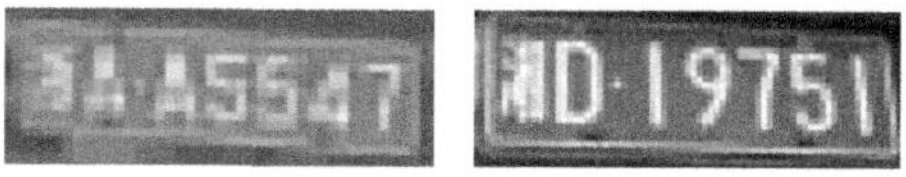

b. DC+2AC texture intensity images

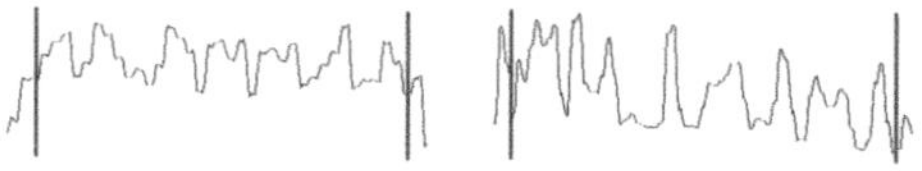

c. Vertical texture intensity projection profile

d. Horizontal texture intensity projection profile

e. Fusion results

Fig. 3. The process of character localization

The bounding boxes of two detected vehicle plate regions are shown in Fig. 3a. Their DC+2AC texture intensity images are shown in Fig. 3b. We can find that the peaks and valleys in the vertical texture intensity projection profiles in Fig. 3c can accurately indicate the position of the character in horizontal direction. Horizontal texture intensity projection profiles Fig. 3d can also be used to finely locate the character position in vertical direction. Fusing the results of both horizontal and vertical directions, the characters of vehicle plate are located accurately, see Fig. 3e.

5 Segmentation and Binarization

It is difficult to directly extract characters in a DC+2AC image, so minimal decompression processing is required. The decompression is done only on the located text region. This helps to reduce processing time in comparison with full decompression. Text has relatively strong contrast with its background, which is a necessary condition for character extraction. The adjusting operation for dynamic range has the effect of increasing contrast in comparison with the background. This operation also makes the boundary between character and its background clearer. The expression for this operation is given in formulation (5).

$$P(x, y) = C(1 + \log(G(x, y))) \tag{5}$$

where P(x, y), G(x, y) and C mean the pixel value applied by dynamic range adjustment, the gray-level at pixel point (x, y) and the constant, respectively. After adjustment, median filtering is carried out to remove noise in the image and OTSU thresholding technique is used for coarse extraction [8].

6 Experimental Results

The proposed algorithm has been evaluated on several real-life video sequences compressed using the MPEG-2 standard. The characters of vehicle plate are detected directly from the compressed video stream. The experimental platform is PC with Intel Celeron 2.0GHz CPU and 1G DDR. The encoder configuration is listed in Table 1. Other parameters are set default values.

Table 1. The configuration of MPEG-2 encoder

Profile	Main=4
Level	High = 6
input picture file format	Yuv=3
number of frames	20
I/P frame distance	1
horizontal_size×vertical_size	912×608, 768×512 416×320, 400×300

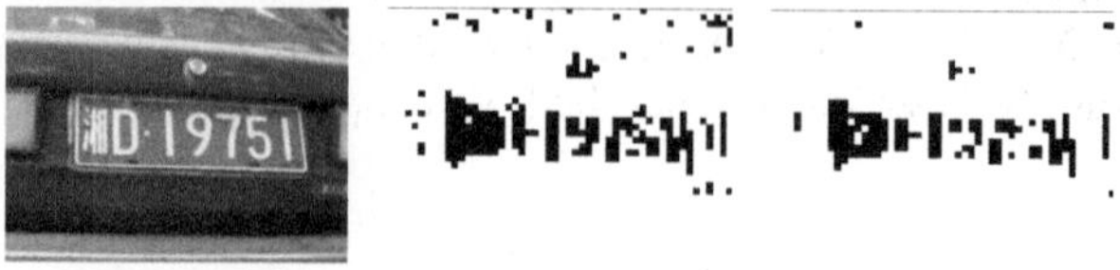

(a) Captured vehicle plate (b)Zhong's method (c)our method

Fig. 4. Results comparison of our texture feature based text detection algorithm with Zhong's method (Frame size: 416×320)

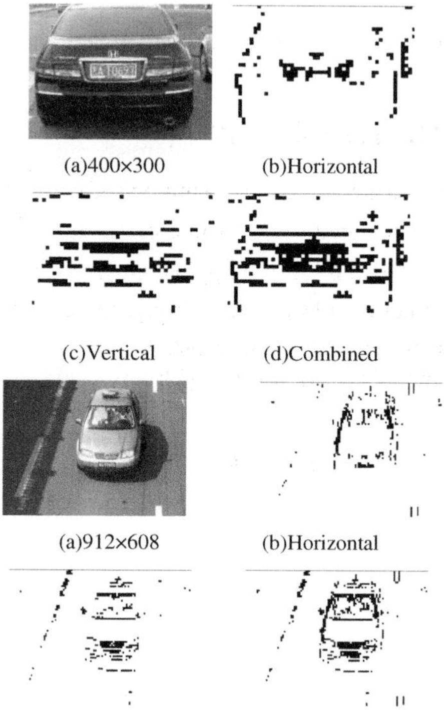

(a)400×300 (b)Horizontal

(c)Vertical (d)Combined

(a)912×608 (b)Horizontal

(c)Vertical (d) Combined

Fig. 5. More experimental results of horizontal and vertical texture energy thresholding

Though in text detection step the texture feature based method is still used, the weighted horizontal texture energy and the proposed adaptive thresholding value seem more effective in identifying text than the Zhong's method. Results comparison of our texture feature based text detection algorithm with Zhong's method is shown in Fig.4. We can see less noise included in the result of our proposed method. For various complex sequences, the proposed algorithm can accurately locate the vehicle plate region. More experimental results refer to Fig. 5. We can find that if we combine the horizontal and vertical thresholding results, the accuracy can be improved. See Fig. 5d.

The processing speed of the proposed algorithm is very fast since it does not require a fully decompressed MPEG video. The DCT information which is used is readily available from the video stream. Furthermore, the processing unit for the algorithm is a DCT block, which is 1/8 of the original image size in each dimension.

7 Conclusion

The paper aims to extract a special type of scene text directly from compressed domain. The contribution of this paper mainly lies in two aspects: 1) The character extraction of vehicle plate is proposed to be conducted in compressed domain. Candidate vehicle plate region is detected directly from the compressed domain using text detection method. 2) DC+2AC texture intensity projection profile is used for character of vehicle plate localization. Because most operations are conducted in compressed domain, the proposed algorithm is fast. However, usually the captured vehicle plate image has different perspective distortions, performing correction using computed warping parameters is needed, which is our next task.

Acknowledgements. The author also thanks the support of Scientific Research Project of Hunan Provincial Education Department (Project No. 09C1233).

References

1. Cui, Y., Huang, Q.: Character extraction of license plates from video. In: Proceedings of Computer Vision and Pattern Recognition, Puerto Rico, pp. 502–507 (1997)
2. Zhong, Y., Zhang, H., Jain, A.K.: Automatic caption localization in compressed video. IEEE Transactions on Pattern Analysis and Machine Intelligence 22(4), 385–392 (2000)
3. Lim, Y.-K., Choi, S.-H., Lee, S.-W.: Text extraction in MPEG compressed video for content-based indexing. In: Proceedings of 15th International Conference on Pattern Recognition, Spain, pp. 409–412 (2000)
4. Gu, L.: Text detection and extraction in MPEG sequences. In: Proceedings of CBMI 2001, Brescia, Italy, pp. 19–21 (2001)
5. Gllavata, J.: Extraction textual information from images and videos for automatic content-based annotation and retrieval
6. Jung, K., Kim, K.I., Jain, A.K.: Text information extraction in images and video: a survey. Pattern Recognition 37, 977–997 (2004)
7. Zhou, Q., Yang, G., Chen, W., Zhang, Z.: A fast and accurate moving object extraction scheme in the MPEG compressed domain. In: ICIG 2007, Chengdu, China, pp. 592–597 (2007)
8. Otsu, N.: A threshold selection method from gray-level histograms. IEEE Transactions on Systems, Man, and Cybernetics smc-9(1), 62–66 (1979)

Study on an Adaptive Fuzzy Controller for High-Frequency Electromagnetic Water Disposal*

Guohui Zeng, Minnian Cao, and Yuchen Chen

College of Electrical and Electronic Engineering
Shanghai University of Engineering Science
Shanghai, P.R. China
guohuizeng0012@sina.cn

Abstract. Existed high-frequency electromagnetic water disposal device with fixed output frequency from the measured data cannot achieve the desired effect. To obtain better effect, the output frequency of high-frequency electromagnetic water disposal device should be adjusted according to changes in temperature and water conductivity. As the recycled cooling water system is a complex nonlinear system, it is difficult to descript mathematically their impact on the output frequency. In this paper, an adaptive fuzzy controller is presented which can adjust the output frequency according to changes in temperature and water conductivity. The simulation and experimental results prove its effectiveness and superiority.

Keywords: Recycled cooling water system, Adaptive fuzzy controller, High-frequency electromagnetic, Water disposal.

1 Introduction

With the development of industry production, more and more recycled cooling water systems are used in power plant boilers, central air conditioning et al to exchange heat. The main problem in recycled cooling water system (RCWS) is descaling and sterilization. Water disposal is becoming an important and emergent problem increasingly. With the characteristic of being safe, economical, high-efficient and environmental, water disposal with high-frequency electromagnetic (HFEM) is a good physical method with full promise and attention [1][2]. Effects of HFEM device for water disposal currently available in different operation environment varied widely; mainly due to its output frequency is fixed and can not be adjusted with changes of temperature and water conductivity [3][4][5]. Therefore, in this paper an adaptive fuzzy controller is presented which can adjust the output frequency adaptively according to changes in temperature and water conductivity. It is proved to be effective by simulation and experimental results.

The scheme of water disposal device with an adaptive fuzzy controller is shown as Fig. 1. The presented controller consists of four parts: fuzzification, inference engine, defuzzification and self-adjusting set [6][7].

* This work is supported by SUES DSF Project A24000821 to G. Zeng and SUES SF Project 2008XZ02 to M. Cao.

J. Zhang (Ed.): ICAIC 2011, Part V, CCIS 228, pp. 574–580, 2011.

2 Adaptive Fuzzy Controller for Water Disposal

A. Fuzzification

In this controller, there are two input variables, namely temperature and water conductivity, whose fuzzification can be done by triangular membership functions. Temperature (indicated by T) was classified as "very low", "low", "medium" ,"high" and "very high" five fuzzy sets, respectively expressed by NL, NM, ZE, PM, PL. The center of each fuzzy set is taken to be 5 ℃, 25 ℃, 40 ℃, 55 ℃, 75 ℃, as shown in Fig. 2(a).

Conductivity (indicated by C) can also be classified as "very low", "low", "medium", "high" and "very high" five fuzzy sets, respectively expressed by NL, NM, ZE, PM, PL. The center of each fuzzy set is taken to be 200 $\mu s/cm$, 600 $\mu s/cm$, 900 $\mu s/cm$, 1200 $\mu s/cm$, 1600 $\mu s/cm$, as shown in Fig. 2(b).

To simplify the control calculation, normalize the values of output frequency using (1) before fuzzification.

$$f_i = (f_x - f_o)/(f_b/2) \tag{1}$$

Where f_o is the center frequency and f_b is the resonance bandwidth of RCWS resonance frequency band that can be informed by frequency scanning and calculation, f_x is any one in the range of frequencies [$f_0 - (f_b/2)$, $f_0 + (f_b/2)$].

Output frequency can be classified as "very low", "low", "medium", "high" and "very high" five fuzzy sets in the range of [-1,1] after normalization, respectively expressed by NL, NM, ZE, PM, PL. The center of each fuzzy set is taken as -1, -0.4, 0, 0.4, and 1, as shown in Fig. 2(c).

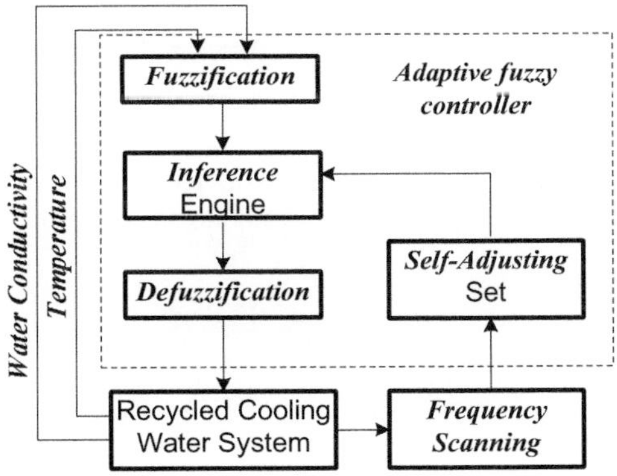

Fig. 1. The scheme of water disposal device with an adaptive fuzzy controller

B. Fuzzy Inference

Fuzzy rule base is a collection of IF-THEN rules. In this design, the initial set of the fuzzy rules such as "IF temperature is NL and conductivity is NL THEN frequency is NL" is based on a test record from a water disposal equipment company, see Table 1.

Inference engine is the most important part of a fuzzy controller. In this control, min-max inference with the multi-rule, two-input single-output model based on Mandani algorithm is adopted. Fig. 3 shows the inference process of two rules with two inputs and one output.

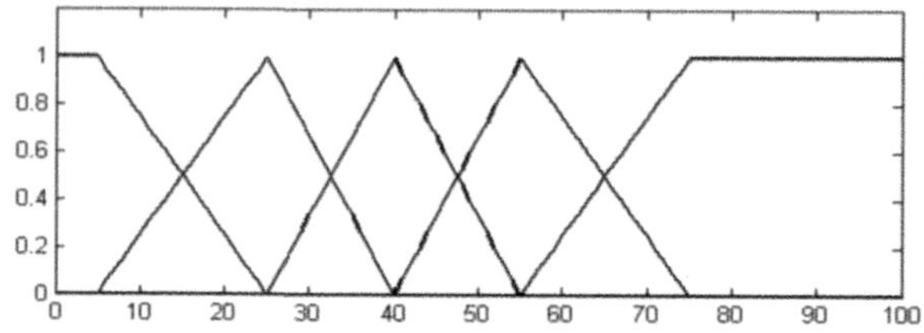

(a)

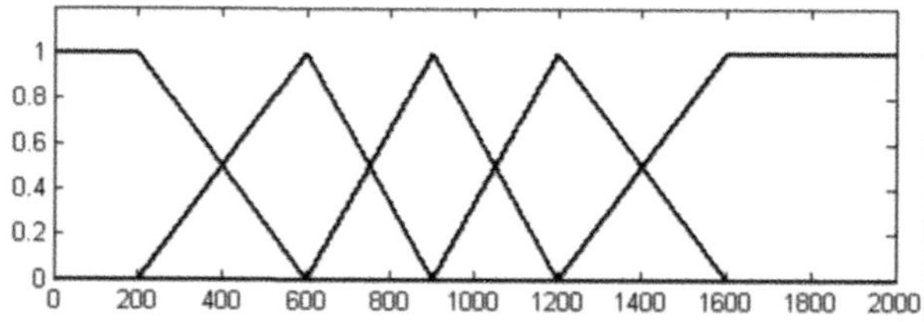

(b)

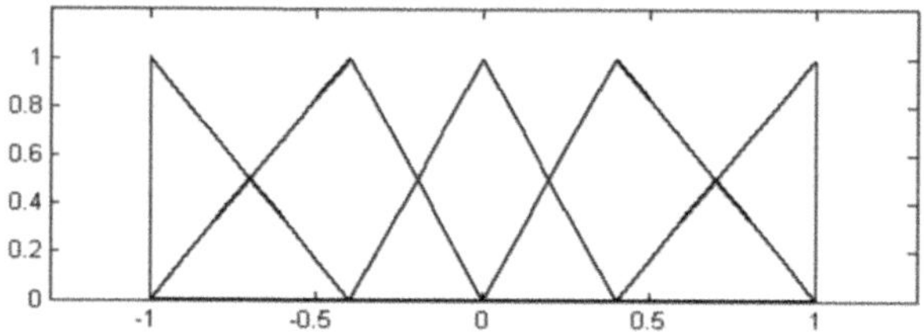

(c)

Fig. 2. Fuzzy control membership functions of (a) temperature (b) conductivity (c) output frequency

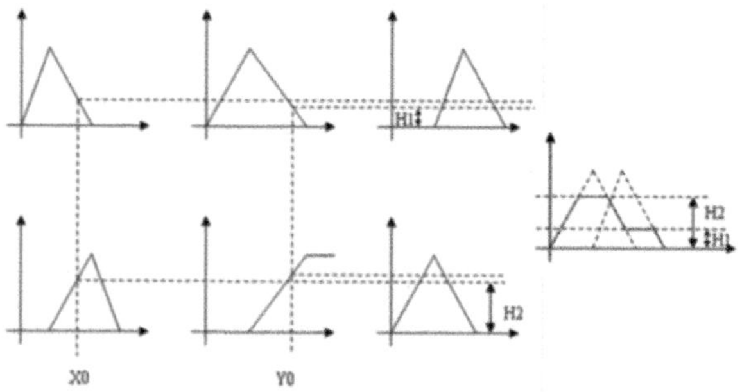

Fig. 3. Fuzzy inference process

C. Defuzzification

The output of this fuzzy controller is a fuzzy subset of control. To get a nonfuzzy value of control, defuzzfication is necessary, which can be performed by height method. The nonfuzzy value of control output can be gained with the height method simply and quickly using (2):

$$y = \sum_{l=1}^{M} \bar{y}^{l} \omega_l \Big/ \sum_{l=1}^{M} \omega_l \tag{2}$$

Since the fuzzy output is also a normalized value and can not be directly used as output frequency, it should be restored to the original range using (3):

$$f_{out} = f_i \times (f_b/2) + f_o \tag{3}$$

D. Self-adjusting Set

One of the most important merits of the presented controller is that it can optimize the fuzzy inference rules on-line to improve the system performance according to the operating conditions. As we know, the RCWS will be constantly changing due to many reasons, so the range of output frequency is not fixed. It must be regularly scanned to obtain the latest resonance frequency band, thus we can adjust the parameters of the system to track the center of resonance frequency. To adjust the parameters of fuzzy controller, f_0' and f_b' when temperature and conductivity are ZE are needed. Minor fluctuations in temperature and conductivity have little effect on the resonance frequency bandwidth, so $f_b' = f_b$. And f_0' can be calculated by (4):

$$f_0' = f_0 - y^* \times (f_b'/2) \tag{4}$$

Therefore, parameters obtained by scanning can be used to adjust the fuzzy controller output.

Table 1. Fuzzy Rules Base

Output Frequency		Temperature(T)				
		NL	NM	ZE	PM	PL
Conductivity(C)	NL	NL	0.9NL	0.7NL	NM	NM
	NM	NL	NM	0.9NM	ZE	PM
	ZE	NL	NM	ZE	PM	PL
	PM	0.8NM	0.9ZE	PM	PM	PL
	PL	0.9NM	PM	0.7PL	0.9PL	PL

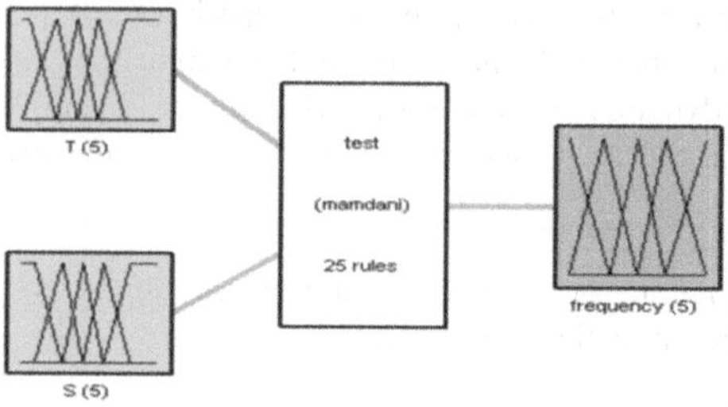

Fig. 4. The presented fuzzy controller model in MATLAB/Simulink

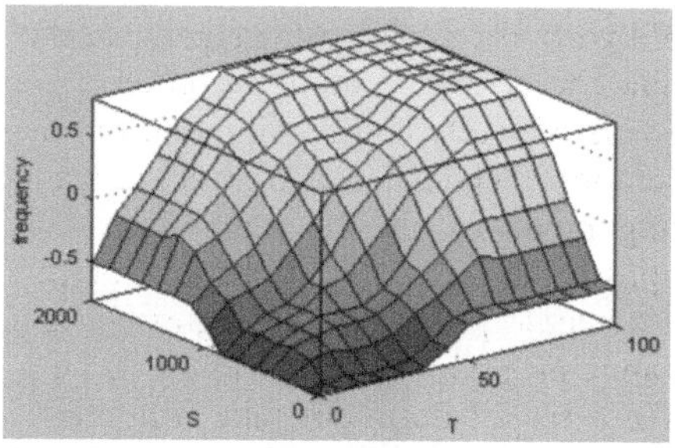

Fig. 5. The control output surface for one example

3 Simulation and Experimental Result

The presented controller can be modeled and simulated in MATLAB/Simulink, shown as Fig. 4. The control output surface for one example is shown in Fig.5. Fig. 6 shows the output frequency and the temperature curve, (a) at 900 $\mu s/cm$ of the water conductivity, (b) at 1400 $\mu s/cm$. Fig. 7 shows the output frequency and the conductivity curve, (a) at 40 ℃ of the temperature, (b) at 70 ℃.

Further more, field test was done using a water disposal device with this controller, from March 17, 2010 to June 16, 2010 in a RCWS. Water conductivity was measured and recorded every 3 days except weekend. The recorded results were curved in Fig. 8.

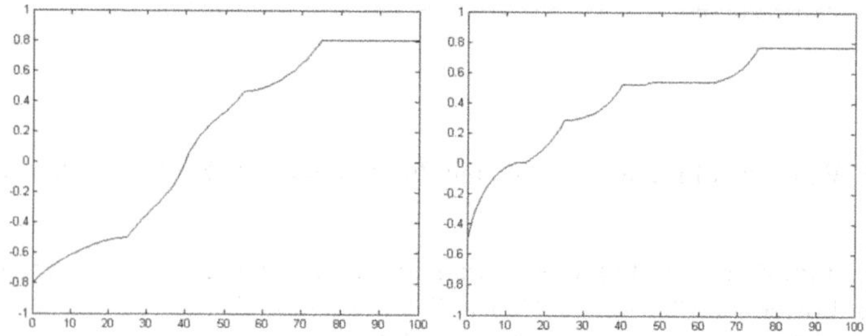

(a) (b)

Fig. 6. Illustration of output frequency versus temperature. (a) at 900 $\mu s/cm$ of the conductivity (b) at 1400 $\mu s/cm$ of the conductivity.

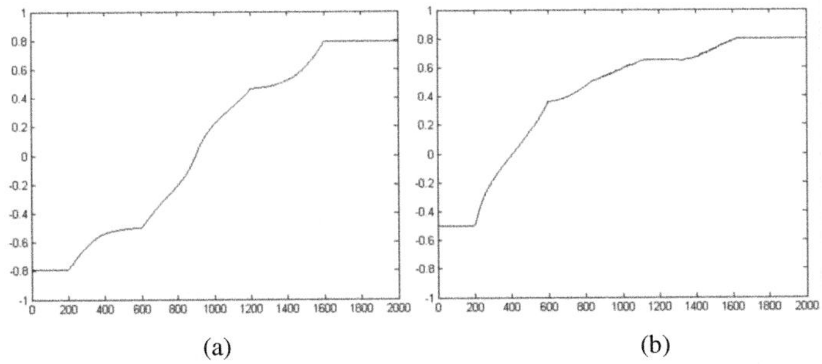

Fig. 7. Illustration of output frequency versus conductivity.(a) at 40 ℃ of the temperature (b) at 70 ℃of the temperature

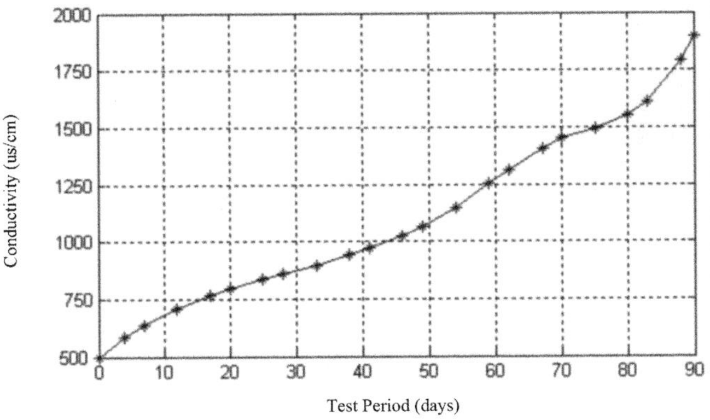

Fig. 8. Illustration of water conductivity versus time tested with adaptive fuzzy controller

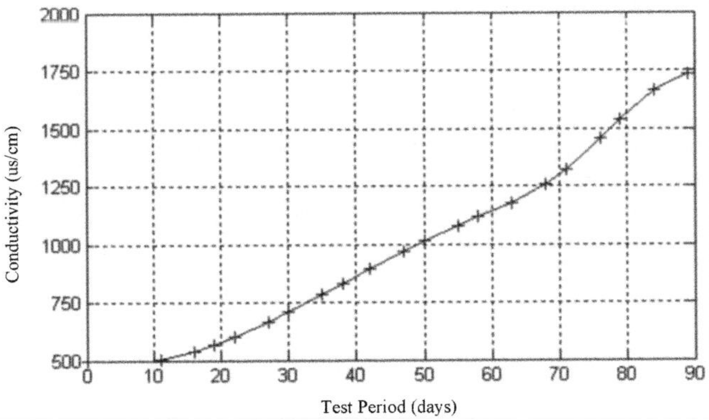

Fig. 9. Illustration of water conductivity versus time tested with fixed output frequency(12MHz)

From the conductivity curve, we can see that the tested RCWS has been descaled well after 3 months operation, which resulted in an increased conductivity from 500 $\mu s/cm$ to 1890 $\mu s/cm$. The tested pipe wall eliminated most of scale layer and formed a black iron oxide protective film.

As a comparative test, the same type device with fixed output frequency (12MHz) was installed at the same site. The result recorded every 3 days form June 19, 2010 to September 18, 2010 was curved in Fig. 9.

4 Conclusions

An adaptive fuzzy controller for HFEM water disposal is presented in this paper. The simulation shows that output frequency of water disposal device with this controller can vary with temperature and water conductivity to achieve better effect. The result of field tests shows that recycled cooling water through the HFEM water disposal device with the presented controller can get higher conductivity and less scale than that with fixed frequency.

References

1. Fang, D.: Boiler Descaling and Water Disposal. The Yuhang Publishing Company, Beijing (1990)
2. Wu, J., Ye, B.: The application of high-frequency electromagnetic water disposal technique on anti-scaling and antisepsis of heating system. Complete Erosion Control (8), 39–40 (2003)
3. Wang, P.: The application mechanism of high-frequency electromagnetic field descaling technique in circulation water system. Power Construction (9), 43–48 (2001)
4. Xu, X., Luo, R.: The test research of descaling with high-frequency electromagnetic field. Trans. of XJTU (1), 124–126 (1997)
5. Yang, J., He, J., Ma, X.: Water electromagnetic disposal. (9), 228–232 (1999)
6. Wang, L.X.: A Course in Fuzzy System and Control. Prentice-Hall, New Jersey (1997)
7. Feng, D.Q., Xie, S.H.: Fuzzy Intelligent Control, ch. 3, pp. 80–81. Chemistry-Industry, Beijing (1998)

Absorptive Capacity of OFDI Reverse Technology Spillover: An Empirical Analysis on Inter-provincial Panel Data in China

Huan Yang, Yan Chen, Wenzheng Han, and Mingpeng Wang

School of Economics and Management Beijing University of Posts and Telecommunications
Beijing, China
Huan_Yang1@yeah.net

Abstract. Reverse technology spillover of OFDI to technological progress of the home country is a relatively new topic in international economics and also one of the important expectations that the Chinese government launched the'going out'strategy. This paper studies the absorptive capacity of reverse technology spillover effect of China's OFDI using inter-provincial panel data (2003~2008). We find that human capital significantly enhances the total factor productivity, but hardly increases levels of absorptive capacity of reverse spillover through an empirical analysis. In the end, the paper provides a reasonable explain for the result, suggests that it is essential to increase education investment, and proposes the corresponding policy recommendations for the implementation of China's'going out'strategy on the basis of above conclusions.

Keywords: Outward foreign direct investment, reverse technology spillover, absorptive capacity, human capital.

1 Introduction

The application of new growth models and empirical techniques to expanded datasets has stimulated research on how interntional businesses enhance technology and knowledge diffusion between countries (Coe et al., 1997, 2008 ; Keller, 2004). Inspired by work on technology improvement caused by import, export (Gross and Helpman, 1991), attention has gradually moved to technology diffusion associated with foreign direct investment (FDI) in the host countries (Haskel et al. 2002; Keller and Yeaple, 2003) and the extent to which foreign presence improves productivity in host developing countries.

However, the research on international technology diffusion obviously prefer business way and FDI inlows to FDI outflows (outward FDI, OFDI) which bring reverse technology spillover effect. Little attempt has, moreover, been made to undertake quantitative analysis of OFDI-related productivity effects in the developing countries that are making the investments, probably because developing countries are genarally the recipients of foreign capital and few have emerged as investors in the world economy.

J. Zhang (Ed.): ICAIC 2011, Part V, CCIS 228, pp. 581–588, 2011.

With China's accession to the World Trade Organisation (WTO) and the implementation of "going out" strategy, Chinese outward FDI (OFDI) has increased rapidly from 2004. In 2009, Chinese total OFDI is 62.06 billion dollars, which is 20 times than the beginning of entering WTO. However, the research of China as investment sources is not enough. As a country with large population, if the human capital in China matters the absorptive capacity of OFDI reverse technology spillover, little study has been done so far.

This paper aims to examine the effects of China's OFDI on changes in its productivity between 2003 and 2008, through estimating the total factos productivity (TFP). Then add human capital as the key factor to examine its effect on the absorptive capacity of reverse technology spillover.

The paper is organized as follows: Section 2 review the theoretical framework of technology sourcing associated with OFDI and analyzes reverse spillover in China with the introduction of absorptive capacity, especially the human capital; Section 3 describes the econometric model and data description; Section 4 presents the empirical analysis and the results; and Section 5 discusses the results and points out the direction to further the research.

2 Literature Review

A. Reverse Technology Spillover

Dunning (1988, 1994) and Cantwell (1989) suggest a close relationship between foreign direct investment and the intensity of R&D expenditures. The implicit mechanism here is that FDI can overcome geographic constraints and provide an important channel of knowledge dissemination. In the inward FDI technology transfer, adoption of foreign technology by domestic companies can be stimulated through competition effects, demonstrative effects and the mobility of trained labor. However, the transfer effect may be reduced by the availability of technology. This will be endogenously determined, and may even be blocked by foreign investors (Keller, 2004).

Across the report"Global Economic Prospects: technology diffusion in developing countries"(The World Bank, 2008), the technology diffusion in developing countries depends not only on its approach of acquiring foreign technology (through trade, foreign direct investment , international migrants and other network), but also its technology absorption ability (government policy, the quality of institution, human capital stock, R&D and financial system composition, etc.). Speak of China's outward foreign direct investment and its reverse technology spillover, two relevant point was mentioned: on the one hand, the reverse technology spillover, drived by outward foreign direct investment, would influenced by international technology transfer channels; On the other hand, China itself must have certain digestion and absorption ability to realize smoothly the absorption and utilization of reverse technology spillover. While the latter can be measured through the government support, freedom of domestic market and human capital stock and so on.

Empirical investigation of the relationship between productivity growth in industrialized countries and foreign presence of their MNEs tends to support the view that FDI-related R&D spillovers contribute to the productivity growth of home countries. For example, Bracornier and Ekholm (2002) found a positive association in Sweden between the size of outward and inward FDI, R&D expenditure by the host country and technological spillovers absorbed by the home country; Neven and Siotis (1993) similarly found FDI in Europe to be concentrated mainly in innovative sectors. Examining total factor productivity of the US, Japan and 11 European countries for the period 1971–1990, Lichtenberg and Van Pottelsberghe (2001) found that both imports and ODI contribute to productivity growth in these economies. Inward FDI appears to be insignificant in technology transfer; the possible explanation is that foreign investors may try to minimize technology diffusion to host country companies in order to maintain their firm-specific advantages (Blomstrom et al., 1999; Dunning, 1994). Other empirical research also finds evidence supporting the stronger technology sourcing effects associated with ODI (Globerman et al., 2000; Singh, 2004).

While a number of studies have estimated ODI-related technology diffusion in developed countries, little attention has been paid to productivity changes induced by ODI in developing countries. This is probably because developing countries are, in general, recipients of foreign capital and, by and large, receive foreign technology through importing, licensing and inward FDI. Few of them have shown a strong demand for the technology and even fewer have engaged in technology sourcing. China has recently emerged as an active investor with an increasing interest in technology in leading developed countries. Despite the growing evidence in qualitative case studies (Child and Rodriguez, 2005; Liu and Tian, 2008), tracking the motives and capabilities of ChineseMNEs in technology sourcing has not been prevalent. Empirical analysis of the technology sourcing effects on productivity change in China has also not been available.

B. Absorptive Capacity of Reverse Technology Spillover

Researchers have used the absorptive capacity construct to explain various organizational phenomena. Several studies of technology spillovers via FDI have explored the hypothesis that the incidence of externalities is dependent on absorptive capacity (Cohen and Levinthal, 1989). Depending upon data availability and the context of the investigation, two basic approaches are usually adopted. One is to divide the plants in the sample according to some perceived proxies for absorptive capacity, and compare the degree of spillovers across the subsamples. Thus Kokko, Tansini and Zejan (1996) divide their sample of Uruguayan manufacturing plants by the size of their technology gap vis-a-vis foreign-owned firms, and find that spillovers are present when the technology gaps are 'moderate'. Girma and Wakelin (2001) stratify microdata for the UK electronics industry according to size and skill intensity, and report that smaller plants or plants with a low share of skilled workers in the workforce lack the necessary absorptive capacity to benefit from FDI. But they also report that large establishments with higher skill intensity do not benefit from FDI, as they presumably operate near the technological frontier. This last point is echoed in the work of Haskel, Pereira and Slaughter (2002). The authors pool all industries in the same UK microdata and the sample is split by employment, total factor productivity (TFP) and skill-intensity quartiles. But in contrast to Girma and Wakelin

(2001), they also conclude that plants further away from the technology frontier gain most from foreign presence in their sector. This would appear to suggest that low absorptive capacity is not a hindrance to learning from foreign technology.

Through the theoretical framework above, we realize that reverse technology spillover will not automatically improve the TFP in home countries. The effect depends on the absorptive capacity in domestic. However, correlative research on the factors affect China's OFDI reverse technology spillover is still not enough. This paper will use LP model (Pottelsberghe and Lichtenberg, 2001) to examine the factors of reverse spillover on the basis of theoretical framework.

C. Human Capital

Grossman and Helpman (1991) identify various mechanisms of technology diffusion through international trade, and Coe et al. (1997) summarize the main mechanisms as usage of high technology embodied in intermediate goods or capital equipment, learning through international trade, and application through imitation. These mechanisms may also be related to domestic human capital as an important factor in understanding the foreign high technologies embodied in import goods and absorbing them for domestic use.

Using the technology-driven endogenous growth setup, many empirically research have test the relationaships among degree of openess, human capital accumulation, international technology diffusion, and economic growth (Grossman and Helpman, 1991; Rivera-Batiz and Romer, 1991; Aghion and Howitt, 1992). As for the role of human capital, Coe et al. (1997) find that the effect of human capital on economic growth in developing countries is substantial, they fail to show that the benefits from international R&D spillovers are greater when the labor force in developing countries is more educated, which is stressed as "absorptive capacity" by Keller (1996).

3 Data Description and Model

A. Data Description

1) Total Factor Productivity (TFP)

In the Cobb-Douglas production function $Y=AK^{\alpha}L^{\beta}$, "A" is the technical level in a country, which can represent TFP, so:

$$TFP=A=K^{\alpha}L^{\beta}/Y \tag{1}$$

Choose 2003-2008, the data originates calendar year " Chinese statistic almanac". α and β both value for 0.5 according to a large number of literatures about Chinese situation. We value K with the perpetual inventory method (Coldsmith, 1951):

$$K_t = I_t/P_t+(1-\delta_t)K_{t-1} \tag{2}$$

K_t is the existing fixed capital at the time "t", I_t is gross capital formation during "t", P_t is fixed assets price index, δ_t is depreciation rate at "t", usually valued 5%,

K_{t-1} is the existing fixed capital at last year. All data came from the China Statistical Yearbook. So the TFP of provinces can be calculated by equation (1).

2) R&D stock S_{it}^{d}

R&D stock in each province can be generated with the method of perpetual inventory:

$$\ln S_{it}^{d}=(1-\delta)S_{it-1}^{d}+RD_{it} \tag{3}$$

S_{it}^{d} is the R&D stock, δ is depreciation rate of R&D. Due to the diffusion from new knowledge to old knowledge, the domain specificity of knowledge is decline. So the depreciation rate of R&D must be higher than that of fixed capital, which is valued as 5%. In China, the actual useful year of technology usually 14, we set the depreciation rate of R&D as the reciprocal of 14, mean 7.14%. RD_{it} is the expenditure stream based on the year 2003, which can be obtain from "China statistical yearbook on science and technology". S_{it-1}^{d} is the expenditure at last year of each province. Then the R&D stock from 2003 to 2008 can be calculated with the equation (4) (Griliches, 1980):

$$S_{it}^{d}=RD_{it}/(g+\delta) \tag{4}$$

3) R&D Reverse Spillover Affected by OFDI oversea

According to the equation generated from LP model, the R&D reverse spillover $SPILL_{it}$ affected by OFDI can be calculated as follow:

$$SPILL_{it}=\sum_{j\neq i}\frac{OFDI_{ijt}}{Y_{jt}}S_{jt}^{f} \tag{5}$$

In this equation, three variables are involved. OFDI stock of China's provinces come from 《Statistical Bulletin of China's Outward Foreign Direct Investment》. In view of the technology-seeking OFDI usually aim at developed countries. So we select a group of developed countries of G7, Hong Kong, macau, Singapore, Korea, and Japan as host countries.Using the GDP of destination countries, and related data above, the R&D reverse spillover affected by OFDI in each province can be obtained from 2003 to 2008.

B. Model

On the basis of the theoretical and empirical discussion, $SPILL_{it}$ in equation (5) represents the R&D stock spilled from OFDI (Zhao, 2006; Bai, 2009). Refer to the influencing factors of spillover effect, Kokko (1994) used cross terms of foreign-invested enterprises' assets and technology gap indicators as explanatory variable to analyze by regression.Qiu (2005) and Zhao, Wang (2006) also used cross terms of related factors and technology spillover to examine the absorptive capacity of China. Thereupon the model be established as follows:

$$\ln TFP_t = a_0 + a_1 \ln S_{it}^d + a_2 \ln SPILL_{it}^f + \varepsilon_t \qquad M1$$

$$\ln TFP_t = a_0 + a_1 \ln S_{it}^d + a_2 \ln SPILL_{it}^f + a_3 H + a_4 H^* \ln SPILL_{it}^f + \varepsilon_t \qquad M2$$

Model 1 is the basic model of reverse technology spillover of OFDI, S_{it}^d stands for the R&D stock in "i" province at "t" year, $SPILL_{it}^f$ is the reverse technology spillover of OFDI in "i" province at "t" year. Model 2 includes human capital "H", which represented by the average education background (Liu, 1995), to examine its influence to TFP and its cross effet with SPILL to TFP.

4 Empirical Findings

Using panel data to research, we test the model with Hausman specification test to test evaluates the significance of an estimator versus an alternative estimator. It helps to evaluate if a statistical model corresponds to the data. The following table gives the results of Model 1 under fixed effects:

Table 1.

Effects Test	Statistic	d.f	Prob.
Cross-section F	100.330692	(28,143)	0.0000
Cross-section Chi-square	526.781767	28	0.0000

The original hypothesis of this test is that fixed effects are unnecessary. From the table, it reject the null hypothesis at a 5% level. So we found it is appropriate to bring in the fixed effects, and fixed effects model could be used.

We use EVIEWS6.0 to analyze Model 1 and 2 by regression. Concrete results shown in the table below:

Table 2.

	Model 1	Model 2
S_d	0.196***	0.188***
SPILL	0.029***	0.067*
H		0.073**
H*SPILL		0.005
R^2	0.979563	0.980836
adjusted R^2	0.975275	0.976487
F-statistic	228.4663	225.5219
Prob(F-statistic)	0.000000	0.000000

*Correlation is significant at the 0.1 level.

**Correlation is significant at the 0.05 level.

*** Correlation is significant at the 0.01 level.

From the table, it can be seen that the goodness of fit and F-Statistics of the models are both well.

5 Discussion and Conclusion

As shown in our empirical analysis, we can find that R&D stock in domestic and the reverse R&D stock from OFDI can both significantly enhance the level of TFP, as the results of Model 1. In addition, the contribution that R&D stock from OFDI made to TFP is 0.029. It means that with a 1% increase in the size of reverse R&D stock from OFDI, there is a 0.029% increase in total factor productivity. The results proved the existence of reverse technology spillover from OFDI.

This conforms well to China's ongoing outward-oriented development strategy and its recent policy shift towards encouraging more companies to engage in overseas investment. An explicitmotivation for both inward and outward internationalization is the great potential for technology spillovers to domestic industry. Technology sourcing through OFDI seems desirable since foreign technology can be specifically targeted and thus can be adopted more effectively. More and more companies are actively participating in technology sourcing OFDI. With the growing international experience of Chinese companies, this trend is expected to continue.

An important finding is that technical efficiency changes have benefitted more from OFDI-related technology diffusion than domestic R&D investment during the period of our study. As is shown in the result, the coefficient of domestic R&D stock is 0.196, which is significantly higher than the coefficient of reverse R&D diffusion from OFDI. It shows that the contribution of domestic R&D is more important than R&D spillover from abroad. In view of this, in order to improve technical level of China, more investment to domestic R&D should be encouraged together with promoting the technology-seeking OFDI. However, neither domestic R&D nor OFDI-related R&D capital stocks made a large contribution to either technological or efficiency change over the study period.

As an way of international technology spillovers, OFDI has a positive effect on the growth of total factor productivity, so some good companies should fully utilize the opportunities of technology globalization, to integrate global resources and access to advanced foreign technology through establishing R &D institutions abroad or acquisition some of core technology, then passed to the parent company with reverse technology spillover effect, so technology improvement, technological upgrading and industrial upgrading of other domestic enterprises will be driven and promoted, reflected at the national level, national technological progress is promoted in China.

Model 2 examine the influence of human capital stock to Chinese absorptive capacity of reverse technology spillover. Given closer analysis, we can see if considered purely from the human capital stock, the positive effect is significant at a level of 5%. It means that the increase of human capital stock can improve TFP, which is in accordance with in accordance with that human capital can promote endogenous technology growth in new growth theory.

From another point of view, the cross term of human capital and spillover from R&D abroad do not significantly affect TFP, which show that present human captical stock do not positive increase the absorptive capacity. Part of the explanation may be seen as "threshold effect". For starters, Grossman and Helpman (1991) found the existance of "threshold effect" when they studied international tradetechnology spillover. They pointed there is little stimulus of international business to the economic development when the developing level has not yet reached the threshold. Once it reached, the stimulus effect will be significant. Thereafter, many scholars discovered the existence of "threshold effect" in the study of international technology spillover. Borensztein et al. (1998) used human capital as proxy variable of technical absorptive capacity to analyze FDI technology spillover, found that the combination of FDI and human capital stock in host countries significantely improved economic development, and the influence is highly more than pure capital accumulation effect. Meanwhile, the technology spillover of FDI had a critical level, namely, when human capital in host countries is rich enough, host country economy can absorb the technology overflow of FDI.

Overall, the findings from this paper suggest that it is essential to increase education investment to improve the quality of human capital, so that promote the absorb of reverse technology spillover. We believe that this study has important implications for research on China's OFDI. It confirms and extends previous research suggesting the importance of learning from OFDI, and offers a more theoretically grounded model of such learning. It also adds to the absorptive capacity literature by examining human capital. For further study, we can also add other components such as economic openness, economic structure, industry structure, institution factors, and so on, to find better method to utilize OFDI-related R&D.

References

1. Coe, D., Helpman, E., Hoffmaister, A.: North-South R&D spillovers. Economic Journal 107, 134–149 (1997)
2. Dahlman, C., Nelson, R.: Social absorption capability national innovation systems and economic development. In: Social Capability and Long-term Growth. Macmillan Press, Basingstoke (1995)
3. Grossman, G., Helpman, E.: Innovation and growth in the global economy, pp. 384–429. MIT Press, Cambridge (1991)
4. Grossman, G., Helpman, E.: Comparative advantage and long-run growth. American Economic Review 80(4), 796–815 (1991)
5. Kokko, A., Tansini, R., Zejan, M.: Productivity spillovers from FDI in the Uruguayan manufacturing sector. Journal of Development Studies 32, 602–611 (1996)
6. Kokko, A.: Journal of Development Economics 43(2), 279–293 (1994)
7. Laces, R.E.: On the Mechanics Of Economic Development. Journal Monetary Economic 22, 342–369 (1988)
8. Levine, R.: Financial Development and Economic Growth Views and Agenda. Journal of Economic Literature 35(2), 688–726 (1997)
9. Naruls, R., Marin, A.: FDI spillovers, absorptive capacities and human capital development: Evidence from Argentina. MERIT Research Memorandum series, vol. 16 (2003)
10. Romer, P.M.: Increasing Returns and Long Run Growth. Journal of Political Economy 194(5), 1003–1034 (1986)

Application and Study on Artificial Intelligence Technoligy in Traffic Singal Control System

Baozhong Liu and Yajin Sun

College of Computer Science and Engineering,Wuhan Institute of Technoligy Wuhan Institute of Technoligy 693ChuXiong Road, WuChang, Wuhan, China 430073
baozhongliu221@sina.cn

Abstract. With scientist and technology developing , especially quickly developing computer technology, artificial intelligence technology has knew about by the people in gradually ,and applies in pratice.

In the article ,we will introduction artificial intelligence principle and apply in traffic singal control,solve some traffic problem in actual.With the development of intelligent transportation system,urban traffic singal control has been growing into one of the most important aspect.As complexity of transportation,the tradition method are not able to achieve the problem of traffic singal control .we will introduction a self-learning system base on RBF neural network. The system can simulate the traffic police's experience.According to the queue length in each intersection,the system can give out both the singal cycle and the spilt of intersection.Furthermore,it can evaluate the effect of the control with the changing of the traffic ,and adjust the singal.Simulate results reveal that the system can much more prefectly control the actual traffic condition and improve the passing ability of intersection.

Keywords: Artificial intelligence, RBF, Self-learning system.

I Introduction

Artificial intelligence were came into existence1956 .It has rapid progress in more than 50 years. Because artificial intelligence is apply very widely and has very big potential in research and developing .Absorb more and more scientific and technical worker who come into the field.

Artificial intelligence is relevant with computer of scientifical, contoral theory, information theory, neurophysiology, psychology and linguistics. It is a comprehensiveness science. Artificial intelligence is research how make manufacture intelligent machine and intelligent system, imitate ability of the man intelligent action , extend the man intelligent scientific.

Computer is be ability which imitate the man intelligence action. Now, it is two main approach in artifical intelligence research. The first, psychologist hope that they may acknowledged how handle information in brain. They believe that the brain is matter basis with intelligent action. For discover nature of intelligent, we must know the man brain structure. Actually, the brain have more than 10 million nerve basic, and in phase ,it is very difficult to carrying out an experiment that imitate the man

J. Zhang (Ed.): ICAIC 2011, Part V, CCIS 228, pp. 589–595, 2011.

brian. So it is very difficult to accomplish the taste .But the scientist would be build "information handle intelligent theory" that will be as a long time of purpose. The view must be take seriously. The second apporach, the computer scientisc think that imitate the brian aspect achieve artificial intelligence. To computer program runing ,achieve the man intelligent action similar in efficially, as a purpose. The view is actual, so it is absorb the most people research.

Resaerch artificial intelligence must be achieved by computer .so may be speaking "artificial intelligence main purpose is clear know theory that achieve artificial intelligence ,it is change more intelligent ,more wise and more usefull.

2 Urban Traffic Singal Control System Research Direction

A great quantity of date indicate it is main three aspect to urban traffic singal control research, following:

(1) It may be improve the actual traffic singal control system function to articficial intelligence apply into urban traffic singal control system.

Lately, more and more people begin to research articficial intelligence technology in traffic aspect application. It has three aspect of important in traffic singal control : collect date of traffic, analyse and handle date, judge and control. From collect date of traffic, until achieve control process, key is improve information efficiently.

(2) Discrete time, rolling horzion approach research traffic singal control system.

Rolling horzion approach is checking the lately date, evaluate state of system now. It is find a most approach in rolling horzion period. The time of rolling horzion must be enough ,it has two parts, the first part is carry out the apporach of the rolling horzion method, the second part is checking the approach with function of terminal price. It is right ,continute to operate ,else stop.

(3) Disperse system is a research direction in future.

Disperse system is efficiently approach in practise .when a part of the system is bad ,it may be operate to the system ,only the part is not core. It increase reliability of entire system.

3 Base Fuzzy Control and Neural Network Theory

3.1 Theory of the Fuzzy Control

Fuzzy control is computer control that is base on fuzzy set theory , variable of fuzzy language and fuzzy logic reasoning, as following:
Fig. 1。

Fuzzy control is nonlinear control .Fuzzy controller is the most part in the system Fuzzy control is a intelligence control ,it is a important and valid method. Especially fuzzy control link up with neural network and GA, that have became a very important style ,it show the bigger latent capacity.

Fuzzy controller is make up 4 portions, as following :

(1) .Fuzzy

The portion is change parameter of accurate import into fuzzy parameter. The import parameter including :inference import and output or state of the system.

<1>. It is deal with the parameter of accurate import, which is change fuzzy parameter.we usually handle e and ec ;e=r-y,ec=de/dt. r is import parameter ,y are parameter that is output from the system. e is inaccuracy , ec is variable of inaccuracy.

<2>.The import parameter will become to the number that turn into self range.

<3>.The accurate parameter is change to the fuzzy parameter.

(2) .Knowledge base

There are knowledge of application range and purpose of the control in the knowledge base. Including :every function , constant and the level of fuzzy space.There are some rule of control in the knowledge base,it is show the expert experience and knowledge.

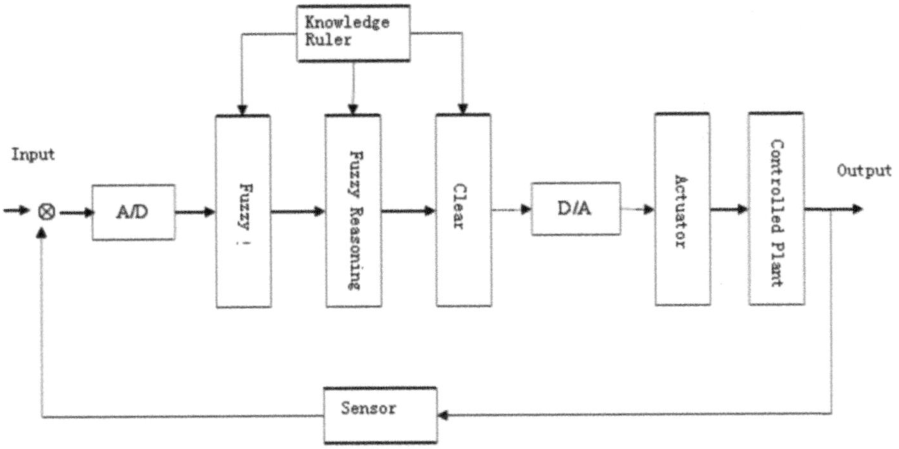

Fig. 1. variable of fuzzy language and fuzzy logic reasoning.

(3) . Fuzzy reasoning

Fuzzy reasoning is core in the fuzzy controller , it can imitate the man capbility of fuzzy concept reasoning .The fuzzy reasoning is depending on fuzzy logical relation and reasoning rule .There is four step:

<1>.Match degree

It is compare the actual to precondition of the fuzzy rule,getting match degree that is the actual to every precondition .

<2>.Simulate strength

Use the fuzzy by and or ,unite match degree of precondition MF in the rule,get simulate strength.

<3>.Effective post condition MF

It is made simulate strength to the post condition of rule . Get valid post condition MF.

<4>.Total output MF

Sum up to all valid post condition MF, get total output MF.

(4) Clearing

It is change the parameter of fuzzy reasoning control into the parameter that can use practise control in the step.

3.2 Base Neural Network Control Theory

There is a Mcculloch-Pitts model of artificial neuron that is abstract to the brain neuron.as following

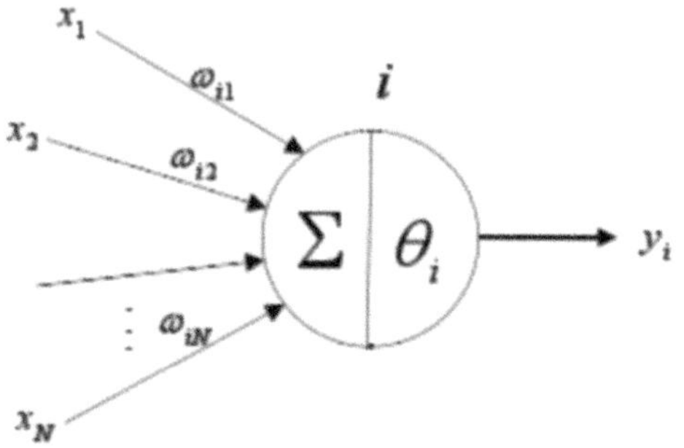

To n neuron,x1,x2,-----,xn is information that take over by the neuron. wi1,wi2,----win as connect strength,is called right .Use the method that is link up with all import singal ,as total efficient ,called pure import.According to different operation style , pure import show some style .It is the easiest to the weighted sum of line.The action will change state of i neuron.the state function （g （) ） is import function (yi) according to i neuron. Θi as i neuron threshold.

$$net_i = \sum_{j=1}^{N} \omega_{ij} x_j - \theta_i$$

$$y(i) = g(net_i)$$

Studying rule is amending connect strength in neuron or the weighted factor algorithm. The structure of knowledge is suit to the surrounding environment alter. There are supervision and no supervision algorithm.

4 Self-learsing System of RBF Neural Network in Traffic Singal Control

4.1 Problem Introduction

Every intersection has four direction, include east, south, west and north. every direction has three method ,include right, left and move straight. following:

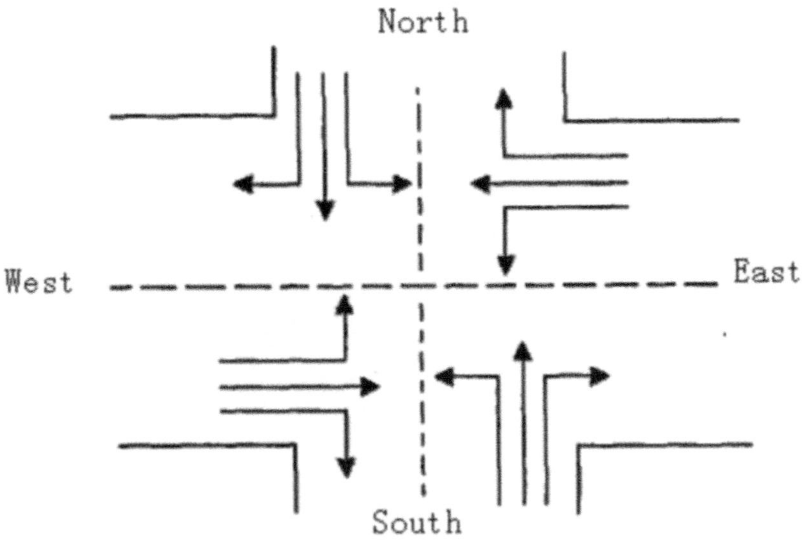

If a policeman who stay in intersection command traffic .He may be reason command traffic according to the four direction queue of automobile length and rate of flow .Rate of flow of automobile is random in most time,so we are not settle same long time in four direction. So it is very importment to imitate the policeman handle with real traffic enviorment.we may be make a function ,four direct long automobile as import parameter,be relative with singal period and every phase's ration of yellow singal as output function.It is a neural network.. Training the function according to the policeman handle real enviorment.

4.2 Base on Neural Network Singal Intersection Model

According to four phase intersection singal control problem descrip ,the neural network s structure ,following: L as import function , L is long automobile,.The period and every phase ration of yellow singal as output function.

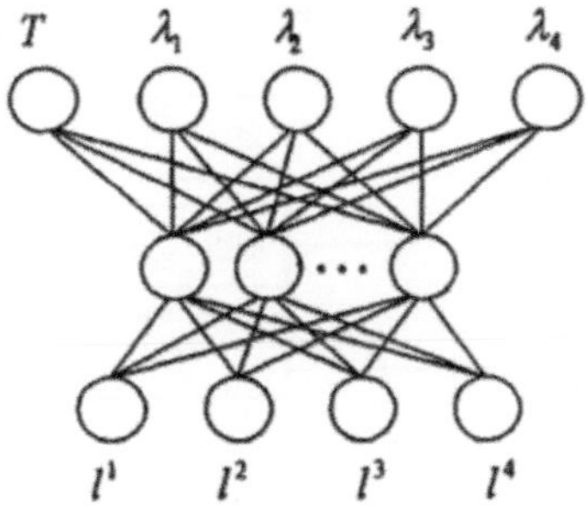

4.3 RBF Neural Network

RBF(Radial Basis function)structure of picture,following:

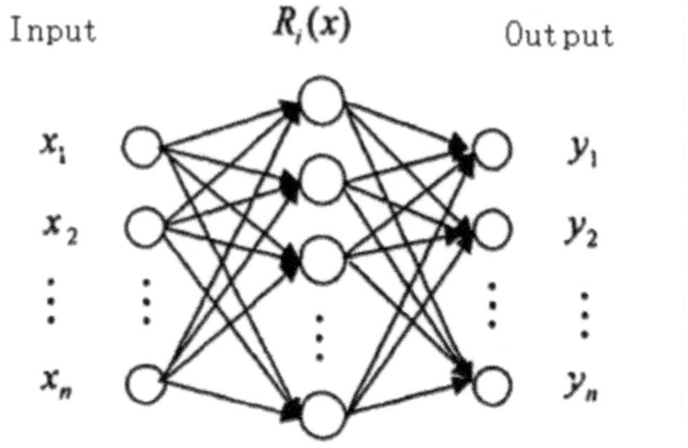

R(x) as base fuction, the base function is Gauss function.

$$R_i(x) = \exp[-\frac{\| x - c_i \|^2}{2\sigma_i^2}] \qquad i=1, 2, \cdots, m$$

x is n phase import function;Ci is base function's center,is vector .
σ_i is variable, it decise width of the function center.
m is element number.
When import singal near range of base function center ,the base function is bring about the biggest output .

4.4 Neural Network Self-learning Progress

For guarantee whenever the system can operate, the system has two RBF neural network controller. whenever the one is operating, the other is learning. Acoording to control efficient evaluate rule decision whether or not learning.

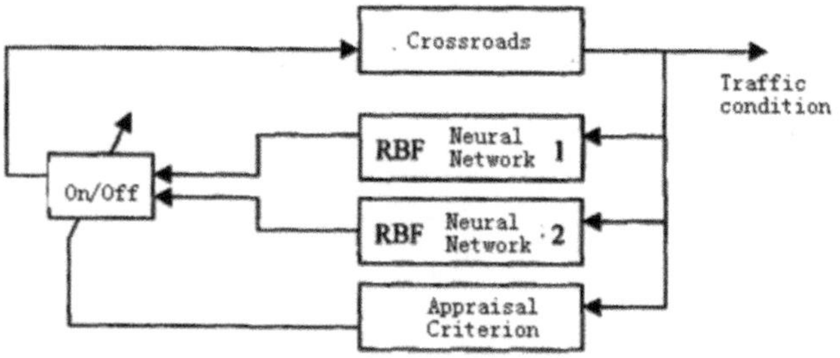

5 Sum Up

With social developing ,more and more artificial intelligence technology will be apply in every aspect of our life.It will be bring into many convenience in our diary life. Artificial intelligence is important direct in research .In future ,from one aspect saying , artificial intelligence represent developing level of the country information. Training the function according to the policeman handle

References

1. Li, h.: Artificial intelligence developing histroy. Information technology (2007)
2. Yang, h.: Research of urban traffic singal control based on artificial intellgence (2007)
3. Liu, K., Cai, B.: Traffic Control Method for Isolated Intersection Based on Fuzzy Control and Simulink Simulation Road project (2007)
4. Wang, J.-y., Peng, W., Wang, K.-p., Cai, H., Xing, Y.-m., Guo, D.-w., Zhou, C.-g.: Application of reinforcement learning and neural network in traffic signal control. Computer engineering and application (2007)

Topographical Change Detection from UAV Imagery Using M-DSM Method*

Wenling Xuan

School of Earth and Space Sciences, Peking University,
3403, YiFuEr Bldg., No.5 YiHeYuan Rd. HaiDian District, BeiJing 100871
wenlingxuan222@sina.com

Abstract. Many megacities in China have a particular interest in detecting change, caused by construction, disaster etc., as soon as possible, on a media which allows quick and immediate processing, of a limited and small area, to an accuracy appropriate to data use and scale and within a modest budget. Topographical change detection projects using imagery obtained from low altitude UAV (Unmanned Aerial Vehicle) system and processed by newly developed software dedicated to the creation of quick and accurate DSM (Digital Surface Models) are being carried out in several large cities in China. This paper outlines the specific UAV system, with its hardware and the special wide field of view camera system as well as the software developed for the processing of this imagery. M-DSM method workflow and Technical route of employing UAV system to detect illegal buildings are provided.

Keywords: UAV system, M-DSM method, topographical change detection, UAV imagery, photogrammetry, GIS.

1 Introduction

Changes of topographical details, caused by human development of the environment, natural disaster, etc. and the recording of these changes is a frequent requirement by city authorities. The access to updated digital database with information changed through building construction and demolition, is essential for city authorities to know the current status, to put corresponding policies in practice, and to plan for future development etc. In China, with its vast and rapid advancements in urban and rural development, there is an immediate and frequent requirement to know of these changes and to consider them for management and planning purposes.

Several photogrammetric and remote sensing methods can be employed to monitor and detect geometrical change in objects and these include conventional photogrammetry with planimetry and height comparisons, photogrammetry by comparison of old map data superimposed on new imagery, comparison of DSMs generated by LiDAR point clouds and the detection of radiometric change in multi-spectral imagery by remote sensing methods etc. In this paper, we provide a 3-dimensional change detection method which is

* This work is partially supported by Key Lab of Earth Exploration & Information Techniques of Ministry of Education in Chengdu University of Technology.

J. Zhang (Ed.): ICAIC 2011, Part V, CCIS 228, pp. 596–605, 2011.

characterized by comparing DSMs created from low altitude UAV imagery obtained in 2 different epochs. This specific method is named M-DSM change detection method.

This paper is organized as follows: section I states the fact that the provision of timely data on construction change by employing a suitable, cost-efficient and simple operation model providing such data, is in great demand in present China. In section 2, the intention of developing a cost-efficient UAV remote sensing system to replace manned aerial platforms in some circumstances, like fast topographical change detection, is explained. A complete hardware and software solution developed for a low altitude and low platform speed UAV system, using highly overlapped optical imagery is discussed. In section 3, new software utilizing the special features of such UAV imagery to finally detect change by comparing DSMs generated from stereo imagery is mentioned and the workflow of M-DSM method is illustrated. The example of applying M-DSM method in the airport area of Wuhan city to detect building construction and demolition is given. Field operation and DEM accuracy aspects of the integrated UAV system is analyzed in Section 4. Further precision analysis and comments on such UAV systems is given in section 5.

2 UAV System

A. Why Develop UAV Platform for Topographical Change Detection?

The detection of change as recorded on imagery obtained from metric sensors is a common process within digital photogrammetry and remote sensing software packets. However, meeting the timely and simple operation requirement for such a remote sensing platform is a challenge. In the past, fixed wing aircraft were used as airborne sensor platform and they required a professional crew of pilot, navigator and sensor operator, required flight permission from the local air traffic control authorities and could only operate in fine weather. Two challenges have been identified in an analysis of the solution required to present timely data: to introduce another airborne platform and camera system and to improve the data processing by the development of a dedicated software packet and workflow.

In China and as in many other countries, UAV platforms equipped with image sensors for civilian use are ideally suited to capture images of relatively small areas and at a favourable cost. UAVs do not require the same highly skilled operation team of pilot/navigator/sensor operator and this leads to easier staff recruitment, lower costs and easier mission disposition. UAVs are not under the same strict operational control by external authorities as fixed wing aircraft are and these advantages have inspired many organizations, including Peking University and Chinese Academy of Survey and Mapping, to investigate, analyze the characteristics of available UAVs, their shortcomings, their payload, operation and practical use and define a suitable platform and image acquisition system. Their final analysis has resulted in a palette of products with some newly designed and constructed components in which several UAV platforms, cameras, can be combined to form the most suitable image acquisition system for the task, the area and the conditions of the site.

B. UAV Platform Features

Two UAV platform technologies have been defined and developed; inflatable blimps (air ships) and fixed-wing miniature aircraft, each with 3 different models and characteristics, all being powered by benzine piston engines and operated by remote control from a portable control system transported in a vehicle which accompanies the mission and crew from site to site.

1) Blimps
Three models, twin propellers, are designated as FT070, FT100, FT140. The blimp is helium inflated with a nylon outer skin. Helium bottles for inflating the blimp are readily available throughout the country and do not have to be transported with the system. Operation is with a 5- man crew. Launching is from a flat clear area, ca 30m x 40m, accessible by vehicle with unlimited vertical clearance.

Main characteristics of the blimps are:

Fig. 1. Blimp FT140

Length:	12.4m – 18.32m
Transit flights:	60km/hr – 72km/hr;
Image flights:	36km/hr (10m/sec) ground speed
Flying height:	max. 2,000m above ground
Payload:	6kg – 15kg
Flight duration:	2hrs – 3hrs
Operating conditions:	10.7 - 13.8 m/s max. wind speed

2) Fixed wing miniature aircraft
Three models, single propeller, are designated as GY01, GY02, GY03. The aircraft is operated by a 3- man crew and launching is from a flat, clear area, ca. 100m x 30m, accessible by vehicle, with unlimited vertical clearance and no height obstructions within 200m.

Main characteristics of the aircraft are:

Fig. 2. Fixed wing UAV GY03

Length: 2.0m – 2.8m
Travel speed: Transit flights:120km/hr – 130km/hr;
Image flights: 36km/hr (10m/sec) ground speed
Flying height: max. 3,500m above ground
Payload: 2kg – 5kg
Flight duration: 1hr – 2hrs, 40km – 80km
Operating conditions: 5.4m/s – 7.9m/s max. wind speed.

C. Navigation Components

The navigational and flight control components of computers, large LCD display, time and speed displays, radio communication to the UAV, steering controls, battery power supply, UAV fuel and maintenance tools etc. are provided in a compact and safe form and are easily transported in a common van without special adaption. See fig. 3.

Fig. 3. Navigational and control units in van

The perimeter of the area of interest is to be defined on a large scale digital, topographic map and even visualized on Google Earth. Flight planning software, as part of the system, determines the parallel flight lines to be covered by the UAV. Camera exposure points are defined as a function of UAV location, UAV flying height and image overlap and the camera shutter is released automatically when this position is reached during flight. Flight planning is performed to allow a tolerance in navigation of +/- 10m from true position on the flight line. The camera system mount does not enable drift caused by wind to be compensated in flight and so generous image overlap of nominal 80% forward and 50% side overlap is planned to overcome this operational aspect. See fig. 4.

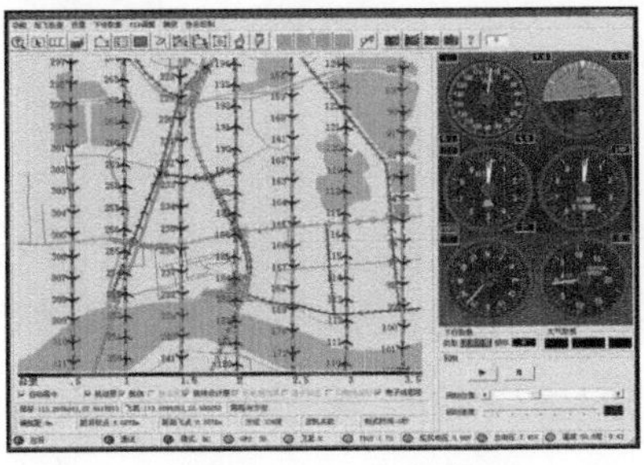

Fig. 4. Flight planning and navigating software interface

The position of the UAV is determined by real-time GPS. The GPS is only used for navigation and does not provide exact camera position at this time. Navigational data is available to the camera operator at the control system and any flight corrections can be implemented directly and online through radio connection.

D. Camera Component

Three combinations of the basic camera unit are available, designated as LAC01, LAC02 and LAC04 and constructed from 1, 2 and 4 exchangeable off-the-shelf camera units respectively.

Light weight, standard, high quality， digital， true color camera units have been chosen to reduce overall system cost, provide reliable availability of component supply to guarantee an upgrade path as camera improvements are performed by their manufacturers and benefit the UAV system. The most innovative and interesting camera system consists of 4 camera units, diametrically and divergently opposed to each other on a cross axis at 90°so as to produce a wide angle image coverage. Imagery is stored on board and not transmitted to the ground because of noise interference to the signal. The camera shutters of each individual camera unit are tested to determine their mechanical delay. This information, as well as the 16 msec time interval required for the UAV to travel the distance caused by the physical separation of the 2 cameras recording imagery in the direction of flight, provide a time adjustment to electronically and automatically release the shutter over the exact area on the ground at the desired location. The image produced covers a typical butterfly pattern, with less divergence in the flight direction as in the cross flight direction so that the raw image extends further across line. The four separate image components are each rectified and joined at the common overlaps and cut to maximum size to create a virtual image with central projection. See fig. 5.

Calibration of the camera unit is done with imagery from the actual area of interest taken during the flight and an individual error model is applied to every image of the project. Camera self calibration is calculated by using 3 dimensional coordinates of sufficient GCPs from site and the calculation is included in a routine of the SW module MAP-AT.

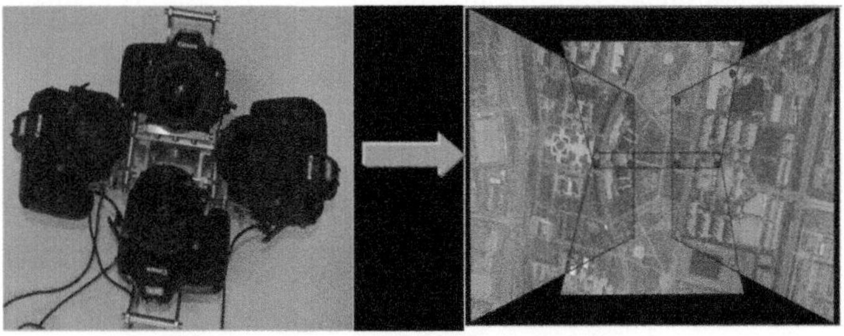

Fig. 5. wide field of view camera system LAC and its typical imagery

E. UAV Software Package

This digital photogrammetry software suite has been specifically designed, using proven algorithms and much automation, to specifically process this UAV imagery to produce a DSM as quick as possible and suitable for the determination of height differences as the identifier of topographic change. One of the specific challenges of UAV image processing is the irregular divergence from the vertical with which the imagery is taken. The SW has been designed to accommodate and process oblique imagery with a far greater angle of divergence than tests to date have revealed. These processes, based on the correlation of redundant points in many images, aid the automatic image correlation. The digital photogrammetry software suite consists of the following 6 modules:

- *MAP-AT 2.0+*

This module is for aerial block triangulation and adjustment and camera self-calibration of the UAV virtual imagery. It uses GPS values for approximation camera position only to aid convergence but will be extended to include GPS camera principal point coordinates within the adjustment and hence reduce GCPs.

- *MAP-DEM*

This module generates and allows the editing of the DEM and builds the difference between two DEMs. This TIN terrain form, is later used for the production of digital orthophoto maps, DOM.

- *MAP-DOM*

This module creates individual orthophotos and builds a mosaic over the total area of interest. The DOM is used as a medium to record the areas of change.

- *MAP-DSM*

The design of this module produces image correlation of every pixel of interest of the imagery. Pixels which do not contribute to the value of the DSM are eliminated by filtering. This dense carpet of points defines planimetric positional change as well as height changes to a high degree of definition and the module is designed for fast convergence and speed.

The height differences between different DSMs, called M-DSMs and based on images from different epochs, is the basis of the change detection. A threshold of height difference is able to be set as a parameter depending on the site characteristics.

- *MAP-DLG*

This is the module which contains the vector information recording the new situation in the field and enables data insertion, manipulations with different GIS systems.

3 City Building Change Detection by M-Dsm Method

A. Application Scenario

With the fast urbanization in China, more illegal buildings are being erected in cities. For example, in the new airport area of Wuhan city, some residential buildings were about to be demolished and a lot of dwellers rushed to erect new buildings for the purpose of getting demolition compensation, causing unnecessary trouble to the local city planning administration. Although city topographical change detection by remote sensing methods with manned aerial vehicles has often been used, say in Japan, Wuhan city authorities were advised on the possibilities offered by high resolution imagery from a UAV platform from which a DSM with accuracy of +/- 30cm could be generated to identify building changes of +/- 1m. This method proved to be very effective in Wuhan.

B. M-DSM Method

Building change detection using UAV imagery is developed from the practical city management requirements. This method takes the advantages of easy, flexible and low-cost UAV remote sensing, using DSMs from low altitude UAV imagery obtained in 2 different epochs. The algorithm is DSM of epoch2 minus DSM of epoch1 which derives the DSM differences. The M-DSM principle is illustrated by the diagram below:

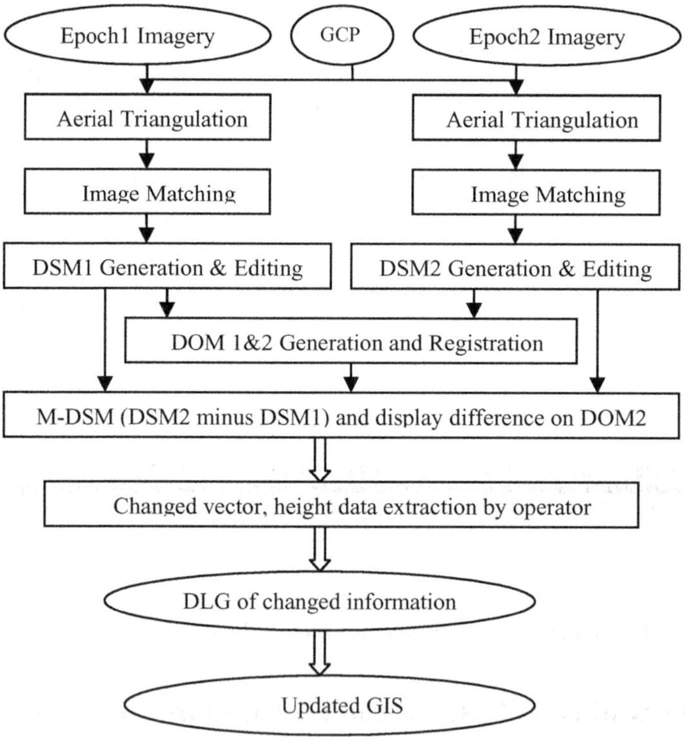

Fig. 6. M-DSM method workflow

C. The Technical Design

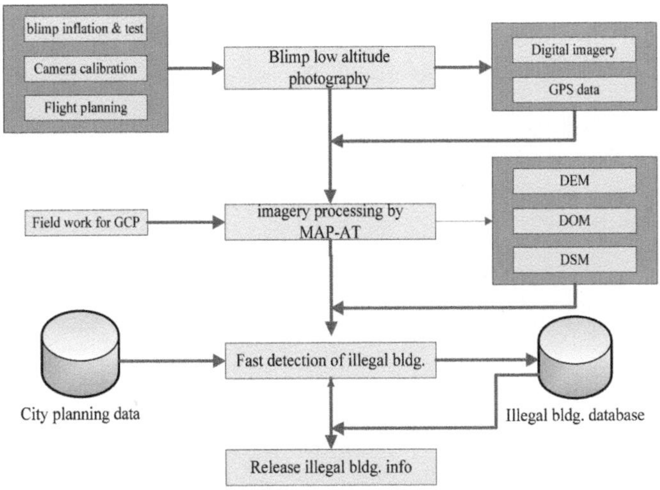

Fig. 7. Technical route of employing UAV system to detect illegal building

4 The Analysis of the Integrated UAV System

A. Field Operation Aspect

The overall concept of a UAV platform, wide field of view camera and stabilizer, in each of several variations, with processing software for the imagery, forms a complete and well integrated system. Parameters which influence the primary choice of UAV system in capturing imagery for change detection purposes are:

- Size of project area
- Choice of camera system to achieve most coverage with least images to enable the technical specification of accuracy, deadline of delivery etc. to be met
- Expected weather conditions.

Low altitude of the UAV, slow UAV ground speed, fast camera shutter speed, large image overlap, tolerance in maintaining the desired flight line etc. are practical measures which are considered in the operational planning and project execution.

B. DEM Accuracy Aspect

The accuracy of many photogrammetric projects depends on the results of the aerial triangulation (AT) and much care and good methods are required to achieve the best possible results of the AT adjustment. The special conditions of UAV virtual imagery created from original oblique imagery, with deviation of the vertical axis exceeding normal acceptance levels, a large number of images with a relative small ground coverage, calibrated non-metric cameras, imagery unable to be compensated for drift etc. are balanced to a certain extent with sufficient and well spaced GCPs and numerous redundant observations.

Experimental AT results show less accuracy than should be expected.

Whereas photogrammetry theory would show after AT work:

Sigma null, XY = +/- 0.3 GSD and
Sigma null, Z = +/- 0.5 GSD (1)

The results obtained after the AT of the UAV imagery with the LAC04 camera have been

Sigma null, XYZ = +/- 2 GSD (2)

Related to LAC04 imagery from an altitude above ground level of 600m and a GSD of 0.20m, accuracy after AT is:

Sigma null, XYZ of +/- 0.4m (3)

This initial AT error would have an effect on the accuracy of the DEM and so Sigma null, Z DEM would be in the order of +/- 0.5m (4)

More significant is the reliability of the differences of the epoch DEMs to detect change and does a height error of 0.5m reduce the ability to detect change from UAV imagery? This aspect still has to be investigated and will be conducted on the basis of the projects completed and the experienced gained to date. Notwithstanding the above, efforts are being made to improve the AT results. Work is being done to include GPS measurements of the camera principal point in the AT adjustment and to record better stereo imagery with improved camera stabilization.

5 Conclusion

A complete and low cost UAV photogrammetry system has been developed and includes hardware and software and a proven work flow for fast city topographical change detection. Comparing with other change detection method, M-DSM method takes full advantages of automatic geometric measuring and calculation capability in digital photogrammetry system, leading to fast and automatic operation and results provision. Experience has been gained which have proven the UAV system merits of the easy operation, flexibility, ability to perform projects frequently and in a short working cycle, with non-professional flying personnel etc.. The resulting change detection information is sufficient and has been used by city authorities and contributed to their city management and decision making.

Operational and accuracy improvements can be gained from modifications to existing components such as the camera stabilizer and the inclusion of the GPS position of the camera principal point in the adjustment. A further component which may aid the automatic detection of change is radiometric change detection, say of vegetation, water condition, ground cover etc. recorded at the two different epochs. This would require a different camera sensor, say a multi-spectral/IR sensor and may be a future addition to the overall system. Such imagery could be processed by existing commercial remote sensing software packets

New developments to support the UAV platform and add new applications could be the miniaturization of airborne LiDAR, with IMU and GPS, to be small, light enough to be carried by a UAV platform. An operational and logistics improvement can be considered using PPP (Precise Point Positioning) post processing of GPS results so that precise satellite orbits, satellite clock correction and modern error modelling eliminate the requirement to have a DGPS unit on the ground in the area of interest and for the duration of the flight.

Acknowledgment. This paper has acquired the fund support from Key Lab of Earth Exploration & Information Techniques of Ministry of Education in Chengdu University of Technology.

References

1. Xuan, W., Wang, D., Sun, J.: Make Wide-Angled Aerial Camera by Combining Small-Sized Area Array CCD Sensors. ISPRS XXth Commission I, 501–506 (2004)
2. Cui, H., Lin, Z.: Non-metric CCD Camera Calibration for Low Altitude Photogrammetric Mapping. In: 8th International Conference on Electronic Measurement & Instruments, pp. 270–278 (2007)
3. Wang, C., Lin, Z.: A research on aero-triangulation in UAVRS Image. Science of Surveying and Mapping 132(14), 41–43 (2007); ISSN 1009-2307

Based on the Theory of Principal-Agent Model of Enterprise Outsourcing Services Platform's Game Complexity Study

Dong Xinjian[1,2] and Ma Junhai[1]

[1] School of Management, Tianjin University,
Tianjin 300072, China
[2] Shandong PEONY Technology cultural Development Co., Ltd.,
Shandong 250100, China
Dongxinjian1978@tom.com

Abstract. This paper discusses the process of outsourcing in the service platform for the principal-agent relationship between enterprises and ASP and establishes an analysis theoretical framework of the relationship between the enterprise and ASP. Under conditions of asymmetric information, the game between the enterprise and ASP is studied under conditions of asymmetric information, and how the enterprise designs appropriate incentives to prompt ASP to provide high level services and to maximize the benefit of enterprise and ASP so that the enterprise and ASP could have a victory for both sides are also studied.

Keywords: Principal-agent theory, ASP, game complexity.

1 Introduction

Since the late 20th century, with the information technology widely applied in economic, social and other fields, e-commerce has become a compulsory election for enterprises to increase efficiency, improve management and service levels. But some enterprises are in face of some restrictions in e-commerce constructions, such as the lack of IT personnel, inadequate infrastructure, limited financial capacities. With the social development and progress, enterprises pay more attention to their core competencies and outsource some low value-added common professional services to some professional services organizations so as to get the most profitable. In 1999, the first application service provider appears in the world. Application Service Providers (ASP) is third-party companies which provide customers with application services from the data center based on a central server implementation of the management application software through the wide area network by hiring. ASP provides application services to customers while the customer gives a payment according to usage or subscriptions, simultaneity, ASP manages centralized and maintains applications and data. Thus, a principal-agent relationship is formed between ASP and the client. ASP pattern has a huge economic benefit in improving the organizational development, operating and maintaining IT applied systems. However, many potential risks which can not be ignored exist in ASP pattern. These risks may be due

J. Zhang (Ed.): ICAIC 2011, Part V, CCIS 228, pp. 606–613, 2011.

to some uncertainty, the information asymmetry between customers and providers or the properties of the application itself (such as data security). Therefore, principal-agent problem exists between the client and ASP. Principal-agent theory was widely studied in the late 60's of the twentieth century. Jensen&Meckling [1]think principal-agent relationship is the contractual relationship that a person or persons (principals) entrust a person or persons (agents) to engage in certain activities according to the principals' interests, and accept certain agents' corresponding decision-making authority." In the relationship, the principal want to maximize the choice of action of the agent in accordance with the principal's interests, but because of uncertainty, information asymmetries, transaction costs and other factors, the agent's choice can not be directly observed by principal, while only the results of actions can be observed, the results depend on the agent's actions and exogenous random factors, which will inevitably lead to "agency problems", that is the action of the agent does not always avail to the principal interests, in some cases, the agent will damage the interests of principal, this is called moral hazard[2]. This problem also exists between the enterprise and ASP, so domestic and foreign scholars have done a lot of research in recent years. FAN Zhi-ping, WANG Yan [3] for the IT outsourcing decision problem between a user and single developer, that paper proposed a game analysis approach based on some hypotheses. This approach was to solve a two-people non-cooperative game problem so as to obtain outsourcing strategy of the user and developing strategy of the developer. The research results of this paper provide decision basis for making IT outsourcing decision. Benoit A. Aubert Suzane Rivard [4]combined transaction costs and incomplete contract theory, assuming that the process of outsourcing asset specificity, uncertainty, business and technical capacity to influence the level of service outsourcing. They also had proposed a IT outsourcing decision-making model. Taking ASP outsourcing model for mobile information work in the product strategy as an example, ZHANG Hao, ZHI Yong [5] studied the dynamics multi-objective game problem between the ASP and user, and used catastrophe progression method to normalize the relevant targets, then, solved the game model of system dynamics as a tool, thus simplified the problem as a static single-target model. Time-varying game equilibrium strategy under the condition of different indicators for the proceeds was solved using simulation method, so that it provided a reference for decision-making.

Scholars from different angles and different approach had done a wide-ranging and in-depth study about the relationship between the enterprise and ASP, but so far there has not been principal-agent theory to analyze the relationship between the two. This paper takes business process outsourcing in the service platform for enterprise and ASP as the research object and uses the principal-agent theory for quantitative research the game relationship between the two, moreover, establishes an analysis theoretical framework of the relationship between the enterprise and ASP. Under the conditions of asymmetric information, this paper studies how the enterprise designs possible mechanisms which can be implemented to maximize the expected utility of the contract and to maximize the benefit of enterprise and ASP.

2 Principal-Agent Theory Analytical Framework for Enterprise and ASP

Model analysis method is to build the incentive constraint sets between enterprise and ASP under certain constraint conditions, ultimately, the problem reduces to an optimal programming, then, Lagrange method can be used to solve the optimal programming problem. In order to build models and simplify the analysis, we refer to Jean-Jacques Laffont and David Martimort of Multitask incentive problems analysis, as well as Hal-Varian relevant research results and make the following assumptions.

A. Basic Assumption

Participation is enterprises (principal) and ASP (agent), both of which pursue their own maximum benefit.

Assumption 1: Enterprise is risk-neutral, so we can use a linear function to represent the utility enterprises.

Assumption 2: ASP's utility function is u, and ASP is risk averse, $u^{'} > 0$， $u^{''} < 0$.

Assumption 3: ASP's behavior and the state of nature co-deciding the revenue v, v=v(α, θ), $\partial v / \partial \alpha > 0$, $\partial v / \partial \theta > 0$. θ is a random variable, and α is decided by the ASP's level of effort. A is the set of all actions for the ASP. Only v can be observed by the enterprise, this assumption reflects the asymmetry of information. Enterprises can not judge the action by ASP, because the information of an important variable α is a disadvantage relative to the ASP for enterprises.

Assumption 4: F(v, α) and f(v, α) are the distribution function and density function of v. Assume that F(v, α)and f(v, α) are differentiable on α and f(v, α)= $F^{'}(v, \alpha)$.

Assumption 5: s(v) is a reward that company paid for ASP, and s(v) will change with the changes in v, rather than paying a fixed remuneration to ASP, $s^{'}(v)>0$. c is a negative effect about ASP action(cost) and c=c(α, θ), $c^{'}>0$, $c^{''}>0$.

B. Model Formulation

If the principal wants to make sure the contract can be executed, which is the agent agree, the expected utility of the agent's adoption of the task should not be less than the utility value (opportunity cost) of other tasks, such is individual rationality constraint (Hereinafter referred to as IR). The contract does not include α as it is not be observed by the client, then the expected action of the client must accord with the benefits of the agent, that is the incentive compatibility constraint, such is incentive compatibility constraint (Hereinafter referred to as IC). The expection of each principal can only be realized by the maximum of the principal's utility. Between the enterprise and ASP relationship, enterprises as principals design the method of payment compensation for ASP to stimulate ASP to improve the service level, then bring about enterprise's expected utility maximization. With "the parameters method of distribution function" denotes a generalized model of the above situation as follows:

$$\max_{\alpha, s(v)} \int V(v - s(v)) f(v, \alpha) dv$$
$$\alpha + \beta = \chi. \tag{1}$$

ASP as an agent, if the enterprise accepted the contract, its individual rationality constraint is as follows:

$$\text{IR} \quad \int U[s(v)-c)]f(v,\alpha)dv \geq \bar{U} \tag{2}$$

In the (2), $\bar{U}$ is reservation utility for the ASP. ASP accepts the contract to obtain the expected utility can not be less than the maximum expected utility obtained if ASP does not accept the contract. When the ASP accepts the contract, he chooses their own actions to make their expected utility maximized, then its incentive compatibility constraint is as follows:

$$\text{IC} \quad \int U[s(v)-c)]f(v,\alpha)dv \geq \int U[s(v)-c)]f(v,\alpha^{'})dv \quad \forall \alpha^{'} \in A \tag{3}$$

$$\alpha + \beta = \chi.$$

Inequality (3) shows that only action α the ASP chooses to obtain the utility is greater than the expected utility to choose action $\alpha^{'}$.

In the case of asymmetric information, the relationship between the enterprise and ASP is a typical three-stage dynamic game of incomplete information and it is shown in Figure 1.

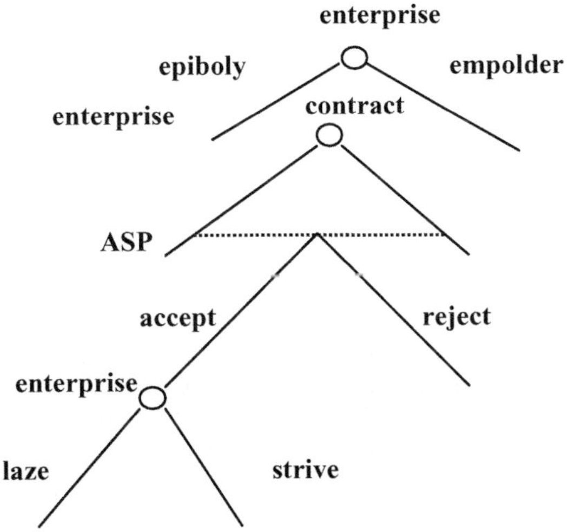

Fig. 1. The game of enterprise and ASP

According to the enterprise's reward program, ASP chooses the best one to maximize their benefits. Firstly, the enterprise designs a contract; secondly, ASP chooses whether to accept the contract designed for enterprise. If the ASP rejects the contract, the exogenous reservation utility can be gained; thirdly, after accepting the contract, the ASP can choose an action on the basis of the contract. According to proof of Mirrlees & Holmstrom, when dealing with incentive compatibility constraint

(3), we can use "first-order approach", which the first-order conditions of (3) replaces the conditions of (3) to solve it. Then the model becomes:

$$\max_{\alpha} \int V[v(\alpha,\theta)-c(\alpha,\theta)]f(v,\alpha)dv \qquad (4)$$
$$\alpha + \beta = \chi.$$

$$\text{s.t. (IR)} \int U[s(v)-c)]f(v,\alpha)dv \geq \bar{U} \qquad (5)$$
$$\alpha + \beta = \chi.$$

$$\text{(IC)} \int U[s(v)-c]f_{\alpha}(v,\alpha)dv = 0 \qquad (6)$$
$$\alpha + \beta = \chi.$$

where $f_{\alpha}(v,\alpha) = \partial f(v,\alpha)/\partial\alpha$.

C. The Inherent Complexity of the Model

Proposition 2.3.1 : In the case of asymmetric information, enterprises can observe benefit v(α, θ) to speculate ASP's action α and judge ASP service level which is high or low, then adjust the compensations for ASP to reward hard-working or punish laziness. ASP service level is high and reward raises; otherwise, reward lowers.

Proof: Supposing ASP service level is divided into α_H (high) and α_L (low), enterprise certainly hopes that a higher level of ASP to provide services, at this point the model incentive compatibility constraint is as follows:

$$\text{IR} \quad \int U[s(v)-c]f_H(v,\alpha_H)dv > \int U[s(v)-c]f_L(v,\alpha_L)dv \qquad (7)$$

$f_H(v,\alpha_H)$ and $f_L(v,\alpha_L)$ are the probability density of v that ASP level of service for α_H and α_L , respectively, so that$\eta$and$\delta$are Lagrange multipliers of the individual rationality constraint and incentive compatibility constraint.

Constructing Lagrange function:

$$\begin{aligned} L[s(v)] = & \int V[v(\alpha,\theta)-c(\alpha,\theta)]f(v,\alpha)dv \\ & +\eta[\int U[s(v)-c)]f(v,\alpha)dv - \bar{U}] \\ & +\delta[\int U[s(v)-c]f_H(v,\alpha_H)dv \\ & -\int U[s(v)-c]f_L(v,\alpha_L)dv] \\ & \alpha + \beta = \chi. \end{aligned} \qquad (8)$$

The optimal first-order conditions are:

$$\frac{V'[v-s(v)]}{U'[s(v)-c]} = \eta + \delta\left[1 - \frac{f_L(v,\alpha_L)}{f_H(v,\alpha_H)}\right] \qquad (9)$$
$$\alpha + \beta = \chi.$$

In the equation (9), if $\delta = 0$, the equation becomes an optimal risk-sharing. Namely, business gives ASP fixed remuneration so the ASP does not bear any risk. However, this situation violates the incentive compatibility constraint, because under the fixed remuneration the ASP will definitely not be taken a high level of service conduct. Otherwise, the ASP's cost will increase. Therefore, under asymmetric information, δ is strictly positive number, $\delta > 0$, this means that compensation contracts will change as v changes.

f_L / f_H is called the likelihood ratio. If $f_L / f_H > 1$, the probability of v_0 when the ASP chooses low-level services is greater than it chooses high-level ones; If $f_L / f_H < 1$, the probability of v_0 when the ASP chooses high-level services is greater than it chooses low-level ones; If $f_L / f_H = 1$, the probability of v_0 when the ASP chooses α_H and α_L is same, thus, it is difficult for enterprises to make accurate judgments.

Under the conditions of information symmetry, when the ASP level of service α is fixed, Pareto-optimal risk-sharing can be achieved. The changes of the rewards that enterprises pay to ASP will be in the same direction with the changes of the enterprises benefiting from the ASP. At this point the model does not contain incentive compatibility constraint:

$$\max_{s(v)} \int V[v - s(v)] f(v, \alpha) dv \qquad (10)$$
$$\alpha + \beta = \chi.$$

$$\text{s.t. (IR)} \quad \int U[s(v) - c)] f(v, \alpha) dv \geq \bar{U} \qquad (11)$$
$$\alpha + \beta = \chi.$$

Constructing Lagrange function:

$$L[s(v)] = \int V[v - s(v)] f(v, \alpha) dv + \eta \left[\int U[s(v) - c] f(v, \alpha) dv - \bar{U} \right] \qquad (12)$$
$$\alpha + \beta = \chi.$$

The optimal first-order conditions are:

$$\eta = \frac{V'[v - s(v)]}{U'[s(v) - c]} \qquad (13)$$
$$\alpha + \beta = \chi.$$

Here, because the individual rationality constraint equation conditions are met, η is strictly the normal constant. The optimal condition shows the ratio of marginal utility for enterprise and ASP is a constant.

Set up $b_{(13)}(v)$ for the type (13) determined the most risk-sharing contracts, $b_{(9)}(v)$ for the type (9) decided to contract. Comparing of (13) and (9), then:

If $f_L/f_H \geq 1$, $b_{(9)}(v) \leq b_{(13)}(v)$; If $f_L/f_H < 1$, $b_{(9)}(v) > b_{(13)}(v)$. Thus, the proposition is proved.

The analysis above shows that the enterprise benefits become higher as a result of the services provided by ASP, and the higher the service level of ASP is, the more reward should be given to ASP. Otherwise, the ASP got less reward as a punishment by the enterprise.

3 Conclusion

In this paper, after ASP gets the outsourcing service platform, the problem of the enterprise, how to adjust compensation to stimulate higher level of service ASP, are analyzed. Symmetric information is only used in the theory for comparison because it is rare in the reality. In the case of asymmetric information, the enterprise gives an estimate to ASP according to the benefits due to their service level. And the remuneration is adjusted combined with the development of the information industry to improve the ASP's positivity and service level. The enterprise may bear more risks because of the mistakable judgment under the condition of asymmetric information. In order to improve ASP service platform outsourcing model's implementation better, and pushing forward the development of productive forces and the promotion of economic growth, the following aspects should be focused on efforts: Firstly, to further improve the market economy system, give full efforts to the price mechanism of the signal function and regulatory function. Secondly, the government should increase support for the information industry so that the overall progress of the industry can contribute to raising the level of ASP services. Thirdly, formulating appropriate laws and regulations should be provided a good regulatory framework for enterprise and ASP. Particularly, protection of intellectual property and maintenance of enterprise data security should are necessary. Fourthly, enterprise is required to enable the sharing business philosophy and culture of the ASP, with this soft environment giving ASP behavior of incentives and constraints, so that the enterprise and ASP could have a victory for both sides.

References

1. Jensen, W.C., Meckling, W.H.: Theory of the Firm Managerial Behavior, Agency Costs and Ownership Structure. Journal of Financial Economics 3(4), 308 (1976)
2. Tian, K.: A Study of R&D Outsourcing: A Framework on the Theory of Principle-Agent. Journal of Hunan University of Science & Technology Journal of Beijing Normal University(Social Science Edition) 1, 90–93 (2007)
3. Fan, Z.-p., Wang, Y.: A Game Analysis Approach to IT Outsourcing Decision Problems. Journal of Industrial Engineering and Engineering Management 3, 5–7 (2002)
4. Aubert, B.A., Rivard, S., Patry, M.: A transaction cost model of IT outsourcing. Information & Management 21, 921–932 (2004)

5. Zhang, H., Zhi, Y.: A Study on dynamic Multiobjective Game with Information in ASP Mode. China Management Informationization 11(16), 85–87 (2008)
6. Bajari, P., Steven, T.: Incentives versus transaction costs: a theory of procurement contracts. Rand Journal of Economics 21, 387–407 (2001)
7. Wu, X.-b.: Major Problems in Entrusting-Acting Relationship under Message Asymmetry. Journal of Beijing Normal University(Social Science Edition) 5, 112–116 (2005)

A Study on Competitiveness Evaluation of the Electronic and Information Industrial Cluster in Jinan

Dong Xinjian[1,2] and Ma Junhai[1]

[1] School of Management, Tianjin University, Tianjin 300072, China
[2] Shandong PEONY Technology cultural Development Co., Ltd., Shandong 250100, China
Dongxinjian1978@tom.com

Abstract. In this paper, GEM model, a quantitative analysis tool to analysis and research competitiveness of industrial clusters, was used. Considering the characteristics of Jinan electronics and information industry, the competitiveness evaluation model of the electronic information industrial clusters in Jinan was built. Then the competitiveness was evaluated, and the enhanced solutions were proposed.

Keywords: Industrial Cluster, Competitiveness, GEM Model.

1 Introduction

The purpose of researching the competitiveness of industrial clusters do not just to objectively describe the actual results of industry clusters' competition, and more important but also to find factors which could determine or affect the competitiveness[1]. Building the right Evaluation Index System is a key component of evaluating the competitiveness of industry clusters. In this paper, GEM model is used to build the evaluation index system of the competitiveness of electronic information industrial clusters. Considering the characteristics of Jinan electronics and information industry, and in accordance with the classification principles, the Evaluation Index System was built. , It is divided into 3 factors, 6 major categories, 22 indicators. The result showed this kind of classification was reasonable and measured the competitiveness of electronic information industrial clusters very well.

2 Type Style and Fonts

Padmore and Hervey Gibson (1998) improved Porter's "diamond model" and proposed an analytical model of the competitiveness of enterprise clusters --GEM model. GEM model identified six factors that affect the cluster. The factors are "Establishment", "Resource", "Supporting Industries", "Enterprise Structure, Competition and Strategy", "Local Market", "External Market". These six factors are classified to three pairs: the first pair called "Foundation" includes "Establishment" and "Resource"; the second called "Enterprise" includes "Supporting Industries" and "Enterprise Structure, Competition and Strategy"; the last pair called "Market" includes "Local Market" and "External Market"[2].

J. Zhang (Ed.): ICAIC 2011, Part V, CCIS 228, pp. 614–618, 2011.

3 Ease of Use

The Evaluation Index System of the competitiveness of electronic information industrial cluster in Jinan City (Table 1) was built in this paper. Based on the research at home and abroad and according to other industries' Evaluation Index Systems [3, 6], 22 indexes were screened eventually. They composed the Evaluation Index System reflecting the features of electronic information industrial cluster in Jinan City. They are divided into 3 factor pairs, totally 6 classify indexes. The result showed this kind of classification was reasonable and measured the competitiveness of electronic information industrial clusters very well.

First, calculates sample means of each indicator according the scores made by experts (Table 2).

Second, determinates the weight of each indicator. Determining the indicators' weight is one of the key problems in judging if the evaluation result is reasonable [6]. Generally, the weight should be determined based on the indicators' relative importance to the evaluation result. There are many methods to weight. In this paper AHP was used. Considering there are too many indicators to compare every pair, those interviewed were just asked to evaluate every factor. Then the judge matrixes were instructed on the basis of scores' means. The calculation about the judge matrixes were completed by MATLAB, and the result is shown in Tab.2.

In the next step, every factor pair's score is as the follows:

Foundation:

$$\text{PAIR SCORE} = (A+B)/2 = (6.18+6.29)/2 = 6.24 \tag{1}$$

Enterprise:

$$\text{PAIR SCORE}=(E+F)/2=(7.30+7.61)/2=7.46 \tag{2}$$

Market:

$$\text{PAIR SCORE}= (C+D)/2= (5.80+5.15)/2=5.66 \tag{3}$$

Finally, calculate the GEM score:

$$\text{GEM}=2.5\times[(A+B)\times(C+D) \times(E+F)]2/3=499.9\approx 400 \tag{4}$$

Where the score of foundation is the evaluation of establishment and resource, the score of enterprise is the evaluation of the supporting industries, enterprise structure competition and strategy, the score of market is the evaluation the local market and external market.

The GEM score of the electronic information industry cluster in Jinan is 400 points. The total score indicates that the competitiveness of the electronic and information industrial cluster in Jinan has achieved above the national average, but the overall competitive advantage is not obvious. This is consistent with the reality of the situation. In the second place, the score of "External Market" is the highest. That is mainly reflected in high scores of the "external market's prospects for development" and the "external market demand".

In addition, the score of "Capital Resource" is high too. Jinan City is the capital of Shandong Province. As the center of politics, culture and economy, Jinan City owns strong economic strength (Fig.1). The economy of Jinan City developed rapidly. Since

2001 to 2008, the average annual GDP chain index of Jinan City is 15.0, 4.9 percent points than China's GDP. It is in the higher level than average. The economic Advantage of Jinan City is very prominent. Above all of these, the rapid economic growth of Jinan City provided enough capital resources for the development of electronic and information industry.

4 Copyright Forms and Reprint Orders

These can be seen through the specific analysis of each factor of the electronic and information industrial cluster:

(1) The government should enact and improve rules and regulations.
In the cores of valuation indexes of "facilities", the core of the "government policies, laws and regulations," is lower than others. With the development of the electronic and

Table 1. Score and Weight

Target	Rule	Indicator	
The competitiveness of the electronic and information industrial cluster in Jinan	Foundation	Resource	A_1 Market location
			A_2 Capital Resource
			A_3 The situation of human
		Establishment	B_4 Traffic facilities
			B_5 Policies
			B_6The IT application standard
			B_7 Business environment
			B_8 The number of college and
			B_9 living environment
			B_{10}The situation of intermediary
	Enterprise	Supporting Industries	C_{11} Allied industries
			C_{12} The situation of cooperation
		Enterprise Structure, competition and Strategy	D_{13} enterprise system
			D_{14} Enterprise Size
			D_{15} The advantage of Brand
			D_{16} The situation of cooperation
	Market	Local Market	E_{17} The competition among local
			E_{18} The development prospects of
			E_{19} The demand of local market
		External Market	F_{20} The development prospects of
			F_{21} rampart in external market
			F_{22} the demand of external market

Table 2. Score and Weight

Index	Mean	Weight	Score
A_1	6.51	0.072	6.18
A_2	7.01	0.279	
A_3	5.79	0.649	
B_4	6.53	0.032	6.22
B_5	5.71	0.265	
B_6	6.54	0.316	
B_7	6.68	0.074	
B_8	6.03	0.142	
B_9	7.41	0.085	
B_{10}	5.32	0.085	
C_{11}	5.03	0.259	5.15
C_{12}	5.14	0.075	
D_{13}	5.68	0.441	5.80
D_{14}	5.71	0.260	
D_{15}	6.61	0.170	
D_{16}	5.33	0.129	
E_{17}	7.01	0.339	7.30
E_{18}	7.45	0.589	
E_{19}	7.37	0.072	
F_{20}	8.01	0.218	7.61
F_{21}	6.54	0.110	
F_{22}	7.65	0.672	

information industrial cluster in Jinan, Jinan government is working to give policy support to it. But because of that the electronic and information industry emerged in the 1970s, the government has less experience. To sum up the rules and regulations set by the government should be texted and improved.

(2) Paying attention to the absence of intermediary service organizations.
"The situation of intermediary service organizations" has the low score. The intermediary service organizations include accounting firms, law firms, executive search firms, marketing, leasing companies, consultative bodies, business association and the service organizations supporting to the high-tech companies' operation. While

the world is generally associated, the development of a cluster requires the support of improved social service system. If the role of intermediary service organizations does not play fully, the industrial cluster effect will be certainly restrained. Therefore, the absence of the intermediary service organizations should be paid attention to.

(3) Strengthening the cooperation among the enterprises and between enterprises and the research institutions in the industry clusters.

Of the above 6 factors, "Supporting Industries" and "Enterprise Structure, competition and Strategy" got the lower score. The reason is that there are three factors holding back the development of the electronic and information industrial cluster in Jinan. First is the clusters' relatively simple structure of the enterprise, second is the low cooperation among enterprises, the last one is the cooperation between enterprises and research institutions. Thus, adjusting the industrial structure and strengthening the cooperation among enterprises and between enterprises and research institutions are the two efficient means for improving the competitiveness of the electronic and information industrial cluster in Jinan.

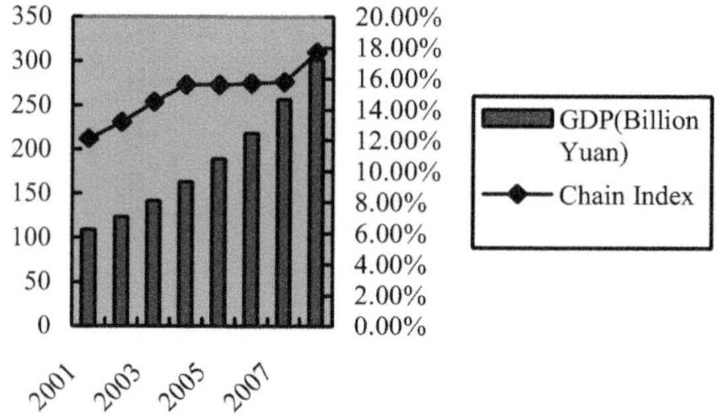

Fig. 1. GDP and Chain Index of Jinan City (2001-2008)

References

1. Jin, P.: Economics of competitiveness. Guangdong Economic Press, Guangzhou (2004)
2. Tim, P., Hervey, G.: Modeling system of innovation: A framework for industrial cluster analysis in regions. Research Policy (1998)
3. Wu, N.: On Bio-techindustrial Cluster and Its Accessment Model of Competitive Power. Journal of Hubei University of Technology (12), 65–67 (2008)
4. Jiang, L., Wu, R., Liu, H., Li, H.: Systematical analysis of industrial cluster competitiveness and the evaluation index. Economic Geography (1), 37–40 (2006)
5. Saaty, T.L.: Decision making with dependence and feedback. RWS Publication, Pittsburgh (1996)
6. Jin, J., Wei, Y.: Generalized intelligent assessment methods for complexsystems and applications. Science Press, Beijing (2008)

Medical Eye Image Detection System

Hongjin Yan[1,2] and Yonglin Zhang[1]

[1] Institute of Optoelectronic Engineering, Jinan University, Guangdong, Guangzhou 510632, P.R. China
[2] Department of Biomedical Engineering, GuangZhou Medical University, Guangdong, Guangzhou 510182, P.R. China
HongjinYan1972@126.com

Abstract. In order to detect the eye for medical needs, an eye detection system was designed, and an algorithm detecting the pupil was proposed. The system included the image acquisition device, the stimulatory light source and the software edited with Visual C++6.0. The stimulatory light source was controlled by a personal computer using serial communication. The algorithm was based on pupil geometry and gray features, and used mathematical morphology operations. Firstly the images were taken by the CCD camera, then transferred to the computer, and processed by the algorithm. Finally the location, size and dynamical characteristics of the pupil were presented. It is pointed that the system can acquisite clear images of bilateral eyes synchronously under medical conditions, and process them fast. The results show that the system can effectively detect and evaluate the eye by the pupil size and the change of pupillary response to light. It is an objective, accurate, quantitative method to detect the pupil, so it provides physicians with reliable diagnostic clues.

Keywords: Pupil, Image Acquisition, Stimulatory Light, Morphology, Hough Transformation.

1 Introduction

Eyes are the most important human senses. The characteristics of eyes is closely linked with health status of people, and especially the pupil eigenvalues of the eye perform crucial roles in diagnosis of ophthalmology and neurological diseases[1,2]. The pupillary light reflex causes the pupil to reduce its area in response to a bright light, thereby providing the earliest control of light intensity reaching the retina. Because it is easy to elicit and observe, the light reflex has been a handy clinical tool with which to evaluate damage to the CNS(central nervous system) and to assess the level of consciousness in humans[3]. The modulation of pupillary diameter might provide an indirect mean to assess of the integrity of neuronal pathways controlling pupil size[4]. At present the traditional method of pupil detection is to observe the pupil and give qualitative descriptions, or use simple measuring tools for manual inspections, thus, the result is subjective, inaccurate. It is dependent on the examiners' skill and environmental conditions. To tackle these issues, this study orients at providing an objective, accurate, quantitative approach to detect pupil. The current study about it is rarely. In this work, an image acquisition system was constructed, and series of eye

J. Zhang (Ed.): ICAIC 2011, Part V, CCIS 228, pp. 619–627, 2011.

images were captured under controlled conditions. Then, these images were processed by the proposed method, so the eigenvalues of pupil were presented, including the center, radius, area of pupil and the dynamical characteristics with time. During the whole procedure, image processing was very important. The eye image of this research was different from that of iris recognition, because it was taken under medical conditions, such as the intensity-changed light source, different wavelengths of light, pathological eye, etc. Therefore, the image acquisition system must meet medical requirements, and those additional disturbances should be taken into account in image processing. In order to detect pupil fast and precisely, reducing of search range and extracting of pupil edge were crucial. It was carried out by the proposed method, based on pupil geometry and gray feature, and using mathematical morphology operators.

Mathematical morphology is a tool for extracting image components that are useful for representation and description. The technique was originally developed by Matheron and Serra[5] at the Ecole des Mines in Paris. It is a set-theoretic method of image analysis providing a quantitative description of geometrical structures.

The paper is organized as follows. Section 2 describes the system hadware, and the system software is introduced in section 3 subsequently. In section 4, the proposed method of pupil detection is presented in detail, followed by experimental results in section 5. Finally, conclusion is made in section 6.

2 System Hardware

The eye image detection system includes the photoelectric subsystem, image acquisition subsystem and stimulatory light source. The photoelectric subsystem takes images under different light; the image acquisition subsystem simultaneously captures two-way or any one-way image; The stimulatory light source is used to observe and analyze the pupillary light reflex, The intensity and time of stimulation can be controlled. The system block diagram is shown in figure 1.

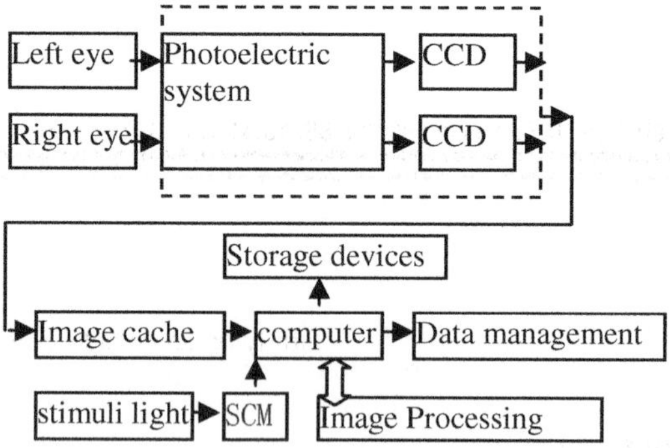

Fig. 1. System block diagram

A. Photoelectric Subsystem

The light source is made up of several infrared(IR) light-emitting diodes(LEDs) and visible LEDs, they are uniformly distributed along the circumference around the camera so that the arrangement was conducive to a clear image acquisition. At the same time bright spots can be uniformly distributed within a region of the image so as to provide convenience in subsequent image processing. The back-ground light is from the ordinary visible LEDs. The light intensity is adjusted by the control circuit so as to obtain a clear image without dizziness, and the different light can be free to switch, Figure 2 shows the principle taking the eye image.

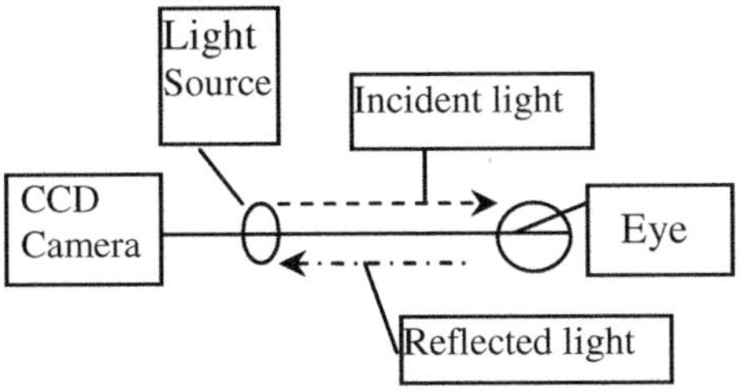

Fig. 2. Principle taking eye image

B. Image Acquisition Subsystem

The image acquisition subsystem consists of CCD cameras, an image acquisition card, a computer and so on. The measured object was illuminated by the light source, so its image is received by CCD camera, and then transferred to the image acquisition card. The Image acquisition card converts the analog signals into digital signals through the analog-to-digital converter, and then the signals pass through the PCI bus after video decoding. Finally, they enter into the computer for processing.

C. Stimulatory Light Source

This section includes stimulatory light source, single-chip microcomputer(SCM) and control circuits. It is divided into left and right stimulus parts, because they are used to stimulate the left and right eyes. The personal computer sent commands to the SCM using serial communication, and then the SCM adjusted the stimulatory light souce by a controlling circuit. The stimulation parameters, including duration, frequency, intensity and mode. There were unilateral stimulation(to single eye) and bilateral stimulation(to both eyes) in stimulatory modes.

3 System Software

The system software is based on windows XP/2003/Vista operating system platform and in the help of Visual C + +6.0 development tools.

The whole image system included image acquisition module, processing and analysis module, control module and so on. The relevant information and data were stored in the database, and the function of each module was as follows.

(1) image acquisition module: acquisition of the eye image. A program was implemented to automatically capture, record and save the images.
(2) light control module: choice of light intensity, the switching time and the stimulation pattern.
(3) Image pre-processing module: adjustment of the image brightness and contrast, image cropping, noise reduction, binarization.
(4) image processing and analysis modules: the image segmentation, the feature region extraction, measurement and analysis of data.
(5) helping module: to provide users with the description of the system performance and operating instructions.

4 Pupil Detection

The eye image acquisited by the system is divided into two categories. One is infrared, and the other is visible-light. Firstly we analysized the infrared image, then the visible-light image. The eye image has a certain gray distribution: Gray value of iris is smaller than that of sclera, as well as gray value of pupil is smaller than that of iris. The pupil region is where gray value is the smallest in the whole image. Gray value within the pupil changes least, but the gray value between pupil and iris or between iris and sclera changes largerly. Although spots of light source, eyelashes and other factors also influence this gray distribution, these impacts are very small because of their small proportion. Therefore, we can determine the pupil boundary according to the characteristics that gray value of pupil is low and uniform. Moreover, the pupil approximates a circle, with circular geometric feature. The method of mathematical morphology is well suited to detect the target with certain geometrical characteristics. Thus we proposed the algorithm using mathematical morphology operators, which is based on gray level and geometric characteristics of the pupil. Finally we obtained the location and size of the pupil by it. Detailed steps are as follows.

Firstly, depending on the gray scale features of pupil, the rough center of pupil was determined by gray projection. Then, based on the rough center, the sub-image that contains the pupil was intercepted with a rectangular window. After that, according to the geometrical characteristics that the pupil approximats a circle, spots were removed by reconstruction with a disc-shaped structural element, Furthermore, the interference of eyelashes was eliminated by the morphological operations. Finally, the location and size of the pupil were obtained after edge extraction and circle hough transformation.

A. Interception of Sub-images and Image Enhancement

The eye is generally located at the middle of the captured image, therefore, a rectangular region centered on the whole image was intercepted, which was called eye sub-image, shown as figure 3(a). The size of eye sub-image is slightly larger than the eye. After this processing, the low gray interference around the eye was eliminated. Subsequently, we determined the rough center of pupil according to the gray scale distribution of pupil. Based on the rough center of pupil, a sub-image was again intercepted, which contained pupil and was called pupil sub-image(see figure 3(b)). The follow-up image processing and calculations were based on this sub-image. Because the interception reduced the search range during image processing, it would improve speed of image processing greatly.

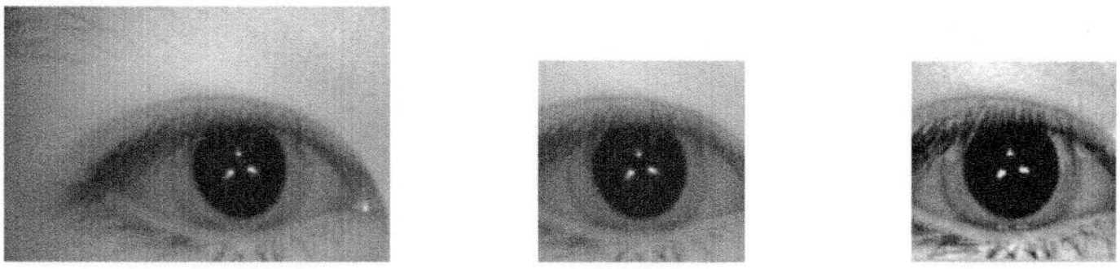

a- eye sub-image　　b- Pupil sub-image　　c- Enhanced image

Fig. 3. Interception and enhancement of image

To highlight the interesting part in the image, top hat and bottom hat transformation were applied[6]. The contrast of the target in the whole image was enhanced so that the edge was more distinct. It was shown in Figure 3(c).

B. Elimination of Light Spots and Eyelashes

Because light spots and eyelashes might influence the pupil gray distribution, and cause interference in image processing, they should be removed from the image. There are some methods commonly used to remove spots, such as filling method and region growing[7,8]. However, these methods need tedious calculation. When opening by reconstruction used in morphology was applied to remove spots, it not only eliminated spots, but also restored the shape of the pupil accurately. The structuring elements used in opening by reconstruction operation should be similar to the target shape. The pupil approximates a circle, so a disc-shaped structuring element was choosen in this test. If the structuring element is B, and the image f was treated by this operation, the symbol was denoted by $R_f\,(f \ominus B)$, The result after this opration is shown in Figure 4(a). This process is not only to achieve the purpose of removing spots, but also to maintain the real outline of the pupil. Compared figure 4(a) with figure 3(b), it showed that the spots disappeared. It is also judged that the pupil edge did not expand and still maintained real from the iris context of figure 4(a).

After the reconstruction operation eliminated spots, there was still interference of eyelashes in the image. Removal of eyelashes also might use mathematical morphology operations. Set Figure 4(a) as A, then select the appropriate structuring element B according to the features that the eyelashes are longer, more intensive and wider. After Implementation of the operation as $((A \ominus B) \oplus B)$ B , the upper and lower eyelashes could be removed, and the result was shown as figure 4(b).

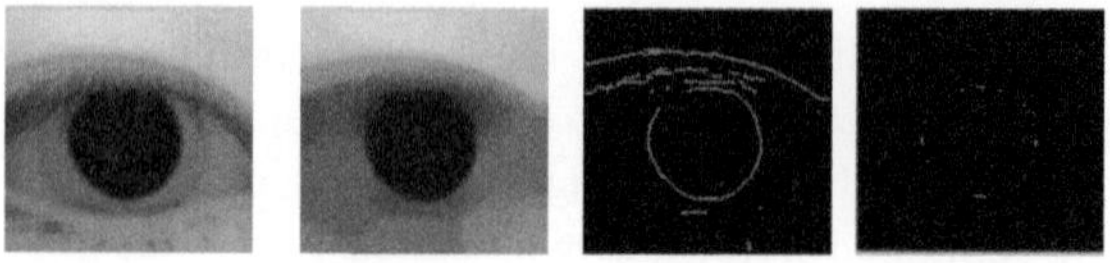

a b c d

Fig. 4. Exclution of disturbances and dection of edge

C. Extraction and Detection of Edge

The sobel operator is usually used in image processing, particularly in edge detection algorithms. The sobel operator is based on convolving the image with a small, separable, and integer valued filter in horizontal and vertical direction and is therefore relatively inexpensive in terms of computations.

In this paper, the boundary between iris and pupil is called pupil edge, as well as the boundary between iris and sclera is called iris edge. In order to obtain center, radius, area and other features of pupil, pupil edge should be extracted. In this study, although the iris edge in the eye image taken under infrared light is fuzzy, the pupil edge is obvious, and its gray gradient is large. Under visible light, both the pupil edge and iris edge are significant, and their gray gradients are large. So the pupil edge can be extracted using sobel operator.

The Hough transformation is a technique which can be used to isolate features of a particular shape within an image. the classical Hough transformation is most commonly used for the detection of regular curves such as lines, circles, ellipses, etc.

In the case of circles, the parametric equation is $(x - a)^2 + (y - b)^2 = r^2$, where a and b are the coordinates of the center of the circle, and r is the radius. Circular Hough transformation is defined as following[9].

$$H(a,b,r) = \sum_{i=1}^{n} h(x_i, y_i, a, b, r) \tag{1}$$

where

$$h(x_i, y_i, a, b, r) = \begin{cases} = 1 \ldots\ldots g(x_i, y_i, a, b, r) = 0 \\ = 0 \ldots\ldots g(x_i, y_i, a, b, r) \neq 0 \end{cases} \tag{2}$$

with

$$g(x_i, y_i, a, b, r) = (x_i - a)^2 + (y_i - b)^2 - r^2. \tag{3}$$

By Hough transformation, the point (x, y) in image space corresponds to the point (a, b, r) in parameter space. The cumulative array M (a, b, r) is established so that M is added up when a, b, r change. M(a,b,r)=M(a,b,r)+1. The maximum corresponding to M is the result of Hough operation.

$$(a_0, b_0, r_0) = (a, b, r) \mid \max\left(\bigcup_{a,b,r} H(a,b,r)\right) \tag{4}$$

Furthermore, the pupil area is shown as the formula $S = \pi r_0^2$. Figure 4(c),(d) shows the result of edge detection and Hough transformation.

5 Experimental Results

The system captured some eye images in clinical practice, and these images were processed by the proposed method. In the end, the pupil eigenvalues were presented. This work was done on a computer with 2.4GHz processor and 1G RAM. Images' resolution was 720 ×576.

Figure 5 shows the test results. According to the test values (a_0, b_0, r_0), the pupil center (a_0, b_0) and the pupil boundary were marked out in these images. The pupil center was marked with "+", while the pupil boundary was marked with a circle that the center was (a_0, b_0), the radius was r_0. Figure 5(a), (b), (c) are three different infrared eye images, The interference of spots and eyelash existed in the three figures. The interference in figure 5(a) was greater. Particularlly, there were longer and thicker eyelashes in figure 5(a), However, this algorithm could still accurately locate the pupil center and boundary in these images. The pathological eye was shown in figure 5(c). Figure 5(d) gived results to detect the visible-light image. The above results show that this algorithm can detect the pupil under different light and medical conditions, and give the pupil eigenvalues. Its anti-interference is strong, and its calculation is greatly reduced, so detection speed is faster.

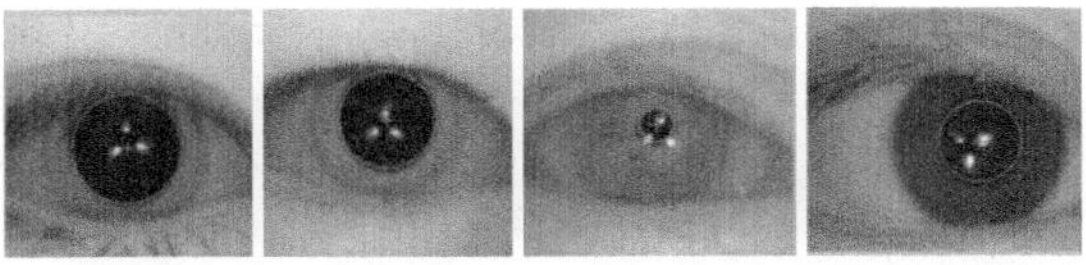

a b c d

Fig. 5. Pupil marking graph

While the eye was stimulated by light, a series of images were obtained by the image acquisition system. After these images were processed by the above approach, the pupil area was gained, and then a scatter diagram was drawn that the pupil area changed with time(see figure 6). The figure shows that the pupil reduced in size with time in optical stimulation, and then increased gradually after withdrawal of stimulus. This increasing showed a recovery trend. In figure 6, the unit of pupil area is pixels square, and the time unit is seconds

Figure 7 gives several bilateral eye images taken by the system in the stimulation mode at the different time: t=0ms, t=1400ms, t=2200ms. It shows that when the left eye was stimulated, the bilateral pupils changed dynamically and were related. The bilateral pupils reduced simultaneously after stimulus, although the unilateral was stimulated.

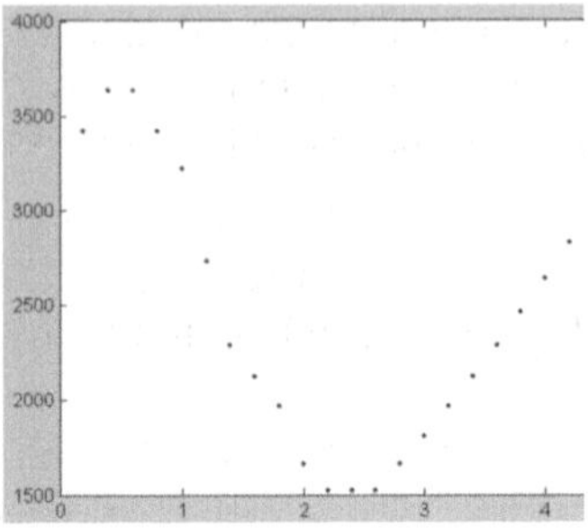

Fig. 6. Pupil area scatter vs. time

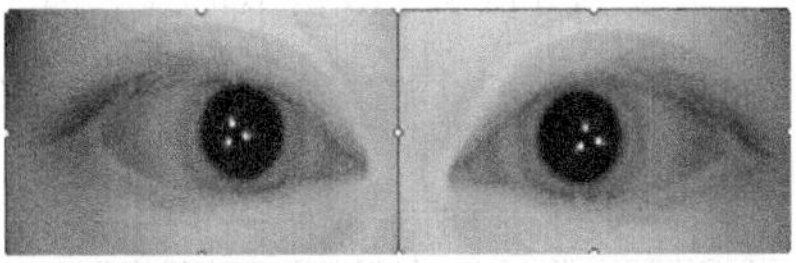

(a) t=0ms right eye left eye

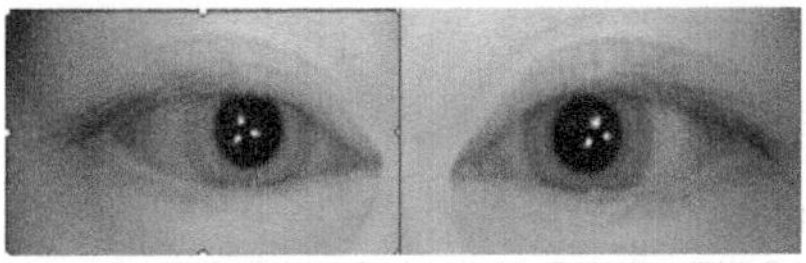

(b) t=1400ms right eye left eye

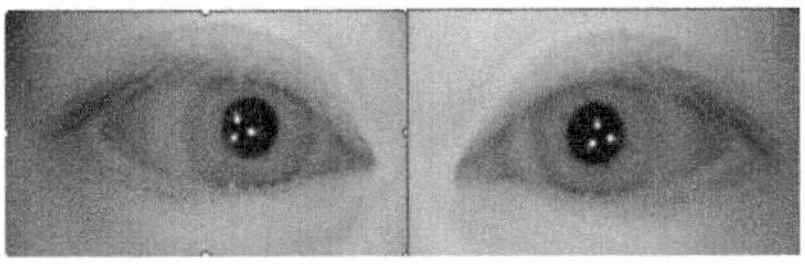

(c) t=2200ms right eye left eye

Fig. 7. Pupil size vs. time

6 Conclusion

This paper described the eye image detection system, including the hardware and software, and introduced a method to detect the pupil in detail. It was applied to detect the eye and evaluate the dynamical characteristics of the pupil in clinical practice. The results show that this system can detect the pupil under infrared light or visible light and medical conditions, and analysize the change of the pupil under the optical stimulation. Results are objective, accurate and quantitative, and satisfy medical requirements. It is better than the qualitative descriptions, or using simple manual test tools. The system can provide doctors with important clues in disease diagnosis.

References

1. Verney, S.P., Granholm, E., Marshall, S.P.: Pupillary responses on the visual backward masking task reflect general cognitive ability. Psychophysiol. 52(1), 23–36 (2004)
2. Stergiou, V., Fotiou, D., Tsiptsios, D., et al.: Pupillometric findings in patients with Parkinson's disease and cognitive disorder. Psychophysiology 72, 97–101 (2009)
3. Pong, M., Fuchs, A.F.: Characteristics of the Pupillary Light Reflex in the Macaque Monkey: Metrics. J. Neurophysiol. 84(2), 953–963 (2000)
4. Meeker, M., Du, R., Bacchetti, P., Privitera, C.M., Larson, M.D., Holland, M.C., Manley, G.: Pupil Examination: Validity and Clinical Utility of an Automated Pupillometer. Journal of Neuroscience Nursing 37, 34–40 (2005)
5. Serra, J.: Image Analysis and Mathematical Morphology. Academic Press, London (1982)
6. Gonzalez, R.C., Woods, R.E., Eddins, S.L.: Digital Image Processing Using MATLAB, p. 373. Pearson Education, Inc., London (2004)
7. Gonzalez, R.C.: Digital Image Processing, pp. 519–530. Publishing House of Electronics Industry, Beijing (2006)
8. Yuan, W.Q., Bai, G., Feng, Q.: New Iris Image Pre-processing Method. Opto-Electronic Engineering 36(4), 133–139 (2009)
9. Tian, Q.C., Pan, Q., Zhang, H.C.: Application of Hough Transform on IRIS Segmentation. Application research of computers 22(1), 249–254 (2005)

Content Adaptive Framelet Image Watermarking

Runhai Jiao

School of Control and Computer Engineering,
North China Electric Power University,
Beijing, China
RunhaiJiao113@yeah.net

Abstract. In this paper, we propose a new adaptive image watermarking algorithm based on framelet and image texture. The host image is firstly decomposed by framelet. Secondly, the texture factors and edge factors of lowest frequency subband(HH) coefficients are calculated, and the HH coefficients are sorted by texture factor in descending order. Thirdly, the scrambled watermark bit is embedded into the HH coefficient one by one, and the embedding strength is adaptive adjusted by edge factor. Experimental results show that the proposed algorithm gains good performance in robustness and invisibility with respect to traditional image processing. Moreover, the watermark invisibility of adaptive embedding is better than that of non-adaptive embedding.

Keywords: mage watermarking, framelet, adaptive watermarking, robustness, invisibility.

1 Introduction

The internet and multimedia technology have gained a rapid development in the past decades, which make the digital multimedia productions(such as image, video and audio) are easy to distribute, duplicate and modify. Therefore, effective copyright protection techniques for digital media are urgently needed. Watermarking is one kind of popular techniques that permits the owner embedding a digital marking into the host media to verify its ownership. Digital watermarking has several characteristics such as invisibility, unambiguous, robustness and security [1]. Among these characteristics, invisibility and robustness are the basic but the most important.

The watermarking algorithms can be divided into pixel domain algorithms [2,3] and frequency domain algorithms[4-6]. Experimental results show that the watermark embedded in pixel domain is fragile to be destroyed by various kinds of image processing attacks. So researchers pay more attention to study frequency domain algorithm. One of the key points of frequency domain watermarking algorithm is to choose a proper transform. Wavelet, as the typical multi-resolution analysis(MRA) transform, has gained better performance than DCT in many signal processing areas. Therefore, wavelet domain watermarking algorithms prevail in the recently years.

Framelet is very similar to wavelet but has some important differences. Framelet has two or more high frequency filter banks, which produces more subbands in decomposition. This can achieve better time-frequency localization ability. Moreover,

J. Zhang (Ed.): ICAIC 2011, Part V, CCIS 228, pp. 628–637, 2011.

the filter banks of framelet are not orthogonal, which means that there exists redundancy between framelet subbands. That is, the coefficients of different subbands are correlated, and modification on coefficients of one band can be compensated by coefficients of other subbands. This can help to improve the invisibility of the watermark.

In this paper, we propose an adaptive image watermarking method in framelet domain. In our method, the watermark embedding position and strength are decided by the image texture complexity. The rests of the paper are organized as follows. Section 2 analyzes the characteristic of framelet transform. Section 3 illustrates the adaptive embedding strategy of our algorithm. Section 4 describes the watermark embedding and extracting algorithm. Experiments are given in section 5, and the conclusion is drawn is section 6.

2 Framelet Transorm Analysis

A.Ron, Z.shen and I. Daubechies gave the definition of MRA-based framelet[7-8]. From its definition, we can see that framelet has at least one mother wavelet function. Wavelet can be thought as a special kind of framelet. Considering the computation complexity, in our algorithm the framelet adopted has one scaling function $\phi(t)$ and two mother wavelet functions $\psi_1(t)$ and $\psi_2(t)$, which can be represented by the following low pass filter(scaling function) $h_0(n)$ and high pass filters(wavelet function) $h_1(n)$ and $h_2(n)$:

$$\begin{aligned} \phi(t) &= \sqrt{2}\sum_n h_0(n)\phi(2t-n) \\ \psi_i(t) &= \sqrt{2}\sum_n h_i(n)\phi(2t-n), i=1,2 \end{aligned} \tag{1}$$

Mallat proposes the framelet pyramid decomposition method for two dimension signal[9].

$$\begin{aligned} C_j &= H_c H_r C_{j-1} \\ D_j^i &= G_{l,c} G_{m,r} C_{j-1}, i=1,\dots,4; l=1,2; m=1,2 \\ D_j^{i'} &= H_c G_{m,r} C_{j-1}, i'=5,6; m=1,2 \\ D_j^{i''} &= G_{l,c} H_r C_{j-1}, i''=7,8; l=1,2 \end{aligned} \tag{2}$$

Where $C_j, D_j^i, D_j^{i'}, D_j^{i''}$ are the low frequency part and high frequency parts of image C_{j-1}.

One-level framelet decomposition on image generates nine subbands, and Fig.1 lists the coefficients map of Lena after one-level decomposition. From Fig. 1 we can see that HH subband is the approximate part of the image, and the other subbands give the detail information of the image at different directions(horizontal, vertical, diagonal). Each subband size is 1/4 of the original image size. Continuous decomposition on HH will generates the pyramid structure of framelet decomposition.

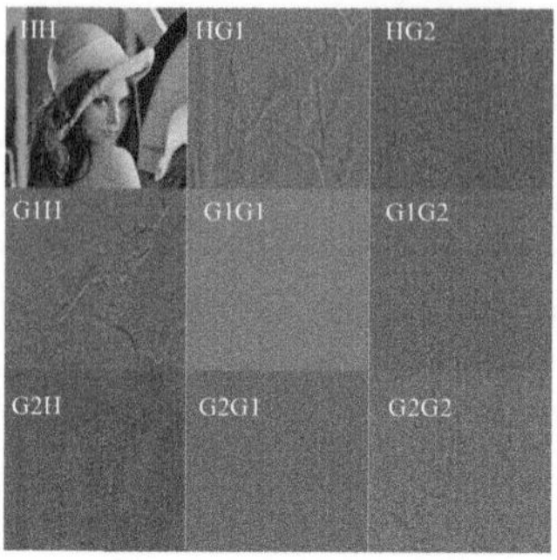

Fig. 1. One-level framelet decomposition on Lena

The subbands of multi-level decomposition constitute a multi-resolution analysis of the original image. Each subband is the representation of the original image in specific frequency range and specific direction. Most of the image energy is compacted into the lowest frequency subband(HH). Usually, the coefficients in HH subband have big magnitudes. By the contrary, coefficients of the other subbands are smaller.

3 Adaptive Embedding Strategy

Usually, frequency watermark embedding is to modify the coefficients magnitude to signify the watermark information. This can be thought as adding noise to the host image, which makes the host image quality degrade. In order to improve the watermark robustness, people hope to make large modification on coefficient magnitude. However, this will make the host image quality go down quickly, and even to the extent that the adding noise is visible to human eye. Therefore, there should be tradeoff between watermark robustness and invisibility. Fig. 2 gives the common watermark embedding procedure. There are two key techniques of watermarking. The first is to choose which coefficients for embedding. The second is how to control the embedding strength.

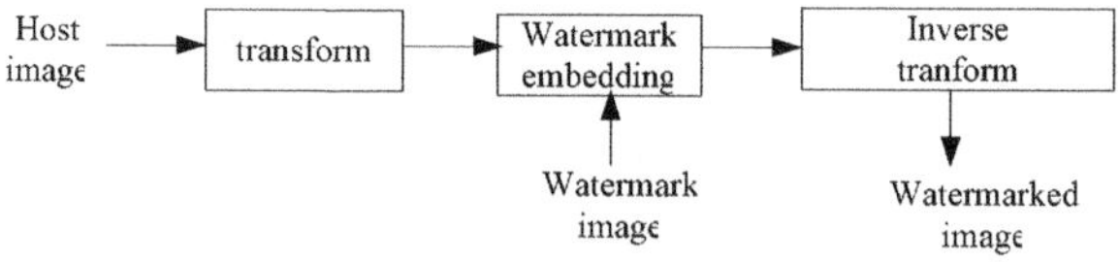

Fig. 2. Watermark embedding Procedure

For the first question, the lowest frequency subband coefficients are usually chosen for embedding[10]. But this kind of method has disadvantage. According to human visual system characteristic, human eye's sensibility to different image regions distortion are different. The sooth region distortion is most sensitive to human eye, and the texture region distortion is not sensitive. The edge region distortion sensibility is between them. Therefore, watermark is better embedded into the texture region or edge region. After multi-level framelet decomposition, one coefficient in HH is corresponding to a series of coefficient blocks(quadric-tree structure) in high frequency subbands, and a square block of pixels in original image [11].

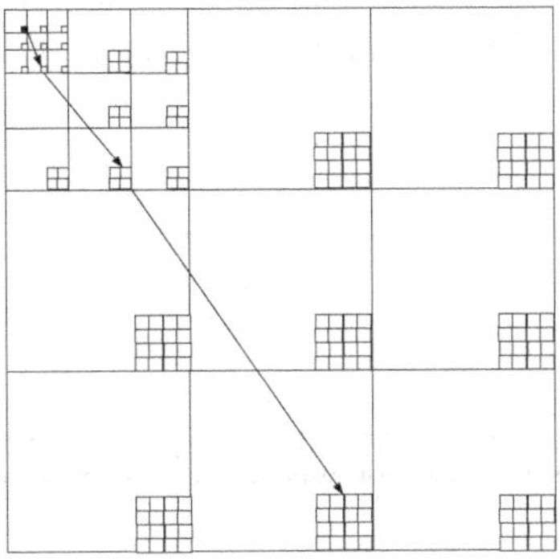

Fig. 3. Quadric-tree structure

As in Fig.3, after 3-level framelet transform, one HH coefficient is corresponding to 8×(1+2×2+4×4)=168 high frequency coefficients, and 64 image pixels. In this paper, we calculate the texture factor f_T and edge factor f_E by the formula (3) and (4). Moreover, f_T and f_E are regularized to [0,1].

$$f_T = \sum_{l=1}^{3}\sum_{d=1}^{3}\sum_{r=2^l \bullet i}^{2^l \bullet (i+1)}\sum_{c=2^l \bullet j}^{2^l \bullet (j+1)} \left| c_{l,d,r,c} \right| \tag{3}$$

$$f_E = \sum_{r=2^{l+1} \bullet i}^{2^{l+1} \bullet (i+1)}\sum_{c=2^{l+1} \bullet j}^{2^{l+1} \bullet (j+1)} \left| p_{r,c} \right| \tag{4}$$

Where l is decomposition level, d is direction, (i,j) is the coefficient position in HH subband, and (r,c) is the corresponding coefficient position, $c_{l,d,r,c}$ is corresponding coefficient value, and $p_{r,c}$ is the corresponding pixel value.

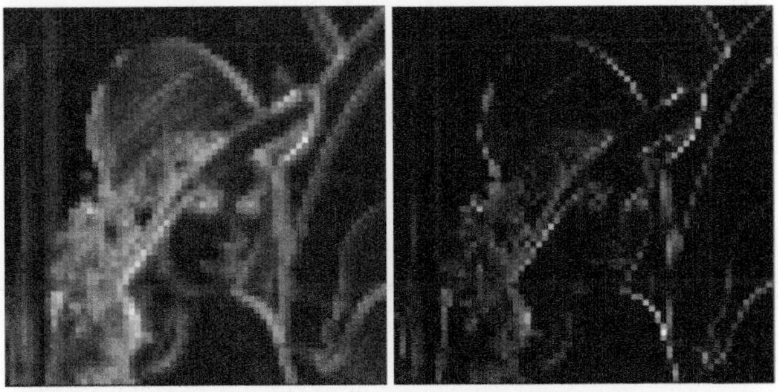

(a) texture (b) edge

Fig. 4. Texture factor map and edge factor map

Usually, smooth region has small f_T, and texture region or edge region has big f_T. As to f_E, edge region usually has big f_E than that of texture region. In Fig. 4, we map f_T and f_E into [0,255] and show them as grayscale image. It is clear that these two factors can distinguish smooth region, edge region and texture region perfectly. In our algorithm, we calculate f_T and f_E of each HH coefficient, and sort HH coefficients in descending order according to f_T. From the beginning, we select the coefficient one by one to embed watermark. Because edge coefficient's distortion is more sensitive to human eyes than texture coefficient, we adaptively adjust the embedding strength according to f_E..

For the second question, we choose framelet in our algorithm. Framelet breaks out the restriction of orthogonality. There are correlations between the framelet filters, and the coefficients in one subband have correlation with coefficients in other subband. This means that changes on one coefficient can be compensated by its related coefficients in reconstruction stage, which produces less noise to host image. This characteristic is helpful for watermarking.

4 Our Proposed Method

A. Watermark Embedding

In order to improve the security and robustness of the watermark, this paper use Arnold transform to scramble the watermark. The watermark is embedded into the host image by using the following formula

$$C = C_i(1+\alpha\beta W_i) \tag{5}$$

where C_i is the framelet coefficient,. $W_i \in \{1,-1\}$ is the pixel value of the binary watermark image, and α is the parameter controlling watermark embedding strength, which is manually input into the algorithm, β is set according to edge factor f_E. In our algorithm, based on typical image experiments, β is set as following

$$\beta = \begin{cases} 0.75 & f_E > 0.2 \\ 1.2 & f_E < 0.1 \\ 1 & others \end{cases} \tag{6}$$

The embedding procedure is given below:

(1) Perform 3-level framelet transform
(2) Calculate HH coefficients' f_T and sort coefficients in descending order
(3) Calculate HH coefficients' f_E
(4) Scramble watermark using Arnold transform
(5) For each watermark bit

a) Choose one HH coefficient in order
b) Calculate its corresponding β according(6)

c) Embedding the watermark bit according (5)

(6) Perform 3-level reverse framelet transform.

B. Watermark Extracting

Generally, the procedure of watermark extracting is the inverse process of embedding. Three-level framelet decomposition is performed on watermarked image and original host image. We calculate f_T for each coefficient in HH subband of host image, and decide the watermark embedding position. The watermark bit is extracted by

$$W_i' = C_i' - C_i \tag{7}$$

Where C_i' is the framelet coefficient of the watermarked image, and C_i is the corresponding framelet coefficient of the original image.

5 Experimental Results

Our proposed algorithm is implemented in Visual C++6.0. The framelet used in our algorithm has one low pass filter and two high pass filters. The filter coefficients are given below:

$$\begin{aligned} H &= \sqrt{2}\begin{pmatrix} -0.03243267,-0.20234149,0.27938597,-0.10133703, \\ 0.27938597,-0.20234149,-0.03243267 \end{pmatrix} \\ G1 &= \sqrt{2}\begin{pmatrix} 0.03294782,0.20555539,-0.32060556,0,0.32060556, \\ -0.20555539,-0.03294782 \end{pmatrix} \\ G2 &= \sqrt{2}\begin{pmatrix} -0.00580348,-0.03620686,0.25881349,0.56632033, \\ 0.25881349,-0.03620686,-0.00580348 \end{pmatrix} \end{aligned} \tag{8}$$

A. Attack Experiment

In our experiment, we choose 512×512×8 bit gray image Lena as host image and 32×32 binary image as watermark, which are illustrated in Fig. 5.

Host image

Watermrak image

Fig. 5. Host image and watermark image

Four kinds of common attacks were chosen to evaluate algorithm performance. We choose PSNR(Peak Signal Noise Ratio) to evaluate the watermarked image's quality, and Normalized Correlation(NC) to evaluate the correlation between the original watermark and the extracted watermark.

Test 1: free attack

(a) (b)

Fig. 6. (a) watermarked image PSNR=37.77dB (b) extracted watermark NC=1

Fig. 7. Gaussian noise attack

Fig. 8. JPEG compression attack

Test 2: Gaussian noise attack, and the parameter is 0.01.The watermarked image PSNR is 19.97dB, and the NC of extracted watermark is 0.950. Images are listed in Fig 7.

Test 3: JPEG compression attack, compression parameter is 90%. The watermarked image PSNR is 31.60dB, and the NC of extracted watermark is 0.996. Images are listed in Fig. 8.

Fig. 9. Image cropping attack

Fig. 10. Gaussian filtering attack

Test 4: Image cropping attack. The watermarked image PSNR is 9.16dB, and the NC of extracted watermark is 0.920. Images are listed in Fig. 9.

Test 5: Gaussian low pass filter, and parameters are [7,7] and 0.8. The watermarked image PSNR is 32.32dB, and the NC of extracted watermark is 0.998. Images are listed in Fig. 10.

From the experimental results we can see that the host image still keep good visual quality after watermark embedding, which proves the invisibility of our watermark. Under several common attacks, the extracted watermarks are clear enough for recognition. This proves the watermark's robustness.

B. Adaptive vs. Non-adaptive Embedding

In our algorithm, watermark bits are adaptively embedded into texture regions and edge regions of the host image. For comparison, we embed watermark bits just according to the magnitude order of HH coefficients. That is, the largest magnitude coefficient is first chosen for embedding, and the second larger is the next, and so on. The embedding formula is as follows:

$$C = C_i(1+\alpha W_i) \tag{9}$$

All parameters have the same meaning as in formula (5). The value of α is 0.031, and PSNR of the watermarked image(in Fig. 11) is 38.75dB.

If we apply Gaussian noise attack as in Test 2, the NC of the extracted watermark is 0.950. Under this condition, we can think that the robustness of adaptive algorithm is equal to that of non-adaptive algorithm. But from Fig.11 we can see that there exist visible noises as marked in the figure. Moreover, most of the noise points exist in smooth area, which are sensitive for human eyes.

Fig. 11. Watermarked image of non-adaptive embedding

6 Conclusion

A content adaptive image watermark algorithm based on framelet and image texture is proposed in this paper. Experimental results show that our watermarking embedding achieves good invisibility. At the same time, it has good robustness, and can resists common watermarking attacks such as Gaussian noise, JPEG compression and image cropping etc.

Acknowledgment. This research is sponsored by PhD Youth Teacher Funds of North China Electric Power University, and here shows special thanks for its support.

References

1. Bender, W., Gruhl, D., Morimoto, N., Lu, A.: Techniques for data hiding. IBM Sys. J. 35(3-4), 313–336 (1996)
2. Bender, W., Gruhl, D., Morimoto, N., et al.: Techniques for data hiding. IBM System Journal 35(3&4), 313–336 (1996)
3. Tirkel, A.Z., et al.: Electronic watermark. In: Proceeding of Digital image computing Technology and Applications, Macquarie University, pp. 666–673 (1993)
4. Lin, W.-H., Horng, S.-J., et al.: An Efficient Watermarking Method Based on Significant Difference of Wavelet Coefficient Quantization. IEEE Transaction on Multimedia 10(5), 746–757 (2008)
5. Yang, S., Lu, Z., Zou, F.: A robust adaptive image watermarking algorithm. In: Proc. IEEE ICME, pp. 1723–1726 (2004)
6. Lien, B.K., Lin, W.H.: A watermarking method based on maximum distance wavelet tree quantization. In: Proc. 19th Conf. Computer Vision, Graphics and Image Processing, pp. 269–276 (2006)

7. Ron, A., Shen, Z.w.: Affine system in $L^2(R^d)$:The analysis of analysis operator. J.Funct. Anal., 408–447 (1997)
8. Daubechies, I., Han, B., Ron, A., Shen, Z.: Framelets :MRA_based construction of wavelet frame. Appl. Comput.Anal., pp.1–46 (2003)
9. Mallat: Theory formultiresolution signal decomposition: the wavelet representation. IEEE Trans on Pattern Analysis and Machine Intelligence 11(7), 674–693 (1989)
10. Corvi, M., Niochiotti, G.: Wavelet-based Image Watermarking for Copyright Protection. In: Scandinavian Conference on Image Analysis, SCIA 1997, Lappeenranta, Finland (June 1997)
11. Shapiro, J.M.: Embedded image coding using zerotrees of wavelet coefficients. IEEE Transactions on Signal Processing 41(12), 3445–3462 (1993)

Pairwise Balanced Design of Order $6n + 4$

Bagchi Satya, Basu Manjusri, and Ghosh Debabrata Kumar

Department of Mathematics,
National Institute of Technology, Durgapur, Burdwan, W.B., India, Pin-713209
bagchisatya@ymail.com
http://www.nitdgp.ac.in

Abstract. In Steiner triple system Bose constructed $2-(v,3,1)$ designs for $v = 6n+3$ and later on Skolem constructed the same for $v = 6n+1$. In literature we can find also a pairwise balanced design (PBD) for $v = 6n+5$. In this paper, we construct a PBD for $v = 6n+4$, $n \in N$.

Keywords: Steiner triple system, Latin square, support, t-design, t-wise balanced design.

1 Introduction

A pairwise balanced design (PBD) is an ordered pair (S, B), where S is a finite set of symbols, and B is a collection of subsets of S called blocks, such that each pair of distinct elements of S occurs together in exactly one block of B. In this paper, we develop a new PBD $2-(6n+4, \{4,3\}, 1)$ for $2n+1$ blocks of size 4 and $3.{}^{2n+1}C_2$, the rest of the blocks of size 3 where $n \in N$.

1.1 Latin Square

A latin square of order n is an $n \times n$ array, each cell of which contains exactly one of the symbols in $\{1, 2, \ldots, n\}$, such that each row and each column of the array contains each of the symbols in $\{1, 2, \ldots, n\}$ exactly once.

A latin square is said to be idempotent if cell (i,i) contains symbol i for $1 \leq i \leq n$.

A latin square is said to be commutative if cell (i,j) and (j,i) contain the same symbol, for all $1 \leq i, j \leq n$.

1.2 Quasi-group

A quasi-group of order n is a pair $(Q, \circ)$, where Q is a set of size n and " $\circ$ " is a binary operation on Q such that for every pair of elements $a, b \in Q$, the equations $a \circ x = b$ and $y \circ a = b$ have unique solutions. As far as we are concerned a quasi-group is just a latin square with a headline and a sideline.

Support: Let c be a binary codeword of length n. The set of positions in which c has non-zero entries is called support of c.

J. Zhang (Ed.): ICAIC 2011, Part V, CCIS 228, pp. 638–642, 2011.

1.3 Design [1]

Let C be a binary code of length n. Let S_k be the set of codewords in C of weight k. We say that S_k holds a $t-(v,k,\lambda)$ design if the supports of codewords in S_k from the blocks of a $t-(v,k,\lambda)$ design, if for any t-set $T \subset \{1,2,\ldots,n\}$ there are exactly λ codewords of weight $k \in C$ with with 1's in the positions given by T.

A $t-(v,k,\lambda)$ design is defined as a Steiner system and denoted by $S(t,k,v)$.

1.4 Steiner Triple System

A Steiner triple system (STS) $2-(v,3,1)$ is an ordered pair (S,T), where S is a finite set of points or symbols, and T is a set of 3-element subsets of S called triples, such that each pair of distinct elements of S occurs together in exactly one triple of T. The order of a Steiner triple system (S,T) is the size of the set S, denoted by $|S| = v$.

A Steiner triple system of order v exists if and only if $v \equiv 1 \; or \; 3 \; (mod \; 6)$.

1.5 t-Wise Balanced Design

Let V be a of v points, $0 \leq t < v$ an integer and K is the set of integers k_i such that $t < k_i < v$. A t-wise balanced design $D = (V,B)$ consists of a family of subsets, called blocks of v, such that each size k of a block lies in K and each t-elements subset of v is contained in the same number of λ blocks. A t-wise balanced design with these parameters is a $t-(v,K,\lambda)$ design, shortly a tBD.

2 Preliminaries

2.1 The Bose Construction ($v \equiv 3(mod\ 6)$) [2,3]

Let $v = 6n+3$ and let $(Q,\circ)$ be an idempotent commutative quasi-group of order $2n+1$, where $Q = \{1,2,\ldots,2n+1\}$. Let $S = Q \times \{1,2,3\}$, and define T to contain the following two types of triples.

Type 1: For $1 \leq i \leq 2n+1$, $\{(i,1),(i,2),(i,3)\} \in T$.

Type 2: For $1 \leq i \leq 2n+1$, $\{(i,1),(j,1),(i \circ j,2)\}, \{(i,2),(j,2),(i \circ j,3)\}, \{(i,3),(j,3),(i \circ j,1)\} \in T$.

2.2 The Skolem Construction ($v \equiv 1(mod\ 6)$) [2,4]

Let $v = 6n+1$ and let $(Q,\circ)$ be a half idempotent commutative quasi-group of order $2n$, where $Q = \{1,2,\ldots,2n\}$. Let $S = \{\infty\} \cup (Q \times \{1,2,3\}$, and define T as follows:

Type 1: For $1 \leq i \leq n$, $\{(i,1),(i,2),(i,3)\} \in T$.

Type 2: For $1 \leq i \leq n$, $\{\infty,(n+i,1),(i,2)\}$, $\{\infty,(n+i,2),(i,3)\}$, $\{\infty,(n+i,3),(i,1)\} \in T$.

Type 3: For $1 \leq i \leq 2n$, $\{(i,1),(j,1),(i \circ j,2)\}$, $\{(i,2),(j,2),(i \circ j,3)\}$, $\{(i,3),(j,3),(i \circ j,1)\} \in T$.

Then (S,T) is a Steiner triple system of order $6n+1$.

2.3 The $6n+5$ Construction [2]

Let $(Q,\circ)$ be an idempotent commutative quasi-group of order $2n+1$, where $Q = \{1,2,\ldots,2n+1\}$ and let α be the permutation $(1)(2,3,\ldots,2n+1)$. Let $S = \{\infty_1,\infty_2\} \cup (Q \times \{1,2,3\}$ and let B contain the following blocks:

Type 1: $\{\infty_1,\infty_2,(1,1),(1,2),(1,3)\}$.

Type 2: $\{\infty_1,(2i,1),(2i,2)\}$, $\{\infty_1,(2i,3),((2i)\alpha,1)\}$, $\{\infty_1,((2i)\alpha,2),((2i)\alpha,3)\}$, $\{\infty_2,(2i,1),((2i)\alpha^{-1},3)\}$, for $1 \leq i \leq n$ and

Type 3: $\{(i,1),(j,1),(i \circ j,2)\}$, $\{(i,2),(j,2),(i \circ j,3)\}$, $\{(i,3),(j,3),((i \circ j)\alpha,1)\}$ for $1 \leq i \leq 2n+1$.

Then (S,B) is a $PBD(6n+5)$ with exactly one block of size 5 and the rest of size 3.

3 Main Result

The $6n+4$ **Construction:** Let $(Q,\circ)$ be an idempotent commutative quasi-group of order $2n+1$, where $Q = \{1,2,\ldots,2n+1\}$. Now we construct (S,B) where $S = \{\infty\} \cup \{Q \times \{1,2,3\}\}$ and B contains the following blocks:

Type 1: $\{\infty,(i,1),(i,2),(i,3)\}$, for $1 \leq i \leq 2n+1$.

Type 2: $\{(i,1),(j,1),(i \circ j,2)\}, \{(i,2),(j,2),(i \circ j,3)\}, \{(i,3),(j,3),(i \circ j,1)\}$ for $1 \leq i < j \leq 2n+1$.

Thus (S,B) is a PBD with $2n+1$ blocks of size 4 and the rest $3.{}^{2n+1}C_2$ blocks of size 3.

Hence (S,B) is a PBD $2-(6n+4,\{4,3\},1)$.

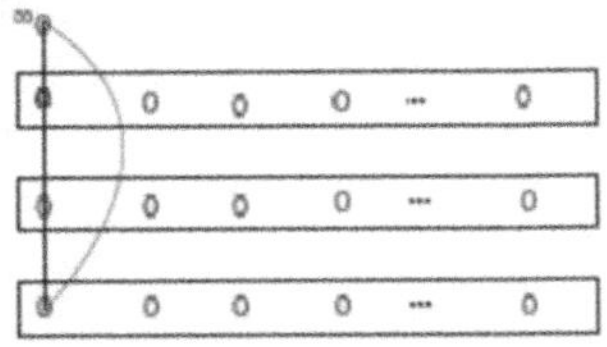

Type 1

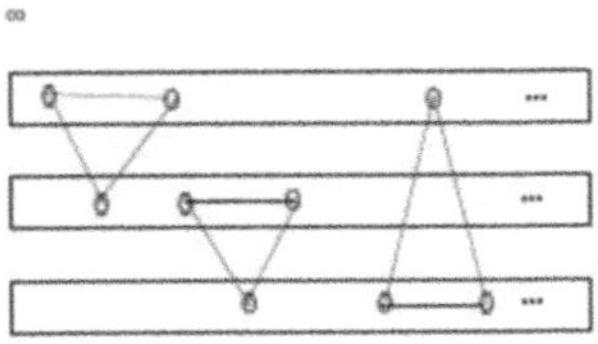

Type 2

So, the number of blocks of the PBD $2-(6n+4, \{4,3\}, 1)$ are $(2n+1) + (3.^{2n+1}C_2) = (2n+1)(3n+1)$.

Example 1: For $n = 1$ i.e. $v = 6.1 + 4 = 10$

o	1	2	3
1	1	3	2
2	3	2	1
3	2	1	3

$S = \{\infty, 1, 2, \ldots, 9\}$ and $B = T_1 \cup T_2$, where

$T_1 = \{\infty, 1, 4, 7\}, \{\infty, 2, 5, 8\}, \{\infty, 3, 6, 9\}$.

$T_2 = \{1, 2, 6\}$, $\{1, 3, 5\}$, $\{2, 3, 4\}$, $\{4, 5, 9\}$, $\{4, 6, 8\}$, $\{5, 6, 7\}$, $\{7, 8, 3\}$, $\{7, 9, 2\}$, $\{8, 9, 1\}$.

So (S,B) is a PBD $2-(10, \{4,3\}, 1)$ with total 12 blocks in which $2.1+1(=3)$ blocks of size 4 and $3.^3C_2(=9)$ blocks of size 3.

Example 2: For $n = 2$ i.e. $v = 6.2 + 4 = 16$

o	1	2	3	4	5
1	1	5	2	3	4
2	5	2	4	1	3
3	2	4	3	5	1
4	3	1	5	4	2
5	4	3	1	2	5

$S = \{\infty, 1, 2, \ldots, 15\}$ and $B = T_1 \cup T_2$, where

$T_1 = \{\infty, 1, 6, 11\}, \{\infty, 2, 7, 12\}, \{\infty, 3, 8, 13\}, \{\infty, 4, 9, 14\}, \{\infty, 5, 10, 15\}.$

$T_2 = \{1, 2, 10\}, \{1, 3, 7\}, \{1, 4, 8\}, \{1, 5, 9\}, \{2, 3, 9\}, \{2, 4, 6\}, \{2, 5, 8\}, \{3, 4, 10\}, \{3, 5, 6\}, \{4, 5, 7\}, \{6, 7, 15\}, \{6, 8, 12\}, \{6, 9, 13\}, \{6, 10, 14\}, \{7, 8, 14\}, \{7, 9, 11\}, \{7, 10, 13\}, \{8, 9, 15\}, \{8, 10, 11\}, \{9, 10, 12\}, \{11, 12, 5\}, \{11, 13, 2\}, \{11, 14, 4\}, \{11, 15, 4\}, \{12, 13, 4\}, \{12, 14, 1\}, \{12, 15, 3\}, \{13, 14, 5\}, \{13, 15, 1\}, \{14, 15, 2\}.$

So (S,B) is a PBD $2-(16, \{4, 3\}, 1)$ with total 35 blocks in which $2.2+1(= 5)$ blocks of size 4 and $3.{}^5C_2(= 30)$ blocks of size 3.

4 Conclusion

There already exists STS for $v = 6n + 1$, $v = 6n + 3$, and PBD for $v = 6n + 5$. In this paper, we construct PBD for $v = 6n+4$. The PBD for $v = 6n$ and $v = 6n+2$ are still now open.

References

1. Basu, M., Rahaman Md., M., Bagchi, S.: On a new code, $[2^n, n, 2^{n-1}]$. Discrete Applied Mathematics 175, 402–405 (2009)
2. Lindner, C.C., Rodger, C.A.: Design Thery. CRC Press, Boca Raton (2009)
3. Bose, R.C.: On the construction of balanced incomplete block design. Ann. Eugenics 9, 353–399 (1939)
4. Skolem, T.: Some remarks on the triple systems of Steiner. Math. Scand. 6, 280–373 (1958)

Efficient Load Balancing Algorithm Using Complete Graph

Lakshmi Kanta Dey, Debashis Ghosh, and Satya Bagchi

Department of Mathematics,
National Institute of Technology, Durgapur
Durgapur-713209, West Bengal, India
bagchisatya@ymail.com
http://www.nitdgp.ac.in

Abstract. For load balancing process in a distributed network consisting of v nodes it requires $O(v^2)$ communication complexity. Here we present a new synchronous dynamic distributed load balancing algorithm on a network having $v = k^2$ nodes, where $k(\geq 2)$ is any positive integer. Our algorithm needs a complete graph of k nodes for an efficient communication system. Each node receives workload information from all other nodes with two-round message interchange without redundancy and hence it requires $O(v\sqrt{v})$ communication complexity.

Keywords: Complete graph, distributed network, communication complexity.

1 Introduction

In the past few decades the Load balancing problem has received a good attention by researchers. Various approaches has adapted to reduce the communication complexity In general, for a distributed network, very often it happens that loads are changes in the nodes, some nodes are loaded more than the other nodes. So, it is reasonable to consider whether the workload can be distributed among the nodes such that the utility of each node increases as well as the response time decreases. Here the distributed system consists of a set of processors those are connected by a point to point communication network which guarantees the delivery of messages. The distributed manner is prescribed in a network in such a way that every node can be informed about the workload information of the other nodes, which should be of latest, otherwise, outdated information may effect the inconsistency of the system.

In order to decrease communication overhead for obtaining workload information from all the nodes in the network, information should be distributed by an arrangement of all nodes. These nodes are kept into a number of sets each of which consisting of same number of nodes, which exchange their information after an interval among themselves. After rearranging the nodes, the information are exchanged once again among themselves.

J. Zhang (Ed.): ICAIC 2011, Part V, CCIS 228, pp. 643–646, 2011.

In [2] CWA is employed for load balancing on hypercube network for which the communication complexity is $O(v^2)$ and the corresponding communication path is $O(log_2 v)$. Lot of analogous works have been done for communication complexity of $O(v^2)$.

O. Lee et. al. [4] designed the network topology consisting of v nodes and $v.k$ links, where each node is linked to $2k$ nodes, with $v = k^2 + k + 1$ and k is prime. On this network, each node receives information from k adjacent nodes and then sends these informations to other $k-1$ adjacent nodes periodically. So, each node receives workload information for $k^2 + k = v - 1$ with two round of message interchange. This algorithm needs $O(v^{\frac{3}{2}})$ communication complexity.

Here we are trying to establish a network in which number of nodes k can be generalized. Then we develop an Algorithm for $v = k^2$ nodes, where $k(\geq 2)$ is any positive integer and we show that in this case also the communication complexity is $O(v\sqrt{v})$. We present a new synchronous dynamic distributed load balancing algorithm on a network having $v = k^2$ nodes, where $k(\geq 2)$ is any positive integer. Our algorithm needs a complete graph of k nodes for an efficient communication system. Each node receives workload information from all other nodes with two-round message interchange without redundancy and hence it requires $O(v\sqrt{v})$ communication complexity.

2 Preliminaries

In this section we state the following well known definitions for developing the main result.

Simple Graph: Let $G = (V, E)$ be a graph, in which V denotes the vertex set and E denotes the edge set. Then G is said to be a simple graph if every pair vertices may be connected by only one edge, i.e., there is no parallel edge or loop.

Complete Graph: Let $G = (V, E)$ be a simple graph, then G is said to be a Complete graph if every pair of vertices are connected. The complete graph on n vertices has $\frac{n(n-1)}{2}$ edges, and is denoted by K_n. The disjoint union of a set of complete graphs forms a strongly regular graph.

3 Main Result

The distributed network has been modeled as a complete graph $G = (V, E)$, where each vertex is associated with nodes and each edge represents a path between nodes. An edge e_{ij} denote the flow from i^{th} node to j^{th} node. In this paper, we develop an Algorithm for finding the workload information process of the network containing $v = k^2$ nodes, where $k \geq 2$, integer.

3.1 Algorithm

We generate an Algorithm for the load balancing distributed network containing $v = k^2$ nodes, where k is any integer ≥ 2.

Step-I: Select a number k and compute $v = k^2$.

Step-II: Form k disjoint sets $G_1, G_2, G_3, ..., G_k$ each of which containing k nodes, where $G_i = \{v_{0i,1}, v_{0i,2}, v_{0i,3}, ..., v_{0i,k}\}$ and $v_{0i,j}$ denotes the jth node of G_i for $i = 1, 2, 3, ..., k$ at the initial stage i.e. at time T_0.

Step-III: $v_{0i,j} \leftrightarrow v_{0i,p}, \forall j, p = 1, 2, 3, ..., k;\ j \neq p$.
In this step each node within a set G_i will simultaneously send the workload information to its remaining nodes without any redundancy and at the same time interval t i.e. at T_t the process will be going on for each G_i, for $i = 1, 2, 3, ..., k$.
Now rearranging the elements of the sets $G_i, i = 1, 2, 3, ..., k$ and construct the new set G_i^* such that $G_i^* = \{v_{t1,i}, v_{t2,i}, v_{t3,i}, ..., v_{tk,i}\}$.

Step-IV: At time T_t we shall follow the same process for G_i^* as that of G_i in ***Step-III***, as follows:
Again $v_{tj,i} \leftrightarrow v_{tq,i},\ j \neq q\ \ and\ j, q = 1, 2, 3, ..., k$. After this second round message interchanging, i.e. at time T_{2t} all k^2 nodes obtain the workload information from all others nodes.

3.2 Theorem

From the above Algorithm, we can conclude the following fact and present this as a Theorem.

Theorem 1. According to the Algorithm, each node will obtain the workload information from all other nodes at T_{2t}.

Proof: Let T_0 be the initial time of the process. At this time, we distribute all k^2 nodes equally into k sets, say $G_i, i = 1, 2, 3, ..., k$ and at this level, all the nodes of G_i set for $i = 1, 2, 3, ..., k$; denoted by $v_{0i,j}$, do not have workload information of other nodes. Now, within the time t each node $v_{0i,j}$ from the set $G_i, i = 1, 2, 3, ..., k$ will exchange the workload information from the remaining nodes of G_i. Within the same time interval this process will be executed in the set $G_i, \forall i$. So, at time T_t, all nodes will receive the workload information of each other within the set $G_i, \forall i$ and consequently, the nodes are represented as $v_{ti,j}$.
After this stage, we rearrange the nodes to generate the new set G_i^* from the previous set G_i. Then the nodes $v_{ti,j}$ will become $v_{tj,i}, \forall i, j \in \{1, 2, ..., k\}$, which are regarded as j^{th} element of the new set G_i^*, $for\ i = 1, 2, 3, ..., k$. Then over the next time interval t, every nodes $v_{tj,i}$ of G_i^* will exchange the workload information among themselves $\forall i = 1, 2, 3, ..., k$. Hence at time T_{2t} all k^2 nodes will receive the workload information from each other.
Here the distributed network has been developed by the help of complete graph. For each set G_i initially forms a complete graph for $i = 1, 2, 3, ..., k$.

So, the corresponding adjacency matrix for this graph will be all 1's, i.e. $J_{k \times k}$. Finally the union of all these k disjoint complete graphs forms a strongly regular graph, with parameters $(k^2, k-1, k-2, 0)$.

4 Conclusion

This approach can certainly be considered as much simpler and improved version of the previously known methods for handling the efficient load balancing Algorithms in the literature. It also covers more number of nodes in general. To carry out this algorithm, each node should receive the workload information from all other nodes without redundancy after two rounds of message exchanging. Hence the Communication complexity is $O(v\sqrt{v})$, as opposed to the $O(v^2)$ communication complexity.

References

1. Ford, L., Fulkerson, D.: Flow in Network. Princeton University Press, Princeton (1962)
2. Wu, M.: On runtime parallel scheduling for processor load balancing. IEEE Transactions on Parallel and Distributed System 8(2), 173–185 (1997)
3. Liu, C.L.: Introduction to Combinatorial Mathematics, pp. 359–383. McGraw-Hill, NY (1968)
4. Lee, O., Yoo, S., Park, B., Chung, I.: The design and analysis of an efficient load balancing algorithm employing the symmetric balanced incomplete block design. Information Sciences 176, 2148–2160 (2006)

Communication Model in Agents System on Multi-modal Logic S5n

Takashi Matsuhisa

Department of Natural Sciences, Ibaraki National College of Technology
Nakane 866, Hitachinaka-shi, Ibaraki 312-8508, Japan
takashimatsuhisa@hotmail.com

Abstract. This article investigates a foundation of communication in multi-agents system. A communication logic based on the multi-modal logic **S5n** is presented from multi-modal logical point of view. We introduce models for the logic as knowledge revision processes on agents' knowledge by communication through messages, and the completeness theorem is shown: the communication logic is determined by the class of all the models.

Keywords: Communication logic, Completeness theorem, Messages, Multi-modal logic **S5n**, Knowledge revision.

AMS 2000 Mathematics Subject Classification: Primary 03B45, Secondary 91A35.

Journal of Economic Literature Classification: C62, C78.

1 Introduction

This paper aims to treat communication model through messages among agents from multi-modal logical points of view.

Recently, researchers in AI, and computer science, economics become entertained lively concerns about relationships between knowledge and actions. The most interest to us is the situation involving the knowledge of a group of agents rather than of just a single agent. At what point does agents sufficiently know to stop gathering information and make decisions? There are also concerns about the complexity of computing knowledge. In investigating the situation, a communication process among them plays important role to reach consensus among all agents.

In this paper formal model of communication is treated: We present a communication logic based on the multi-modal logic **S5n**, and we introduce models for the logic as revision processes on agents' knowledge by communication through messages, and show that he completeness theorem; i.e., the communication logic is determined by the models.

This paper is organised as follows. Section 2 recalls the multi-modal logic **S5n** together with the semantics models. In Section 3 we present the communication logic **CS5n** based on the logic **S5n**, and introduce the communication model

J. Zhang (Ed.): ICAIC 2011, Part V, CCIS 228, pp. 647–654, 2011.

as semantics for the logic **CS5n**. Section 4 gives the proof of the completeness theorem of the logic **CS5n**. In Section 5 we conclude with remarks on 'agreeing to disagree theorem' on the finite models for the logic **CS5n**.

2 Multi-modal Logic

2.1 Syntax

Let N be a finite set of *agents* $\{1, 2, \cdots, k, \cdots, n\}$, and T the time horizontal line $\{0, 1, 2, \cdots t, \cdots\}$. The *language* is founded on as follows: The *sentences* of the language form the least set containing each *atomic* sentence $\mathbf{P}_m (m = 0, 1, 2, \dots)$ and closed under the following operations: nullary operators for *falsity* $\bot$ and for *truth* $\top$; unary and binary syntactic operations for *negation* $\neg$, *conditionality* $\rightarrow$ and *conjunction* $\wedge$, *disjunction* $\vee$, respectively; two unary operations for *modalities* $\Box_i$, $\Diamond_i$ together with countably many binary operations $\Box_i^t, \odot_i^t$ for $i \in N, t \in T$. Other such operations are defined in terms of those in usual ways.

The intended interpretation of $\Box_i\varphi$ is the sentence that 'agent i knows a sentence φ,' the sentence $\Diamond_i\varphi$ is interpreted as the sentence that 'i thinks a sentence φ is possible,' the sentence $\odot_i^t(\varphi|\psi)$ is meant to express propositions of the *messages* that i sent at time t and $\Box_i^t(\varphi|\psi)$ for $t \in T$ is meant that agent i knows φ when he/she receives ψ on communication through messages at time t.

2.2 Multi-modal Logic S5n

A *multi-modal logic* L is a set of sentences containing all truth-functional tautologies and closed under substitution and modus ponens. A multi-modal logic L' is an *extension* of L if $L \subseteq L'$. A sentence φ in a modal logic L is a *theorem* of L, written by $\vdash_L \varphi$. Other proof-theoretical notions such as *L-deducibility, L-consistency, L-maximality* are defined in usual ways. (See, Chellas [2].)

A *normal system* of multi-modal logic L is a multi-modal logic containing $(\mathrm{Df}_\Diamond)$ and is closed under the rules of inference $(\mathrm{RK}_{\Box_i})$.

$(\mathrm{Df}_\Diamond)$ $\Diamond_i\varphi \longleftrightarrow \neg\Box_i\neg\varphi$;

$(\mathrm{RK}_{\Box_i})$ $\dfrac{(\varphi_1 \wedge \varphi_2 \wedge \cdots \wedge \varphi_k) \longrightarrow \psi}{(\Box_i\varphi_1 \wedge \Box_i\varphi_2 \wedge \cdots \wedge \Box_i\varphi_k) \longrightarrow \Box_i\psi}$ $(k \geqq 0)$

The *logic* **S5n** is the minimal normal system of multi-modal logic containing the following schemas:

(T) $\Box_i\varphi \longrightarrow \varphi$

(5) $\Diamond_i\varphi \longrightarrow \Box_i\Diamond_i\varphi$

The logic **S5n** contains the schema:

(4) $\Box_i\varphi \longrightarrow \Box_i\Box_i\varphi$

2.3 Knowledge Structure, Model and Truth

Let Ω be a non-empty set called a *state space* and 2^{Ω} the field of all subsets of Ω. Each member of 2^{Ω} is called an *event* and each element of Ω called a *state*. A *knowledge structure* is a tuple $\langle \Omega, (K_i)_{i\in N}, (\Pi_i)_{i\in N}\rangle$ in which Ω is a state space and $K_i : 2^{\Omega} \to 2^{\Omega}$ is agent i's *knowledge operator*. The interpretation of the event K_iE is that 'i knows E.' Π_i is i's *possibility operator* on 2^{Ω} defined by $\Pi_i E = \Omega \setminus K_i(\Omega \setminus E)$ for every E in 2^{Ω}. The interpretation of $\Pi_i E$ is that 'E is possible for i.'

A *model on* a knowledge structure is a tuple $\mathcal{M} = \langle \Omega, (K_i)_{i\in N}, (\Pi_i)_{i\in N}, V\rangle$ in which $\langle \Omega, (K_i)_{i\in N}, (\Pi_i)_{i\in N}\rangle$ is a knowledge structure and a mapping V assigns either **true** or **false** to every $\omega \in \Omega$ and to every atomic sentence $\mathbf{P}_m$. The model $\mathcal{M}$ is called *finite* if Ω is a finite set.

Definition 1. By $\models^{\mathcal{M}}_{\omega} \varphi$, we mean that a sentence φ is *true* at a state ω in a model $\mathcal{M}$. *Truth at a state* ω in $\mathcal{M}$ is defined by the inductive way as usuals:

1. $\models^{\mathcal{M}}_{\omega} \mathbf{P}_m$ if and only if $V(\omega, \mathbf{P}_m) = \mathbf{true}$, for $m = 0, 1, 2, \ldots$;
2. $\models^{\mathcal{M}}_{\omega} \top$, and not $\models^{\mathcal{M}}_{\omega} \bot$;
3. $\models^{\mathcal{M}}_{\omega} \neg\varphi$ if and only if not $\models^{\mathcal{M}}_{\omega} \varphi$;
4. $\models^{\mathcal{M}}_{\omega} \varphi \longrightarrow \psi$ if and only if $\models^{\mathcal{M}}_{\omega} \varphi$ implies $\models^{\mathcal{M}}_{\omega} \psi$;
5. $\models^{\mathcal{M}}_{\omega} \varphi \wedge \psi$ if and only if $\models^{\mathcal{M}}_{\omega} \varphi$ and $\models^{\mathcal{M}}_{\omega} \psi$;
6. $\models^{\mathcal{M}}_{\omega} \varphi \vee \psi$ if and only if $\models^{\mathcal{M}}_{\omega} \varphi$ or $\models^{\mathcal{M}}_{\omega} \psi$;
7. $\models^{\mathcal{M}}_{\omega} \Box_i\varphi$ if and only if $\omega \in K_i(||\varphi||^{\mathcal{M}})$, for $i \in N$;
8. $\models^{\mathcal{M}}_{\omega} \Diamond_i\varphi$ if and only if $\omega \in \Pi_i(||\varphi||^{\mathcal{M}})$, for $i \in N$:

Where $||\varphi||^{\mathcal{M}}$ denotes the set of all the states in $\mathcal{M}$ at which φ is true; this is called the *truth set of* φ.

We say that a sentence φ is *true in the model* $\mathcal{M}$ and write $\models^{\mathcal{M}} \varphi$ if $\models^{\mathcal{M}}_{\omega} \varphi$ for every state ω in $\mathcal{M}$. A sentence is said to be *valid in* a knowledge structure if it is true in every model on the knowledge structure. Let Γ be a set of sentences. We say that $\mathcal{M}$ is a *model for* Γ if every member of Γ is true in $\mathcal{M}$. A knowledge structure is said to be *for* Γ if every member of Γ is valid in it. Let $\mathbf{C}$ be a class of models on a knowledge structure. A multi-modal logic L is *sound with respect to* $\mathbf{C}$ if every member of $\mathbf{C}$ is a model for L. It is *complete with respect to* $\mathbf{C}$ if every sentence valid in all members of $\mathbf{C}$ is a theorem of L. A multi-modal logic L is said to be *determined by* $\boldsymbol{C}$ if L is sound and complete with respect to $\boldsymbol{C}$. It is said to have the *finite model property* if it is sound and complete with respect to the class of all finite models in $\mathbf{C}$. We denote by $\mathbf{C}_F$ the class of all finite models in $\mathbf{C}$.

Theorem 1. A normal system of modal logic L has the finite model property; i.e., $\vdash_L \varphi$ if and only if $\models^{\mathcal{M}} \varphi$ for all $\mathcal{M} \in \mathbf{C}_F$. In particular, **S5n** has the finite model property.

Proof. Chellas [2].

3 Communication

We assume that the players communicate by sending *messages*. A *protocol* is a mapping $\Pr : T \to N \times N, t \mapsto (s(t), r(t))$ such that $s(t) \neq r(t)$. Here t stands for *time* and $s(t)$ and $r(t)$ are, respectively, the *sender* and the *recipient* of the communication which takes place at time t. We consider the protocol as the directed graph whose vertices are the set of all players N and such that there is an edge (or an arc) from i to j if and only if there are infinitely many t such that $s(t) = i$ and $r(t) = j$.

A protocol is said to be *fair* if the graph is strongly connected; in words, every player in this protocol communicates directly or indirectly with every other player infinitely many often. It is said to contain a *cycle* if there are players $i_1, i_2, \ldots, i_k$ with $k \geq 3$ such that for all $m < k$, i_m communicates directly with i_{m+1}, and such that i_k communicates directly with i_1. The communication is assumed to proceed in *rounds*[1] The protocol is said to be *acyclic* if the graph has no cycle. (C.f.: [4], [3])

3.1 Communication Logic CS5n

Into the language of the logic **S5n** we introduce sentences of the forms $\odot_i^t(\varphi|\psi)$ and $\Box_i^t(\varphi|\psi)$ for $i \in N, t \in T$. The sentence $\odot_i^t(\varphi|\psi)$ is meant to express propositions of the *messages* that i sent at time t and $\Box_i^t(\varphi|\psi)$ expresses propositions of the form that that agent i knows φ when he/she receives ψ on communication through messages at time t.

A *normal system of communication logic* based on the logic **S5n** according to a fair protocol $Pr : T \to N \times N$ is an extension of **S5n** is closed under the rules of inference (RCE$_\mu$) and (RCK$_\mu$)): For $\mu = \odot, \Box$ and each $t \in T$,

$$\text{(RCE}_\mu) \quad \frac{\psi_1 \longleftrightarrow \psi_2}{\mu_i^t(\varphi|\psi_1) \longleftrightarrow \mu_i^t(\varphi|\psi_2)};$$

$$\text{(RCK}_\mu) \quad \frac{\varphi_1 \wedge \varphi_2 \wedge \cdots \wedge \varphi_k \longrightarrow \varphi}{\mu_i^t(\varphi_1|\psi) \wedge \mu_i^t(\varphi_1|\psi) \wedge \cdots \wedge \mu_i(\varphi_k|\psi) \longrightarrow \mu_i^t(\varphi|\psi)} \quad (k \geqq 0)$$

Definition 2. The *communication logic* **CS5n** is the smallest normal system of communication logic based on the logic **S5n** which contains the schemas: For every $t \in T$ with $t \geqq 1$,

(CDef) $\Box_i \varphi \longleftrightarrow \Box_i^0(\varphi|\psi)$
(UN) $\Box_i^t(\varphi|\bot) \longrightarrow \bot$;
(CS5) $\mu(*|\psi)$ gives a unary modal operator for **S5n** for $\mu = \odot_i^t, \Box_i^t$;
(MESDef)

$$\Box_j^{t+1}(\varphi|\psi) \longleftrightarrow (\Box_j^t \wedge \odot_i^t)(\varphi|\psi) \quad \text{for } (i,j) = \Pr(t);$$

$$\Box_j^{t+1}(\varphi|\psi) \longleftrightarrow \Box_j^t(\varphi|\psi) \quad \text{for } (i,j) \neq \Pr(t);$$

(COH) $\Box_i^t(\varphi|\psi) \longrightarrow \Box_i^{t+1}(\varphi|\psi)$ for $\psi \neq \bot$;
(COHMES) $\odot_i^t(\varphi|\psi) \longrightarrow \Box_i^t(\varphi|\psi)$

[1] There exists a time m such that for all t, $\Pr(t) = \Pr(t+m)$. The *period* of the protocol is the minimal number of all m such that for every t, $\Pr(t+m) = \Pr(t)$.

3.2 Communication Model

Let Z be a set of decisions, which set is common for all agents. By a *system design function* (decision function) we mean a mapping f of $2^{\Omega} \times 2^{\Omega}$ into the set of decisions Z. We refer the following property of the function f: Let X be an event.

DUC (Disjoint Union Consistency): For every pair of disjoint events S and T, if $f(X;S) = f(X;T) = d$ then $f(X;S \cup T) = d$;

Remark 1. If f is intended to be a posterior probability, we assume given a probability measure μ on a state-space Ω which is common for all agents; precisely, for some event X of Ω, $f(X;\cdot)$ is given by $f(X;\cdot) = \mu(X|\cdot)$. This function satisfies (DUC).

Definition 3. Let $\mathcal{M} = \langle \Omega, (K_i)_{i\in N}, (\Pi_i)_{i\in N}, V\rangle$ be a model for the logic **S5n**. A *communication model* $\mathcal{C}$ is a triple

$$\langle \Omega, (K_i)_{i\in N}, (\Pi_i)_{i\in N}, V, \mathrm{Pr}, f, (Q_i^t), (W_i^t) \mid (i,t) \in N \times T\rangle$$

in which $\mathrm{Pr}(t) = (s(t), r(t))$ is a fair protocol such that for every t, $r(t) = s(t+1)$, f is a system design function , and $Q_i^t(*|*)$ is the mapping of $2^{\Omega} \times \Omega$ into 2^{Ω} for i at time t that is defined inductively in the following way: We assume given a mapping $Q_i^0(X;\omega) := \Pi_i(\{\omega\})$. Suppose Q_i^t is defined. Let

$$W_i^t(X;\omega) = \{\xi \in \Omega \mid f(X; Q_i^t(X;\{\xi\})) = f(X; Q_i^t(X;\{\omega\}))\}.$$

This is interpreted as the message that i sends to j at time t. Then $Q_j^{t+1}(X|\omega)$ is defined as follows:

- If j is not a recipient of a message at time $t+1$, then $Q_j^{t+1}(X;\omega) = Q_j^t(X;\omega)$.
- If j is a recipient of a message at time $t+1$, then $Q_j^{t+1}(X;\omega) = Q_j^t(X;\omega) \cap W_{s(t)}^t(X;\omega)$.

Truth at a state ω in $\mathcal{C}$ is defined by the usual way. In particular,

$$\models_{\omega}^{\mathcal{C}} \Box_i^t(\varphi|\psi) \quad \text{if and only if} \quad Q_i^t(||\psi||^{\mathcal{C}};\omega) \subseteq ||\varphi||^{\mathcal{C}}$$
$$\models_{\omega}^{\mathcal{C}} \odot_i^t(\varphi|\psi) \quad \text{if and only if} \quad W_i^t(||\psi||^{\mathcal{C}};\omega) \subseteq ||\varphi||^{\mathcal{C}}$$

We can show that

Theorem 2. *The logic* **CS5n** *is determined by all communication models.*

4 Proof of Theorem 2

The proof will be given in the same line as described in Chellas [2].

4.1 Proof Sets

Let L be a system of communication logic based on the logic **S5n** according to a fair protocol Pr. We recall the *Lindenbaum's lemma*:

Lemma 1. *Every L-consistency set of sentences has an L-maximal extension.*

This is because L includes the ordinary propositional logic.

We call the extension an *L-maximally consistent set.* As a consequence, *a sentence in L is deducible from a set of sentences Γ if and only if it belongs to every L-maximally consistent set of Γ*, and thus

Theorem 3. *Let L be a system of communication logic on the logic* **S5n** *according Pr. A sentence is a theorem of L if and only if it is a member of every L-maximally consistent set of sentences.*

We denote by $|\varphi|_L$ the class of L-maximally consistent sets of sentences containing the sentence φ; this is called the *proof set* of φ. We note that

Corollary 1. *Let L be a system of communication logic on the logic* **S5n** *according Pr.*

(i) *A sentence φ is a theorem of L if and only if $|\varphi|_L = \Omega_L$;*
(ii) *A sentence $\varphi \to \psi$ is a theorem of L if and only if $|\varphi|_L \subseteq |\psi|_L$.*

4.2 Canonical Model

Let L be a system of communication logic on the logic **S5n** according Pr.

Definition 4. The *canonical model* for L is the model

$$\mathcal{C}_L = \langle N, \Omega_L, (K_{Li}), (\Pi_{Li}), V_L, (Q_{Li}^{\ t}), (W_{Li}^{\ t}) \mid (i,t) \in N \times T \rangle$$

for L consisting of:

1. Ω_L is the set of all the L-maximally consistent sets of sentences;
2. $K_{Li} : 2^{\Omega_L} \to 2^{\Omega_L}$ is given by

$$\omega \in K_{Li}(E) \text{ if and only if there exists } \varphi \in L \text{ such that } |\varphi|_L = E \text{ and } \Box_i\varphi \in \omega$$

3. $\Pi_{Li} : 2^{\Omega_L} \to 2^{\Omega_L}$ is given by

$$\omega \in \Pi_{Li}(E) \text{ if and only if there exists } \varphi \in L \text{ such that } |\varphi|_L = E \text{ and } \Diamond_i\varphi \in \omega$$

4. V_L is the mapping defined by $V_L(\omega, \mathbf{P}_m) = \{\omega \in \Omega^L \mid \mathbf{P}_m \in \omega\}$ for $m = 0, 1, 2, \dots$.
5. $Q_{Li}^{\ t} : 2^{\Omega_L} \times \Omega_L \to 2^{\Omega_L}$ is given as follows: For every $\omega \in \Omega_L$,

$$\Box_i^t(\phi|\psi) \in \omega \quad \text{if and only if} \quad Q_{Li}^{\ t}(|\psi|_L; \omega) \subseteq |\varphi|_L$$

6. $W_{Li}^{\ t} : 2^{\Omega_L} \times \Omega_L \to 2^{\Omega_L}$ is given as follows:

$$W_i^t(|\psi|_L; \omega) = \{\xi \in \Omega_L \mid f(|\psi|_L; Q_i^t(|\psi|_L; \{\xi\})) = f(|\psi|_L; Q_i^t(|\psi|_L; \{\omega\}))\}.$$

We can easily observe that

Proposition 1. *The canonical model $\mathcal{C}_L$ is a model for a system L of communication on the multi-modal logic* **S5n**. □

The important result about a canonical model is the following:

Basic theorem 1. *Let $\mathcal{C}_L$ be the canonical model for a system of communication logic L. Then for every sentence φ, $\models^{\mathcal{C}_L} \varphi$ if and only if $\vdash_L \varphi$. In other words, $||\varphi||^{\mathcal{C}_L} = |\varphi|_{\mathcal{C}_L}$.*

Proof. By induction on the complexity of φ. We treat only that φ is $\Box_i\psi$ and φ is $\Box_i^t(\psi|\chi)$. As an inductive hypothesis we assume that $||\psi||^{\mathcal{C}_L} = |\psi|_{\mathcal{C}_L}$ and $||\chi||^{\mathcal{C}_L} = |\chi|_{\mathcal{C}_L}$. Then

$$\begin{aligned}
\models^{\mathcal{C}_L} \Box_i\varphi \quad &\text{if and only if} \quad \omega \in K_{Li}(||\psi||^{\mathcal{C}_L}) \text{ for every } \omega \in \Omega_L, \\
&\qquad \text{by the definition of validity ;} \\
&\text{if and only if} \quad \omega \in K_{Li}(|\psi|^{\mathcal{C}_L}) \text{ for every } \omega \in \Omega_L, \\
&\qquad \text{by the inductive hypothesis as above;} \\
&\text{if and only if} \quad \vdash_L \Box_i\psi \\
&\qquad \text{by the definition of canonical model.} \\
\models^{\mathcal{C}_L} \Box_i^t(\psi|\chi) \quad &\text{if and only if} \quad Q_{Li}^{\ t}(||\chi||^{\mathcal{C}_L}; \omega) \subseteq ||\psi||^{\mathcal{C}_L} \text{ for every } \omega \in \Omega_L, \\
&\qquad \text{by the definition of validity ;} \\
&\text{if and only if} \quad Q_{Li}^{\ t}(|\chi|_L; \omega) \subseteq |\psi|_L \text{ for every } \omega \in \Omega_L, \\
&\qquad \text{by the inductive hypothesis as above;} \\
&\text{if and only if} \quad \vdash_L \Box_i^t(\psi|\chi) \\
&\qquad \text{by the definition of canonical model.}
\end{aligned}$$

Proof of Theorem 2. Soundness has been already observed in Proposition 1. The completeness will be shown as follows: Suppose that some sentence φ is not a theorem in L. In view of Basic theorem 1, it follows that φ is not valid for a canonical model $\mathcal{C}_L$. □

5 Concluding Remarks

For a finite model for the logic **CS5n** we can show the agreeing to disagree theorem ([1], [3]) that every agents can be reached consensus after long run communication: If Pr is acyclic then for sufficient large $t \in T$ and if f satisfies (**PUD**), $f(||\varphi||^{\mathcal{C}}; Q_i^t(||\varphi||^{\mathcal{C}}; \{\omega\})) = f(||\varphi||^{\mathcal{C}}; Q_j^t(X; \{\omega\}))$ for any $\omega \in \Omega$.@

References

1. Aumann, R.J.: Agreeing to disagree. Annals of Statistics 4, 1236–1239 (1976)
2. Chellas, B.F.: Modal Logic: An introduction. Cambridge University Press, Cambridge (1980)
3. Krasucki, P.: Protocol forcing consensus. Journal of Economic Theory 70, 266–272 (1996)
4. Parikh, R., Krasucki, P.: Communication, consensus, and knowledge. Journal of Economic Theory 52, 178–189 (1990)

Answer Set Programming Modulo Theories

Yisong Wang[1] and Mingyi Zhang[2]

[1] Department of Computer Science, Guizhou University, Guiyang, China
[2] Guizhou Academy of Sciences, Guiyang, China

Abstract. In the paper, we propose a formalism of answer set programming with background theories, for which the modulo answer set semantics is presented. It is shown that the formalism (under the given semantics) is a generalization of normal logic programs under answer set semantics, and it can capture the SPT-PDB answer set semantics of logic programs with aggregate functions. It is also proved that the formalism coincides with the one defined by Shen and You in default manner whenever the background theory is propositional logic.

Keywords: Logic Programs, Answer Sets, Satisfiability Modulo Theories, Aggregate Functions.

1 Introduction

Logic programming based on answer set semantics, called *answer set programming* (ASP in short), has been considered a promising paradigm for declarative problem solving. The general idea is to encode a problem by a logic program such that the answer sets of the program correspond to the solutions to the problem. With the development of efficient ASP solvers such as CLASP, CMODELS, SMODELS, and DLV, ASP has been applied to a number of practical domains [1].

It is well-known that applications in artificial intelligence and formal methods for hardware and software development have greatly benefited from the recent advances in SAT which has efficient implementations and various applications. With respect to some background theory which fixes the interpretations of certain predicate and function symbols, such as theory of integers, theory of equality with uninterpreted functions (EUF in short) and theory of arrays, the satisfiability of formulas, called *Satisfiability Modulo Theories* (SMT in short) has caused extensive study. The SMT solvers or techniques have been integrated into interactive theorem provers for high-order logic, extended static checkers, verification system, formal CASE environments, model checkers, certifying compilers and unit test generators [2].

Due to the declarative nature of ASP and the utilities of modulo theories, it is of both theoretical and practical interest to integrate background theories into ASP following the spirit of SMT. For this purpose, we propose a formalism of answer set programming with background theories, for which modulo answer set semantics is defined. The formalism under the modulo answer set semantics is a generalization of normal logic programs. We show that the formalism can

J. Zhang (Ed.): ICAIC 2011, Part V, CCIS 228, pp. 655–661, 2011.

capture logic programs with aggregate functions under the SPT-PDB semantics [7,9]. The modulo answer set semantics coincides with the one defined in default manner whenever the background theory is propositional logic [8]. Due to space limitation, the technical proofs of propositions is omitted. A full version is available at `http://210.40.7.234/~wys/ICTMF-11.pdf`.

2 ASP Modulo Theories: Syntax and Semantics

We assume an underlying first-order language $\mathcal{L}$. We also assume that the syntax and semantics of background theories are well-defined, such as *difference logic*, the *theory of integers, rationals and reals*, the *theory of equality with uninterpreted functions* (*EUF* in short) and so on [2]. As mentioned in [6], even though the satisfiability of two theories $\mathcal{T}_1$ and $\mathcal{T}_2$ are decidable respectively, the satisfiability of their combination $\mathcal{T}_1 \oplus \mathcal{T}_2$ may be undecidable. In what follows, we assume the satisfiability of ground combined theories are always decidable, in addition that the signatures of background theories are pairwise disjoint, except for those from $\mathcal{L}$.

A *(modulo) rule* r is of the form

$$h \leftarrow \varphi \tag{1}$$

where h is a ground atom over $\mathcal{L}$ and φ is a ground formula. Intuitively, it means that "if φ holds then h does". h and φ are called its *head* written *Head*(r), and *body* written *Body*(r) respectively. A *(modulo) program* P is a finite set of modulo rules. By $\mathbf{T}_P$ we denote the combined background theories of program P, and by HB_P we mean the set of Herbrand base (atoms) of $\mathcal{L}$. While a *normal logic program* is defined as a finite set of rules of the form

$$A \leftarrow B_1, \ldots, B_m, not\, B_{m+1}, \ldots, not\, B_n, (n \geq m \geq 0) \tag{2}$$

where A, B_i $(1 \leq i \leq n)$ are atoms of a propositional language [4].

Let P be a modulo program and $M \subseteq HB_P$. The *inverse reduct* of P relative to M, written $IR(P, M)$, is the modulo program obtained from P by replacing every rule "$h \leftarrow \varphi$" of P with "$h \leftarrow \varphi[\overline{M}/\bot]$" where $\overline{M} = HB_P \setminus M$ and $\varphi[M/\bot]$ stands for the formula obtained from φ by replacing every atom $a \in M$ with $\bot$. In what follows we identify a finite set M of formulas as the conjunction of the formulas in M when it is clear from its context. Given a modulo program P and $M \subseteq HB_P$, we define an operator $\delta_P : 2^{HB_P} \longrightarrow 2^{HB_P}$ as follows:

$$\delta_P(M) = \{h | (h \leftarrow \varphi) \in P \ s.t.\ M \vdash_{\mathbf{T}_P} \varphi\}$$

where $M \vdash_{\mathbf{T}_P} \varphi$ stands for $M \cup \mathbf{T}_P \vdash \varphi$. It is evident that δ_P is monotonic, i.e,. if $M_1 \subseteq M_2$ then $\delta_P(M_1) \subseteq \delta_P(M_2)$. Thus the least fixpoint of δ_P always exists, denoted by $lfp(\delta_P)$, which can be iteratively constructed:

$$\delta_P^0 = \emptyset;$$
$$\delta_P^{n+1} = \delta_P(\delta_P^n).$$

We say that M is a *(modulo) answer set* of P if $M = lfp(\delta_{IR(P,M)})$.

Example 1. Let's consider the following modulo programs:

- Let $P_1 = \{a \leftarrow \neg b, \quad b \leftarrow \neg a\}$. Suppose $M_1 = \{a\}$, we have that $IR(P, M_1) = \{a \leftarrow \neg\bot, \quad b \leftarrow \neg a\}$ and $lfp(\delta_{IR(P,M_1)}) = \{a\}$. So that M_1 is a modulo answer set of P_1. It is similar to check that $\{b\}$ is the other modulo answer set of P_1.
- Let $P_2 = \{a \leftarrow \neg a\}$. Suppose $M_1 = \emptyset$ and $M_2 = \{a\}$. We have that $IR(P_2, M_1) = \{a \leftarrow \neg\bot\}$ and $IR(P_2, M_2) = P_2$. It is not difficult to verify that $lfp(\delta_{IR(P_2,M_1)}) = \{a\}$ and $lfp(\delta_{IR(P_2,M_2)}) = \emptyset$. Thus neither M_1 is a modulo answer set of P_2, nor is M_2.
- Let P_3 consists of

$$p \leftarrow \neg q \wedge (x \geq 1 \wedge y > 1 \supset x + y \geq 2),$$
$$q \leftarrow \neg p \wedge (x + y < 2).$$

We note that $\mathbf{T}_{P_3}$ is the theory of integers. It is easy to see that P has a unique answer set $\{p\}$.

It is easy to see that, in terms of the modulo answer set semantics, modulo programs are nonmonotonic, e.g., we know that $\{a\}$ is a modulo answer set of the modulo program P_1 in the above example, but $\{a, b\}$ is not a modulo answer set of P_1. The below proposition shows that modulo answer set semantics is a generalization of standard answer set semantics of normal logic programs.

Proposition 1. *Let P be a logic program and $I \subseteq HB_P$. We have that I is an answer set of P if and only if I is a modulo answer set of P' where P' is obtained from P by replacing each rule of the form (2) by*

$$A \leftarrow \bigwedge_{1 \leq i \leq m} B_i \wedge \bigwedge_{m+1 \leq j \leq n} \neg B_j.$$

3 An Application: Programs with Aggregate Functions

3.1 Programs with Aggregates Functions

As aggregate functions, such as *sum, count, max* etc., are often used in database query language, they are introduced into answer set programming for practice [3]. An *aggregate atom* is an expression of the form $f\{S\} \prec$ g where g is a constant (called *guard*), $f \in \{sum,\ count,\ max,\ min,\ times\}$, and S is a set whose elements have the form $\langle \mathbf{c} : \psi \rangle$ where $\mathbf{c}$ is a list of constants and ψ is a ground conjunction of literals (i.e., an atom or its default negation: p or $not\, p$), and $\prec \in \{<, \leq, >, \geq, =\}$ that are interpreted in a standard way. For example,

$$max\{\langle 1 : r(2), not\, a(2,1) \rangle, \langle 2 : r(2), not\, a(2,2) \rangle\} > 1 \qquad (3)$$

is an aggregate atoms.

A *(logic) program with aggregate functions* consists of a finite set of rules of the form (2) where each $B_i (1 \leq i \leq n)$ may be a (classical) atom or an aggregate atom. Let M be a set of (classical) atoms and $f(S) \prec \mathrm{g}$. Let $\alpha = (f(S) \prec \mathrm{g})$ be an aggregate atom. A set of atoms I *satisfies* α, denoted $I \models \alpha$, whenever

$$f\{c_i | \langle c_i, \psi_i \rangle \in S \;\&\; I \models \psi_i\} \prec \mathrm{g}.$$

In particular, if $\{c_i | \langle c_i, \psi_i \rangle \in S \;\&\; I \models \psi_i\}$ is an empty set then we define the truth value of α under I is $\bot$ (since the values of $min\{\}$ and $max\{\}$ are undefined) unless $f \in$ {times, sum, count}, where we define $sum\{\} = count\{\} = 0$ and $times\{\} = 1$. For instance, the aggregate atom (3) is satisfied by $I_1 = \{r(2)\}$ but not by $I_2 = \{r(2), a(2,2)\}$ and its truth value under any interpretation I is $\bot$ if $r(2) \notin I$. The answer set semantics for logic programs with aggregate functions are different in literature. We consider the SPT-PDB semantics only.

Pelov, Denecker and Bruynooghe defined the answer set semantics of logic program with aggregate atoms by translating logic program with aggregate functions into ones with nested expressions [7]. Son, Pontelli and Tu proposed an answer set semantics for logic program with arbitrary constraint atoms [9], which can be applied to logic programs with aggregate atoms by regarding every aggregate atom A as an abstract constraint atom (D_A, C_A) and giving its semantics by conditional satisfaction where *not*-atoms are translated into their complements of the atoms. The two definitions are equivalent.

Example 2. Let's consider the logic program with aggregate functions [5].

Let $P_1 = \{p(1), \quad p(0) \leftarrow sum\{x : p(x)\} = 1\}$. The corresponding propositional formula in terms of Ferraris transformation is

$$p(1) \wedge [(p(0) \vee p(1)) \wedge (\neg p(0) \vee p(1)) \supset p(0)]$$

and the corresponding logic program with abstract constraint atoms is

$$\{p(1); \qquad p(0) \leftarrow (\{p(0), p(1)\}, \{\{p(1)\}, \{p(0), p(1)\}\})\}.$$

It is not difficult to verify that the unique answer set of P_1 is $\{p(0), p(1)\}$ in terms of SPT-PDB semantics.

3.2 Translating Aggregate Atoms into Modulo Theories

Let $\alpha = (f\{S\} \prec \mathrm{g})$ be an aggregate atom where $S = \langle c_1 : \psi_1 \rangle, \ldots, \langle c_k : \psi_k \rangle$. If $k \geq 1$ then we define $\gamma(\alpha)$ as follows:

$$\begin{cases} \left[\bigwedge_{1 \le i \le k} (x_i = ite(\psi_i, 1, 0))\right] \supset (sum\{x_1, \ldots, x_k\} \prec \mathrm{g}), & \text{if } f = count; \\ \left[\bigwedge_{1 \le i \le k} (x_i = ite(\psi_i, c_i, +\infty))\right] \supset (min\{x_1, \ldots, x_k\} \prec \mathrm{g}), & \text{if } f = min; \\ \left[\bigwedge_{1 \le i \le k} (x_i = ite(\psi_i, c_i, -\infty))\right] \supset (max\{x_1, \ldots, x_k\} \prec \mathrm{g}), & \text{if } f = max; \\ \left[\bigwedge_{1 \le i \le k} (x_i = ite(\psi_i, c_i, 0))\right] \supset (sum\{x_1, \ldots, x_k\} \prec \mathrm{g}), & \text{if } f = sum; \\ \left[\bigwedge_{1 \le i \le k} (x_i = ite(\psi_i, c_i, 1))\right] \supset (times\{x_1, \ldots, x_k\} \prec \mathrm{g}), & \text{if } f = times. \end{cases}$$

where $1 \le i \le k$, $x_i(1 \le i \le k)$ are distinct fresh variables (free constants), and $x = ite(\psi, c, c')$ stands for the formula

$$(\psi \supset (x = c)) \wedge (\neg\psi \supset (x = c')).$$

If $k = 0$ then we define $\gamma(\alpha)$ as follows:

$$\begin{cases} 0 \prec \mathrm{g}, \text{ if } f = count \text{ or } f = sum; \\ \bot, \quad\;\; \text{if } f = min \text{ or } f = max; \\ 1 \prec \mathrm{g}, \text{ if } f = times. \end{cases}$$

We let $\pi(\alpha)$ be the formula obtained from $\gamma(\alpha)$ by replacing $min\{x_1, \dots, x_k\} \prec \mathrm{g}$ and $max\{x_1, \dots, x_k\} \prec \mathrm{g}$ with the following formulas

$$\left[\left(y_1 = x_1 \wedge \bigwedge_{2 \le i \le k} (y_i = ite(y_{i-1} \le x_i, y_{i-1}, x_i))\right) \supset (y_k \prec \mathrm{g})\right]$$

and

$$\left[\left(y_1 = x_1 \wedge \bigwedge_{2 \le i \le k} (y_i = ite(y_{i-1} \ge x_i, y_{i-1}, x_i))\right) \supset (y_k \prec \mathrm{g})\right]$$

respectively, where $y_i(1 \le i \le k)$ are distinct fresh variables (free constants).

It is easy to see that $\pi(\alpha)$ is an SMT formula for the aggregate atoms since $sum\{x_1, \dots, x_k\} \prec \mathrm{g}$ is identical to $(x_1 + \ldots + x_k) \prec \mathrm{g}$, and $times\{x_1, \dots, x_k\} \prec \mathrm{g}$ is identical to $(x_1 \times \ldots \times x_k) \prec \mathrm{g}$ that are formulas over integers, rationales and reals. Please note that, $\pi(\alpha)$ is a formula built over integers (rationales or reals). And if it is not *times* aggregate atom then $\pi(\alpha)$ is a formula of LIA (resp. LRA) – the theory of Ints (resp. Reals). Further more, if $k \le 2$ for the *sum* and *count* aggregate atoms then $\pi(\alpha)$ is a formula of IDL (the theory of difference logic)[1].

For a logic program with aggregate functions P, we let $\pi(P)$ be the modulo program obtained from P by replacing every aggregate atom α by $\pi(\alpha)$ and *not* by $\neg$ respectively.

Example 3 (Continued from Example 2). We have that $\pi(P_1)$ consists of

$$p(1),$$
$$p(0) \leftarrow ((x_1 = ite(p(0), 0, 0)) \wedge (x_2 = ite(p(1), 1, 0)) \supset (x_1 + x_2 = 1)).$$

The background theory $\mathbf{T}_{\pi(P_1)}$ is difference logic. Let $M = \{p(0), p(1)\}$. We have that $IR(\pi(P_1), M) = \pi(P_1)$ since $HB_{\pi(P_1)} = M$ implies $\overline{M} = \emptyset$,
$\delta^1_{IR(\pi(P_1),M)} = \{p(1)\}$,
$\delta^2_{IR(\pi(P_1),M)} = \{p(0), p(1)\}$ (this is because
$p(1) \wedge (x_1 = ite(p(0), 0, 0)) \wedge (x_2 = ite(p(1), 1, 0) \vdash_{\mathbf{T}_{\pi(P_1)}} x_1 = 0 \wedge x_2 = 1))$ and
$\delta^3_{IR(\pi(P_1),M)} = lfp(\delta_{IR(\pi(P_1),M)}) = M$. Thus M is a (modulo) answer set of $\pi(P_1)$. It is easy to verify that $\pi(P_1)$ has no other modulo answer sets.

[1] http://combination.cs.uiowa.edu/smtlib/

Proposition 2. *Let P be a logic program with aggregate functions and M a set of atoms. M is an SPT-PDB answer set of P if and only if M is a modulo answer set of $\pi(P)$.*

4 Related Work

4.1 The Default Approach

Shen and You proposed an answer set semantics for *general logic programs* whose rules is has the form (1) but φ is a propositional formula [8]. Let P be a logic program and $I \subseteq HB_P$, we denote $I^\neg$ the set $\{\neg p | p \in HB_P \setminus I\}$. Let P be a logic program and W a theory. The *deductive set* $Th(P, W)$ is the smallest set of formulas satisfying (1) $W \subseteq Th(P, W)$, and (2) for each rule r of the form (1) in P, if $Th(P, W) \models \varphi$ then $h \in Th(P, W)$. We say that I is an *answer set* of P if $Th(P, I^\neg)$ is consistent and for each $p \in I$, $Th(P, I^\neg) \models p$. It has been shown that I is an answer set of P if and only if $Th(P, I^\neg) = I \cup I^\neg$ (cf. Theorem 1 of [8]). The semantics defined above for modulo programs is a generalization to theirs in the sense that,

Proposition 3. *Let P be a general logic program and M is a set of atoms. Then M is an answer set of P if and only if M is a modulo answer set of P.*

4.2 Niemelä's Proposal

Niemelä proposed an integration of ASP and SMT where ASP rules are extended with constraints supported by SMT solvers[2] of the form

$$A \leftarrow B_1, \ldots, B_m, not\, B_{m+1}, \ldots, not\, B_n, T_{n+1}, \ldots, T_k (k \geq n \geq m \geq 0) \quad (4)$$

where $A, B_i (1 \leq i \leq n)$ are defined as before, and $\{T_j | n+1 \leq j \leq k\}$ is a set of ground theory literals (which can contain free constants).

An interpretation of a program P is a pair (M, I) where M is a set of atoms and I is a set of ground theory atoms such that $T \wedge I \wedge \overline{I}$ is consistent where $\overline{I} = \{\neg t | t$ is a theory atom in P but $t \notin I\}$. An interpretation (M, I) of a program P is an *answer set* of P iff (i) $(M, I) \models P$ and (ii) M is the least model of $P^{(M,I)}$ where $P^{(M,I)}$ is the *reduct* of P (relative to (M, I)) which is obtained from P by

- deleting each rule of the form (4) if $B_i \in M$ for some i $(m+1 \leq i \leq n)$ or $I \not\models T_j$ for some j $(n+1 \leq j \leq k)$;
- deleting $not\, B_i$ and T_j from the remaining rules.

For example, consider the program P, with the background theory of integers, consisting of

$$\leftarrow not\, s, \qquad s \leftarrow x > z,$$
$$p \leftarrow x \leq y, \qquad p \leftarrow q, \qquad q \leftarrow p, y \leq z.$$

[2] http://www.cs.uni-potsdam.de/lpnmr09/downloads/niemela-lpnmr09.pdf

Consider the interpretation $\mathcal{M}_1 = (M, I)$ where $M = \{s\}$ and $I = \{x > z\}$. $(x > z) \wedge \neg(x \leq y) \wedge \neg(y \leq z)$ is T-consistent. $\mathcal{M}_1 \models P$. We have that $P^{\mathcal{M}_1} = \{s \leftarrow, \quad p \leftarrow q\}$ whose least model is $\{s\}$. Thus $\mathcal{M}_1$ is an answer set of P. It is easy to see that $(\{s, p, q\}, \{x > z, y \leq z\})$ is not an answer set of P. This approach is totally different from ours.

5 Conclusion and Future Work

A formalism of answer set programming with background theories was proposed in the spirit of SMT, for which the modulo answer set semantics was given. It has been shown that the formalism (under the proposed semantics) is a generalization of normal logic programs under standard answer set semantics [4], and a generalization for general logic program [8]. The formalism can be used to define the semantics of logic programs with aggregate functions. In that case it coincides with SPT-PDB answer set semantic of logic program with functions [7,9]. It is a challenging task to design a prototype for the formalism.

Acknowledgment. This work was partially supported by the Natural Science Foundation of China under grant 60963009, the Fund of Guizhou Science and Technology: 2008[2119], the Fund of Education Department of Guizhou Province: 2008[011] and Scientific Research Fund for talents recruiting of Guizhou University 2007[042].

References

1. Baral, C.: Knowledge Representation, Reasoning and Declarative Problem Solving. Cambridge University Press, New York (2003)
2. Seshia, S.A., Barrett, C., Sebastiani, R., Tinelli, C.: Handbook of Satisfiability. Frontiers in Artificial Intelligence and Applications, Satisfiability Modulo Theories, ch. 12, vol. 185, pp. 737–797. IOS Press, Amsterdam (2008)
3. Faber, W., Pfeifer, G., Leone, N., Dell'Armi, T., Ielpa, G.: Design and implementation of aggregate functions in the dlv system. TPLP 8(5-6), 545–580 (2008)
4. Gelfond, M., Lifschitz, V.: The stable model semantics for logic programming. In: Proceedings of the Fifth International Conference and Symposium on Logic Programming, pp. 1070–1080. MIT Press, Seattle (1988)
5. Lee, J., Meng, Y.: On Reductive Semantics of Aggregates in Answer Set Programming. In: Erdem, E., Lin, F., Schaub, T. (eds.) LPNMR 2009. LNCS, vol. 5753, pp. 182–195. Springer, Heidelberg (2009)
6. Nelson, G., Oppen, D.C.: Simplification by cooperating decision procedures. ACM Trans. Program. Lang. Syst. 1(2), 245–257 (1979)
7. Pelov, N., Denecker, M., Bruynooghe, M.: Translation of aggregate programs to normal logic programs. In: ASP 2003, pp. 29–42 (2003)
8. Shen, Y.-D., You, J.-H.: A Default Approach to Semantics of Logic Programs with Constraint Atoms. In: Erdem, E., Lin, F., Schaub, T. (eds.) LPNMR 2009. LNCS, vol. 5753, pp. 277–289. Springer, Heidelberg (2009)
9. Son, T.C., Pontelli, E., Tu, P.H.: Answer sets for logic programs with arbitrary abstract constraint atoms. Journal of Artificial Intelligence Research 29, 353–389 (2007)

In the Product-Innovation Discovered Needs Important Discussion

Xiao Wang–qun

Department of Industrial Design, Anhui University of Technology,
Ma'anshan 243002, China
xiaowangqun999@sina.com

Abstract. The final purpose of product-innovation is to meet gradually upgrading and endless human needs. In order to invigorate product-innovation, we should persist in looking for human needs. This paper specifies the notation and meaning of product-innovation. Moreover according to the theory of needs-layer, human needs can be divided to two kinds: real needs and potential needs, through which it deeply analyzes and researches human needs. Furthermore, many facts and examples are employed to expound and prove the source of product-innovation in detail.

Keywords: product-innomation, the reality needs, latent needs.

1 Introduction

Making a general survey of the world, science and technology is developed at a full speed, the process of the economic globalization is accelerated day by day, the commercial competition of the internationalization is fierce day by day, and the existence of enterprises faces greater challenge. Compare with famous trans-corporation of the world, our enterprises develop time is short, the whole is in small scale, the adaptability to changes is low, the competitiveness is weak. How should our enterprises meet to the changing fast market and advancing technology, How to dissolve the harm that double-edged sword globalization might bring. All these problems are badly in need of solution. The answer may be different, but one of the vital part that can't be avoided is how we see product-innovation, how to think about designing, how to design.

2 Product-Innovation's Meaning and Intension

Product-innovation is developing new product using such a series of means as scientific research and technical development. These new products can meet consumers' needs well; can bring more business benefit and more social benefit. The meaning of product-innovation include technological innovation and designing innovation , the high-level technological innovation has nature invented; the low-level technological innovation includes improvement of the spare part of the products, changing the produce of the technological process , try to lower costs by technology etc. It is mainly technological innovation that the traditional product-innovation meaning, namely

J. Zhang (Ed.): ICAIC 2011, Part V, CCIS 228, pp. 662–665, 2011.

product-innovation is go on with technological development and technological improvement. In modern society, the enormous transition has taken place in the social demand, so has taken place in economic mode, Because of these transition single product mode of technological innovation can't meet the need of economic development, so we must pay attention to creating new mode for design which consider the user's needs as direction.

The products are ties that enterprises contact consumer, and product-innovation is a source of enterprise's life motive force. The competition between enterprises mostly shows by the competition of products especially the new products. The key of enterprises survival and development is to hold the pulse of the times to go on product-innovation; this is also the source of enterprise's life motive force. With the raise of people's living standard and life sampling, Human needs from products have changed from pursuit of meeting quality to direction of quantity, even the communication of feelings, this gives product-innovation more meaning and extension. Therefore, product-innovation is not only transforming the product appearance or technical development now, but a complex system engineering, a course with various kinds of mutual information, including economic, technology, culture and society etc.

3 Analysis and Reasearch about Human Needs

A. The Theory of Human Needs-layer

Maslow divided human needs to five kinds: Physiological needs——this most original and basic needs, for example when we hungry we need food, thirsty need drink, cool need clothes, inhabiting want the residence, cure sickly and so on. This is most motive needs too, if it can't meet the needs, we lives will be in danger. So physiological needs are the strongest demands and lacks least.

Safety needs——this needs means the body is not in danger, avoid the harm from unemployment , property , food or inhabiting . After the physiology needs satisfying, the individual hope to avoid the harm from cold, hot, poison gas, calamity, and pain; require the job to have guaranteed to stabilize, work under security, and not to be worry about future life.

Emotion needs—— after above two kinds needs to satisfy, people will produce the emotion , friendship and belong to needs, for example wish eagerly to keep a good relation with family , friend , colleague , superior ,etc. offer to others and get friendliness and help from others, and finally make ourselves belong to someone collective.

Esteem needs——People are animals of the society, in a collective, wish eagerly others respect one's own character, acknowledge one's own talent. For example, the superior affirm our work and the position promoting.

Self-fulfillment needs——these are the highest needs. That means one hopes his own work assort with his own ability; hopes his own ability and talent get maximum full play, and then become the person that himself hopes.

These five needs are our human being shared needs, and gradually go up. Everyone's needs are very complex and work on our action every time. The purpose of design is to meet to physiological and mental needs, which become the motivity of design. Continually produced and satisfied the human needs drive the design to go ahead, meanwhile influence and restrict the content and fashion of design. It reflects that need level rise step by step constantly that mankind design from simple and practical to require products innovation move towards. Namely has explained this kind of priority level relation that the mankind needs to meet more or less.

B. The Real Needs and Potential Needs of Human Needs

Human needs can be divided to two kinds: real needs and potential needs. The mean of real needs is that the physical product or intellectual product mainly people need urgently at present but which the realistic society can't satisfy. For example, in fervent summertime, an electric fan or the air conditioner just become people real needs. The potential needs of human needs refer to the physical product or intellectual product people has not experienced yet in the realistic society but which the society will hope to enjoy in the future. For instance, when the telephone has been just invented, the videophone is people's potential demand.

Real needs and potential demands are dynamic development. After people's reality needs satisfying constantly, the potential demands will turn into new reality needs progressively. Thus keep the dynamic equilibrium of the realistic needs and potential demands, Move in circles and endless.

There are very great differences on the basis of different crowds between realistic needs and potential demands. For example, Peasants of the outlying mountain area that struggles in the poverty line, their reality needs just feeding themselves, the potential demands are a fairly comfortable life. In already marching toward city residents who has even stridden the fairly comfortable life, their current demands are higher than the formers' potential demand, which are more high-quality life. The potential demands are that material life and cultural life are got and satisfied more greatly. And for example, there are very differences between children real needs and potential demands and old people's, and the disabled person's realistic needs and potential demands, etc.

4 The Resource of Product-Innovation Is Looking for Needs

From remote antiquity to modern times, mankind always through constant innovation and make progress to meet corresponding period various kinds of needs of people. In Stone Age, people created stone zax for hunting or being apt to cut things; in handicraft stage, people have created various kinds of handicrafts in order to pursue aesthetically etc. needing; and in industrial stage, people have invented the plane, train and steamer in order to travel conveniently more fast; in information age, people have invented computer and internet for working efficiently and so on. All these indicate that the human innovative history is a history of meeting mankind's needs in fact. Therefore, we can make the conclusion that needs is the motivity of innovation from converse thinking. No needs, no innovation. The innovation will be unworthiness and inanition without considering the needs of people.

The final purpose of product-innovation is to meet the needs of people. People's needs always exists following the mankind exist, and reality needs which are born with us is stored in existing in the historical long river of the human development forever.

People's needs can't be invented and created, but can only be found and excavated. According to needs' contents included and concrete demand to carry on products innovation, this will make products innovation have more realistic value and meaning. For instance, in 1989, Haier is on the top to take the lead in putting out the cool room totally out of the common of designing on the domestic market, refrigerator that the freezer is on the bottom, this refrigerator is under the industry designers of Haier through an analysis of user's occupation mode to design, place the cool room of the application on above which use more, that have reduced the number of times that the user bends over every day greatly. That's a very representative example. In that time, the design of such abnormal rule has not merely got consumers' approval, but also became the mainstream of Japan and Korea to design. And for Konka, when all televisions are developed to the large screen, Konka has put out the small screen TV——Konka Xiao Hua Xian TV, this is satisfying the children and bachelordom, because they need such TV small-scale, with fashionable appearance. This one is a typical example, which carries on products innovation from users needs. Idiot camera of Japan, have exactly found the masses of housewives' reality needs, and then innovate the camera, thus the idiot camera obtained a so unprecedented and unexpected good result of the whole world.

Human invention and creation achievement of each period enumerated above, all is found and innovated by unconscious or consciously. All kinds of crowds' realistic needs and potential demands are not the same and unusually complicated, especially in this material, which is great and abundant and economy, develops extremely uneven era world. So we must study and excavate thoroughly to different crowds, fully find this kind of crowd's realistic needs and potential demands, thus carry on products innovation on the premise of meeting their needs to the maximum extent. Only this can meet their real needs. We know our people's needs are endless, so are the findings of demands, and the product-innovation which is on the base of findings of needs is also endless. So we can say: the resource of product-innovation is looking for needs.

5 Close

The research and discussion about products innovation have already become the hot topics of a great deal of disciplines in our country gradually, and have drawn much concrete products innovation methods. The author of this text mainly passes and varies an angle, stand in the source of products innovation, has thought again to products innovation, thus draw such conclusion that "finding the needs is the source of products innovation". Finding the needs is the motive force of products achieve. Therefore finding the needs to merge each other with the concrete products innovation method will make products innovation radiate out immortal vigor.

References

1. Bian, S.-R.: Product Innovation Design. Publishing House of Beijing University of Science and Technology (2002)
2. Jing, R.-H.: Overcome mistake and press close to market–grasping the method of industrial design of Chinese corporation from the development of Haier group. The Science and Technology of Family Wirling (1997)

Comparative Study on Land Quality Comprehensive Evaluation of Different Landforms in Henan Province

Qu Chenxiao[1], Meng Qingxiang[2], Lin Yan[3], and Cai Congfa[1,*]

[1] College of Resources and Environment,
Huazhong Agricultural University Wuhan 430070, China
cfcai@public.wh.hb.cn
[2] College of Resources Environment,
Henan Agricultural University, Zhengzhou 450002, China
[3] College of Urban and Environment Science,
Central China Normal University, Wuhan 430079, China
1436731004@qq.com

Abstract. According to the characteristics of land consolidation project, this paper established the quality indicators of land consolidation from micro-topography, soil, water, landscape and other factors. The weights of these indexes were reckoned by AHP, and then the land quality based on the land consolidation project about Baiyang town and Guqiao township were evaluated, which both are hilly area in Henan province. The results showed that different landforms had been changed significantly before and after land consolidation in the level of spatial distribution, size and overall quality. Land consolidation had a great impact on land quality, and project area in Baiyang town was up to 75.7% and Guqiao township to 37.3%.

Keywords: landforms, land consolidation, Land Quality, Comprehensive Evaluation, Henan Province.

1 Introduction

"Land Consolidation", originating in Germany Since the mid-16th century and through years of practice in the Netherlands, Finland and other European countries, has accumulated a wealth of experience, and gradually form a complete system [1]. Land consolidation in its true sense was formally proposed and operated till the late 90s of last century in china, marked by establishing Land Consolidation and Rehabilitation Center of National Land Agency in 1998 (renamed Ministry of Land and Resources in 1999) and launching the first land consolidation project in 2001. Land consolidation was being generally launched nationwide with billions of RMB invested[2-5]. The important factors including land use type, soil physical and chemical properties, road accessibility, drainage and irrigation conditions would been improved through measures of land leveling, farmland capital construction, land regularization and parcel merging[6-9],and the same as the effective area and quality of arable land after Land Consolidation project, Bindraban [10] pointed out that the

[*] Corresponding author.

J. Zhang (Ed.): ICAIC 2011, Part V, CCIS 228, pp. 666–675, 2011.

interaction of LQI determined land quality, and also had scale effect. Ryder[11] and Leibig & Doran[12] found that there were a high degree of agreement between farmers view and experiment test results at farm and site scales when they studied in the U.S.; Jinliang Ouyang[13] also reached a similar conclusion that they emphasized the importance of the behavior and views of peasant households in the evaluation of land quality.

Compared with its foreign counterparts, land quality research was started rather late in our country, and a comprehensive study system has not been formed. The studies mainly stayed in the construction of index system, which were not only weak in the mechanism and practical application but also lacked of long-term fixed-point monitoring data for assessment. From the standpoint of the department, Ministry of Land and Resources proposed agricultural land regulations [14].

2 Project Outline, Research Method and Date

2.1 Project Outline

Project I is located in Baiyang township Yiyang county in the southeast of Henan Province, involving 3 administrative villages of Tugou, Honggou and Longwo. Micro-topography is hill, and agrotype is red clay which is adhesive and with poor tilth and poor nutrient. Agriculture production is undeveloped and irrigation system leaves much to be desired. Layout of the field road, with large occupying area, is very unreasonable and most roads are poorly maintained, which hindered the development of agricultural production. The project is a key provincial investment for land consolidation with a construction period of 1 year and was completed in 2006. Construction scale was 153.00hm2, including new land 32.84hm2 which accounts for 21.46%.

Project II, is located in Guqiao Township belonging to Changge County in Henan Province, Involving 2 administrative villages of Caiwang and Xuwangzhao. The whole region is on the transition belt between the mountainous area in west of Henan and the plain regions in southeast of Henan. Northwest was the landscape High-Low southeast, showing a slow tilt changes. Groundwater is the main source of industrial and agricultural water use for the scarce surface water resource. Natural vegetation in the field characterized with secondary nature, is mainly covered with weed. Irrigation system is imperfect. Agro type is Fluvo-aquic soil with weak structure, permeability and poor capacity of water and nutrient. The project also is a key national investment for land consolidation with a construction period of 1 year and was completed in 2006. The construction scale was 214.59hm2, including new land 24.6hm2 which accounts for 11.24%.

2.2 Data Information Collection and Processing

Data information was obtained through the investigation of relevant units and local people, field observation, soil samples collection and laboratory analysis. Text information consisted of project feasibility research report, planning and design report and final acceptance report. Graphic information included 1:5000 land use maps before and after land consolidation and topographic maps of the project area.

After digitizing by MAPGIS, ARC/INFO, the village road maps before and after land consolidation was constructed on the basis of present land consolidation maps, the planning maps and the implemented land use maps; the irrigation guaranty ration maps and the parcel merging maps were drafted combining with the fieldwork and household surveys; the slope maps were created from the topographical maps; and the soil depth, the soil texture and soil organic matter dot distribution maps were drafted. Then the land use structure changes before and after consolidation was analyzed and land use degree were calculated, finally the changes in land quality were worked out.

3 Land Quality Evaluation on Land Consolidation Project

3.1 Establishment of Evaluation Index System

Before and after land consolidation project was due to differences in land quality caused by project construction. The differences of land quality before and after land consolidation were due to project construction. land consolidation works included land leveling, irrigation and water conservation, the field path and the shelter forest construction. The land leveling project generally included the cubic meter of earth and stone to excavate, the backfill, the transportation, the land leveling and so on. The topographical condition was meaningful to farmland quality evaluation for the macroscopic region. To the hilly area land arrangements project, the land leveling project would change the micro-relief landform inevitably in project area, and then influenced land quality.

According to the land quality evaluation indicator selection principle and above-mentioned analysis and combining with characteristic of land arrangements project, this study selected 4 factors, 11 indices, and established a three levels land quality evaluation indicator system (Table 1).

Table 1. The evaluation index system of land quality

Object	Item	index
Land Quality Impact Assessment of Land Consolidation A	Micro-topography factors B1	Flatness of land C1
		Production convenient degree C2
		Entire degree of field block gauge C3
	Soil factors B2	Topsoil thickness C4
		Soil fertility level C5
		Land utilization ratio C6
	Water Factors B3	The degree of the Water source guarantee C7
		The degree of irrigation guaranteesC8
		The degree of draining water guarantees C9
	Landscape factors B4	Landscape productivity C10
		Landscape aesthetic degree C11

The first layer was object, land quality impact assessment of land consolidation. The second layer was item, including micro-topography, soil, water, landscape and other factors, and the third layer was the index level (index), that was what each of the criterion level indexes.

3.2 Evaluation Index Classification

In land quality evaluation of the land arrangements' target description, some evaluation indicators such as topsoil thickness and so on could be described

Table 2. Index Classification

Evaluation Index	Index description method	Index characteristics and Classification standards	Rank
Flatness of land	May describe the land leveling situation, flatness of land is high, and high ease of mechanization and irrigation.	Slope <10,levelling degree is high[69]	1
		Slope 10—30,levelling degree is little high[69]	2
		Slope 30—50，leveling degree is mid	3
		Slope >50,levelling degree is bad	4
Production convenient degree	Reflect the production convenient situation before and after arrangement qualitative description	convenient	1
		inconvenient	2
Entire degree of field block gauge	Reflect the fragmented plots before and after arrangement breaking situation qualitative description	regular	1
		Irregular	2
Topsoil thickness	Soil indicators can be measured, its value is too small affecting crop growth quantitative description	25—40cm	1
		15—25cm	2
		Less than 15cm	3
Soil fertility level	Reflect the degree of soil fertility improvement after arrangement qualitative description	Fertile soil	1
		The soil fertility level is little high	2
		The fertility level is little low	3
		The soil is barren, the fertility level is low	4
Land utilization ratio	Reflect the degree of the land utilization before and after arrangement quantitative description	>90%, Land utilization ratio Is high	1
		85%—90%，little high	2
		80%—85%，middle	3
		〈80%，little low	4
The degree of the Water source guarantee	Reflect the guarantee situation of the irrigation water source before and after arrangement qualitative description	The irrigation water source is sufficient	1
		The irrigation water source is limit	2
		No irrigation water source	3
The degree of irrigation guarantee	The conditions of irrigation can be reflected by data, the more too close to 100%, the guaranteed degree is higher quantitative description with the percentage quota	The guarantee degree of irrigation is more than 90%	1
		75%—90%	2
		Less than 75%	3
The degree of draining water guarantees	Drainage conditions can be described by the combined effect of topography and drainage	Have perfecting facilities of draining away water, can discharge in time	1
		The drainage facility is ordinary, the	2

quantificationally, and some evaluation indicators such as production convenience and so on needed to be given the qualitative description. The evaluation Index Classification for details was shown in Table 2.

3.3 Determination of Evaluation Index Weight

The determination of the weight of farmland quality evaluation factors, should been given according to the actual state of the farmland in project area and Combined with actual agricultural production affection extent of the evaluation factors , the weight of the evaluation factors were determined with the hierarchical analysis method. May also use regression analysis, principal component analysis, Delphi method and so on.

3.4 Determination of the Evaluation Factors Index Score

In the selected land arrangement of land quality evaluation index system, some factors could be partitioned quantitatively, some factors could only be partitioned qualitatively。 In order to operating conveniently, four levels of evaluation factors were assigned with 100-10.

Table 3. Factor level-Score contraposition

	100	90	80	70	60	50	40	30	20	10
Flatness of land	1		2		3					
Production convenient degree	1		2							
Entire degree of field block gauge	1				2					
Topsoil thickness	1		2		3					
Soil fertility level	1		2		3		4			
Land utilization ratio	1		2		3		4			
The degree of the Water source guarantee	1		2		3					
The degree of irrigation guarantee	1		2		3					
The degree of draining water guarantees	1		2		3					
Landscape productivity	1		2		3		4			
Landscape aesthetic degrees	1		2		3					

3.5 The Standardization of Evaluation Factors Index

The study on land quality influence of land consolidation is aimed at revealing the implementation influence type and degree to project area. There were few studies about land arrangements' land quality influence, and there was not the unification evaluation system at present. Simultaneously, the different evaluation system's evaluating indicator's standardization used different method. This study based on status background value of the project area before the land arrangements, and then compared with index values after land consolidation. The land quality target value before land consolidation defaults 1.

Index standardized calculation method: $F = X_{after} / X_{before}$

After index value standardization, the F value was greater than 1, showing land quality index value enhanced; The F value was equal to 1, which explained that the land quality index value did not change; The F value was less than 1, which explained that the land quality index value reduces.

3.6 Determination of Evaluation Model

Comprehensive evaluation model: $E = \sum_{i=1}^{n} F_i * W_i$

In formula: E was the comprehensive evaluation value of land quality; Fi was standard valuesof the index i; Wi was the weight of the index i.

4 Results of Evaluation

According to the above methods, through field surveys and data analysis of the project areas, firstly assign and standardize the land quality indicators, and then assign different Index Weight, finally calculate the comprehensive evaluation results

4.1 Evaluation Results of Land Quality in Guqiao Township of Changge

4.1.1 Factors

Micro-topography factors: Generally, land consolidation region in Guqiao township of Changge was plain, but micro-topography undulating. The land consolidation projects were mainly about land leveling, and re-planning the layout of the plots. After land consolidation finished, flatness of land was improved, and the size and layout of the field plot was reasonable, in favor of the lighting for field crops, the facility for agricultural mechanization and irrigation drainage and increasing the effect of the wind defense.

Soil factors: the project of the land formation effectively improved the soil plow layer thickness, soil ability to retain water and nutrients, and the content of soil organic matter. Meanwhile, the land consolidation project would change brick and tile kilns, waste ponds and other terrain into arable land, which increased the effective area of cultivated land , improved the efficiency of the project area land, and finally enhanced the quality of the all project area of the land.

Water factors: The length of low-pressure pipeline buried in project area was 15.75km; the drainage channel was 10.547km, and flood control standard was up to against once-in-50 years flood. Through the construction of water conservancy projects, the margin of the water level of the project area was effectively improved, irrigation guarantee rate was improved to 100%, and the flood control capacity has also been enhanced.

Landscape factors: Through land consolidation, the project area was changed into high-quality farmland which had high flatness, high soil ability to retain water and fertilizer, high level of mechanization, appropriate sale, highly efficient irrigation and drainage. The food production capacity and production capacity in the project area were also improved. Scientific planning of terraces, rational distribution of agricultural and farmland protection forest road construction had improved the infrastructure of the project area. And it also made the terraces flat, the drainage interlinked; the roads connected into a network of forests, changed the former messy land utilization and promoted the agricultural landscape aesthetic degree.

Table 4. The evaluation index system of land quality and Standard value (I)

object	item	indictor	Before consolidation	After consolidation	Standard
Results of the quality assessment of consolidated land A	Micro-topographic factors B1	Flatness C_1	80	100	1.25
		Facilitation C_2	80	100	1.25
		Regularity C_3	60	100	1.67
	Soil factors B2	Topsoil thickness C_4	80	100	1.25
		Fertility C_5	60	80	1.33
		Land utilization C_6	80	100	1.25
	Water factors B3	Water level of assurance C_7	60	100	1.67
		Irrigation level of assurance C_8	60	100	1.67
		Drainage level of assurance C_9	60	100	1.67
	Landscape factors B4	Landscape productivity C_{10}	80	100	1.25
		Landscape aesthetics C_{11}	50	90	1.8

4.1.2 Results of Comprehensive Assessment

As is shown in Table 5, after land consolidation, quality factors of the project land have shown increasing trend, especially to water factors which had increased as much as 66.7%. The quality of integrated land had significantly been improved as much as 37.3%. It indicated that the management of the project areas had an obvious positive effect on land quality.

Table 5. The result of Comprehensive evaluation (I)

	Category Index				The result of Comprehensive evaluation
	Micro-topographic factors	Soil factors	Water factors	Landscape factors	
After consolidation	1.318	1.282	1.667	1.433	1.373

4.2 Results of the Quality Assessment of Land Consolidation in Baiyang Town, Yiyang County

4.2.1 Factors

Micro-topography factors: the land consolidation project, located in Yiyang county Poplar town, was hilly and micro-terrain showed fluctuations. Through land consolidation, the terraces were constructed , the former slope become smooth piece of paddy field and the layout of agricultural land and roads production were re-planned and adjusted. After land consolidation, flatness of land was improved and land size and layout became reasonable, which contribute field crops to lighting; and new roads were built to facilitate the farmers for mechanized production and to improve the ease of agricultural production.

Table 6. The evaluation index system of land quality and Standard value

object	item	index	Before finishing	After finishing	Standard
Results of the quality assessment of land consolidation A	Micro-topographic factors B1	Flatness C1	40	100	2.5
		Facilitation C2	80	100	1.25
		Regularity C3	60	100	1.67
	Soil factors B2	Topsoil thickness C4	60	80	1.33
		Fertility C5	40	60	1.5
		Land utilization C6	80	100	1.25
	Water factors B3	Water level of assurance C7	60	100	1.67
		Irrigation level of assurance C8	60	100	1.67
		Drainage level of assurance C9	100	100	1
	Landscape factors B4	Landscape productivity C10	40	80	2
		Landscape aesthetics C11	50	90	1.8

Soil factors: by converting steep land into terraced fields, the risk of soil erosion reduced, the soil ability to retain water and nutrients was improved effectively, which enhanced the soil organic matter content. Meanwhile, the land such as mud flat or waste land were flatted into arable land , increasing the effective area of cultivated land, improving the project area's land use internsityand enhancing the overall quality of the project area of the land.

Water factors: two stations and two Pumping wells were newly built, which made the project area with sufficient water to meet irrigation requirements and to improve the water level of assurance. Meanwhile, the new agricultural drainage, which was 7.09 km long, made each plot irrigated better, increased drought tolerance, and improved irrigation guarantee rate. Natural conditions of drainage were very good in project area, so there were no constructions on the drainage.

Landscape factors: After land consolidation finished, the original steep slopes were transformed into well-structured trapezoidal field, and there was a new farmland shelterbelt. Project area of slope cropland was changed into land which had high formation and the modest size. These enhanced the capacity of the soil retain water and nutrients and drought, increased the level of mechanization of agricultural production, and greatly improved the food production capacity and production capacity in the project area. Construction of terraces, the re-layout of the road and the construction of farmland, made terraces regular, network distribution uniform, and windbreak structure good. These also changed the disordered land-utilization before consolidation. The project area of the landscape features was substantially improved.

4.2.2 Results of Comprehensive Assessment

As shown in Table 7, after land consolidation, quality factors of the project land had shown increasing trend, especially in micro-topography and landscape which were as high as 99.3% and 93.3%. The quality of integrated land had significantly improved, as much as 75.7%. It indicated that the management of the project areas had a obvious positive effect on land quality.

Table 7. The result of Comprehensive evaluation

	Category Index				The result of Comprehensive evaluation
	Micro-topographic factors	Soil factors	Water factors	Landscape factors	
After finished	1.993	1.379	1.589	1.933	1.757

5 Conclusion and Discussion

5.1 Conclusion

It's one of the main objectives of land consolidation to improve land productivity. Empirical research showed that: with the implementation of land consolidation and analysis of various factors, the project significantly improved the land quality, factors of which had shown increasing trend. In Guqiao township of Change, the plain area, water factors increased particularly by as much as 66.7%, and the land quality raised by as much as 37.3%; and In Baiyang township of Yiyang, the hilly region, especially factors of micro-topography and landscape increased by as much as 99.3% and 93.3% respectively, and the land quality improved as much as 75.7%.

5.2 Discussion

Studies on land quality of land consolidation belonged to a new work, with the characteristics of the relatively small range of subjects, the data of quantitative indicators being difficult to obtain, big variations in land quality before and after land consolidation and so on. The land quality evaluation method used by this paper had better reflected on the influence extent over land consolidation to land quality, but qualitative indexes held quite large proportion, making the final evaluation result strong subjective because of relatively small scope of the evaluation, the data of quantitative indexes difficult to obtain and land quality with large fluctuations of the objects. Therefore, land quality research of land consolidation should be paid much attention to avoid evaluation subjectively and establish a more scientific system of land quality indicators, to ensure an objective evaluation results.

Due to its nonlinearity, integrity and complexity, land quality assessment and investigation mathematical methods and modern technology to carry out[15]. ①land quality evaluation method in land consolidation: Based on principal component analysis, modified weighting method, comprehensive evaluation method and multi-angle single-factor evaluation method, summarize relative standards for land quality evaluation and then establish operational land quality assessment appropriate for land consolidation; ② land quality investigation indicators and methods in land consolidation: Based on the analysis of land quality property influenced by land consolidation, construct the land quality index system combined with "3S" technology and ground investigation and explore technological methods of land quality in land consolidation by use of "3S" technologies.

References

1. Jun, W., Li, Y., Ming, L., Gang, Z.: The overview of Land Consolidation. Areal Research and Development 2, 8–11 (2003)
2. Xinshe, L.: General theory of land consolidation strategies. Transactions of the Chinese Society of Agricultural Engineering 18(1), 1–5 (2002)
3. Xiangjun, G.: Theory and Practice of Land Consolidation. Geilogical Publishing House, Beijing (2003)
4. Wu, z., Liu, m., Davis, J.: Land consolidation and productivity in Chinese household crop production. China Economic Review 16, 28–49 (2005)
5. Planning Division in Ministry of Land and Resources, Land Consolidation and Rehabilitation Center. Planning examples for land development and rehabilitation. Geilogical Publishing House, Beijing (2001)
6. Crecente, R., Alvarez, C., Fra, U.: Economic, Social and Environmental Impact of Land Consolidation in Galicia. Land Use Policy (19), 135–147 (2002)
7. Coelho, J.C.: A systems approach for the estimation of the effects of land consolidation projects (LPCs): a model and its application. Agricultural Systems 68(3), 179–195 (2001)
8. Ishii, A.: The methods to consolidate scattered tenanted lots into large rice paddy lots by the land consolidation projects in Japan. Paddy Water Environ. (3), 225–233 (2005)
9. Planning department of country land administration ministry, et al. Land consolidation reference at home and broad, pp. 291–380. Chinese land publishing house, Beijing (1998)
10. Bindraban, P.S., Stoorvogel, J.J., Jansen, D.M.: Land quality indicators for sustainable land management: proposed method for yield gap and soil nutrient balance. Agriculture, Ecosystems and Environments 81, 102–103 (2000)
11. Reder: Land evaluation for steepland agriculture in the Dominican republic. The Geographical Journal 160(l), 74–86 (1994)
12. Leibig, M.A., Doran, J.W.: Evaluation of Point-scale assessments of soil quality. Journal of Soil and Water Conservation 54(2), 510–518 (1999)
13. Jinliang, O., Zhenrong, Y., Fengrong, Z.: The soil quality changes and peasant behavior analysis based on eco-economic regions. Acta Ecologica Sinica 23(6), 1147–1155 (2003)
14. Ministry of Land and Resources(PRC). Regulations for gradation on agricultural land. Standards Press of China, Beijing (2003)
15. Jun, W.: Implications for Land Consolidation Research of China from Land Quality Development. Areal Research and Development 25(6), 108–111 (2006)

Author Index

GPSR Compliance

The European Union's (EU) General Product Safety Regulation (GPSR) is a set of rules that requires consumer products to be safe and our obligations to ensure this.

If you have any concerns about our products, you can contact us on ProductSafety@springernature.com

In case Publisher is established outside the EU, the EU authorized representative is:

Springer Nature Customer Service Center GmbH
Europaplatz 3
69115 Heidelberg, Germany

Batch number: 10838023

Printed by Printforce, the Netherlands